MODELS OF QUANTUM MATTER

AF844888

Models of Quantum Matter

A First Course on Integrability and the Bethe Ansatz

Hans-Peter Eckle

Humboldt Study Centre
Ulm University

OXFORD
UNIVERSITY PRESS

OXFORD
UNIVERSITY PRESS

Great Clarendon Street, Oxford, OX2 6DP,
United Kingdom

Oxford University Press is a department of the University of Oxford.
It furthers the University's objective of excellence in research, scholarship,
and education by publishing worldwide. Oxford is a registered trade mark of
Oxford University Press in the UK and in certain other countries.

© Hans-Peter Eckle 2019

The moral rights of the author have been asserted.

First Edition published in 2019

First published in paperback in 2026

All rights reserved. No part of this publication may be reproduced, stored in a retrieval system, transmitted, used for text and data mining, or used for training artificial intelligence, in any form or by any means, without the prior permission in writing of Oxford University Press, or as expressly permitted by law, by licence or under terms agreed with the appropriate reprographics rights organization. Enquiries concerning reproduction outside the scope of the above should be sent to the Rights Department, Oxford University Press, at the address above.

You must not circulate this work in any other form
and you must impose this same condition on any acquirer.

Links to third party websites are provided by Oxford in good faith and
for information only. Oxford disclaims any responsibility for the materials
contained in any third party website referenced in this work.

Published in the United States of America by Oxford University Press
546 Fifth Avenue, New York, NY 10036, United States of America

British Library Cataloguing in Publication Data
Data available

Library of Congress Control Number: 2019930263

ISBN 9780199678839 (hbk)
ISBN 9780197919613 (pbk)

DOI: 10.1093/oso/9780199678839.001.0001

The manufacturer's authorized representative in the EU for product safety is
Oxford University Press España S.A. of Parque Empresarial San Fernando de Henares,
Avenida de Castilla, 2 – 28830 Madrid (www.oup.es/en or product.safety@oup.com).
OUP España S.A. also acts as importer into Spain of products made by the manufacturer.

Für Irene

Preface

Why study Bethe ansatz? The Bethe ansatz provides one of the very few methodologies to calculate the physical properties of models for strongly interacting quantum matter *non-perturbatively*. Arguably it is the only such method we have which is exact. This means, once we have set up the model, there are no approximations or further assumptions necessary: we can exactly compute physically relevant properties of the model. There is, furthermore, an infinite set of conserved quantities: the quantum mechanical model is integrable.

This makes the search for quantum models which are amenable to an exact solution by the Bethe ansatz methodology so important and rewarding. Even if—as is sometimes, but *certainly* not always, the case—the model with an exact solution is not the most physically relevant one, the exact solution will provide important benchmarks for other models that may occasionally be closer to physical reality, but do not admit an exact solution. Hence, for a plethora of cases, the Bethe ansatz solution provides valuable insight into the physics of strongly interacting quantum matter.

Since Hans Bethe provided the eponymous method to solve the Heisenberg quantum spin chain, many more models of low-dimensional quantum systems have been found to be integrable by the Bethe ansatz.

Moreover, these models and their integrability have been and continue to play an influential role in many subfields of physics, which include classical and quantum statistical mechanics, quantum field theory, and quantum many-particle and condensed matter physics, the latter in recent times especially in connection with modern developments in physics on the nanometre scale and in low dimensions. Quantum optics has also benefited from studying integrable models, especially in investigations of ultracold Bosonic and Fermionic quantum gases and Bose–Einstein condensates in artificial crystals of light, the so-called optical lattices. Recently in string theory and cosmology there is a hype of activity involving conjectures of Bethe ansatz integrability in the framework of the celebrated anti-de-Sitter space/conformal field theory (AdS/CFT) correspondence.

Of course, Bethe ansatz and integrability are discussed in Mathematical Physics, but there is also an ongoing cross-fertilization with various subfields of pure Mathematics.

Some prominent examples of integrable models include: various variants of the Heisenberg quantum spin chain whose physical realizations are probed by neutron scattering; the Hubbard model and its variants which inter alia have been discussed in connection with high-temperature superconductivity; the Kondo model which has recently seen a renaissance because of the development of tunable quantum dots; interacting Bose and Fermi gases which can now be produced in very pure and tunable form in optical lattices.

But, what is the Bethe ansatz? In its original form, devised by Bethe, it is an ingeniously guessed form for the wave function of a one-dimensional quantum system. However, why this wave function is correct and even exact remained an open question which has only been answered much later through the algebraic form of the Bethe ansatz. This method enables us to construct an integrable quantum model in one dimension from a two-dimensional statistical mechanical model. The construction reveals the reason for quantum integrability and delivers the infinite set of conserved quantities together with the wave function which Bethe guessed. In our exposition of the Bethe ansatz methodology, we shall therefore start, somewhat unhistorically, but more systematically, with the algebraic Bethe ansatz.

Who is this book for? Interestingly, a look at Richard Feynman's last blackboards (Paz, 1989) reveals that he may well have been interested. In one of his last publications (1988), Feynman in fact wrote:

> 'I got really fascinated by these $(1 + 1)$-dimensional models that are solved by the Bethe ansatz and how mysteriously they jump out at you and work and you don't know why. I am trying to understand all this better.' Cited after Batchelor (2007).

In view of the exciting developments in Bethe ansatz of the last two decades, Richard Feynman's fascination would certainly have continued.

The most helpful prerequisites for present readers are a good grounding in quantum mechanics, statistical mechanics, and the basics of quantum many-particle theory, especially second quantization. However, we shall comprehensively discuss the necessary tools and background in part I of the book. Through this approach, the book should be smoothly accessible for Master's students who look for an area of specialization as well as for beginning graduate students. Moreover, to paraphrase Paul Halmos (in the preface of his book on Measure Theory (Halmos, 1978)), the novice to the Bethe ansatz methodology should *not* be discouraged if she or he finds that she or he does not have the prerequisites to read the preliminaries. After all, as Max Born reminds us, where would quantum physics be if Werner Heisenberg had been discouraged that he did not know what a matrix was when he developed the matrix form of quantum mechanics?

The book grew out of lecture notes the author prepared for an invited graduate lecture series at the Indian Institute of Science in Bangalore in 1995, summer school courses at the University of Jyväskylä in Central Finland in 1997 (on *Bethe Ansatz Methods in Many–Body Physics*) and 1999 (on *Conformal Invariance in Statistical Physics*), a graduate course at the same University which, together with an amiable group of students, made the extremely cold Finnish winter of 1999 actually an enjoyable experience, and postgraduate courses at the University of New South Wales in Sydney in 2000, and within the Mathematics–Physics $\mathcal{MP}^2$ Platform at Göteborg University in 2009, as well as summer school lectures in Turkey: 2013 in Turunç, 2014 in Izmir and 2013 in Ireland in Dungarven under the auspices of the School of Theoretical Physics of the Dublin Institute for Advanced Studies.

Acknowledgements

The author's *working knowledge* of the topics and methods on which this book focuses was shaped through collaborations with several people who are gratefully acknowledged in chronological order: Ferenc Woynarovich, Tuong Truong, Chris Hamer, Rudolf Römer, Bill Sutherland, Alexander Punnoose, Henrik Johannesson, Charles Stafford, Tim Byrnes, Robert Bursill, Anders Sandvik, and Johan Nilsson. Gratefully acknowledging discussions with many others, especially my thesis advisor Dieter Schotte and with Ingo Peschel and Karol Penson, I shall not attempt to mention all of them for fear that someone's name might escape my attention.

I learned coordinate Bethe ansatz from and during a fruitful collaboration with Ferenc Woynarovich. We both were initiated to the algebraic Bethe ansatz by Tuong Truong.

I also often enjoyed stimulating discussions with Holger Frahm as well as mutual visits with Fabian Essler, Andreas Klümper, and Andreas Schadschneider.

My deeply felt thanks go to Henrik Johannesson for his kind invitations to prepare and deliver lectures in Göteborg and, very especially, for our long-standing collaboration and friendship.

Great thanks are also due to Henrik again and Alexander Stolin for organizing and co–lecturing the course on 'Integrable Models and Quantum Groups' within the Mathematics–Physics $\mathcal{MP}^2$ Platform at Göteborg University.

Jussi Timonen's initiative is highly appreciated for suggesting and encouraging my lectures in Jyväskylä.

Valuable discussions with Balazs Hetényi, especially in connection with the co-organization of the summer schools in Turkey, and with Daniel Braak, are gratefully acknowledged. Many thanks also go to Karsten Seifert who volunteered to undertake the altruistic task of proof-reading parts of this book at various stages of its formation.

Many thanks are due to Natan Andrei, Nikolay Bogoliubov, and the late Anatoli Izergin for their hospitality at Rutgers and in Sankt Petersburg, respectively, and the sharing of their knowledge in stimulating personal discussions and in the written form of lecture notes, and also to Vladimir Korepin for encouragement on many occasions.

Many thanks are also due to Peter Horsch, Joachim Stolze, and Johannes Voit, and the late Heinz Barentzen and Hellmut Keiter for discussions and hospitality in Stuttgart, Bayreuth, and Dortmund, respectively, and the sharing of pertinent sets of lecture notes.

The team at Oxford University Press cannot be praised highly enough, especially their friendly patience and cheerful spirit, which appear to be inexhaustible. The incentive of April Warman got this project under way and helped it along in its early infant steps. Sönke Adlung supplied fresh guidance to help the project along whenever there seemed to be an impasse. Ania Wronksi never tired of providing motivation and momentum for the book project and, of course, all the necessary information.

Contents

1 Introduction

A journey of a thousand miles begins with a single step.

– Lao Tzu

This brief introductory chapter's purpose is to direct you, the reader, quickly to those places in the book where you can find general introductory information that may be helpful for an overview of and the orientation within the book. It is deliberately kept short to avoid redundancies.

The general motivation for the writing of the book and the main targeted readerships as well as the levels of sophistication assumed and detail aimed at in the presentation of the various parts of the book are outlined in the preface. There, we also attempt a delineation of an assessment of the relevance of the book's topics for current and potential future research.

The list of contents, by its nature in the form of key words and key phrases, provides a more comprehensive orientation of all the topics treated and their mutual dependence.

The book is divided into six major parts and, including this short introductory chapter, into twenty chapters. Each part and each chapter begins with a description detailing their respective subject matter. The part descriptions supply the bigger picture, while the outlines at the beginnings of each chapter point more specifically to the topics treated. Wherever this seemed to be helpful, we have attempted to supply further signposts about what we have achieved and where we plan to go from there.

At various places, we also remark on the depth with which the topics are treated, what may have been left out or will only be mentioned in passing, the relation to other parts and chapters and the book's intentions as a whole, and where to find alternative and further specialized treatments of these topics.

In order to supply a rough overview, let us briefly summarize the major parts of the book and their interrelationship.

Part I ranges from the fundamental concepts and tools required for an understanding of strongly interacting quantum matter to the fundamental models that represent the physical systems of strongly interacting quantum matter. In this book, we want to investigate selected aspects of these models with a particular emphasis on the uses of the exact methodology of the Bethe ansatz and of quantum integrability.

Part II is devoted to the quantum inverse scattering method and the algebraic Bethe ansatz that demonstrate the quantum integrability of certain one-dimensional strongly interacting quantum models and provide their exact solution. Our approach makes decisive use of the intimate connection between these models and two-dimensional

Models of Quantum Matter: A First Course on Integrability and the Bethe Ansatz. Hans-Peter Eckle,
Oxford University Press (2019). © Hans-Peter Eckle. DOI: 10.1093/oso/9780199678839.001.0001

models of classical statistical mechanics. The concrete models we shall be enlisting in this part are the Heisenberg quantum spin chain as one-dimensional quantum model and the six-vertex model as two-dimensional classical statistical model.

In part III, we introduce the coordinate Bethe ansatz, the original approach Bethe used to solve the Heisenberg quantum spin chain. Since this approach does not allow us to understand why the models are quantum integrable, we shall address it only after we discuss the algebraic Bethe ansatz and the quantum inverse scattering method. The coordinate Bethe ansatz approach is, however, still extremely useful. We demonstrate this again for the Heisenberg quantum spin chain and also for a gas of Bosons interacting via δ-function potentials in one dimension.

Part IV is concerned with strongly interacting quantum models where the fundamental constituents have internal degrees of freedom. Our examples, the one-dimensional gas of Fermions interacting via δ-function potentials, the one-dimensional Hubbard model, and the Kondo model of a magnetic impurity interacting with conduction electrons, are all electronic models where there is only one internal degree of freedom in addition to the particle degree of freedom, which is electronic spin. We shall find that these models can be solved by two interconnected Bethe ansätze. The method is thus called *nested* Bethe ansatz.

Thus far, the Bethe ansatz methods discussed were mainly investigating the low-lying and therefore zero temperature properties of the quantum models. In part V we examine how to extend the Bethe ansatz to finite temperatures. Again, our quantum models of choice will be the Heisenberg quantum spin chain and also the Bose gas interacting via δ-function potentials in one dimension.

In part VI, the final part of the book, we relinquish another assumption we made or had to make in order to find solutions of the Bethe ansatz equations, equations which generally hold for a finite system. So far, we usually considered the thermodynamic limit, the limit of an infinite system size. This limit allowed us to rewrite the Bethe ansatz equations as linear integral equations for certain densities whose solutions characterized solutions of the Bethe ansatz equations, but only, of course, for the thermodynamic limit. The Bethe ansatz for finite systems attempts to find corrections to the Bethe ansatz solutions and physical quantities, e.g. the ground state energy of the thermodynamic limit that takes into account the finiteness of a system. Again, we inquire into how this can be achieved using the Heisenberg quantum spin chain as our exemplary model system.

The focus of this book is on selected concepts, methods, and mathematical techniques in the area of strongly interacting quantum matter systems, especially the various Bethe ansatz techniques discussed. We hope that these techniques will prove useful in future research in the area of strongly interacting quantum matter. We also hope that some physical insight will be gained from the models of quantum matter used as examples to demonstrate the concepts and techniques and will provide guidance for the understanding of other systems not treated here.

For the most part, we shall use natural units where the speed of light, Boltzmann's, and Planck's constants are

$$c = k \equiv k_B = \hbar = 1, \tag{1.1}$$

except when including the constants explicitly will render the results more transparent.

Part 1

Methods and Models in the Theory of Quantum Matter

Cannot we be content with experiment alone? No, that is impossible; that would be a complete misunderstanding of the true character of science. The man of science must work with method. Science is built up of facts, as a house is built of stones; but an accumulation of facts is no more a science than a heap of stones is a house.

– Henri Poincaré (1854–1912)

This first part of the book presents an overview of the most important methods indispensable for an understanding of the theory of strongly interacting quantum matter. Moreover, we introduce a selection of quantum mechanical many-particle models and the related concepts that form the background of the theory of quantum matter, especially in view of the quantum integrable models, whose exact Bethe ansatz solutions are discussed in later parts of the book. These methods and models are relevant also in many other parts of theoretical and mathematical physics. It is therefore recommended that readers review this material to judge how familiar they are with it.

These chapters, however, do not treat and do not attempt to treat their topics in a fully comprehensive manner. There is always a lot more that could be covered. In fact, there is a vast literature specifically devoted to these topics. Nevertheless, we attempt as clear and comprehensible a treatment as possible of the aspects we cover with the intention to render those aspects that we do cover self-contained. Where a self-contained treatment is beyond the limitations of this book, we provide appropriate hints to the literature specially devoted to these topics.

More specifically, in chapter 2, basic facts are reviewed from the quantum mechanics of many-particle systems, in particular leading from the Hilbert spaces representing quantum many-particle systems to a discussion of second quantization, which is the language most useful to formulate the models of strongly interacting quantum matter.

Moreover, in chapter 3 we address the quantum mechanical theory of angular momentum, especially for many quantum particles, which is indispensable for an understanding of the magnetic properties of the models of strongly interacting quantum matter. These magnetic properties will be at the centre of much of our discussion of quantum integrable models and their exact Bethe ansatz solutions.

Quantum many-particle theory is, of course, resting on the foundations of equilibrium statistical mechanics, especially quantum statistical mechanics. But classical statistical mechanics also will be necessary to appreciate the developments of quantum models that are integrable by the Bethe ansatz method. An examination of the methods and results of equilibrium statistical mechanics, both classical and quantum, will therefore be a useful addition in this first part of the book, and which we take up in chapter 4.

Among the most fascinating phenomena of many-particle systems, again classical and quantum, phase transitions and critical phenomena occupy a prominent place. Their theoretical description is challenging and requires an arsenal of sophisticated and innovative methods that are outlined in chapter 5, where we also analyse the approach to the thermodynamic limit of systems of finite size.

There is an intimate connection between quantum field theory and (classical) statistical mechanics on which much of the quantum inverse scattering method and the algebraic Bethe ansatz is founded. Chapter 6 offers an introduction to this immensely useful connection, which will also play a central role in the subsequent chapter.

Bethe ansatz calculations for finite systems, being rather more involved than those in the thermodynamic limit, produce results that can be directly compared to predictions based on the conformal symmetry of two-dimensional classical statistical mechanics. In order to fully appreciate this connection, chapter 7 offers an introduction into basic aspects of the conformal symmetry of critical systems.

While the chapters described so far were mainly concerned with methods useful for a thorough appreciation of the Bethe ansatz methodologies examined in later parts of the book, chapter 8 of this first part introduces the physical background of a selection of models of strongly interacting quantum matter together with methods to investigate and understand them. The selection criteria have been whether appropriate versions of the models exhibit quantum integrability and are solvable by Bethe ansatz. The quantum many-particle models considered range from the Bose fluid to models of itinerant as well as localized magnetism and to the Fermi liquid and ultimately to models of strong light–matter interaction.

2

Quantum Many-Particle Systems and Second Quantization

One of the principal objects of theoretical research is to find the point of view from which the subject appears in the greatest simplicity.

Josiah Willard Gibbs (1839–1903)

This chapter reviews some aspects of the quantum mechanics of systems composed of many particles (many-body or many-particle systems), which will prove useful for the later developments in this book. We mainly concentrate on the foundations of quantum many-particle physics leading to the formalism of *second quantization* as a convenient language for the formulation of the properties of the many-particle systems of quantum matter.

Many-particle quantum systems can be described by a many-particle Schrödinger equation, whose corresponding wave function depends on the configuration of the particles, e.g. their positions $\mathbf{r}_i$ and possibly further quantum numbers, e.g. the spin quantum numbers σ_i of the particles. In practice this approach is very cumbersome even for quite modest numbers of particles, let alone for the macroscopic numbers of particles of quantum statistical mechanics and condensed matter physics. *Second quantization* is a formulation or language of many-particle quantum mechanics that helps to minimize the technical complications of practical calculations for many-particle systems.

It is also the appropriate language of other branches of theoretical physics, most notably quantum field theory (see, for example Lancaster and Blundell, 2014).

Two examples illustrate the usefulness of second quantization. In quantum field theory as well as many applications of many-particle physics and condensed matter physics, the number of particles is variable, i.e. particles can be created as well as destroyed.[1] Moreover, the Schrödinger equation, and hence the wave function, for a

[1] As we shall see, the concept of creating and annihilating particles is a rather abstract one, especially for Fermions, i.e. particles for which the Pauli exclusion principle applies. Richard Feynman, in his Nobel prize acceptance speech (Feynman, 1965), alludes to this in a personal reminiscence: 'I remember that when someone had started to teach me about creation and annihilation operators, that this operator creates an electron, I said, "how do you create an electron? It disagrees with the conservation of charge", and in that way, I blocked my mind from learning a very practical scheme of calculation.'

Models of Quantum Matter: A First Course on Integrability and the Bethe Ansatz. Hans-Peter Eckle,
Oxford University Press (2019). © Hans-Peter Eckle. DOI: 10.1093/oso/9780199678839.001.0001

system of $N = 10^{24}$ particles is different from those of a system of $N = 10^{24} - 1$ particles. Yet, we expect both systems to exhibit the same macroscopic physics. In the language of second quantization, which is specifically adapted to accommodate variable numbers of particles, we can cope easily with both situations as detailed work in this chapter and in chapter 4 on equilibrium statistical mechanics shows, as well as throughout many other chapters of this book.

Second quantization is a standard topic of quantum many-particle theory and treatments can be found in many books wholly or partly devoted to this topic. A classic reference devoted to the method of second quantization is Berezin (1966). Some standard references are the corresponding chapters of Abrikosov *et al.* (1975), Fetter and Walecka (2003), Mahan (2000), and Negele and Orland (1998). More recent work includes Nazarov and Danon (2013), Altland and Simons (2010), and Coleman (2015). Of course, all of these works treat many more topics in the theory of quantum many-particle physics.

In particular, applications of the method of second quantization can be found in the cited works and in later chapters of this book, where we shall make ample use of the language of second quantization. In this chapter, however, we concentrate on the formalism, demonstrating its power with only a few elementary examples.

Section 2.1 of this chapter constructs the Hilbert spaces, the tensor product spaces, and the Fock spaces, appropriate for the states of the many-particle system and then section 2.2, selects from these Hilbert spaces the symmetrized many-particle states of the many-particle Hilbert space that describe Bosons, where any number of particles can occupy the same quantum state and the antisymmetrized many-particle states that describe Fermions, where at most one particle can occupy the same quantum state.

This construction of Hilbert spaces is more general: for any quantum system composed of subsystems, a Hilbert space can be constructed in the way we describe. Therefore, we initially keep the discussion more general before we focus again on Hilbert spaces composed of (many) quantum particles.

For the following few sections, we focus on the Bosonic case, before eventually also discussing Fermions. In section 2.3, we introduce creation and annihilation operators for Bosons. These operators are the main objects in which the formalism of second quantization is expressed.

The creation and annihilation operators can be expressed in different orthonormal and complete bases. The transformations of the creation and annihilation operators between different abstract orthonormal and complete bases are derived in section 2.4, while in section 2.5 the creation and annihilation operators in the position basis, then called quantum field operators, are introduced as one of the most important examples.

Section 2.6 is devoted to the introduction of one-particle operators, section 2.7 to two-particle operators in the formalism of second quantization.

An elementary introduction of second quantization starts from the time-dependent single–particle Schrödinger equation of basic quantum mechanics. How this can be achieved is demonstrated for Bosons in section 2.8 and for Fermions in section 2.10.

Section 2.9 finally returns to Fermions, introducing creation and annihilation operators for the Fermionic case.

The penultimate section of this chapter, 2.11 demonstrates explicitly the equivalence of the many-particle wave function and the formalism of second quantization, while the final section 2.12 of this chapter touches on the issue of the correct ordering of creation and annihilation operators, i.e. the normal ordering.

As a prelude to the formalism, we start with an elementary exercise: the quantum treatment of the single particle one-dimensional harmonic oscillator in terms of creation and annihilation operators, sometimes also called ladder operators. In this exercise we are reminded of important notions that help illuminate the more formal treatment of the rest of this chapter. Moreover, the algebraic structure encountered here will reappear many times in this and in later chapters. Hence, it may be quite a good idea to go through this elementary exercise to gain confidence for the more involved later developments.

EXERCISE 2.1 Quantum mechanical harmonic oscillator The Hamiltonian of the one-dimensional harmonic oscillator of a mass m and frequency ω is given in terms of the position operator x and the momentum operator $p = -i\frac{d}{dx}$ satisfying the canonical commutation relation

$$[x, p] = i \tag{2.1}$$

as

$$\mathcal{H} = \frac{1}{2m}p^2 + \frac{m\omega^2}{2}x^2. \tag{2.2}$$

Among the many ways to solve the Schrödinger equation of the harmonic oscillator, a particularly elegant, and fruitful, way deconstructs the Hamiltonian (2.2) into operators, the creation and annihilation operator, respectively

$$a^\dagger = \sqrt{\frac{m\omega}{2}}\left(x - i\frac{1}{m\omega}p\right) \quad \text{and} \quad a = \sqrt{\frac{m\omega}{2}}\left(x + i\frac{1}{m\omega}p\right). \tag{2.3}$$

- Show, using the ladder operators a and $a^\dagger$, that the canonical commutation relation $[x, p] = i$ becomes

$$[a, a^\dagger] = 1 \tag{2.4}$$

and the Hamiltonian (2.2)

$$\mathcal{H} = \omega\left(a^\dagger a + \frac{1}{2}\right). \tag{2.5}$$

- Furthermore, show that, if λ is the eigenvalue corresponding to the normalized eigenstate $|\lambda\rangle$ of the operator $\Lambda = a^\dagger a$, then

$$a|\lambda\rangle = c_\lambda|\lambda - 1\rangle \tag{2.6}$$

$$a^\dagger|\lambda\rangle = d_\lambda|\lambda + 1\rangle. \tag{2.7}$$

- Calculate the coefficients c_λ and d_λ.
- Prove that $\lambda \geq 0$ and that $\lambda = 0$ must be an eigenvalue. What is, hence, the spectrum of eigenvalues of Λ and H, respectively? Show that this implies for the ground state $a|0\rangle = 0$.
- Use the representation (2.3) of the ladder operators as differential operators to solve the Schrödinger equation for the ground state corresponding to $\lambda = 0$, i.e. determine the ground state wave function $\langle x|0\rangle$.
- Determine the wave function of the first excited state by applying the creation operator $a^\dagger$ once to the ground state wave function.

  The results obtained thus far suggest that the operator Λ can be interpreted as an operator counting the number of excitations of the harmonic oscillator. In order to make this even more suggestive let us change the notation and replace Λ by n and $|\lambda\rangle$ by $|n\rangle$. Furthermore, as we shall see in more detail in this chapter, these results suggest an interpretation of the excitations of the harmonic oscillator as *particles* or *quasiparticles*. The state with no particles $|0\rangle$ then corresponds to the vacuum state.
- Show with the help of (2.6) and (2.7) that the properly normalized state of n excitations or n particles is

$$|n\rangle = \frac{1}{\sqrt{n!}}\left(a^\dagger\right)^n|0\rangle. \tag{2.8}$$

- Finally, to appreciate how fruitful this algebraic treatment of the harmonic oscillator is, calculate the expectation values of the first few powers of the position operator in the state $|n\rangle$, let us say x, x^2, x^3, and x^4.
- Hint: Prove first that the number operator $n = a^\dagger a$, and hence the Hamiltonian $\mathcal{H} = \omega\left(a^\dagger a + \frac{1}{2}\right)$, are Hermitian operators and that therefore the corresponding eigenstates, which are non–degenerate (why?), are orthogonal.

....

With this exercise at the back of our minds, we can now start to develop the formalism of second quantization by first constructing a Hilbert space appropriate for a quantum many-particle system.

2.1 Many-particle Hilbert spaces

The formalism of 'second quantization'[2] provides an elegant and economic way to describe a physical system containing a great, possibly indeterminate, number of

[2] The name is a trifle unfortunate and originates from the interpretation of the algebra of ladder operators (see exercise 2.1). These operators and their corresponding quantum excitations can be viewed as discrete 'quantized' units. It must be emphasized, however, that 'second' quantization is a representation of quantum mechanics particularly suitable for problems involving many particles. It is *not* a quantizing an already quantized theory. However, to be able to distinguish representations, we shall in places also have to use the equally unfortunate epithet 'first' quantization.

particles. It describes particles as quanta of a quantum field and is, hence, at the heart of the modern understanding of quantum mechanics and quantum field theory.

We assume that the solution of the quantum problem for one particle is known, i.e. we assume that

- there is a one-particle Hilbert space (quantum state space) $\mathscr{H}_1$, with, especially, the scalar product $\langle\phi|\psi\rangle$ of states $|\psi\rangle, |\phi\rangle \in \mathscr{H}_1$ from this Hilbert space;
- this means in particular, that the one-particle Schrödinger eigenvalue problem has been solved for the single particle Hamiltonian $\mathcal{H}$:

$$\mathcal{H}|\lambda\rangle = \epsilon_\lambda |\lambda\rangle, \tag{2.9}$$

  where $|\lambda\rangle \in \mathscr{H}_1$ is a normalized eigenstate and ϵ_λ the corresponding eigenvalue;
- furthermore, the time evolution of the particle is determined by the unitary operator (ignoring the possibility of an explicitly time-dependent Hamiltonian)

$$U(t) = \mathrm{e}^{-i\mathcal{H}t}; \tag{2.10}$$

- lastly, observables such as position $\mathbf{r}$, momentum $\mathbf{p}$, angular momentum $\mathbf{L}$, etc., of the single particle problem have been determined.

Second quantization is a formalism permitting to construct quantities that correspond to a system composed of an arbitrary, indeterminate number of such quantum particles under the assumption that the statements above for a single particle hold true.

The basis of the formalism consists in constructing Hilbert spaces and states for an arbitrary number of particles from the Hilbert space and states of a single particle.

2.1.1 Composite Hilbert space of two systems A and B

As mentioned in the introduction to this chapter, we begin with a more general point of view. Assume that there are two quantum systems A and B which may but need not be individual quantum particles. For instance, system A could denote a microscopic system, while system B could represent a macroscopic measurement apparatus. Their respective Hilbert spaces are $\mathscr{H}^A$ and $\mathscr{H}^B$. We are interested in the composite quantum system AB.

We can construct a Hilbert space for the composite quantum system in two different ways. Both ways begin by forming a space of all ordered pairs of states taken from the Hilbert spaces $\mathscr{H}^A$ and $\mathscr{H}^B$

$$\mathscr{M} \equiv \mathscr{H}^A \times \mathscr{H}^B = \left\{F|F = (f^A, f^B), f^A \in \mathscr{H}^A, f^B \in \mathscr{H}^B\right\} \tag{2.11}$$

which can be made into a composite Hilbert space by choosing a scalar product in two different ways.

2.1.1.1 *Tensor product Hilbert space*

The first construction is achieved through the introduction of a tensor product Hilbert space $\mathscr{H}^{AB}$ of two Hilbert spaces $\mathscr{H}^A$ and $\mathscr{H}^B$, whose dimensions $\dim \mathscr{H}^A$ and $\dim \mathscr{H}^B$ need not be the same,

$$\mathscr{H}^{AB} = \mathscr{H}^A \otimes \mathscr{H}^B \tag{2.12}$$

with dimension $\dim \mathscr{H}^{AB} = (\dim \mathscr{H}^A)(\dim \mathscr{H}^B)$. With respect to the composite Hilbert space $\mathscr{H}^{AB}$, the Hilbert spaces $\mathscr{H}^A$ and $\mathscr{H}^B$ are called factor spaces.

It should be emphasized that already the Hilbert space of a single particle can be a tensor product space as the example of the spin-orbit Hilbert space of a single particle shows where

$$\mathscr{H}_{\text{spin-orbit}} = \mathscr{H}_{\text{orbit}} \otimes \mathscr{C}^2 \tag{2.13}$$

is the tensor product of the orbital Hilbert space $\mathscr{H}_{\text{orbit}}$ of the particle with its two-dimensional spin state space $\mathscr{C}^2$.

Now the construction of the tensor product Hilbert space $\mathscr{H}^{AB}$ proceeds as follows. For each pair of states $f^A \equiv |\psi^A\rangle \in \mathscr{H}^A$ and $f^B \equiv |\varphi^B\rangle \in \mathscr{H}^B$ there is a (formal) product state F for which different notations are in use

$$F \equiv |\psi^{AB}\rangle \equiv |\psi^A\rangle \otimes |\varphi^B\rangle \equiv |\psi^A\rangle|\varphi^B\rangle \equiv |\psi^A, \varphi^B\rangle \equiv |\psi, \varphi\rangle. \tag{2.14}$$

These notations, going from left to right, emphasize less and less that the states belong to, in general, different Hilbert spaces. Hence, their use requires more and more caution and a clear understanding of their meaning in particular situations.

The composite states are linear in each of their factors separately, i.e.

$$|\psi^A\rangle \otimes |\left(\lambda|\varphi_1^B\rangle + \mu|\varphi_2^B\rangle\right) = \lambda|\psi^A\rangle \otimes |\varphi_1^B\rangle + \mu|\psi^A\rangle \otimes |\varphi_2^B\rangle, \tag{2.15}$$

$$\left(\lambda|\psi_1^A\rangle + \mu|\psi_2\rangle\right) \otimes |\varphi^B\rangle = \lambda|\psi_1^A\rangle \otimes |\varphi^B\rangle + \mu|\psi_2^A\rangle \otimes |\varphi^B\rangle \tag{2.16}$$

with complex numbers λ and μ.

The scalar product between composite states is formed in a space-wise manner by

$$\langle\psi^A|\langle\varphi^B| \cdot |\xi^A\rangle|\zeta^B\rangle = \langle\psi^A|\xi^A\rangle\langle\varphi^B|\zeta^B\rangle. \tag{2.17}$$

In order to obtain a composite Hilbert space $\mathscr{H}^{AB}$ large enough to contain states that cannot be written as pure product states, i.e. states of the form

$$|\psi_1^A\rangle \otimes |\phi_1^B\rangle + |\psi_2^A\rangle \otimes |\phi_2^B\rangle \tag{2.18}$$

we need to consider the linear span (also called linear hull) of the space $\mathscr{M}$. This is achieved by taking bases of the factor Hilbert spaces $\mathscr{H}^A$ and $\mathscr{H}^B$, e.g. $\{|n^A\rangle\}$ and $\{|m^B\rangle\}$, to form a basis of the composite Hilbert space $\mathscr{H}^{AB}$

$$\{|n^A\rangle \otimes |m^B\rangle\}, \tag{2.19}$$

a so-called (tensor) product basis, in the sense that all states of the form

$$|\psi^{AB}\rangle = \sum_{n,m} c_{n,m}|n^A\rangle \otimes |m^B\rangle \tag{2.20}$$

with complex numbers $c_{n,m}$ as expansion coefficients, i.e. the linear span of $\mathscr{M}$, define the composite Hilbert space $\mathscr{H}^{AB}$.

Since an orthonormal basis in $\mathscr{H}^{AB}$ satisfies, using (2.17),

$$\langle n^A, m^B|p^A, q^B\rangle = \langle n^A|p^A\rangle\langle m^B|q^B\rangle = \delta_{np}\delta_{mq}, \tag{2.21}$$

that is, the basis of the composite Hilbert space is orthonormal if the bases of the factor states are orthonormal, we obtain for the scalar product of two states of the form (2.20)

$$\langle \psi_1^{AB}|\psi_2^{AB}\rangle = \sum_{n,m} c_{nm}^{(1)*} c_{nm}^{(2)}, \tag{2.22}$$

which completes the construction of the composite Hilbert space $\mathscr{H}^{AB}$.

Using the notion of a product Hilbert space $\mathscr{H}^{AB}$, we are now in a position to introduce in a formal way an important notion at the heart of many investigations in quantum physics. It is the notion of *quantum entanglement*, which goes back to important publications by Einstein, Podolsky, and Rosen 1935, and especially by Schrödinger (1935). A composite state in $\mathscr{H}^{AB}$ is called entangled if it cannot be represented as a product state but only as a superposition of product states, as in (2.20).

Examples of composite states that cannot be written as product states are the so-called *Bell states*

$$|\Phi_\pm^{AB}\rangle = \frac{1}{\sqrt{2}}\left(|0^A, 0^B\rangle \pm |1^A, 1^B\rangle\right) \tag{2.23}$$

$$|\Psi_\pm^{AB}\rangle = \frac{1}{\sqrt{2}}\left(|0^A, 1^B\rangle \pm |1^A, 0^B\rangle\right), \tag{2.24}$$

which are states in the composite Hilbert space

$$\mathscr{H}^{AB} = \mathscr{H}_2^A \otimes \mathscr{H}_2^B \tag{2.25}$$

where the factor states describe two-level systems or so-called qubits. These states are maximally entangled states. Details about maximally entangled states and entanglement measures in general can be found in Audretsch (2007).

2.1.1.2 Composite Hilbert space as a direct sum

However, a different way is also possible to construct a composite Hilbert space from the Hilbert spaces $\mathscr{H}^A$ and $\mathscr{H}^B$ of the two quantum systems A and B. The starting point is again the space $\mathscr{M}$ of ordered pairs of states given in (2.11).

However, now we define the scalar product of two states $F = (f^A, f^B) \equiv |\psi^{AB}\rangle \equiv (|\psi^A\rangle, |\psi^B\rangle)$ and $G = (g^A, g^B) \equiv |\varphi^{AB}\rangle \equiv (|\varphi^A\rangle, |\varphi^B\rangle)$ from $\mathscr{M}$ as

$$F \cdot G = f^A \cdot g^A + f^B \cdot g^B \equiv \langle\psi^{AB}|\varphi^{AB}\rangle \equiv \langle\psi^A|\varphi^A\rangle + \langle\psi^B|\varphi^B\rangle. \tag{2.26}$$

This definition of a scalar product for the space $\mathscr{M}$ makes it into a composite Hilbert space $\tilde{\mathscr{H}}^{AB}$ different from the tensor product Hilbert space $\mathscr{H}^{AB}$ we constructed in the previous section. This composite Hilbert space is called a direct sum space and is denoted

$$\tilde{\mathscr{H}}^{AB} = \mathscr{H}^A \oplus \mathscr{H}^B. \tag{2.27}$$

In order to avoid possible confusion, we mention that multiplication by a complex number λ is to be understood component-wise

$$\lambda F = \left(\lambda f^A, \lambda f^B\right) = \lambda|\psi^{AB}\rangle = \left(\lambda|\psi^A\rangle, \lambda|\psi^B\rangle\right), \tag{2.28}$$

in contrast to the tensor product Hilbert space $\mathscr{H}^{AB}$ where

$$\lambda|\psi^{AB}\rangle = \left(\lambda|\psi^A\rangle\right)|\psi^B\rangle = |\psi^A\rangle\left(\lambda|\psi^B\rangle\right) = \lambda|\psi^A\rangle|\psi^B\rangle. \tag{2.29}$$

We encounter this important kind of space later under the name of Fock space when we discuss situations involving particle production and destruction and, hence, need a Hilbert space with a variable number of particles.

There we shall also need the notion of orthogonality for states from different Hilbert spaces that can be conveniently defined using a direct sum of these spaces. We extend the vectors $f^A \in \mathscr{H}^A$ and $g^B \in \mathscr{H}^B$ to the space $\mathscr{H} = \mathscr{H}^A \oplus \mathscr{H}^B$ by defining

$$F_0 \equiv \left(f^A, 0\right), \tag{2.30}$$

$$G_0 \equiv \left(0, g^B\right). \tag{2.31}$$

Calculating the scalar product (2.26) of these two special vectors of $\mathscr{H}$, we discover the important result

$$F_0 \cdot G_0 = f^A \cdot 0 + 0 \cdot g^B = 0, \tag{2.32}$$

namely, that states from different subspaces of a direct sum Hilbert space are orthogonal to each other.

2.1.2 Case of many distinguishable particles

The construction of a composite Hilbert space consisting of the Hilbert spaces of two subsystems, discussed in the previous section, can easily be generalized to the case of many subsystems. From now on we focus on the case of many subsystems where each subsystem is a single quantum particle. In constructing the many-particle Hilbert space, let us assume for the time being that the particles are *distinguishable*. Of course, in quantum physics, we encounter many situations where particles are indeed distinguishable, e.g. the electron and proton forming a hydrogen atom are distinguishable quantum particles.

The Hilbert space of a system composed of *exactly* N particles is given by the tensor product Hilbert space of N copies of the single-particle Hilbert space $\mathscr{H}_1$ with dimension $\dim \mathscr{H}_1$:

$$\mathscr{H}_1^{\otimes N} \equiv \prod_{i=1}^{N} \otimes \mathscr{H}_1 \equiv \underbrace{\mathscr{H}_1 \otimes \mathscr{H}_1 \otimes \cdots \otimes \mathscr{H}_1}_{N} \tag{2.33}$$

The states of this Hilbert space are formed from linear combinations of product states of the form (cp. (2.14))

$$|\varphi_1, \varphi_2, \ldots, \varphi_N\rangle = |\varphi_1\rangle|\varphi_2\rangle \ldots |\varphi_N\rangle, \qquad |\varphi_i\rangle \in \mathscr{H}_1. \tag{2.34}$$

Such a state vector represents a state of the system where the first particle is in state $|\varphi_1\rangle$, the second in the state $|\varphi_2\rangle, \ldots$, and the Nth in the state $|\varphi_N\rangle$. Crucially, we note that the states of different particles may coincide, i.e. that the possibility

$$|\varphi_i\rangle = |\varphi_j\rangle \tag{2.35}$$

for particles $i \neq j$ must be taken into account.

The scalar product in $\mathscr{H}_1^{\otimes N}$ is defined by generalizing (2.17)

$$\langle \varphi_1, \varphi_2, \ldots, \varphi_N | \psi_1, \psi_2, \ldots, \psi_N\rangle = \langle \varphi_1|\psi_1\rangle\langle\varphi_2|\psi_2\rangle \cdots, \langle\varphi_N|\psi_N\rangle. \tag{2.36}$$

In order to be able to consider a state with zero particles, we formally introduce a one-dimensional Hilbert space $\mathscr{V}$, the *vacuum* space,

$$\mathscr{V} \equiv \mathscr{H}_1^{\otimes 0} \tag{2.37}$$

which contains only one normalized basis state, the zero particle or vacuum state $|0\rangle$, i.e. $\langle 0|0\rangle = 1$.

To each basis of $\mathscr{H}_1$, there corresponds a basis of $\mathscr{H}_1^{\otimes N}$. Explicitly this statement means that, if a set of orthonormal vectors $|k\rangle$ ($k = 1, \ldots, \dim \mathscr{H}_1 \equiv M$) spans the one-particle Hilbert space $\mathscr{H}_1$, then the set of vectors

$$|k_1, k_2, \ldots, k_M\rangle = |k_1\rangle|k_2\rangle \ldots |k_M\rangle \tag{2.38}$$

forms an orthonormal basis of $\mathscr{H}_1^{\otimes n}$, i.e. with complex numbers c_k, we have for each single-particle state $|\varphi\rangle$

$$|\varphi\rangle = \sum_{k=1}^{M} c_k |k\rangle. \tag{2.39}$$

The number of single-particle basis states $|k\rangle$, i.e. $\dim \mathscr{H}_1 = M$, depends on the concrete single-particle problem at hand and may be finite or infinite, or there can be a continuous dependence, e.g. for a free particle described by a plane wave of continuous momentum p. In the latter case, the sum in (2.39) will be replaced by an integral over the appropriate range of the continuous variable k.

If the number of particles of the system is indeterminate, the states of the system are obtained by superposition of states that correspond to each possible value of N. The state space is the direct sum of all state spaces $\mathscr{H}_1^{\otimes N}$ for $N = 0, 1, 2, \ldots$:

$$\mathscr{H} = \mathscr{H}_1^{\otimes 0} \oplus \mathscr{H}_1^{\otimes 1} \oplus \mathscr{H}_1^{\otimes 2} \oplus \ldots \equiv \sum_{N=0}^{\infty} \mathscr{H}_1^{\otimes N}. \tag{2.40}$$

The state space $\mathscr{H}$ is defined by its subspaces $\mathscr{H}_1^{\otimes N}$. More precisely, if $|\varphi^{(N)}\rangle$ denotes an arbitrary state vector of $\mathscr{H}_1^{\otimes N}$, then a state in $\mathscr{H}$ is defined as an infinite series of states with particle numbers $N = 0, 1, 2, \ldots$ which we can formally write as a sum (cp. section 2.1.1.2)

$$\left(|\varphi^{(0)}\rangle, |\varphi^{(1)}\rangle, \ldots, |\varphi^{(N)}\rangle, \ldots\right) \equiv |\varphi^{(0)}\rangle + |\varphi^{(1)}\rangle + \ldots + |\varphi^{(N)}\rangle + \ldots \tag{2.41}$$

and the scalar product of this state with another state $|\psi^{(0)}\rangle + |\psi^{(1)}\rangle + \ldots + |\psi^{(N)}\rangle + \ldots$ of $\mathscr{H}$ is given by

$$\sum_{N=0}^{\infty} \langle \varphi^{(N)} | \psi^{(N)} \rangle \tag{2.42}$$

where the scalar products $\langle \varphi^{(N)} | \psi^{(N)} \rangle$ are calculated according to equation (2.36). The vacuum state for $N = 0$ is here denoted by $|\varphi^{(0)}\rangle = \alpha|0\rangle$, for some complex number α.

Using the notation of (2.38), the set of states of the form

$$|0\rangle, |k\rangle, |k_1, k_2\rangle, \ldots, |k_1, k_2, \ldots, k_N\rangle, \ldots \tag{2.43}$$

is a complete orthonormal basis of $\mathscr{H}$, i.e.

$$\begin{aligned} \langle 0|0\rangle &= 1, \\ \langle k|k'\rangle &= \delta_{kk'}, \\ \langle k_1, k_2|k_1', k_2'\rangle &= \delta_{k_1 k_1'} \delta_{k_2 k_2'}, \\ &\ldots \end{aligned} \tag{2.44}$$

and

$$\begin{aligned} |0\rangle\langle 0| &= \mathbb{I}, \\ \sum_k |k\rangle\langle k| &= \mathbb{I}, \\ \sum_{k_1,k_2} |k_1, k_2\rangle\langle k_1, k_2| &= \mathbb{I}, \\ &\ldots . \end{aligned} \tag{2.45}$$

2.1.3 Case of many indistinguishable particles

At the microscopic level, quantum particles of the same kind, e.g. two electrons or two Helium atoms, are indistinguishable, except for possible inner degrees of freedom of the particles, e.g. spin.

For a system of N indistinguishable particles, also called identical particles, the Hamiltonian $\mathcal{H}$ depends on the dynamical and internal degrees of freedom of all particles. However, indistinguishability implies that the Hamiltonian cannot change if the degrees of freedom of two arbitrary particles i and j are exchanged

$$\mathcal{H} = \mathcal{H}(1, 2, \ldots, i, \ldots, j, \ldots, N) = \mathcal{H}(1, 2, \ldots, j, \ldots, i, \ldots, N), \tag{2.46}$$

where the labels $1, 2, \ldots, i, \ldots, j, \ldots, N$ represent the degrees of freedom of the particles, e.g. their momenta, positions, and internal degrees of freedom, e.g. spin.

The corresponding many-particle state

$$|\varphi\rangle = |\varphi_1, \varphi_2, \ldots, \varphi_N\rangle = |\varphi_1\rangle|\varphi_2\rangle \ldots |\varphi_N\rangle \tag{2.47}$$

of this assembly of N indistinguishable quantum particles is, under exchange of two arbitrary particles, either totally symmetric or totally antisymmetric. This can be seen by acting with the operator exchanging two arbitrary particles, i and j, on a state of N identical particles

$$\mathcal{T}_{ij}|\varphi_1, \varphi_2, \ldots, \varphi_i, \ldots, \varphi_j, \ldots, \varphi_N\rangle = |\varphi_1, \varphi_2, \ldots, \varphi_j, \ldots, \varphi_i, \ldots, \varphi_N\rangle \tag{2.48}$$

$$= \eta_{ij}|\varphi_1, \varphi_2, \ldots, \varphi_i, \ldots, \varphi_j, \ldots, \varphi_N\rangle, \tag{2.49}$$

which produces a state that may not be physically different from the original state, so that there can only be a phase factor

$$\eta_{ij} = e^{i\alpha_{ij}}. \tag{2.50}$$

Moreover, because of

$$\mathcal{T}_{ji} = \mathcal{T}_{ij} \tag{2.51}$$

we have

$$\mathcal{T}_{ij}^2 = 1 \tag{2.52}$$

and, hence

$$\eta_{ij} = \pm 1. \tag{2.53}$$

Lastly, the phase factor must be the same for all pairs (i, j) of particles because the exchange operators, and thus the phases, satisfy

$$\mathcal{T}_{ij}\mathcal{T}_{jk} = \mathcal{T}_{ki}\mathcal{T}_{ij} \qquad \text{and} \qquad \eta_{ij}\eta_{jk} = \eta_{ki}\eta_{ij}, \tag{2.54}$$

hence

$$\eta_{jk} = \eta_{ki} = \eta = \pm 1 \tag{2.55}$$

the phase factors are the same for all pairs of particles because the choice of particles i, j, and k has been arbitrary.

The validity of (2.54) is easiest seen graphically (Figure 2.1).

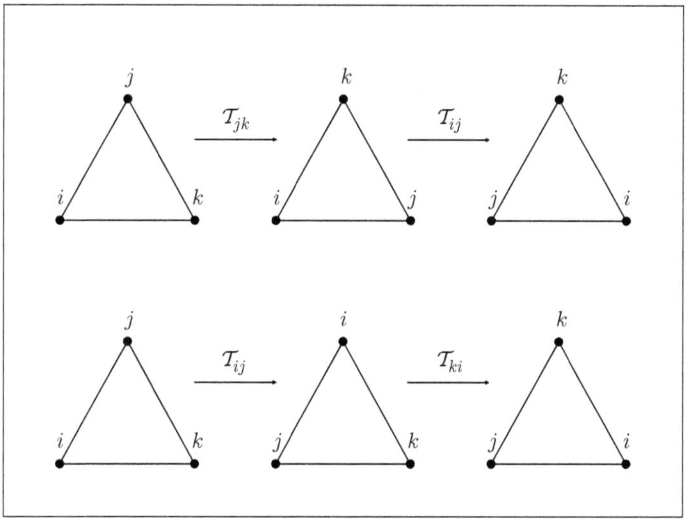

Figure 2.1 *Action of two pairs of particle exchange operators leading to the same final state.*

These results imply that there are only two types of quantum particles:

- $\eta = +1$ particles whose states are symmetrical under particle exchange,
- $\eta = -1$ particles whose states are antisymmetrical under particle exchange.

2.1.3.1 Pauli exclusion principle

As an immediate consequence of the restriction of the eigenvalues of the transposition operator to only two allowed values $\eta = \pm 1$, (2.55), we obtain an important result that is connected to Wolfgang Pauli's name: the exclusion principle in its form for many-particle systems.

While the case $\eta = +1$ imposes no restriction on the many-particle state

$$|\varphi\rangle = |\varphi_1, \varphi_2, \ldots, \varphi_i, \ldots, \varphi_j, \ldots, \varphi_N\rangle = |\varphi_1\rangle|\varphi_2\rangle \ldots |\varphi_i\rangle \ldots, |\varphi_j\rangle \ldots, |\varphi_N\rangle \tag{2.56}$$

if two single-particle states in it are equal, e.g. $|\varphi_i\rangle = |\varphi_j\rangle$, the case $\eta = -1$ implies that the many-particle state $|\varphi\rangle$ vanishes if two single-particle states in it are equal.

Many-particle states with $\eta = +1$ under exchange of single-particle states describe systems of Bose particles, or Bosons. Many-particle states with $\eta = -1$ under exchange of single-particle states describe systems of Fermi particles, or Fermions. These two possibilities are the only two possibilities for quantum particles under ordinary circumstances (with an exception). The two kinds of particles show clearly distinct properties.

Thus we can state the famous Pauli exclusion principle:

- The probability for a many-particle state to contain two or more Fermions in the same single-particle state vanishes.

- In other words, it is impossible that two Fermions in a many-Fermion state coincide in all quantum variables, e.g. in position and spin quantum number.
- There can be at most one Fermion in each single-particle state of a many-Fermion state.

Such a many-particle state is, however, quite possible for Bosons, as we have seen. A many-particle Bosonic state can contain an arbitrary number of Bosons, described by equal single-particle states.

A few further remarks are in order at this stage.

- In another important contribution, Wolfgang Pauli (1940) proved by his famous spin-statistics theorem that Bosons carry integer spin, $s = 0, 1, 2 \ldots$, while Fermions carry half-integer spin, $s = 1/2, 3/2, \ldots$.
- The following sections discuss the drastic consequences of this distinction. Furthermore, the two types of particle exhibit very different statistical behaviour as we discuss in chapter 4. This fact is also the reason for the name of the famous theorem Pauli proved.
- Note that the spin-statistics theorem depends on the spatial dimension. It presupposes three (or more) space dimensions. In lower dimensions, the behaviour of particles under exchange may offer much richer possibilities. This has been demonstrated by Leinaas and Myrheim (1977) and has been a major research area of theoretical physics of low-dimensional systems ever since. The postulated particles of fractional statistics, called anyons, play a major role in the physics of the quantum Hall effect. For recent reviews of anyons in relation to the quantum Hall effect, see Stern (2008) and Stern (2010).

EXERCISE 2.2 Consequence of the indistinguishability of particles For this exercise we first need to show that any permutation $\mathcal{P}$ of the permutation group S_N is a unitary operator, i.e.

$$\mathcal{P}^{\dagger} = \mathcal{P}^{-1}. \tag{2.57}$$

Reminder: Permutations are defined as arrangements of the numbers $(1, 2, \ldots, N)$ in a particular sequence and written as $\mathcal{P} = (p_1, p_2, \ldots, p_N)$. The number of arrangements of N numbers $(1, 2, \ldots, N)$ is $a(N) = N!$. The permutations of N numbers form a group S_N with $\mathcal{I} = (1, 2, \ldots, N)$ as identity element. A pair of numbers p_i and p_j in the permutation $\mathcal{P}$ for which $p_i > p_j$ form an inversion. A permutation with an even number of inversions is called even, a permutation with an odd number of inversions is called odd. Each permutation is thus characterized by

$$\chi(\mathcal{P}) = e^{\mathcal{J}(\mathcal{P})} = \pm 1 \tag{2.58}$$

where $\mathcal{J}(\mathcal{P})$ is the number of inversions and $\chi(\mathcal{P}) = +1$ for an even, $\chi(\mathcal{P}) = -1$ for an odd permutation.

The indistinguishability of quantum particles means that there can be *no* observable $\mathcal{O}$ of the many-particle system that can be used to distinguish the particles.

Express this requirement as a relation between the observable $\mathcal{O}$ and an arbitrary permutation $\mathcal{P} \in S_N$. You may want to start from the expressions for the expectation values of the observable $\mathcal{O}$ in an arbitrary many-particle state $|\psi\rangle$ and the permuted state $\mathcal{P}|\psi\rangle$.

2.2 Occupation number representation: Bosons and Fermions

The states in $\mathscr{H}_1^{\otimes N}$ and $\mathscr{H} = \sum_N \mathscr{H}_1^{\otimes N}$ behave in complicated ways under the exchange of pairs of particles. To describe this behaviour would require the full representation theory of the permutation group S_N. However, as we have seen, for indistinguishable particles, we only need to consider many-particle states that are symmetric or antisymmetric under particle exchange, which corresponds to two simple one-dimensional representations of the permutation group. Only those states which remain either completely unchanged or only change by an overall sign, corresponding to $\eta = \pm 1$, are physically relevant.

In order to project out the physically relevant basis states, we introduce symmetrization $\mathcal{S}$ and antisymmetrization $\mathcal{A}$ operators for a fixed number N of particles. The symmetrization operator $\mathcal{S}$ produces states which automatically satisfy $\eta = 1$ by permuting the basis states to form a *permanent*

$$\mathcal{S}|k_1, k_2, \ldots, k_N\rangle = \frac{1}{\sqrt{N!N_1! \cdots N_M!}} \sum_{\mathcal{P}} |k_{i_1}, k_{i_2}, \ldots, k_{i_N}\rangle \tag{2.59}$$

where the sum is taken over the $N!$ permutations of the basis states $|k_{i_1}\rangle, \ldots, |k_{i_N}\rangle$. The factor $1/\sqrt{N!}$ normalizes the sum consisting of $N!$ terms.

As discussed, not all single-particle basis states in the many-particle basis state must be different (cf. the discussion with respect to 2.35). In order not to overcount basis states for which this happens, we introduce the factors $N_l!$ for $l = 1, \ldots, M$ with $M \leq N$, the numbers N_l counting the basis states in $|k_1, k_2, \ldots, k_N\rangle$ which are equal. $M = N$ corresponds to all basis states being different, i.e. all $N_l = 1$. The numbers N_l, which here appear for normalization purposes, will obtain a physical interpretation as occupation numbers of the corresponding states and, thus, play an important role in the following developments.

Similarly, we introduce an antisymmetrization operator $\mathcal{A}$

$$\mathcal{A}|k_1, k_2, \ldots, k_N\rangle = \frac{1}{\sqrt{N!}} \sum_{\mathcal{P}} (-1)^{\mathcal{P}} |k_{i_1}, k_{i_2}, \ldots, k_{i_N}\rangle, \tag{2.60}$$

where the sum is over all $N!$ permutations $\mathcal{P}$ of 1, 2, ..., N. As opposed to the symmetrized state, (2.59), we can disregard multiple single-particle states because their contribution to (2.60) vanishes by construction as is required from an antisymmetrical many-particle state.

The real space wave function $\langle \mathbf{r}_1, \mathbf{r}_2, \ldots, \mathbf{r}_N | \mathcal{A} | k_1, k_2, \ldots, k_N \rangle$ then becomes a determinant, called the Slater determinant,

$$\langle \mathbf{r}_1, \mathbf{r}_2, \ldots, \mathbf{r}_N | \mathcal{A} | k_1, k_2, \ldots, k_N \rangle = \frac{1}{\sqrt{N!}} \begin{vmatrix} \varphi_1(\mathbf{r}_1) & \varphi_2(\mathbf{r}_1) & \cdots & \varphi_N(\mathbf{r}_1) \\ \varphi_1(\mathbf{r}_2) & \varphi_2(\mathbf{r}_2) & \cdots & \varphi_N(\mathbf{r}_2) \\ \vdots & \vdots & \ddots & \vdots \\ \varphi_1(\mathbf{r}_N) & \varphi_2(\mathbf{r}_N) & \cdots & \varphi_N(\mathbf{r}_N) \end{vmatrix} \tag{2.61}$$

where $\varphi_i(\mathbf{r}_j) = \langle \mathbf{r}_j | k_i \rangle$.

2.2.1 Fock spaces for Bosons and Fermions

We now come back to our goal to deal with systems where the number of particles is indeterminate, potentially even infinite. In order to achieve this, we recall that the notion of a composite Hilbert space as a direct sum introduced in section 2.1.1.2 allowed us to construct a composite Hilbert space $\mathscr{H}$ for an indeterminate number of particles as a direct sum of composite tensor product Hilbert spaces $\mathscr{H}_1^{\otimes N}$ each for a fixed number $N = 0, 1, 2, \ldots$ of particles (cf. 2.40)

$$\mathscr{H} = \mathscr{H}_1^{\otimes 0} \oplus \mathscr{H}_1^{\otimes 1} \oplus \mathscr{H}_1^{\otimes 2} \oplus \ldots \equiv \sum_{N=0}^{\infty} \mathscr{H}_1^{\otimes N}. \tag{2.62}$$

This Hilbert space can be symmetrized or antisymmetrized by performing the corresponding operation on all many-particle basis states for a fixed number N of particles, as described in section 2.1.2. The Hilbert spaces $\mathcal{S}\mathscr{H}$ and $\mathcal{A}\mathscr{H}$ constructed in this way are subspaces of $\mathscr{H}$, called Fock spaces, describing Bosonic and Fermionic many-particle spaces where the numbers of Bosons and Fermions are indeterminate.

The many-particle states (2.59) and (2.60) for $N = 0, 1, 2, \ldots$ form bases of the respective Fock spaces, which can be characterized by the sequence of occupation numbers N_l and written as

$$|\phi\rangle = (\mathcal{S}/\mathcal{A})|k_1, k_2, \ldots, k_N\rangle \equiv |N_1, N_2, \ldots\rangle = |\{N\}\rangle \tag{2.63}$$

with $N_l = 0, 1, 2, \ldots$ for Bosons ($\mathcal{S}$), and $N_l = 0, 1$ for Fermions ($\mathcal{A}$). The occupation numbers N_l completely characterize the state $|\phi\rangle$. Viewed as a state in the Fock space, the state $|\phi\rangle$ has an infinite number of occupation number levels labelled by $l = 1, 2, \ldots \infty$. Nevertheless, each state $|\phi\rangle$ contains a fixed number of particles

$$N = \sum_{l=1}^{\infty} N_l. \tag{2.64}$$

However, an important new feature the construction of the Fock spaces generates is that the total number of particles in these spaces is not fixed any more as it was in each tensor product many-particle Hilbert space. It is now rather a dynamic variable whose corresponding operator we obtain in the following section.

The states $|\phi\rangle$ form a complete orthonormal set of many-particle states

$$\langle N_1, N_2, \ldots | N_1', N_2', \ldots \rangle = \delta_{N_1 N_1'} \delta_{N_2 N_2'} \cdots \tag{2.65}$$

and

$$\sum_{N_1, N_2, \ldots} |N_1, N_2, \ldots\rangle\langle N_1', N_2', \ldots| = \mathbb{I}. \tag{2.66}$$

EXERCISE 2.3 Properties of symmetrization and antisymmetrization operators Show that the symmetrization and antisymmetrization operators have the following properties:

- $\mathcal{S}$ and $\mathcal{A}$ are Hermitian operators.
- $\mathcal{S}$ and $\mathcal{A}$ are idempotent, i.e. $\mathcal{S}^2 = \mathcal{S}$ and $\mathcal{A}^2 = \mathcal{A}$.
- $\mathcal{S}$ and $\mathcal{A}$ are orthogonal, i.e. $\mathcal{S}\mathcal{A} = \mathcal{A}\mathcal{S} = 0$.

In the following, we treat the cases of Bosons and Fermions separately, beginning in the next section with the Bose case.

2.3 Creation and annihilation operators for Bosons

This section concentrates on the Bosonic case while Fermions are discussed in section 2.9.

On the Fock space $\mathcal{S}\mathcal{H}$, we define an operator $a_i^\dagger$ acting on occupation level i of the state $|\phi\rangle$ as

$$a_i^\dagger |\phi\rangle = \sqrt{N_i + 1}|\phi'\rangle = \sqrt{N_i + 1}|N_1, \ldots, N_i + 1, \ldots\rangle. \tag{2.67}$$

This operator, called a creation operator, creates a particle in the occupation level i of the many-particle state $|\phi\rangle$. It is important to note that, while $|\phi\rangle$ is a many-particle state with N particles, the state $|\phi'\rangle$ is a many-particle state with $N + 1$ particles. However, both states are states in the direct sum Hilbert or Fock space $\mathcal{S}\mathcal{H}$. The Fock space $\mathcal{S}\mathcal{H}$

has been constructed expressly for the purpose of accommodating the action of creation operators $a_i^\dagger$ for $i = 1, 2, \ldots$.

Similarly, we define an operator a_i

$$a_i|\phi\rangle = \sqrt{N_i}|\phi''\rangle = \sqrt{N_i}|N_1, \ldots, N_i - 1, \ldots, N_l\rangle \tag{2.68}$$

which annihilates a particle in the occupation level i of the many-particle state $|\phi\rangle$, and hence is called an annihilation operator. From these definitions it is easy to demonstrate that $a_i^\dagger$ and a_i are, as the notation suggests, mutually adjoint operators. Hence, the two definitions are not independent of each other and one would have sufficed since it implies the other.

The creation and annihilation operators $a_i{}^\dagger$ and a_i permit a convenient representation of the many-particle state $|\phi\rangle$. By repeated application of the annihilation operator a_i for $i = 1, 2, \ldots$ all particles in a given state $|\phi\rangle$ can be eliminated one by one and we arrive at an empty, no–particle, or vacuum, state

$$|vac\rangle = |0, 0, 0, 0, \ldots\rangle \in \mathcal{SH}. \tag{2.69}$$

This state has formally to be distinguished from the empty state $|0\rangle \in \mathcal{H}^{\otimes 0}$. One more application of an arbitrary annihilation operator a_i

$$a_i|vac\rangle = a_i|0, 0, 0, 0, \ldots\rangle = 0 \tag{2.70}$$

annihilates the vacuum state altogether.

On the other hand, applying creation operators $a_i{}^\dagger$ for $i = 1, 2, \ldots$ on the vacuum state $|vac\rangle$ we can built up any many-particle state $|\phi\rangle$ one particle at a time. For example, a many–particle basis state with just one particle has the form

$$|1_\alpha\rangle = a_\alpha^\dagger|vac\rangle = |0, 0, \ldots, 0, N_\alpha = 1, 0, \ldots\rangle. \tag{2.71}$$

We shall call such a many-particle state a one-particle state to contrast it with the single-particle states $|k\rangle$. Later, we will shall also need two-particle states

$$|1_\alpha 1_\beta\rangle = a_\alpha^\dagger a_\beta^\dagger|vac\rangle = |0, 0, \ldots, 0, N_\alpha = 1, 0, \ldots, 0, N_\beta = 1, 0, \ldots\rangle, \tag{2.72}$$

$$|2_\alpha\rangle = \frac{1}{\sqrt{2!}}\left(a_\alpha^\dagger\right)^2 |vac\rangle = |0, 0, \ldots, 0, N_\alpha = 2, 0, \ldots\rangle. \tag{2.73}$$

As shown in previous sections, Bosonic, as well as Fermionic, many-particle states are characterized by their behaviour under the exchange of pairs of particles. Building up a general many-particle basis state by repeated application of creation operators will have to be consistent with this symmetry under pairwise exchange of particles. This is guaranteed for Bosons by the commutation relations

$$[a_i, a_j] = [a_i^\dagger, a_j^\dagger] = 0, \qquad [a_i, a_j^\dagger] = \delta_{ij} \tag{2.74}$$

which can be verified by evaluating the action of the commutators on an arbitrary many-particle state $|\phi\rangle$.

The general many-particle basis state in the Fock space $S\mathcal{H}$ can now be written with the help of the creation operators $a_i^\dagger$ for $i = 1, 2, \ldots$

$$|\phi\rangle = |N_1, N_2, \ldots, N_\alpha, \ldots\rangle \tag{2.75}$$

$$= \frac{1}{\sqrt{N_1!N_2!\ldots N_\alpha!\ldots}} \left(a_1^\dagger\right)^{N_1} \left(a_2^\dagger\right)^{N_2} \ldots \left(a_\alpha^\dagger\right)^{N_\alpha} \ldots |vac\rangle \tag{2.76}$$

or, due to the commutation relations, in any other permuted order of the creation operators acting on the vacuum state.

From the definitions (2.67) and (2.68) we also obtain

$$a_i^\dagger a_i |\phi\rangle = N_i |\phi\rangle \tag{2.77}$$

and

$$a_i a_i^\dagger |\phi\rangle = (N_i + 1)|\phi\rangle. \tag{2.78}$$

This justifies the interpretation of the operator

$$n_i = a_i^\dagger a_i \tag{2.79}$$

as the operator that counts the number of particles in the occupation level i of the many–particle state $|\phi\rangle$, the particle number operator. The total particle number operator is thus

$$\mathcal{N} = \sum_{i=1}^{\infty} n_i = \sum_{i=1}^{\infty} a_i^\dagger a_i. \tag{2.80}$$

EXERCISE 2.4 Bogoliubov transformation for Bosons A canonical transformation of the form

$$b = ua + va^\dagger, \tag{2.81}$$

$$b^\dagger = ua^\dagger + va, \tag{2.82}$$

where u and v are real, is called a Bogoliubov transformation (after the mathematician and theoretical physicist Nikolai Bogoliubov).

Derive the condition under which the Bose commutation relations are preserved.

The following exercises furnish us with useful formulas for calculations involving Bosonic creation and annihilation operators.

However, let us start with a useful property of operator-valued functions, which are defined by a power series

$$f(\mathcal{O}) = \sum_{n=0}^{\infty} a(n)\mathcal{O}^n \tag{2.83}$$

with complex numbers $a(n)$.

EXERCISE 2.5 Operator relation Show that for operators $\mathcal{O}$ and $\mathcal{P}$ and an operator-valued function $f(\mathcal{O})$, defined by the power series (2.83), the equation

$$\mathcal{P}^{-1} f(\mathcal{O})\mathcal{P} = f(\mathcal{P}^{-1}\mathcal{O}\mathcal{P}) \tag{2.84}$$

holds.

This equation is particularly useful for the exponential function

$$f(\mathcal{O}) = e^{\mathcal{O}}, \tag{2.85}$$

for which it becomes

$$\mathcal{P}^{-1} e^{\mathcal{O}} \mathcal{P} = e^{(\mathcal{P}^{-1}\mathcal{O}\mathcal{P})}. \tag{2.86}$$

The result of this exercise will prove useful for exercise 2.7.

Here are the first set of exercises that expound important relations for Bosonic creation and annihilation operators.

EXERCISE 2.6 Functions of Bosonic creation and annihilation operators A function $f(b, b^\dagger)$ of the Bosonic operators b and $b^\dagger$ is defined by its power series with respect to the two arguments, i.e.

$$f(b, b^\dagger) = \sum_{n,m=0}^{\infty} a(n, m) b^n (b^\dagger)^m \tag{2.87}$$

with complex numbers $a(n, m)$.

- Show that the following generalized commutator relations

$$bf(b, b^\dagger) - f(b, b^\dagger)b = \frac{\partial f(b, b^\dagger)}{\partial b^\dagger}, \tag{2.88}$$

$$b^\dagger f(b,b^\dagger) - f(b,b^\dagger)b^\dagger = -\frac{\partial f(b,b^\dagger)}{\partial b}. \tag{2.89}$$

are satisfied.

- Starting from the elementary commutation relation

$$bb^\dagger - b^\dagger b = 1 \tag{2.90}$$

prove first, by complete induction, the special case of, e.g., (2.88)

$$b(b^\dagger)^n - (b^\dagger)^n b = n(b^\dagger)^{n-1}. \tag{2.91}$$

- Show that the relations (2.88) and (2.89) imply

$$e^{-\alpha b^\dagger} b e^{\alpha b^\dagger} = b + \alpha, \tag{2.92}$$

$$e^{-\alpha b} b^\dagger e^{\alpha b} = b^\dagger - \alpha \tag{2.93}$$

for any complex number α.

More generally, you may want to show that

$$U^\dagger b U = b + \alpha \tag{2.94}$$

$$U^\dagger b^\dagger U = b^\dagger + \alpha^* \tag{2.95}$$

with the unitary operator

$$U = e^{\alpha b^\dagger - \alpha^* b} = e^{-|\alpha|/2} e^{\alpha b^\dagger} e^{-\alpha^* b}. \tag{2.96}$$

EXERCISE 2.7 Factorization of the exponential function For this exercise the result of exercise 2.5 will prove useful.

Show that the exponential function of a sum of a creation and an annihilation operator can be factorized

$$e^{\alpha b^\dagger + \beta b} = e^{\alpha b^\dagger} e^{\beta b} e^{\frac{\alpha\beta}{2}} = e^{\beta b} e^{\alpha b^\dagger} e^{-\frac{\alpha\beta}{2}}. \tag{2.97}$$

EXERCISE 2.8 Commutation with an exponential of $b^\dagger b$ Derive the formulas

$$e^{\alpha b^\dagger b} b e^{-\alpha b^\dagger b} = e^{-\alpha} b \tag{2.98}$$

$$e^{\alpha b^\dagger b} b^\dagger e^{-\alpha b^\dagger b} = e^{\alpha} b^\dagger \tag{2.99}$$

which can also be written in the form of commutation relations

$$b e^{-\alpha b^\dagger b} = e^{-\alpha} e^{-\alpha b^\dagger b} b \tag{2.100}$$

$$b^\dagger e^{-\alpha b^\dagger b} = e^{\alpha} e^{-\alpha b^\dagger b} b^\dagger. \tag{2.101}$$

....

2.4 Basis transformation

The states $|1_\alpha\rangle$ we constructed in (2.71), special cases of the many-particle states $|\phi\rangle = |N_1, N_2, \ldots\rangle$, are one-particle states in the many-particle Fock space. These states are in themselves orthonormal and complete

$$\langle 1_\alpha | 1_\beta \rangle = \delta_{\alpha\beta}, \tag{2.102}$$

$$\sum_\alpha |1_\alpha\rangle\langle 1_\alpha| = \mathbb{I}, \tag{2.103}$$

and hence, they form a many-particle basis $\{|1_\alpha\}$. They are useful for a number of purposes where the fully general many-particle state $|\phi\rangle$ would obscure the argument.

The creation and annihilation operators have been defined using the general many-particle state $|\phi\rangle$, but an equivalent definition makes use of only the one-particle states

$$a_\alpha^\dagger |0\rangle = |1_\alpha\rangle, \tag{2.104}$$

$$a_\alpha |1_\beta\rangle = \delta_{\alpha\beta}|0\rangle. \tag{2.105}$$

We now want to see how they transform under a basis transformation $\{|1_\alpha\rangle\} \to \{|1_i\}$.

Different basis states $\{|1_\alpha\rangle\}$ and $\{|1_i\rangle\}$ are mutually related via the completeness relations

$$|1_\alpha\rangle = \sum_i |1_i\rangle\langle 1_i | 1_\alpha\rangle, \tag{2.106}$$

$$|1_i\rangle = \sum_\alpha |1_\alpha\rangle\langle 1_\alpha | 1_i\rangle. \tag{2.107}$$

The creation and annihilation operators which use the specific one-particle basis states $\{|1_\alpha\rangle\}$, are transformed by the unitary transformations

$$a_i^\dagger = \sum_\alpha \langle 1_\alpha | 1_i\rangle a_\alpha^\dagger, \tag{2.108}$$

$$a_i = \sum_\alpha \langle 1_i | 1_\alpha\rangle a_\alpha, \tag{2.109}$$

such that in the new basis the corresponding relations

$$|1_i\rangle = a_i^\dagger |0\rangle, \tag{2.110}$$

$$|0\rangle = a_i |1_i\rangle \tag{2.111}$$

hold as expected.

The inverse transformations are

$$a_\alpha^\dagger = \sum_i \langle 1_i | 1_\alpha \rangle a_i^\dagger, \tag{2.112}$$

$$a_\alpha = \sum_i \langle 1_\alpha | 1_i \rangle a_i. \tag{2.113}$$

These relations for the creation and annihilation operators, together with the completeness relations in the different bases, are mutually consistent.

In the next section we exploit a particularly important case of basis transformation, which involves one-particle and two-particle basis states of well-defined position $\mathbf{r}$

$$\{|1_\alpha\rangle\} = \{|\mathbf{r}\rangle\}, \tag{2.114}$$

where orthonormality and completeness take the form

$$\langle \mathbf{r} | \mathbf{r}' \rangle = \delta(\mathbf{r} - \mathbf{r}'), \tag{2.115}$$

$$\int d^3r \, |\mathbf{r}\rangle\langle \mathbf{r}| = \mathbb{I} \tag{2.116}$$

The resulting operators are called quantum field creation and annihilation operators.

2.5 Quantum field operators

So far, the states have been states of an abstract Hilbert space, and, consequently, the creation and annihilation operators have been acting in abstract Hilbert spaces.

We now use for the following developments of investigating quantum field operators, as prepared in Section 2.4, the transformation to the particular position basis $\{|\mathbf{r}\rangle\}$, i.e. we deal with the creation and annihilation of particles at definite points $\mathbf{r}$ in physical space. This can be achieved by introducing *quantum field operators* for creating and annihilating particles at position $\mathbf{r}$ using the following transformations, which are special cases of the general transformations discussed earlier

$$a_\alpha^\dagger \to \psi^\dagger(\mathbf{r}) = \sum_\alpha \langle 1_\alpha | \mathbf{r} \rangle a_\alpha^\dagger = \sum_\alpha u_\alpha^*(\mathbf{r}) a_\alpha^\dagger, \tag{2.117}$$

and

$$a_\alpha \to \psi(\mathbf{r}) = \sum_\alpha \langle \mathbf{r} | 1_\alpha \rangle a_\alpha = \sum_\alpha u_\alpha(\mathbf{r}) a_\alpha, \tag{2.118}$$

where $u_\alpha(\mathbf{r})$ is the wave function of the one-particle state $|1_\alpha\rangle$. The quantum field operators create and annihilate, respectively, a particle at position $\mathbf{r}$

$$\psi^\dagger(\mathbf{r})|0\rangle = |\mathbf{r}\rangle, \tag{2.119}$$

$$\psi(\mathbf{r})|\mathbf{r}'\rangle = \delta(\mathbf{r}-\mathbf{r}')|0\rangle. \tag{2.120}$$

The inverse transformations are

$$a_\alpha^\dagger = \int d^3r\,\langle\mathbf{r}|1_\alpha\rangle\psi^\dagger(\mathbf{r}) = \int d^3r\,u_\alpha(\mathbf{r})\psi^\dagger(\mathbf{r}) \tag{2.121}$$

and

$$a_\alpha = \int d^3r\,\langle 1_\alpha|\mathbf{r}\rangle\psi_\sigma(\mathbf{r}) = \int d^3r\,u_\alpha^*(\mathbf{r})\psi(\mathbf{r}). \tag{2.122}$$

The number of particles at point $\mathbf{r}$ in a small volume d^3r is

$$\psi^\dagger(\mathbf{r})\psi(\mathbf{r})d^3r = \rho(\mathbf{r})d^3r, \tag{2.123}$$

and, hence, the total number of particles

$$\mathcal{N} = \int d^3r\,\rho(\mathbf{r}) = \int d^3r\,\psi^\dagger(\mathbf{r})\psi(\mathbf{r}). \tag{2.124}$$

The commutators become

$$\left[\psi^\dagger(\mathbf{r}), \psi^\dagger(\mathbf{r}')\right] = 0, \tag{2.125}$$

$$\left[\psi(\mathbf{r}), \psi(\mathbf{r}')\right] = 0, \tag{2.126}$$

$$\left[\psi(\mathbf{r}), \psi^\dagger(\mathbf{r}')\right] = \delta(\mathbf{r}-\mathbf{r}'). \tag{2.127}$$

We close this section with an elementary example.

Examples 1 *Free particle in a box* A simple but important example is provided by choosing the abstract one-particle basis $\{1_i\}$ in (2.108) and (2.109) to be the basis of momentum eigenstates $\{|\mathbf{k}\rangle\}$ with

$$\mathbf{p}_i|\mathbf{k}\rangle = \mathbf{k}|\mathbf{k}\rangle, \tag{2.128}$$

$$\sum_{\mathbf{k}} |\mathbf{k}\rangle\langle\mathbf{k}| = \mathbb{I}, \tag{2.129}$$

$$\langle\mathbf{k}|\mathbf{k}'\rangle = \delta_{\mathbf{k}\mathbf{k}'}. \tag{2.130}$$

We obtain

$$a_{\mathbf{k}}^{\dagger} = \sum_{\alpha} \langle 1_{\alpha} | \mathbf{k} \rangle a_{\alpha}^{\dagger} = \sum_{\alpha} \int d^3 r \, \langle 1_{\alpha} | \mathbf{r} \rangle \langle \mathbf{r} | \mathbf{k} \rangle a_{\alpha}^{\dagger} = \int d^3 r \, \langle \mathbf{r} | \mathbf{k} \rangle \psi^{\dagger}(\mathbf{r}), \tag{2.131}$$

$$a_{\mathbf{k}} = \sum_{\alpha} \langle \mathbf{k} | 1_{\alpha} \rangle a_{\alpha} = \sum_{\alpha} \int d^3 r \, \langle \mathbf{k} | \mathbf{r} \rangle \langle \mathbf{r} | 1_{\alpha} \rangle a_{\alpha} = \int d^3 r \, \langle \mathbf{k} | \mathbf{r} \rangle \psi(\mathbf{r}), \tag{2.132}$$

where we have used (2.117) and (2.118). The functions

$$\langle \mathbf{r} | \mathbf{k} \rangle = (\langle \mathbf{k} | \mathbf{r} \rangle)^* = u_{\mathbf{k}}(\mathbf{r}) = e^{i \mathbf{k} \cdot \mathbf{r}} \tag{2.133}$$

where the wave vector $\mathbf{k}$ is given by

$$\mathbf{k} = \frac{2\pi}{L} \begin{pmatrix} n_1 \\ n_2 \\ n_3 \end{pmatrix}, \quad n_i \in \mathbb{N}. \tag{2.134}$$

describe a free particle confined to a cubic volume $V = L^3$ and subject to periodic boundary conditions. The final result for creation and annihilation operators is then given by

$$a_{\mathbf{k}}^{\dagger} = \int d^3 r \, e^{i \mathbf{k} \cdot \mathbf{r}} \psi^{\dagger}(\mathbf{r}), \tag{2.135}$$

and

$$a_{\mathbf{k}} = \int d^3 r \, e^{-i \mathbf{k} \cdot \mathbf{r}} \psi(\mathbf{r}), \tag{2.136}$$

respectively, with the inverse relations

$$\psi^{\dagger}(\mathbf{r}) = \frac{1}{V} \sum_{\mathbf{k}} e^{-i \mathbf{k} \cdot \mathbf{r}} a_{\mathbf{k}}^{\dagger}, \tag{2.137}$$

$$\psi(\mathbf{r}) = \frac{1}{V} \sum_{\mathbf{k}} e^{i \mathbf{k} \cdot \mathbf{r}} a_{\mathbf{k}}. \tag{2.138}$$

We will now express arbitrary operators in terms of creation and annihilation operators. In doing so, we distinguish operators acting on one, two, ..., N particles. The most important operators in practical situations are the one- and two-particle operators, to which the next two sections are devoted.

2.6 One-particle operators

The simplest operators to consider are additive one-particle operators. Recall that in first quantization an additive one-particle operator $\mathbf{O}_1$ is given formally by its dependence on the dynamical variables of the particles, e.g. position operator $\mathbf{r}$ and momentum operator

$\mathbf{p}$. The additive one-particle operator is given by the sum over operators $\mathbf{O}_i^{(1)}$, each of which acts only on one particle

$$\mathbf{O}_1 = \sum_i \mathbf{O}_i^{(1)} \tag{2.139}$$

An example is the Hamiltonian of non-interacting particles in an external potential

$$\mathcal{H} = \sum_i \left(\frac{\mathbf{p}_i^2}{2m} + V(\mathbf{r}_i) \right) = \sum_i \mathcal{H}_i. \tag{2.140}$$

In order to make contact with the occupation number representation introduced in the previous discussion, we need to drop the particle index i of the operators $\mathbf{O}_i^{(1)}$ and derive their action first on the abstract basis states discussed in sections 2.2 and especially 2.4, which are basis states where occupation levels are populated with particles (cf. (2.75)), especially one or two particles (cf. (2.71), (2.72), and (2.73)).

Let us assume now that there is a one-particle basis $\{|1_\alpha\rangle\}$ in which the operators $\mathbf{O}^{(1)}$ are diagonal with eigenvalues ω_α

$$\mathbf{O}^{(1)}|1_\alpha\rangle = \omega_\alpha |1_\alpha\rangle, \tag{2.141}$$

as is, for instance, the case for momentum eigenstates of the momentum operator $\mathbf{p}$ (see the example 2.2), then the one-particle operator becomes

$$\mathbf{O}_1 = \sum_\alpha \omega_\alpha N_\alpha = \sum_\alpha \omega_\alpha a_\alpha^\dagger a_\alpha. \tag{2.142}$$

In any other one-particle basis $\{|1_i\rangle\}$ whose creation and annihilation operators are related to those of the one-particle basis $\{|1_\alpha\rangle\}$ by (2.108) and (2.109) the second quantized form of the general operator (2.139) is then

$$\mathbf{O}_1 = \sum_{ij} \langle 1_i|\mathbf{O}^{(1)}|1_j\rangle a_i^\dagger a_j. \tag{2.143}$$

In the real space or position basis $\{|\mathbf{r}\rangle\}$, the matrix elements in this expression can be written as

$$\langle 1_i|\mathbf{O}^{(1)}|1_j\rangle = \int d^3r\, d^3r'\, u_i^*(\mathbf{r})\langle \mathbf{r}|\mathbf{O}^{(1)}|\mathbf{r}'\rangle u_j(\mathbf{r}'), \tag{2.144}$$

which reduces for operators $\mathbf{O}^{(1)}$ diagonal in this representation to

$$\langle 1_i|\mathbf{O}^{(1)}|1_j\rangle = \int d^3r\, u_i^*(\mathbf{r})\mathbf{O}^{(1)}(\mathbf{r},\mathbf{p})u_j(\mathbf{r}), \tag{2.145}$$

Let us look at a few concrete examples.

2.6.1 Examples of one-particle operators

Examples 2 *Position operator* The simplest example is arguably the position operator, i.e. $\mathbf{O}_1 = \mathbf{R}$ and $\mathbf{O}^{(1)} = \hat{\mathbf{r}}$.[3] We use quantum field operators $\psi^\dagger(\mathbf{r})$ and $\psi(\mathbf{r})$ and the basis of position eigenstates $\{|\mathbf{r}\rangle\}$ in the general expression (2.143) to obtain

$$\mathbf{R} = \int d^3r\, d^3r'\, \langle\mathbf{r}|\hat{\mathbf{r}}|\mathbf{r}'\rangle\psi^\dagger(\mathbf{r})\psi(\mathbf{r}') = \int d^3r\, \mathbf{r}\psi^\dagger(\mathbf{r})\psi(\mathbf{r}), \tag{2.146}$$

where we used the property of the position eigenstates $\langle\mathbf{r}|\hat{\mathbf{r}}|\mathbf{r}'\rangle = \mathbf{r}'\langle\mathbf{r}|\mathbf{r}'\rangle = \mathbf{r}'\delta(\mathbf{r}-\mathbf{r}')$.

Examples 3 *Linear momentum operator* In the case of linear momentum $\mathbf{O}_1 = \mathbf{P}$ and $\mathbf{O}^{(1)} = \mathbf{p}$ we obtain for arbitrary one-particle states $\{|1_i\rangle\}$

$$\mathbf{P} = \sum_{ij}\langle 1_i|\mathbf{p}|1_j\rangle a_i^\dagger a_j, \tag{2.147}$$

which can be rewritten using the completeness relation for position states $\int d^3r\, |\mathbf{r}\rangle\langle\mathbf{r}| = \mathbb{I}$ as

$$\mathbf{P} = \sum_{ij}\int d^3r\, d^3r'\, a_i^\dagger\langle 1_i|\mathbf{r}\rangle\langle\mathbf{r}|\mathbf{p}|\mathbf{r}'\rangle\langle\mathbf{r}'|1_j\rangle a_j = \int d^3r\, d^3r'\, \psi^\dagger(\mathbf{r})\langle\mathbf{r}|\mathbf{p}|\mathbf{r}'\rangle\psi(\mathbf{r}'). \tag{2.148}$$

Next, we insert the completeness relation $\sum_{\mathbf{k}}|\mathbf{k}\rangle\langle\mathbf{k}| = \mathbb{I}$ for the basis of momentum eigenstates $\mathbf{p}_i|\mathbf{k}\rangle = \mathbf{k}|\mathbf{k}\rangle$, which yields the final result that the momentum operator is diagonal in this basis (as was to be expected)

$$\mathbf{P} = \sum_{\mathbf{k}\mathbf{k}'}\int d^3r\, d^3r'\, \psi^\dagger(\mathbf{r})\langle\mathbf{r}|\mathbf{k}\rangle\langle\mathbf{k}|\mathbf{p}|\mathbf{k}'\rangle\langle\mathbf{k}'|\mathbf{r}'\rangle\psi(\mathbf{r}') = \sum_{\mathbf{k}}\mathbf{k}\, a_{\mathbf{k}}^\dagger a_{\mathbf{k}}, \tag{2.149}$$

where $\langle\mathbf{r}|\mathbf{k}\rangle = e^{i\mathbf{k}\cdot\mathbf{r}}$ are plane waves and (2.135) and (2.136) have been used.

We could have obtained this result more directly by choosing as one-particle basis states $\{|1_i\rangle\}$ the basis of momentum eigenstates $\{|\mathbf{k}\rangle\}$ from the outset.

[3] Where, for once, we use the symbol ^ to distinguish an operator.

However, it is also often useful to stay in the position basis and use the momentum operator in position representation $\mathbf{p} = -i\nabla$, so that we can write (2.148) as

$$\mathbf{P} = \int d^3r\, \psi^\dagger(\mathbf{r})\, (-i\nabla)\, \psi(\mathbf{r}). \tag{2.150}$$

Examples 4 *Kinetic energy* Similarly the kinetic energy of a particle of mass m

$$\mathbf{O}^{(1)} = \frac{\mathbf{p}^2}{2m} \tag{2.151}$$

yields the kinetic energy of N identical non-interacting particles

$$\mathcal{H}_0 = \sum_{i=1}^{N} \frac{\mathbf{p}_i^2}{2m} = \frac{1}{2m} \sum_{ij} \langle u_i|\mathbf{p}^2|u_j\rangle a_i^\dagger a_j, \tag{2.152}$$

which in the basis of momentum eigenstates becomes again diagonal

$$\mathcal{H}_0 = \sum_{\mathbf{k}} \frac{k^2}{2m} a_{\mathbf{k}}^\dagger a_{\mathbf{k}}. \tag{2.153}$$

In the position basis, we have the useful expression

$$\mathcal{H}_0 = \int d^3r\, \psi^\dagger(\mathbf{r}) \left(-\frac{1}{2m}\nabla^2\right) \psi(\mathbf{r}), \tag{2.154}$$

which can be readily generalized to particles in an external potential $U(\mathbf{r})$

$$\mathcal{H} = \int d^3r\, \psi^\dagger(\mathbf{r}) \left(-\frac{1}{2m}\nabla^2 + U(\mathbf{r})\right) \psi(\mathbf{r}). \tag{2.155}$$

Examples 5 *Particle density* We obtain the particle density from (2.123) together with (2.117) and (2.118), choosing as basis $\{|\mathbf{k}\rangle\}$ for $\{|1_\alpha\rangle\}$

$$\rho(\mathbf{r}) = \psi^\dagger(\mathbf{r})\psi(\mathbf{r}) = \sum_{\mathbf{k}\mathbf{k}'} \langle \mathbf{k}|\mathbf{r}\rangle \langle \mathbf{r}|\mathbf{k}'\rangle a_{\mathbf{k}}^\dagger a_{\mathbf{k}'}, \tag{2.156}$$

whose Fourier transform is

$$\rho(\mathbf{q}) = \int d^3r\, e^{-i\mathbf{q}\cdot\mathbf{r}} \rho(\mathbf{r}) = \sum_{\mathbf{k}\mathbf{k}'} \left(\int d^3r\, e^{i(\mathbf{k}'-(\mathbf{k}+\mathbf{q}))\cdot\mathbf{r}}\right) a_{\mathbf{k}}^\dagger a_{\mathbf{k}'}, \tag{2.157}$$

so that

$$\rho(\mathbf{q}) = \sum_{\mathbf{k}} a_{\mathbf{k}}^\dagger a_{\mathbf{k}+\mathbf{q}} = \sum_{\mathbf{k}} a_{\mathbf{k}-\mathbf{q}/2}^\dagger a_{\mathbf{k}+\mathbf{q}/2}. \tag{2.158}$$

EXERCISE 2.9 Current In the same way it can be shown that the Fourier transform of the particle current

$$\mathbf{j}(\mathbf{r}) = -\frac{i}{2m}\left[\psi^\dagger(\mathbf{r})\left(\nabla\psi(\mathbf{r})\right) - \left(\nabla\psi^\dagger(\mathbf{r})\right)\psi(\mathbf{r})\right] \tag{2.159}$$

is given by

$$\mathbf{j}(\mathbf{q}) = \frac{1}{m}\sum_{\mathbf{k}}\left(\mathbf{k}+\frac{\mathbf{q}}{2}\right)a^\dagger_{\mathbf{k}}a_{\mathbf{k}+\mathbf{q}}. \tag{2.160}$$

2.7 Two-particle operators

A general two-particle operator is

$$\mathbf{O}_2 = \sum_{ij}\mathbf{O}^{(2)}_{ij}, \tag{2.161}$$

where the sum runs over all pairs of particles. The second quantized form in the abstract two-particle basis (2.72) and (2.73) becomes thus

$$\mathbf{O}_2 = \sum_{ijmn}\langle 1_i 1_j|\mathbf{O}^{(2)}|1_m 1_n\rangle a^\dagger_i a^\dagger_j a_n a_m. \tag{2.162}$$

Note the reversed order of the indices n and m of the pair of annihilation operators as compared to their order in the matrix elements. This is a convention we adopt that is useful especially for Fermions where, as we shall see below in sections 2.9 and 2.11, and, especially, 2.12, changing the order of operators produces minus signs.

In the position basis, using (2.117) and (2.118), the two-particle operator becomes

$$\mathbf{O}_2 = \int d^3r_1\, d^3r_1'\, d^3r_2\, d^3r_2'\, \psi^\dagger(\mathbf{r}_1)\psi^\dagger(\mathbf{r}_1')\langle\mathbf{r}_1,\mathbf{r}_1'|\mathbf{O}^{(2)}|\mathbf{r}_2,\mathbf{r}_2'\rangle\psi(\mathbf{r}_2')\psi(\mathbf{r}_2). \tag{2.163}$$

Note that, as a consequence of the earlier ordering of the creation and annihilation operators in the general two-particle operator (2.162), the sequence of positions in the quantum field operators is $\mathbf{r}_1, \mathbf{r}_1', \mathbf{r}_2', \mathbf{r}_2$.

If $\mathbf{O}^{(2)}$ is diagonal in the position space basis, using (2.117) and (2.118), this expression further reduces further to

$$\mathbf{O}_2 = \int d^3r\, d^3r'\, \psi^\dagger(\mathbf{r})\psi^\dagger(\mathbf{r}')\mathbf{O}^{(2)}(\mathbf{r},\mathbf{r}')\psi(\mathbf{r}')\psi(\mathbf{r}). \tag{2.164}$$

An important special case of this last expression is obtained if

$$\mathbf{O}^{(2)}(\mathbf{r}, \mathbf{r}') = \mathbf{O}^{(2)}(\mathbf{r} - \mathbf{r}') \tag{2.165}$$

as is, for instance, the case for many pairwise interaction potentials between particles. In order to exploit it, we start again from the general expression (2.162) for the two-particle operator $\mathbf{O}_2$ but choose as two-particle basis states $\{|1_i 1_j\rangle\}$ the momentum eigenstates $\{|\mathbf{k}_1 \mathbf{k}_2\rangle\}$

$$\mathbf{O}_2 = \sum_{\mathbf{k}_1 \mathbf{k}_2 \mathbf{k}_1' \mathbf{k}_2'} \langle \mathbf{k}_1 \mathbf{k}_2 | \mathbf{O}^{(2)} | \mathbf{k}_1' \mathbf{k}_2' \rangle a^\dagger_{\mathbf{k}_1} a^\dagger_{\mathbf{k}_2} a_{\mathbf{k}_2'} a_{\mathbf{k}_1'} . \tag{2.166}$$

Evaluating the matrix element in this expression in the plane wave representation when (2.165) holds, we finally obtain

$$\mathbf{O}_2 = \sum_{\mathbf{k}_1 \mathbf{k}_2 \mathbf{q}} \mathbf{O}^{(2)}(\mathbf{q}) a^\dagger_{\mathbf{k}_1} a^\dagger_{\mathbf{k}_2} a_{\mathbf{k}_2 + \mathbf{q}} a_{\mathbf{k}_1 - \mathbf{q}} = \sum_{\mathbf{k}_1 \mathbf{k}_2 \mathbf{q}} \mathbf{O}^{(2)}(\mathbf{q}) a^\dagger_{\mathbf{k}_1 + \mathbf{q}} a^\dagger_{\mathbf{k}_2 - \mathbf{q}} a_{\mathbf{k}_2} a_{\mathbf{k}_1}, \tag{2.167}$$

where $\mathbf{O}^{(2)}(\mathbf{q})$ is the Fourier transform of $\mathbf{O}^{(2)}(\mathbf{r})$

$$\mathbf{O}^{(2)}(\mathbf{q}) = \int d^3 r \, \mathbf{O}^{(2)}(\mathbf{r}) e^{-i\mathbf{q}\cdot\mathbf{r}}. \tag{2.168}$$

....

EXERCISE 2.10 Two-particle operator in momentum representation Complete the steps leading from (2.166) to (2.167).

EXERCISE 2.11 Fourier transform of the 1/*r*-interaction in three, two, and one dimension Calculate the Fourier transform of the interaction potential $1/r$ for three

$$V(\mathbf{q}) = \int_{-\infty}^{\infty} dz \, e^{-iq_z z} \int_{-\infty}^{\infty} dy \, e^{-iq_y y} \int_{-\infty}^{\infty} dx \, \frac{e^{-iq_x x}}{\sqrt{x^2 + y^2 + z^2}}, \tag{2.169}$$

two

$$V(\mathbf{q}) = \int_{-\infty}^{\infty} dy \, e^{-iq_y y} \int_{-\infty}^{\infty} dx \, \frac{e^{-iq_x x}}{\sqrt{x^2 + y^2}}, \tag{2.170}$$

and one dimension

$$V(q) = \int_{-\infty}^{\infty} dx \, \frac{e^{-iqx}}{x}. \tag{2.171}$$

For the first integral, we can use spherical coordinates instead of Cartesian ones. In this approach we must regularize the integral by introducing a convergence factor which, in the final result, can be set to an appropriate value.

This approach becomes more difficult in two and one dimensions. In two dimensions, the introduction of polar coordinates leads to the appearance of a special function, the Bessel function $\mathcal{J}_0$, with an appropriate argument.

An interesting approach to calculate the integral in any number of dimensions is to stay with Cartesian coordinates and evaluate the integrals 'from the inside out', i.e. start with the integration over x. A careful perusal of Gradshteyn and Ryzhik (1980) shows that the requisite integrals are all tabulated there, providing, en passant, the result for two dimensions.

However, the one-dimensional case presents a difficulty because the integral diverges. That being said, we can introduce on physical grounds a regularization starting with the innermost integral of the three-dimensional case and setting $a = \sqrt{y^2 + z^2}$ as the width of the quasi-one-dimensional system. Moreover, it is possible to obtain an expression for the regularized integral in the limit $q = |q_x| \to 0$.

2.8 Second quantization of the Schrödinger equation: Bosonic case

In elementary quantum mechanics, the time-dependent Schrödinger equation for a particle in a potential $V(\mathbf{r})$ is

$$i\frac{\partial \psi}{\partial t} = -\frac{1}{2m}\nabla^2\psi + V(\mathbf{r})\psi \tag{2.172}$$

where $\psi(\mathbf{r}, t)$ is the Schrödinger wave function or the Schrödinger field. The stationary eigenvalue equation

$$-\frac{1}{2m}\nabla^2\psi_n + V(\mathbf{r})\psi_n = E_n\psi_n \tag{2.173}$$

has solutions $\psi_n(\mathbf{r})$, which are eigenfunctions with energy eigenvalues E_n. The general solution of the time-dependent Schrödinger equation is thus

$$\psi(\mathbf{r}, t) = \sum_n a_n(t)\psi_n(\mathbf{r}) \tag{2.174}$$

where the $a_n(t)$ satisfy

$$\dot{a}_n(t) = -iE_n a_n(t). \tag{2.175}$$

The energy, or, more precisely, the expectation value of the Hamilton operator, can now be written as

$$E = \langle H \rangle = \langle \psi | H | \psi \rangle = \int d^3r\, \psi^*(\mathbf{r}, t)\left[-\frac{1}{2m}\nabla^2 + V(\mathbf{r})\right]\psi(\mathbf{r}, t) \tag{2.176}$$

$$= \sum_n E_n a_n^* a_n = \sum_n \omega_n a_n^* a_n. \tag{2.177}$$

This is the energy of a system of harmonic oscillators with eigenfrequencies $\omega_n = E_n$.

Thus, it is consistent to interpret these result in terms of a second quantized Hamiltonian

$$H = \sum_n E_n a_n^\dagger a_n, \tag{2.178}$$

where the creation and annihilation operators satisfy the commutation relations (2.74).

The equation of motion for the creation operators a_n, its Heisenberg equation of motion, is

$$\frac{da_n}{dt} = i\,[H, a_n] = -iE_n a_n. \tag{2.179}$$

In section 2.10, we briefly review the second quantization of the Schrödinger for the case of Fermions.

2.9 Creation and annihilation operators for Fermions

As shown in section 2.1.3, for Fermions, i.e. particles that obey the Pauli exclusion principle, the wave function is totally antisymmetric, i.e. exchange of particles leads to a change of sign of the wave function. To achieve this, we have introduced an antisymmetrization operator $\mathcal{A}$.

Section 2.1.3 also showed that the only possible states in an occupation number representation for Fermions are given by

$$|\phi\rangle = |N_1, N_2, \ldots, N_l, \ldots\rangle = |\{N\}\rangle, \tag{2.180}$$

with either $N_l = 1$ or $N_l = 0$, $l = 1, 2, \ldots$, which expresses the exclusion principle and a fixed number of particles $\sum_{l=1}^{\infty} N_l = N$.

The definition of creation and annihilation operators acting on the occupation number states $|\{N\}\rangle$ is somewhat more delicate than in the Bose case. The Fermionic creation and annihilation operators are more abstract entities than their Bosonic counterparts. They cannot be introduced in any way with recourse to position and momentum operators. This is a consequence of the fact that Fermions are genuinely quantum mechanical objects. Using the notation c_i and $c_i^\dagger$ to distinguish from the Bose case, we define

$$c_i|\{N\}\rangle = \Theta_i\sqrt{N_i}|\ldots(1 - N_i)\ldots\rangle \tag{2.181}$$

$$c_i^\dagger|\{N\}\rangle = \Theta_i\sqrt{1 - N_i}|\ldots(1 - N_i)\ldots\rangle \tag{2.182}$$

with phase factors Θ_i. We could have also introduced phase factors for the definition of the Bose creation and annihilation operators, but choosing them as unity would have been consistently possible. In the Fermi case, however, the phase factors are of crucial importance and cannot be chosen as unity.

A consistent choice is the following. First, order the sequence of occupation number levels l in $|\phi\rangle$ in a fixed way that cannot be changed any more. In the definition of the creation and annihilation operators c_i and $c_i{}^\dagger$, if the number ν_i of levels that are occupied by a Fermion is even for the levels which are predecessors of the level i, then the phase is $\Theta_i = +1$, and if it is odd, then put $\Theta = -1$. Obviously, we can write

$$\Theta_i = (-1)^{\sum_{j=1}^{i-1}} \equiv (-1)^{\nu_i}. \tag{2.183}$$

From the definitions (2.181) and (2.182) of the creation and annihilation operators, we see that

$$(c_i)^2|\{N\}\rangle = (c_i^\dagger)^2|\{N\}\rangle = 0, \tag{2.184}$$

which again expresses the Pauli exclusion principle.

For, without limiting generality, $i < j$, the definitions (2.181) and (2.182) of the creation and annihilation operators imply the anti-commutation relations

$$\{c_i, c_j\} = c_i c_j + c_j c_i = 0, \tag{2.185}$$

$$\{c_i^\dagger, c_j^\dagger\} = c_i^\dagger c_j^\dagger + c_j^\dagger c_i^\dagger = 0, \tag{2.186}$$

and

$$\{c_i, c_j^\dagger\} = c_i c_j^\dagger + c_j^\dagger c_i = \delta_{ij}. \tag{2.187}$$

EXERCISE 2.12 Fermion anti-commutation relations Derive the Fermion anti-commutation relations using the definitions (2.181) and (2.182) of the creation and annihilation operators.

With the help of the creation and annihilation operators for Fermions c_i and $c_i^\dagger$ and the definition for vacuum state

$$|vac\rangle = |0, 0, \ldots, 0, \ldots\rangle, \tag{2.188}$$

a general N-particle Fermi state can be now be written as

$$|\{N\}\rangle = \left(c_1^\dagger\right)^{N_1}\left(c_2^\dagger\right)^{N_2}\ldots|vac\rangle. \tag{2.189}$$

The creation and annihilation operators for Fermions satisfy

$$c_i^\dagger c_i|\{N\}\rangle = N_i|\{N\}\rangle \tag{2.190}$$

and

$$c_i c_i^\dagger|\{N\}\rangle = (1-N_i)|\{N\}\rangle, \tag{2.191}$$

which again allows the interpretation of

$$n_i = c_i^\dagger c_i \tag{2.192}$$

as the particle number operator and

$$\mathcal{N} = \sum_{i=1}^{\infty} n_i = \sum_{i=1}^{\infty} c_i^\dagger c_i \tag{2.193}$$

as the operator of total particle number.

If proper care is taken concerning the order of creation and annihilation operators, all results for Bosons remain valid for Fermions.

EXERCISE 2.13 Bogoliubov transformation for Fermions Consider a system consisting of two Fermions, c_1 and c_2. Which condition must the complex parameters u and v satisfy such that the Bogoliubov transformation

$$c_1 = uc_1 + vc_2^\dagger, \tag{2.194}$$

$$c_2^\dagger = -vc_1 + uc_2^\dagger, \tag{2.195}$$

preserves the canonical Fermi anti-commutation relations?

EXERCISE 2.14 Band splitting A system of N electrons in one dimension, disregarding the spin of the electrons, in a periodic potential is described by the Hamilton operator

$$\mathcal{H} = \sum_{j=1}^{N}\left(-\frac{1}{2m}\frac{d^2}{dx_j^2} + V_0\cos\left(\frac{2\pi}{a}x_j\right)\right) = \sum_{j=1}^{N}\mathcal{H}_j = \sum_{j=1}^{N}\mathcal{H}(x_j). \tag{2.196}$$

Assume V_0 to be a small perturbation to the Hamiltonian of free, i.e. non-interacting, electrons.

a) Express $\mathcal{H}$ in the occupation number representation. Which are the appropriate one-particle basis and one-particle wave functions for this problem?
b) Restrict the Hamiltonian $\hat{\mathcal{H}}$ in occupation number representation to a small interval

$$I = \left(\frac{K}{2} - \delta, \frac{K}{2} + \delta\right) \tag{2.197}$$

with $K = \frac{2\pi}{a}$ and $\delta \ll K$, i.e.

$$\hat{\mathcal{H}} = \sum_{k \in I} \mathcal{H}_k \tag{2.198}$$

where

$$\mathcal{H}_k = \epsilon_k c_k^\dagger c_k + \epsilon_q c_q^\dagger c_q + \frac{V_0}{2} c_k^\dagger c_q + \frac{V_0}{2} c_q^\dagger c_k \tag{2.199}$$

with $q = k - K$.
Diagonalize $\mathcal{H}_k$ using the appropriate choice of parameters λ_k in the transformation

$$b_k = c_k \cos \lambda_k + c_q \sin \lambda_k, \tag{2.200}$$

$$b_q = -c_k \sin \lambda_k + c_q \cos \lambda_k. \tag{2.201}$$

Which eigenvalues does $\mathcal{H}_k$ thus have?
c) The one-particle eigenvalue spectrum ϵ_k of free electrons has a parabolic form. The periodic perturbation potential in (2.196) leads to a deviation from this parabolic form, which is called a band splitting. Calculate this band splitting at $k = \frac{K}{2}$. In order to do this, expand the eigenvalues of $\mathcal{H}_k$ up to and including second order in δ.

Similar to the case of Bosonic creation and annihilation operators, we now summarize useful formulas for calculations involving Fermionic creation and annihilation operators as exercises. As we shall see, these formulas have a form simpler than their Bosonic counterparts. The reason for this is that for Fermionic creation and annihilation operators, we have

$$c^n = \left(c^\dagger\right)^n = 0 \qquad \text{for} \qquad n \geq 2. \tag{2.202}$$

Using this fact, we can furthermore restrict the functions of Fermionic operators to the two cases

$$f(c) = f(0) + f'(0)c, \tag{2.203}$$

$$f(c^\dagger) = f(0) + f'(0)c^\dagger, \tag{2.204}$$

which yields as immediate consequence the special cases

$$e^{\alpha c} = 1 + \alpha c, \tag{2.205}$$

$$e^{\alpha c^\dagger} = 1 + \alpha c^\dagger \tag{2.206}$$

for any complex number α.

EXERCISE 2.15 Functions of Fermionic creation and annihilation operators As a consequence of (2.203) and (2.204), there are only two non-trivial cases to show

$$e^{-\alpha c^\dagger} c e^{\alpha c^\dagger} = c - \alpha^2 c^\dagger - 2\alpha s, \tag{2.207}$$
$$e^{-\alpha c} c^\dagger e^{\alpha c} = c^\dagger - \alpha^2 c + 2\alpha s \tag{2.208}$$

with the operator

$$s = c^\dagger c - \frac{1}{2}. \tag{2.209}$$

However, the exponential $e^{\alpha c^\dagger c}$ gives rise to interesting relations that are analogous to the ones in the Bosonic case and can be derived in an completely analogous way. They are

$$e^{\alpha c^\dagger c} c e^{-\alpha c^\dagger c} = e^{-\alpha} c, \tag{2.210}$$
$$e^{\alpha c^\dagger c} c^\dagger e^{-\alpha c^\dagger c} = e^{\alpha} c^\dagger. \tag{2.211}$$

2.10 Second quantization of the Schrödinger equation: Fermionic case

The same line of argument as in section 2.8, although via using the anti-commutation relations (2.185–2.187), leads to the formally same results.

In particular, a second quantized Hamiltonian can be identified

$$H = \sum_n E_n c_n^\dagger c_n \tag{2.212}$$

where the creation and annihilation operators satisfy the anti-commutation relations (2.185–2.187).

The equation of motion for the annihilation operators c_n, its Heisenberg equation of motion, again turns out to be

$$\frac{dc_n}{dt} = i\,[H, c_n] = -iE_n c_n. \tag{2.213}$$

EXERCISE 2.16 Repeat explicitly the steps in section 2.8 for the case of Fermions. Especially ascertain the validity of the Heisenberg equation of motion (2.213) in the Fermionic case.

2.11 Second quantization formalism and the many-particle wave function

In this penultimate section of part I, we demonstrate that the second quantization formalism is indeed equivalent to the Schrödinger equation of a many-particle system. In order to keep the equations manageable, we restrict this demonstration to the case of two particles. The considerations are valid for both the Bosonic and the Fermionic cases. In order to express this notationally, we use $b^\dagger$ and b for creation and annihilation operators. The general two-particle state is created from the vacuum state by applying two creation operators

$$|\phi_2\rangle = \sum_{i,j} c_{i,j} b_i^\dagger b_j^\dagger |0\rangle. \tag{2.214}$$

We rewrite this state in terms of quantum field operators

$$b_i^\dagger = \int d^3r\, \phi_i(\mathbf{r}) \psi^\dagger(\mathbf{r}), \tag{2.215}$$

$$b_i = \int d^3r\, \phi_i(\mathbf{r}) \psi(\mathbf{r}) \tag{2.216}$$

to obtain

$$|\phi_2\rangle = \sum_{i,j} c_{i,j} \int d^3r\, \phi_i(\mathbf{r})\psi^\dagger(\mathbf{r}) \int d^3r'\, \phi_j(\mathbf{r}')\psi^\dagger(\mathbf{r}')|0\rangle. \tag{2.217}$$

Introducing the function

$$f(\mathbf{r},\mathbf{r}') = \sum_{i,j} c_{i,j}\phi_i(\mathbf{r})\phi_j(\mathbf{r}'), \tag{2.218}$$

the two-particle state becomes

$$|\phi_2\rangle = \int d^3r \int d^3r'\, f(\mathbf{r},\mathbf{r}')\psi^\dagger(\mathbf{r})\psi^\dagger(\mathbf{r}')|0\rangle \tag{2.219}$$

$$= \int d^3r \int d^3r'\, f(\mathbf{r}',\mathbf{r})\psi^\dagger(\mathbf{r}')\psi^\dagger(\mathbf{r})|0\rangle \tag{2.220}$$

$$= \int d^3r \int d^3r'\, f(\mathbf{r}',\mathbf{r})\left(\pm\psi^\dagger(\mathbf{r})\psi^\dagger(\mathbf{r}')\right)|0\rangle, \tag{2.221}$$

where, going from the first line to the second, we have exchanged the variables $\mathbf{r}$ and $\mathbf{r}'$, and then, going from the second line to the third, we used the Bose (upper sign) and Fermi (lower sign) commutation and anti-commutation relations, respectively. Comparing the first and last lines, we find

$$f(\mathbf{r},\mathbf{r}') = \pm f(\mathbf{r}',\mathbf{r}) \tag{2.222}$$

for Bosons (upper sign) and Fermions (lower sign), respectively.

As we have seen, in sections 2.6 and 2.7 respectively, the second quantized Hamiltonian in the position basis of a many-particle system in an external potential $U(\mathbf{r})$ and with two-particle interactions $V(\mathbf{r},\mathbf{r}')$ can be written as

$$\mathcal{H} = \int d^3r\, \psi^\dagger(\mathbf{r})\left(-\frac{1}{2m}\nabla^2 + U(\mathbf{r})\right)\psi(\mathbf{r}) + \frac{1}{2}\int d^3r \int d^3r'\, \psi^\dagger(\mathbf{r})\psi^\dagger(\mathbf{r}')V(\mathbf{r},\mathbf{r}')\psi(\mathbf{r}')\psi(\mathbf{r}) \tag{2.223}$$

$$= \mathcal{H}_1 + \mathcal{H}_2 \equiv \mathcal{H}_1 + \mathcal{V} \tag{2.224}$$

where the factor 1/2 avoids double counting of the interactions and the two-particle interaction potential could, e.g. be the Coulomb potential

$$V(\mathbf{r},\mathbf{r}') = V(\mathbf{r}-\mathbf{r}') = \frac{e^2}{|\mathbf{r}-\mathbf{r}'|}. \tag{2.225}$$

The Schrödinger equation

$$\mathcal{H}|\phi_2\rangle = E|\phi_2\rangle \tag{2.226}$$

for the two-particle state can now be rewritten as the Schrödinger equation for the two-particle wave function $f(\mathbf{r}_1,\mathbf{r}_2)$

$$\left(-\frac{1}{2m}\nabla_1^2 + U(\mathbf{r}_1) - \frac{1}{2m}\nabla_2^2 + U(\mathbf{r}_2) + V(\mathbf{r}_1,\mathbf{r}_2)\right) f(\mathbf{r}_1,\mathbf{r}_2) = Ef(\mathbf{r}_1,\mathbf{r}_2). \tag{2.227}$$

In order to verify this result, we have to use the commutation or anti-commutation relations for quantum field operators of Bosons and Fermions, respectively. Moreover, we need a twofold partial integration in the integrals containing the single–particle Hamiltonian $\mathcal{H}_1$.

2.12 Normal ordering

In section 2.11 and in section 2.7 on two-particle operators, we have always written the field and particle creation and annihilation operators in such a way that all creation operators appeared to the left of the annihilation operators. This convention is called normal ordering. Adopting this convention ensures that the vacuum expectation value vanishes

$$\langle 0|\mathcal{V}|0\rangle = 0 \tag{2.228}$$

with $\mathcal{V}$ given in (2.224) and (2.223) where we have also observed the convention $\mathbf{r},\mathbf{r}',\mathbf{r}',\mathbf{r}$ for the order of arguments of the quantum field creation and annihilation operators.

EXERCISE 2.17 Normal ordering of two-particle potential operator: wrong choice Show that the apparently obvious choice for the second quantized form of a two-particle potential operator

$$\mathcal{V} = \frac{1}{2}\int d^3r\, d^3r'\; V(\mathbf{r},\mathbf{r}')\rho(\mathbf{r})\rho(\mathbf{r}') \tag{2.229}$$

where

$$\rho(\mathbf{r}) = \psi^\dagger(\mathbf{r})\psi(\mathbf{r}) \tag{2.230}$$

leads to an unwanted self-energy term

$$\frac{1}{2}\int d^3r\, V(\mathbf{r},\mathbf{r})\psi^\dagger(\mathbf{r})\psi(\mathbf{r}). \tag{2.231}$$

Therefore, observing normal ordering of the creation and annihilation operators from the start avoids unnecessary arguments about why certain terms are to be disregarded later on.

The following exercise confirms this lesson from the positive side.

EXERCISE 2.18 Correct normal ordering Show that the normal ordered two-particle potential operator

$$\mathcal{V} = \frac{1}{2}\int d^3r\, d^3r'\ \psi^\dagger(\mathbf{r})\psi^\dagger(\mathbf{r}')V(\mathbf{r},\mathbf{r}')\psi(\mathbf{r}')\psi(\mathbf{r}) \tag{2.232}$$

for a symmetric two-particle potential $V(\mathbf{r},\mathbf{r}') = V(\mathbf{r}',\mathbf{r})$ applied to a many-particle state

$$|\mathbf{r}_1,\mathbf{r}_2,\ldots,\mathbf{r}_N\rangle = \psi^\dagger(\mathbf{r}_1),\psi^\dagger(\mathbf{r}_2),\ldots,\psi^\dagger(\mathbf{r}_N)|0\rangle \tag{2.233}$$

reproduces the correct (first quantized) expression

$$\mathcal{V}|\mathbf{r}_1,\mathbf{r}_2,\ldots,\mathbf{r}_N\rangle = \frac{1}{2}\sum_{n\neq m} V(\mathbf{r}_n,\mathbf{r}_m)|\mathbf{r}_1,\mathbf{r}_2,\ldots,\mathbf{r}_N\rangle. \tag{2.234}$$

Furthermore, show that the two-particle potential operator of the previous exercise does not reproduce this result.

....

This chapter has introduced and reviewed the basics of the theory of quantum many-particle systems, especially the method of second quantization. We will use these concepts and results throughout the book.

Of course, the theory of quantum many-particle systems contains much more than the method of second quantization. The books mentioned at the beginning of this chapter provide more comprehensive expositions of this theory.

3

Angular Momentum

No, it's quite impossible for the electron to have a spin. I have thought of that myself, and if the electron did have a spin, the speed of the surface of the electron would be greater than the velocity of light. So, it's quite impossible.

Hendrik Antoon Lorentz (1853–1928) to Uhlenbeck and Goudsmit (fortunately Ehrenfest had already submitted their paper)

The first model solved by Bethe ansatz, by Hans Bethe himself in 1931, is a model of atoms arranged in a one-dimensional lattice whose magnetic moments or, for short, quantum spins, interact—the so-called Heisenberg model for magnetism.

Therefore, this chapter briefly reviews the concept of angular momentum in quantum mechanics, especially the coupling of spin-$\frac{1}{2}$ operators for several quantum spins. We shall be needing this when we discuss the properties of quantum many–particle Hamiltonians, for example, the Heisenberg quantum spin chain Hamiltonian in parts II and III.

Section 3.1 briefly develops the theory of angular momentum for a single quantum particle, first discussing its general theory, then specialized to case of spin-$\frac{1}{2}$, which has no classical counterpart but is a special case of an internal property of quantum particles. Section 3.2 is devoted to a discussion of the angular momentum of several quantum particles.

3.1 Angular momentum of a single quantum particle

3.1.1 General theory

The operator[1] L for angular momentum in quantum mechanics is defined in analogy to the corresponding relation of classical mechanics

$$\mathbf{L} = \mathbf{r} \times \mathbf{p} \tag{3.1}$$

[1] Again we shall not distinguish, for instance by using hats (ˆ), between quantum mechanical operators and classical dynamical variables except in cases where such a distinction reduces the chances of possible confusion.

Models of Quantum Matter: A First Course on Integrability and the Bethe Ansatz. Hans-Peter Eckle, Oxford University Press (2019). © Hans-Peter Eckle. DOI: 10.1093/oso/9780199678839.001.0001

where position[2] and (linear) momentum

$$\mathbf{r} = (x, y, z)^T = (r_1, r_2, r_3)^T, \qquad \mathbf{p} = \left(p_x, p_y, p_z\right)^T = (p_1, p_2, p_3)^T \tag{3.2}$$

are now identified with operators **r** and **p** with non-vanishing commutators which, in components, are given by

$$[r_j, p_k] = i\delta_{jk}. \tag{3.3}$$

From these commutators follow the commutators for the components L_1, L_2, and L_3 of the angular momentum operator **L**

$$\left[L_j, L_k\right] = i\epsilon_{jkl}L_l, \tag{3.4}$$

where ϵ_{jkl} are the components of the totally antisymmetric Levi–Civita tensor

$$\epsilon_{123} = 1, \quad \epsilon_{jkl} = -\epsilon_{kjl} = -\epsilon_{jlk} = -\epsilon_{lkj}. \tag{3.5}$$

EXERCISE 3.1 Angular momentum commutator It is a good exercise to verify (3.4) with the help of the commutation relations (3.3).

It follows from (3.4) and the corresponding Heisenberg uncertainty relation that the components L_j of the angular momentum operator **L** cannot be sharply measured simultaneously. However, for the square of the angular momentum

$$\mathbf{L} \cdot \mathbf{L} = L^2 = L_1^2 + L_2^2 + L_3^2 \tag{3.6}$$

we find that all its commutators with the components of **L** vanish

$$\left[L^2, L_j\right] = 0, \quad j = 1, 2, 3, \tag{3.7}$$

which means that the square L^2, and *one* component L_j of the angular momentum **L**, can be measured simultaneously with arbitrary precision. This component is conventionally chosen to be L_3. Moreover, we can always find simultaneous eigenstates of L^2 and L_3. In order to show (3.7), it is helpful to use the general relation for commutators involving three operators *A*, *B*, and *C*

[2] In this chapter, we use several conventions since all of them are appropriate in certain situations. We label quantities at times by their Cartesian component, e.g. p_x for the x-component of momentum, at other times by a numerical label i with $i = 1, 2, 3$, e.g. σ_3 for the 3–component instead of σ_z for the z-component of the vector of Pauli matrices. Moreover, it will sometimes be convenient to resort to upper indices, like in σ_i^z where there is also a lower index i to distinguish between the spin of different particles.

$$[AB, C] = A\,[B, C] + [A, C]\,B. \tag{3.8}$$

In order to investigate the eigenvalues and eigenfunctions of the angular momentum operators further, it will prove useful to introduce two operators with the help of the components of the angular momentum operator **L**

$$L^+ = L_1 + iL_2, \tag{3.9}$$

and

$$L^- = L_1 - iL_2, \tag{3.10}$$

which, together with L_3, satisfy the commutation relations

$$\left[L_3, L^+\right] = L^+, \tag{3.11}$$

$$\left[L_3, L^-\right] = -L^-, \tag{3.12}$$

and

$$\left[L^+, L^-\right] = 2L_3. \tag{3.13}$$

The operators $L^{\pm}$ are constructed in analogy to the harmonic oscillator creation and annihilation operators. It is instructive to demonstrate how the operators $L^{\pm}$ can be motivated by the following consideration. Assume we have an eigenstate $|m\rangle$ of the operator L_3

$$L_3|m\rangle = m|m\rangle. \tag{3.14}$$

We now want to find an operator L_S which produces a state $|\lambda\rangle = L_S|m\rangle$ with a shifted eigenvalue, i.e.

$$L_3|\lambda\rangle = L_3\left(L_S|m\rangle\right) = (m+\lambda)\left(L_S|m\rangle\right). \tag{3.15}$$

Multiplying (3.14) by L_S and subtracting it from (3.15) gives

$$[L_3, L_S]\,|m\rangle = \lambda L_S|m\rangle \tag{3.16}$$

or, as an operator equation

$$[L_3, L_S] = \lambda L_S. \tag{3.17}$$

In order to satisfy (3.17), it suffices to construct the shift operator L_S as a linear combination of the operators L_1 and L_2

$$L_S = L_1 + aL_2. \tag{3.18}$$

so that from (3.17)

$$[L_3, L_1] + a\,[L_3, L_2] = iL_2 - iaL_1 = \lambda L_1 + \lambda a L_2 \tag{3.19}$$

or

$$i = a\lambda \qquad \text{and} \qquad -ia = \lambda \tag{3.20}$$

must hold. The latter two equations are only compatible if $\lambda = \pm 1$. The shift operator can thus be identified with the operators L^+ and L^- of (3.9) and (3.10), creating states where the eigenvalue m of an eigenstate $|m\rangle$ of L_3 is shifted by $\pm$ 1.

In terms of the operators L^+, L^-, and L_3, the square of the angular momentum operator can be expressed in three ways, which will prove useful for the following considerations

$$L^2 = \left(L_3^2 + \frac{1}{2}\left(L_+L^- + L^-L^+\right)\right) \tag{3.21}$$

$$= \left(L_3^2 + L_3 + L^-L^+\right) \tag{3.22}$$

$$= \left(L_3^2 - L_3 + L^+L^-\right). \tag{3.23}$$

We consider a simultaneous eigenstate $|\Psi\rangle$ of L^2 and L_3 with eigenvalues C^2 and m, respectively, i.e.

$$L^2|\Psi\rangle = C^2|\Psi\rangle, \tag{3.24}$$

and

$$L_3|\Psi\rangle = m|\Psi\rangle, \tag{3.25}$$

which, as we have seen, is always possible. But what happens if we act with $L^\pm$ on this state? Let us thus consider the states

$$|\Phi^\pm\rangle = L^\pm|\Psi\rangle. \tag{3.26}$$

We already know that the eigenvalue of L_3 gets shifted

$$L_3|\Phi^\pm\rangle = (m \pm 1)|\Phi^\pm\rangle. \tag{3.27}$$

Furthermore, since L^2 commutes with all component of **L**, we obtain

$$L^2|\Phi^\pm\rangle = L^2L^\pm|\Psi\rangle = L^\pm L^2|\Psi\rangle = C^2\left(L^\pm|\Psi\rangle\right) = C^2|\Phi^\pm\rangle. \tag{3.28}$$

In summary, the operator L^+ thus increases and the operator L^- decreases the eigenvalue of the third component L_3 of the angular momentum by one, while leaving the eigenvalue of the square of the angular momentum L^2 unchanged. Because of the properties just discussed, the angular momentum operators L^+ and L^- are also called ladder operators, with L^+ being the raising, and L^- the lowering, operator.

More generally, we have, by iteration,

$$L_3\left(\left(L^\pm\right)^n|\Psi\rangle\right) = (m \pm n)\left(\left(L^\pm\right)^n|\Psi\rangle\right). \tag{3.29}$$

The square of the third component of the angular momentum cannot exceed the square of the angular momentum, i.e.

$$(m+n)^2 \leq C^2, \tag{3.30}$$

which can also be seen from the expectation values in the state $|\Psi\rangle$

$$\langle L^2\rangle \equiv \langle\Psi|L^2|\Psi\rangle = \langle L_1^2\rangle + \langle L_2^2\rangle + \langle L_3^2\rangle, \tag{3.31}$$

i.e.

$$C^2 = \langle L_1^2\rangle + \langle L_2^2\rangle + m^2 \geq m^2. \tag{3.32}$$

Thus, m must be bounded from above and below, i.e. there must be a largest value $m_{\max}$ and a smallest value $m_{\max}$ with

$$m_{\max} \leq m \leq m_{\min} \tag{3.33}$$

such that

$$L_3|m_{\max}\rangle = m_{\max}|m_{\max}\rangle \quad \text{and} \quad L^+|m_{\max}\rangle = 0, \tag{3.34}$$

$$L_3|m_{\min}\rangle = m_{\min}|m_{\min}\rangle \quad \text{and} \quad L^-|m_{\min}\rangle = 0. \tag{3.35}$$

Choosing $|m_{\max}\rangle$ and $|m_{\min}\rangle$ as simultaneous eigenstates of L^2, we find with the help of (3.22) and (3.23)

$$L^2|m_{\max}\rangle = C^2|m_{\max}\rangle = \left(L^-L^+ + L_3^2 + L_3\right)|m_{\max}\rangle \tag{3.36}$$

$$= m_{\max}\left(m_{\max}+1\right)|m_{\max}\rangle \tag{3.37}$$

and

$$L^2|m_{\min}\rangle = C^2|m_{\min}\rangle = \left(L^+L^- + L_3^2 - L_3\right)|m_{\min}\rangle \tag{3.38}$$

$$= m_{\min}(m_{\min} - 1)|m_{\min}\rangle, \tag{3.39}$$

which are only compatible if $m_{\min} = -m_{\max}$.

It is conventional to denote $m_{\max}$ by the letter l and the corresponding simultaneous eigenstate of the square L^2 and the third component L_3 of the angular momentum operator by $|l, l\rangle$.

This state thus satisfies

$$L^+|l, l\rangle = 0, \tag{3.40}$$

$$L_3|l, l\rangle = l|l, l\rangle. \tag{3.41}$$

and

$$L^2|l, l\rangle = C^2|l, l\rangle = l(l+1)|l, l\rangle, \tag{3.42}$$

i.e. the eigenvalue of the square L^2 of the angular momentum is $C^2 = l(l+1)$. Moreover, we can conclude from the action of $L^\pm$ (cf. (3.29)) that the possible values of m extend from $-l$ to l in integer steps and that l can only take integer or half-odd integer values.

We conclude that the states of the system are labelled, for any fixed value of l, by

$$|l, m\rangle, \quad 2l = 0, 1, 2, \ldots, \quad -l \leq m \leq l. \tag{3.43}$$

The corresponding state space, denoted by $\mathscr{V}_l$, has dimension $\dim \mathscr{V}_l = 2l + 1$. Another name for the eigenvalue of L_3 of a state $|\Phi\rangle$, often used in connection with quantum groups, is the *weight* of $|\Phi\rangle$. We call any state $|\Phi\rangle$ satisfying

$$L^-|\Phi\rangle = 0 \tag{3.44}$$

a *lowest weight state*.

For the states $|l, m\rangle$ we have just constructed, the weight is given by m, also called the azimuthal quantum number, and the lowest weight state is $|l, -l\rangle$.

So far, the states $|l, m\rangle$ have not been normalized. Normalizing them, we find the eigenvalues of the operators L^+ and L^- as well. If we write

$$L^\pm|l, m\rangle = N^\pm_{l,m}|l, m \pm 1\rangle, \tag{3.45}$$

now assuming that both states $|l, m\rangle$ and $|l, m \pm 1\rangle$ are normalized, we find with the help of the relations (3.21–3.23) that

$$N^{\pm}_{l,m} = [l(l+1) - m(m \pm 1)]^{\frac{1}{2}}\,. \tag{3.46}$$

Our results can be interpreted in terms the Heisenberg uncertainty relation. The maximally possible eigenvalue of L_3 is l and, hence, the maximally possible value of L_3^2 is l^2. However, the eigenvalue of L^2 is $l(l+1)$, larger than l^2. This implies that the angular momentum operator $\mathbf{L}$ can never align with certainty with L_3 and the uncertainty principle is satisfied. If $\mathbf{L}$ could be aligned with L_3, then $L_1 = L_2 = 0$ and we would have simultaneously sharp values of all three components of the angular momentum operator, in contradiction to Heisenberg's uncertainty relations, which follow from (3.4).

3.1.2 Spin

The half-odd integer values of angular momentum of the form $l = \frac{2n+1}{2}$, with n a non-negative integer, which we have found as a possibility in quantum mechanics in the previous section, have no correspondence in terms of classical angular momentum of the orbital form (3.1) given at the beginning of the previous section. However, they are realized in nature and correspond to the internal structure of the quantum particle and are referred to as *spin*. In this connection, the symbol $\mathbf{S}$ is used instead of $\mathbf{L}$.

As also discussed extensively in chapter 2, there is a fundamental distinction between particles with half-odd integer spin, called *Fermions*, which obey the Pauli exclusion principle, and those with integer spin,[3] called *Bosons*, which do not obey the Pauli exclusion principle. This distinction is rather deep and manifests itself also in different thermodynamic behaviour of Bosons and Fermions. Where the former have a many-particle wave function that is symmetric with respect to the exchange of particles, and they obey Bose–Einstein statistics, the latter have an antisymmetric many-particle wave function with respect to particle exchange and they obey Fermi–Dirac statistics. Some of the many-particle aspects of this distinction have been discussed in chapter 2, and the corresponding aspects of statistical mechanics and thermodynamics will be discussed in chapter 4. This chapter, focuses on the angular momentum, especially of Fermions with spin-$\frac{1}{2}$.

For the special case $l \equiv s = \frac{1}{2}$, on which we focus from now on, we introduce the operator $\mathbf{S}$ whose components satisfy (3.4), i.e.

$$[S_x, S_y] = iS_z, \qquad [S_y, S_z] = iS_x, \qquad [S_z, S_x] = iS_y. \tag{3.47}$$

The weight of a state $|\chi\rangle$, as previously introduced, measures the eigenvalue of the spin operator of the state, e.g. in the 3– or z–direction, of the quantum particle

$$S_3|\chi\rangle = m_S|\chi\rangle = \pm\frac{1}{2}|\chi\rangle \tag{3.48}$$

[3] Nevertheless, there is no interpretation of quantum particles with integer spin in terms of $\mathbf{L} = \mathbf{r} \times \mathbf{p}$!

and

$$\mathbf{S}^2|\chi\rangle = s(s+1)|\chi\rangle = \frac{3}{4}|\chi\rangle \tag{3.49}$$

with

$$-s \leq m_s \leq s, \qquad \text{i.e.} \qquad m_S = \pm\frac{1}{2}, \tag{3.50}$$

where s (which is $s = \frac{1}{2}$ in this case) is often simply denoted as the spin of the quantum particle.

Corresponding to $m_S = \pm\frac{1}{2}$, we introduce a two-dimensional basis for a spin-$\frac{1}{2}$ particle. The general element of the space spawned by this basis is called a spinor and can be formally written as

$$|\chi\rangle = \begin{pmatrix} |\chi_+\rangle \\ |\chi_-\rangle \end{pmatrix}, \tag{3.51}$$

and whose components are in general complex numbers and which is in general a function of position and time $|\chi\rangle = |\chi(\mathbf{r}, t)\rangle$. Moreover, $||\chi_\pm\rangle|^2$ is interpreted as the probability to find that the particle in the state $|\chi\rangle$ has $m_S = \pm\frac{1}{2}$.

Fixing the coordinate system, we have two possible states: the spin state with $m_S = \frac{1}{2}$

$$\left|\frac{1}{2}, \frac{1}{2}\right\rangle = |\uparrow\rangle = \begin{pmatrix} 1 \\ 0 \end{pmatrix}, \tag{3.52}$$

and the spin state with $m_S = -\frac{1}{2}$

$$\left|\frac{1}{2}, -\frac{1}{2}\right\rangle = |\downarrow\rangle = \begin{pmatrix} 0 \\ 1 \end{pmatrix}, \tag{3.53}$$

which we can choose as basis states. The different notations depend on the context we want to emphasize.

In order to find realizations for the spin operators S_i, we introduce first an operator

$$S^+ \equiv S_1 + iS_2, \tag{3.54}$$

a special case of (3.9), which increases m_S by one. In analogy to the general case, this operator satisfies

$$S^+|\uparrow\rangle = 0 \qquad \text{and} \qquad S^+|\downarrow\rangle = S^+\begin{pmatrix} 0 \\ 1 \end{pmatrix} = |\uparrow\rangle = \begin{pmatrix} 1 \\ 0 \end{pmatrix}. \tag{3.55}$$

In the same way, the operator S^-, a special case of (3.10), decreases m_S by one

$$S^- = S_1 - iS_2 \tag{3.56}$$

i.e.

$$S^-|\uparrow\rangle = S^-\begin{pmatrix} 1 \\ 0 \end{pmatrix} = |\downarrow\rangle = \begin{pmatrix} 0 \\ 1 \end{pmatrix} \quad \text{and} \quad S^-|\downarrow\rangle = 0. \tag{3.57}$$

In the expressions above, we have used that (3.46) in these cases gives

$$N^+_{\frac{1}{2},\frac{1}{2}} = N^-_{\frac{1}{2},-\frac{1}{2}} = 0 \quad \text{and} \quad N^+_{\frac{1}{2},-\frac{1}{2}} = N^-_{\frac{1}{2},\frac{1}{2}} = 1. \tag{3.58}$$

From these results, we deduce that

$$\left(S^+ + S^-\right)\begin{pmatrix} 1 \\ 0 \end{pmatrix} = 2S_1\begin{pmatrix} 1 \\ 0 \end{pmatrix} = \begin{pmatrix} 0 \\ 1 \end{pmatrix} \tag{3.59}$$

and

$$\left(S^+ + S^-\right)\begin{pmatrix} 0 \\ 1 \end{pmatrix} = 2S_1\begin{pmatrix} 0 \\ 1 \end{pmatrix} = \begin{pmatrix} 1 \\ 0 \end{pmatrix}. \tag{3.60}$$

Both of these equations can be realized by S_1 in the form of the 2×2 matrix

$$S_1 = \frac{1}{2}\begin{pmatrix} 0 & 1 \\ 1 & 0 \end{pmatrix}. \tag{3.61}$$

Repeating the arguments above for $S^+ - S^- = 2iS_2$, we find

$$S_2 = \frac{1}{2}\begin{pmatrix} 0 & -i \\ i & 0 \end{pmatrix}. \tag{3.62}$$

Finally, the matrix for S_3 produces the eigenvalues $m_S = \pm\frac{1}{2}$

$$S_3 = \frac{1}{2}\begin{pmatrix} 1 & 0 \\ 0 & -1 \end{pmatrix}, \quad S_3\left|\frac{1}{2}, \pm\frac{1}{2}\right\rangle = \pm\frac{1}{2}\left|\frac{1}{2}, \pm\frac{1}{2}\right\rangle. \tag{3.63}$$

Here are some elementary exercises to get acquainted with spin matrices and spinors.

EXERCISE 3.2 Spin operators in matrix form Verify that the spin operators in matrix form that we just derived satisfy the commutation relations (3.47).

Show that they also satisfy the relations

$$S_1 S_2 = -S_2 S_1 = \frac{i}{2} S_3, \qquad S_3 S_1 = -S_1 S_3 = \frac{i}{2} S_2, \qquad S_2 S_3 = -S_3 S_2 = \frac{i}{2} S_1 \quad (3.64)$$

and that hence the anti-commutators of the spin matrices vanish.

Moreover, show that the squares of the spin matrices are proportional to the 2×2 unit matrix.

Summarizing the matrix realizations of the spin operator $\mathbf{S}$ we introduce the vector of the Pauli spin matrices $\boldsymbol{\sigma}$

$$\mathbf{S} = \frac{1}{2} \boldsymbol{\sigma} \quad (3.65)$$

or, in components

$$S_1 = \frac{1}{2} \begin{pmatrix} 0 & 1 \\ 1 & 0 \end{pmatrix} = \frac{1}{2} \sigma_1, \; S_2 = \frac{1}{2} \begin{pmatrix} 0 & -i \\ i & 0 \end{pmatrix} = \frac{1}{2} \sigma_2, \; S_3 = \frac{1}{2} \begin{pmatrix} 1 & 0 \\ 0 & -1 \end{pmatrix} = \frac{1}{2} \sigma_3, \quad (3.66)$$

which satisfy the relations (3.4) and (3.47), respectively.

EXERCISE 3.3 A useful relation Prove that for two vectors $\mathbf{a}$ and $\mathbf{b}$

$$(\mathbf{a} \cdot \boldsymbol{\sigma})\,(\mathbf{b} \cdot \boldsymbol{\sigma}) = \mathbf{a} \cdot \mathbf{b} + i\,(\mathbf{a} \times \mathbf{b}) \cdot \boldsymbol{\sigma} \quad (3.67)$$

EXERCISE 3.4 Unit vector corresponding to a spinor We define a unit vector $\mathbf{e}$ in the spinor state $|\chi\rangle$ by

$$\mathbf{e} = \langle \boldsymbol{\sigma} \rangle_\chi = \langle \chi | \boldsymbol{\sigma} | \chi \rangle. \quad (3.68)$$

Calculate $\mathbf{e}$ for the spinor

$$|\zeta\rangle = \frac{1}{\sqrt{2}} \begin{pmatrix} 1 \\ 1 \end{pmatrix} \quad (3.69)$$

and show how, once you have the unit vector $\mathbf{e}$, you can recover the spinor $|\zeta\rangle$.

EXERCISE 3.5 A simple spin Hamiltonian Calculate the energy eigenvalues and the eigenstates of the Hamiltonian

$$\mathcal{H} = B\cos\theta \cdot \sigma_x + B\sin\theta \cdot \sigma_y. \tag{3.70}$$

EXERCISE 3.6 Two-level system A two-level system can be described by the Hamiltonian

$$\mathcal{H}_0 = \sigma_x. \tag{3.71}$$

Add a term

$$\mathcal{V} = B\sigma_z \tag{3.72}$$

and calculate the energy eigenvalues of $\mathcal{H} = \mathcal{H}_0 + \mathcal{V}$ by direct diagonalization.

For small field $B \ll 1$ you may consider $\mathcal{V}$ as a perturbation. Calculate the corrections to the energy eigenvalues of $\mathcal{H}_0$ up to second order in perturbation theory and compare with the exact result.

EXERCISE 3.7 Useful relations for spin-$\frac{1}{2}$ operators The spin-1/2 raising and lowering operators in terms of Pauli matrices are

$$\sigma^{\pm} = \frac{1}{2}\left(\sigma^x \pm i\sigma^y\right) \tag{3.73}$$

Calculate all products between pairs of these two operators and their commutation and anti-commutation relations. Furthermore, calculate the commutation and anti-commutation relations between the raising and lowering operators and the Pauli operators σ^x, σ^y, and σ^z.

Show that a function f that can be expanded into a power series can, as an operator function of the product of the spin-$\frac{1}{2}$ raising and lowering operators, only have the form

$$f(\sigma^+\sigma^-) = f(0) + [f(1) - f(0)]\,\sigma^+\sigma^-, \tag{3.74}$$

$$f(\sigma^-\sigma^+) = f(0) + [f(1) - f(0)]\,\sigma^-\sigma^+. \tag{3.75}$$

Moreover, show that the following relations hold

$$e^{\frac{\alpha\sigma_z}{2}}\sigma^+ e^{\frac{\alpha\sigma_z}{2}} = e^{\alpha}\sigma^+, \tag{3.76}$$

$$e^{\frac{\alpha\sigma_z}{2}}\sigma^- e^{\frac{\alpha\sigma_z}{2}} = e^{\alpha}\sigma^-. \tag{3.77}$$

In order to conclude this section on the spin of one particle, we now demonstrate that the spin-$\frac{1}{2}$ degree of freedom has a most remarkable property. Consider a new coordinate system where spin space has been rotated by an angle θ about the 2-axis. This rotation transforms S_3 into

$$S_3' = S_3 \cos\theta + S_1 \sin\theta = \frac{1}{2}\cos\theta \begin{pmatrix} 1 & 0 \\ 0 & -1 \end{pmatrix} + \frac{1}{2}\sin\theta \begin{pmatrix} 0 & 1 \\ 1 & 0 \end{pmatrix} \tag{3.78}$$

$$= \frac{1}{2}\begin{pmatrix} \cos\theta & \sin\theta \\ \sin\theta & -\cos\theta \end{pmatrix}. \tag{3.79}$$

The transformed spin operator S_3' must have the same eigenvalues $m_S = \pm\frac{1}{2}$ in the rotated system as the original spin operator S_3 had in the unrotated system. In the rotated coordinate system the eigenvector $|\uparrow'\rangle$ for eigenvalue $\left(+\frac{1}{2}\right)$, i.e.

$$S_3'|\uparrow'\rangle = +\frac{1}{2}|\uparrow'\rangle \tag{3.80}$$

is

$$|\uparrow'\rangle = \begin{pmatrix} \cos\frac{\theta}{2} \\ \sin\frac{\theta}{2} \end{pmatrix} = \cos\frac{\theta}{2}|\uparrow\rangle + \sin\frac{\theta}{2}|\downarrow\rangle \tag{3.81}$$

and for eigenvalue $\left(-\frac{1}{2}\right)$, i.e.

$$S_3'|\downarrow'\rangle = -\frac{1}{2}|\downarrow'\rangle \tag{3.82}$$

we have

$$|\downarrow'\rangle = \begin{pmatrix} -\sin\frac{\theta}{2} \\ \cos\frac{\theta}{2} \end{pmatrix} = -\sin\frac{\theta}{2}|\uparrow\rangle + \cos\frac{\theta}{2}|\downarrow\rangle. \tag{3.83}$$

Therefore, if we measure the spin to be in the spin up state in the rotated coordinate system, we shall, upon measuring, find it in the original coordinate system in the spin up state with probability $\cos^2\frac{\theta}{2}$ and in the spin down state with probability $\sin^2\frac{\theta}{2}$.

Now, if we rotate through an angle of 2π, a remarkable situation arises. A particle we measure to be in the spin up state in the rotated coordinate system will be with probability 1 in the down state with respect to the original coordinate system. In other words, a rotation of the universe by 2π causes all spin-$\frac{1}{2}$ particles to assume unitarily transformed eigenstates, a distinguishing feature between intrinsic spin and the usual notion of angular momentum.

3.2 Angular momentum of several quantum particles

We begin our discussion of the coupling of the angular momentum of several quantum particles by first considering the case of spin $s = l$ before we specialize to $s = \frac{1}{2}$, which is the case most important for us. In order to gain a clear understanding, in this section we shall use a slightly cumbersome but precise notation that we will later abandon again (in most cases), except when it may be rendered necessary by the demands of a special situation.

The total spin vector of a system of N identical particles each of spin $s = l$ is

$$\mathbf{S}(N) = \sum_{i=1}^{N} \mathbf{S}_i \tag{3.84}$$

where

$$\mathbf{S}_i = I \otimes I \otimes \ldots \otimes \underbrace{\mathbf{S}}_{i} \otimes \ldots \otimes I \tag{3.85}$$

acts non-trivially only on the spin state of particle i. The unit operators I in the tensor product act on the spins of all particles except the spin of particle i. The total state space is written as the N-fold tensor product[4] of the state space $\mathscr{V}_l$ of a single particle's spin state space

$$\mathscr{V} = \underbrace{\mathscr{V}_l \otimes \mathscr{V}_l \otimes \ldots \mathscr{V}_l}_{N}, \tag{3.86}$$

which is the vector space whose basis states are N-fold product states of the single–particle basis states $|l, m_i\rangle$

$$|l, \mathbf{m}\rangle = |l, m_1\rangle \otimes |l, m_2\rangle \otimes \ldots \otimes |l, m_N\rangle. \tag{3.87}$$

The components of the total spin operators $S^\alpha, \alpha = 1, 2, 3$ acting on the spin of particle i[5] are given by

$$S^\alpha(N) = \sum_{i=1}^{N} \left(I \otimes I \otimes \ldots \otimes \underbrace{S^\alpha}_{i} \otimes \ldots \otimes I \right). \tag{3.88}$$

Hence, $S^\alpha(N)$ can be realized by a $(2l{+}1)^N \times (2l{+}1)^N$ matrix. The many-particle spin operators $S^\alpha(N)$ also obey the commutation relations (3.4).

4 For more on tensor products, see section 10.7.1.
5 Cf. footnote 2.

The action of the total third component $S_3(N)$ of the spin operator on the state (3.87) above, for example, is

$$S^3(N)|l,\mathbf{m}\rangle = \left(\sum_{i=1}^{N} m_i\right)|l,\mathbf{m}\rangle \tag{3.89}$$

i.e. the total third component of the spin of the many-particle system is simply the sum of the individual third components of spin.

However, already for the square of the total spin the situation is more complicated. For just two particles, to see the essential feature, we have

$$\mathbf{S}(2)\cdot\mathbf{S}(2) = (\mathbf{S}\otimes I + I\otimes\mathbf{S})\,(\mathbf{S}\otimes I + I\otimes\mathbf{S}) = S^2\otimes I + I\otimes S^2 + 2\mathbf{S}\otimes\mathbf{S} \tag{3.90}$$

which is usually written in the notation

$$\mathbf{S}(2)\cdot\mathbf{S}(2) \equiv (\mathbf{S}_1+\mathbf{S}_2)^2 = S_1^2 + S_2^2 + 2\mathbf{S}_1\cdot\mathbf{S}_2 \tag{3.91}$$

$$= 2\frac{1}{2}\left(\frac{1}{2}+1\right) + 2\mathbf{S}_1\cdot\mathbf{S}_2 = \frac{3}{2} + 2\mathbf{S}_1\cdot\mathbf{S}_2 \qquad \text{for} \qquad s=\frac{1}{2} \tag{3.92}$$

with the understanding that the spin operators only act on particle 1 or 2, respectively. The latter is the usual notation in physics, where the tensor product character of the expressions is usually suppressed, but (at most) a sentence added '…operator with index i only acts on particle or spin i …'. We shall, however, adopt a clear policy here: we will flip back and forth between both notations as we see fit for the situation we are describing.

The allowed values for the square of the spin of two coupled particles, one with spin l, the other with spin k, assuming $l \geq k$, are derived from the general result for the state spaces, which we state here without proof

$$\mathscr{V}_l \otimes \mathscr{V}_k = \bigoplus_{i=l-k}^{l+k} \mathscr{V}_i, \tag{3.93}$$

and which means that, if the squares of the spins of the two particles are given by $l(l+1)$ and $k(k+1)$ respectively, these particles may be coupled so that the allowed values of the spin for the two particle system fall in the range

$$\{i(i+1): \quad i = l-k, l-k+1, \ldots l+k-1, l+k\}. \tag{3.94}$$

Later, we will need to couple two and more spin-$\frac{1}{2}$ particles. Therefore, we state that the allowed values for the square of the spin of two coupled spin-$\frac{1}{2}$ particles are determined by the following relation for the state space of the two particles

$$\mathcal{V}_{\frac{1}{2}} \otimes \mathcal{V}_{\frac{1}{2}} = \mathcal{V}_1 \oplus \mathcal{V}_0, \tag{3.95}$$

which expresses the intuitive expectation that two spin-$\frac{1}{2}$ particles can couple to form a compound of either spin-1 or spin-0. Indeed, the general formula gives that one particle with spin $s = \frac{1}{2}$ and another with $r = \frac{1}{2}$ are coupled to form a two-particle system with spin $S = |s - r| = 0$ and $S = s + r = 1$, the corresponding azimuthal quantum numbers being $m_S = 0$ for $S = 0$ and $m_S = 0, \pm 1$ for $S = 1$, the former called *singlet*, the latter *triplet*. A corresponding basis is given by

$$|\chi_s\rangle = \frac{1}{\sqrt{2}} (|\uparrow\rangle \otimes |\downarrow\rangle - |\downarrow\rangle \otimes |\uparrow\rangle) \tag{3.96}$$

for the spin $S = 0$ space, the singlet, and

$$|\chi_t^{(+)}\rangle = |\uparrow\rangle \otimes |\uparrow\rangle \tag{3.97}$$

$$|\chi_t^{(0)}\rangle = \frac{1}{\sqrt{2}} (|\uparrow\rangle \otimes |\downarrow\rangle + |\downarrow\rangle \otimes |\uparrow\rangle) \tag{3.98}$$

$$|\chi_t^{(-)}\rangle = |\downarrow\rangle \otimes |\downarrow\rangle \tag{3.99}$$

for the spin $S = 1$ space, the triplet.[6]

We conclude this chapter with some exercises.

EXERCISE 3.8 Permutation operator Show that the operator

$$\mathcal{P}_{12} = \frac{1}{2} [(\boldsymbol{\sigma} \otimes I)(I \otimes \boldsymbol{\sigma}) + (I \otimes I)] = \frac{1}{2} (\boldsymbol{\sigma}_1 \cdot \boldsymbol{\sigma}_2 + 1) \tag{3.102}$$

exchanges the spin variables of a two-particle state, i.e.

$$\mathcal{P}_{12}\psi(\sigma_1, \sigma_2) = \psi(\sigma_2, \sigma_1) \tag{3.103}$$

[6] In the exercises below, we shall often use notations more frequently used in physics in the same vain as in (3.91), e.g.

$$|\uparrow_1\uparrow_2\rangle = |\uparrow_1\rangle|\uparrow_2\rangle = \begin{pmatrix} 1 \\ 0 \end{pmatrix}_1 \begin{pmatrix} 1 \\ 0 \end{pmatrix}_2 \tag{3.100}$$

for

$$|\uparrow\rangle \otimes |\uparrow\rangle. \tag{3.101}$$

by applying it to the four spin basis states

$$|\uparrow_1\uparrow_2\rangle = |\uparrow_1\rangle|\uparrow_2\rangle, \quad |\uparrow_1\downarrow_2\rangle = |\uparrow_1\rangle|\downarrow_2\rangle, \quad |\downarrow_1\uparrow_2\rangle = |\downarrow_1\rangle|\uparrow_2\rangle, \quad |\downarrow_1\downarrow_2\rangle = |\downarrow_1\rangle|\downarrow_2\rangle, \tag{3.104}$$

where

$$|\uparrow\rangle = \begin{pmatrix} 1 \\ 0 \end{pmatrix} \quad \text{and} \quad |\downarrow\rangle = \begin{pmatrix} 0 \\ 1 \end{pmatrix}. \tag{3.105}$$

EXERCISE 3.9 Two interacting spins The analysis of atomic or molecular systems with two electrons, e.g. the helium atom, suggests that a system consisting of two interacting spins can be described by the Hamiltonian

$$\mathcal{H} = -\mathcal{J}\boldsymbol{\sigma}_1 \cdot \boldsymbol{\sigma}_2 = -\mathcal{J}\left(\sigma_1^x\sigma_2^x + \sigma_1^y\sigma_2^y + \sigma_1^z\sigma_2^z\right) \tag{3.106}$$

$$= -\mathcal{J}\left((\sigma^x \otimes \mathbb{I})(\mathbb{I} \otimes \sigma^x) + (\sigma^y \otimes \mathbb{I})(\mathbb{I} \otimes \sigma^y) + (\sigma^z \otimes \mathbb{I})(\mathbb{I} \otimes \sigma^z)\right) \tag{3.107}$$

a so-called Heisenberg Hamiltonian. The interaction constant $\mathcal{J}$ is customarily called exchange interaction or exchange integral, reflecting its origin in atomic physics. A basis for the Hilbert space of this Hamiltonian is again given by (3.104) with (3.105). The action of the operators in the Heisenberg Hamiltonian on the states of this basis is given by

$$\sigma_1^x\sigma_2^x|\uparrow_1\uparrow_2\rangle = \left(\sigma_1^x|\uparrow_1\rangle\right)\left(\sigma_2^x|\uparrow_2\rangle\right), \tag{3.108}$$

i.e. the operator with index i acts only on particle i.

a) Write $\mathcal{H}$ as a 4 × 4-matrix in this basis. Determine the energy eigenvalues of $\mathcal{H}$ and their degree of degeneracy.
b) Write the eigenstates of $\mathcal{H}$ as linear combinations of the basis states.
c) Determine the eigenvalues of the z-component of the total spin operator $\sigma^z = \sigma_1^z + \sigma_2^z$ in the eigenstates of $\mathcal{H}$.

EXERCISE 3.10 Many interacting spins: spin waves in the one-dimensional Heisenberg model A chain of N spins with lattice spacing a where only nearest neighbours interact with each other can be described by the Heisenberg Hamiltonian

$$\mathcal{H} = -\mathcal{J}\sum_{j=1}^{N} \boldsymbol{\sigma}_j \cdot \boldsymbol{\sigma}_{j+1} \tag{3.109}$$

where the exchange interaction is $\mathcal{J} > 0$. We assume periodic boundary conditions

$$\boldsymbol{\sigma}_{N+1} = \boldsymbol{\sigma}_1. \tag{3.110}$$

a) Show that $\mathcal{H}$ commutes with the third component of the total spin operator

$$\sigma^3 = \sum_{j=1}^{N} \sigma_j^3. \tag{3.111}$$

It will prove useful to express $\mathcal{H}$ through the spin raising and lowering operators $\sigma_i^{\pm} = \sigma_i^1 \pm i\sigma_i^2$ and the third component of the spin operator σ_i^3 and to first calculate the commutators $\left[\sigma_i^3, \sigma_i^{\pm}\right]$.

b) The ground state of this model is

$$|0\rangle \equiv \prod_{j=1}^{N} |\uparrow\rangle_j = \prod_{j=1}^{N} \begin{pmatrix} 1 \\ 0 \end{pmatrix}_j, \tag{3.112}$$

i.e. all spins are in eigenstates of σ_i^3 with eigenvalue $\left(\frac{1}{2}\right)$.
Calculate the ground state energy E_0 from

$$\mathcal{H}|0\rangle = E_0|0\rangle. \tag{3.113}$$

The generalization for many spins of the operator (3.102) introduced in exercise 3.8

$$\boldsymbol{\sigma}_i \cdot \boldsymbol{\sigma}_{i+1} = 2P_{i,i+1} - 1 \tag{3.114}$$

will simplify calculations.

c) Introduce the state with a down spin at lattice site m

$$|m\rangle = \sigma_m^-|0\rangle \tag{3.115}$$

and show that this state satisfies

$$\mathcal{H}|m\rangle = E_0|m\rangle + 2\mathcal{J}\left(2|m\rangle - |m+1\rangle - |m-1\rangle.\right) \tag{3.116}$$

The energy eigenstates, i.e. the eigenstates of $\mathcal{H}$, can now be calculated by forming the linear combination of one-spin states

$$|k\rangle = \sum_{m=1}^{N} e^{ikma}|m\rangle, \tag{3.117}$$

the so-called spin waves.
Calculate the spin wave energies E_k.
Which values are allowed for the wave number k?
Calculate the eigenvalue of σ^3 in the spin wave state $|k\rangle$.

EXERCISE 3.11 Spin operators for electrons Spin operators for electrons at a lattice site $\mathbf{R_i}$ are defined as

$$\mathbf{S_i} = \frac{1}{2}\sum_{\sigma\sigma'} c^{\dagger}_{\mathbf{i}\sigma} (\boldsymbol{\sigma})_{\sigma\sigma'} c_{\mathbf{i}\sigma'} \tag{3.118}$$

using the electron creation and annihilation operators $c^{\dagger}_{\mathbf{i}\sigma}$ and $c_{\mathbf{i}\sigma}$ and the vector of Pauli spin matrices $\boldsymbol{\sigma} = (\sigma^x, \sigma^y, \sigma^z)^T$.

Show that these operators indeed satisfy the commutation relations of angular momentum operators.

Furthermore, show that the spin raising and lowering operators and the z-component of the spin operators are given by

$$S^{+}_{\mathbf{i}} = S^{x}_{\mathbf{i}} + iS^{y}_{\mathbf{i}} = c^{\dagger}_{\mathbf{i}\uparrow} c_{\mathbf{i}\downarrow}, \tag{3.119}$$

$$S^{-}_{\mathbf{i}} = S^{x}_{\mathbf{i}} - iS^{y}_{\mathbf{i}} = c^{\dagger}_{\mathbf{i}\downarrow} c_{\mathbf{i}\uparrow} = \left(S^{+}_{\mathbf{i}}\right)^{\dagger}, \tag{3.120}$$

$$S^{z}_{\mathbf{i}} = \frac{1}{2}\left(c^{\dagger}_{\mathbf{i}\uparrow} c_{\mathbf{i}\uparrow} - c^{\dagger}_{\mathbf{i}\downarrow} c_{\mathbf{i}\downarrow}\right) = \frac{1}{2}\left(n_{\mathbf{i}\uparrow} - n_{\mathbf{i}\downarrow}\right). \tag{3.121}$$

....

4

Equilibrium Statistical Mechanics

I am conscious of being only an individual struggling weakly against the stream of time. But it still remains in my power to contribute in such a way that, when the theory of gases is again revived, not too much will have to be rediscovered.

Ludwig Boltzmann (1844–1906)

David Goodstein (1985) made the macabre remark, that 'Ludwig Boltzmann, who spent much of his life studying statistical mechanics, died in 1906, by his own hand. Paul Ehrenfest, a student of Boltzmann's, carrying on the same work, died similarly in 1933. Now it is our turn to study statistical mechanics.'

Ludwig Boltzmann's pessimism about his own times, expressed in this chapter's epigraph, has given way to an almost unanimous acceptance of the atomistic and statistical understanding of the physics of large agglomerations of particles. With this modern background, David Goodstein's remark only appears to be frightening and unpleasant.

Therefore, this chapter reviews the basic notions of equilibrium statistical mechanics without making an attempt either at a complete coverage or at delving into fundamental questions or advanced applications of this fascinating subject.

More detailed expositions of statistical mechanics which, in particular, contain many illuminating applications of the concepts described briefly in this chapter, can be found in many textbooks, of which a short selection includes Chandler (1987), Sethna (2006), Schwabl (2006), and Reichl (2016). An approach to the subject mainly via problems is given in the classic book by Kubo (1988).

The first section of this chapter, section 4.1, sets forth the fundamental postulate of equilibrium statistical mechanics and outlines from this fundamental assumption the structure of the theory. The various statistical ensembles concretize the fundamental postulate via a specification of the ways systems can be partitioned.

The *microcanonical* ensemble (section 4.2) draws conclusions for an isolated system. It leads to the introduction of the concept of entropy (section 4.3) and the second law of thermodynamics (section 4.4), as well as to a definition of temperature (section 4.5), pressure, and chemical potential (section 4.6), and, finally, the formulation of the first law of thermodynamics (section 4.7).

Models of Quantum Matter: A First Course on Integrability and the Bethe Ansatz. Hans-Peter Eckle, Oxford University Press (2019). © Hans-Peter Eckle. DOI: 10.1093/oso/9780199678839.001.0001

The *canonical* ensemble (section 4.8) partitions the whole system into the system of interest and an energy reservoir. It is especially versatile for calculational purposes via the introduction of the important concept of the *partition function* (section 4.9).

As the last of the many possible ensembles, we introduce here the *grand canonical ensemble* (section 4.10), where the system of interest also exchanges particles with a reservoir system. This ensemble will be especially useful for quantum statistical mechanics.

However, before concentrating on quantum statistics, we insert a brief section on an important generalization due to Gibbs, aptly called the Gibbs entropy (section 4.11), connecting the thermostatistical entropy concept with the modern information concept.

In order to discuss the basics of quantum statistics, we first introduce the density matrix or density operator formalism (section 4.12), which we concretize for the statistical ensembles previously introduced.

One of the simplest cases in quantum statistics are the non-interacting quantum gases (section 4.13) in analogy to the classical ideal gas. These systems are nonetheless of great current research interest as the basis of the ongoing studies of ultracold quantum gases.

An introduction to the Ising model will render many of the subsequent discussions meaningful. Moreover, some acquaintance with other classical lattice models (section 4.14) of statistical mechanics will also prove helpful for our discussion in part II of the algebraic Bethe ansatz.

As a preparation for the following chapter on *phase transitions and critical phenomena*, we treat interacting magnetic moments (section 4.15) in a mean-field theory approximation, thus introducing in a simple setting many of the notions of phase transitions and critical phenomena.

The final two sections of this chapter are devoted to a first introduction, in section 4.16, of the transfer matrix and its application to exactly solve the Ising model in one dimension (section 4.17).

4.1 Fundamental postulate of equilibrium statistical mechanics

Equilibrium statistical mechanics and, more generally, statistical mechanics studies the macroscopic properties of huge assemblies of interacting subsystems. To achieve this goal, statistical mechanics attempts to extract from the laws of microscopic physics a description of phenomena on a macroscopic scale. Due to the enormous number of subsystems, any detailed knowledge of all subsystems would be so overwhelming as to render it meaningless. In a famous essay, Philip Anderson (1972) expressed this by saying 'More is different' which points to the emergence of many key concepts of macroscopic systems, which are not visible in the underlying laws of physics of the subsystems, but only when large collections of subsystems are considered. Take one of the simplest examples: temperature. There is no sensible way to ascribe a temperature to a single electron.

The *fundamental postulate* of *equilibrium* statistical mechanics is the postulate of *equal a priori probabilities*:

> **An isolated system in equilibrium is equally likely to be found in any of the microstates accessible to it.**

This statement introduces a number of concepts that need further clarification. Some of the concepts will become more refined as we proceed in this chapter.

System refers to that part of the universe in which we are currently interested. We assume that the subsystems of this system are so weakly coupled to any outside systems that these couplings are negligible compared to the couplings of the subsystems within the system of interest.

This is further specified by the term *isolated*, an idealization that means that all external influences, e.g. energy or particle fluxes between the system of interest and the rest of the universe, are eliminated.

For a system in *equilibrium* all transient behaviour has decayed and therefore all macroscopic quantities are independent of time. To achieve this idealized situation one 'simply' has to wait long enough. Often it is practically impossible to wait long enough and one has to study the non-equilibrium situation directly. For example, systems, such as glasses, with a complex energy landscape can remain for a long time in a deep energy minimum before relaxing to the equilibrium. However, problems of non-equilibrium statistical mechanics will not concern us in for the most part in this book.

A *microstate* is a complete specification of all subsystems of the system. In principle, the dynamics of the whole system could be mapped by following the dynamics of all microstates. In most cases, this is not possible in practice and the very reason we develop the methods of statistical mechanics is to deal with this impossibility. For example, for a system of classical particles, a microstate specifies the positions and velocities or momenta of all particles. For a quantum mechanical system, the quantum state of a quantum mechanical many-particle system must be specified to obtain the corresponding microstate. This can be done, for example, by specifying the many-particle wave function. Chapter 2 on *quantum many-particle systems and second quantization* has been devoted already to methods to deal with large numbers of quantum particles. In this chapter, we return to these issues in some more detail in sections 4.12 on the *density matrix* and 4.13 on *non-interacting quantum gases*.

The word *likely* refers to the fact that statistical mechanics deals with probabilities to account for the overwhelmingly many possible microstates. The isolated system is found with equal probability in any of these microstates.

Finally, not all microstates may be *accessible* to a system because macroscopic constraints, e.g. a fixed energy or a fixed number of particles, have been imposed on the system.

Equilibrium statistical mechanics is based on statistical *ensembles*. An ensemble is a collection of microstates with a probability measure, i.e. a probability attached to each microstate. We imagine a huge number of repetitions of essentially the same system. The term *same* here means *macroscopically* the same, but possibly differing in many microscopic details. The probability of a microstate is then the fraction of the states of the ensemble that represent the same microstate. This so-called ensemble average, introduced by Gibbs, is far superior conceptually to taking time averages. Experimentally, of course, preparing many macroscopically nearly identical copies of a system can be very difficult. It is much easier to observe a single system over time and then take time averages of its properties after large enough intervals of time. In most cases, the *ergodicity* hypothesis, which states that the system, given enough time, comes sufficiently close to each microstate, warrants that measurements on a single system averaged over time can be calculated by ensemble averages. However, there exists no general proof of the validity of the ergodicity hypothesis. On the contrary, we can devise examples of non–ergodic behaviour of systems. Again, we shall not follow this exciting problem at the foundation of statistical mechanics further.

The postulate reduces the task of finding the probability for a macroscopic configuration of an isolated system to counting the number of microstates that are consistent with the macroscopic configuration. The simplest formulation for this concept is based on quantum mechanics, where discrete states are considered and, hence, discrete energy states have to be counted that lead to the same macroscopic energy.

Three ensembles play a prominent role and will be discussed in the subsequent sections. The first of these, the microcanonical ensemble, is based directly on the fundmental postulate discussed so far in this section. The other two ensembles, the canonical and the grand canonical, are obtained by dispensing with the requirement of a totally isolated system allowing for energy exchange in the canonical ensemble or energy and particle exchange in the grand canonical ensemble with an extremely huge reservoir of energy or particles that is not noticeably changed by the exchanges with our system of interest. The overall system, i.e. the system we are interested in together with the huge reservoir, is, again, an isolated system for which we assume the microcanonical ensemble to be applicable. The fundamental idea to extract information about a macroscopic system therefore consists, as we shall see several times in the following developments, of partitioning the isolated system into two macroscopic subsystems allowing for specified exchange of certain quantities while keeping the others still fixed. This concept even works well within the microcanonical ensemble where all quantities are kept fixed in the completely isolated system, which, however, we may imagine as consisting of two subsystems.

To each of these (and other) ensembles there is an associated thermodynamic potential, internal energy, Helmholtz free energy, and grand canonical potential, respectively, for the three cases discussed next.

4.2 Microcanonical ensemble

Consider, as an idealization, an isolated system. Isolation means that, by definition, there is no exchange whatever with the outside world. Therefore, all quantities—energy, particle number, momentum, angular momentum—are fixed. Since, in practice, we often think of a closed container, momentum and angular momentum are not conserved quantities and thus are not taken into account here. In astrophysical applications of statistical mechanics this may be different. The isolated system is described by the microcanonical ensemble. The fundamental assumption of the microcanonical ensemble, as we have no further information about the system, is that all states with fixed energy $E = U$, fixed particle number N, fixed volume V, and possibly other fixed quantities like, e.g. the magnetization $\mathbf{M}$ in an appropriate magnetic system, are equally probable. Thus, the probability of each state is equal to the inverse of the number of microstates $W = W(U, V, N, \ldots)$ consistent with the constraints of fixed energy U, particle number N, volume V. In other words, the microcanonical ensemble is characterized via the counting of microstates of the isolated system.

It should be noted that all fixed quantities in the microcanonical ensemble are *extensive* or *additive*, i.e. they are quantities that depend on the size of the system. If we add together, for instance, two identical systems, the extensive quantities all double. Later, we shall introduce intensive quantities that do not depend on the system size. For both extensive and intensive quantities we shall also use the name *thermodynamic variable.*

4.3 Entropy

The entropy S is a measure for the uncertainty about the microstate and is given, according to Boltzmann,[1] through W as (where from now on we suppress the dots indicating further possible dependencies on extensive thermodynamic variables)

$$S = S(U, V, N) = k \ln W(U, V, N) = \ln W(U, V, N) \tag{4.1}$$

introducing for a moment Boltzmann's constant k to obtain the correct physical dimension and then immediately setting $k = 1$. This definition ensures that the entropy is additive, because two non-interacting systems A and B with numbers of microstates $W_A = W(U_A, V_A, N_A)$ and $W_B = W(U_B, V_B, N_B)$ have a combined number of microstates

$$W(U_A, U_B, V_A, V_B, N_A, N_B) = W(U_A, V_A, N_A) W(U_B, V_B, N_B), \tag{4.2}$$

and, hence

$$S(U_A, U_B, V_A, V_B, N_A, N_B) = S(U_A, V_A, N_A) + S(U_B, V_B, N_B). \tag{4.3}$$

The additivity property of the entropy S (4.3) implies that S is a monotonous function of its variables. Therefore, S can be inverted to obtain the thermodynamic potential connected with the microcanonical ensemble the internal energy $U = U(S, V, N)$ of

[1] This formula is carved on Ludwig Boltzmann's gravestone in Vienna's Zentralfriedhof (Central Cemetery). See, e.g. the entry on Boltzmann's entropy formula in http://en.wikipedia.org/wiki/

the system. By its construction, the internal energy depends on the entropy and all other specified extensive thermodynamic variables, e.g. volume V and particle number N.

4.4 Second law of thermodynamics

Now, we bring the two non-interacting systems A and B together in such a way that they can exchange energy but assume that the changes in the energy levels of the individual systems are negligible. Under this assumption, the energy of the combined system remains unchanged

$$U_{\text{total}} = U_A + U_B, \tag{4.4}$$

but this energy can be redistributed such that system A has an energy $U \leq U_{\text{total}}$ while system B has the remaining energy $U_{\text{total}} - U$ where, for simplicity, we assume that the energy distribution is such that the energies are permissible as discrete quantum mechanical energy levels of their respective systems.

The number of available states of the combined system becomes[2]

$$W(U_{\text{total}}) = \sum_U W_A(U) W_B(U_{\text{total}} - U) \tag{4.5}$$

$$= \sum_U \exp\left(S_A(U) + S_B(U_{\text{total}} - U)\right). \tag{4.6}$$

In practice, for large systems the discrete energy levels U of the above sums become continuous and the sums can be replaced by integrals.

Since the combined system is again an isolated system, we can employ the micro-canonical ensemble, i.e. the probability for the combined system to be in any of its states is equally distributed with

$$p = \frac{1}{W(U_{\text{total}})}. \tag{4.7}$$

As can be inferred from (4.5), the states of the originally separated systems A and B are a subset of the states of the combined system. Therefore, the entropy of the combined system cannot be smaller than the entropy of the original separated systems

$$S(U_{\text{total}}) \equiv \ln W(U_{\text{total}}) \geq S_A(U_A) + S_B(U_B). \tag{4.8}$$

This is the second law of thermodynamics, which states the entropy is never decreasing for equilibrium systems.

[2] Here and in the following we suppress the variables V and N until we shall need them again explicitly.

4.5 Temperature

The temperature of the isolated statistical system can now be defined as

$$\frac{1}{T} = \frac{\partial S}{\partial U} \tag{4.9}$$

by the following argument.

Consider again a system consisting of two subsystems A and B that may only exchange energy via a weak link. The energy of the combined system is $E = E_A + E_B$. The volumes and particle numbers of the subsystems remain unchanged. The energy will typically flow in the direction that increases the number of microstates available to the two systems. Therefore, the number of microstates of the combined system

$$W_{A+B}(E) = W_A(E_A)W_B(E_B) \tag{4.10}$$

assumes a maximum when the combined system reaches equilibrium and, hence

$$\frac{\partial}{\partial E}W_{A+B}(E) = \frac{\partial W_A(E_A)}{\partial E_A}W_B(E_B) + W_A(E_A)\frac{\partial W_B(E_B)}{\partial E_B} = 0, \tag{4.11}$$

which can be rewritten, using the entropy formula (4.1) as

$$\frac{\partial S_A(E_A)}{\partial E_A} = \frac{\partial S_B(E - E_A)}{\partial E_A} \quad \text{i.e.} \quad \frac{1}{T_A} = \frac{1}{T_B}. \tag{4.12}$$

Using the temperature definition (4.9), this argument shows that the energy will flow from the system with the higher temperature to that with the lower with the result that the temperatures in both systems will change, and the net flow of energy will terminate when both subsystems reach the same temperature. Thus, the systems reach the most probable distribution of energy, i.e. the one that corresponds to the maximal number of microstates. As a caveat, we mention that to confirm the temperature definition (4.9) is correct we need to go through the calculations leading to (4.9) for a concrete system, e.g. the ideal gas.

4.6 Pressure and chemical potential

In the microcanonical ensemble volume V and number of particles N, and perhaps other extensive quantities, have been kept fixed. The following two exercises intend to show how, in close analogy to temperature in the previous section, pressure and chemical potential can be derived.

EXERCISE 4.1 Pressure Repeat the arguments of section 4.5 for two systems that are in contact through a moveable partition such that the overall volume remains fixed but may be distributed freely between the systems. Energy and particles may not be exchanged between the systems. Show that $\partial S/\partial V$ must be the same for both systems when, in equilibrium, the volumes of the systems cease to change and that then

$$\frac{\partial S}{\partial V} = \frac{p}{T} \tag{4.13}$$

where p is identified as the pressure.

EXERCISE 4.2 Chemical potential In the same way, show that for two systems that are only allowed to exchange particles the chemical potential defined as

$$\frac{\mu}{T} = -\frac{\partial S}{\partial N} \tag{4.14}$$

has to be the same in equilibrium.

4.7 First law of thermodynamics

The results of the last two sections can be combined to write a general entropy change as

$$dS(U, V, N) = \frac{\partial S}{\partial U}dU + \frac{\partial S}{\partial V}dV + \frac{\partial S}{\partial N}dN = \frac{1}{T}dU + \frac{p}{T}dV - \frac{\mu}{T}dN. \tag{4.15}$$

Rearranging this, we obtain the first law of thermodynamics for equilibrium systems

$$dU = TdS - pdV + \mu dN. \tag{4.16}$$

The following sections derive further ensembles from the microcanonical ensemble. The main idea is always to consider a part of a huge reservoir system as a subsystem where the system is treated entirely as an isolated system, according to the microcanonical ensemble, but our interest will be focused on what happens in the subsystem when certain quantities are exchanged between it and the reservoir. The reservoir is assumed to be so huge that the exchange with its subsystem does not change the reservoir noticeably.

4.8 Canonical ensemble

The canonical ensemble describes a system S that exchanges energy with a reservoir, in this context also called a heat bath, R. The assumptions of the canonical ensemble are, thus,

1. The reservoir R is much larger than the system S, to be described.
2. Furthermore, the interaction between the reservoir R and the system S is weak, i.e. at any given time the energy in the system is much larger than the energy in transit between S and R.
3. Finally we assume that we can consider the combined system of S and R as isolated from the outside world and, hence, use the microcanonical ensemble to describe the combined system.

Because the interaction between S and R is assumed to be weak, we can treat the states of S and R separately, and, moreover, we may view the states of S + R as composed of a state of S and a state of R each. For the energy of a general state $E_{\mathrm{R+S}}$ this means that energy contributions from the interaction between S and R, $E_{\mathrm{R,\,S}}$ say, can be neglected and we have

$$E_{\mathrm{R+S}} = E_{\mathrm{R}} + E_{\mathrm{S}} + E_{\mathrm{R,S}} \approx E_{\mathrm{R}} + E_{\mathrm{S}} \tag{4.17}$$

as an excellent approximation.

The probability of a given microstate j of S with energy $E_{\mathrm{S}} = E_j$ is proportional to the number of states of R compatible with the microstate j of S, i.e. to the ratio

$$P_j \propto \frac{W_{\mathrm{R}}}{W_{\mathrm{R+S}}} = \exp\left(S_{\mathrm{R}} - S_{\mathrm{R+S}}\right), \tag{4.18}$$

where the entropies are

$$S_{\mathrm{R}} = S_{\mathrm{R}}(E_{\mathrm{R}}), \tag{4.19}$$

and

$$S_{\mathrm{R+S}} = S_{\mathrm{R+S}}(E_{\mathrm{R+S}}) = S_{\mathrm{R+S}}(E_{\mathrm{R}} + E_{\mathrm{S}}) = S_{\mathrm{R+S}}(E_{\mathrm{R}} + E_j) \approx S_{\mathrm{R}}(E_{\mathrm{R}} + E_j). \tag{4.20}$$

As the energy of the reservoir does not change perceivably, P_j is a function only of the energy E_j of microstate j. Given the relative size of S and R, it is reasonable to attempt a Taylor expansion in the exponent of (4.18) in powers of E_j.

$$P_j \propto \frac{W_{\mathrm{R}}}{W_{\mathrm{R+S}}} = \exp\left(S_{\mathrm{R}} - S_{\mathrm{R+S}}\right) = \exp\left(\frac{1}{k}\left[-E_j\frac{\partial S_{\mathrm{R}}}{\partial E_{\mathrm{R}}} + \ldots\right]\right). \tag{4.21}$$

Recalling the temperature definition (4.9), we obtain

$$P_j \propto \exp\left(-\frac{E_j}{T}\right). \tag{4.22}$$

This derivation makes it clear that the temperature is determined by the heat reservoir or heat bath. Moreover, it became also obvious that the system S need not be macroscopic. It can even be as small as, e.g. a single particle whose motion is studied when coupled harmonically to a heat bath of harmonic oscillators (Ford *et al.*, 1965).

The next term in the Taylor expansion may be neglected: it is proportional to the change of temperature of the reservoir R as a result of the loss of energy E_j. Since the reservoir R is sufficiently big compared to the system S, this can be made arbitrarily small.

The proportionality (4.22) can be written as an equality

$$P_j = \frac{1}{Z} \exp\left(-\frac{E_j}{T}\right), \tag{4.23}$$

where

$$Z = \sum_j \exp\left(-\frac{E_j}{T}\right) \tag{4.24}$$

normalizes the total probability to unity.

4.9 Partition function

The normalization Z is called the partition function.[3] Despite its humble origin as a mere factor to normalize the probability distribution (4.22) to unity, as probability distributions have to be normalized, the partition function is arguably the central quantity of statistical mechanics. As we shall see shortly, knowledge of the partition function suffices to compute all the macroscopic thermodynamic quantities of a system.

This function over all discrete energy states can be rewritten using the microcanonical number of states for a given energy $W(E)$, as a sum over energies

$$Z = \sum_E W(E) \exp\left(-\frac{E}{T}\right) = \sum_E \exp\left(-\frac{1}{T}(E - TS(E))\right). \tag{4.25}$$

The dominant contribution to the latter sum comes from the maximal value of the summand which is, by differentiating it with respect to the energy E, the temperature definition in the microcanonical ensemble (4.9) for the value $E = U$ at this maximum. We obtain

[3] The German expression 'Zustandssumme' may be more revealing, as its literal translation is 'sum of states' (with the appropriate weight).

$$Z = \exp\left(-\frac{1}{T}(U - TS(U))\right) = \exp\left(-\frac{F(T)}{T}\right) \tag{4.26}$$

which defines the free energy

$$F(T, V, N) = U - TS \tag{4.27}$$

of the system S.

As anticipated, we see that knowledge of the partition function is indeed the key to the thermodynamics of a statistical mechanics system.

The variables of the free energy $F = F(T, V, N)$ are determined from the observation that F and the internal energy U are formally related by a Legendre transformation

$$F = U - \frac{\partial U}{\partial S}S = U - TS \tag{4.28}$$

from which the differential relation follows

$$dF = dU - TdS - SdT = -SdT - pdV + \mu dN \tag{4.29}$$

using the first law of thermodynamics (4.16).

For a system of N non-interacting subsystems, e.g. the non-interacting atoms of an ideal gas, the partition function becomes particularly simple. The energy E_j of a microstate j is then a sum of energies $E_j(n)$ of the subsystems in this particular microstate j

$$E_j = \sum_{n=1}^{N} E_j(n). \tag{4.30}$$

The partition function in the non-interacting case takes then the form

$$Z = \sum_j \exp\left(-\frac{1}{T}\sum_{n=1}^{N} E_j(n)\right) = \sum_j \prod_{n=1}^{N} \exp\left(-\frac{E_j(n)}{T}\right) \tag{4.31}$$

In the last expression, sum and product can be interchanged. The summation is then only over the microstate j_n of subsystem n, no longer over the full macrostate j of the system S

$$Z = \prod_{n=1}^{N}\left(\sum_{j_n} \exp\left(-\frac{E_{j_n}(n)}{T}\right)\right). \tag{4.32}$$

The expression in brackets is the partition function of one subsystem. For identical subsystems, we can write

$$Z_1 = \left(\sum_{j_n} \exp\left(-\frac{E_{j_n}(n)}{T} \right) \right), \tag{4.33}$$

which gives the final result

$$Z = \prod_{n=1}^{N} Z_1 = (Z_1)^N. \tag{4.34}$$

An example of non-interacting subsystems is an ideal paramagnet consisting of N magnetic moments in an external magnetic field $\mathbf{B}$ which do not mutually interact. In the following two exercises, we consider two models describing paramagnetic behaviour.

EXERCISE 4.3 Paramagnetism of non-interacting magnetic moments: isotropic model Assume a collection of magnetic moments μ with a fixed length which can point in any direction, i.e. μ moves on the surface of a sphere of length $|\mu|$. The magnetic moments do not interact with each other but with a magnetic field $\mathbf{B}$ such that the energy of the magnetic moment in the field is

$$E = -\boldsymbol{\mu} \cdot \mathbf{B}. \tag{4.35}$$

a) Calculate the partition function Z_1 of a single magnetic moment from an appropriate integral over the unit sphere and, from this result, the partition function Z_N of the collection of N magnetic moments.
b) The Gibbs free energy is a function of T, B, and N (rather than of T, p, and N as, e.g. for a gas, cp. exercise 4.6) and can be calculated from

$$G(T, B, N) = -T \ln Z_N. \tag{4.36}$$

Use this expression to calculate the magnetization M of the collection of non-interacting magnetic moments. There are two ways to obtain the magnetization M: using (4.36) with the expression for $Z_1 = Z_1(T, H)$ obtained in a) or by directly differentiating with respect to the magnetic field B the surface integral representation of Z_1 that is the starting point in a).

c) Calculate the magnetic susceptibility

$$\chi = \frac{\partial M}{\partial B}, \tag{4.37}$$

the entropy

$$S = -\left(\frac{\partial G}{\partial T}\right)_{B,N}, \tag{4.38}$$

and the specific heat at constant magnetic field

$$c_B = T\left(\frac{\partial S}{\partial T}\right)_B. \tag{4.39}$$

EXERCISE 4.4 Paramagnetism of non-interacting magnetic moments: strongly anisotropic model Now consider the case, which corresponds to a quantum mechanical spin-1/2 system, where the magnetic moment can only be parallel or antiparallel to the magnetic field that can be chosen, e.g. in the z-direction $\mathbf{B} = B\mathbf{e}_z$. Calculate the partition function, the Gibbs free energy, the magnetization, and the magnetic susceptibility. Calculate also the entropy of the system from the Gibbs free energy, especially for the case of vanishing magnetic field $B = 0$, and vanishing temperature $T \to 0$, and interpret your result. Finally, calculate the specific heat at constant magnetic field c_B. What happens if you lower the magnetic field from B to $B/2$ adiabatically, i.e. slowly enough to keep the entropy constant?

EXERCISE 4.5 Behaviour at low and high temperatures Calculate the behaviour of the magnetization of the models of non-interacting magnetic moments of the previous two exercises for low and high temperatures. Interpret your results.

4.10 Grand canonical ensemble

When a system exchanges not only energy but also particles with the reservoir, it is described by the grand canonical ensemble. Consider a reservoir R of which the system S is a small subvolume V partitioned off from the total, such that the subvolume may still exchange energy and particles with the whole. Otherwise, the same assumptions as described for the canonical ensemble hold. Again, the probability of a state of S can be written as the ratio of the number of states of R compatible with the chosen state of S and the total number of states for the whole:

$$P(j, N, V) \propto \frac{W_\mathrm{R}}{W_\mathrm{R} + W_\mathrm{S}} = \exp\left(\frac{S_\mathrm{R} - S_\mathrm{R+S}}{k}\right). \tag{4.40}$$

As before, the entropy of R will be expanded, now simultaneously in powers of $E_{j,N}$ and N, i.e. the energy, which now itself depends on the particle number, and the particle number of S.

$$S_\mathrm{R} = S_\mathrm{R+S} - E_{j,N}\frac{\partial S_\mathrm{R}}{\partial E_\mathrm{R}} - N\frac{\partial S_\mathrm{R}}{\partial N_\mathrm{R}}. \tag{4.41}$$

The coefficients in this expansion we know from our treatment of the microcanonical ensemble that applies to the reservoir. They are from (4.9)and (4.14), respectively, the inverse temperature

$$\frac{\partial S_R}{\partial E_R} = \frac{1}{T}, \tag{4.42}$$

and the chemical potential

$$\frac{\partial S_R}{\partial N_R} = -\frac{\mu}{T} \tag{4.43}$$

of the bath, which is now a heat bath and a reservoir of particles.

Again, these identifications explain why the second order terms may be neglected: since they correspond to the change in these thermodynamic parameters as a result of the withdrawal of N particles with energy $E_{j,N}$ from R, these second order terms vanish in the limit of a large bath $\mathcal{R}$. As a result, we have for the probability of finding N particles, in state j:

$$P(j, N, V) \propto \exp\left(-\frac{1}{T}(E_{j,N} - \mu N)\right). \tag{4.44}$$

In principle, other ensembles are possible. One can, e.g. keep the pressure constant rather than the volume, if the separator between S and R is a moveable piston rather than a fixed wall. The treatment is analogous. The most commonly considered ensembles are, however, the three described just above. Furthermore, for a mixture of different particles one needs to introduce a chemical potential μ_i for each particle species i.

In principle, while the microcanonical, the canonical, and the grand canonical ensemble describe different situations, the relations between the extensive functions like energy, entropy, particle number, and volume can be derived from all these cases, as can be explicitly verified. Therefore, the selection of the appropriate ensemble is usually guided by convenience rather than the physical situation. However, when one is interested in fluctuations of a particular quantity, or in the variation of one quantity as the result of variation of another, one should take into account precisely which quantities are to be kept constant.

For completeness sake, we note the grand canonical partition function

$$\mathcal{Z}(T, \mu, V) = \sum_{j,N} \exp\left(-\frac{1}{T}(E_{j,N} - \mu N)\right) = \exp\left(-\frac{\Phi}{T}\right) \tag{4.45}$$

and the grand canonical potential

$$\Phi = F - \mu N \tag{4.46}$$

the latter being a function of the variables T, μ, and V which, as usual, we obtain from observing the differential

$$d\Phi = dF - \mu dN - N d\mu = -SdT - pdV - Nd\mu. \tag{4.47}$$

EXERCISE 4.6 Gibbs ensemble Repeat the steps in this section but, instead of allowing for particle exchange between system and reservoir, introduce a moveable piston instead of a fixed wall as separator. This ensemble is called the Gibbs ensemble, the corresponding thermodynamic potential, the Gibbs potential G, which is commonly called the Gibbs free energy in analogy to the Helmholtz free energy F.

The potentials connect the statistical mechanical description to thermodynamics. A good place to explore the riches of the thermodynamic formalism is Callen (1985) or the more focused and concise first edition of this classic book (Callen, 1960).

4.11 Gibbs entropy

The discussion of ensemble theory closes with an exercise that elucidates an important generalization of Gibbs' entropy notion, which has been rediscovered in the context of information theory: it is called Shannon entropy for classical systems, and von Neumann entropy for quantum systems. According to Gibbs, the entropy of a system whose states have, assuming some preparation of the system by imposing certain constraints, a probability distribution $p(n)$ is given by

$$S = -k \sum_n p(n) \ln p(n). \tag{4.48}$$

This entropy definition allows a unified treatment of all ensembles through a variational principle. The entropy (4.48) which is subject to various constraints, has to be maximized. The different ensembles are then distinguished by the constraint. The exploitation of this maximal entropy principle is the content of the following exercises.

EXERCISE 4.7 Gibbs entropy

a) *Microcanonical ensemble*
Even if we have no further information about the system, the probability distribution must still satisfy $\sum_n p(n) = 1$. Use a Lagrange multiplier to add the constraint $\sum_n p(n) = 1$ to the Gibbs entropy (4.48). Show that the entropy is maximized by the microcanonical ensemble in which all states, restricted to fixed energy E, are equally likely. Further, show that in this case the Gibbs entropy (4.48) coincides with the Boltzmann entropy (4.1).

b) *Canonical ensemble*
Next, show that the Gibbs entropy is maximized by the canonical ensemble for fixed average energy $\langle E \rangle = U = \sum_n p(n) E_n$. Moreover, show that the corresponding Lagrange multiplier is proportional to $\beta = 1/T$, the inverse temperature. Confirm that maximizing the entropy corresponds to minimizing the free energy.

c) *Grand canonical ensemble*
Show that at fixed average energy $\langle E\rangle = U$ and average particle number $\langle N\rangle$, the Gibbs entropy is maximized by the grand canonical ensemble. Interpret the Lagrange multipliers in this case.

EXERCISE 4.8 Lattice gas and grand canonical ensemble
Consider a lattice gas on a two-dimensional square lattice of size $M = L \times L$. The number of particles is $N < L^2$. Particles sit on lattice sites N_i, numbered in, e.g. a row-by-row manner, which are either occupied ($N_i = 1$) or unoccupied ($N_i = 0$) with energy E_i (e.g. $E_i = \epsilon$ for occupied, $E_i = 0$ for unoccupied sites). Determine the probability distribution p_i that the particles sit on certain lattice sites i maximizing the Gibbs entropy

$$S = -k \sum_{i=1}^{M} p_i \ln p_i \tag{4.49}$$

given an average particle number $N = \sum_{i=1}^{M} N_i p_i$ and average energy $E = \sum_{i=1}^{M} E_i p_i$. Use the method of Lagrange multipliers and interpret the physical meaning of the Lagrange multipliers.

4.12 Density matrix

The expert will be fully aware of the debt this section owes to Richard Feynman's 1972 masterful exposition.

An important tool in quantum statistical mechanics is the *density operator* or the *density matrix*, also called *statistical operator* or *statistical matrix*. In the spirit of decomposing the large system, in the last consequence the universe, into the system on which our interest focuses and, as a reservoir, the rest of the universe, we write a quantum state $|\psi\rangle$ of the universe, i.e. our system of interest plus the rest of the universe, as a product state of the complete orthonormal sets $|\phi_i\rangle$ of the system and $|\chi_u\rangle$ of the rest of the universe

$$|\psi\rangle = \sum_{i,u} c_{iu} |\phi_i\rangle |\chi_u\rangle. \tag{4.50}$$

An observable of the system represented by the Hermitian operator $\mathcal{A}$ acts only on system states $|\phi_i\rangle$. Its expectation value with respect to the quantum state $|\psi\rangle$, using the orthonormality condition for the reservoir $\langle \chi_v | \chi_u \rangle = \delta_{uv}$, defines the density matrix ρ_{ij}

$$\langle \psi | \mathcal{A} | \psi \rangle = \sum_{ij} \langle \phi_i | \mathcal{A} | \phi_j \rangle \sum_u c_{iu}^* c_{ju} = \sum_{ij} \langle \phi_i | \mathcal{A} | \phi_j \rangle \rho_{ij}. \tag{4.51}$$

The density operator is then defined as the operator ρ with the matrix elements

$$\rho_{ij} = \langle \phi_j | \rho | \phi_i \rangle, \tag{4.52}$$

so that (4.51) can be written as

$$\langle\psi|\mathcal{A}|\psi\rangle = \langle\mathcal{A}\rangle = \sum_i \langle\phi_i|\mathcal{A}\rho|\phi_i\rangle = \mathrm{Tr}(\mathcal{A}\rho) = \mathrm{Tr}(\rho\mathcal{A}). \tag{4.53}$$

Obviously, from(4.51), ρ is a Hermitian operator and can, therefore, be diagonalized with real eigenvalues p_i and a complete orthonormal set $\{|i\rangle\}$, hence

$$\rho_{ij} = p_i\delta_{ij} \quad \text{or} \quad \rho = \sum_i p_i|i\rangle\langle i|. \tag{4.54}$$

Choosing special operators $\mathcal{A}$, we obtain from (4.53) for the identity operator $\mathcal{A} = 1$,

$$\sum_i p_i = 1, \tag{4.55}$$

and for the projection operator onto the state $|j\rangle$, $\mathcal{A} = |j\rangle\langle j|$,

$$p_j = |\langle j|\psi\rangle|^2 \geq 0 \quad \text{for all} \quad j. \tag{4.56}$$

These last two results show that p_i can be interpreted as the probability that the system is in the quantum state $|i\rangle$. Moreover, in the basis $\{|i\rangle\}$ that diagonalizes ρ the expectation value of an observable $\mathcal{A}$ is given by

$$\langle\mathcal{A}\rangle = \sum_i p_i\langle i|\mathcal{A}|i\rangle, \tag{4.57}$$

i.e. the weighted average of the expectation values $\langle i|\mathcal{A}|i\rangle$ of the observable in the quantum states $|i\rangle$. However, as our derivation showed, this is only true in the basis $\{|i\rangle\}$ that diagonalizes ρ and not for any other arbitrary basis.

We recover the elementary formulation of quantum mechanics in the special case where all but one p_i vanish. This corresponds to what Feynman (1972) calls 'acting as if the system in which we are interested comprised the whole universe'. A system with one $p_i = 1$ and all others vanishing is in a *pure* state. The density operator can be used to test whether a system is in a pure state, for which a necessary and sufficient condition is that $\rho^2 = \rho$, i.e. ρ is then a projection operator.

Generally, if several p_i are non-zero, the system is in a *mixed* state. The density operator then incorporates both the quantum and statistical information necessary to describe the system.

In order to work with the density operator instead of with the wave functions, we need an equation that deals with the time evolution of the density operator in analogy to the Schrödinger equation for the wave function which, for a system sufficiently weakly coupled to the rest of the universe, is governed by the system Hamiltonian $\mathcal{H}$

$$-i\frac{\partial}{\partial t}|i\rangle = \mathcal{H}|i\rangle. \tag{4.58}$$

The equation for the time evolution of the density operator in its diagonal basis is therefore

$$-i\frac{\partial}{\partial t}\rho = \sum_i p_i\left(-i\frac{\partial}{\partial t}|i\rangle\langle i|\right) = \sum_i p_i\left(\mathcal{H}|i\rangle\langle i| - |i\rangle\langle i|\mathcal{H}\right) = [\mathcal{H}, \rho] \tag{4.59}$$

where $[\mathcal{H}, \rho]$, as usual, denotes the commutator.

Finally, we must ask how our previous results for the various ensembles connect to the notion of the density operator. In other words, how is the behaviour of the rest of the universe contained in the density operator ρ? The answer to this question depends on the physical situation and we briefly examine it for the cases of the microcanonical, the canonical, and the grand canonical ensembles.

When the system has reached equilibrium, the density operator becomes stationary, viz. $\partial\rho/\partial t = 0$ and hence $[\mathcal{H}, \rho] = 0$. This is certainly satisfied if $\rho = \rho(\mathcal{H}, \{\Lambda_i\})$ where $\{\Lambda_i\}$ are further conserved quantities, i.e. $[\mathcal{H}, \Lambda_i] = 0$.

A very important result follows from the assumption that $\rho = \rho(\mathcal{H}, \{\Lambda_i\})$ is the only way to satisfy $[\mathcal{H}, \rho] = 0$: $\mathcal{H}$ and ρ are diagonalized by the same basis of eigenstates, namely the energy eigenstates $|n\rangle$ with eigenenergies E_n, i.e. $\mathcal{H}|n\rangle = E_n|n\rangle$ and $\Lambda_i|n\rangle = \lambda_{n,\,i}|n\rangle$

$$\rho = \sum_n p_n|n\rangle\langle n| \quad \text{and} \quad \langle\mathcal{A}\rangle = \sum_n p_n\langle n|\mathcal{A}|n\rangle. \tag{4.60}$$

Now we are in a position to write down the density operators for the various ensembles. We just state the results that can be arrived at in close analogy to the corresponding results for the ensembles directly (as previously discussed).

4.12.1 Density operator for the microcanonical ensemble

In the case of the microcanonical ensemble, the energy and all other conserved quantities are fixed

$$\mathcal{H}|n\rangle = E_n|n\rangle \qquad \text{and} \qquad \Lambda_i|n\rangle = \lambda_{n,i}|n\rangle \tag{4.61}$$

so that the matrix elements of H and all Λ_i and, thus, ρ itself between the energy eigenstates $|n\rangle$, are all proportional to unit matrices. Not all states may be accessible, so that an explicit representation of ρ is

$$\rho = \begin{pmatrix} \frac{1}{W}\begin{pmatrix} 1 & 0 & 0 & 0 \\ 0 & 1 & 0 & 0 \\ 0 & 0 & \ddots & 0 \\ 0 & 0 & 0 & 1 \end{pmatrix} & (0) \\ (0) & (0) \end{pmatrix} \tag{4.62}$$

where the upper left block of size $W \times W$ represents the W accessible states. Clearly, this can be expressed simpler by just giving the probabilities

$$p_n = \begin{cases} \frac{1}{W} & \text{for the } W \text{ accessible states} \\ 0 & \text{otherwise.} \end{cases} \tag{4.63}$$

4.12.2 Density operator for the canonical ensemble

The probabilities are

$$p_n = \frac{1}{\sum_n e^{-\beta E_n}} e^{-\beta E_n}, \tag{4.64}$$

which we can rewrite in terms of the density operator as

$$\rho = \frac{1}{\mathrm{Tr}\, e^{-\beta \mathcal{H}}} e^{-\beta \mathcal{H}}. \tag{4.65}$$

We read off the partition function

$$Z = \mathrm{Tr}\, e^{-\beta \mathcal{H}} \tag{4.66}$$

and the entropy becomes

$$S = -k\mathrm{Tr}\,(\rho \ln \rho). \tag{4.67}$$

4.12.3 Density operator for the grand canonical ensemble

Similarly, the probabilities for the grand canonical ensemble are

$$p_n = \frac{1}{\sum_n e^{-\beta(E_{n,N} - \mu N)}} e^{-\beta(E_{n,N} - \mu N)}, \tag{4.68}$$

which, written for the density operator

$$\rho = \frac{1}{\mathrm{Tr}\, e^{-\beta(\mathcal{H} - \mu \mathcal{N})}} e^{-\beta(\mathcal{H} - \mu \mathcal{N})}, \tag{4.69}$$

where $\mathcal{N}$ is the number operator, entails for the partition function

$$\mathcal{Z} = \mathrm{Tr}\, e^{-\beta(\mathcal{H}-\mu\mathcal{N})}. \tag{4.70}$$

4.13 Non-interacting quantum gases

Let us briefly summarize the results obtained in chapter 2 for the special case of non-interacting quantum particles. Non-interacting quantum gases are sometimes also called *ideal* or *perfect* quantum gases. However, some authors distinguish in their terminology between perfect gases (where the particles are indeed assumed to be non-interacting) and ideal gases (as the limit of high temperature and low density of a gas of, in general, interacting particles). In the latter case, the gas of particles behaves ideally because the interactions between particles and effects of their finite volume can be neglected. In order to avoid possible confusions, we shall use in this section only the designation *non-interacting* quantum gas.

A non-relativistic non-interacting quantum gas, i.e. a gas consisting of N non-interacting identical non-relativistic quantum particles, is described by the Hamiltonian

$$\mathcal{H} = \sum_i \epsilon_i a_i^\dagger a_i, \tag{4.71}$$

subject to the condition that the total number operator

$$\mathcal{N} = \sum_i a_i^\dagger a_i, \tag{4.72}$$

yields

$$N = \langle \{N\} | \mathcal{N} | \{N\} \rangle, \tag{4.73}$$

where

$$|\{N\}\rangle = |N_1, N_2, \ldots\rangle \tag{4.74}$$

are the occupation number states introduced in section 2.2.1. The energies ϵ_i depend only on particle i, e.g.

$$\epsilon_i \to \epsilon_{\mathbf{k}} = \frac{k^2}{2m} = \frac{p^2}{2m} \tag{4.75}$$

in the momentum representation (cp. (2.153) in section 2.6.1).

The occupation numbers in one state for Bosons and Fermions are, respectively,

$$N_i = \langle N_i | a_i^\dagger a_i | N_i \rangle = 0, 1, 2, \ldots, \tag{4.76}$$

$$N_i = \langle N_i | a_i^\dagger a_i | N_i \rangle = 0, 1. \tag{4.77}$$

Furthermore, the creation and annihilation operators obey commutation relations

$$\left[a_i, a_j\right] = \left[a_i^\dagger, a_j^\dagger\right] = 0 \quad \text{and} \quad \left[a_i, a_j^\dagger\right] = \delta_{ij} \tag{4.78}$$

for Bosons, and anti-commutation relations

$$\left\{a_i, a_j\right\} = \left\{a_i^\dagger, a_j^\dagger\right\} = 0 \quad \text{and} \quad \left\{a_i, a_j^\dagger\right\} = \delta_{ij} \tag{4.79}$$

for Fermions.

EXERCISE 4.9 Simple example of Maxwell–Boltzmann, Bose–Einstein, and Fermi–Dirac statistics Consider a gas of only two particles, each of which can be in one of three states $s = 1, 2, 3$. These states can be either classical states in the case of Maxwell–Boltzmann statistics or quantum states in the cases of Bose–Einstein and Fermi–Dirac statistics.

What different characteristics do the particles therefore have in the different cases and what does this imply for the occupation of the different states with particles?

Enumerate the possible states of the gas according to Maxwell–Boltzmann, Bose–Einstein, and Fermi–Dirac statistics and calculate the ratio between the probability that the particles are found in the same state and the probability that they are found in different states. Interpret this ratio.

4.13.1 Non-interacting Bose gas

In the case of non-interacting Bosons, the Hamiltonian (4.71) together with the total number operator (4.72) are used in the grand canonical partition function (4.70)

$$\mathcal{Z} = \mathrm{Tr}\left(e^{-\beta(\mathcal{H}-\mu\mathcal{N})}\right), \tag{4.80}$$

which is evaluated in the occupation number states (4.74) subject to (4.76)

$$\mathcal{Z} = \sum_{\{N_i\}} \left\langle N_1, N_2, \ldots \left| e^{-\beta(\mathcal{H}-\mu\mathcal{N})} \right| N_1, N_2, \ldots \right\rangle \tag{4.81}$$

$$= \sum_{\{N_i\}} \prod_i e^{-\beta(\epsilon_i-\mu)n_i} = \prod_i \left(\sum_{N_i} e^{-\beta(\epsilon_i-\mu)n_i} \right) \tag{4.82}$$

$$= \prod_i \frac{1}{1 - e^{-\beta(\epsilon_i-\mu)}}. \tag{4.83}$$

The following exercise demonstrates that we can now calculate all thermodynamic quantities in which we might be interested, starting from the partition function we just obtained.

EXERCISE 4.10 Average particle number of a non-interacting Bose gas Using the general expression for an observable (4.53), calculate the average number of particles for the non-interacting Bose gas, i.e.

$$\langle \mathcal{N} \rangle = \mathrm{Tr}\,(\mathcal{N}\rho)\,. \tag{4.84}$$

Recalling the grand canonical ensemble defined in (4.45), i.e.

$$\Phi(T,\mu,V) = -\frac{1}{\beta}\ln \mathcal{Z} = -T\ln \mathcal{Z} \tag{4.85}$$

and remembering that the average number of particles $N = \langle \mathcal{N} \rangle$ and the chemical potential μ are conjugate variables, this calculation can be somewhat cut short.

The result of this exercise

$$N = \sum_i \frac{1}{e^{\beta(\epsilon_i-\mu)}-1} = \sum_i n_{\mathrm{BE}}(\epsilon_i) \tag{4.86}$$

leads to the introduction of the Bose–Einstein distribution function

$$n_{\mathrm{BE}}(\epsilon_i) = \frac{1}{e^{\beta(\epsilon_i-\mu)}-1}. \tag{4.87}$$

4.13.2 Bose–Einstein condensation

The results of the previous section can now be used to elucidate an intriguing macroscopic quantum phenomenon that recently has been at the centre of intense research (although it was predicted by Satyendra Nath Bose and Albert Einstein in the 1920s). Bose's (1924)[4] work focused on the statistics of photons and reproduced Planck's law for black-body radiation without any resort to classical physics (as had still been the case in Planck's original derivation). Einstein, who immediately realized the importance of Bose's contribution, not only translated and submitted Bose's paper, but also followed up on it by considering a gas of non-interacting, massive Bosons, and concluded that, below a certain temperature, a non-zero fraction of the total number of particles would occupy the lowest-energy single-particle state (Einstein, 1924, 1925).

[4] An English translation and some additional, mainly historical, remarks are in (Theimer and Ram, 1976).

Bose–Einstein condensation is the phenomenon that, at sufficiently low temperatures, the dominant fraction of Bose particles condenses into the lowest, and hence macroscopically occupied, state, i.e. the ground state of the non-interacting Bose gas.

In order to understand how this happens, we shall further investigate the expression for the average particle number (4.86), which we derived in section 4.13.1.

We assume the non-interacting Bose particles to be contained in a box of volume V but otherwise free (cp. example 1 in section 2.5 and also in particular section 2.6.1). In momentum space representation, (4.86) becomes

$$N = \sum_{\mathbf{p}} \frac{1}{e^{\beta\left(\frac{p^2}{2m}-\mu\right)} - 1} \tag{4.88}$$

where we used the energies of free particles in momentum space (4.75).

Moreover, in order to investigate the behaviour of the non-interacting Bose gas, we use the grand canonical potential, which has been obtained as an intermediate result of the previous exercise

$$\Phi = \frac{1}{\beta} \sum_i \ln\left(1 - e^{-\beta(\epsilon_i - \mu)}\right) = \frac{1}{\beta} \sum_{\mathbf{p}} \ln\left(1 - e^{-\beta\left(\frac{p^2}{2m}-\mu\right)}\right). \tag{4.89}$$

In a first attempt to further understand these expressions, we approximate the discrete set of momentum modes (cp. example 1) by a continuum of momentum modes. The latter would be correct for completely free particles and thus, we expect it to be a good approximation for a sufficiently large volume V. The approximation allows us to replace discrete sums by integrals

$$\sum_{\mathbf{p}} \rightarrow gV \int \frac{d^3p}{(2\pi)^3} \tag{4.90}$$

where the factor $g = 2s + 1$ allows for spin degeneracy.

We thus obtain for N and Φ

$$\rho = \frac{N}{V} = g \int \frac{d^3p}{(2\pi)^3} \frac{ze^{-\beta\frac{p^2}{2m}}}{1 - ze^{-\beta\frac{p^2}{2m}}}, \tag{4.91}$$

$$\Phi = \frac{gV}{\beta} \int \frac{d^3p}{(2\pi)^3} \ln\left(1 - ze^{-\beta\frac{p^2}{2m}}\right) \tag{4.92}$$

where we introduced $z = e^{\beta\mu}$, sometimes called the fugacity. Instead of the grand canonical potential Φ itself, we are more interested in the internal energy U

$$U = \frac{\partial\,(\beta\Phi)}{\partial\beta} + \mu N \tag{4.93}$$

$$= gV \int \frac{d^3p}{(2\pi)^3}\left(\frac{p^2}{2m}\right)\frac{ze^{-\beta\frac{p^2}{2m}}}{1 - ze^{-\beta\frac{p^2}{2m}}}. \tag{4.94}$$

EXERCISE 4.11 Density and internal energy of the non-interacting Bose gas Evaluate the integrals (4.91) for the density and (4.94) for the internal energy of the non-interacting Bose gas.

The substitution $x^2 = \beta\frac{p^2}{2m}$ will be helpful. Furthermore, the expansion

$$\frac{y}{1-y} = y + y^2 + y^3 + \ldots \tag{4.95}$$

for

$$y = ze^{-x^2} \tag{4.96}$$

and, finally, the standard integrals

$$\int_{-\infty}^{\infty} dx\, e^{-ax^2} = \sqrt{\frac{\pi}{a}} \tag{4.97}$$

$$\int_{-\infty}^{\infty} dx\, x^2 e^{-ax^2} = -\frac{d}{da}\int_{-\infty}^{\infty} dx\, e^{-ax^2} = \frac{1}{2}a^{-3/2}\sqrt{\pi} \tag{4.98}$$

$$\int_{-\infty}^{\infty} dx\, x^2 e^{-ax^2} = \frac{d^2}{da^2}\int_{-\infty}^{\infty} dx\, e^{-ax^2} = \frac{3}{4}a^{-5/2}\sqrt{\pi} \tag{4.99}$$

can be usefully employed.

You will encounter the polylogarithm

$$\mathrm{Li}_r(z) = \zeta_r(z) = \sum_{k=1}^{\infty}\frac{z^k}{k^r} = z + \frac{z^2}{2^r} + \frac{z^3}{3^r} + \ldots, \tag{4.100}$$

a generalization of the Riemann ζ function, for the fugacity $z = e^{\beta\mu}$ and some values of r.

The results of the exercise are, for the density

$$\rho = g\left(\frac{mT}{2\pi}\right)^{3/2}\zeta_{3/2}(z) \tag{4.101}$$

and for the internal energy

$$U = \frac{3}{2} g V T \left(\frac{mT}{2\pi} \right)^{3/2} \zeta_{5/2}(z). \tag{4.102}$$

In both expressions appears the quantity

$$\lambda_{\text{th}} = \sqrt{\frac{2\pi}{mT}}, \tag{4.103}$$

which bears the name thermal de Broglie wavelength and which appears in many of the subsequent expressions.

These two expressions, especially the two functions $\zeta_{3/2}(z)$ and $\zeta_{5/2}(z)$, now allow us to discuss the behaviour of the non-interacting Bose gas in various temperature regimes.

For high enough temperatures or low densities, or both, the function

$$\zeta_{3/2}(z) \propto \frac{\rho}{T^{3/2}} \tag{4.104}$$

and, hence, z is small. Thus, also $\zeta_{5/2}(z)$ is small and both are approximately $\zeta_{3/2}(z) \approx z\, \zeta_{5/2}(z) \approx z$. Their ratio approaches

$$\frac{\zeta_{5/2}(z)}{\zeta_{3/2}(z)} \approx 1 \tag{4.105}$$

and, thus, the ratio of U and ρ implies

$$\frac{U}{V} \approx \frac{3}{2} \rho T, \tag{4.106}$$

which is the result for a classical non-interacting or ideal gas. Note that we obtained this result here as a high temperature and low density approximation.

If we now lower the temperature for fixed density ρ, $\zeta_{3/2}(z)$ and hence z will increase and ultimately reach the value $z = 1$. Since the function $\zeta_{3/2}(z)$ diverges for $z > 1$, our analysis of the non-interacting Bose gas breaks down, although it yielded the correct low temperature behaviour.

The reason for this breakdown lies at the very beginning of our analysis when we converted the discrete sum (4.88) into an integral. This conversion is not justified for low temperatures. However, we can determine a critical temperature T_c when the breakdown occurs at $z = 1$

$$T_c = \frac{2\pi}{m} \left(\frac{\rho}{g \zeta_{3/2}(1)} \right)^{2/3}. \tag{4.107}$$

Looking again at the sum for N in (4.86), when we examine the summand

$$\left(z^{-1}e^{\beta\epsilon_i}-1\right)^{-1} \tag{4.108}$$

we see that for high temperatures $z^{-1} = e^{\beta\mu} = e^{\mu/T}$ is large and dominates the contribution of the lowest energies ϵ_i in $e^{-\beta\epsilon_i}$, which accordingly contribute only little to the sum. This was the reason we could approximate the sum by an integral leading to good results for high temperatures.

This approximation increasingly loses its validity when the temperature is lowered, as z^{-1} becomes smaller and smaller, and stops altogether to be viable for $T < T_c$. We cannot any longer neglect the discreteness of the lowest energy levels. Critically, the $\mathbf{k} = 0$, i.e. the ground state, contribution

$$N_0 = \frac{1}{e^{-\beta\mu}-1} = \frac{1}{z^{-1}-1} \tag{4.109}$$

dominates the sum (4.88) for $T < T_c$ from which we deduce the chemical potential

$$\mu = -T\ln\left(1+\frac{1}{N_0}\right) \approx -\frac{T}{N_0}, \tag{4.110}$$

which is also small for low temperatures. We may write

$$N - N_0 \approx \sum_{\mathbf{k}>0} \frac{1}{e^{\beta p^2/2m}-1} \tag{4.111}$$

where we have neglected $z^{-1} \approx 1$ and replace now the remaining sum by an integral

$$N - N_0 \approx gV \int \frac{d^3p}{(2\pi)^3} \frac{1}{e^{\beta p^2/2m}-1} = gV\left(\frac{mT}{2\pi}\right)^{3/2} \zeta_{3/2}(1) = N\left(\frac{T}{T_c}\right)^{3/2}. \tag{4.112}$$

This expression, valid for $T < T_c$, shows that there is a sizable fraction of all Bose particles in the ground state, i.e. in the condensate N_0 at low temperatures

$$N_0 = N\left(1-\left(\frac{T}{T_c}\right)^{3/2}\right). \tag{4.113}$$

For $T \to 0$, all Bose particles condense into the ground state. It must, however, be emphasized that this condensation takes place in momentum space.

4.13.3 Non-interacting Fermi gas

The treatment of the non-interacting Fermi gas proceeds in exactly the same way as the non-interacting Bose gas up to the point where we have to accommodate the Pauli exclusion principle, i.e. that there are only the occupation numbers $N_i = 0$ and $N_i = 1$. So, we start again with the grand canonical partition function

$$\mathcal{Z} = \mathrm{Tr}\left(e^{-\beta(\mathcal{H}-\mu\mathcal{N})}\right), \tag{4.114}$$

which is evaluated in the occupation number states (4.74) now subject to (4.77)

$$\mathcal{Z} = \sum_{\{N_i\}} \left\langle N_1, N_2, \dots \left| e^{-\beta(\mathcal{H}-\mu\mathcal{N})} \right| N_1, N_2, \dots \right\rangle \tag{4.115}$$

$$= \sum_{\{N_i\}} \prod_i e^{-\beta(\epsilon_i-\mu)n_i} = \prod_i \left(\sum_{N_i=0}^{1} e^{-\beta(\epsilon_i-\mu)n_i} \right) \tag{4.116}$$

$$= \prod_i \left(1 + e^{-\beta(\epsilon_i-\mu)}\right). \tag{4.117}$$

The corresponding grand canonical potential is given by

$$\Phi = -\frac{1}{\beta} \ln \mathcal{Z} = -\frac{1}{\beta} \sum_i \left(1 + e^{-\beta(\epsilon_i-\mu)}\right) \tag{4.118}$$

and the average number of particles

$$N = \langle \mathcal{N} \rangle = -\frac{\partial \Phi}{\partial \mu} = \sum_i \frac{1}{e^{\beta(\epsilon_i-\mu)} + 1} \equiv \sum_i n_{\mathrm{FD}}(\epsilon_i) \tag{4.119}$$

with $n_{\mathrm{FD}}(\epsilon_i)$ the Fermi–Dirac distribution function.

These expressions can be further evaluated by again replacing sums by integrals and using the momentum representation, especially

$$\epsilon_i \to \epsilon_{\mathbf{p}} = \frac{p^2}{2m}. \tag{4.120}$$

Contrary to the Bose case, in the Fermi case (due to the Pauli exclusion principle) we cannot encounter a situation where the ground state is occupied by a sizable fraction of particles.

We obtain after the replacement of sums by integrals for the grand canonical partition function and the particle density

$$\Phi = -\frac{gV}{\beta} \int \frac{d^3p}{(2\pi)^3} \ln\left(1 + e^{-\beta(\frac{p^2}{2m} - \mu)}\right), \tag{4.121}$$

$$\rho = \frac{N}{V} = -\frac{1}{V}\frac{\partial \Phi}{\partial \mu} = g \int \frac{d^3p}{(2\pi)^3} \frac{1}{e^{\beta(\frac{p^2}{2m} - \mu)} + 1}, \tag{4.122}$$

where again $g = 2s + 1$ is the spin degeneracy, e.g. $g = 2$ for electrons.

In order to gain further insight into the physics of the non-interacting Fermi gas, we investigate separately the cases of vanishing temperature $T = 0$ and finite temperature $T > 0$.

We first observe that the Fermi–Dirac distribution defined in (4.119) becomes a step function for $T = 0$, depending on the chemical potential $\mu(T = 0) \equiv \mu_0$, which is conventionally referred to as the *Fermi energy* ϵ_F

$$n_{\text{FD}}(\epsilon) = \begin{cases} 1 & \text{for} \quad \epsilon < \mu_0 \\ 0 & \text{for} \quad \epsilon > \mu_0 \end{cases}. \tag{4.123}$$

The particle density at $T = 0$ becomes

$$\rho(0) = g \int_0^{p_0} dp \, \frac{4\pi p^2}{(2\pi)^3} = g \frac{1}{(2\pi)^3} \frac{4\pi}{3} p_0^3, \tag{4.124}$$

where the Fermi momentum $p_0 \equiv p_{\text{F}}$ is defined via the chemical potential $p_0^2/2m = \mu_0$ at $T = 0$, i.e. the Fermi energy. Similarly, we can define a Fermi temperature $T_{\text{F}} = \mu_0/k$.

At this point, we can also introduce the notion of density of states $f(\epsilon)d\epsilon$, i.e. the number of states per energy interval $d\epsilon$, by

$$g \frac{4\pi}{(2\pi)^3} p^2 \, dp = g \frac{4\pi\sqrt{2}}{(2\pi)^3} m^{3/2} \sqrt{\epsilon} \, d\epsilon \equiv f(\epsilon) d\epsilon. \tag{4.125}$$

Due to the property of the Fermi–Dirac distribution function to be a step function, other thermodynamic quantities of the non-interacting Fermi gas can also be calculated easily. Again, instead of the grand canonical potential Φ, we calculate the internal energy, given for arbitrary temperature, by

$$U(T) = \frac{\partial(\beta\Phi)}{\partial\beta} + \mu N = g \int \frac{d^3p}{(2\pi)^3} \left(\frac{p^2}{2m}\right) \frac{1}{e^{\beta\left(\frac{p^2}{2m} - \mu\right)} + 1} \tag{4.126}$$

$$= g \int \frac{d^3p}{(2\pi)^3} \left(\frac{p^2}{2m}\right) n_{\text{FD}}\left(\frac{p^2}{2m}\right). \tag{4.127}$$

EXERCISE 4.12 Internal energy at $T = 0$ of the non–interacting Fermi gas Evaluate (4.126) for $T = 0$ and show how $U(0)$ and the particle density $\rho(0)$ (4.124) relate to the Fermi energy $\mu(0) = \mu_0$.

The case of finite temperatures, $T > 0$, involves the evaluation of more complicated integrals as we have to take into account occupied states beyond the Fermi energy, i.e. the full Fermi–Dirac distribution function.

EXERCISE 4.13 Non-interacting Fermi gas at finite temperatures $T > 0$ Rewrite the expressions for the internal energy U (4.126) and the particle density ρ (4.122) in terms of the density of states $f(\epsilon)$.

In both expressions, you will find an integral of the form

$$I = \int_0^\infty d\epsilon \, \frac{\kappa(\epsilon)}{e^{\beta(\epsilon-\mu)} + 1} \tag{4.128}$$

with different functions $\kappa(\epsilon)$ for ρ and U. These integrals can be split up by separating the integration interval into $(0, \mu)$ and (μ, ∞), where the first of the resulting integrals can be further split up by observing that

$$\frac{1}{e^x + 1} = 1 - \frac{1}{e^{-x} + 1}. \tag{4.129}$$

Moreover, in most cases of practical interest (e.g. in metals, but even in neutron stars) the Fermi energy $\mu_0 = \epsilon_F \gg T$, which can be used to further approximate the integrals for ρ and U.

We use the results

$$\rho = \left(\frac{2a}{3}\right)\mu^{3/2} + a\frac{\pi^2}{12}\mu^{-1/2} \tag{4.130}$$

$$u = \frac{U}{V} = \left(\frac{2a}{5}\right)\mu^{5/2} + a\frac{\pi^2}{4}\mu^{1/2} \tag{4.131}$$

with the constant

$$a = g\frac{4\sqrt{2}\pi m^{3/2}}{(2\pi)^3} \tag{4.132}$$

of the exercise above to calculate the contribution of the non-interacting Fermi gas to the specific heat c_V at constant volume V at low temperatures.

Using the grand canonical potential means that there is a reservoir of particles and the number of particles is not fixed, but only given by an average $N = \langle N \rangle$. On the other hand, the chemical potential is a fixed quantity. Now, we reverse this situation, as is often the case in condensed matter experiments, and keep the number of particles in a volume and, hence, the particle density ρ, fixed. In such situations, the chemical potential will become a function of temperature $\mu = \mu(T)$. The particle density ρ in (4.130) must then be equal to the particle density at $T = 0$, given in (4.124), which, expressed in terms of μ_0, becomes

$$\rho = \frac{2a}{3}\mu_0^{3/2}. \tag{4.133}$$

Comparing (4.133) and (4.130), we consider the second term in (4.130) as a small correction term where we are allowed to make the replacement $\mu \approx \mu_0$. Thus, we may write

$$\mu \approx \mu_0 \left(1 - \frac{\pi^2}{12}\left(\frac{T}{\mu_0^2}\right)\right) \tag{4.134}$$

to lowest order in the temperature T.

Inserting this expression into (4.131) and keeping terms up to order T^2 gives

$$u \approx \frac{2a}{5}\mu_0^{5/2} - a\mu_0^{5/2}\frac{\pi^2}{12}\left(\frac{T}{\mu_0}\right)^2 + \frac{a\pi^2}{4}T^2\sqrt{\mu_0} = u_0 + \frac{a\pi^2}{6}\sqrt{\mu_0}T^2 \tag{4.135}$$

$$\equiv u_0 + \gamma T^2 \tag{4.136}$$

and, thus, for the specific heat

$$c_V = \frac{\partial u}{\partial T} = 2\gamma T. \tag{4.137}$$

The non-interacting Fermi gas is a good starting point for the theory of metals, which are investigated in chapter 8.

4.13.4 Ultracold atomic gases

Methods to produce temperatures low enough to test the predictions for the non-interacting Bose gas just outlined became available only in the mid-1990s. They depended on the two developments of laser cooling and magnetic trapping of atoms. Laser cooling uses the radiation pressure of a laser beam to slow down atoms and thereby to cool and to localize them. Magnetic trapping techniques allow us to confine atoms without a material container. Combining these techniques permits to unite a large number of atoms into a macroscopic condensate evincing quantum properties similar to those predicted by Bose–Einstein condensation.

The simplest many-particle systems are the non-interacting Bose and Fermi gases and we might expect that they are relevant to the description of dilute weakly interacting, degenerate quantum systems. In condensed matter, most systems are dense, and some of the many-particle theories that were developed for dilute systems were never tested by experiment. It is only recently that the appropriate temperature and density conditions have been realized in ultracold atomic gases thanks to advances in evaporative and laser cooling techniques. Trapped ultracold atomic gases have become exciting experimental systems that enable the study of the effect of tunable interactions on dilute quantum Bose and Fermi gases. Phenomena that are now accessible to experiments include the Bose–Einstein Condensation (BEC), interference of matter waves, vortices, Fermi condensates, BCS superconductivity, and BCS-to-BEC crossover. Chapter 8 provides a discussion of quantum many-particle systems and focuses on how to deal with their interactions. While most of the systems considered there are electronic, i.e. Fermionic or quantum spin systems, section 8.1 is devoted to introducing to weakly interacting Bose systems.

4.14 Classical lattice models in statistical mechanics

This book is mainly concerned with quantum mechanical models in one spatial dimension. We shall, however, exploit a deep connection between *classical* lattice models in statistical mechanics and quantum statistical mechanical lattice models (see below in part II). 'Classical' in this regard refers to the description of the dynamical variables of these models in terms of numbers, as opposed to operators in a quantum setting. This connection requires that we take a brief look at a few classical statistical mechanical models. There is a vast literature on the subject and we recommend Baxter (1982) as the most apposite for the purposes of this book, as will become amply clear.

These classical statistical mechanical models, for the most part classical spin models, in all their variations have played an important role in elucidating the theory of phase transitions and critical phenomena a, as discussed in chapter 5. Although they are, in some sense, toy models chosen for their simplicity rather than their realism, many of them accurately describe the essential features of real systems undergoing phase transitions .

4.14.1 Classical spin models

First we must review a few basic notions of classical spin models in statistical mechanics. These models can be defined in arbitrary spatial dimension D. However, we shall restrict our discussion mostly to two-dimensional classical spin models, because almost all exact solutions of classical spin models are for $D = 2$ and because the mentioned connection between classical and quantum models, though valid for arbitrary D, relates exactly solvable classical models in $D = 2$ and quantum integrable models in $D = 1$.

Figure 4.1 shows a typical example of a classical spin model. On a two-dimensional lattice Λ of, e.g. $N = L_1 \times L_2$ lattice sites, there is a classical variable σ_i ($i \in \Lambda$), called

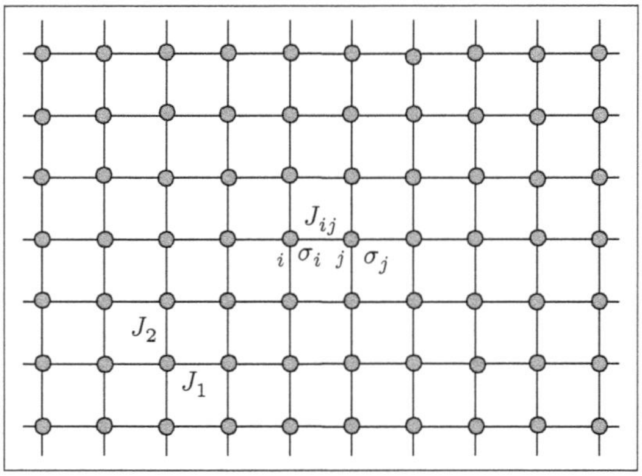

Figure 4.1 *A typical two-dimensional spin model with nearest-neigbour interactions $\mathcal{J}_{ij}$. There is a classical variable σ_l connected with each site l of the lattice. More precisely, the index labelling a lattice site j would be written as a vector index with a horizontal and a vertical component $j = (j_1, j_2)$. A typical case is indicated where there is an interaction $\mathcal{J}_1$ along horizontal and a, in general different, interaction $\mathcal{J}_2$ along vertical nearest-neighbour bonds $\langle ij \rangle$.*

'spin' or spin variable, assigned to each lattice site. The spins on different lattice sites interact via interactions $\mathcal{J}_{ij}$, which can, in the general case, have an arbitrary range. In figure 4.1 only nearest-neighbour interactions are indicated, which act along the (nearest-neighbour) horizontal and vertical bonds. Interactions across the diagonal would already constitute next-nearest neighbour interactions.

The possible values of the spin variables σ_i and the interactions $\mathcal{J}_{ij}$ determine the energy

$$E(\sigma_1, \ldots, \sigma_N) \tag{4.138}$$

of a given configuration of spins $\{\sigma_i\}$.

4.14.2 The Ising model

For the Ising class of models, we have two values

$$\sigma_i = \pm 1 \tag{4.139}$$

for the spin variables, and, the energy of a configuration of spins is

$$E(\sigma_1, \ldots, \sigma_N) = \sum_{ij} \mathcal{J}_{ij} \sigma_i \sigma_j - \sum_i H_i \sigma_i, \tag{4.140}$$

where we have introduced local magnetic fields H_i, which may be different at different lattice sites. We restrict the further discussion to the case of a homogeneous magnetic field $H_i \equiv H$ for all lattice sites i and to nearest-neighbour sites $\langle ij \rangle$ indicated by the brackets.

Under these restrictions and, moreover, for vanishing magnetic field $H = 0$, the Ising model was solved exactly by Lars Onsager in two dimensions, whose famous solution (Onsager, 1944; Kaufman, 1949; Kaufman and Onsager, 1949) is valid even for anisotropic interactions $\mathcal{J}_1 \neq \mathcal{J}_2$, i.e. different interaction strengths in the horizontal and vertical directions. *Exact solvability* refers to the fact that the partition function

$$Z_N = \sum_{\{\sigma\}} e^{-E(\{\sigma\})} = \sum_{\{\sigma_1,\ldots,\sigma_N\}} \exp\left[\sum_{\langle ij \rangle} K_{ij}\sigma_i\sigma_j - h\sum_i^N \sigma_i\right] \tag{4.141}$$

can be calculated exactly, i.e. without the necessity for any approximation, for $h = 0$. Here we introduced

$$K_{ij} \equiv \beta \mathcal{J}_{ij} \quad \text{and} \quad h \equiv \beta H \quad \text{with} \quad \beta = \frac{1}{T}. \tag{4.142}$$

From the partition function, the free energy

$$F = -T \ln Z_N, \tag{4.143}$$

and the free energy per spin, also sometimes called free energy density

$$f = -T \lim_{N\to\infty} \frac{1}{N} \ln Z_N \tag{4.144}$$

and all other thermodynamic quantities can be calculated. Thus, once the hard part of exactly calculating the partition function is achieved, the model's thermodynamics is completely known. This is always the case in statistical mechanics, alas, there are only very few models whose partition functions are known exactly.

In section 4.15, we treat the Ising model in the so-called mean-field approximation, where the influence on one spin of its neighbouring spins is treated in an averaged way. Hence, we assume that every spin sees a field, the mean-field, which is produced by all its neighbours. This approximation thereby neglects, as we shall see, the fluctuations of the spin variables.

Before we embark on this important approximative scheme, we briefly introduce a sample of the many lattice models discussed and that produce important insights in statistical physics.

4.14.3 Lattice gas

Despite the 'magnetic' language we have been using, the Ising model can be used to model many other physical situations. For example, the spin variable can be reinterpreted as an occupation variable of a lattice gas model or a model of a binary mixture

$$n_i = \frac{\sigma_i + 1}{2} = \begin{cases} 1 & \text{if lattice site } i \text{ is occupied} \\ 0 & \text{otherwise.} \end{cases} \tag{4.145}$$

There are many possible variations for spin models. Besides having interactions of longer range, or random interactions, the latter describing so-called spin glasses, or living on various lattices (triangular, hexagonal, etc.), the spin variables can also take other values. The next sections briefly introduce a number of those with different values for the spin variables.

4.14.4 The classical planar XY and Heisenberg models

These two classes of models relax the restriction of the Ising model to only two discrete spin values $\sigma = \pm 1$. Instead, they consider spin variables σ_i, which are two component vectors for the classical planar or XY model, or three component (or even, more generally, D component) vectors for the classical Heisenberg model. In order to remain physically meaningful, there must exist a restriction on the length of the spins. This restriction is mostly chosen to be

$$\boldsymbol{\sigma}_i^2 = 1 \quad \text{for all} \quad i. \tag{4.146}$$

Consequently, due to the greater variability of orientations of the spins of these models, they can exhibit qualitatively different phase transitions from the phase transitions described by the Ising model.

Choosing the usual nearest-neighbour exchange interaction, the partition function becomes

$$Z_N = \sum_{\{\boldsymbol{\sigma}_i\}} \exp\left(\beta\left[\sum_{\langle ij\rangle} \mathcal{J}_{ij}\boldsymbol{\sigma}_i\boldsymbol{\sigma}_j + \sum_i \mathbf{H}_i \cdot \boldsymbol{\sigma}_i\right]\right). \tag{4.147}$$

The XY model can also be described by a complex variable ψ_i, which is equivalent to the two-component vector σ_i of the original definition of the XY model. The interaction energy is then written as

$$E_{ij} = \mathcal{J}_{ij}\left(\psi_i^*\psi_j + \psi_i\psi_j^*\right). \tag{4.148}$$

4.14.5 The Potts model

If we allow the spin variables to take one of q possible values

$$\sigma_i = 1, \ldots, q \tag{4.149}$$

and choose the energy of a configuration to be

$$E(\sigma_1, \ldots, \sigma_N) = \sum_{ij} \mathcal{J}_{ij} \delta_{\sigma_i, \sigma_j}, \tag{4.150}$$

we arrive at the so-called q-state Potts model.

In the Potts model, as in the Heisenberg model, the alignment of neighbouring spins may be favourable or unfavourable, depending on the sign of the interaction $\mathcal{J}_{ij}$, but only, in contrast to the Heisenberg model, if the spins are identical.

4.14.6 The Gaussian and spherical models

In another attempt to introduce models that are more tractable than the Ising model the Gaussian and the spherical models relax the restriction (4.146) on the length of the spins σ_i, which were introduced for the XY and Heisenberg models rather than considering spin vectors. Both models work with spins that take arbitrary real values $-\infty < \sigma_i < \infty$ instead of just $\sigma_i = \pm 1$ as in the Ising model.

Despite its use for perturbative calculations, the Gaussian model is deeply flawed because it cannot be avoided that, below a certain temperature, the energy can become arbitrarily negative by letting the spins grow without limit. In order to remedy this unphysical behaviour of the Gaussian model, in the spherical model an overall restriction

$$\frac{1}{N} \sum_i \sigma_i^2 = 1 \tag{4.151}$$

on the length of the spins, as opposed to a restriction on the length of each spin individually, as in (4.146), is introduced.

The partition function of the spherical model can thus be written

$$Z_N = \int_{-\infty}^{\infty} \cdots \int_{-\infty}^{\infty} d\sigma_1 \ldots d\sigma_N \exp\left(\beta \sum_{\langle ij \rangle} \sigma_i \sigma_j + H \sum_i \sigma_i\right) \delta\left(N - \sum_i \sigma_i\right) \tag{4.152}$$

where the delta-function enforces (4.151).

The spherical model, then, is one of the very few models in statistical mechanics whose partition function can be calculated in three space dimensions. However, we can argue that the model is somewhat unphysical because, due to (4.151), interactions between all spins of the lattice are introduced, no matter how far apart they are.

4.15 Interacting magnetic moments: mean-field theory

Using the Ising model we introduced in the previous section, we shall now discuss the interaction of magnetic moments in an important particular approximation, the so-called mean-field approximation which we shall present here in its simplest form. While non-interacting magnetic moments (cp. exercises 4.3, 4.4, and 4.5) did not exhibit transitions from a disordered to an ordered arrangement of the moments as the temperature is lowered (analogous to a gas of non-interacting particles—an ideal gas—which remains a gas at low temperatures and does not condensate), we shall see that the inclusion of interactions between the magnetic moments does lead to such a transition. We thus encounter a first example of a phase transitions, an important phenomenon in interacting many-particle systems, which we shall discuss in further detail in chapter 5.

The energy E or Hamiltonian $\mathcal{H}$ of a configuration $\{s_i\}$ of N interacting magnetic moments or spins of the Ising model in a constant external magnetic field h is

$$E(\{s_i\}) \equiv \mathcal{H}(\{s_i\}) = -\frac{1}{2}\sum_{ij} \mathcal{J}_{ij} s_i s_j - h \sum_i s_i \tag{4.153}$$

where the magnetic moments are represented by 'classical' spin variables $s_i = \pm 1$ located on the lattice sites of a D-dimensional lattice. The spins interact via exchange couplings

$$\mathcal{J}_{ij} = \begin{cases} \mathcal{J} > 0 & \text{if } i \text{ and } j \text{ are nearest neighbours} \\ 0 & \text{otherwise} \end{cases} \quad . \tag{4.154}$$

Again we shall use the bracket notation $\langle ij \rangle$ to indicate nearest-neighbour summation only. The factor 1/2 is included to account for the double counting of bonds between lattice sites i and j. It could be avoided if we explicitly agreed to count each bond only once. Since we are considering an Ising model where only nearest neighbours are coupled, the summation is over the z nearest neighbours of the spin at lattice site i. The coordination number z depends on the lattice we are considering and its dimensionality. For simple cubic lattices in D dimensions, the coordination number is $z = 2D$.

Let us rewrite the spin at each site, using its site-independent mean value $m = \langle s \rangle$

$$s_i = \langle s_i \rangle + (s_i - \langle s_i \rangle) = m + (s_i - \langle s_i \rangle) = m + \delta s_i \tag{4.155}$$

and insert it into the energy function of the Ising model (4.153)

$$\mathcal{H}(\{s_i\}) = -\mathcal{J} \sum_i \left(z m s_i - \frac{z}{2} m^2 \right) - h \sum_i s_i - \frac{\mathcal{J}}{2} \sum_{\langle ij \rangle} \delta s_i \delta s_j . \tag{4.156}$$

In the mean-field approximation we are going to neglect the last term in this expression, which is tantamount to the assumption that fluctuations around the mean are small

$$\frac{\langle \delta s_i \delta s_j \rangle}{\langle s_i \rangle \langle s_i \rangle} = \frac{\langle \delta s_i \delta s_j \rangle}{m^2} \ll 1. \tag{4.157}$$

We consider now an arbitrary spin at lattice site i. All contributions in the energy (4.153) on this chosen spin at lattice site i are

$$E(s_i) = -hs_i - \mathcal{J}s_i \sum_j s_j = s_i \left(-h - \mathcal{J} \sum_j s_j \right) = -s_i h' \tag{4.158}$$

where the sum is over all z nearest-neighbour sites j of site i. We observe that the magnetic field acting on the spin at lattice site i consists of two parts: the external magnetic field h and an internal field generated by all other spins, i.e. the spin s_i is subject to the field

$$h' = h + \mathcal{J} \sum_j s_j. \tag{4.159}$$

The approximation now consists of replacing the spin at lattice site j by its site-independent mean value $m = \langle s \rangle$ to determine the mean or effective field acting on the spin at lattice site i, which consequently is also site-independent

$$h_{\text{eff}} = h + \mathcal{J}z\langle s \rangle = h + \mathcal{J}zm. \tag{4.160}$$

Hence, the name mean-field approximation.

The energy (4.156), fluctuations neglected, then becomes the mean-field energy

$$E_{\text{MF}}(\{s_i\}) = -h_{\text{eff}} \sum_i s_i + \frac{1}{2} N\mathcal{J}zm^2, \tag{4.161}$$

which has, up to an additive term, the form of the energy of an ideal paramagnet in a magnetic field h_{eff} (cp. exercise 4.4).

The mean-field partition function is

$$Z_{\text{MF}} = \sum_{\{s_i\}} e^{-\beta E_{\text{MF}}(\{s_i\})} = e^{-N\mathcal{J}zm^2/2\beta} \left(2 \cosh \beta h_{\text{eff}}\right)^N \tag{4.162}$$

from which we can derive a mean-field free energy

$$F_{\text{MF}}(T, h, m) = -T \ln Z_{\text{MF}} = \frac{1}{2} N\mathcal{J}zm^2 - NT \ln \left(2 \cosh \beta h_{\text{eff}}\right). \tag{4.163}$$

The mean value $m = \langle s \rangle$ is determined by the implicit equation

$$m = \frac{1}{Z_{\text{MF}}} \sum_{\{s_i\}} s_i e^{-\beta E_{\text{MF}}(\{s_i\})} = \tanh\left(\beta h_{\text{eff}}\right) = \tanh\left(\frac{h + \mathcal{J}zm}{T}\right). \tag{4.164}$$

It is instructive to compare these results with those of exercise 4.4.

The same result can be obtained by minimizing the mean-field free energy (4.163) with respect to m keeping T and h constant: $\partial F_{\text{MF}}/\partial m = 0$. However, inserting (4.164) into (4.163) we observe that $F_{\mathrm{MF}}(T, h, m) = F_{\mathrm{MF}}(T, h, m(h))$ is really a function of the intensive thermodynamic variables temperature T and external magnetic field h, and therefore (4.163) should be considered the Gibbs free energy $G(T, h)$. (We have, moreover, suppressed the dependence on the extensive variable, the number N of spins.) Therefore, calculating (4.164) from $\partial F_{\mathrm{MF}}/\partial m = 0$ is to be considered as a minimization of $F_{\mathrm{MF}}(m)$ with respect to m, rather than a thermodynamic relation.

The total magnetization

$$M = Nm = N \tanh\left(\frac{h_{\mathrm{eff}}}{T}\right) \approx N\frac{h_{\mathrm{eff}}}{T} \tag{4.165}$$

can be linearized for $h_{\mathrm{eff}}/T \ll 1$. The effective field is also given implicitly

$$h_{\mathrm{eff}} = h + \mathcal{J}zm = h + \mathcal{J}z \tanh\frac{h_{\mathrm{eff}}}{T} \approx h + \frac{h_{\mathrm{eff}}}{T} \tag{4.166}$$

which yields

$$h_{\mathrm{eff}} = \frac{h}{1 - \frac{\mathcal{J}z}{T}} \tag{4.167}$$

so that the magnetization is given for small fields by

$$M = N\frac{h}{T - \mathcal{J}z} \tag{4.168}$$

and the magnetic susceptibility

$$\chi = \frac{\partial M}{\partial h} = \frac{N}{T - \mathcal{J}z}. \tag{4.169}$$

4.15.1 Mean-field critical behaviour

For non-interacting spins (cp. exercise 4.4) we had the Curie law

$$\chi \propto 1/T \tag{4.170}$$

while we now obtain the Curie–Weiss law

$$\chi \propto 1/(T - T_c) \tag{4.171}$$

the susceptibility diverges at the critical temperature

$$T_c = \mathcal{J}z. \tag{4.172}$$

These results reflect an important qualitative difference between the non-interacting and the interacting magnetic systems, even in the approximations we employed in the latter case, the mean-field approximation for small fields.

However, if the temperature becomes too low, the small field approximation $h_{\text{eff}}/T \ll 1$ employed can no longer be maintained. Most importantly, for $T < T_c \equiv \mathcal{J}z$, we obtain a finite magnetization, which does not disappear for vanishing external magnetic field, $h \to 0$. This *spontaneous* magnetization, which vanishes continuously for $T \to T_c$, can be calculated from the mean-field results we obtained so far. For $h = 0$ and T close to but smaller than T_c, we consider the implicit equation (4.164) for m

$$m = \tanh \frac{\mathcal{J}zm}{T} = \tanh \left(\frac{T_c}{T} m \right). \tag{4.173}$$

Obviously, there is always a trivial solution $m = 0$ for arbitrary T on which we shall comment in a moment. For $T < T_c$, however, we have

$$\frac{d}{dm} \tanh \left(\frac{T_c}{T} m \right) \bigg|_{m=0} = \frac{T_c}{T} > 1 \tag{4.174}$$

and two more solutions of (4.173) emerge from $m = 0$ which, of course, remains a solution.

To find the temperature dependence of the non-zero solutions, we expand the right-hand side of (4.164)

$$m = \tanh \frac{\mathcal{J}zm}{T} = \tanh \left(\frac{T_c}{T} m \right) \approx \frac{T_c}{T} m - \frac{1}{3} \left(\frac{T_c}{T} \right)^3 m^3 + \ldots \tag{4.175}$$

which may be solved for m. The other two, nontrivial solutions are determined by

$$m^2 = 3 \left(\frac{T}{T_c} \right)^3 \left(\frac{T_c}{T} - 1 \right). \tag{4.176}$$

Defining the reduced temperature $t = (T - T_c)/T_c = T/T_c - 1$ we find to linear order in t

$$m^2 \approx -3t. \tag{4.177}$$

For $T < T_c$, i.e. $t < 0$ the magnetization behaves as

$$m \approx \pm (3|t|)^{\frac{1}{2}}. \tag{4.178}$$

This expression shows the typical behaviour of an order parameter at a continuous phase transition: it vanishes continuously at the critical temperature, but its slope $\partial m / \partial t$

diverges for $T \to T_c$. Chapter 5 discusses phase transitions in more detail and also properly defines the terminology used (especially in the last sentence). Before we close this chapter, let us fit the results we obtained into a framework that will prove very useful in chapter 5.

While $m = 0$ is always a solution, it is the only solution, as we have seen, for $T > T_c$. For $T < T_c$ it can be shown that $m = 0$ corresponds to an unstable situation that can be destroyed easily such that the system assumes one of the finite solutions with $|m| > 0$. This quite general phenomenon is called *spontaneous symmetry breaking*.

We next calculate the magnetic susceptibility

$$\chi = \left.\frac{\partial m}{\partial h}\right|_{h=0} \tag{4.179}$$

from the implicit expression for the magnetization (4.164), for which we again use an expansion analogous to (4.175), keeping, however, only a linear term in the external magnetic field h to capture the leading order behaviour.

$$m = \tanh \frac{h + \mathcal{J}zm}{T} = \tanh\left(\frac{h}{T} + \frac{T_c}{T}m\right) \approx \frac{h}{T} + \frac{T_c}{T}m - \frac{1}{3}\left(\frac{T_c}{T}\right)^3 m^3 + \dots . \tag{4.180}$$

Taking the derivative with respect to h on both sides of this equation yields

$$\chi = \frac{1}{T} + \chi\frac{T_c}{T} + \chi m^2 \left(\frac{T_c}{T}\right)^3 . \tag{4.181}$$

Hence, the magnetic susceptibility diverges if the critical temperature T_c is approached from higher or lower temperatures. Using (4.177), we obtain for these divergent behaviours of the susceptibility

$$T > T_c \qquad \chi \approx \frac{\beta}{t}, \tag{4.182}$$

$$T < T_c \qquad \chi \approx \frac{\beta}{2|t|}. \tag{4.183}$$

4.15.2 Correlation function and correlation length

One of the central concepts in statistical and condensed matter physics is the *correlation length* ξ. It measures the distance over which degrees of freedom, e.g. spins in the Ising model, are correlated and is fundamental to the understanding of phase transitions. Beyond the correlation length, the degrees of freedom are essentially statistically independent. For the Ising model, the (two-point) correlation function can be defined as

$$\mathcal{G}(i,j) = \langle s_i s_j \rangle - \langle s_i \rangle \langle s_j \rangle = \langle (s_i - \langle s_i \rangle)(s_j - \langle s_j \rangle) \rangle = \langle \delta s_i \delta s_j \rangle. \tag{4.184}$$

$\mathcal{G}(i,j)$ measures the correlations between the fluctuations δs_i. A finite $\mathcal{G}(i,j)$ indicates that the fluctuations of the spins about their mean values are correlated. In our analysis so far, we have, however, explicitly neglected these fluctuations (see 4.157), i.e. we assumed $\mathcal{G}(i,j) \approx 0$.

Beyond the mean-field approximation as outlined, we nevertheless expect that there are correlations in a statistical mechanical system and hence that for some, possibly large, distance R the correlation function behaves as

$$\mathcal{G}(R) \propto e^{-R/\xi}. \tag{4.185}$$

This relation introduces the correlation length ξ. We see that the correlation length ξ sets the scale over which correlations decrease significantly.

In a magnetic system like the one described by the Ising model, the overall magnetization for $T > T_c$ vanishes, $m = 0$. However, there still may be clusters of magnetic moments for which the magnetization does not vanish, see e.g. figure 5.1. In an intuitive picture, the correlation length is a measure of the linear size of the largest clusters of correlated magnetic moments. As the temperature decreases towards the critical temperature T_c, these clusters increase and finally diverge in size to produce a macroscopic magnetization. Thus, we expect the correlation length to diverge at the critical temperature.

There are refinements of mean-field theory that include the effects of two-point correlations described by $\mathcal{G}(i,j)$ and predict that the correlation length diverges as

$$\xi \propto |t|^{-1/2}. \tag{4.186}$$

However, these mean-field theories are still deficient and do not describe the details of the behaviour of systems at a phase transition correctly because they still neglect higher order correlations, i.e. three-point, four-point, etc. correlations. This indicates that near a critical point, correlations and fluctuations on all length scales become important. We shall have a closer look into this in chapter 5.

4.15.3 Summary

In conclusion, mean-field theory, in the simple version presented here, and with some extensions to include two-point correlations (which we only mentioned in passing), predicts

- a critical point at $h = 0$: $T_c = \mathcal{J}z$,
- singular critical behaviour.[5]

[5] There is no danger of confusing the inverse temperature $\beta = \frac{1}{T}$ and the critical exponent, traditionally also called β.

$$m \approx m_-|t|^\beta \qquad (T < T_c) \tag{4.187}$$

$$\chi \approx \chi_\pm|t|^{-\gamma} \tag{4.188}$$

$$\xi \approx \xi_\pm|t|^{-\nu} \tag{4.189}$$

β, γ, and ν are known as *critical exponents* for which mean-field theory predicts the values $\beta = 1/2$, $\gamma = 1$, and $\nu = 1/2$.

A useful definition of the critical exponent, κ say, of the singular critical quantity K is

$$\kappa = \lim_{|t|\to 0} \frac{\ln K}{\ln|t|}. \tag{4.190}$$

Besides those mentioned, there are more critical exponents. For example, the zero magnetic field heat capacity behaves near criticality as

$$c_h \propto |t|^{-\alpha}. \tag{4.191}$$

Mean-field theory predicts for this exponent the value $\alpha = 0$, which corresponds to a discontinuity.

It is instructive to look at a different approach to mean-filed theory that makes use of the symmetries of a system and guesses a general form of the mean-field free energy consistent with those symmetries. Chapter 5 returns to this so-called *Landau theory*.

4.15.4 Simple Landau theory

Let us, however, at this stage construct a thermodynamic potential, the Gibbs free energy $G(T, h, N)$, in such a way as to reproduce the above results of our mean-field approach, especially in the vicinity of $T = T_c$. This will exemplify the more general development of the section 5.3.

We have obtained the effective field

$$h_{\text{eff}} = h' = h + \mathcal{J}zm = T\tanh^{-1} m. \tag{4.192}$$

The free energy per site $g(T, h) = G(T, h, N)/N$satisfies

$$h = \left(\frac{\partial g}{\partial m}\right)_T = T\tanh^{-1} m - \mathcal{J}zm, \tag{4.193}$$

from which we obtain by integration with respect to m, making use of (4.192),

$$g(T, h) = g_0(T) + mh + \frac{1}{2}\mathcal{J}zm^2 + \frac{T}{2}\ln(1 - m^2). \tag{4.194}$$

Some algebraic manipulation, using (4.192), and an appropriate choice of the integration constant $g_0(T)$ shows that this expression is

$$g(T, h) = mh + \frac{1}{2}\mathcal{J}zm^2 - T\ln\left(2\cosh\left(\frac{h_{\text{eff}}}{T}\right)\right) = mh + f_{\text{MF}}(T, h, m) \qquad (4.195)$$

where $Nf_{\text{MF}}(T, h, m) = F_{\text{MF}}(T, h, m)$ is the mean-field free energy (4.163). It is now apparent that we have to interpret this free energy as

$$F_{\text{MF}}(T, h, m, N) = F_{\text{MF}}(T, h(m), m, N) = F_{\text{MF}}(T, m, N) \qquad (4.196)$$

and that g and f_{MF} are a pair of thermodynamic potentials related by a Legendre transformation.

An expression for $F_{\text{MF}}(T, h(m), m, N)$ and hence, via (4.195), G can also be derived via a different, also quite instructive, route. The mean energy of the Ising energy (4.153) in the mean-field approximation, i.e. $\delta s_i = 0$, is given by

$$\langle E\rangle = -\mathcal{J}\sum_{\langle ij\rangle}\langle s_i s_j\rangle - h\sum_i\langle s_i\rangle \approx -\frac{1}{2}N\mathcal{J}zm^2 - Nhm. \qquad (4.197)$$

We identify this expression as the mean-field internal energy $U(m) = -\frac{1}{2}N\mathcal{J}zm^2 - Nhm$. To a given magnetization, $m = N_+ - N_-$ and total number of spins $N = N_+ + N_-$, there correspond

$$W(M) = \frac{N!}{N_+!N_-!} = \frac{N!}{\left[\frac{1}{2}(N+M)\right]!\left[\frac{1}{2}(N-M)\right]!} \qquad (4.198)$$

configurations. Using the usual Stirling approximation, this leads to an entropy

$$S(m) = \ln W(M) = -N\left[\frac{1+m}{2}\ln\left(\frac{1+m}{2}\right) + \frac{1-m}{2}\ln\left(\frac{1-m}{2}\right)\right]. \qquad (4.199)$$

The mean-field free energy per site $f(m) = \frac{1}{N}F(m)$ combining these expressions

$$\begin{aligned} f(m) &= \frac{1}{N}\left(U(m) - TS(m)\right) \\ &= -\frac{1}{2}\mathcal{J}zm^2 - hm + T\left[\frac{1+m}{2}\ln\left(\frac{1+m}{2}\right) + \frac{1-m}{2}\ln\left(\frac{1-m}{2}\right)\right] \end{aligned} \qquad (4.200)$$

can be shown to reproduce the mean-field free energy (4.163), i.e. $Nf(m) = F_{\text{MF}}$.

EXERCISE 4.14 Consistency of approaches Convince yourself of the last statement by a direct calculation using the mean-field relation (4.164) between magnetization m and external field h.

The form of F_{MF} (4.200) is very instructive when discussing some key concepts of the physics of phase transitions. Especially for $h = 0$ we observe

1. (4.200) is symmetric in m: $f(-m) = f(m)$;
2. For $T > T_c = \mathcal{J}z$ there is a single minimum at $m = 0$, which is thus the equilibrium state;
3. At $T = T_c$

$$\left.\frac{\partial^2 f}{\partial m^2}\right|_{m=0} = 0; \tag{4.201}$$

4. For $T < T_c$ two symmetric minima emerge and the minimum at $m = 0$ becomes a maximum and thus unstable;
5. As mentioned earlier, for $T < T_c$ the symmetry is *spontaneously broken* as the system has to select one of the minima as its equilibrium state.

This spontaneous symmetry breaking can be achieved by the minutest of magnetic fields, as will become clear in exercise 4.15.

EXERCISE 4.15 Finite magnetic field Discuss the free energy (4.200) for a non-vanishing external magnetic field $h \neq 0$.

The picture becomes even easier to comprehend when we restrict ourselves to small m, which is certainly justified, as we have seen, close to the critical temperature.

EXERCISE 4.16 Small magnetization close to the critical temperature Discuss the free energy (4.200) for small m, i.e. close to the critical temperature, by expanding it in powers of m for both cases, vanishing and non-vanishing external magnetic field h.

4.16 Transfer matrix

The transfer matrix technique is the central technique used in the algebraic Bethe ansatz, one of the core parts (part II) of this book. It is therefore valuable to introduce this technique in its simplest setting: the one-dimensional Ising model.

Assume a one-dimensional lattice of N sites wrapped around to form a ring on whose lattice sites we have Ising spins $\sigma_i = \pm 1$. The ring geometry implies that the periodic boundary condition $\sigma_{N+1} = \sigma_1$ holds. The interaction energy between the spin on lattice site i and the spin on lattice site $i+1$ in the Ising model is assumed to be

$$\mathcal{H}_0(\sigma_i, \sigma_{i+1}) = \mathcal{J}\sigma_i\sigma_{i+1} \tag{4.202}$$

and hence the total energy of a configuration of spins

$$\mathcal{H} = \sum_{i=1}^{N} \mathcal{H}_0(\sigma_i, \sigma_{i+1}). \tag{4.203}$$

In order to introduce the notion of the transfer matrix, the exact form of $\mathcal{H}_0(\sigma_i, \sigma_{i+1})$ is, however, not important. The partition function of the system with a general $\mathcal{H}_0(\sigma_i, \sigma_{i+1})$ is

$$Z_N = \sum_{\{\sigma_i\}} \exp\left(-\beta \sum_{i=1}^{N} \mathcal{H}_0(\sigma_i, \sigma_{i+1})\right) \tag{4.204}$$

$$= \sum_{\{\sigma_i\}} \prod_{i=1}^{N} \exp\Big(- \beta\mathcal{H}_0(\sigma_i, \sigma_{i+1})\Big) \tag{4.205}$$

$$= \prod_{i=1}^{N} \sum_{\sigma_i=\pm 1, \sigma_{i+1}=\pm 1} \exp\Big(- \beta\mathcal{H}_0(\sigma_i, \sigma_{i+1})\Big). \tag{4.206}$$

The partition function written in the last form suggests introducing the 2×2 matrix with matrix elements

$$T = (T_{ij}) = \Big(\exp\Big(- \beta\mathcal{H}_0(\sigma_i, \sigma_{i+1})\Big)\Big) = \begin{pmatrix} e^{-\beta\mathcal{H}_0(+1,+1)} & e^{-\beta\mathcal{H}_0(+1,-1)} \\ e^{-\beta\mathcal{H}_0(-1,+1)} & e^{-\beta\mathcal{H}_0(-1,-1)} \end{pmatrix}. \tag{4.207}$$

This matrix is called the transfer matrix T. The interaction Hamiltonian $\mathcal{H}_0(\sigma, \sigma')$ is symmetrical in its arguments. This implies that the transfer matrix T is a real symmetrical matrix with positive matrix elements.

Physically the elements of the transfer matrix represent all possible Boltzmann weights of two coupled spins. Multiplying two transfer matrices together gives a new matrix T^2, which represents all Boltzmann weights of a short chain of three spins where the inner spin has been summed over. The resulting matrix depends on the spins of the first and third spin. Further multiplications produce matrices T^m, where more and more inner spins have been summed over, the resulting matrices depending only on the first and last spin. In this way, we produce the expression (4.206) for the partition function one multiplication at a time. Finally, once we have reached a chain of length N, recalling the periodic boundary condition $\sigma_{N+1} = \sigma_1$, we have to take the trace of the matrix to obtain (4.206), but now expressed through the transfer matrix, i.e.

$$Z_N = \mathrm{Tr}\left(T^N\right). \tag{4.208}$$

This method can be easily generalized to the case of a Hamiltonian

$$\mathcal{H} = \mathcal{H}_0(\sigma_i, \sigma_j) + V(\sigma_i) \tag{4.209}$$

as is, for instance, the case when there is an applied external field. All we have to do is to rewrite this expression as

$$\mathcal{H} = \mathcal{H}_0(\sigma_i, \sigma_j) + \frac{1}{2}\left(V(\sigma_i) + V(\sigma_j)\right). \tag{4.210}$$

This approach is useful as it can be employed in solving the one-dimensional Ising model.

4.17 Exact solution of the one-dimensional Ising model

Consider first the case with no external magnetic field, $B = 0$. For a coupling

$$\mathcal{J} = -\epsilon \quad \text{with} \quad \epsilon > 0 \tag{4.211}$$

the transfer matrix (4.207) becomes

$$T = \begin{pmatrix} e^{\beta\epsilon} & e^{-\beta\epsilon} \\ e^{-\beta\epsilon} & e^{\beta\epsilon} \end{pmatrix}. \tag{4.212}$$

Since T is symmetric and positive, it can be diagonalized to yield real and positive eigenvalues. Moreover, the trace of a matrix is invariant under similarity transformations, and thus specially under orthogonal transformations $\mathbf{O}$

$$\mathrm{Tr}\left(\mathbf{O}^{-1} T \mathbf{O}\right) = \mathrm{Tr}\, T, \tag{4.213}$$

which diagonalize the transfer matrix T.

The two eigenvalues of the transfer matrix (4.212) are

$$\lambda_0 = 2\cosh \beta\epsilon \tag{4.214}$$

and

$$\lambda_1 = 2\sinh\beta\epsilon \tag{4.215}$$

with $\lambda_0 > \lambda_1$.

In the limit of long chains, $N \to \infty$, we therefore obtain from the partition function

$$\lim_{N\to\infty}\frac{1}{N}\ln\left(\mathrm{Tr}\,T^N\right) = \lim_{N\to\infty}\frac{1}{N}\ln\left(\lambda_0^N + \lambda_1^N\right) \tag{4.216}$$

$$= \lim_{N\to\infty}\frac{1}{N}\ln\left(\lambda_0^N\left[1 + \left(\frac{\lambda_1}{\lambda_0}\right)^N\right]\right) \tag{4.217}$$

$$= \ln\lambda_0 \tag{4.218}$$

$$= \ln(2\cosh\beta\epsilon). \tag{4.219}$$

The partition function itself is

$$Z_N = 2^N\cosh^N(\beta\epsilon) \tag{4.220}$$

from which we obtain the free energy per site

$$f = -\frac{1}{\beta}\ln\left[2\cosh\beta\epsilon\right] \tag{4.221}$$

and the internal energy per site

$$u = u(\beta) = \frac{\partial(\beta f)}{\partial\beta} = -\epsilon\tanh\beta\epsilon. \tag{4.222}$$

This smooth function of inverse temperature β will not give rise to any anomalous behaviour of the specific heat $c_V = \partial u/\partial T$, explicitly

$$c_V = \frac{\partial u}{\partial T} = -\beta^2\frac{\partial u}{\partial\beta} = \frac{\beta^2\epsilon^2}{\cosh^2\beta\epsilon}, \tag{4.223}$$

see figure 4.2. The Ising model in one dimension, thus, does not exhibit a phase transition. The peak exhibited by the specific heat is an example of a more general phenomenon called a Schottky anomaly (for an elementary discussion of Schottky anomalies, see, e.g. MacDonald, 2006). We have encountered Schottky anomalies, which are a typical phenomenon of two-level systems already in exercises 4.3–4.5.

This rather uninteresting, although exact, result (from the point of view of phase transitions) does not change if we consider the model in an external magnetic field.

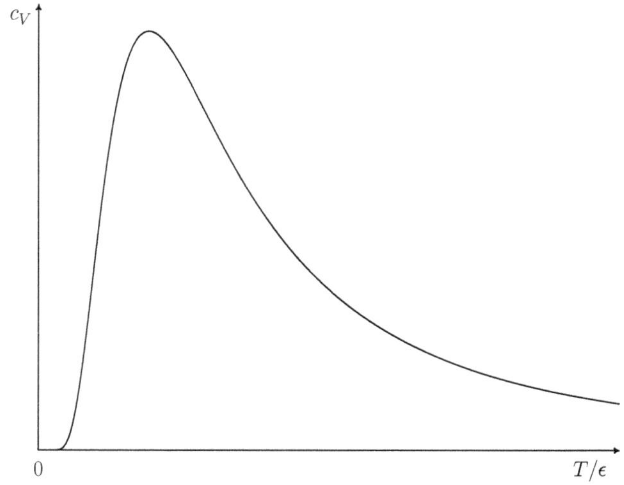

Figure 4.2 *The specific heat of the One-dimensional Ising model.*

EXERCISE 4.17 One-dimensional Ising model in a magnetic field Set up and diagonalize the transfer matrix for the One-dimensional Ising model in a magnetic field H. Consider the larger of the two eigenvalues of the transfer matrix to calculate first the free energy per site and then the average spin

$$\langle \sigma \rangle = -\left(\frac{\partial f}{\partial H}\right). \tag{4.224}$$

What behaviour does $\langle \sigma \rangle$ exhibit for vanishing magnetic field $H \to 0$ and arbitrary inverse temperature β? Does the One-dimensional Ising model in a magnetic field ever become ferromagnetically ordered? Is the model thus capable of a phase transition?

An extension of the transfer matrix method allows the calculation of correlation functions of the One-dimensional Ising model. Here, to conclude this section, we only quote the result for the correlation length

$$\xi = \frac{a}{\ln(\lambda_0/\lambda_1)} \tag{4.225}$$

where a is the lattice spacing, introduced here explicitly to obtain the correct dimension of length.

In chapter 5, we extend the study of statistical mechanics in the particularly interesting direction of an investigation of phase transitions and critical phenomena, a topic introduced in section 4.15 on the mean-field theory of the Ising model.

5

Phase Transitions, Critical Phenomena, and Finite-Size Scaling

One should not search for anything behind the phenomena. They themselves are the message.

Johann Wolfgang von Goethe (1749–1832)

Phase transitions and critical phenomena are ubiquitous in nature and have played an indispensable part in human history. They are among the most striking features of interacting many-particle systems. The interacting 'particles' may be atoms or molecules of a gas, liquid, or solid, but, arguably, also human beings interacting in an economy (see, e.g. Voit, 2005; or Sornette, 2003) or in other societal contexts (see, e.g. Solé, 2011). Examples of phase transitions and critical phenomena abound and are all around us; e.g. steam, water, and ice, and their multitude of phases and the changes between them, which are crucial, not least, for global weather; or the metals and their mixtures (alloys) which gave their names to whole human epochs, e.g. the bronze and iron ages. Binney *et al.* (1992) suggested that:

> Life without the endless movement of H_2O from one phase to another would be a sorry thing. ...Phase transitions are central to life on Earth and understanding them is one of the prime tasks of condensed–matter physicists.

Beyond the human realm, phase transitions and critical phenomena also play an important role, e.g. in the development of the early universe (see Kolb and Turner, 1989) The characteristic of phase transitions is that very small changes in some external parameter, e.g. temperature for simple gas-liquid transitions or falling house prices for the global economy, can cause the system as a whole to undergo dramatic, even catastrophic, changes from one phase to another.

These catastrophic changes manifest themselves in certain thermodynamic quantities, e.g. in discontinuities in the case of first order, discontinuous or abrupt phase transitions,[1]

[1] The density of the fluid in a typical liquid-gas phase transition may change by three orders of magnitude at the transition temperature.

Models of Quantum Matter: A First Course on Integrability and the Bethe Ansatz. Hans-Peter Eckle, Oxford University Press (2019). © Hans-Peter Eckle. DOI: 10.1093/oso/9780199678839.001.0001

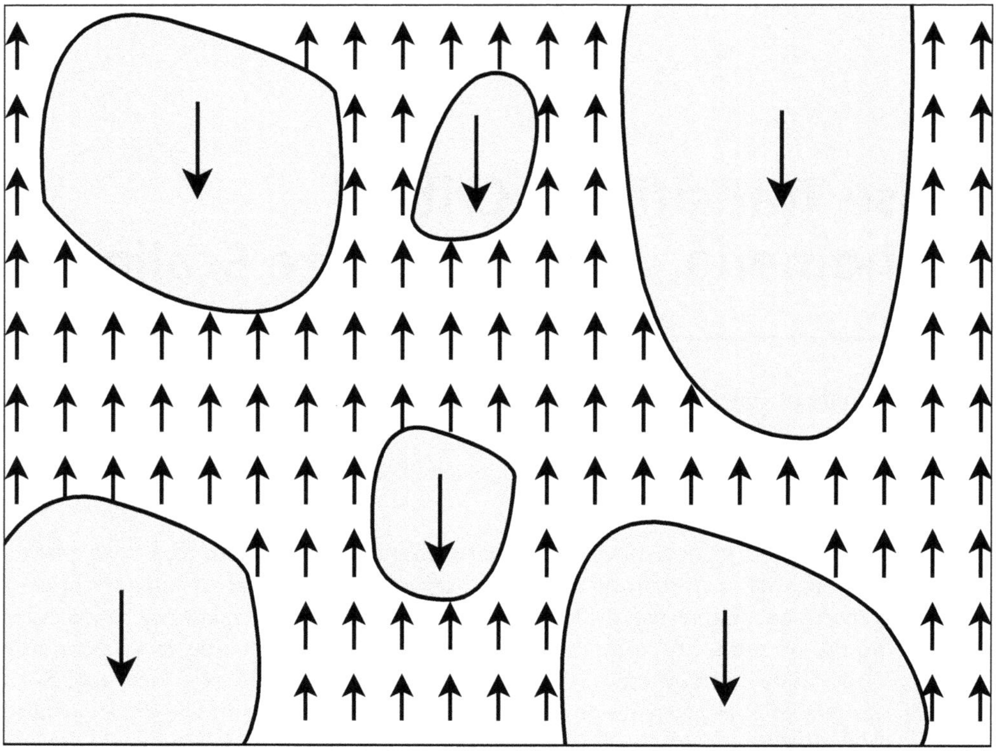

Figure 5.1 *Clusters of down spins forming in a sea of up spins. Clusters form in all sizes up to lengths of the order of the correlation length* ξ.

or power law singularities in the case of second order or continuous phase transitions, i.e. phase transitions at so-called critical points.[2] Here, we concentrate mostly on the latter case of continuous phase transitions, also called *critical phenomena*, where, near the critical point or points in the phase diagram, the quantities of paramount interest are the critical exponents describing the power law behaviour. For an elementary treatment of abrupt phase transitions, see e.g. Sethna (2006).

This chapter on phase transitions and critical phenomena introduces the most important concepts of this field of study to an extent that developments in later chapters, especially the chapter on *conformal symmetry in statistical mechanics* (chapter 7), are properly grounded and their significance can be fully appreciated. While the subsequent section 5.1 surveys the basic notions of phases, phase diagrams, and phase transitions, section 5.2 focuses on the discussion of critical behaviour at a second order phase transition.

[2] Paul Ehrenfest's classification of phase transitions, into orders has, with time, not proved to be the most useful.

Section 5.3 on Landau–Ginzburg theory and section 5.4 on scaling prepare for section 5.5, which provides an elementary introduction to the concepts of the renormalization group.

Furthermore, it discusses how phase transitions and critical phenomena are foreshadowed in finite systems in section 5.6 on finite-size scaling. In order to further build up a working knowledge of the field of phase transitions and critical phenomena, we recommend, e.g. Ma (1976), Binney *et al.* (1992), Goldenfeld (1992), Cardy (1996), Kardar (2007*a*) and Kardar (2007*b*), and Nishimori and Ortiz (2011). Finally, section 5.7 of this chapter is devoted to an introduction to the field of quantum phase transitions where quantum fluctuations take over the role of thermal fluctuations. The standard reference in this field is Sachdev (2011).

5.1 Phases, phase diagrams, and phase transition

As discussed in chapter 4, statistical mechanics attempts to explain the thermodynamic behaviour of systems of many particles on a macroscopic length scale from microscopic considerations, i.e. from the atomic perspective. In other words, the term *macroscopic* refers to systems that can be described by thermodynamic potentials, e.g. the Helmholtz free energy F.

An elementary definition, going back to Gibbs, states that a *phase* is a macroscopic region of space of a thermodynamic system throughout which all its physical properties are uniform. Or, to use an expression synonymous in this respect, homogeneous. The larger system may overall be inhomogeneous. In this respect, *uniform* or *homogeneous* means that all physical properties in the macroscopic region of the system in question vary only smoothly. The partition function or, equivalently, the appropriate thermodynamic potential (e.g. the Helmholtz free energy F or the Gibbs free energy G) must hence be smooth functions within a thermodynamic phase. When crossing the boundaries of the region, the phase boundaries in real space, the partition function, and hence, the physical properties, change dramatically, they exhibit non-analytic behaviour. As an example think of the difference in density between the liquid and gaseous phase of a fluid like water.

Instead of looking at different parts of a spatially inhomogeneous system, it is more instructive to consider a homogeneous system. By varying external parameters like pressure or temperature, abrupt transitions can be induced at sharp values of the parameters between the various phases of which the system is capable. For example, water can be evaporated by increasing the temperature or by lowering the pressure. At the values of the external parameters where the transition occurs, parts of the system will coexist in different phases (sometimes more than two—the triple point of a liquid is an example where solid, liquid, and gas coexist).

In this way, a diagram of the transitions from one phase to another can be constructed: a *phase diagram* where the external parameters serve as axes and where the loci of external parameter values at which the transitions occur delineate the phase boundaries in the parameter space. A familiar example of such a phase diagram is the phase diagram depicting the curves of the phase boundaries between steam, water, and ice, depending on the applied pressure and temperature.

Examining such a diagram, we observe that there are boundaries across which the physical properties indeed change dramatically, e.g. the boundary where either water evaporates or steam condenses, depending on the direction in which the boundary is crossed. Such phase transitions are usually accompanied by a certain amount of *latent heat*, which is either released or has to be expended. They are called *first order*, *discontinuous*, or *abrupt* phase transitions.

If one follows the water-steam phase boundary towards higher temperatures and higher pressures, the amount of latent heat decreases until it completely vanishes at a point (p_c, T_c), i.e. the critical point. At this point the distinction between water and steam ceases to exist. This corresponds to another kind of phase transition, called *second order* or *continuous*. It is an example of a *critical phenomenon*.

A terminology for phase transitions was introduced by Paul Ehrenfest.

> At an nth order phase transition the nth derivative of a thermodynamic potential with respect to one of its natural variables is discontinuous while all lower derivatives, i.e. all derivatives of order $n' < n$, are continuous.

Some examples are:

1. Transitions for $T < T_c$ where a liquid evaporates at constant pressure. Consider the Gibbs free energy $dG = -SdT$ and, hence, $S = -(\partial G/\partial T)_p$. There is a discontinuity, a jump, ΔS in the entropy and, hence, a latent heat $\Delta Q = T\Delta S$. This phase transition is of first order.
2. The fluid-superfluid transition of ^{4}He at constant pressure. The entropy is continuous, but has a diverging derivative with respect to temperature at T_c, the specific heat $C_p = T(\partial S/\partial T)_p$ diverges logarithmically. This phase transition is, hence, of second order.
3. A superfluid film of ^{4}He on graphite. According to the *Berezinskii–Kosterlitz–Thouless* theory, this system has a phase transition, yet the Gibbs free energy G has arbitrary many continuous derivatives at $T = T_c$. This phase transition is, hence, of infinite order.

A phenomenological way of looking at phase transitions is provided by examining the equations of state with which a thermodynamic system can be described on the macroscopic level. The most basic and well-known equation of state is the ideal gas law $pV = NT$ describing an (idealized) gas where N classical point particles are considered and which do not interact among themselves but only with the walls of their container of volume V, thereby generating a pressure p. The ideal gas, however, does not exhibit a phase transition.

There have been several attempts to modify the ideal gas law in a such way as to obtain a transition from a gas phase to a condensed liquid phase. Two of these attempts are the subject of the following two exercises.

EXERCISE 5.1 Van der Waals equation of state The van der Waals equation of state, an equation to describe real fluids by introducing phenomenological parameters to account for the interactions between particles, is

$$\left(p + \frac{aN^2}{V^2}\right)(V - Nb) = NT, \tag{5.1}$$

where N is the number of particles in the fluid. The phenomenological parameters a and b describe the interaction between particles and the volume excluded by the presence of the particles, respectively.

Determine the critical isotherm of the $p(V)$ diagram, i.e. the isotherm separating the region where there is no phase transition from gas to liquid phase from the region where there are two phases, a high density liquid and a low density gaseous phase.

EXERCISE 5.2 Dieterici equation of state Another phenomenological attempt to describe a real gas by accounting phenomenologically for interactions is the Dieterici equation of state

$$p(V - b) = NTe^{-a/VT}, \tag{5.2}$$

where the physical meaning of the constants a and b is similar to the one for the van der Waals equation.

Repeat the previous exercise for this equation of state.

As we shall see, the singularities expressed in the power laws at continuous phase transitions correspond to singularities in the free energy or, equivalently, the partition function of the systems. It is, however, clear from the definition of the partition function being, as the sum of regular functions, itself a regular function that any finite system cannot have a singular partition function. Only in the thermodynamic limit of an infinite system can singularities occur. How the singularities emerge from the regular partition function of the finite system is an important question studied in the theory of finite-size scaling (see section 5.6).

The study of phase transitions is, hence, concerned with the origin and characterization of the various singularities in the free energy or, equivalently, the partition function of systems in the thermodynamic limit. This is a very complex problem and no simple solutions are known which would work for all conceivable cases.

However, a considerable simplification in the description of phase transitions and critical phenomena occurs due to *universality*. Let us describe what that means qualitatively via an example. On the microscopic level, consider magnetic moments on the lattice sites of a crystal. These magnetic moments are formed from the interacting electrons of the atoms constituting the crystal and, hence, strongly interact with each other. Depending on the details of these interactions, magnetic moments may order. This ordering of the magnetic moments is, thus, a complex cooperative phenomenon.

However, in many cases, a very simple description suffices. We assume that the magnetic moments can be represented by spin degrees of freedom σ, which retain the quantum mechanical behaviour of the electrons only in a very rudimentary form: the spin variables are assumed to take only two values, conventionally chosen to be $\sigma = \pm 1$. Furthermore, the interactions are captured as so-called exchange interactions $\mathcal{J}_{ij}$ between the spins on lattice sites i and j. This is the model (or rather class of models), i.e. the Ising model, introduced in section 4.14.2 and whose mean-field description is outlined in section 4.15.

This model, however crude it appears to be, captures the critical physics of many real magnetic systems and even quite different systems like a binary mixture (cp. section 4.14.3) quite well. This is a manifestation of *universality*, the phenomenon that the critical behaviour of physical systems does not (or only very weakly) depend on the microscopic details of the system. Hence, their critical behaviour can be described by simple models that only have to incorporate roughly the microscopic details. Various models like these, differing in essential ingredients, e.g. the number of spin orientations in the Ising versus the Potts model (cp. section 4.14.5), lead to different *classes* of universality.

5.2 Critical behaviour

5.2.1 Order parameter

Since there is more than one equilibrium phase on a coexistence line, we seek to identify a thermodynamic function, the *order parameter*, which is different for the coexisting phases. It is not always easy to find a thermodynamic function that serves this purpose. Moreover, the choice of the function may not be unique, i.e. more than one function may be possible.

As mentioned in section 4.15.1, for a magnetic system the spontaneous magnetization can serve as order parameter. It is defined in the limit of vanishing magnetic field H as

$$m(T) = \frac{1}{V} \lim_{H \to 0} M(H, T) \tag{5.3}$$

where V is the macroscopic volume of the sample. In the vicinity of the critical temperature T_c the spontaneous magnetization has a power law behaviour

$$m(T) \propto \begin{cases} 0 & : \quad T > T_c \\ |t|^\beta & : \quad T < T_c \end{cases} \tag{5.4}$$

with the reduced temperature

$$t = \frac{T_c - T}{T_c}. \tag{5.5}$$

The critical exponent β, called a thermal exponent, characterizes the singular behaviour of the order parameter along the coexistence line in the vicinity of the critical point $T = T_c$.[3]

Along the critical isotherm $T = T_c$, the magnetization exhibits the singular behaviour

$$m(T = T_c, H) \propto H^{1/\delta} \tag{5.6}$$

with the field exponent δ.

For a general order parameter, we shall use the notation ϕ or Φ, the latter in case of a multicomponent order parameter.

5.2.2 Static response function

The response of the order parameter to its conjugate field gives rise to a further exponent γ:

$$\chi_\pm(T, H = 0) = \left.\frac{\partial m(H, T)}{\partial H}\right|_{H=0} \propto |t|^{-\gamma_\pm} \tag{5.7}$$

where we have distinguished, for generality's sake, the two exponents γ_+ and γ_- for the two sides of the phase transition $T \to T_c+$ and $T \to T_c-$. In most cases, however, the same exponent governs both sides, i.e. $\gamma_+ = \gamma_- = \gamma$.

5.2.3 Long-range correlations

The divergence of response functions implies that critical fluctuations are correlated over long distances. We demonstrate that this is indeed the case for a particular model: the Ising ferromagnet with the Hamiltonian

$$\mathcal{H} = -\mathcal{J}\sum_{\langle ij\rangle} \sigma_i\sigma_j - H\sum_i \sigma_i = -\mathcal{J}\sum_{\langle ij\rangle} \sigma_i\sigma_j - HM \tag{5.8}$$

where the Ising spin variables $\sigma_i = \pm 1$ are summed over nearest neighbours $\langle ij\rangle$ and the exchange coupling constant is $\mathcal{J} > 0$. The magnetization is given by

$$M = \sum_i \sigma_i \tag{5.9}$$

[3] Using $|t|$, i.e. the modulus, in (5.4) is, of course, not necessary in this case of an order parameter, but for the critical behaviour of other quantities we may find its use helpful to write formulas in a more compact way.

and the partition function, which depends on the magnetic field H, is

$$Z(H) = \sum_{\{\sigma\}} e^{-\beta\mathcal{H}} \tag{5.10}$$

where the sum incorporates all configurations (microstates) $\{\sigma\}$ of the N spins.

The average magnetization and the susceptibility follow from the partition function as

$$\langle M\rangle = \frac{\partial \ln Z}{\partial(\beta H)} = \frac{1}{Z}\sum_{\{\sigma\}} M e^{-\beta\mathcal{H}}, \tag{5.11}$$

$$\chi = \frac{\partial\langle M\rangle}{\partial H} = \beta\left(\frac{1}{Z}\sum_{\{\sigma\}} M^2 e^{-\beta\mathcal{H}} - \frac{1}{Z^2}\left(\sum_{\{\sigma\}} M e^{-\beta\mathcal{H}}\right)^2\right). \tag{5.12}$$

The susceptibility is related to the variance of the magnetization by

$$T\chi = \mathrm{var}(M) \equiv \langle M^2\rangle - \langle M\rangle^2. \tag{5.13}$$

Next, we reinterpret this result in the continuum limit, since we are only interested in long-range fluctuations and ignore short-range fluctuations on a microscopic scale. In order to do so, we introduce a coarse-grained local magnetization $m(\mathbf{r})$ which we obtain by summing over all spins in a small volume centred at position $\mathbf{r}$ that is, however, large enough to be macroscopic. Any fluctuations of the spins at a shorter length scale are averaged out. Then

$$M = \int d^3r\, m(\mathbf{r}) \tag{5.14}$$

is the magnetization of the sample. The susceptibility can then be rewritten as

$$T\chi = \int d^3r\, d^3r' \Big(\langle m(\mathbf{r})m(\mathbf{r}')\rangle - \langle m(\mathbf{r})\rangle\langle m(\mathbf{r}')\rangle\Big). \tag{5.15}$$

For a homogeneous, i.e. translationally invariant, system the average magnetization is a constant in space

$$\langle m(\mathbf{r})\rangle = m. \tag{5.16}$$

Furthermore, we can introduce the correlation function $\mathcal{G}(\mathbf{r},\mathbf{r}')$ which, for a translationally invariant system, depends only on the spatial separation $\mathbf{r}-\mathbf{r}'$

$$\mathcal{G}(\mathbf{r}-\mathbf{r}') = \langle m(\mathbf{r})m(\mathbf{r}')\rangle. \tag{5.17}$$

Introducing the connected correlation function

$$\mathcal{G}_c(\mathbf{r}) = \langle m(\mathbf{r})m(0)\rangle_c = \langle m(\mathbf{r})m(0)\rangle - m^2 \tag{5.18}$$

we obtain

$$\chi = \beta V \int d^3r\, \langle m(\mathbf{r})m(0)\rangle_c = \beta V \int d^3r\, \mathcal{G}_c(\mathbf{r}). \tag{5.19}$$

Since above T_c, $m = 0$, the connected correlation function $\mathcal{G}_c(\mathbf{r}) \equiv \mathcal{G}(\mathbf{r})$. However, below T_c, $\mathcal{G}_c(\mathbf{r})$ measures only how the local fluctuations of the order parameter in one part of the system affect those of another part. The large contribution of the partially ordered state has been subtracted.

There will be a typical distance ξ over which such influences of one part of the system on a distant one are appreciable. This distance is called the *correlation length*, discussed in section 4.15.2.

For a rough quantitative estimate, we assume the connected correlation function to fall off exponentially[4]

$$\mathcal{G}_c(\mathbf{r}) \propto e^{-|\mathbf{r}|/\xi} \quad \text{for} \quad |\mathbf{r}| > \xi. \tag{5.20}$$

Within the correlation length, i.e. within a region $|\mathbf{r}| < \xi$, a reasonable estimate assumes that $g \approx m^2$ is an upper limit for $\mathcal{G}_c(\mathbf{r})$, i.e. $\mathcal{G}_c(\mathbf{r}) < g$ for $|\mathbf{r}| < \xi$ and, hence, the susceptibility can be estimated roughly from

$$T\chi/V < g\xi^3. \tag{5.21}$$

The divergence of the susceptibility χ therefore implies the divergence of the correlation length ξ, which explains, e.g. the observation of critical opalescence at a liquid-gas critical point that can be measured by light scattering experiments.

For a magnetic system, where the correlation length is accessible to neutron scattering experiments, we have accordingly

$$\xi_\pm(t, H=0) \propto |t|^{-\nu_\pm} \tag{5.22}$$

[4] As we shall see later, a more sophisticated expectation is $\mathcal{G}_c(\mathbf{r}) = e^{r/\xi}/r$. This would, however, not change our semi-quantitative argument here.

where again, in many cases $\nu_+ = \nu_- = \nu$. At the critical point, $t = 0$, the correlation length diverges as a function of the magnetic field

$$\xi(t = 0, H) \propto H^{-\nu_H}. \tag{5.23}$$

5.3 Landau–Ginzburg theory

As discussed in the previous sections, microscopic details, e.g. the lattice structure, are unimportant for the qualitative behaviour of systems at critical points. Therefore, a phenomenological approach, similar but more general and hence more versatile than the mean-field theory outlined in section 4.15, may be applicable and shed further insight on critical phenomena. Such an approach is provided in the *Landau–Ginzburg* theory (Landau and Ginzburg, 1950).[5] It is an attempt to build a generic framework for mean-field theories, like the one outlined in section 4.15 on the basis of the Ising model.

Above, we introduced the coarse-grained magnetization variable $m(\mathbf{r})$ as order parameter, which we now generalize to be a generic order parameter $\Phi(\mathbf{r})$ with n components ϕ_i, where also the position space we consider has d dimensions.

5.3.1 Examples of systems with multicomponent order parameters

Before we delve into the construction of the Landau–Ginzburg Hamiltonian and the consequences to be drawn from this object, we first explore examples of the multicomponent order parameters employed in systems exhibiting phase transitions and critical behaviour.

5.3.1.1 One-component order parameters

A one-component order parameter is a scalar function or field. A typical example is the liquid-gas phase transition where

$$\phi(\mathbf{r}) = \rho(\mathbf{r}) - \rho_{\text{Gas}}(\mathbf{r}) \tag{5.24}$$

is the difference in local density between liquid and gas phase where both $\rho(\mathbf{r})$ and $\rho_{\text{Gas}}(\mathbf{r})$ are averaged over a macroscopically large volume centred at position $\mathbf{r}$, which is, however, small compared to the sample volume, i.e. $\phi(\mathbf{r})$ is an example of a coarse-grained variable.

Similarly, we have a scalar order parameter

$$\phi(\mathbf{r}) = x_1(\mathbf{r}) - x_2(\mathbf{r}) \tag{5.25}$$

[5] An English translation of this original paper by Landau and Ginzburg, who develop their theory for the case of superconductivity, can be found in ter Haar (1965).

for the coarse-grained, molar densities $x_1(\mathbf{r})$ and $x_2(\mathbf{r})$ with $x_1(\mathbf{r}) + x_2(\mathbf{r}) = 1$ of a binary mixture, e.g. a mixture composed of two fluids whose miscibility depends on temperature.

For an uniaxial magnetic system, e.g. the one described by the Ising model, the magnetization $m(\mathbf{r})$ is yet another example of a one-component order parameter. Its value signifies the transition between the disordered paramagnetic and the ordered ferromagnetic phase.

5.3.1.2 *Two-component order parameters*

Helium, more precisely ^{4}He, undergoes a transition at about 2 Kelvin from a phase where it is a normal viscous liquid, which is called helium I (He I), to a phase that contains a fraction that exhibits no viscosity at all and that grows as the temperature is lowered further. This superfluid condensate is called helium II (He II) and is a Bose condensate formed in a system of interacting Bosons. The natural choice of order parameter to describe the phase transition between He I and He II, i.e. between normal fluid and superfluid, is the wave function or quantum amplitude $\psi(\mathbf{r})$ to find a particle in the condensed state at position $\mathbf{r}$. This quantum amplitude is a delocalized quantity since the thermal de Broglie wavelength of the He particles in the condensate is macroscopic in size. Hence, $\psi(\mathbf{r})$ is a naturally coarse-grained variable. Obviously, since $\psi(\mathbf{r})$ is a complex function, the order parameter has two components.

A similar situation arises for the conductor to superconductor phase transition where pairs of electrons, the Cooper pairs, which develop mediated through phonons at low temperatures, form a Bose condensate.

Another example are planar magnets, described, for example, by the XY model (cp. 4.14.4) where the magnetization, a two-component vector, serves as order parameter.

5.3.1.3 *Three-component order parameters*

Classical isotropic magnets, as described by the classical Heisenberg model (cp. 4.14.4), fall into the class of models whose order parameters have three components, and which is again given by the magnetization vector for the isotropic magnet.

5.3.1.4 *More exotic order parameters*

More exotic order parameters are possible. The helium isotope ^{3}He is a spin-1/2 Fermion. However, a liquid formed by this isotope is possible by applying suitably high pressure. In such a liquid, at very low temperatures, the ^{3}He atoms form pairs that rotate around their common centre of mass with total angular momentum taking values of $\mathcal{J} = 0$, 1, or 2. This results in $\sum_{\mathcal{J}=0}^{2}(2\mathcal{J} + 1) = 9$ angular momentum configurations, each of which can form a Bose condensate with a complex wave function. Hence, the order parameter has 9 complex or 18 real components.

Another esoteric order parameter consists of a second-rank tensor, which is appropriate to describe phase transitions in liquid crystals.

5.3.2 Construction of the Landau–Ginzburg Hamiltonian

In order to construct a phenomenological Hamiltonian, we need to employ symmetry arguments gleaned from microscopic models and combine them with our phenomenological knowledge of critical behaviour (see section 5.2) and mean-field theory (e.g. of the Ising model as in section 4.15.1). This will determine the generic form a phenomenological coarse-grained or effective Hamiltonian $\mathcal{H}$, which is capable of extending our picture of phase transitions and critical phenomena further. It is clear that such a Hamiltonian and the corresponding theory cannot be considered as derived from first principles. It is, as it were, a designed Hamiltonian, which has to prove its mettle against experimental observations of phase transitions.

It will turn out that the effective Hamiltonian, which for convenience we write as $\beta\mathcal{H} = \mathcal{H}/T$, is not a Hamiltonian or even a free energy in the traditional sense. There are in general no canonical relations between the order parameter fields and derivatives with respect to those fields. Moreover, the coupling parameters in $\beta\mathcal{H}$ will depend on the temperature and on external fields.

The first requirement or assumption that goes into the design of the effective Hamiltonian is *locality*: the order parameter fields defining the effective Hamiltonian should mirror the local short-range interactions of an underlying microscopic theory, i.e. the effective Hamiltonian is a functional of the order parameter $\Phi(\mathbf{r})$ (returning a number, the effective Hamiltonian, for a function, the order parameter)

$$\beta\mathcal{H}\left[\mathbf{\Phi}(\mathbf{r})\right] = \int d^d r \,\widehat{\mathcal{H}} \tag{5.26}$$

with a Hamiltonian density $\widehat{\mathcal{H}}$. This Hamiltonian density must be itself a local function of the order parameter fields and their generalized gradients

$$\widehat{\mathcal{H}} = \widehat{\mathcal{H}}\left(\mathbf{\Phi}(\mathbf{r}), \nabla\mathbf{\Phi}(\mathbf{r})\right) \tag{5.27}$$

where ∇ acts on each component of $\Phi(\mathbf{r})$. In principle, $\widehat{\mathcal{H}}$ can also explicitly depend on the position $\mathbf{r}$ when the system is not uniform, i.e. when there is no translational invariance of the microscopic system. We shall not further pursue this latter possibility.

The symmetry of the underlying microscopic model further restricts $\widehat{\mathcal{H}}$. A symmetry under spin reversal $\sigma_i \to -\sigma_i$ as for the Ising model in the absence of a magnetic field, or a rotational invariance as for the classical Heisenberg model, mandates that a power series expansion of the Hamiltonian density in the order parameter field and its generalized gradients depends only on even powers

$$\widehat{\mathcal{H}} = \frac{1}{2}a\mathbf{\Phi}^2 + \frac{1}{4}b\mathbf{\Phi}^4 + \frac{1}{6}c\mathbf{\Phi}^6 + \ldots + \frac{1}{2}A(\nabla\mathbf{\Phi})^2 + \frac{1}{2}B(\nabla^2\mathbf{\Phi})^2 + \ldots - \mathbf{h}\cdot\mathbf{\Phi} \tag{5.28}$$

except for the linear dependence coupling the order parameter to an external field $\mathbf{h}$ (e.g. a magnetic field as in the magnetic case). Here

$$(\nabla\boldsymbol{\Phi}(\mathbf{r}))^2 \equiv \sum_{i=1}^{n}\sum_{\alpha=1}^{d}\left(\frac{\partial\phi_i}{\partial r_\alpha}\right)^2. \tag{5.29}$$

The order parameter field Φ can be any of the particular order parameters for which we gave examples in section 5.3.1.

However, for a more complete specification of the Hamiltonian density (5.28), we need to find constraints for the phenomenological coefficients a, b, c,... and $A, B, \ldots$, which depend on the microscopic interactions and are, thus, *a priori* unknown. Generally, these coefficients must be analytic functions of temperature and other external parameters, e.g. pressure. Further constraints are furnished by the demands that the Hamiltonian density remain bounded and stable. For example, in the absence of an external field $\mathbf{h}$, this implies that the coefficient for the highest power of Φ, which we retain in the expansion, must be positive such that the Hamiltonian density remains bounded from below. This also guarantees that the probability (Boltzmann weight)

$$\exp\Big(-\beta\mathcal{H}\left[\boldsymbol{\Phi}(\mathbf{r})\right]\Big) \tag{5.30}$$

for a configuration Φ remains normalizable.

Under these quite general symmetry assumptions, the coarse-grained Hamiltonian assumes the form of the *Landau–Ginzburg* Hamiltonian

$$\beta\mathcal{H} = \int d^3r\left[\frac{1}{2}a\boldsymbol{\Phi}^2 + \frac{1}{4}b\boldsymbol{\Phi}^4 + \ldots + \frac{1}{2}A(\nabla\boldsymbol{\Phi})^2 + \ldots - \mathbf{h}\cdot\boldsymbol{\Phi}\right]. \tag{5.31}$$

This functional of the coarse-grained order parameter field Φ constitutes a considerable simplification compared to the original microscopic problem.

We now attempt to determine approximately the dependence of some of the parameters in the Landau–Ginzburg Hamiltonian on the external parameters, in particular, the temperature, in a simplified version of the Landau–Ginzburg approach, the *Landau* mean-field theory. Moreover, we introduce plausible assumptions for other parameters, especially for where to terminate the expansion in (5.28). Within this approach we shall also succeed in determining the values of some critical exponents.

5.3.3 Landau mean-field theory

The Landau–Ginzburg functional (5.31) can be used to express the partition function as a path integral (cf. section 6.1.1 for more details on the Feynman path integral)

$$Z = \int \mathcal{D}\boldsymbol{\Phi}(\mathbf{r})\exp\{-\beta\mathcal{H}\left[\boldsymbol{\Phi}(\mathbf{r})\right]\}. \tag{5.32}$$

It is, in general, not easy to evaluate this path integral. We shall restrict ourselves to a saddle–point approximation that will lead us back to results for the critical exponents with which we are already familiar from our discussion of mean-field theory for the Ising magnet in section 4.15. However, with the Landau–Ginzburg functional (5.31) and the partition function path integral (5.32), we have gained a point of departure for implementing refinements that go beyond the simple mean-field theory discussed in section 4.15.

The saddle-point approximation we shall adopt consists in retaining only the terms explicitly shown in (5.31), assuming that the parameter $A > 0$, and replacing the integrand in (5.32) by its maximum. A uniform order parameter, i.e. one that satisfies $\nabla \Phi(\mathbf{r}) = 0$, maximizes (5.32) if it is also aligned with the external field, i.e. $\Phi(\mathbf{r}) = \Phi \mathbf{e}_h$ where Φ is a constant and $\mathbf{e}_h$ is a unit vector along the external field direction.

The uniform order parameter that maximizes the partition function (5.32) is then determined by the minimum of

$$f(\phi) = \frac{1}{2}a\Phi^2 + \frac{1}{4}b\Phi^4 - h\Phi. \tag{5.33}$$

For small magnitudes of the order parameter Φ, i.e. in the vicinity of the critical point, $f(\Phi)$ is dominated by its first term and especially the sign of the parameter a.

- If $a > 0$, the quartic term in (5.33) becomes unimportant near the critical point, i.e. for small Φ and, hence, the minimum is located at $\phi = h/a$, which vanishes for vanishing external field h. The corresponding susceptibility $\chi = \partial\Phi/\partial h = 1/a$ diverges for $a \to 0$.
- If $a < 0$, the quartic term becomes important and can no longer be neglected. Moreover, only if in the parameter of the quartic term $b > 0$, there will be a minimum at a finite value of the order parameter Φ. In fact, there are now two non–zero minima in addition to a maximum at $\Phi = 0$.

These properties of the Landau–Ginzburg Hamiltonian in the saddle-point approximation can be matched with the behaviour near a critical point of, e.g. a magnetic system, if we identify the temperature behaviour of the phenomenological parameters to be

$$a(t) = a_1 t + \mathcal{O}(t^2) \tag{5.34}$$

$$b(t) = b_0 + b_1 t + \mathcal{O}(t^2) \tag{5.35}$$

where the reduced temperature is $t = (T - T_c)/T_c$. The positive constants $a_1, b_0, \ldots$ will depend on the microscopic details of a concrete material and are, still, unknown. However, the temperature dependence, generic for the critical behaviour, is now explicit in the above Taylor expansions in the reduced temperature t.

We now recover the critical point behaviour, i.e. the critical exponents, which, as already anticipated, will be the same as for our previous mean-field calculations for the

Ising model in section 4.15. For example, the exponent of the order parameter is found by determining the minimum of (5.33) from $\partial f/\partial\Phi = 0$ as $\beta = 1/2$.

We shall not repeat these explicit calculations here. Moreover, we shall also not further pursue more refined developments which the Landau–Ginzburg Hamiltonian (5.31) offers. Two final exercises will suffice for our purposes.

EXERCISE 5.3 Landau theory of a continuous magnetic phase transition The Landau free energy to describe a continuous magnetic phase transition is an expansion in terms of the magnetization M

$$F(T, M) = F_0 + a(T - T_c)M^2 + bM^4, \tag{5.36}$$

where $a, b > 0$ are constants, T_c is the critical temperature. F_0 is the non-singular part of the free energy which, for simplicity, can be assumed to be temperature independent in the vicinity of T_c.

a) Discuss $F(T, M)$ as a function of M by sketching it for three isotherms: i. $T > T_c$, ii. $T = T_c$, and iii. $T < T_c$.
b) Derive the magnetic equation of state $H = H(T, M)$ and sketch isotherms for the same three cases as in a) in a (H, M)–diagram.
c) Sketch the magnetization $M = M(T)$ for a fixed value of H for $H = 0$ and $H \neq 0$.
d) Calculate the isothermal susceptibility

$$\chi_T = \left(\frac{\partial M}{\partial H}\right)_T \tag{5.37}$$

for $H = 0$ in the two cases $T > T_c$ and $T < T_c$.
e) For $H = 0$, calculate also the specific heat $c_H(T)$. How large is the jump in the specific heat, $c_H(T, H = 0)$, at $T = T_c$? Hint: Construct the Gibbs potential $G(T, H)$ for $H = 0$.

EXERCISE 5.4 Landau theory for abrupt phase transition Within the framework of Landau theory, the simple Gibbs potential for the order parameter ψ

$$G(t) - G_0(T_0) \equiv \mathscr{G}(t) = A\psi^2 + B\psi^4 + C\psi^6 \tag{5.38}$$

describes an abrupt phase transition. The constant parameters are $B < 0$ and $C > 0$, while the parameter $A = at$ $(a > 0)$ varies with the distance $t = (T - T_0)/T_0$ to the temperature T_0 of the phase transition.

Show that for $A = B^2/4C$, this Gibbs potential has three minima with the same energy. Calculate the transition temperature T_0 as a function of the constant parameters in the Gibbs potential. Further, calculate the discontinuity $\Delta\psi$ in the order parameter at the phase transition.

5.4 Scaling

The various partition functions and their corresponding thermodynamic potentials discussed in chapter 4 contain the thermodynamic behaviour of a macroscopic system. They also play a major role in the mean-field theory of the Ising model in section 4.15. Adopting two assumptions, this section shows how the thermodynamic potentials can be used to further understand critical point behaviour of macroscopic systems. These assumptions are that the thermodynamic potentials can be split up into two additive parts, one regular and one singular at the critical point, and that the singular part of the thermodynamic potentials is a generalized homogeneous function of some set of thermodynamic variables.

We begin by introducing the notion of homogeneous functions.

5.4.1 Homogeneous functions

A function $f(x)$ of one (real) variable x is homogeneous if, for an arbitrary number $\lambda \neq 0$, called a scale factor, we have

$$f(\lambda x) = \nu(\lambda) f(x). \tag{5.39}$$

EXERCISE 5.5 Homogeneous function of one variable Show that in (5.39) $\nu(\lambda) = \lambda^p$ where $p = d\nu/d\lambda|_{\lambda=1}$. Moreover, show that a homogeneous function is always of power form $f(x) = cx^p$ with $c = f(1)$.

In view of the result of exercise 5.5, the homogeneous function is sometimes also said to be of degree p.

The notion of a homogeneous function can be extended to functions of more than one variable, e.g. for two variables we have $f(\lambda x, \lambda y) = \nu(\lambda) f(x, y)$.

EXERCISE 5.6 Homogeneous function of two variable Show that for a homogeneous function of two variables also $\nu(\lambda) = \lambda^p$ with $p = d\nu/d\lambda|_{\lambda=1}$. Moreover, show that any homogeneous function of two variables can always be written as $f(x, y) = x^p g_1(y/x) = y^p g_2(x/y)$, where $g_1(z) = f(1, z)$ and $g_2(z) = f(z, 1)$ are functions of one variable only.

For the case of two (or more) variables, a generalized homogeneous function can be defined by

$$f(\lambda^a x, \lambda^b y) = \lambda f(x, y) \tag{5.40}$$

with a and b arbitrary constants and λ a scale factor. It is often useful to replace $\lambda \to \lambda^d$ to obtain the alternative definition of the generalized homogeneous functions

$$f(\lambda^{a'} x, \lambda^{b'} y) = \lambda^d f(x, y) \tag{5.41}$$

with $a' = ad$ and $b' = bd$.

By putting, e.g. $\lambda = y^{-1/b}$ in (5.40), we obtain

$$f(x, y) = y^{1/b} g\left(\frac{x}{y^{a/b}}\right) \tag{5.42}$$

where $g(z) = f(z, 1)$.

EXERCISE 5.7 Generalized homogeneous function Show that the Legendre transformation

$$g(u, y) = f(x, y) - xu(x, y) \quad \text{where} \quad u(x, y) = \left(\frac{\partial f}{\partial x}\right)_y \tag{5.43}$$

of a generalized homogeneous function $f = f(x, y)$ of two variables is also a homogeneous function. How do the parameters a and b defining the generalized homogeneity of $f(x, y)$ change for $g(u, y)$.

This implies that all thermodynamic potentials are generalized homogeneous functions if one of them can be shown to be a generalized homogeneous function.

The generalized homogeneous functions have proved very useful in the theory of critical phenomena. As indicated at the beginning of this section, the first ingredient of the scaling hypothesis consists of positing that thermodynamic potentials behave as generalized homogeneous functions near critical points of the phase diagram.

5.4.2 The scaling hypothesis

The power law dependencies of thermodynamic quantities at critical points indicate scaling behaviour and, thus, suggest that the thermodynamic potentials near critical points may have a generalized homogeneous form. Using this hypothesis, the idea that critical behaviour is invariant under rescaling then leads to relations between the critical exponents. In order to be concrete, we shall now investigate the consequences of this hypothesis for the Gibbs potential $G(T, H)$ of a magnetic system.

We further assume, also as previously indicated, that the Gibbs potential can be split up into a regular and a singular part.

$$G(T, H) = G_{\mathbf{reg}}(T, H) + G_{\mathbf{sing}}(T, H) \tag{5.44}$$

where the regular part at $T = T_c$ does not show any discontinuous or diverging behaviour

$$\lim_{T \to T_c} G_{\mathbf{reg}}(T, H) = G_{\mathbf{reg}}(T_c, H) \tag{5.45}$$

and will not be of further interest to us, while the singular part

$$G_{\mathbf{sing}}(T, H) = G\left(\frac{T - T_c}{T_c}, H\right) = G(t, H) \tag{5.46}$$

(where we drop the subscript from now on) describes the behaviour close to the critical point. This singular thermodynamic potential was assumed by Widom (1965) to be a generalized homogeneous function, i.e.

$$G(\lambda^{a_t} t, \lambda^{a_H} H) = \lambda G(t, H). \tag{5.47}$$

This Widom, or static scaling, hypothesis includes no assumptions about the values of the scaling parameters a_t and a_H. However, it allows us to relate all critical exponents to these two numbers, which are also called scaling dimensions.

EXERCISE 5.8 Static scaling hypothesis Show with the help of the static scaling hypothesis (5.47) that the critical exponent β of the magnetization for $H = 0$ and $t \to 0$

$$M \propto (-t)^{\beta} \tag{5.48}$$

can be expressed in terms of the scaling parameters a_t and a_H as

$$\beta = \frac{1 - a_H}{a_t} \tag{5.49}$$

and the critical exponent δ of the magnetization for $t = 0$ and $H \to 0$

$$M \propto H^{1/\delta} \tag{5.50}$$

in terms of the scaling parameters and a_H alone as

$$\delta = \frac{a_H}{1 - a_H}. \tag{5.51}$$

The scaling hypothesis can be used to show that the critical exponents for the susceptibility $\chi \propto |t|^{-\gamma_\pm}$ approaching the critical temperature from above, γ_+, and from below, γ_- are the same: $\gamma_- = \gamma_+$. Moreover, they are related to scaling parameters by

$$\gamma' = \gamma = \frac{2a_H - 1}{a_t}. \tag{5.52}$$

These results follow from $\chi(t, H) = -\partial^2 G(t, H)/\partial^2 H$ and the scaling hypothesis (5.47). Moreover, the specific heat exponent α is related to the scaling dimensions by

$$\alpha = 2 - \frac{1}{a_t}, \tag{5.53}$$

which follows from $c(t, H) = -T\partial^2 G(t, H)/\partial T^2$ and the scaling hypothesis (5.47).

We have now expressed four critical exponents (α, β, γ, and δ) in terms of two scaling dimensions a_t and a_H. By eliminating the two scaling dimensions a_t and a_H, we can obtain relations between the critical exponents directly, which are valid as a consequence of the scaling hypothesis (5.47). These relations, as inequalities rather than equalities, can be obtained by purely thermodynamic reasoning (see, e.g. Stanley (1971), an early classic in the field that appeared just before the breakthrough of the renormalization group approach to critical phenomena).

EXERCISE 5.9 Scaling relations As a simple exercise, extract the relations between the critical exponents α, β, γ, and δ by eliminating the scaling dimensions a_t and a_H from the results mentioned.

5.4.3 Correlation function and hyperscaling

The scaling hypothesis does not yet suffice to obtain all critical exponents in terms of only two parameters: the scaling dimensions a_t and a_H. We have already encountered one more critical exponent describing the divergence of the correlation length $\xi \propto t^{-\nu}$ close to the critical temperature. There is a further critical exponent η associated with the correlation function. Since the correlation length is closely connected with the correlation function, we expect that the critical behaviour of these two quantities is also closely connected.

In addition to the scaling hypothesis, which furnished relations between the four exponents of the previous section, a second scaling hypothesis, the hyperscaling hypothesis, is needed to account for the two further exponents ν and η. As we shall see, the hyperscaling hypothesis will involve the dimension d of space of the system.

The hyperscaling hypothesis assumes that the correlation function has the scaling form

$$\mathcal{G}_c(\mathbf{r}, t) = \frac{1}{r^{d-2+\eta}} \Psi_{\pm}\left(\frac{r}{\xi(t)}\right) \tag{5.54}$$

with the critical exponent η and the correlation length, which diverges at the critical temperature $t \to 0$: $\xi(t) \propto t^{-\nu}$ with the critical exponent ν. The scaling functions $\Psi_{\pm}(z)$ accommodate possible different behaviour above and below the critical temperature T_c. They remain finite for $z \to 0$, i.e. when ξ diverges at criticality, while they behave as e^{-z} for large distances away from criticality. At criticality, the correlation function thus behaves as

$$\mathcal{G}_c(\mathbf{r}, t = 0) \propto \frac{1}{r^{d-2+\eta}}, \tag{5.55}$$

which is found experimentally and corroborated (with $\eta = 0$) in the Gaussian approximation of the Landau–Ginzburg theory (for the details of these results we refer to the literature given at the beginning of this chapter).

In order to connect the exponents ν and η to the other exponents, in fact first to the exponent γ, at this point we need some input from the general theory of linear response which we cannot explicate within the limits of this book (see, however, e.g. chapter 9 of Coleman (2015)). The local magnetic susceptibility is defined as

$$\chi(\mathbf{r}-\mathbf{r}') = \frac{\partial m(\mathbf{r})}{\partial H(\mathbf{r}')}|_{H=0} \tag{5.56}$$

and its Fourier transform, according to linear response theory, is identical to the Fourier transformed correlation function

$$\chi(\mathbf{q}) = \mathcal{G}_c(\mathbf{q}). \tag{5.57}$$

The $\mathbf{q} = 0$-component, i.e. the uniform part, of this susceptibility is thus related to the correlation function by

$$\chi(t) = \chi(\mathbf{q}=0,t) = \int d^d r\, \mathcal{G}_c(\mathbf{r},t). \tag{5.58}$$

Inserting the scaling form of the correlation function (5.54) and observing that a change of variables $x/\xi = z$ pulls out the relevant dependence on the correlation length ξ

$$\chi \propto \xi^{2-\eta} \tag{5.59}$$

while resulting in a finite integral on both sides of the critical point. Comparing with the critical behaviour of the correlation length $\xi \propto |t|^{-\nu}$, we obtain

$$\gamma = \nu(2-\eta), \tag{5.60}$$

which is known as Fisher's scaling law.

Since, at criticality, the correlation length ξ is assumed to be the only relevant length scale, the Gibbs potential per volume $g = G(t,h)/V$ should scale with the correlation length as

$$g \propto \xi^{-d}, \tag{5.61}$$

which, after differentiating twice with respect to temperature, implies $c_H \propto |t|^{\nu d-2}$ and, thus,

$$\alpha = 2 - \nu d, \tag{5.62}$$

known as Josephson's scaling law. In order to derive this scaling law, which involves the dimensionality of space d, we needed to invoke the additional scaling assumption (5.61). Josephson's scaling law is therefore also called hyperscaling.

As an important consequence of these scaling ideas we have the dilatation symmetry: under a change of scale $\mathbf{r} \to \lambda\mathbf{r}$, the critical correlation function (5.55) is obviously invariant

$$\mathcal{G}_c(\lambda\mathbf{r}) = \lambda^p \mathcal{G}_c(\mathbf{r}). \tag{5.63}$$

The critical system exhibits scale invariance or self-similarity.

Conformal invariance will be a direct generalization of scale invariance and the correlation functions will be the quantities to investigate in order to reveal the consequences of this larger symmetry. Before we deal with conformal invariance (see chapter 7), we shall have a look at the renormalization group. We shall restrict our attention to the, arguably, more intuitive real-space renormalization, which is, however, much less systematic and general than momentum-space renormalization. A treatment that adopts systematically the latter methodology can be found, e.g. in Herbut (2007).

5.5 Renormalization group

As emphasized in the previous sections, the quantities of major interest for critical phenomena are the critical exponents of the various power laws. The renormalization group is a powerful methodology to achieve the goal of calculating the critical exponents with an accuracy going beyond mean-field theory results.

The following sections provide an overview of the renormalization group, restricting us to the real-space version, which is both more intuitive and less technically involved. Even the real-space renormalization group is covered by only two examples, which are still reasonably simple, and by briefly outlining the general procedure to provide the flavour of this set of ideas.

This is almost invariably so, even in books predominantly devoted to the renormalization group: in the words of John Cardy (1996), '...the misleading impression [is] that the renormalization group is a kind of universal machine through which any problem may be processed, producing neat tables of critical exponents at the other end. This is quite false. It cannot be stressed too strongly that the renormalization group is merely a framework, a set of ideas, which has to be adapted to the nature of the problem at hand.'

5.5.1 Conceptual steps of the renormalization group

As discussed in the previous sections, the successful prediction of identities between the critical exponents has strongly corroborated the assumption that the correlation length ξ becomes the only important length scale as a system approaches criticality, and that microscopic length scales become correspondingly irrelevant. The critical behaviour is dominated by fluctuations that are statistically self-similar up to the scale set by the correlation length ξ.

The fundamental idea of the renormalization group is to take advantage of this self-similarity of the fluctuations by successively eliminating degrees of freedom of the system at length scales $x \ll \xi$. In this way, eventually only uncorrelated degrees of freedom at length scales comparable to the correlation length ξ remain. Conceptually, the renormalization group rests on three steps. However, it must be emphasized that these steps cannot be applied in a simple mechanical way but need careful consideration for each particular system.

In the previous sections, we have already come across some of the fundamental ideas behind the renormalization group which enter the three-step procedure. The first step consists of a a coarse-graining of the system, where the minimum length scale (initially, for example, the microscopic length scale of the lattice constant a) is changed by a factor $b > 1$ to $ba > a$. All degrees of freedom between a and ba are averaged over. Next, in the second step, all lengths scales are rescaled and, finally, in step three, the degrees of freedom are renormalized.

The following sections demonstrate how this three-step procedure works using two examples, in some detail for the one- (and more cursorily for the two-) dimensional Ising model. Finally, it outlines the general procedure for classical Ising-like spin models in order to summarize the terminology of the real-space renormalization group.

5.5.2 Renormalization of the one-dimensional Ising model

The classical one-dimensional Ising model is arguably the simplest model with which to demonstrate the procedure of the real-space renormalization group. We write its Hamiltonian in a form where the inverse temperature is absorbed into the Hamiltonian via the reduced coupling constant $K = \beta \mathcal{J}$

$$H = -\beta \mathcal{H} = K \sum_{i=1}^{N} \sigma_i \sigma_{i+1} \tag{5.64}$$

where we have set the magnetic field $h = 0$. Moreover, we shall assume periodic boundary conditions such that $\sigma_{N+1} = \sigma_1$ and, for simplicity, we assume N to be even. We want to evaluate the partition function

$$Z_N(K) = \sum_{\{\sigma\}} e^{K \sum_{i=1}^{N} \sigma_i \sigma_{i+1}} = \sum_{\{\sigma\}} \prod_{i=1}^{N} e^{K \sigma_i \sigma_{i+1}} \tag{5.65}$$

of the system described by the Hamiltonian (5.64) where the sum runs over all possible configurations of the N spins $\{\sigma = \pm 1\}$. The basic idea of the real-space renormalization group transformation, as outlined in section 5.5.1, is to perform the summation over a subset of the spins instead of over all spins while holding the other subset of spins fixed. For the one-dimensional chain of spins we are now considering, the simplest way to implement this is to hold every second spin, let us say the spins σ_{2j-1} on the odd sites, fixed and sum over the spins σ_{2j} on the even sites ($j = 1, 2, \ldots, N$). This corresponds to a rescaling of the original lattice by a factor $b = 2$. This procedure does not follow the letter, but rather the spirit, of what we said in the previous section. It does not average over regions or blocks of spins, but achieves an averaging by a decimation procedure.

We thus decompose the configuration sum in (5.65) in in the following way

$$Z_N(K) = \left(\sum_{\sigma_1=\pm 1} \cdots \sum_{\sigma_{2j-1}=\pm 1} \cdots \right) \left(\sum_{\sigma_2=\pm 1} \cdots \sum_{\sigma_{2j}=\pm 1} \cdots \prod_{i=1}^{N/2} e^{K\sigma_{2i}(\sigma_{2i-1}+\sigma_{2i+1})} \right) \tag{5.66}$$

$$= \left(\sum_{\sigma_1=\pm 1} \cdots \sum_{\sigma_{2j-1}=\pm 1} \cdots \right) \prod_{i=1}^{N/2} \sum_{\sigma_{2i}=\pm 1} e^{K\sigma_{2i}(\sigma_{2i-1}+\sigma_{2i+1})} \tag{5.67}$$

$$= \left(\sum_{\sigma_1=\pm 1} \cdots \sum_{\sigma_{2j-1}=\pm 1} \cdots \right) \prod_{i=1}^{N/2} 2\cosh K(\sigma_{2i-1} + \sigma_{2i+1}). \tag{5.68}$$

Next, we introduce a new coupling constant K' such that the partition function can be written in a form that closely resembles the original partition function. In order to achieve this goal, we write

$$2\cosh K(\sigma + \sigma') = Ae^{K'\sigma\sigma'}, \tag{5.69}$$

which, when evaluated for the possible values of the two spins σ and σ', implies for the renormalization group transformation

$$K'(K) = \frac{1}{2} \ln (\cosh 2K), \tag{5.70}$$

which is a non-linear transformation to a renormalized coupling constant. For the multiplicative factor, we obtain

$$A(K) = 2\sqrt{\cosh 2K}. \tag{5.71}$$

Hence, using the expression (5.68) for the partition function of the original one-dimensional Ising model, we realize that it can be rewritten apart from a multiplicative factor as the partition function of a one-dimensional Ising model on a chain of $N/2$ sites with a modified coupling constant K'

$$Z_N(K) = A^{N/2} Z_{N/2}(K'). \tag{5.72}$$

In the thermodynamic limit, $N \to \infty$, the free energy per site now has the form of a recursion or scaling relation

$$f(K) = \lim_{N\to\infty}\left(\frac{1}{N}\ln Z_N\right) = \lim_{N\to\infty}\left\{\frac{1}{N}\ln\left(A^{N/2}Z_{N/2}(K')\right)\right\} = g(K) + \frac{1}{2}f(K') \quad (5.73)$$

where

$$g(K) = \frac{1}{2}\ln A = \frac{1}{4}\ln(4\cosh 2K) = \frac{1}{2}K'(K) + \frac{1}{2}\ln 2. \quad (5.74)$$

Recall that the free energy (5.73) gets multiplied by minus the inverse temperature in accordance with our choice to absorb minus the inverse temperature $-\beta$ in the Hamiltonian and hence the coupling constant (cp. (5.64)). The free energy per site in the conventional definitions would have been

$$\beta f = -\frac{1}{N}\ln Z_N. \quad (5.75)$$

EXERCISE 5.10 Numerical iteration Show that for all positive initial values of the coupling constant $K > 0$ we obtain upon iteration of the renormalization group transformation (5.70) that

$$K' < K. \quad (5.76)$$

More explicitly, show that

$$\lim_{n\to\infty} K^{(n)} \to 0 \quad (5.77)$$

for the non-linear recursion

$$K^{(n)} = \frac{1}{2}\ln\left(\cosh 2K^{(n-1)}\right) \quad (5.78)$$

with $K^{(0)} = K > 0$, while the inverse transformation

$$K = \frac{1}{2}\cosh^{-1}\left(e^{2K'}\right) \quad (5.79)$$

leads to larger and larger couplings such that

$$\lim_{n\to\infty} K^{(n)} \to \infty. \quad (5.80)$$

Neither transformation has any other fixed points.

In order to use the recursion relation (5.73), which we rewrite in the form

$$f(K) = \frac{1}{2}\left(f(K') + K' + \ln 2\right) \quad (5.81)$$

in practice, we need an initial value for the free energy $f(K')$. Assume that, for small K', e.g. $K' = 0.01$, we can neglect the interaction between the spins. Then the partition function for the almost free spins is simply

$$Z_N \approx 2^N \tag{5.82}$$

and hence

$$f(0.01) \approx \ln 2. \tag{5.83}$$

Calculate numerically the first ten iterations of the recursion using these initial values and compare with the exact result of Ising (1925)

$$Z_N = (2\cosh(K))^N. \tag{5.84}$$

This exercise is adapted from Maris and Kadanoff (1978).

The result of exercise 5.10 shows that the renormalized Ising model is mapped towards higher and higher temperatures as we iterate the renormalization group transformation (5.70). Therefore, we expect that the model will become more and more disordered and hence the correlation length will decrease under the recursion.

In order to check this, consider the correlation function of the original and the renormalized system. The new spins $\{\sigma'\}$ are a subset of the original spins $\{\sigma\}$. Therefore, we have e.g.

$$\langle\sigma_1\sigma_{2j+1}\rangle_H = \langle\sigma'_1\sigma'_j\rangle_{H'} \tag{5.85}$$

where the expectation values are taken with respect to the original Hamiltonian H on the left-hand side and with respect to the renormalized Hamiltonian H' on the right-hand side. This relation for the correlation function can also be written as

$$g(2r, K) = g(r, K'), \tag{5.86}$$

which implies, assuming exponential decay for g, for the correlation length

$$\xi(K') = \frac{1}{2}\xi(K). \tag{5.87}$$

The correlation length decreases in the renormalization step, which again shows that the one-dimensional Ising model does not become critical at a finite temperature.

EXERCISE 5.11 Correlation function renormalization Convince yourself through a direct calculation of (5.85) and (5.87).

EXERCISE 5.12 Renormalization of the one-dimensional Ising model in a magnetic field Repeat the renormalization group calculations of this section for the one-dimensional Ising model in a magnetic field, i.e. for the model Hamiltonian

$$H = -\beta\mathcal{H} = \sum_{i=1}^{N} (K\sigma_i\sigma_{i+1} + h\sigma_i) \tag{5.88}$$

where $K = \beta\mathcal{J}$ and $h = \beta h_{\mathrm{m}}$ with the magnetic field h_{m}.

Nauenberg (1975) showed that the scaling equation (5.73) can be solved iteratively and leads to the exact solution (Ising, 1925)

$$f(K) = \ln(2\cosh K). \tag{5.89}$$

The first step in his analysis is to simplify the renormalization group transformation (5.70) to

$$\zeta' = \zeta^2 \tag{5.90}$$

using

$$\zeta = \tanh K. \tag{5.91}$$

This renormalization group transformation is simpler, although still non-linear, than the original transformation (5.70).

EXERCISE 5.13 Transformed renormalization group transformation Show that (5.70) and (5.90) are indeed equivalent under the transformation (5.91).

However, show also that the alternative transformation

$$\zeta = \coth K. \tag{5.92}$$

would be possible. The next exercise shows why this transformation does not work.

EXERCISE 5.14 Iterative solution of the scaling relation Use the scaling relation (5.73) rewritten as

$$f(K) = \frac{1}{2}f(K') + g(K) = \frac{1}{2}f(K') + \frac{1}{2}K' + \frac{1}{2}\ln 2 \tag{5.93}$$

$$= \frac{1}{2}f(K^{(1)}) + \frac{1}{2}K^{(1)} + \frac{1}{2}\ln 2, \tag{5.94}$$

which can be viewed as the first step of an iteration process and continue iterating.

Write down first the general nth iteration of the scaling relation and then take the limit $n \to \infty$. We know that $\lim_{n\to\infty} K^{(n)} = 0$. What does this imply for $f(K^{(n)})$ for large n?

Next we need to insert $K^{(n)}$ in terms of the transformed variable ζ into the scaling relation iterated infinitely many times. Evaluating the infinite product we will obtain recovers the exact result (5.89) for the free energy of the one-dimensional Ising model.

The following identity will prove useful:

$$\frac{1}{1-x} = \prod_{n=0}^{\infty} \left(\frac{1+x^{2^n}}{1-x^{2^n}} \right)^{\frac{1}{2^{n+1}}}, \tag{5.95}$$

which can be proven using

$$1 - x^{2^{n+1}} = \left(1 - x^{2^n}\right)\left(1 + x^{2^n}\right). \tag{5.96}$$

....

It is well known that the one-dimensional Ising model does not exhibit a phase transition at a finite temperature. It is therefore somewhat disappointing in the physical phenomena it displays, as Ising (1925) already realized. Moreover, it can be solved exactly. Therefore, the renormalization group considerations in this section have largely been an exercise to get an understanding of the real-space renormalization group procedure of arguably the simplest possible example where all steps are still elementary and easily controllable. Of course, as we shall see in the next section, a few interesting complications of more involved situations have not even arisen for the one-dimensional Ising model. The next section demonstrates how the essentially same procedure can immediately lead to a more complicated and interesting situation when we consider the same model in two dimensions.

5.5.3 Renormalization of the two-dimensional Ising model

We consider the Ising model in two dimensions on a square lattice in the absence of a magnetic field where classical spins $\sigma = \pm 1$ are located on the lattice sites and interact with their four nearest neighbours. In order to avoid unnecessary complications, we assume that the interactions are the same in the horizontal and vertical direction. The Hamiltonian of this model has the form

$$H = -\beta\mathcal{H} = K \sum_{\langle ij \rangle} \sigma_i \sigma_j \tag{5.97}$$

where $\langle ij \rangle$ indicates summation over all pairs of nearest-neighbour spins on the two-dimensional lattice.

This model also has an exact solution which, however, is much more involved than the exact solution of its one-dimensional counterpart (see Onsager, 1944; Kaufman, 1949; Kaufman and Onsager, 1949; and Schultz *et al.*, 1964).

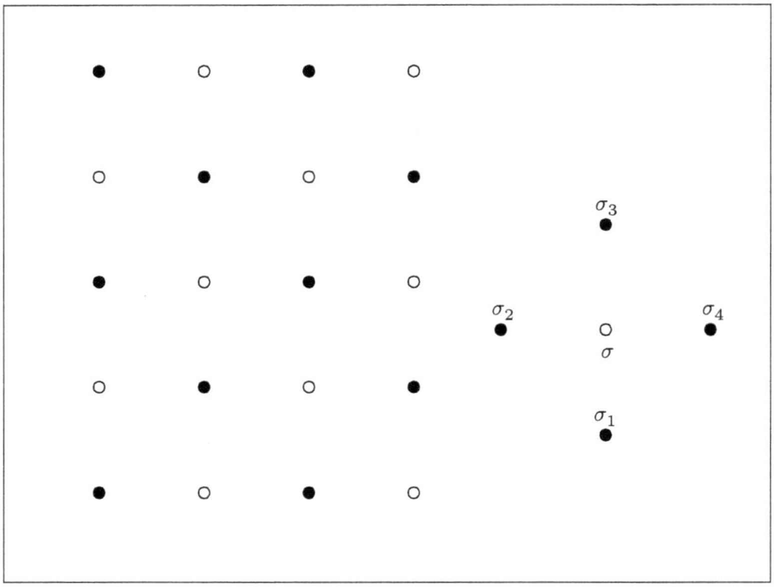

Figure 5.2 *Left: decomposition of the square lattice into two isomorphic sub-lattices, both of which are rotated by $\pi/4$ with respect to the original lattice. Right: a single square or plaquette centred at a central spin σ and his four nearest neigbours $\{\sigma_1, \sigma_2, \sigma_3, \sigma_4\}$ of the original lattice. The central spin of such a plaquette will be summed over in the first step of the renormalization group procedure. On the renormalized lattice, e.g. 'black' spins 1 and 2 are nearest neighbours, while, e.g. 'black' spins 1 and 3 are next–nearest neighbours.*

As in the case of the one-dimensional model, we divide the lattice into two sub-lattices like a checkerboard and then sum over the spins on one sub-lattice (see the left side of figure 5.2). More precisely, we pick an arbitrary spin σ of one of the two sub-lattices (e.g. as on the right side of figure 5.2, a spin on the 'white' sub-lattice) which interacts by nearest-neigbour interaction with four spins, let us call them $\{\sigma_1, \sigma_2, \sigma_3, \sigma_4\}$, all of which belong to the other sub-lattice (the 'black' sub-lattice, see the right side of figure 5.2). The partition function can then be rewritten as

$$Z_N(K) = \sum_{\{\sigma\}} \exp\left(K \sum_{\langle ij \rangle} \sigma_i \sigma_j\right) \tag{5.98}$$

$$= \sum_{\{\sigma^p\}} \prod_p \left(e^{K(\sigma_1^p+\sigma_2^p+\sigma_3^p+\sigma_4^p)} + e^{-K(\sigma_1^p+\sigma_2^p+\sigma_3^p+\sigma_4^p)}\right). \tag{5.99}$$

The product in the second expression is over all plaquettes p defined by the sites of one sub-lattice, the summation is over the spins on the corners of all plaquettes, each of which belongs to the other (dual) lattice (see again figure 5.2). Hence, summing over this central spin σ gives a typical term in the partition function

$$e^{K(\sigma_1+\sigma_2+\sigma_3+\sigma_4)} + e^{-K(\sigma_1+\sigma_2+\sigma_3+\sigma_4)} = 2\cosh\{K(\sigma_1+\sigma_2+\sigma_3+\sigma_4)\}. \quad (5.100)$$

In analogy to the one-dimensional case, we can rewrite (5.100) in the same way as in (5.69), i.e. introduce a renormalized coupling constant K' and an overall multiplicative factor $A(K)$. In order to achieve this goal, we would have to satisfy equations for all nonequivalent combinations of $\sigma_i = \pm 1$ $(i = 1, 2, 3, 4)$ in (5.100). There are four such combinations, which cannot in general be satisfied with only two parameters, K' and A. Hence, here the simple emulation of the one-dimensional case ends. On the other hand, we shall be rewarded by physically richer results for the two-dimensional Ising model.

As a matter of fact, we have to introduce four new parameters in order to write the partition function again as a sum over Boltzmann factors with spin-spin interactions in the exponential. We find that (5.100) can be rewritten as

$$\begin{aligned} 2\cosh K(\sigma_1+\sigma_2+\sigma_3+\sigma_4) = A\exp\big(&K'(\sigma_1\sigma_2+\sigma_2\sigma_3+\sigma_3\sigma_4+\sigma_4\sigma_1) \\ &+ L'(\sigma_1\sigma_3+\sigma_2\sigma_4) + M'(\sigma_1\sigma_2\sigma_3\sigma_4)\big), \end{aligned} \quad (5.101)$$

i.e. the renormalization of the lattice has introduced three types of spin-spin interaction. The usual nearest-neighbour spin-spin interaction on the decimated lattice with coupling constant K', but in addition a next-nearest-neighbour interaction with coupling constant L' and a four-spin interaction around a plaquette of the decimated lattice with coupling constant M' (see again figure 5.2 to identify the types of interaction on the decimated 'black' sub-lattice).

Those three coupling constants and the multiplicative factor A are given explicitly by

$$A = 2(\cosh 4K)^{1/8}(\cosh 2K)^{1/2}, \quad (5.102)$$

$$K' = L' = \frac{1}{8}\ln(\cosh 4K), \quad (5.103)$$

$$M' = \frac{1}{8}\ln(\cosh 4K) - \frac{1}{2}\ln(\cosh 2K). \quad (5.104)$$

Compared to the one-dimensional case, this is a considerably more complicated situation. Not only have we acquired next-nearest neighbour and four-spin interactions, but it has also become impossible to continue the renormalization group procedure with an exact calculation.

EXERCISE 5.15 Renormalization group transformation of the two-dimensional Ising model Derive the renormalization group equations of the two-dimensional Ising model (5.102–5.104).

The partition function, which, as we recall, may not be changed by the renormalization procedure, after the first iteration, takes the form

$$Z_N = A(K)^{N/2} \sum_{\{\sigma\}} \exp\left(2K' \sum_{\langle nm\rangle} \sigma_n\sigma_m + L' \sum_{[kl]} \sigma_k\sigma_l + M' \sum_{p} \sigma_q\sigma_r\sigma_s\sigma_t\right) \tag{5.105}$$

$$= A(K)^{N/2} Z_{N/2}(K', L', M') \tag{5.106}$$

where we have dropped again the superscript p for plaquette on the spin variables. The notation $[kl]$ stands for next-nearest-neighbour pairs, while p signifies the sum over all plaquettes. The sum over $\{\sigma\}$ is now only to be taken over half the lattice sites of one sub-lattice. The factor of 2 is due to the fact that each nearest-neighbour pair appears in two plaquettes in the sum over plaquettes in (5.99).

The free energy becomes

$$f(K', L', M') = \frac{1}{2} \ln A(K) + \ln Z_{N/2}(K', L', M') \tag{5.107}$$

$$= \frac{1}{2} \ln\left(2(\cosh 4K)^{1/8}(\cosh 2K)^{1/2}\right) + \frac{1}{2} f(K', L', M'). \tag{5.108}$$

In order to make progress in this situation, we have to treat the coupling constants that emerged from the renormalization group procedure in an approximate way. The simplest possibility would be to ignore the next–nearest and four-spin coupling. However, the recursion relations

$$K' = \frac{1}{8} \ln(\cosh 4K), \tag{5.109}$$

$$f(K') = 2f(K) - \ln\left(2(\cosh 4K)^{1/8}(\cosh 2K)^{1/2}\right) \tag{5.110}$$

again have no fixed point and lead to an analytic free energy $f(K)$. The situation is thus quite similar to the one-dimensional Ising model and exhibits again no critical point and, hence, no continuous phase transition. However, unlike for the one-dimensional Ising model, where also the exact solution exhibits no critical behaviour, we know from the exact solution that the Ising model in two dimensions *does* have a critical point.

In order to do better, we therefore have to include the other couplings L' and M' or, at least, one of them in some approximate way. We restrict ourselves to a discussion of only one of many possibilities, albeit a still relatively simple one. More sophisticated approximation schemes can be found in Niemeijer and van Leeuwen (1976).

Since $K > 0$ by definition of the model, according to (5.103) also $K' > 0$ and $L' > 0$, while M' in (5.104) may be (and in fact is) negative. Therefore M' would counteract the ferromagnetic tendency of the model to align spins. Hence, as an approximation, we may choose to neglect M'. Furthermore, we may approximate the effect of K' and L' by an averaged K', which is chosen so that the energy of a (renormalized) lattice of $N/2$ spins with interactions K' and L' and hence N nearest and N next-nearest bonds

$$-N\left(2K' + L'\right) \to -NK' = -N\frac{3}{8}\ln\left(\cosh 4K\right) \tag{5.111}$$

is equal to the energy of the averaged interaction K' with N nearest-neighbour bonds. The renormalized Hamiltonian is now of the same form as the original Hamiltonian of the two-dimensional Ising model, i.e. has only nearest-neighbour interactions.

In this approximation, the recursion relation of the renormalization has a fixed point

$$K' = K^* = \frac{3}{8}\ln\left(\cosh 4K^*\right) \tag{5.112}$$

and maps K to smaller values $K' < K$ and eventually to zero for $K < K^*$ and maps K to larger $K' > K$ and eventually to infinity for $K > K^*$. We recall that for the one-dimensional Ising model K was always mapped to smaller values for any $K > 0$ (see exercise 5.10). The numerical value of the recursion (5.112) is $K^* = K_c \approx 0.50698$.

We are now also in a position to calculate a numerical value for a critical exponent. With the recursion relation (5.112) and assuming that the free energy (5.110) exhibits

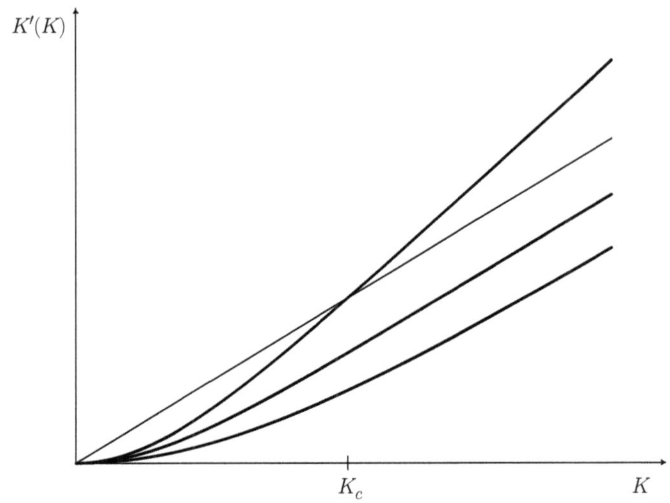

Figure 5.3 *Recursion relations (5.112) and (5.109) for the two-dimensional Ising model. For comparison, the lowest curve is the recursion relation (5.70) of the one-dimensional Ising model and the line $K' = K$. Only the approximation leading to the recursion relation (5.112) has a critical coupling constant K_c.*

singular behaviour $f_{\mathbf{sing}}(K) = a|K - K_c|^{2-\alpha}$, we can obtain the singular behaviour of the specific heat $c \propto |T - T_c|^{-\alpha} \propto |K - K_c|^{-\alpha}$ and the corresponding critical exponent α. On the right-hand side of (5.110) we have the singular behaviour

$$2f_{\mathbf{sing}}(K) = 2a|K - K_c|^{2-\alpha} \tag{5.113}$$

while, on the left-hand side, we have

$$f_{\mathbf{sing}}(K') = a|K' - K_c|^{2-\alpha}. \tag{5.114}$$

Expanding (5.111) to first order

$$K' = K_c + (K - K_c)\frac{dK'(K)}{dK}|_{K=K_c} \tag{5.115}$$

the free energy recursion relation (5.110) becomes

$$a|\,(K - K_c)\,\frac{dK'(K)}{dK}|_{K=K_c}|^{2-\alpha} = 2a|K - K_c|^{2-\alpha} \tag{5.116}$$

and, finally, the critical exponent of the specific heat

$$\alpha = 2 - \frac{\ln 2}{\ln\left(\frac{dK'(K_c)}{dK}\right)}. \tag{5.117}$$

This yields an numerical value of $\alpha \approx 0.1308$, which has to be compared to the value $\alpha = 0$ of the exact solution of the two-dimensional Ising model corresponding to a logarithmic singularity at T_c.

We now briefly survey how the examples discussed in this and the previous section can be used to obtain a general real-space renormalization procedure.

5.5.4 General real-space renormalization group procedure

Specifically, as starting point we use a generalized Ising spin system in d spatial dimensions, parametrized by a set of coupling constants, which, for convenience, we combine in an infinite dimensional vector

$$\mathbf{K} = (K_1, K_2, K_3, \ldots). \tag{5.118}$$

We assume the general form of the Hamiltonian

$$H = -\beta\mathcal{H} = K_1 \sum_i \sigma_i + K_2 \sum_{\langle ij \rangle} \sigma_i\sigma_j + K_2 \sum_{[ij]} \sigma_i\sigma_j + K_4 \sum_{\langle ijk \rangle} \sigma_i\sigma_j\sigma_k + \ldots. \tag{5.119}$$

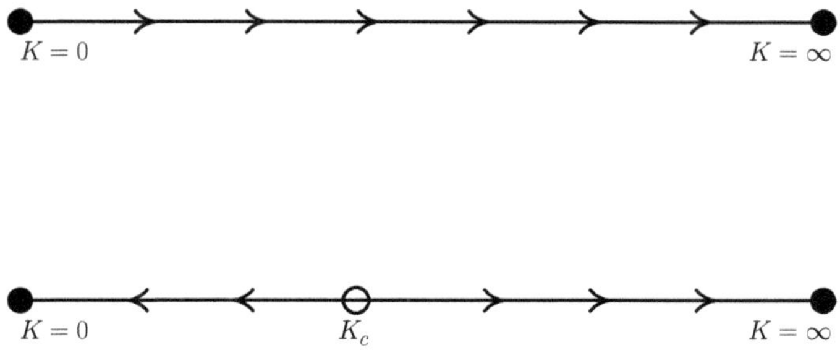

Figure 5.4 *Renormalization group flows for the one- and two-dimensional Ising models. While the one-dimensional Ising model has only $K = 0$ as unstable and $K = \infty$ as stable fixed point, the two-dimensional model has also a fixed point at a finite value $K = K^* = K_c$, which corresponds to a critical coupling.*

Thus, for the usual Ising model in a magnetic field, we have $K_1 = -h$ and $K_2 = K$ while all other components of the vector $\mathbf{K}$ are initially zero, but, as we have seen in the previous section, further not-zero components can emerge after each renormalization step.

Similar to our examples of the one- and two-dimensional Ising models in sections 5.5.2 and 5.5.3, the task now is to construct, in a renormalization step, a transformation of the coupling constants

$$\mathbf{K} \to \mathbf{K}' = \mathcal{R}_b(\mathbf{K}). \tag{5.120}$$

The index b indicates the spacial scale factor $b > 1$. In the decimation schemes in the examples of the one- and two-dimensional Ising models, the scale factor has been $b = 2$ for the one- and $b = \sqrt{2}$ for the two-dimensional case. How to construct such a renormalization group transformation depends on the model and will, in general, not be straightforward.

In order to further elucidate the general procedure, we assume that such a construction has been successfully carried out and that the transformation has the following characteristics.

- The transformation $\mathcal{R}_b$ is a smooth, differentiable function of its arguments $\mathbf{K}$.
- Spatial distances rescale under the transformation $\mathcal{R}_b$ as $\mathbf{r} \to \mathbf{r}' = \mathbf{r}/b$ while the number of spins is reduced to $N' = N/b^d$.
- The physical requirement that the partition function remains unchanged except for a rescaling factor implies that the free energies of the system corresponding to the two sets of coupling constants $\mathbf{K}$ and $\mathbf{K}'$ are related by

$$\frac{1}{b^d} f(\mathbf{K}') = f(\mathbf{K}) + g(\mathbf{K}). \tag{5.121}$$

Moreover, the correlation length behaves as

$$b\xi(\mathbf{K}') = \xi(\mathbf{K}). \tag{5.122}$$

- The renormalization group transformation $\mathcal{R}_b$ is a *semi-group* with respect to the spatial scale factor b, which means that two successive transformations with scale factors b and b' result again in a transformation

$$\mathcal{R}_b\mathcal{R}_{b'} = \mathcal{R}_{bb'}. \tag{5.123}$$

However, there is no inverse transformation.[6]

5.5.4.1 *Fixed points and criticality*

As shown in the examples, the fixed points of the renormalization group transformation or recursion relation $\mathcal{R}_b$ determine the critical behaviour of a system. Suppose we have found a fixed point $\mathbf{K}^*$, i.e.

$$\mathbf{K}^* = \mathcal{R}_b(\mathbf{K}^*). \tag{5.124}$$

Then the correlation length at the fixed point satisfies

$$\xi^* = b\xi^*. \tag{5.125}$$

Since $b > 1$, this equation can have only two solutions. Either

$$\xi^* = 0 \quad \text{or} \quad \xi^* = \infty. \tag{5.126}$$

The first possibility corresponds to maximal disorder since the system exhibits no correlations over an appreciable distance as is expected for a system at high temperature $T \to \infty$. The second possibility is more interesting. It corresponds to either a completely ordered system at $T = 0$ (i.e. $\mathbf{K}^* = \infty$) or to a critical point if the correlation length diverges for a finite $\mathbf{K}^* = \mathbf{K}_c$.

5.5.4.2 *Relevant and irrelevant scaling fields*

In order to learn more, we have to examine the renormalization group recursion relation $\mathcal{R}_b$ in more detail in the neighbourhood of the fixed point

$$\mathbf{K} = \mathbf{K}^* + \boldsymbol{\epsilon}, \tag{5.127}$$

[6] The name renormalization *group* is therefore, strictly speaking, a misnomer.

which implies the linear relationship

$$\mathbf{K}' = \mathbf{K}^* + \mathcal{T}\boldsymbol{\epsilon} + \mathcal{O}(\epsilon^2) \tag{5.128}$$

with the matrix

$$\mathcal{T}_{\alpha\beta} = \left(\frac{\partial\left(\mathcal{R}_b\right)_\alpha}{\partial K_\beta}\right)_{\mathbf{K}=\mathbf{K}^*} = \left(\frac{\partial K'_\alpha}{\partial K_\beta}\right)_{\mathbf{K}=\mathbf{K}^*}, \tag{5.129}$$

which, in general, is not a symmetric matrix.

Hence, we need to consider right *and* left eigenvectors, $\mathbf{d}_p$ and $\mathbf{g}_p$, respectively, such that the right and left eigenvalue problems are

$$\mathcal{T}\mathbf{d}_p = \lambda_p \mathbf{d}_p \quad \text{and} \quad \mathbf{g}_p \mathcal{T} = \lambda_p \mathbf{g}_p, \tag{5.130}$$

which we can normalize and which satisfy $\mathbf{g}_p \cdot \mathbf{d}_q = \delta_{pq}$ if the eigenvalues are non-degenerate $\lambda_q \neq \lambda_p$.

The vector ϵ in (5.127), as indeed any arbitrary vector, can be developed in terms of the set of right eigenvectors $\mathbf{d}_p$

$$\boldsymbol{\epsilon} = \sum_p a_p \mathbf{d}_p = \sum_p \left(\mathbf{g}_p \cdot \epsilon\right) \mathbf{d}_p \tag{5.131}$$

and, hence, (5.127) becomes

$$\mathbf{K} = \mathbf{K}^* + \sum_p \left(\mathbf{g}_p \cdot \epsilon\right) \mathbf{d}_p \equiv \mathbf{K}^* + \sum_p u_p \mathbf{d}_p \tag{5.132}$$

which defines the *linear scaling fields* u_p. Applying the linear recursion matrix (5.129) to this equation yields for (5.128)

$$\mathbf{K}' = \mathbf{K}^* + \sum_p \lambda_p(b) u_p \mathbf{d}_p + \mathcal{O}(\epsilon^2). \tag{5.133}$$

Thus, we can write the linearized renormalization group transformation in terms of the linear scaling fields

$$u_p \to u'_p = \lambda_p(b) u_p. \tag{5.134}$$

The semi-group property (5.123) of the renormalization group transformation translates into the functional equation

$$\lambda_p(b)\lambda_p(b') = \lambda_p(bb') \tag{5.135}$$

whose solution is

$$\lambda_p(b) = b^{y_p}. \tag{5.136}$$

The renormalization group eigenvalues y_p determine the behaviour of the linear scaling fields u_p under repeated renormalization group transformations. For $y_p > 0$, the corresponding linear scaling field is amplified and called relevant, while for $y_p < 0$ the linear scaling field decreases eventually to zero and is therefore called irrelevant. Linear scaling fields with $y_p = 0$ remain unaffected and are called marginal.

In terms of the linear scaling fields, criticality now means that all relevant scaling fields are zero. This corresponds to the experience that, in order to reach criticality, we need to adjust some of the thermodynamic variables of the macroscopic system to particular values. For example, as seen in the Ising spin systems, at the ferromagnetic critical point, that the magnetic field is zero, while the temperature attains its critical value T_c. In this example, we expect that the critical point can be described by two relevant scaling fields

$$u_1 \equiv u_t \propto (T - T_c)/T_c \quad \text{and} \quad u_2 \equiv u_h \propto h. \tag{5.137}$$

We now have the following picture in the space spanned by the linear scaling fields u_p. A finite fixed point of the renormalization group transformation corresponds to a critical point of the system. If the system is close to a fixed point, relevant scaling fields will be amplified and, hence, drive the system away from the critical point, whereas irrelevant scaling fields will return the system to the critical point. The effect of marginal fields cannot be decided on the basis of the linear analysis to which we have restricted our discussion. An example where a marginal field becomes important is the non-universal critical behaviour of the eight-vertex, or Baxter, model which is an example of a model exhibiting critical exponents continuously varying with a model parameter rather than critical exponents which are mere numbers Kadanoff and Brown (1979). For more on vertex models in the context of exact solutions, see part II.

5.5.4.3 Universality

The picture of a critical point, which has now emerged due the renormalization group, allows us now to make the notion of *universality* more concrete. The class of physical systems that differ from each other only in irrelevant fields form a universality class of systems since all irrelevant fields flow to same fixed point under iteration of the renormalization group. These irrelevant fields of coupling constants, flowing to the same particular fixed point, form the so-called *basin of attraction* of this fixed point. In chapter 7, we shall return to the important question of how to classify all possible universality classes for the case of two dimensions. This question cannot be answered within the renormalization group framework alone. We need to assume an additional symmetry: conformal invariance of two-dimensional critical systems.

5.5.4.4 Scaling and critical exponents

This section demonstrates how scaling and critical exponents emerge within the framework of the renormalization group as we have developed it so far.

If there is only one relevant scaling field u, we can write the correlation length (5.122) as

$$\xi(K^* + b^y u) = \xi(K^* + u)/b, \tag{5.138}$$

which implies for $\xi \propto (T - T_c)^{-\nu} \propto (K - K^*)^{-\nu}$ that

$$\left(b^y u\right)^{-\nu} = u^{-\nu}/b \tag{5.139}$$

and, hence,

$$y = 1/\nu. \tag{5.140}$$

Assuming two relevant ($u_t \propto t$ and $u_h \propto h$) scaling fields and one irrelevant (u_i) scaling field, the recursion relation (5.121) for the free energy becomes after n iterations

$$f_{\text{sing}}(u_t, u_h, u_i) \approx b^{-nd} f_{\text{sing}}(b^{ny_t} u_t, b^{ny_h} u_h, b^{-n|y_i|} u_i) \tag{5.141}$$

where we have only written down the singular part of the free energy, assuming especially that the contribution g in (5.121) is analytic.

EXERCISE 5.16 Specific heat exponent Show that the singular part of the free energy (5.141) leads to the relation $\alpha = 2 - d/y_t$ or, using (5.140), $\alpha = 2 - d\nu$, which we recognize as the hyperscaling relation of section 5.4.3.

This section showed how the real-space version of the renormalization group works for two examples, the one- and two-dimensional Ising models, and gave a brief outline of how it works in principle. Recalling the quotes from John Cardy at the beginning of this section, it is clear that real-space renormalization is not a tool that can be applied to a system of interest in a straightforward mechanical way. The books mentioned at the beginning of this chapter and further references they provide will have to be consulted to acquire a working knowledge of this exciting field.

5.6 Finite-size scaling

Strictly speaking, in finite systems, i.e. *all* conceivable systems, there can be no phase transitions. The partition function for a finite system is always an analytic function, and, thus, are all its derivatives. Therefore no abrupt changes like discontinuities or divergencies can occur. How can this apparent contradiction with the observed phase transitions be resolved?

Ferdinand and Fisher's (1969) study of finite two-dimensional Ising models provides an answer. These authors demonstrated, *inter alia*, that very small systems indeed do not show singular behaviour in their thermodynamic quantities, like the specific heat per spin $c = C/nm$, where $N = nm$ is the size of the system. However, as the system size grows, increasingly pronounced peaks develop. If the system size grows further, the rounded peak will eventually become indistinguishable from a true singularity. This behaviour of finite systems is sketched in figure 5.5.

The question finite-size scaling tries to answer is how the infinite system size or bulk behaviour is approached when the actual finite size of the system accessible to calculations changes. Of particular interest is the determination of the critical point parameters of the bulk system from the finite-size data.

Chapter 7, especially section 7.10, shows that the conformal symmetry of a bulk critical system in two dimensions allows not only to predict, via a conformal transformation to a finite geometry, its finite-size behaviour, but that this finite-size behaviour also contains the parameters of the bulk critical system. Therefore, reversing the argument, finite-size calculations of two-dimensional critical systems can be used to read off these critical parameters directly. Moreover, the connection between two-dimensional classical

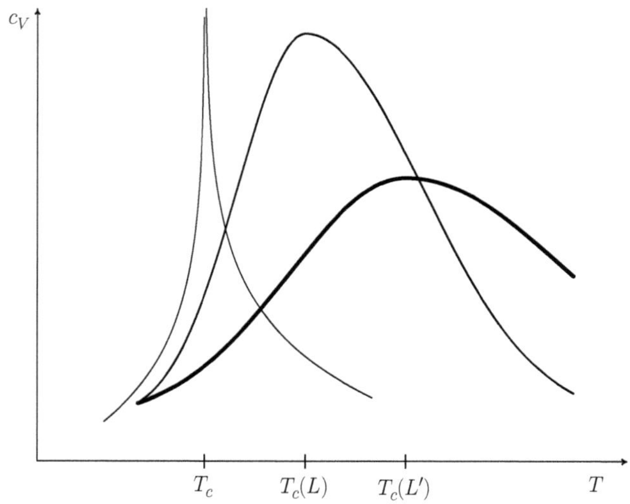

Figure 5.5 *Schematic sketch of the specific heat as a function of temperature for two lattices of sizes L (thin line) and* $L' < L$ *(thick line). The finite-size scaling of the shift in the temperature* $L^{-\lambda}$*, the rounding, and the maximal height* $L^{\alpha/\nu}$ *of the specific heat are shown, if somewhat exaggerated. For comparison, the diverging specific heat in the thermodynamic limit with critical temperature* T_c *is also given (very thin line).*

systems of statistical mechanics and one-dimensional quantum systems, elucidated in chapter 6, will open a further possibility to extract critical parameters, e.g. through exact calculations on one-dimensional quantum systems using, as key ingredient, the Bethe ansatz method. How such calculations can be performed in detail is the topic of part VI.

In finite systems, singularities, such as discontinuities or divergencies, associated with phase transitions of the infinite system, are rounded. The goal of the theory of finite-size scaling is to provide a theory which allows to extrapolate from such data for finite-size systems to the behaviour of the infinite system, i.e. of the system in the thermodynamic limit. The theory makes use of scaling laws, introduced by Fisher (Fisher, 1971; Fisher and Barber, 1972), which involve the size of the system. We present here this 'naive' approach to finite-size scaling, leaving out the fine print that involves considering renormalization group methods Brézin (1982, 1986). However, we briefly discuss the ideas of *phenomenological renormalization* Nightingale (1976, 1982).

In finite-size systems, we distinguish two cases:

- genuinely finite systems, e.g. a finite cube of volume L^3 or a finite square of area L^2; and
- systems which are finite in some spatial directions, but still infinite in the others, e.g. a slab of infinite 'volume' $L \times \infty \times \infty$, a prism of 'volume' $L \times L \times \infty$ or a strip of infinite 'area' $L \times \infty$.

As we shall see, the former do not possess a sharp phase transition, while the latter, being infinite systems, exhibit phase transitions, which are, however, different from the corresponding thermodynamic limit $L \to \infty$ in the finite dimensions.

5.6.1 Fisher's finite-size scaling theory

The previous sections of this chapter have discussed the critical behaviour of a large system, where the number N of particles is practically infinite, $N \to \infty$, with fixed particle density $n = N/V$, i.e. in the thermodynamic limit, is determined by the fluctuations of the order parameter. The correlation length ξ_∞ as a function of temperature T of the critical fluctuations of the order parameter diverges with a power law as the critical temperature T_c is approached

$$\xi_\infty(t) = \xi_0 \, |t|^{-\nu} \tag{5.142}$$

with the reduced temperatute

$$t = \frac{T - T_c}{T_c} \to 0. \tag{5.143}$$

Finite systems *cannot* exhibit singularities. First of all, the correlation length is clearly bounded by the system size $L \propto V^{\frac{1}{3}}$ itself. Secondly, the partition function

$$Z = \sum_n \exp\left(-\beta E_n\right) \tag{5.144}$$

is an analytic function of the temperature $T = 1/\beta$ as long as the summation is over a finite number of degrees of freedom which are indicated here by the quantum numbers n and the corresponding energies E_n of the system since all Boltzmann weights $\exp\left(-\beta E_n\right)$ are analytic functions. Only if the summation is over infinitely many such terms, i.e. in the thermodynamic limit, non-analytic singular behaviour can emerge Yang and Lee (1952).

Thus, strictly speaking, all finite systems, and therefore necessarily all real experiments, do not exhibit singular behaviour. The divergencies of certain thermodynamic quantities predicted in the thermodynamic limit therefore only exist as more or less pronounced, however rounded, maxima or peaks. Also other singularities are smeared out. For macroscopic particle numbers ($N \approx 10^{23}$) measurements are, however, practically indistinguishable from the singular behaviour of the thermodynamic limit. In view of this, we can ask whether finite-size effects play any role and whether they should concern us at all.

However, although we may be primarily interested in the properties of the infinite system, it is important that we have a sound and theoretically underpinned understanding of finite-size effects and the methods to extrapolate the behaviour of the infinite system from the results obtained for a finite system. This statement is certainly true for numerical approaches like exact diagonalization or (quantum) Monte Carlo techniques (for a detailed discussion of finite-size effects concerning the latter, see Landau and Binder (2015)) where the systems under consideration are necessarily finite. Moreover, for systems confined in slap, prism, or strip geometries (as mentioned previously), which are finite in some of their dimensions, conformal symmetry offers predictions where critical properties of the thermodynamic limit can be read off from results obtained for a finite system. These developments are discussed in detail in chapter 7, especially in section 7.10.

The basic idea of finite-size scaling rests on two plausible assumptions which are known as Fisher's finite-size scaling hypotheses Fisher (1971); Fisher and Barber (1972). As we have already seen in this chapter, the critical behaviour of a system in the thermodynamic limit is determined by the correlation length $\xi_\infty(t)$. Since, in a finite system, there cannot be correlations on a scale larger than the linear system size L itself, the first basic assumption is to express thermodynamic quantities of the finite system in terms of the ratio

$$x = \frac{L}{\xi_\infty(t)} \tag{5.145}$$

between this characteristic length L of the system and the correlation length $\xi_\infty(t)$ of the corresponding system in the thermodynamic limit.

Thus, a thermodynamic quantity $G_L(t)$, such as the specific heat or a susceptibility, of the finite system is assumed to be essentially determined by the variable

$$x \equiv \frac{L}{\xi_\infty(t)} = \frac{1}{\xi_0}\left(|t|L^{1/\nu}\right)^\nu \tag{5.146}$$

if the finite system approaches a critical point. The fundamental tenet of finite-size scaling is thus that, near T_c and for a sufficiently large system size L, the thermodynamic quantity $G_L(t)$ will be a function of the single variable x. Furthermore, the second plausible ingredient of Fisher's finite-size scaling hypotheses is to demand that the finite system allows a consistent thermodynamic limit.

If $G(t) \equiv G_\infty(t)$ is the quantity of interest in the thermodynamic limit, then we demand

$$G(t) \equiv G_\infty(t) = \lim_{L\to\infty} G_L(t). \tag{5.147}$$

The finite-size quantity G_L will therefore be related to its bulk counterpart $G_\infty(t)$ by a scaling function f

$$\frac{G_L(t)}{G_\infty(t)} = f\left(\frac{L}{\xi_\infty(t)}\right) = f(x). \tag{5.148}$$

This reveals a first condition which the, as yet *unknown*, scaling function $f(x)$ needs to satisfy. For $L \to \infty$ and fixed $t \neq 0$, we need to have $\lim_{x\to\infty} f(x) = 1$.

Thus far we have not learned very much about the actual finite-size behaviour of $G_L(t)$. In order to make progress on this point, we need an assumption for the size-dependence of $G_L(t)$ for which a scaling form

$$G_L(t) = L^p f\left(\frac{L}{\xi_\infty(t)}\right) = L^p g\left(tL^{1/\nu}\right) \tag{5.149}$$

is assumed. The exponent p which governs the finite-size behaviour of $G_L(t)$ can be determined by comparison with the bulk behaviour of $G_\infty(t)$.

Assuming that, in the thermodynamic limit, $G_\infty(t)$ has the critical power law behaviour

$$G_\infty(t) = G_0\,|t|^{-g} \tag{5.150}$$

with some critical exponent g, then the finite-size quantity G_L (5.149) assumes the correct thermodynamic limit behaviour, if $\nu p = g$. We obtain this result if we assume that the scaling function $f(x)$ in (5.149) behaves as $f(x) \propto x^{-p}$. Moreover, we use the expression (5.142) for the correlation length in the thermodynamic limit. Thus, we obtain the result

$$G_L(t) = L^{g/\nu} f\left(\frac{L}{\xi_\infty(t)}\right) = L^{g/\nu} g(tL^{1/\nu}). \tag{5.151}$$

Clearly, we still need to describe in more detail the finite-size situation that is analogous to critical behaviour in the thermodynamic limit. Therefore, the following focuses on some simple consequences of the scaling assumption (5.151) to elucidate this problem.

First, we analyse the scaling function $f(x)$ for small arguments x, keeping the system size L fixed. In other words, we consider the limit $\xi_\infty \to \infty$, i.e. we approach the critical temperature of the infinite system, $t \to 0$. Since the finite-size quantity G_L will be non-singular at the critical point of the bulk system, we can expand it in a Taylor series

$$G_L(t) = A + Bt + Ct^2 + \ldots \tag{5.152}$$

for $t \to 0$. We can obtain this behaviour of G_L if we require that the scaling function behaves as

$$f(x) = a + bx^{1/\nu} + cx^{2/\nu} + \ldots \tag{5.153}$$

for $x = L/\xi_\infty(t) \to 0$. Moreover, if G_L corresponds to a diverging critical bulk quantity, we can determine the shift in the critical temperature between the critical temperature T_c of the infinite system and the temperature of the maximum of G_L.

For fixed system size L, we assume that G_L has a maximum at

$$x_0 = \frac{L}{\xi_\infty(t_{\max})} = \frac{L}{\xi_0 \, |t_{\max}|^{-\nu}}. \tag{5.154}$$

Then, the critical temperature of the bulk system is shifted by $\Delta t = t_{\max}$ related to the system size L by

$$\Delta t = \frac{T_{\max} - T_c}{T_c} \propto L^{-1/\nu} \tag{5.155}$$

where $\lambda = 1/\nu$ is called the shift exponent. The maximum of G_L itself grows according to (5.149) with $L^{g/\nu}$. Figure 5.5 provides a summary in a schematic sketch.

The shift exponent is not the only way to describe, for a finite system, the vicinity of what corresponds to the critical point. There are other characteristic temperatures that can be ascribed to the finite system to describe its approach to a phase transition as the system size increases (cf. Barber (1983)).

Are finite-size effects observable in macroscopic systems? In order to answer this question, we use our finite-size scaling results to estimate the orders of magnitude of susceptibility and shift in the critical temperature for an Ising-like critical point, such as e.g. a liquid-vapour critical point, i.e. for a real macroscopic system Brézin (1986). We identify $G_L(t) = \chi_L(t)$. The critical exponent of the susceptibility of the infinite size system is $g = \gamma \approx 1.24$. Further, we have $1/\nu \approx 1.59$ and, hence, $\gamma/\nu \approx 1.97$. Assuming a number of particles of the order of Avogadro's number $N_0 \approx 10^{24}$, i.e. $L \approx N_0^{1/3} \approx 10^8$, we obtain for the shift in the critical temperature an order of magnitude of 10^{-12} to 10^{-13}. The maximum/peak of the susceptibility is about 10^{16} times higher than the value of the susceptibility far from the critical temperature. It therefore makes perfect sense to talk of a diverging susceptibility of macroscopic, but still finite-size, systems at $T = T_c$ instead of calling it an increase by a factor of 10^{16}. Also, the rounding of the susceptibility curve remains invisible unless one is able to resolve relative temperature changes of 10^{-13} around the critical temperature.

5.6.2 Phenomenological renormalization

A further implication of the finite-size scaling hypothesis is known as phenomenological renormalization (Nightingale, 1976; Nightingale, 1982). The upshot is that phenomenological renormalization eliminates the scaling function f whose general behaviour is still unknown. Phenomenological renormalization relies only on the finite-size quantities of two systems that are connected by a scale transformation.

The scaling behaviour (5.151) for a general critical quantity can be specialized to the correlation function itself, hence

$$\xi_L(t) = Lf\left(\frac{L}{\xi_\infty(t)}\right) \propto L. \tag{5.156}$$

The definition of the correlation length for the finite system needs careful reconsideration Brézin (1982); Barber (1983). It cannot be defined in the same way as the correlation length in the thermodynamic limit. One possibility is to define ξ_L from the second moment of the two-point correlation function $g_L(r)$ of the finite system

$$\xi_L^2 = \frac{\sum_r r^2 g_L(r)}{\sum_r g_L(r)}. \tag{5.157}$$

We shall not further go into these details here but refer to the cited original works.

The basic tenet of phenomenological renormalization posits that a scale transformation $L \to L'$ and $t \to t'$ can be defined by

$$\frac{L}{\xi_\infty(t)} = \frac{L'}{\xi_\infty(t')}. \tag{5.158}$$

With this transformation, using (5.148) with $\xi_L(t) = G_L(t)$ and $\xi_\infty(t) = G_\infty(t)$, we obtain

$$\frac{\xi_{L'}}{\xi_L} = \frac{L'}{L}. \tag{5.159}$$

Note that we have eliminated the scale transformation for the temperature $t \to t'$ and obtained a relation containing only quantities of the finite system. The relation (5.159) is particularly useful to test calculations that have exclusively to rely on finite systems, i.e. when, for example, the thermodynamic limit is not yet known. For an example of its application to determine the critical temperature of an Ising model from data of rather small systems obtained from the transfer matrix of a semi-infinite lattice, a two-dimensional strip, see Nightingale (1982).

5.7 Quantum phase transitions

Quantum phase transitions have attracted considerable attention in condensed matter physics over the last two or three decades and have become an important research area. These phase transitions occur at zero temperature and are driven by some parameter different from temperature, which, in the context of quantum phase transitions, is conventionally called control parameter. Examples of control parameters are pressure, magnetic field, or chemical doping. Rather than by thermal fluctuations, quantum phase transitions are caused by quantum fluctuations, which are a consequence of Heisenberg's uncertainty relation.

Quantum phase transitions can be of first order or continuous. In the following, as throughout most of this chapter, we focus on continuous quantum phase transitions. As their thermal counterparts, continuous quantum phase transitions occur gradually. Interestingly, in the continuous case a quantum phase transition can also influence properties of the system in a finite-temperature sector above the quantum critical point and so provide a new perspective on how quantum phenomena like non-Fermi liquid behaviour or high-temperature superconductivity emerge at low but finite temperature.

Quantum phase transitions occur as a result of competing phases with different ground states where, depending on the control parameter, one or the other ground state configuration is favoured. The quantum phase transition is then the transition as the control parameter is varied between these different ground states and especially the ground state evolving at and in a finite-temperature sector above the critical transition value of the control parameter. This ground state at the critical value of the control parameter is qualitatively different from the simpler ground states away from this critical value on either side of the quantum phase transition. At the quantum critical point, this ground state extends over the whole system. It is a complicated superposition of exponentially many configurations ranging over all length scales, i.e. exhibiting long-range quantum entanglement.

We concentrate on one exemplary model system that exhibits a continuous quantum phase transition, the one-dimensional quantum Ising model in a transverse magnetic field, rather than attempting an introduction to the general theory of quantum phase transitions.

There are excellent reviews and books devoted to giving a more complete perspective on quantum phase transitions, including Sondhi *et al.* (1997), Vojta (2002),Vojta (2000),

Vojta (2003), Carr (2011), and the seminal book by Sachdev (2011), now in its second edition.

5.7.1 Transverse field quantum Ising chain

The Hamiltonian of the quantum Ising chain in a transverse magnetic field is in terms of the Pauli spin-1/2 matrices or operators

$$\boldsymbol{\sigma} = \left(\sigma^x, \sigma^y, \sigma^z\right)^T \tag{5.160}$$

given by

$$\mathcal{H} = -\sum_{i=1}^{N-1} \sigma_i^z \sigma_{i+1}^z - g \sum_{i=1}^{N} \sigma_i^x = \mathcal{H}_0 + g\mathcal{H}_1 \tag{5.161}$$

where we have assumed free or open boundary conditions, i.e. the z-components of the first and last spin, σ_1^z and σ_N^z, do not interact as they would do for periodic boundary conditions representing a chain closed to form a ring. We shall mostly neglect the difference between these two cases of boundary conditions and assume that the corrections are small and will vanish for long chains, i.e. in the thermodynamic limit, $N \to \infty$.

Furthermore, we have

$$[\mathcal{H}_0, \mathcal{H}_1] \neq 0 \tag{5.162}$$

i.e. the two parts of the Hamiltonian (5.161) cannot be diagonalized simultaneously. For a qualitative discussion of the model, the form of the Hamiltonian $\mathcal{H} = \mathcal{H}_0 + g\mathcal{H}_1$ suffices under appropriate assumptions concerning the states corresponding to $\mathcal{H}_0$ and $\mathcal{H}_1$. Here, however, we discuss the model in some more detail and also treat it within a renormalization group approach in section 5.7.2.

This model is a quantum generalization of the Ising model where classical spins $\sigma_i = \pm 1$ were considered. The spin matrices satisfy the commutation and anti-commutation relations of spin-1/2 matrices discussed in chapter 3. The model is a special case of the XY quantum spin chain models treated exactly by Lieb *et al.* (1961) and its exact solution has been investigated in detail by Pfeuty (1970) using the methods of Lieb *et al.* (1961), which are, in turn, based on the Jordan–Wigner transformation (Jordan and Wigner, 1928) of a quantum spin model to a model of interacting spinless Fermions, both in one dimension. Thus, the model has the somewhat rare advantage that exact solutions are available for comparison. We shall return to this class of models also in exercise 13.1.

The form (5.161) of the transverse field quantum Ising chain Hamiltonian contains only one dimensionless control parameter g expressing the fact that the quantum critical behaviour of the model will only depend on this one parameter. However, in terms of physical intuition, it is helpful to write down the Hamiltonian (5.161) using two parameters, the exchange interaction $\mathcal{J}$ and the magnetic field h

$$\mathcal{H} = -\mathcal{J} \sum_{i=1}^{N-1} \sigma_i^z \sigma_{i+1}^z - h \sum_{i=1}^{N} \sigma_i^x \tag{5.163}$$

although it is obvious that only one parameter is really relevant, which can be chosen either as $h/\mathcal{J} = g$ or as $\mathcal{J}/h = \kappa = 1/g$. The former choice is the conventional one. Choosing the coupling $\mathcal{J} > 0$, i.e. an exchange coupling favouring a ferromagnetic ground state, it is natural to choose the control parameter in the range $0 \leq g < \infty$.

As usual, we denote the eigenvectors of the z-component of spin σ^z by

$$|\uparrow\rangle = \begin{pmatrix} 1 \\ 0 \end{pmatrix} \quad \text{and} \quad |\downarrow\rangle = \begin{pmatrix} 0 \\ 1 \end{pmatrix} \tag{5.164}$$

such that

$$\sigma^z|\uparrow\rangle = |\uparrow\rangle, \qquad \sigma^z|\downarrow\rangle = -|\downarrow\rangle. \tag{5.165}$$

The x-component of spin flips the orientation of these states from spin up to spin down and vice versa

$$\sigma^x|\uparrow\rangle = |\downarrow\rangle, \qquad \sigma^x|\downarrow\rangle = |\uparrow\rangle. \tag{5.166}$$

We can determine eigenvectors of σ^x in terms of the states $|\uparrow\rangle$ and $|\downarrow\rangle$, which satisfy

$$\sigma^x|\rightarrow\rangle = |\rightarrow\rangle, \qquad \sigma^x|\leftarrow\rangle = -|\leftarrow\rangle \tag{5.167}$$

and are given explicitly by

$$|\rightarrow\rangle = \frac{1}{\sqrt{2}}(|\uparrow\rangle + |\downarrow\rangle), \tag{5.168}$$

$$|\leftarrow\rangle = \frac{1}{\sqrt{2}}(|\uparrow\rangle - |\downarrow\rangle). \tag{5.169}$$

Thus, the z-component of spin, σ^z, flips the orientation of these spins from right to left, and vice versa.

We are now in a position to obtain a qualitative picture of the ground state phases of the transverse field quantum Ising chain (5.160) with the help of the control parameter g. It will prove most helpful to start with the limiting cases of the control parameter and consider in turn $g \to 0$ and $g \to \infty$.

5.7.1.1 *Ferromagnetic phase*

For small values of g, the spins align in an ordered ferromagnetic phase which is doubly degenerate, i.e. we have for $g \to 0$ the two degenerate ground states

$$|\Uparrow\rangle = |\uparrow\uparrow\ldots\uparrow\rangle = \prod_{i=1}^{N} |\uparrow_i\rangle, \tag{5.170}$$

$$|\Downarrow\rangle = |\downarrow\downarrow\ldots\downarrow\rangle = \prod_{i=1}^{N} |\downarrow_i\rangle \tag{5.171}$$

with ferromagnetic ground state energy

$$\mathcal{H}_0|\Uparrow\rangle = \mathcal{H}_0|\Downarrow\rangle = E^{(0)}_{N,\mathrm{fm}}|\Uparrow\rangle = E^{(0)}_{N}|\Downarrow\rangle, \tag{5.172}$$

where

$$E^{(0)}_{N,\mathrm{fm}} = -(N-1), \tag{5.173}$$

which is simply given by the number of nearest-neighbour pairs $N-1$ in an open chain of N spins.

For $g = 0$ the Hamiltonian (5.161) has a spin-inversion or $\mathbb{Z}_2$ symmetry. Therefore, which of these degenerate states is chosen depends on small external perturbations, i.e. the spin-inversion symmetry has to be broken spontaneously.

In the weak coupling limit, for a small finite control parameter $g \ll 1$, however, deviations from the totally ferromagnetic state $|\Uparrow\rangle$ develop in the form of domain walls that are the excitations with the lowest energy. The domain walls are the elementary excitations or quasiparticles above the ferromagnetic ground state $|\Uparrow\rangle$. A domain wall between sites i and $i+1$ is described by the state

$$|i\rangle = \prod_{n=1}^{i} |\uparrow_n\rangle \prod_{n=i+1}^{N} |\downarrow_n\rangle = |\uparrow\uparrow\ldots\uparrow\uparrow\downarrow\downarrow\ldots\downarrow\rangle, \tag{5.174}$$

which has an excitation energy of

$$\Delta E = 2 \tag{5.175}$$

above the ground state energy (5.173). This is indeed a lower excitation energy than the excitation energy

$$\Delta E = 4 \tag{5.176}$$

of a state with one spin flipped from spin up to spin down at some site i which corresponds to two adjacent domain walls

$$|\text{flip}\rangle = |\uparrow\uparrow\ldots\uparrow\uparrow\downarrow\uparrow\uparrow\ldots\rangle. \tag{5.177}$$

This difference in energy is important if we consider how, for small but finite $g \ll 1$, the second part $g\mathcal{H}_1$ of the Hamiltonian (5.161) acts on a domain wall excitation $|i\rangle$.

Recall than σ_i^x flips a spin at site i, e.g. $\sigma_i^x|\uparrow_i\rangle = |\downarrow_i\rangle$. Therefore $g\mathcal{H}_1$ generates two kinds of states when acting on $|i\rangle$. The first kind are states where, away from the domain wall, a flipped spin, e.g.

$$|j-1,j,i\rangle = |\uparrow_1,\uparrow_2,\ldots,\uparrow_{j-1},\downarrow_j,\uparrow_{j+1},\ldots,\uparrow_{i-1},\uparrow_i,\downarrow_{i+1},\ldots,\downarrow_N\rangle \tag{5.178}$$

is generated with an extra energy cost $\Delta E = 4$. In the quasiparticle language, these correspond to three-particle excitations.

However, at the site of the domain wall, the effect of the flipped spin is to move the domain wall one site to the left, $i \to i-1$, or to the right, $i \to i+1$,

$$|i\rangle = |\uparrow\uparrow\ldots\uparrow\uparrow\downarrow\downarrow\ldots\downarrow\downarrow\rangle \xrightarrow{\sigma_i^x} |i-1\rangle = |\uparrow\uparrow\ldots\uparrow\downarrow\downarrow\downarrow\ldots\downarrow\downarrow\rangle \tag{5.179}$$

$$|i\rangle = |\uparrow\uparrow\ldots\uparrow\uparrow\downarrow\downarrow\ldots\downarrow\downarrow\rangle \xrightarrow{\sigma_{i+1}^x} |i+1\rangle = |\uparrow\uparrow\ldots\uparrow\uparrow\uparrow\downarrow\ldots\downarrow\downarrow\rangle \tag{5.180}$$

at a lower energy cost of $\Delta E = 2$. This also shows why we can address the domain wall excitation as a quasiparticle.

We can thus write as an approximation

$$g\mathcal{H}_1|i\rangle = -g\sum_{n=1}^{N}\sigma_n^x|i\rangle \approx -g\left(|i-1\rangle + |i+1\rangle\right) \tag{5.181}$$

where we neglect the first kind of states described above.

The action of the Hamiltonian (5.161) on the domain wall state $|i\rangle$ is therefore, after subtracting the ground state energy (5.173),

$$\bar{\mathcal{H}}|i\rangle = \left(\mathcal{H} - E_{N,\text{fm}}^{(0)}\right)|i\rangle = 2|i\rangle - g\left(|i-1\rangle + |i+1\rangle\right). \tag{5.182}$$

We can diagonalize this expression in momentum space

$$|j\rangle = \frac{1}{\sqrt{N}}\sum_k e^{-ikj}|k\rangle, \qquad k = \frac{2\pi}{N}n, n \in \mathbb{Z} \tag{5.183}$$

$$|k\rangle = \frac{1}{\sqrt{N}} \sum_{i=j}^{N} e^{ikj} |j\rangle \tag{5.184}$$

to obtain

$$\bar{\mathcal{H}}|k\rangle = 2\,(1 - g\cos k)\,|k\rangle = \epsilon_{\rm fm}(k)|k\rangle. \tag{5.185}$$

The excitation spectrum $\epsilon_{\rm fm}(k)$ has a gap above the ground state energy for $k = 0$

$$\Delta(g) = \epsilon_{\rm fm}(0) = 2(1 - g). \tag{5.186}$$

As a consequence of the spontaneously broken $\mathbb{Z}_2$ symmetry, the spin-spin correlation function for $g = 0$ is given by the finite magnetization per site

$$m_z = \frac{1}{N} \sum_{i=1}^{N} = \langle \sigma_i^z \rangle = \frac{1}{N} \sum_{i=1}^{N} \langle \Uparrow |\sigma_i^z| \Uparrow \rangle = 1 \tag{5.187}$$

and exhibits ferromagnetic order

$$\langle \sigma_i^z \sigma_j^z \rangle = m_z^2 \tag{5.188}$$

for any distance $|i - j|$ of the spins. For small values of the control parameter, $g \ll 1$, the ferromagnetic long-range order for correlations over large distances $|i - j| \to \infty$ is maintained, although the value of the magnetization decreases due to a loss of ferromagnetic order at shorter distances $|i - j|$. This decrease of the order parameter will continue if the control parameter is further increased until we reach a critical value of $g = g_c$, when the last remnant of the long-range order is destroyed. We anticipate this critical behaviour by assuming that the magnetization behaves as an order parameter

$$m_z \propto |g - g_c|^{\beta}. \tag{5.189}$$

However, since this model has an exact solution, the critical exponent β and other exponents can also be determined exactly—a fortunate but rare situation! We shall sketch the exact solution in section 5.7.1.3 and obtain the exact value of the critical exponent β and other critical exponents there. In section 5.7.2 below we also use a real-space renormalization group approach to investigate the critical behaviour of the quantum Ising chain in a transverse field.

In order to shed more light on the critical region, we now consider the opposite limit of large values of the control parameter $g \to \infty$.

5.7.1.2 *Quantum paramagnetic phase*

In the opposite limit, $g \to \infty$, the spins will align, as a consequence of the Zeeman interaction, parallel to the magnetic field $h \propto g$ along the x-axis. The ground state is then formed by a product of eigenstates $|\rightarrow\rangle$ of σ^x given by (5.168)

$$|\Rightarrow\rangle = \prod_{i=1}^{N} |\rightarrow_i\rangle \tag{5.190}$$

$$= \frac{1}{2^{N/2}} \prod_{i=1}^{N} (|\uparrow_i\rangle + |\downarrow_i\rangle) \tag{5.191}$$

$$= \frac{1}{2^{N/2}} (|\uparrow\uparrow\ldots\uparrow\rangle + |\downarrow\uparrow\ldots\uparrow\rangle + \ldots + |\downarrow\downarrow\ldots\downarrow\rangle) \tag{5.192}$$

with paramagnetic ground state energy

$$-g\mathcal{H}_1|\Rightarrow\rangle = E^{(0)}_{N,\mathrm{pm}}|\Rightarrow\rangle \tag{5.193}$$

where

$$E^{(0)}_{N,\mathrm{pm}} = -gN. \tag{5.194}$$

This state is very different from the states (5.170) and (5.171). As the expansion (5.192) in the last line shows, this state is an equal-weight, i.e. coherent, superposition of all 2^N states one can build from a total of N states that are either $|\uparrow_i\rangle$ or $|\downarrow_i\rangle$ resulting in a completely disordered state with respect to the z-axis in spin space and, hence, describes a paramagnetic phase. In contrast to the ferromagnetic states (5.170) and (5.171), the paramagnetic state $|\Rightarrow\rangle$, as can be seen from (5.192), is invariant under the exchange of all states $|\uparrow_i\rangle$ and $|\downarrow_i\rangle$ and does not break spin-inversion symmetry.

Away from the strong coupling limit $g \to \infty$, but still for $1/g \ll 1$, the elementary excitations or quasiparticles above the paramagnetic ground state $|\Rightarrow\rangle$ are spins in the chain flipped from right- to left-oriented with respect to the x-axis, i.e.

$$|i\rangle = |\rightarrow_1\rangle|\rightarrow_2\rangle\ldots|\rightarrow_{i-1}\rangle|\leftarrow_i\rangle|\rightarrow_{i+1}\rangle\ldots|\rightarrow_{N-1}\rangle|\rightarrow_N\rangle \tag{5.195}$$

at a site i. The cost in energy of such a spin flip is

$$\Delta E = 2g. \tag{5.196}$$

We use the same notation, $|i\rangle$, as for the state with one domain wall above the ferromagnetic ground state, but emphasize that the two states are quite different physically.

In the paramagnetic phase, the part $\mathcal{H}_0$ of the Hamiltonian (5.161) moves the flipped state at site i

$$\mathcal{H}_0|i\rangle = -\sum_{n=1}^{N-1} \sigma_n^z \sigma_{n+1}^z |i\rangle \approx -\left(|i-1\rangle + |i+1\rangle\right), \tag{5.197}$$

but also creates states with two further spin flips away from site i, which we neglect because they correspond to higher excitation energies. The neglected states, in quasiparticle language, are again three-particle states.

For the complete Hamiltonian (5.161), we thus get, after subtracting the paramagnetic ground state energy (5.194),

$$\bar{\mathcal{H}}|i\rangle = \left(\mathcal{H} - E^{(0)}_{N,\text{pm}}\right)|i\rangle = 2g|i\rangle - \left(|i-1\rangle + |i+1\rangle\right), \tag{5.198}$$

which becomes again diagonal in momentum space

$$\bar{\mathcal{H}}|k\rangle = 2g\left(1 - \frac{1}{g}\cos k\right)|k\rangle = \epsilon_{\text{pm}}(k)|k\rangle. \tag{5.199}$$

Again the spectrum of the elementary excitations exhibits a gap

$$\Delta(g) = \epsilon_{\text{pm}}(0) = 2(g-1). \tag{5.200}$$

The magnetization of the quantum paramagnetic state $|\Rightarrow\rangle$ vanishes: $m_z = 0$. The spin-spin correlation function vanishes exponentially for large distances $|i-j| \to \infty$ with a correlation length ξ

$$\langle \sigma_i^z \sigma_j^z \rangle \propto \exp\left(-\frac{|i-j|}{\xi}\right). \tag{5.201}$$

The correlation length measures the distance over which spins are still correlated in the paramagnetic phase.

However, it is crucial to realize that the disordered paramagnetic phase does not come about as a result of the thermally activated flipping in time of spins. It is a *quantum* paramagnetic phase of spins in a coherent quantum superposition of the two spin orientations $|\uparrow_i\rangle$ and $|\downarrow_i\rangle$. The spins do not fluctuate due to a thermal environment.

For $g \to \infty$, the correlation length is $\xi = 0$. However, as the control parameter is decreased, the correlation length becomes finite and we eventually expect it to diverge with a critical exponent ν as the critical value of the control parameter is approached from values $g \to g_c+$

$$\xi \propto |g - g_c|^{-\nu}. \tag{5.202}$$

We can thus summarize our observations. The system undergoes a zero temperature, i.e. quantum, phase transition from one of the degenerate ordered ferromagnetic states $|\Uparrow\rangle$ or $|\Downarrow\rangle$ for $g \to 0$ to a disordered paramagnetic state $|\Rightarrow\rangle$ for $g \to \infty$ at some value of the control parameter g. Thus, we expect that there is a critical value $g = g_c$ of the control parameter such that for $g < g_c$ there is long-range order in the z-direction of spin space and the magnetization

$$m_z = \langle \sigma^z \rangle = \frac{1}{N} \sum_{i=1}^{N} \langle \sigma_i^z \rangle \tag{5.203}$$

is finite. For $g > g_c$ this long-range order will be destroyed and m_z will vanish. The exact calculation of the ground state energy of the infinite chain $N \to \infty$ by Pfeuty (1970) shows a non-analyticity for $g = 1$, signifying a phase transition. In other words, the critical control parameter is $g_c = 1$. Section 5.7.2 confirms this result with a real-space renormalization group calculation. Let us have a brief look at how the exact solution of the one-dimensional quantum Ising model in a transverse field works before we then use the insights from the exact solution to discuss further on in some detail how the ground states develop across the quantum phase transition.

5.7.1.3 Sketch of the exact solution

The methods to solve the one-dimensional quantum Ising model in a transverse field exactly will be applied further and in more detail in exercise 13.1 to solve the related, but more general, one-dimensional quantum XY model. Here we just give a brief overview of these methods and state the results. This will be very instructive to understand the quantum critical behaviour of the quantum Ising chain in a transverse field which we have, so far, only partially and qualitatively elucidated.

The main ingredient of the exact solution of the model is the Jordan–Wigner transformation (Jordan and Wigner, 1928). It serves to map the spin degrees of freedom onto spinless Fermions. The idea leading to the Jordan–Wigner transformation comes from the observation of the similarity of spin-1/2 operators and Fermi operators on the same lattice site, namely that both, spin-1/2 operators and Fermi operators, obey anti-commutation relations. Spin-1/2 operators relating to different sites, however, obey commutation relations, i.e. they behave like Bosons. Thus, the question arises whether spin-1/2 operators can be transformed into either fully Fermionic or fully Bosonic operators. In fact, both can be achieved and prove useful. Later, in section 8.6, we employ even two different ways to transform spin operators for general spin magnitude s into Bosons: the Schwinger and Holstein–Primakoff Bosons.

Relatively recently, generalization of the Jordan–Wigner transformation to higher dimensions or higher spin have been discussed; see, for example, Fradkin (1989) for a generalization of the Jordan–Wigner transformation to two, Huerta and Zanelli (1993) to three and higher dimensions, and Batista and Ortiz (2001) for a generalization to higher spin. For a recent review, see Derzhko (2008).

Here, however, we are interested in a transformation of the spin-1/2 operators into Fermi operators on the one-dimensional lattice: the Jordan–Wigner transformation.

Let us see how this can be achieved. For the spin raising and lowering operators

$$\sigma_i^{\pm} = \frac{1}{2}\left(\sigma_i^x \pm i\sigma_i^y\right) \tag{5.204}$$

we have the anti-commutation relations

$$\{\sigma_i^-, \sigma_i^+\} = 1, \qquad \left(\sigma_i^-\right)^2 = \left(\sigma_i^+\right)^2 = 0. \tag{5.205}$$

Spinless lattice Fermion creation and annihilation operators $c_i{}^\dagger$ and c_i, on the other hand, obey the anti-commutation relations

$$\{c_i, c_j\} = \{c_i^\dagger, c_j^\dagger\} = 0, \tag{5.206}$$

$$\{c_i, c_j^\dagger\} = \delta_{ij}. \tag{5.207}$$

We can identify the spin-up state $|\uparrow_i\rangle$ at lattice site i with the unoccupied spinless Fermion state $|0_i\rangle$ at lattice site i and the spin-down state $|\downarrow_i\rangle$ at lattice site i with the occupied spinless Fermion state $|1_i\rangle$ at lattice site i, if we relate the spin-1/2 operator and the Fermion operators in the following way

$$\sigma_i^z = 1 - 2c_i^\dagger c_i = 1 - 2\mathcal{N}_i. \tag{5.208}$$

This relation between the spin operator σ_i^z and the Fermi operators $c_i{}^\dagger$ and c_i is local in (5.208), i.e. it involves only one site i of the lattice.

However, we also need to express the other spin operators in terms of the Fermi operators in order to make the analogy between spins and Fermions useful. This is possible, but unfortunately there are no local relations. Jordan and Wigner (1928) found an expression for the spin raising and lowering operators (5.204) that involve a whole *string* of Fermi operators

$$\sigma_i^+ = \prod_{j=1}^{i-1}\left(1 - 2c_j^\dagger c_j\right) c_i = U_i^\dagger c_i = U_i c_i, \tag{5.209}$$

$$\sigma_i^- = \prod_{j=1}^{i-1}\left(1 - 2c_j^\dagger c_j\right) c_i^\dagger = U_i c_i^\dagger \tag{5.210}$$

and are, hence, non-local operators. The sting operators U_i satisfy

$$U_i = U_i^\dagger = U_i^{-1}. \tag{5.211}$$

Without the string operators, the Jordan–Wigner transformation would give the correct spin and Fermi operator algebras only for the same site $i = j$, but not for different sites $i \neq j$ (see exercise 5.18).

EXERCISE 5.17 String operator and inverse Jordan–Wigner transformation Show that the string operator can also be expressed as an exponential

$$U_i = \exp\left(i\pi \sum_{j=1}^{i-1} c_i^\dagger c_i \right). \tag{5.212}$$

Furthermore, show that the inverse Jordan–Wigner transformation can be expressed in either of the two forms, by

$$c_i^\dagger = \left(\prod_{j=1}^{i-1} \sigma_j^z \right) \sigma_i^-, \tag{5.213}$$

$$c_i = \left(\prod_{j=1}^{i-1} \sigma_j^z \right) \sigma_i^+ \tag{5.214}$$

or, more similar in form to the transformation from spin operators to Fermi operators, by

$$c_i^\dagger = \prod_{j=1}^{i-1} \left(1 - 2\sigma_j^- \sigma_j^+ \right) \sigma_i^-, \tag{5.215}$$

$$c_i = \prod_{j=1}^{i-1} \left(1 - 2\sigma_j^- \sigma_j^+ \right) \sigma_i^+. \tag{5.216}$$

EXERCISE 5.18 Commutation and anti-commutation relations Verify that the transformations (5.208), (5.209–5.210), and (5.213–5.214) (or 5.215–5.216) are consistent with the canonical commutation relations for spin operators

$$\left[\sigma_i^+, \sigma_j^- \right] = \delta_{ij} \sigma_i^z, \qquad \left[\sigma_i^z, \sigma_j^\pm \right] = \pm \delta_{ij} \sigma_i^\pm \tag{5.217}$$

and with canonical anti-commutation relations for Fermi operators

$$\left\{ c_i, c_j^\dagger \right\} = \delta_{ij}, \qquad \{ c_i, c_j \} = \left\{ c_i^\dagger, c_j^\dagger \right\} = 0. \tag{5.218}$$

For the case of the quantum Ising chain in a transverse field, it is advantageous to rotate the spin operators by $\pi/2$ about the y-axis such that

$$\sigma^z \rightarrow \sigma^x \tag{5.219}$$

$$\sigma^x = \sigma^+ + \sigma^- \rightarrow -\sigma^z. \tag{5.220}$$

Hence, the Jordan–Wigner transformation becomes

$$\sigma_i^x = 1 - 2c_i^\dagger c_i \tag{5.221}$$

$$\sigma_i^z = -\prod_{j=1}^{i-1}\left(1 - 2c_i^\dagger c_i\right)\left(c_i + c_i^\dagger\right), \tag{5.222}$$

which transforms the Hamiltonian (5.161) of the quantum Ising chain in a transverse field into a spinless Fermion Hamiltonian

$$\bar{\mathcal{H}} = -\sum_{i=1}^{N-1}\left\{\left(c_i^\dagger - c_i\right)\left(c_{i+1}^\dagger + c_{i+1}\right) - 2gc_i^\dagger c_i\right\} \tag{5.223}$$

where we subtracted the term $-\ gN$ corresponding to the paramagnetic ground state energy $E^{(0)}_{N,\mathrm{pm}}$ and neglected a boundary term $-\ 2gc_N{}^\dagger c_N$.

This Hamiltonian will then be transformed into momentum space. We first obtain

$$\bar{\mathcal{H}} = -\sum_k\left(c_k^\dagger c_{-k}^\dagger e^{-ik} - c_k c_{-k} e^{ik} + (2\cos k - 2g)c_k^\dagger c_k\right). \tag{5.224}$$

Breaking up the integrand into a positive wave number part, $k > 0$, and a negative wave number part, $k < 0$, we can bring this expression into the form

$$\bar{\mathcal{H}} = -\sum_{k>0}\left((-2i\sin k)\left(c_k^\dagger c_{-k}^\dagger - c_{-k}c_k\right) + (2\cos k - 2g)\left(c_k^\dagger c_k + c_{-k}^\dagger c_{-k}\right)\right). \tag{5.225}$$

This form of the Hamiltonian has an interpretation in terms of particle-hole pairs. Let us introduce hole operators for $k > 0$

$$c_{-k} = h_k^\dagger, \tag{5.226}$$

$$c_{-k}^\dagger = h_k, \tag{5.227}$$

such that the Hamiltonian becomes

$$\bar{\mathcal{H}} = -\sum_{k>0}\left((-2i\sin k)\left(c_k^\dagger h_k - h_k^\dagger c_k\right) + (2\cos k - 2g)\left(c_k^\dagger c_k - h_k^\dagger h_k\right)\right). \tag{5.228}$$

Introducing the spinor

$$\psi_k = \begin{pmatrix} c_k \\ h_k \end{pmatrix} \tag{5.229}$$

we can bring the Hamiltonian into the compact bilinear form

$$\bar{\mathcal{H}} = \sum_{k>0} \psi_k^\dagger \mathcal{A}(k) \psi_k \tag{5.230}$$

with the matrix

$$\mathcal{A}(k) = \begin{pmatrix} 2(g - \cos k) & 2i \sin k \\ -2i \sin k & -2(g - \cos k) \end{pmatrix}. \tag{5.231}$$

The exact result for the excitation spectrum is determined by the eigenvalues of the matrix $\mathcal{A}(k)$ and has two branches or bands

$$\epsilon(k) = \pm 2\sqrt{1 - 2g \cos k + g^2}. \tag{5.232}$$

EXERCISE 5.19 Details of the derivation Repeat the derivations of this section keeping track of all the terms that have been neglected along the way and convince yourself that these approximations do not introduce errors that would spoil the exactness of the final result.

The exact solution for the excitation spectrum reproduces the correct limits for weak coupling $g \ll 1$

$$\epsilon(k) \approx 2(1 - g \cos k) \tag{5.233}$$

and for strong coupling $1/g \ll 1$

$$\epsilon(k) \approx 2g\left(1 - \frac{1}{g} \cos k\right) \tag{5.234}$$

from which the exact energy gaps at zero momentum follow

$$\epsilon(0) = \Delta(g) = \begin{cases} 2(1 - g) & g < g_c = 1, \\ 2(g - 1) & g > g_c = 1, \end{cases} \tag{5.235}$$

or combined together in one equation

$$\Delta(g) = 2|1-g|. \tag{5.236}$$

The gap $\Delta(g)$ will play an important role in the following development. A simple qualitative consideration relates $\Delta(g)$ to the correlation length at criticality by evaluating the excitation energy spectrum (5.232) at long distances, i.e. for $k \approx 0$

$$\epsilon(k) = \pm 2\sqrt{1-2g\cos k + g^2} \approx \pm|\Delta|\sqrt{1+\xi^2 k^2} \tag{5.237}$$

where we introduced the correlation length

$$\xi = \frac{2\sqrt{g}}{|\Delta|} \propto |1-g|^{-1} \tag{5.238}$$

as the length related to long-distance excitations. This reveals the critical exponent

$$\nu = 1 \tag{5.239}$$

for the correlation length near the critical point $g = g_c = 1$.

We shall refrain from a survey of the calculation of the correlation function

$$G(n) = \langle \sigma_i^z \sigma_{i+n}^z \rangle. \tag{5.240}$$

These calculations are also based on the exact solution for the excitation spectrum we have just outlined and are themselves exact. A derivation of these calculations would, however, go beyond the scope and space limitations of this exposition. We refer instead to the literature cited at the beginning of and within this section.

5.7.1.4 *Evolution of the ground states across the quantum phase transition*

It is crucial to realize that one cannot, by varying the control parameter g, smoothly connect the ferromagnetic states (5.170) or (5.171) for $g = 0$ to the quantum paramagnetic state (5.192) for $g \to \infty$. These states are distinguished by their spin-inversion or $\mathbb{Z}_2$ symmetry that is broken for (5.170) or (5.171) at $g = 0$, but intact for (5.192) at $g \to \infty$. This indicates the existence of a value in the control parameter $g = g_c$ where, as the exact solution (Pfeuty, 1970) also shows, the ground state exhibits a non-analyticity. This can only strictly happen for an infinite system.

The ferromagnetic moment is measured by the difference of the number of spin-up and spin-down states and also signifies the presence of the broken spin inversion symmetry (cp. 5.203). Eventually, as g is increased, the ferromagnetic moment vanishes and the spin inversion symmetry is restored by the appearance of the ground state similar to $|\Rightarrow\rangle$. As already discussed, for this scenario to happen there must be a non-analytic behaviour of the ground state as a function of the coupling g at a critical point $g = g_c$.

The picture that has emerged so far for the ground state as a function of the control parameter g can now be summarized and described more completely. As discussed, the simple product ground states (5.170) (or 5.171) and (5.192) describe the limits $g \ll g_c$ and $g \gg g_c$, respectively. At the critical point $g = g_c$, the quantum critical ground state is a highly complicated quantum superposition of all 2^N spin configurations. The ground state (5.192) is also such a superposition of all 2^N spin configurations but it contains all spin configurations with an equal amplitude and can thus be written as a product state as in (5.190). By contrast, for the ground state at $g = g_c$ there is no such product form. This, as briefly discussed at the end of section 2.1.1.1, is the distinguishing mark of quantum entangled states, which cannot be expressed in any simple form in terms of local basis states.

The ground-state spin correlation length ξ can be used for a further characterization of the quantum phases. Away from the critical value of the control parameter $g = g_c$, the ground state averaged over a length larger than the correlation length ξ resembles the quantum ferromagnetic ground states (5.170) or (5.171) for $g < g_c$ and the quantum paramagnetic ground state (5.192) when $g > g_c$. However, at length scales smaller than ξ, the complicated quantum entangled ground state has already developed. When ξ diverges as a function of the control parameter

$$\xi \propto |g - g_c|^{-\nu} \tag{5.241}$$

at $g \to g_c$, the quantum ground state which has developed is entangled on all length scales.

5.7.1.5 *Qualitative description of an experimental system modelled by the quantum Ising chain in a transverse field*

A comprehensive analysis of quantum phase transitions of the class described by the quantum Ising chain in a transverse field lies unfortunately outside the scope not only of this chapter, but also this book. In lieu of this, we give a qualitative description of an experimental system whose quantum phase transition has been interpreted with the help of the quantum Ising chain in a transverse field following the exposition in Sachdev and Keimer (2011).

Lithium holmium fluoride ($LiHoF_4$) is an insulating ionic crystal. The magnetic properties of this material are determined by the holmium ions Ho^{3+} that exhibit an unpaired electron, which, despite the complications of the spin-orbit coupling, gives rise at low temperatures to only two spin orientations per site: $|\uparrow_i\rangle$ and $|\downarrow_i\rangle$. In the terminology of quantum computation and quantum information, each holmium ion realizes a *qubit*, a system whose states consist of only two levels. At low temperatures, due to the crystal structure and the corresponding crystal fields, the spins of the holmium ions interact much stronger along one axis than in the plane perpendicular to this axis. $LiHoF_4$ can thus be considered a quasi-one-dimensional material as far as its magnetic behaviour at low temperatures is concerned. In an external magnetic field h perpendicular to this preferred axis, $LiHoF_4$ is a material whose quantum critical behaviour is well described by the one-dimensional quantum Ising model in a transverse field (Bitko *et al.*, 1996).

In the absence of the magnetic field perpendicular to the magnetically preferred axis of $LiHoF_4$, there is a thermally driven, i.e. conventional, continuous phase transition from a paramagnetic disordered phase at temperatures above $T_c \approx 1.5$K to a ferromagnetic ordered phase at $T < T_c$. This behaviour persists when a small magnetic field h is applied perpendicular to the preferred axis. However, as the magnetic field is increased, the critical temperature $T_c = T_c(h)$ decreases. Eventually, for a critical magnetic field $h = h_c$, even at $T = 0$ the ferromagnetic order will be destroyed. In contrast to the phase transition at $T > 0$, the paramagnetic state for $h > h_c$ does not consist of spins thermally fluctuating in real time. There is rather a ground state which is a coherent superposition of spin-up and spin-down states, as we have seen for the quantum Ising chain in a transverse field. At $T = 0$ and $h > h_c$ we have a quantum paramagnetic phase.

The picture that emerges is the following (cp. figure 5.6): for magnetic fields $h < h_c$, the ground state is the ferromagnetic state, for $h > h_c$ the ground state is a quantum paramagnet in the sense just described. For very low temperature above the ferromagnetic state, elementary excitations in the form of domain walls in the totally ferromagnetic state develop. Above the quantum paramagnetic state, elementary excitations develop where spins $|\rightarrow\rangle$ along the axis of the transverse magnetic field flip, i.e. spin states $|\leftarrow\rangle$ appear in the superposition of this excited state.

Both types of elementary excitations do not describe the approach to the quantum critical point well, i.e. for $T \rightarrow 0$ and $h \rightarrow h_c$ more complicated quantum states appear that also persist and fan out into a region of $T > 0$ above and with apex at h_c. These states are quantum entangled states (cp. the remarks at the end of section 2.1.1.1). The research fields of quantum phase transitions and quantum entanglement are thus intimately connected.

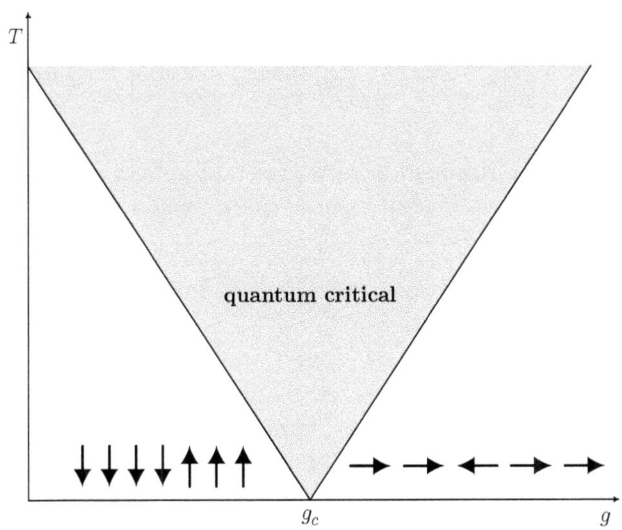

Figure 5.6 *Schematic phase diagram of the one-dimensional quantum Ising model in a transverse magnetic field depending on the temperature T and the dimensionless control parameter g = h/J On and for low temperatures above the line 0 < g < g_c, there are domain wall excitations, while for g_c < g there are excitations consisting of spin-flips. The quantum critical point g = g_c and the quantum critical region fanning out from g = g_c to finite temperatures cannot be adequately described by such simple elementary excitations.*

Another experimental system that has recently attracted interest is the material cobalt niobate ($CoNb_2O_6$) (Coldea *et al.*, 2010), where a quantum phase transition of the the Co^{2+} ions has been observed if a magnetic field is applied perpendicular to the direction of the zig-zag chain formed by the Co^{2+} ions.

There is also an yttrium-doped version of the former material, $LiHo_xY_{1-x}F_4$, for which quantum phase transitions have been observed which are driven by a perpendicular magnetic field h or by the doping concentration x and for which the spin degrees of freedom are governed by more complicated microscopic interactions than the ones described earlier for the transverse field quantum Ising chain.

A comprehensive exposition and guide to the literature of quantum phase transitions in transverse field spin models can be found in the book by Dutta *et al.* (2015) with the eponymous title.

The following section further investigates the quantum Ising chain in a transverse field with a renormalization group approach, which will shed further insight on the quantum critical behaviour of the model.

5.7.2 Real-space renormalization group of the transverse field quantum Ising chain

We follow Nishimori and Ortiz (2011) and begin the real-space renormalization procedure by grouping the Hamiltonian (5.160) in blocks of size $b = 2$, i.e we write it as a sum of two parts

$$\mathcal{H} = \sum_{i \text{ odd}} \mathcal{H}_{\text{block},i} + \sum_{i \text{ even}} \mathcal{H}_{\text{block},i} = \mathcal{H}_{\text{intra}} + \mathcal{H}_{\text{inter}} \tag{5.242}$$

where, in the first part the summation runs over odd indices i, in the second part over even indices i. The block-spin Hamiltonian is thus defined as

$$\mathcal{H}_{\text{block},i} = -\sigma_i^z \sigma_{i+1}^z - g\sigma_i^x = -\sigma^z \otimes \sigma^z - g\sigma^x \otimes \mathbb{I}, \tag{5.243}$$

and is formally the same for both parts of the Hamiltonian. However, if we form blocks of two spins starting with an odd index of the first spin of the block, then we need to distinguish carefully the two parts of the Hamiltonian: $\mathcal{H}_{\text{intra}}$ couples spins within one block, while $\mathcal{H}_{\text{inter}}$ couples the second spin of one block with the first spin of the following block, i.e. it couples spins across blocks. Keeping this distinction in mind will be crucial for the subsequent renormalization group analysis.

This is not the only way to divide the Hamiltonian (5.160) into block Hamiltonians. However, this choice will prove particularly simple to work with.

The block Hamiltonian $\mathcal{H}_{\text{block},i}$ in the basis of eigenstates $\{|\uparrow\uparrow\rangle, |\uparrow\downarrow\rangle, |\downarrow\uparrow\rangle, |\downarrow\downarrow\rangle\}$ of the z-component of the spin operator σ becomes

$$\mathcal{H}_{\text{block},i} = \begin{pmatrix} -1 & 0 & -g & 0 \\ 0 & 1 & 0 & -g \\ -g & 0 & 1 & 0 \\ 0 & -g & 0 & -1 \end{pmatrix} \tag{5.244}$$

where, e.g.

$$|\downarrow\uparrow\rangle = |\downarrow\rangle \otimes |\uparrow\rangle = \begin{pmatrix} 0 \\ 1 \end{pmatrix} \otimes \begin{pmatrix} 1 \\ 0 \end{pmatrix} = \begin{pmatrix} 0 \\ 0 \\ 1 \\ 0 \end{pmatrix}. \tag{5.245}$$

Due to the structure of the matrix of the block Hamiltonian (5.244) the eigenvalue equation

$$\mathcal{H}_{\text{block},i}|\lambda\rangle = \lambda|\lambda\rangle \tag{5.246}$$

decomposes into two equivalent sets of equations for the first and third, and for the second and fourth component of $|\lambda\rangle$, respectively. Writing

$$|\lambda\rangle = \begin{pmatrix} x_1, & x_2, & x_3, & x_4 \end{pmatrix}^T \tag{5.247}$$

we explicitly obtain from the eigenvalue equation (5.246) together with the block Hamiltonian matrix (5.244)

$$(1+\lambda)x_1 + gx_3 = 0 \tag{5.248}$$

$$-gx_1 + (1-\lambda)x_3 = 0 \tag{5.249}$$

and

$$gx_2 + (1+\lambda)x_4 = 0 \tag{5.250}$$

$$(1-\lambda)x_2 - gx_4 = 0. \tag{5.251}$$

Both sets of equations lead to the same eigenvalues, which we can label

$$\lambda_1 = \lambda_2 = -\sqrt{1+g^2} \tag{5.252}$$

$$\lambda_3 = \lambda_4 = \sqrt{1+g^2}. \tag{5.253}$$

The simplest way to determine the eigenstates consists in choosing the trivial solution in turn for one of the sets of eigenvalue equations, i.e. first to choose $x_2 = x_4 = 0$ for $\lambda = \lambda_1$, then to choose $x_1 = x_3 = 0$ for $\lambda = \lambda_2$, and so on for $\lambda = \lambda_3$ and $\lambda = \lambda_4$.

For $\lambda = \lambda_1$, we can satisfy the eigenvalue equation by choosing

$$x_1 = 1 - \lambda_1, \tag{5.254}$$

$$x_2 = g \tag{5.255}$$

such that the unnormalized eigenstate corresponding to the eigenvalue λ_1 becomes

$$|\widehat{1}\rangle = \begin{pmatrix} 1+\sqrt{1+g^2} \\ 0 \\ g \\ 0 \end{pmatrix} = \left(1+\sqrt{1+g^2}\right)|\uparrow\uparrow\rangle + g|\downarrow\uparrow\rangle. \tag{5.256}$$

In the same way, we obtain upon proper normalization the four eigenstates of the block Hamiltonian

$$|1\rangle = \frac{1}{\sqrt{g^2+\left(1+\sqrt{1+g^2}\right)^2}}\left\{\left(1+\sqrt{1+g^2}\right)|\uparrow\uparrow\rangle + g|\downarrow\uparrow\rangle\right\}, \tag{5.257}$$

$$|2\rangle = \frac{1}{\sqrt{g^2+\left(1+\sqrt{1+g^2}\right)^2}}\left\{\left(1+\sqrt{1+g^2}\right)|\downarrow\downarrow\rangle + g|\uparrow\downarrow\rangle\right\}, \tag{5.258}$$

$$|3\rangle = \frac{1}{\sqrt{g^2+\left(1-\sqrt{1+g^2}\right)^2}}\left\{\left(1-\sqrt{1+g^2}\right)|\uparrow\uparrow\rangle + g|\downarrow\uparrow\rangle\right\}, \tag{5.259}$$

$$|4\rangle = \frac{1}{\sqrt{g^2+\left(1-\sqrt{1+g^2}\right)^2}}\left\{\left(1-\sqrt{1+g^2}\right)|\downarrow\downarrow\rangle + g|\uparrow\downarrow\rangle\right\}. \tag{5.260}$$

In the next step of the renormalization group procedure, we neglect the eigenstates $|3\rangle$ and $|4\rangle$ with higher energies $\lambda_{3,4} = \sqrt{g^2+1}$ and keep only the lower lying eigenstates $|1\rangle$ and $|2\rangle$ with lower energies $\lambda_{1,2} = -\sqrt{g^2+1}$ to construct block spin operators and the block Hamiltonian. We expect this approximation to work well as long as we focus on ground state properties.

With the help of the truncated basis $\{|1\rangle, |2\rangle\}$, we can construct spin operators

$$\widehat{\mathbb{I}} = |1\rangle\langle 1| + |2\rangle\langle 2|, \tag{5.261}$$

$$\widehat{\sigma}^x = |1\rangle\langle 2| + |1\rangle\langle 2|, \tag{5.262}$$

$$\widehat{\sigma}^z = |1\rangle\langle 1| - |2\rangle\langle 2|. \tag{5.263}$$

The operator $\widehat{\mathbb{I}}$ is a projection operator

$$\widehat{\mathbb{I}}^2 = \widehat{\mathbb{I}} \tag{5.264}$$

projecting any state of the basis $\{|1\rangle, |2\rangle, |3\rangle, |4\rangle\}$ on the truncated basis $\{|1\rangle, |2\rangle\}$, i.e. for any state

$$|\psi\rangle = \sum_{\alpha=1}^{4} C_\alpha |\alpha\rangle \tag{5.265}$$

we obtain a state in the basis $\{|1\rangle, |2\rangle\}$

$$\widehat{\mathbb{I}}|\psi\rangle = \sum_{\alpha=1}^{4} C_\alpha (|1\rangle\langle 1| + |2\rangle\langle 2|)|\alpha\rangle = C_1|1\rangle + C_2|2\rangle = \sum_{\alpha=1}^{2} C_\alpha |\alpha\rangle. \tag{5.266}$$

The operator $\widehat{\sigma}^x$ exchanges the two states

$$\widehat{\sigma}^x|1\rangle = |2\rangle \qquad \text{and} \qquad \widehat{\sigma}^x|2\rangle = |1\rangle, \tag{5.267}$$

while $\widehat{\sigma}^z$ produces the eigenvalues $\pm$ 1

$$\widehat{\sigma}^z|1\rangle = |1\rangle \qquad \text{and} \qquad \widehat{\sigma}^z|2\rangle = -|2\rangle. \tag{5.268}$$

In other words, $\widehat{\sigma}^x$ and $\widehat{\sigma}^z$ act on the truncated states $\{|1\rangle, |2\rangle\}$ in the same way as σ^x and σ^z acted on the original states $\{|\uparrow\rangle, |\downarrow\rangle\}$.

The original Hamiltonian of the quantum Ising chain in a transverse field was constructed from the operators $\mathbb{I}$, σ^x, and σ^z (cp. (5.161) and (5.243)). The next task of the renormalization group procedure is thus to construct a renormalized Hamiltonian from the operators (5.261–5.263)

In preparation for this goal, we first observe that

$$\widehat{\mathbb{I}}\,\mathcal{H}_{\text{block},i}\,\widehat{\mathbb{I}} = \lambda_1 \widehat{\mathbb{I}}, \tag{5.269}$$

which holds because, remembering $\lambda_1 = \lambda_2$,

$$\mathcal{H}_{\text{block},i}|1\rangle = \lambda_1|1\rangle \qquad \text{and} \qquad \mathcal{H}_{\text{block},i}|2\rangle = \lambda_2|2\rangle = \lambda_1|2\rangle. \tag{5.270}$$

Thus, the intra-block Hamiltonian is diagonal in the truncated basis of $\{|1\rangle, |2\rangle\}$.

However, this is not true for the inter-block Hamiltonian. In order to calculate this Hamiltonian in the truncated basis of $\{|1\rangle, |2\rangle\}$, we need the following expressions where it is crucial to note that the site index i refers to the even site of the block

$$\langle 1|\sigma_i^z|1\rangle = \langle 1|\mathbb{I}\otimes\sigma^z|1\rangle = 1, \tag{5.271}$$

$$\langle 2|\sigma_i^z|2\rangle = \langle 2|\mathbb{I}\otimes\sigma^z|2\rangle = -1, \tag{5.272}$$

$$\langle 1|\sigma_i^z|2\rangle = \langle 2|\sigma_i^z|1\rangle = 0. \tag{5.273}$$

Moreover, we need expressions holding for odd sites of blocks

$$\langle 1|\sigma_{i+1}^z|1\rangle = \langle 1|\sigma^z\otimes\mathbb{I}|1\rangle = \frac{1}{\sqrt{1+g^2}}, \tag{5.274}$$

$$\langle 2|\sigma_{i+1}^z|2\rangle = \langle 2|\sigma^z\otimes\mathbb{I}|2\rangle = -\frac{1}{\sqrt{1+g^2}}, \tag{5.275}$$

$$\langle 1|\sigma_{i+1}^z|2\rangle = \langle 2|\sigma_{i+1}^z|1\rangle = 0. \tag{5.276}$$

Lastly, again for the even site of a block, we have

$$\langle 1|\sigma_i^x|2\rangle = \langle 2|\sigma_i^x|1\rangle = \frac{g}{\sqrt{1+g^2}}, \tag{5.277}$$

$$\langle 1|\sigma_i^x|1\rangle = \langle 2|\sigma_i^x|2\rangle = 0. \tag{5.278}$$

With the help of these matrix elements, we can now calculate the projections of the spin operators of the original Hamiltonian on the truncated basis

$$\widehat{\mathbb{I}}_I\sigma_i^z\widehat{\mathbb{I}}_I = \widehat{\sigma}_I^z, \tag{5.279}$$

$$\widehat{\mathbb{I}}_I\sigma_i^x\widehat{\mathbb{I}}_I = \frac{g}{\sqrt{1+g^2}}\widehat{\sigma}_I^x, \tag{5.280}$$

$$\widehat{\mathbb{I}}_{I+1}\sigma_{i+1}^z\widehat{\mathbb{I}}_{I+1} = \frac{1}{\sqrt{1+g^2}}\widehat{\sigma}_{I+1}^z \tag{5.281}$$

where we have introduced a block site index I in accordance with the calculations of the matrix elements.

The renormalized Hamiltonian after one renormalization step is then the original Hamiltonian projected onto the truncated basis. Putting together the parts of the expressions above, we obtain[7]

[7] We have tacitly assumed that the number of sites N is even. For further renormalization group steps, we may even assume that N takes the form $N = 2^m$.

$$\widehat{\mathbb{P}}\mathcal{H}\widehat{\mathbb{P}} = \widehat{\mathcal{H}} = \lambda_1 \sum_{I=1}^{N/2} \widehat{\mathbb{I}} - \frac{1}{\sqrt{1+g^2}} \sum_{I=1}^{N/2-1} \widehat{\sigma}_I^z \widehat{\sigma}_{I+1}^z - \frac{g^2}{\sqrt{1+g^2}} \sum_{I=1}^{N/2} \widehat{\sigma}_I^x. \tag{5.282}$$

This renormalized Hamiltonian preserves the form of the original Hamiltonian apart from the first term, which is just a constant and can be subtracted off. This first term stems, as we have seen, from the intra-block Hamiltonian, which is diagonal in the truncated basis. The couplings get renormalized but no additional couplings are generated. This feature of the renormalization group procedure is particular to the quantum Ising chain in a transverse field and the choice of blocking which we have adopted.

We may rescale the renormalized Hamiltonian, after having subtracted the constant term, with a factor $\sqrt{1+g^2}$. Therefore, the renormalization group equation, obtained in principle after many more iterations of the renormalization group procedure, has the simple form

$$g' = f(g) = g^2 \tag{5.283}$$

from which we obtain the fixed point equation

$$g^* = \left(g^*\right)^2. \tag{5.284}$$

The two stable fixed points of this equation are

$$g^* = 0, \tag{5.285}$$

which is reached by (5.283) if we start from any value $g < 1$ and

$$g^* = \infty \tag{5.286}$$

reached from any value $g > 1$. The first fixed point corresponds to the ordered ferromagnetic phase, the second to the disordered quantum paramagnetic phase.

The third fixed point of (5.283) separates the attractive regions of $g^* = 0$ and $g^* = \infty$ and is itself instable

$$g^* = 1. \tag{5.287}$$

It is the critical fixed point

$$g^* = g_c = 1. \tag{5.288}$$

As we have already anticipated, this fixed point is the same as the one that the exact calculations of Pfeuty (1970) revealed. In this sense it is exact.

Comparison of the linearized renormalization group equation

$$f(g) - f(g_c) \approx \frac{df}{dg}|_{g=g_c}(g - g_c) = 2g_c(g - g_c) = 2(g - 1) \tag{5.289}$$

with the general result for the correlation length

$$\xi \propto (g - g_c)^{-\nu} \tag{5.290}$$

and the rescaling of the correlation length by blocks of length $b = 2$

$$\xi' = \frac{\xi}{b} \tag{5.291}$$

implies the value of the critical exponent

$$\nu = 1. \tag{5.292}$$

6

Statistical Mechanics and Quantum Field Theory

Every truth has two sides; it is well to look at both, before we commit ourselves to either.

Aesop (c. 620–564 BCE)

This chapter reviews the strong and important connection between statistical mechanics on the one side and quantum mechanics and quantum field theory on the other. The connection is based on the analogy between thermal and quantum fluctuations, an analogy also discussed in section 5.7 on quantum phase transitions. Formally, the connection is expressed through the mathematical equivalence between the partition function in statistical mechanics and the propagator in quantum field theory, which makes both disciplines of theoretical physics virtually the flip sides of the same coin (see e.g. section 1 'Introduction' of Shenker (1982) for an elementary discussion). This has, of course, important consequences, many of which play a significant role in this book.

In section 6.1, we explore the equivalence between statistical mechanics and quantum mechanics or quantum field theory in more general terms using the Feynman path integral and present the example of the equivalence between the classical XY Heisenberg model and the sigma model of quantum field theory. Section 6.2 presents a concrete calculation reviewing the important concepts of the partition function and the transfer matrix. We demonstrate the passage from the quantum mechanics of a single degree of freedom, a zero-dimensional system, to the statistical mechanics of a one-dimensional system represented by classical variables.

6.1 Connection between statistical mechanics and quantum field theory

This section briefly introduces the Feynman path integral and demonstrates its equivalence with the partition function of statistical mechanics discussed in chapter 4. We do this to show that statistical mechanics and quantum field theory, although they describe very different physical situations, are (on a certain level) one and the same theoretical discipline.

Models of Quantum Matter: A First Course on Integrability and the Bethe Ansatz. Hans-Peter Eckle, Oxford University Press (2019). © Hans-Peter Eckle. DOI: 10.1093/oso/9780199678839.001.0001

6.1.1 Feynman path integral

Section 4.9 discussed how the partition function of an interacting many-particle system in statistical mechanics, described by its Hamiltonian $\mathcal{H}(j)$ with energies E_j and given by the sum

$$Z_{\mathrm{SM}} = \sum_j e^{-\beta\mathcal{H}(j)} \tag{6.1}$$

over all microstates j, completely determines the thermodynamics of the system. The microstates j are the possible configurations of the system under consideration. Hence, the partition function (6.1) is a sum over configurations.

In order to be definite, let us consider a spin model on a lattice, where the microstates are given by the possible spin configurations $\{S_i\}$. The partition function is, thus, the sum over all spin configurations $\{S_i\}$

$$Z_{\mathrm{SM}} = \sum_{\{S_i\}} e^{-\beta\mathcal{H}(S_i)} \equiv \left\{\prod_i \sum_{S_i}\right\} e^{-\beta\mathcal{H}(S_i)}. \tag{6.2}$$

Next, let us consider a quantum field theory for a quantum field $y(t, x)$ which, to keep things simple, is supposed to depend only on time t and one spatial dimension x. Specifically, we think of $y(t, x)$ as the displacement field of a string, which, upon quantization, describes acoustic phonons in a one-dimensional solid. The Feynman path integral or sum over histories for the propagator provides a quantum field theory analog of the partition function of a statistical mechanical model. It consists of a sum over all possible paths or histories of the field configuration in a time interval from $t = 0$ to $t = T$ and a space interval from $x = 0$ to $x = L$ where each path is weighted by

$$\exp(iS) \tag{6.3}$$

with the action S corresponding to one particular path

$$S = \int_0^T dt \int_0^L dx\, \mathcal{L}(y, \partial_t y, \partial_x y). \tag{6.4}$$

If we split the continuous time interval into discrete times t_i with spacing $\Delta t = \tau$ and the space interval into discrete points x_i with spacing $\Delta x = a$, we can write an expression that is completely analogous to the partition function (6.2) of the spin system

$$Z_{\mathrm{QFT}} = \lim_{a,\tau\to 0} \left\{\prod_{i=(x_i,t_i)} \int_{-\infty}^{\infty} dy_i\right\} e^{iS} \equiv \int \mathcal{D}y\, e^{iS}. \tag{6.5}$$

The two objects, $Z_{\rm SM}$ and $Z_{\rm QFT}$, are identical in their mathematical structure—they are sums over configurations weighted by a corresponding exponential factor.

Both objects, $Z_{\rm SM}$ and $Z_{\rm QFT}$, respectively, act as *generating functionals* for the respective theory, from which, if known, all the correlation functions (in the case of statistical mechanics) or n–point functions (in the case of quantum field theory) can be derived. All the physics can be extracted from these generating functionals by appropriate mathematical manipulations.

The following section looks at a concrete example to render the foregoing formal discussion more transparent. A much more complete discussion of the ideas briefly outlined here is given by Kogut (1979).

6.1.2 Example: Equivalence of the XY model and the sigma model

Let us look at the example of the classical two-dimensional O(2) Heisenberg model (also known under the names of the classical XY and the planar model), which describes a two-dimensional magnetic system generalizing the Ising model. It is not a quantum mechanical model since the magnetic moments are described by classical variables, not Pauli matrices. Its magnetic moments are confined to a plane. This model has been studied intensively in lattice gauge theory and condensed matter physics. The model exhibits a so-called Berezinskii–Kosterlitz–Thouless phase transition. For more on the physics of this model, we refer to the literature, e.g. to chapter 9 of Nishimori and Ortiz (2011). Here, we restrict our attention to how this model of classical statistical mechanics in two space dimensions gives rise to a quantum field theory in one space and one time dimension.

In the XY model 'spins' $\mathbf{S}(\mathbf{r}) = (S^x(\mathbf{r}), S^y(\mathbf{r}))$ are situated on the N sites $\mathbf{r} = (x, y) = (ai, aj)$ of a two-dimensional square lattice with lattice spacing a. In order to make the continuum limit (which we shall discuss in a moment) more transparent, we note the lattice spacing explicitly. The spins interact via nearest-neighbour interactions of strength K. The Hamiltonian is

$$\mathcal{H}_{\rm SM} = -K \sum_{\langle \mathbf{r}\mathbf{r}' \rangle} \mathbf{S}(\mathbf{r}) \cdot \mathbf{S}(\mathbf{r}') \tag{6.6}$$

in analogy to the Ising Hamiltonian or the quantum mechanical Heisenberg model, where now the two components of the spin $\mathbf{S}(\mathbf{r})$ are real variables that are constrained by

$$\mathbf{S}(\mathbf{r})^2 = (S^x(\mathbf{r}))^2 + (S^y(\mathbf{r}))^2 = 1 \tag{6.7}$$

and the sum is over all adjacent pairs of sites, i.e. all sites which are nearest-neighbours. This Hamiltonian can be rewritten due to the constraint as

$$\mathcal{H}_{\text{SM}} = \frac{1}{2}K \sum_{\langle \mathbf{rr}' \rangle} \left(\mathbf{S}(\mathbf{r}) - \mathbf{S}(\mathbf{r}')\right)^2 \tag{6.8}$$

up to an irrelevant additive constant. The constraint $\mathbf{S}^2(\mathbf{r}) = 1$ can be satisfied by introducing a local phase $\phi(\mathbf{r})$ such that $\mathbf{S}(\mathbf{r}) = (\cos\phi(\mathbf{r}), \sin\phi(\mathbf{r}))$.

If we now let the lattice spacing go to zero, $a \to 0$, we obtain

$$\mathbf{S}(x,y) - \mathbf{S}(x+a,y) = -\partial_x \mathbf{S}(x,y)a + \mathcal{O}(a^2) \tag{6.9}$$

for a horizontal pair of nearest neighbours and

$$\mathbf{S}(x,y) - \mathbf{S}(x,y+a) = -\partial_y \mathbf{S}(x,y)a + \mathcal{O}(a^2) \tag{6.10}$$

for a vertical pair of nearest neighbours. Hence, we obtain the continuum Hamiltonian

$$\mathcal{H}_{\text{SM}} \longrightarrow \frac{1}{2}K \int d^2x \, (\partial_x \mathbf{S}(\mathbf{r}))^2 + \left(\partial_y \mathbf{S}(\mathbf{r})\right)^2 \qquad \text{for} \qquad a \to 0. \tag{6.11}$$

Next, we introduce a transformation, a so-called *Wick rotation*, for one of the spatial directions, say the y direction, which changes it into an imaginary variable

$$y \longrightarrow it, \tag{6.12}$$

where we interpret the variable t now as a time variable. The Hamiltonian thus becomes

$$\mathcal{H}_{\text{SM}} \longrightarrow -iS_{\text{QFT}} \qquad \text{for} \qquad y \to it, \tag{6.13}$$

where

$$\begin{aligned} S_{\text{QFT}} &= \frac{1}{2}K \int dt\, dx \left(\left(\partial_t S^x\right)^2 + \left(\partial_t S^y\right)^2 - \left(\partial_x S^x\right)^2 - \left(\partial_x S^y\right)^2 \right) \\ &= \frac{1}{2}K \int dt\, dx \sum_{\mu} \left(\partial_\mu \mathbf{S} \cdot \partial^\mu \mathbf{S}\right) \end{aligned} \tag{6.14}$$

is the action of the equivalent quantum field theory with the Lagrangian density

$$\mathcal{L}_{\text{QFT}} = \frac{1}{2}K \sum_{\mu} \left(\partial_\mu \mathbf{S} \cdot \partial^\mu \mathbf{S}\right) \qquad \text{with} \qquad \mathbf{S}^2(x,t) = 1. \tag{6.15}$$

This is the so-called O(2) sigma model of quantum field theory.

We have shown the equivalence, in the continuum limit, of a two-dimensional classical statistical mechanical model to a (1 + 1)-dimensional quantum field theory. This equivalence has been exploited by theorists working in lattice gauge field theory who were able to transfer many ideas and techniques from statistical mechanics to their own field of study. Arguably, therefore, Boltzmann and Gibbs discovered the path integral several decades before Dirac, Feynman, and Schwinger.

Let us note that the transformation from the continuum model to the lattice model, however, is not unique; it may be achieved in various ways. For example, instead of the square lattice, a triangular lattice may be employed. Furthermore, the finite-difference approximations of the derivative may be chosen in various ways. This means that many different lattice models correspond to the same continuum field theory.

6.2 Thermal fluctuations and quantum fluctuations

Thermal fluctuations and quantum fluctuations provide a link between quantum mechanics and statistical mechanics.

We begin by considering a quantum mechanical single-particle system. Such a system, described by a single-particle Hamiltonian $\mathcal{H}$, exhibits quantum fluctuations if we have for the vacuum expectation values of an observable $\mathcal{A}$, i.e. for the expectation values in the ground state of the system, that

$$\langle 0|\mathcal{A}^2|0\rangle \neq \langle 0|\mathcal{A}|0\rangle^2. \tag{6.16}$$

In order to establish the connection with thermal fluctuations, we consider the analytical continuation to imaginary time $T = -i\beta$ of the time evolution operator

$$\exp(-i\mathcal{H}T) \to \exp(-\beta\mathcal{H}), \tag{6.17}$$

where $\mathcal{H}$ is a quantum Hamiltonian of a single degree of freedom. This could be, for instance, the Hamiltonian of a qubit, i.e. a two-level system for which a simple Hamiltonian has the form

$$\mathcal{H} = \frac{1}{2}(E_0 + E_1) - \frac{1}{2}(E_1 - E_0)\,\sigma^x - h\sigma^z. \tag{6.18}$$

EXERCISE 6.1 Eigenvalues of the two-level system Calculate the eigenvalues of the Hamiltonian (6.18) of the two-level system and their limits when $h \to 0$.

Furthermore, determine the gap $\Delta(h)$ between the ground state energy and the excited state energy as well as its limit for $h \to 0$.

Let us return to the time evolution operator (6.17). We now subdivide the time interval β of the time evolution operator (6.17) into N equally long segments $\delta\beta$ with $N\delta\beta = \beta$. The matrix element of the time evolution operator between two single-particle states $|a\rangle$ and $|b\rangle$ from a complete orthonormal set $\{|n\rangle\}$ of single-particle states then becomes

$$\langle b|\exp(-\beta\mathcal{H})|a\rangle = \langle b|\exp(-\mathcal{H}\{\delta\beta+\delta\beta+\ldots+\delta\beta\})|a\rangle. \tag{6.19}$$

Next, we insert complete orthonormal sets $\{|n_1\rangle\}$, $\{|n_2\rangle\}$, ..., $\{|n_{N-1}\rangle\}$ of basis states of the single-particle system between each of the exponential factors and obtain

$$\langle b|\exp(-\beta\mathcal{H})|a\rangle = \sum_{n_1}\ldots\sum_{n_{N-1}}\langle b|\exp(-\mathcal{H}\delta\beta)|n_1\rangle\ldots\langle n_{N-1}|\exp(-\mathcal{H}\delta\beta)|a\rangle. \tag{6.20}$$

This is called *Trotter decomposition*, a mathematical construction which has manifold applications, e.g. in quantum Monte Carlo simulations of many-particle problems.

The sums over intermediate states $|n_i\rangle$ in (6.20) closely resemble the sums introduced by Schultz *et al.* (1964) for the solution of the two-dimensional classical Ising model, and which appear in the calculation of the partition function of especially two-dimensional classical statistical mechanics. The mathematical device used in statistical mechanics to express the partition function that makes this resemblance into an explicit equivalence is called the transfer matrix $\mathcal{T}$.

The transfer matrix $\mathcal{T}$ is introduced through

$$\langle n|\mathcal{T}|n'\rangle \equiv e^{-V(n,n')} \equiv \langle n|e^{-\mathcal{H}\delta\beta}|n'\rangle. \tag{6.21}$$

The transfer matrix is thus the equivalent object in statistical mechanics to the time evolution operator in quantum mechanics or to the propagator in quantum field theory. The transfer matrix describes the development of a state, although not in time as the time evolution operator, but the spatial development of a statistical configuration. The latter statement can be seen via the Ising model.

The matrix element (6.20), using the transfer matrix (6.21), now becomes

$$\langle b|e^{-\beta\mathcal{H}}|a\rangle = \sum_{n_1}\ldots\sum_{n_{N-1}} e^{-\{V(b,n_1)+\ldots+V(n_{N-1},a)\}}. \tag{6.22}$$

In this expression, the variables $b, n_1, \ldots, n_{N-1}, a$ are no longer quantum states. In the exemplary case of the Ising model, for instance, these numbers would be $n_i = \pm 1$. In this sense, we are dealing now with a classical statistical mechanics system. The quantum fluctuations in time of the ground state and excited states of the single-particle quantum system with which we started can now be interpreted as statistical, i.e. thermal, fluctuations of the classical variables n_i. The parameter β now acquires the meaning of inverse temperature

$$\beta = \frac{1}{T}. \tag{6.23}$$

Furthermore, this statistical mechanics system is one-dimensional, whereas the single-particle quantum system has spatial dimension zero.

In a last step to obtain an expression for the partition function, we take the trace of (6.22), i.e. we sum over the diagonal elements with $|a\rangle = |b\rangle$

$$Z = \mathrm{Tr}\left(e^{-\beta\mathcal{H}}\right) = \sum_a \langle a|e^{-\beta\mathcal{H}}|a\rangle \tag{6.24}$$

$$= \sum_a \left\{e^{-\{V(a,n_1)+\ldots+V(n_{N-1},a)\}}\right\} = \mathrm{Tr}\left(\mathcal{T}^N\right). \tag{6.25}$$

The calculation of the partition function, and hence, as shown in chapter 4, of all thermodynamic quantities, is now reduced to the diagonalization of the transfer matrix. This will be one of our major tasks, especially in part II of this book.

Here we assume for the moment that the diagonalization of the transfer matrix $\mathcal{T}$ has been achieved and that we know its eigenvalues λ_k $(k = 1, \ldots, N)$. The partition function then reduces to the sum

$$Z = \lambda_1^N + \lambda_2^N + \ldots \lambda_N^N. \tag{6.26}$$

According to the Perron–Frobenius theorem of linear algebra, there is a largest eigenvalue λ_1, such that all other eigenvalues are smaller $\lambda_j < \lambda_1$ for all $j \geq 2$. The theorem holds for matrices with positive elements, as is the case for the transfer matrix $\mathcal{T}$. We can therefore rewrite the partition function as

$$Z = \lambda_1^N\left(1 + \left(\frac{\lambda_2}{\lambda_1}\right)^N + \ldots \left(\frac{\lambda_N}{\lambda_1}\right)^N\right) \to \lambda_1^N \quad \text{for} \quad N \to \infty. \tag{6.27}$$

The partition function is thus determined by the largest eigenvalue λ_1 of the transfer matrix $\mathcal{T}$ in the thermodynamic limit $N \to \infty$. For example, the free energy per site is then given by

$$f = \lim_{N\to\infty}\left(-\frac{1}{\beta N}\ln Z\right) = -\frac{1}{\beta}\ln \lambda_1. \tag{6.28}$$

There are further ways to show and exploit the equivalence of single-particle quantum mechanics and statistical mechanics, which will be of importance for the interpretation of certain results in later chapters. The first concerns the free energy of the statistical system written using its original definition, i.e without the use of the transfer matrix

$$F = -\frac{1}{\beta}\ln Z = -\frac{1}{\beta}\ln\left(\mathrm{Tr}\, e^{-\beta\mathcal{H}}\right). \tag{6.29}$$

Let us assume that we know, e.g. in (6.18) above, the eigenvalues of $\mathcal{H}$, at least the two lowest ones ϵ_0 and ϵ_1, and that these two eigenvalues are non-degenerate, then the free energy in the limit of zero temperature $1/\beta \to 0$

$$F = \lim_{\beta\to\infty}\left(-\frac{1}{\beta}\ln\left(\text{Tr}\, e^{-\beta\mathcal{H}}\right)\right) = \lim_{\beta\to\infty}\left(-\frac{1}{\beta}\ln\left(e^{-\beta\epsilon_0} + e^{-\beta\epsilon_1} + \ldots\right)\right) = \epsilon_0 \tag{6.30}$$

has become the ground state energy of the quantum system.

This is an important connection that, generalized to the analogous relation between the free energy of a two-dimensional statistical mechanical system and the ground state energy of a one-dimensional many-particle system, we will exploit in chapter 7 and again in the interpretation of finite-size Bethe ansatz results in part VI.

Further quantities to consider showing the usefulness of the transfer matrix technique are correlation functions that encapsulate important information about statistical mechanics systems. The simplest correlation function is the one-point correlation function for a variable n_i

$$\langle n_i\rangle = \frac{1}{Z}\sum_{n_1,\ldots,n_N} n_i \exp\left(-\sum_{j=1}^{N} V(n_j, n_{j+1})\right) = \frac{1}{Z}\text{Tr}\left(n_i\mathcal{T}^N\right), \tag{6.31}$$

which often, e.g. for magnetic systems, provides the order parameter at a continuous phase transition. The variable n_i can be written as a diagonal matrix $\mathcal{S}$ with matrix elements

$$\mathcal{S}_{n,n'} = n\delta_{n,n'}. \tag{6.32}$$

The one-point correlation function thus becomes

$$\langle n_i\rangle = \frac{1}{Z}\text{Tr}\left(\mathcal{S}\mathcal{T}^N\right). \tag{6.33}$$

Let us now assume that we find the matrix $\mathcal{D}$ that diagonalizes the transfer matrix $\mathcal{T}$

$$\mathcal{D}^{-1}\mathcal{T}\mathcal{D} = \begin{pmatrix} \lambda_1 & 0 & \ldots & \ldots & 0 \\ 0 & \lambda_2 & \ldots & \ldots & 0 \\ 0 & 0 & \ddots & \ldots & 0 \\ \vdots & \vdots & \vdots & \vdots & \vdots \\ 0 & 0 & 0 & 0 & \lambda_N \end{pmatrix} \equiv \mathcal{T}_D. \tag{6.34}$$

The same transformation matrix $\mathcal{D}$ applied on $\mathcal{S}$ will in general generate a non-diagonal matrix

$$\mathcal{R} = \mathcal{D}^{-1}\mathcal{S}\mathcal{D}. \tag{6.35}$$

Due to the properties of the trace operation, the one-point correlation function can now be written as

$$\langle n_i \rangle = \frac{1}{Z} \mathrm{Tr}\left(\mathcal{R} \mathcal{T}_D^N \right). \tag{6.36}$$

In order to evaluate this expression in the thermodynamic limit, $N \to \infty$, we again only need to consider the largest eigenvalue of the transfer matrix. It is therefore only necessary to know the matrix element in the upper left corner of the matrix $\mathcal{R}$, which we call r_{11} and which determines the one-point correlation function

$$\langle n_i \rangle = r_{11}. \tag{6.37}$$

Since this matrix element is connected to the matrix that diagonalizes the transfer matrix and, hence, the transfer matrix itself, we see how central the transfer matrix is to this approach.

Let us next consider the two-point correlation function or correlation function proper

$$\langle n_i n_j \rangle = \frac{1}{Z} \sum_{n_1,\ldots,n_N} n_i n_j \exp\left(-\sum_{l=1}^{N} V(n_l, n_{l+1}) \right) \tag{6.38}$$

$$= \frac{1}{Z} \mathrm{Tr}\left(\mathcal{S} \mathcal{T}^{j-i} \mathcal{S} \mathcal{T}^{N-j+i} \right) \tag{6.39}$$

$$= \frac{1}{Z} \mathrm{Tr}\left(\mathcal{R} \mathcal{T}_D^{j-i} \mathcal{R} \mathcal{T}_D^{N-j+i} \right). \tag{6.40}$$

Arguments similar to the case of the one-point correlation function reveal that only the largest two eigenvalues of the transfer matrix are relevant for the two-point correlation function in the thermodynamic limit, $N \to \infty$. Moreover, we need the four matrix elements in the upper left corner of the non-diagonal matrix $\mathcal{R}$, namely

$$\mathcal{R} = \begin{pmatrix} r_{11} & r_{12} & \cdots \\ r_{21} & r_{22} & \cdots \\ \vdots & \vdots & \ddots \end{pmatrix}. \tag{6.41}$$

The two-point correlation function is then

$$\langle n_i n_j \rangle = r_{11}^2 + r_{12} r_{21} \left(\frac{\lambda_2}{\lambda_1} \right)^{j-i}. \tag{6.42}$$

As expected for a translationally invariant system, the correlation function depends only on the distance $j - i$. Furthermore, since $\lambda_2 < \lambda_1$ it falls off with increasing distance $j - i$.

We can therefore rewrite the correlation function as an decreasing exponential function of $j - i$

$$g(j - i) \equiv \langle n_i n_j \rangle - \langle n_i \rangle^2 \propto \exp\left(-\frac{j - i}{\xi}\right) \tag{6.43}$$

introducing a correlation length

$$\xi = \ln\left(\frac{\lambda_1}{\lambda_2}\right)^{-1}. \tag{6.44}$$

EXERCISE 6.2 One- and two-point correlation functions Perform the calculations explicitly leading to the one- and two-point correlation functions (6.37) and (6.43), respectively. It may be helpful to consider first the case of a two-level system, where the variables $\{n\}$ can only have two values and hence all matrices are 2×2 matrices.

The correlation length is related to another important quantity, which we have encountered before in section 5.7 on quantum phase transitions. We rewrite the two-point correlation function as

$$\langle n_i n_j \rangle \propto e^{-(\epsilon_1 - \epsilon_0)r} \tag{6.45}$$

with $r = j - i$ where we have assumed that the largest two eigenvalues of the transfer matrix are related to the lowest two eigenvalues of the single-particle Hamiltonian, the ground state eigenvalue, and the first excited state eigenvalue, by

$$\lambda_1 = e^{-\delta\beta\epsilon_0}, \tag{6.46}$$

$$\lambda_2 = e^{-\delta\beta\epsilon_1}. \tag{6.47}$$

We see that the correlation length and the energy gap between ground state and first excited state are then related through

$$\delta\beta\xi = \frac{1}{\epsilon_1 - \epsilon_0} = \frac{1}{\Delta}. \tag{6.48}$$

Observe that in order to obtain physical dimensions of the quantities we have used we need to introduce a length scale a, which is typically a lattice spacing. For example, the inverse temperature has to be scaled according to $\beta \to \beta/a$, energies according to $E \to Ea$, and the number of divisions $N \to Na = L$ becomes the fixed length of the statistical mechanical system. This implies

$$\xi = \frac{1}{\delta\beta\Delta} = \frac{N}{\beta\Delta} \to \frac{N}{\beta\Delta}a = \frac{L}{\beta\Delta} \tag{6.49}$$

giving the correlation length the correct dimension of length.

Again, we shall be putting these important relations to good use with the help of the insights of chapter 7 to interpret exact finite-size Bethe ansatz results in part VI.

The relations (6.30), (6.48), and (6.49), respectively, have here been derived for the low-energy properties of a zero-dimensional quantum system and the low-temperature properties of a one-dimensional statistical mechanical system. However, relations (6.30) and (6.48) hold for any dimension, i.e. they relate the low-energy behaviour of a quantum system in d dimensions to the low-temperature behaviour of a statistical mechanical system in $d + 1$ dimensions. In particular, they can be used to connect the critical behaviour of quantum and statistical mechanical systems. The latter statistical system in $d + 1$ dimensions is critical when the correlation length diverges, which translates into a vanishing energy gap for the former quantum system in d dimensions. These connections become especially powerful for $d = 1$ where they can be backed up by conformal symmetry, which is most useful for critical systems in $d + 1 = 2$.

Therefore, chapter 7 explores the conformal symmetry of statistical mechanical systems at criticality with a special focus on two-dimensional critical systems.

7

Conformal Symmetry in Statistical Mechanics

As long as algebra and geometry have been separated, their progress have been slow and their uses limited, but when these two sciences have been united, they have lent each other mutual forces, and have marched together towards perfection.

Joseph Louis Lagrange (1736–1813)

Soon after the advent of Einstein's special theory of relativity, where the invariance of the Maxwell equations under Lorentz transformations is one of the cornerstones of the theory, Cunningham (1910) and Bateman (1910) showed that Maxwell's equations are also invariant under a larger group of transformations: the conformal transformations. The use that has been made of the symmetry of physical systems under conformal transformations was rather limited until the 1970s, when Alexander Polyakov (1970) initiated the study of critical phenomena employing the local invariance under conformal transformations (see section 7.1). Before this, critical phenomena had been studied only under the aspect of their invariance under global scale transformations (cp. chapter 5). However, in dimensions larger than two, the new insights from conformal symmetry were not very dramatic (see sections 7.2 and 7.3).

This situation changed when, in seminal papers, Belavin, Polyakov, and Zamolodchikov 1984*a*; 1984*b*, Dotsenko (1984), and Friedan, Qiu, and Shenker 1984 demonstrated the power of conformal symmetry in two dimensions (see section 7.4) and triggered an immediate storm of activity, to which collection of papers published by Izykson *et al.* (1988) only four years later is testimony. Ever since, conformal field theory has been an extremely active field of theoretical physics. The origins of conformal field theory in two dimensions lie in statistical physics. However, a conformal field theory, in string theory, also describes the dynamics of a string on the world-sheet, the two-dimensional surface that the string traces out while propagating through space-time (see, e.g., Polchinski (2005) chapter 2, or Blumenhagen and Plauschinn (2009)). The conformal symmetry is much larger in two dimensions than in dimensions higher than two. Therefore, conformal field theories in two dimensions possess a particularly rich mathematical, especially algebraic, structure, which makes them an interesting field of research also from a conceptional viewpoint. For more information about these developments in string theory and mathematical physics, we refer to the literature cited in this introductory section.

Models of Quantum Matter: A First Course on Integrability and the Bethe Ansatz. Hans-Peter Eckle,
Oxford University Press (2019). © Hans-Peter Eckle. DOI: 10.1093/oso/9780199678839.001.0001

This chapter provides only an outline of conformal field theory tailored for our immediate needs that are connected to critical models in statistical mechanics and their one-dimensional quantum mechanical counterparts (cp. chapter 6). Conformal field theory not only predicts the critical exponents of two-dimensional statistical mechanics systems but also offers a way to categorize their possible universality classes. Integrable models, either classical statistical models in two dimensions or one-dimensional quantum models, via the correspondence elucidated in section 6.1, often allow the exact calculation of these critical exponents and, thus, provide an independent confirmation of the predictions of conformal field theory. Therefore, this chapter promotes a basic understanding of the notions of conformal field theory to enable us to connect them in later chapters, especially in part VI on the Bethe ansatz for finite systems, with results from classical statistical or quantum integrable models.

The core of our exposition of the theory of conformal symmetry in statistical mechanics appears in sections 7.5–7.9 and which explain the important concepts of correlation functions of order parameter fields, the energy–momentum tensor, conformal Ward identities and the operator product expansion, and the Virasoro algebra. Section 7.10 revisits finite-size corrections from section 7.1 in more detail and shows how they can be used to determine critical parameters, especially the central charge.

Early expositions include Saint-Aubin (1987), a set of lecture notes in French, the 1988 Les Houches summer school lecture notes by Cardy (1990) and Ginsparg (1990), and the treatment, again in French, by Grandati (1992) that contains detailed mathematical derivations. Monographs devoted to the subject are Henkel (1999) and the encyclopedic Di Francesco *et al.* (1997), known as the 'yellow book'. Moreover, Itzykson and Drouffe (1989), Cardy (1996), Mussardo (2010), and Nishimori and Ortiz (2011) all have chapters devoted to conformal symmetry and its consequences for statistical mechanics. Lastly, there is a newer set of lecture notes by Cardy (2010) from the Les Houches summer school in 2008 that update his earlier exposition and include recent developments.

The sources mentioned here represent only a small fraction of the literature on conformal field theory and its applications. It is, furthermore, mainly biased towards conformal symmetry of statistical mechanical systems, which is our focus, rather than the developments of conformal field theory in other fields, e.g. string theory.

7.1 From scale to conformal invariance

Chapter 5 discussed how the behaviour of many physical quantities close to critical points is dominated by power laws that are invariant under global scale transformations. The hypotheses of the renormalization group provide a natural framework to explain scale invariance and allow the calculation of critical exponents of the power laws.

In particular, we found that at a critical point the (two-point) correlation functions of, e.g. a scalar order parameter $\phi(\mathbf{r})$ (e.g. a suitably coarse-grained spin density $\sigma(\mathbf{r})$) relating two points $\mathbf{r}_1$ and $\mathbf{r}_2$ assumes a power law

$$\langle\phi(\mathbf{r}_1)\phi(\mathbf{r}_2)\rangle \propto |\mathbf{r}_1 - \mathbf{r}_2|^{-2x} \tag{7.1}$$

where we have, for later convenience, written the exponent as $2x = d - 2 + \eta$.

Scale invariance of this correlation function can be easily demonstrated. A dilatation, i.e. a transformation by a scalar factor λ, $\mathbf{r} \to \mathbf{r}' = \lambda\mathbf{r}$, multiplies the correlation function by a factor λ^{-2x}. Scale invariance can, thus, be expressed through

$$\langle\phi(\mathbf{r}_1)\phi(\mathbf{r}_2)\rangle = \lambda^{2x}\langle\phi(\lambda\mathbf{r}_1)\phi(\lambda\mathbf{r}_2)\rangle. \tag{7.2}$$

We now wish to extend the invariance to a larger class of transformations, the conformal transformations, where the scale factor may be a function of the spatial coordinate $\lambda = \lambda(\mathbf{r})$. Transformations of this kind have long been known in mathematics and have been used in physics for various purposes. We have already mentioned the invariance of Maxwell's equations under conformal transformations. Other applications occur in electrostatics, where fields and potentials are transformed between different geometries related to each other by conformal transformations.

In the connection of critical phenomena, invariance of critical systems under this class of transformations is again a hypothesis or principle. To quote again John Cardy (1987): 'It should be emphasized that, just as for scale invariance, there is no rigorous proof of conformal invariance in statistical mechanics at a critical point. Instead it should be regarded as a principle, whose numerous consequences can then be compared with exact and numerical results on specific models, and ultimately with experiment.'

The next section outlines the definition and properties of conformal transformations in more detail. Here, we only intend to demonstrate the most salient features of conformal transformations and the consequences for critical systems of their invariance under such transformations.

Let us first look at systems in dimension $d > 2$, because, as discussed later, $d = 2$ is quite special. In $d > 2$ dimensions, all conformal transformations can be constructed from four independent building blocks: spatial translations and rotations, dilatations, and inversions, the latter mapping points within the (unit) circle to points without, and vice versa. Lastly, instead of the inversion, the so-called special conformal transformation is often chosen as the fourth independent building block of general conformal transformations in $d > 2$. This transformation consists of a concatenation of an inversion and a translation, followed by a second inversion.

We have already seen that a dilatation only multiplies the correlation function by a factor λ^{-2x}, see (7.2), which signifies scale invariance. For homogeneous and isotropic systems, a translation or a rotation does not change the correlation function at all. The remaining case of an inversion is more interesting.

EXERCISE 7.1 Inversion Show that the modulus of the Jacobi determinant $|\det(\partial\mathbf{r}'/\partial\mathbf{r})|$ of the inversion $\mathbf{r} \to \mathbf{r}' = \mathbf{r}R/|\mathbf{r}|^2$ at a d-dimensional hypersphere of radius R is given by

$$\left|\det\left(\frac{\partial\mathbf{r}'}{\partial\mathbf{r}}\right)\right| = \left(\frac{R}{|\mathbf{r}|}\right)^{2d}. \tag{7.3}$$

Further, show that this transformation applied to (7.1) leads to

$$|\mathbf{r}_1 - \mathbf{r}_2|^{-2x} = \lambda(\mathbf{r}_1)^x \lambda(\mathbf{r}_2)^x |\mathbf{r}_1' - \mathbf{r}_2'|^{-2x} \tag{7.4}$$

with $\lambda(\mathbf{r}) = |\det\left(\partial \mathbf{r}'/\partial \mathbf{r}\right)|^{1/d}$.

....

The results of the exercise furnishes the general conformal transformation of the correlation function in $d > 2$

$$\langle \phi(\mathbf{r}_1)\phi(\mathbf{r}_2)\rangle = \lambda(\mathbf{r}_1)^x \lambda(\mathbf{r}_2)^x \langle \phi(\mathbf{r}_1')\phi(\mathbf{r}_2')\rangle, \tag{7.5}$$

which expresses conformal invariance of the correlation function. The local scale factor $\lambda(\mathbf{r})$ of the conformal transformation now contains scale invariance for a constant $\lambda(\mathbf{r}) = \lambda$ as a special case.

In the spirit of John Cardy's quote, we now assume that correlation functions of various physical quantities, e.g. of spin or energy densities, and also mixed and many-point correlation functions transform in a way corresponding to (7.5)

$$\langle \phi_1(\mathbf{r}_1)\phi_2(\mathbf{r}_2)\phi_3(\mathbf{r}_3)\ldots\rangle = \lambda(\mathbf{r}_1)^{x_1}\lambda(\mathbf{r}_2)^{x_2}\lambda(\mathbf{r}_3)^{x_3}\ldots\langle \phi_1(\mathbf{r}_1')\phi_2(\mathbf{r}_2')\phi_3(\mathbf{r}_3')\ldots\rangle. \tag{7.6}$$

A statistical mechanical system, for which (7.6) is satisfied, is a conformally invariant system at its critical point.

As an intermediate summary, we note that a system possessing conformal invariance has to exhibit translational invariance (i.e. it must be homogeneous) and rotational invariance (i.e. it must be isotropic) in addition to its local scale invariance. Moreover, the interactions in the system must be of short range only because, otherwise, the scale factors $\lambda(\mathbf{r})$ would cease to be local functions.

Often, due to the origin of many of the ideas of conformal symmetry in quantum field theory, the order parameters $\phi_i(\mathbf{r})$ are also called scaling operators. Here, we shall not go any deeper into these relations except to the extent discussed in chapter 6.

7.1.1 From scale to conformal invariance in two dimensions

Let us now consider the important special case of $d = 2$ dimensions where the conformal symmetry is most powerful. In $d = 2$ we can parametrize the plane with the complex coordinates z and $\bar{z}$ instead of the two-dimensional position vector $\mathbf{r} = (r^1, r^2)$. Of course, we could choose and shall do so occasionally below other parameterizations of the two-dimensional plane, such as real and imaginary part of the complex coordinate, $\Re z$ and $\Im z$, or modulus and argument, $|z|$ and $\arg(z)$.

Then every holomorphic or anti-holomorphic function, $w(z)$ or $u(\bar{z})$, of the complex plane, i.e. every differentiable function with respect to its complex variable that is not a constant (i.e. $dw\,/\,dz \neq 0$ or $du/d\bar{z} \neq 0$), defines a conformal transformation according to $z \to z' = w(z)$ or $\bar{z} \to \bar{z}' = u(\bar{z})$. There are infinitely many conformal transformations, and hence there are enormously rich possibilities in $d = 2$ to solve problems or infer information in one geometry or its conformally transformed geometry.

The transformation of the correlation function in (7.5) in the $d = 2$ case becomes

$$\langle \phi(z_1, \bar{z}_1)\phi(z_2, \bar{z}_2)\rangle = |w'(z_1)|^x |w'(z_2)|^x \langle \phi(z_1', \bar{z}_1')\phi(z_2', \bar{z}_2')\rangle \tag{7.7}$$

where

$$z' = w(z) \quad \text{and} \quad \bar{z}' = \bar{w}(\bar{z}) = \overline{w(z)} \tag{7.8}$$

and we assume z and $\bar{z}$ to represent two independent variables.

EXERCISE 7.2 Scale factor in $d = 2$ Show that the scale factor of the transformations (7.8) is given by $|dw(z)/dz| = |w'(z)|$.

One of the first successful tests of conformal invariance for critical two-dimensional systems was achieved by deducing that the exponential behaviour of the correlation function in a finite geometry is related to the critical exponent of the infinite system Cardy (1984) (see also Privman (1985)). This prediction from the correlation functions in different geometries related by conformal transformations was in complete agreement with and generalized numerical studies of special models.

7.1.2 Finite-size geometry

It is often possible to obtain results for finite two-dimensional systems at their finite-size critical point (in the sense discussed in section 5.6), for instance, using numerical methods). The assumption of conformal invariance of an infinite two-dimensional critical system predicts that the correlation function of a local scalar order parameter satisfies (7.7), which we write as

$$\langle \phi(z_1)\phi(z_2)\rangle = |w'(z_1)|^x |w'(z_2)|^x \langle \phi(z_1')\phi(z_2')\rangle \tag{7.9}$$

where, in contrast to (7.7), we interpret the complex variable z as representing the two variables needed to parametrize the two-dimensional plane. The exponent x is the scaling dimension of the order parameter ϕ. As we have already seen in the case of general dimension d, a simple scale transformation or dilatation $z \to bz$, and hence $w'(z) = b$, results in the scaling law

$$\langle \phi(z_1)\phi(z_2)\rangle = b^{2x} \langle \phi(bz_1)\phi(bz_2)\rangle, \tag{7.10}$$

which is, as in the case of general d, consistent with the expected power law behaviour

$$\langle \phi(z_1)\phi(z_2)\rangle = |z_1 - z_2|^{-2x}. \tag{7.11}$$

The critical exponent is thus $\eta = 2x$.

We now connect this result for an infinite system to a system in a different geometry. This can be achieved using any conformal transformation mapping the two-dimensional plane onto itself. We are especially interested in a transformation such that the resulting geometry is finite with periodic boundary conditions in one direction, while still infinite in the orthogonal direction.

As an example, let us consider the logarithmic transformation

$$z' = w(z) = \frac{L}{2\pi} \ln z, \tag{7.12}$$

which maps the whole complex plane onto the finite area of a strip of width L bounded by $|\Im w| \leq L/2$, with periodic boundary conditions.

Using the result (7.11) for the correlation function in the infinite plane and writing

$$z = e^{2\pi w/L} = e^{2\pi(u+iv)/L} \tag{7.13}$$

where u is the coordinate in the (infinite) direction along the strip, and v the coordinate in the (finite, periodic) direction across the strip, we obtain from (7.9)

$$\langle \phi(u_1, v_1)\phi(u_2, v_2)\rangle_{\text{strip}} = \left(\frac{2\pi}{L}\right)^{2x} \left(2\cosh\frac{2\pi}{L}(u_1 - u_2) - 2\cos\frac{2\pi}{L}(v_1 - v_2)\right)^{-x} \tag{7.14}$$

where the correlation function on the left hand side $\langle \ldots \rangle_{\text{strip}}$ is evaluated in the strip geometry.

EXERCISE 7.3 Correlation function in a strip Perform the steps of the calculation leading to (7.14).

For large distances $u_1 - u_2 \gg L$ along the strip, (7.14) becomes

$$\langle \phi(u_1, v_1)\phi(u_2, v_2)\rangle_{\text{strip}} = \left(\frac{2\pi}{L}\right)^{2x} \exp\left(-\frac{2\pi x}{L}(u_1 - u_2)\right). \tag{7.15}$$

Comparing this result to the general form of the correlation function $e^{-(u_1-u_2)/\xi(L)}$, we can infer for the correlation length of the finite system

$$\xi(L) = \frac{L}{2\pi x} = \frac{L}{\pi\eta}. \tag{7.16}$$

This result has now been obtained in complete generality from the assumed conformal invariance of the correlation function of two-dimensional critical systems. It has been known from perturbative results for special model systems only but was expected to hold universally (Luck, 1982). Its most remarkable feature is that it allows us to determine the critical exponent η of the infinite system from calculations of the correlation length for a finite system.

Later, section 7.10 examines the result for the correlation function in the strip geometry (7.14) more carefully to extract more information from it. In order to do this, we must connect it to the transfer matrix formalism used in sections 4.16 and 4.17, and again in section 6.2.

Let us now explore some properties of conformal transformations for arbitrary dimension $d > 2$ in more detail before we focus on the special case $d = 2$.

7.2 Conformal symmetry

Section 7.1 showed how for correlation functions of order parameters the notion of scale invariance can be successfully generalized to a larger symmetry: conformal invariance. This section briefly outlines the basic tenets of this idea and mentions some of its applications to give an orientation for the remainder of this chapter. Since conformal symmetry and its consequences and applications have grown into a major field of theoretical physics and mathematics, we will restrict our attention in the subsequent sections on those aspects most relevant for our later purposes, i.e. to those parts of the general theory necessary for the applications to two-dimensional critical systems of statistical mechanics and their one-dimensional quantum counterparts.

In mathematics, conformal transformations are defined as those coordinate transformations that leave angles between non-collinear directions invariant. Lengths, and hence volume elements, however, are not necessarily preserved by conformal transformations. In particular, this implies that via a conformal transformation a triangle will undergo a similarity transformation.

As demonstrated in the previous section, the objects of conformal field theory are fields (e.g. order parameter fields) that possess a special space-time symmetry, the conformal symmetry. The fields are assumed to be invariant under conformal transformations.

In order to understand the import of conformal field theory, we thus need to discuss in more detail what a *conformal transformation* is and which consequences invariance under conformal transformations entails for a field theory. Moreover, we need to discuss the applications of conformal field theories in the theory of low-dimensional quantum matter systems.

Before we elaborate the details of conformal transformations , let us first express in one sentence the gist of conformal transformations. Conformal transformations are *local* space dilatations (which are *angle-preserving*). Hence, they are a generalization of scale transformations that are *global* dilatations of space-time characterized by a uniform scale factor.

The major applications of conformal field theory include

1. Classical statistical systems in two space dimensions at their critical point where they display *conformal invariance* in the long-wavelength limit. Examples of such systems are vertex and other Ising-like models.

2. Critical one-dimensional quantum models, e.g. quantum spin chains, or quantum models that can be effectively reduced to one-dimensional form, e.g. the Kondo impurity problem.
3. String theory: a string describes during its time-evolution a closed or open two-dimensional surface in space-time, which is by construction conformally invariant.

We are concerned with the first two applications of conformal field theory, which are intimately connected.

The major results obtainable by exploiting conformal field theory include

1. A classification of critical behaviour (universality classes);
2. Correlation functions at criticality; and
3. Critical point parameters.

7.3 Conformal transformations in dimensions larger than two

We now infer the restrictions on conformal transformations, which follow from their defining property, the invariance of angles between non-collinear directions. Although we are mainly interested in dimensions $d = 2$, we shall gain more insights, as is often the case, from starting with the more general situation of $d = p + q > 2$, i.e. we consider a generalized Minkowskian space-time $\mathbb{R}^{p,q}$ with metric

$$\eta = \begin{pmatrix} -1_p & 0 \\ 0 & 1_q \end{pmatrix} \tag{7.17}$$

where 1_p and 1_q are the p- and q-dimensional unit matrices, respectively.

A general d-dimensional vector in a basis $\{\mathbf{e}_\mu\}$ can be written as

$$\mathbf{r} = r^\mu \mathbf{e}_\mu, \qquad \mu = 1, \dots, d. \tag{7.18}$$

A transformation

$$\mathbf{r} \rightarrow \tilde{\mathbf{r}} = \mathbf{f}(\mathbf{r}) \tag{7.19}$$

maps a point $\mathbf{r}$ in $\mathbb{R}^{p,q}$ to a point $\tilde{\mathbf{r}} = \mathbf{f}(\mathbf{r})$ in $\mathbb{R}^{p,q}$. Locally, this mapping can be represented by a linear transformation $d\mathbf{r} \rightarrow d\tilde{\mathbf{r}} = \Omega d\mathbf{r}$, which in components is written as

$$d\tilde{r}^\mu = \sum_\nu \left.\frac{\partial \tilde{r}^\mu}{\partial r^\nu}\right|_{\tilde{\mathbf{r}}=\mathbf{r}} dr^\nu \equiv \sum_\nu \Omega^\mu_\nu dr^\nu. \tag{7.20}$$

EXERCISE 7.4 Condition for a coordinate transformation to be conformal As mentioned in the previous section, a conformal transformation preserves angles and thus a triangle undergoes a similarity transformation. Consider two non-collinear direction vectors $\mathbf{u}$ and $\mathbf{v}$, which define a triangle with sides $\mathbf{u}$, $\mathbf{v}$, and $\mathbf{u} - \mathbf{v}$ and for which $\tilde{\mathbf{u}} = \Omega\mathbf{u}$ and $\tilde{\mathbf{u}} = \Omega\mathbf{u}$ define a transformed triangle. The condition that the transformation is conformal becomes

$$\tilde{\mathbf{u}}^2 = \lambda^2(\mathbf{r})\mathbf{u}^2, \quad \tilde{\mathbf{v}}^2 = \lambda^2(\mathbf{r})\mathbf{v}^2, \quad (\tilde{\mathbf{u}} - \tilde{\mathbf{v}})^2 = \lambda^2(\mathbf{r})(\mathbf{u} - \mathbf{v})^2 \tag{7.21}$$

with a local scale factor $\lambda(\mathbf{r})$.

Derive from (7.21) the condition

$$\sum_{\nu\mu} \eta_{\mu\nu} \frac{\partial \tilde{r}^\mu}{\partial r^\sigma} \frac{\partial \tilde{r}^\nu}{\partial r^\rho} = \lambda^2(\mathbf{r})\eta_{\sigma\rho} \tag{7.22}$$

for the transformation to be conformal.

The scalar product between two vectors is defined using the metric η

$$\mathbf{u} \cdot \mathbf{v} = \sum_{\mu\nu} \eta_{\mu\nu} u^\mu v^\nu. \tag{7.23}$$

The conformal transformations form a group since performing two angle-preserving transformations one after the other yields again an angle-preserving transformation.

7.3.1 Finite conformal transformations

Let us first consider special finite transformations that satisfy (7.22) and are thus obviously conformal transformations. The first class are the transformations with scale factor $\lambda(\mathbf{r}) = 1$ (the Poincaré transformations forming the Poincaré group). The first of this class are the translations by a constant vector $\mathbf{a}$. In components, that is the transformation

$$r^\mu \to \tilde{r}^\mu = r^\mu + a^\mu. \tag{7.24}$$

The rigid rotations complete the Poincaré transformations

$$r^\mu \to \tilde{r}^\mu = \sum_\nu R^\mu{}_\nu r^\nu \tag{7.25}$$

with the orthogonal rotation matrix satisfying

$$\sum_{\mu\nu} R^\mu{}_\sigma \eta_{\mu\nu} R^\nu{}_\rho = \eta_{\sigma\rho}, \tag{7.26}$$

which is (7.22) with $\lambda(\mathbf{r}) = 1$.

EXERCISE 7.5 Independent parameters of the rotation matrix Using (7.26), show that the number of independent parameters of the rotation matrix in $\mathbb{R}^{p,q}$ is given by $\frac{1}{2}d(d-1)$ where $d = p + q$.

A further simple conformal transformation is the dilatation by a constant scale factor $\lambda \neq 0$

$$r^\mu \to \tilde{r}^\mu = \lambda r^\mu. \tag{7.27}$$

As discussed in the previous section, an interesting transformation is the inversion considered here as an inversion at the unit hypersphere. This transformation is given by

$$r^\mu \to \tilde{r}^\mu = \frac{r^\mu}{\mathbf{r}^2}. \tag{7.28}$$

EXERCISE 7.6 Scale factor of the inversion transformation Determine the Jacobi matrix of the inversion (7.28) and then show that the scale factor of the inversion is given from (7.22) by

$$\lambda^2(\mathbf{r}) = \frac{1}{\mathbf{r}^4}. \tag{7.29}$$

The inversion transformation is a discrete transformation that will make it inconvenient when we discuss infinitesimal transformations. In order to obtain a transformation that can be studied near the identity transformation by considering a small parameter, we replace the pure inversion by a combination of an inversion followed by a translation and a second inversion. This transformation can be written for a translation vector $\mathbf{s} = \sum_\mu s^\mu \mathbf{e}_\mu$ as

$$r^\mu \to \tilde{r}^\mu = \frac{r^\mu + s^\mu \mathbf{r}^2}{1 + 2\mathbf{s}\cdot\mathbf{r} + \mathbf{s}^2\mathbf{r}^2} \tag{7.30}$$

This transformation, called the special conformal transformation, satisfies the requirements we need for an infinitesimal transformation. For $\mathbf{s} = 0$ it reduces to the identity transformation and for $\mathbf{s} \to 0$ it allows the construction of an infinitesimal transformation close to the identity transformation.

EXERCISE 7.7 Special conformal transformation Derive the formula (7.30) for the special conformal transformation by explicitly performing an inversion, followed by a translation, and finally a second inversion.

In summary, all conformal transformations of the components r^μ of the d-dimensional vector $\mathbf{r} = \sum_\mu r^\mu \mathbf{e}_\mu$ ($\mu = 1, \ldots, d = 1, \ldots, p + q$) are combinations of the four basic transformations:

- a translation by a^μ:

$$r^\mu \to \tilde{r}^\mu = r^\mu + a^\mu; \tag{7.31}$$

- a rotation by an orthogonal rotation matrix $R_\nu{}^\mu$:

$$r^\mu \to \tilde{r}^\mu = \sum_\nu R^\mu_\nu r^\nu; \tag{7.32}$$

- a dilatation by a constant scale factor λ:

$$r^\mu \to \tilde{r}^\mu = \lambda r^\mu; \tag{7.33}$$

- and a special conformal transformation involving a vector $\mathbf{s} = \sum_\mu s^\mu \mathbf{e}_\mu$

$$r^\mu \to \tilde{r}^\mu = \frac{r^\mu + s^\mu \mathbf{x}^2}{1 + 2\mathbf{s} \cdot \mathbf{r} + \mathbf{s}^2 \mathbf{r}^2}. \tag{7.34}$$

  A lengthy calculation shows that the scale factor of the special conformal transformation is given by

$$\lambda(\mathbf{r}) = \left(1 + 2\mathbf{s} \cdot \mathbf{r} + \mathbf{s}^2 \mathbf{r}^2\right)^{-1}. \tag{7.35}$$

7.3.2 Infinitesimal conformal transformation

In order to demonstrate the validity of the ultimate statement of the previous section, we next derive conditions for the infinitesimal transformation

$$r^\mu \to \tilde{r}^\mu = r^\mu + \epsilon \xi^\mu(\mathbf{r}) + \mathcal{O}(\epsilon^2), \qquad \epsilon \to 0, \tag{7.36}$$

i.e. the transformation that is infinitesimally close to the identity transformation. The condition for a general coordinate transformation to be conformal, (7.22), then becomes

$$\sum_{\mu\nu} \eta_{\mu\nu} \left(\delta^\mu_\sigma + \epsilon \frac{\partial \xi^\mu}{\partial r^\sigma}\right) \left(\delta^\nu_\rho + \epsilon \frac{\partial \xi^\nu}{\partial r^\rho}\right) = \eta_{\sigma\rho} + \epsilon \zeta(\mathbf{r}) \eta_{\sigma\rho} \tag{7.37}$$

where we have used the expansion

$$\lambda^2(\mathbf{r}) = 1 + \epsilon \zeta(\mathbf{r}) + \mathcal{O}(\epsilon^2), \qquad \epsilon \to 0. \tag{7.38}$$

The constant term of this equation (order ϵ^0) coincides on both sides of the equation. The terms linear in ϵ are

$$\sum_\nu \eta_{\sigma\nu} \frac{\partial \xi^\nu}{\partial r^\rho} + \sum_\mu \eta_{\mu\rho} \frac{\partial \xi^\mu}{\partial r^\sigma} = \frac{\partial \xi_\sigma}{\partial r^\rho} + \frac{\partial \xi_\rho}{\partial r^\sigma} = \zeta(\mathbf{r})\eta_{\sigma\rho}. \tag{7.39}$$

This equation is called the (conformal) Killing–Cartan equation. The functions $\xi^\mu(\mathbf{r})$ and $\zeta(\mathbf{r})$ are yet to be determined. For a given conformal transformation, the functions $\xi^\mu(\mathbf{r})$ are known. Our task consists in determining the general form of the functions $\xi^\mu(\mathbf{r})$. Using Exercise 7.8, let us check that the conformal transformations we identified in the previous section (translation, rigid rotation, dilatation, and special conformal transformation), in their infinitesimal form indeed satisfy (7.39) and determine the function $\zeta(\mathbf{r})$.

EXERCISE 7.8 Infinitesimal conformal transformations Check that the infinitesimal versions of the finite conformal transformations of the previous section satisfy (7.39), i.e. determine the function $\zeta(\mathbf{r})$ in each of the following cases:

- translation by a constant vector ϵb^μ

$$\tilde{r}^\mu = r^\mu + \epsilon b^\mu \qquad \epsilon \to 0; \tag{7.40}$$

- rigid rotation described by

$$R^\mu{}_\nu = \delta^\mu{}_\nu + \epsilon w^\mu{}_\nu \qquad \epsilon \to 0 \tag{7.41}$$

  where the infinitesimal rotation matrix $w^\mu{}_\nu$ is antisymmetric, $w_{\mu\nu} = -w_{\nu\mu}$, which follows from the definition of the (finite) rotation matrix (7.26); the infinitesimal coordinate transformation associated with this infinitesimal rotation is

$$\tilde{r}^\mu = r^\mu + \epsilon \sum_\nu w^\mu{}_\nu r^\nu \qquad \epsilon \to 0; \tag{7.42}$$

- dilatation by $\lambda = 1 + \epsilon\ell$

$$\tilde{r}^\mu = r^\mu + \epsilon \ell r^\mu \qquad \epsilon \to 0; \tag{7.43}$$

- and special conformal transformation with translation vector ϵt^μ

$$\tilde{r}^\mu = r^\mu + \epsilon \left(t^\mu \mathbf{r}^2 - 2(\mathbf{t}\cdot\mathbf{r}) r^\mu \right) \qquad \epsilon \to 0. \tag{7.44}$$

Next, we observe that we can eliminate $\zeta(\mathbf{r})$ from (7.39) by multiplying this equation by $\eta^{\sigma\rho}$ and summing over σ and ρ, i.e. we contract the equation by η. Because

$$\sum_{\sigma\rho} \eta^{\sigma\rho} \eta_{\sigma\rho} = d, \tag{7.45}$$

we obtain

$$2\sum_{\sigma}\frac{\partial \xi^{\sigma}}{\partial r^{\sigma}} \equiv 2(\partial \cdot \xi) = d\zeta(\mathbf{r}) \tag{7.46}$$

and finally

$$\frac{\partial \xi_{\sigma}}{\partial r^{\rho}} + \frac{\partial \xi_{\rho}}{\partial r^{\sigma}} = \frac{2}{d}(\partial \cdot \xi)\eta_{\sigma\rho}. \tag{7.47}$$

We can now manipulate (7.39) and (7.47) to derive equations that single out the dimensions $d = 1$ and $d = 2$ (see exercise 7.9). The first of these equations is

$$(d-1)\partial^2\zeta = 0, \tag{7.48}$$

which can be used in the derivation of a second equation if $d > 1$. This second equation is

$$(d-2)\frac{\partial^2 \zeta}{\partial r^{\mu}\partial r^{\nu}} = 0 \tag{7.49}$$

whose general solution for $d > 2$ can be written as

$$\zeta(\mathbf{r}) = 2\ell - 4\mathbf{t}\cdot\mathbf{r} \tag{7.50}$$

with the constant ℓ and the constant vector $\mathbf{t}$ defined in exercise 7.8. Finally, we obtain the important result

$$d\partial^2\xi_{\mu} = (2-d)\frac{\partial}{\partial r^{\mu}}(\partial \cdot \boldsymbol{\xi}). \tag{7.51}$$

EXERCISE 7.9 Derivation of (7.48–7.51) Derive (7.48), (7.49), and (7.51) by manipulating (7.39) and (7.47) appropriately, i.e. by taking further derivatives and contracting the ensuing equations with the metric η after taking advantage of exchanges in the indices labelling the functions ξ^{μ} and the derivatives $\partial_{\mu} \equiv \partial/\partial r^{\mu}$.

We note that (7.48) and (7.49) tell us that our analysis does not give any information on the infinitesimal scale factor $\zeta(\mathbf{r})$ for dimensions $d \leq 2$. In one dimension, (7.48) does not pose any constraint on the function ζ and thus any sufficiently smooth function is a conformal transformation. However, this statement is somewhat empty and trivial since the defining angle-preserving characteristic of conformal transformations does not make sense in one dimension.

However, (7.51) does provide information on the special case $d = 2$. Here, all solutions of the Laplace equation $\partial^2\xi_{\mu} = 0$ are conformal transformations. An extra section below explores this important special case, which is at the centre of our attention.

One of the equations we encounter in the preceding exercise is

$$2\frac{\partial^2 \xi_\mu}{\partial r^\nu \partial r^\rho} = -4t_\rho \eta_{\mu\nu} - 4t_\nu \eta_{\mu\rho} + 4t_\mu \eta_{\rho\nu} \tag{7.52}$$

where we also used (7.50). This equation is particularly well suited to answer the question of finding the most general conformal transformations in the case $d > 2$. Since the general solution of this equation is at most quadratic in the coordinates

$$\xi^\mu(\mathbf{r}) = A^\mu + \sum_\nu B^\mu_\nu r^\nu + \sum_{\nu\rho} C^\mu_{\nu\rho} r^\nu r^\rho \tag{7.53}$$

where $C^\mu_{\nu\rho} = C^\mu_{\rho\nu}$, this provides the form of the most general conformal transformation in $d > 2$ with the constants being determined in the following exercise.

EXERCISE 7.10 General conformal transformation in $d > 2$ In order to qualify as a conformal transformation, the general solution (7.53) must satisfy (7.39) and (7.47). Show that this requirement leads to the general conformal transformation in $d > 2$ in the form

$$\xi^\mu = b^\mu + \sum_\nu w^\mu_\nu r^\nu + \ell r^\mu + t^\mu \mathbf{r}^2 - 2(\mathbf{t}\cdot\mathbf{r})r^\mu \tag{7.54}$$

where the constants are those defined in exercise 7.8. Furthermore, show that the most general infinitesimal scale factor is

$$\zeta(\mathbf{r}) = 2\ell - 4\mathbf{t}\cdot\mathbf{r}. \tag{7.55}$$

7.3.3 Lie algebra of conformal transformations

The generators of the conformal transformations form a Lie algebra. For an infinitesimal conformal transformation

$$\mathbf{r} \to \tilde{\mathbf{r}} = \mathbf{r} + \epsilon\boldsymbol{\xi}(\mathbf{r}), \tag{7.56}$$

the generator G_ξ is given by

$$\delta f = f(\mathbf{r} - \epsilon\boldsymbol{\xi}(\mathbf{r})) - f(\mathbf{r}) = -\epsilon(\boldsymbol{\xi}(\mathbf{r})\cdot\partial)f(\mathbf{r}) = -i\epsilon G_\xi f, \tag{7.57}$$

i.e. explicitly

$$G_\xi = -i\sum_\mu \xi^\mu(\mathbf{r})\frac{\partial}{\partial r^\mu}. \tag{7.58}$$

EXERCISE 7.11 Generators of the conformal transformations and their commutators Calculate the generators of translation, rotation, dilatation, and special conformal transformation with the help of exercise 7.8 and determine their mutual commutators.

We now turn our attention to the special case of two dimensions, which is our main focus for the purposes of this book.

7.4 Conformal transformations in two dimensions

We are mostly interested in Euclidian conformal field theory, which corresponds to the two-dimensional Euclidian space $\mathbb{R}^{0,2}$ with metric

$$\eta_{\mu\nu} = \delta_{\mu\nu}. \tag{7.59}$$

However, the analysis for a Minkowskian space-time $\mathbb{R}^{1,1}$ is essentially similar.

For a coordinate transformation $(x, y) \to (\tilde{x}, \tilde{y})$ to be conformal, the invariant infinitesimal length element in $\mathbb{R}^{0,2}$ has to satisfy

$$\begin{aligned} ds^2 = d\tilde{x}^2 + d\tilde{y}^2 = &\left\{\left(\frac{\partial \tilde{x}}{\partial x}\right)^2 + \left(\frac{\partial \tilde{y}}{\partial x}\right)^2\right\} dx^2 + \left\{\left(\frac{\partial \tilde{x}}{\partial y}\right)^2 + \left(\frac{\partial \tilde{y}}{\partial y}\right)^2\right\} dy^2 \\ &+ 2\left\{\frac{\partial \tilde{x}}{\partial x}\frac{\partial \tilde{x}}{\partial y} + \frac{\partial \tilde{y}}{\partial x}\frac{\partial \tilde{y}}{\partial y}\right\} dxdy \\ &\equiv \lambda^2(x, y)\left(dx^2 + dy^2\right). \end{aligned} \tag{7.60}$$

For this equation to hold, we must require

$$\left(\frac{\partial \tilde{x}}{\partial x}\right)^2 + \left(\frac{\partial \tilde{y}}{\partial x}\right)^2 = \left(\frac{\partial \tilde{x}}{\partial y}\right)^2 + \left(\frac{\partial \tilde{y}}{\partial y}\right)^2 = \lambda^2(x, y) \tag{7.61}$$

and

$$\frac{\partial \tilde{x}}{\partial x}\frac{\partial \tilde{x}}{\partial y} + \frac{\partial \tilde{y}}{\partial x}\frac{\partial \tilde{y}}{\partial y} = 0. \tag{7.62}$$

In order to satisfy this latter equation, we need to have, for example,

$$\frac{\partial \tilde{y}}{\partial x} = \rho(x, y)\frac{\partial \tilde{x}}{\partial y}, \qquad \frac{\partial \tilde{y}}{\partial y} = -\frac{1}{\rho(x, y)}\frac{\partial \tilde{x}}{\partial x} \tag{7.63}$$

for some function $\rho(x, y)$.

EXERCISE 7.12 Cauchy–Riemann differential equations Show that $\rho(x,y) = \pm 1$ must hold for the transformation $(x,y) \to (\tilde{x},\tilde{y})$ to be conformal and that therefore the Cauchy–Riemann differential equations, either as

$$\frac{\partial \tilde{x}}{\partial x} = \frac{\partial \tilde{y}}{\partial y}, \qquad \frac{\partial \tilde{x}}{\partial y} = -\frac{\partial \tilde{y}}{\partial x} \tag{7.64}$$

or as

$$\frac{\partial \tilde{x}}{\partial x} = -\frac{\partial \tilde{y}}{\partial y}, \qquad \frac{\partial \tilde{x}}{\partial y} = \frac{\partial \tilde{y}}{\partial x} \tag{7.65}$$

must hold.

The result of exercise 7.12 indicates that the complex function $w(x,y) = \tilde{x}(x,y) + i\tilde{y}(x,y)$ is a holomorphic function of the complex variable $z = x + iy$ if we choose alternative (7.64), while, if we choose alternative (7.65), the complex function $\bar{w}(x,y) = \tilde{x}(x,y) + i\tilde{y}(x,y)$ is an anti-holomorphic function of the complex variable $\bar{z} = x - iy$.

We can thus write $(z,\bar{z}) \to (w(z,\bar{z}), \bar{w}(z,\bar{z}))$ for the coordinate transformation. The condition (7.60) for the transformation to be conformal becomes then in this complex notation

$$dw\, d\bar{w} = \lambda^2(z,\bar{z}) dz\, d\bar{z}, \tag{7.66}$$

which is satisfied if

$$\frac{\partial w}{\partial z}\frac{\partial \bar{w}}{\partial z} = 0 \quad \text{and} \quad \frac{\partial w}{\partial \bar{z}}\frac{\partial \bar{w}}{\partial \bar{z}} = 0. \tag{7.67}$$

Apart from the trivial solution of these equations where all four factors vanish, there are two possibilities to satisfy them. The first consists in

$$\frac{\partial \bar{w}}{\partial z} = 0 \Leftrightarrow \frac{\partial w}{\partial \bar{z}} = 0, \tag{7.68}$$

which implies that $w = w(z)$ is a holomorphic function of z alone and that $\bar{w} = \bar{w}(\bar{z}) \equiv \overline{w(z)}$ is a anti-holomorphic function of $\bar{z}$ alone. The second possibility is

$$\frac{\partial w}{\partial z} = 0 \Leftrightarrow \frac{\partial \bar{w}}{\partial \bar{z}} = 0, \tag{7.69}$$

which implies $w = w(\bar{z})$ is an anti-holomorphic function of $\bar{z}$ and $\bar{w} = \bar{w}(z)$ is a holomorphic function. Obviously the two possibilities differ only in the symbols we use for the functions and, hence, we shall restrict ourselves to the first alternative without any loss of generality.

We note the important result that in two dimensions any holomorphic or anti-holomorphic function, $z \to w(z)$ or $\bar{z} \to \bar{w}(\bar{z})$, defines a conformal transformation in some open set of the complex plane.

EXERCISE 7.13 Möbius transformation Show that the conformal transformations of translation, rotation, dilatation, and special conformal transformation are all contained in the Möbius (or fractional-linear) transformation

$$z \to w(z) = \frac{Az + B}{Cz + D} \tag{7.70}$$

where the complex constants satisfy $AD - BC = 1$.

7.4.1 Lie algebra in two dimensions

As in the case of arbitrary dimension $d > 2$, briefly outlined in section 7.3.3, we can associate a Lie algebra to the group of conformal transformations in two dimensions. In contrast to the general case, we find that in two dimensions there are infinitely many generators of this Lie algebra. For an infinitesimal holomorphic transformation

$$z \to w(z) = z + \epsilon\xi(z), \tag{7.71}$$

the function $\xi(z)$ may be expanded in a Laurent series

$$\xi(z) = \sum_{n=-\infty}^{\infty} \xi_n z^{n+1}. \tag{7.72}$$

The infinitesimal conformal transformation (7.71) acts on an arbitrary function as

$$\delta_\epsilon f = f(z - \epsilon\xi(z)) - f(z) = \epsilon \frac{df}{d\epsilon}|_{\epsilon=0} = -\epsilon\xi(z)\frac{\partial f(z)}{\partial z} \equiv -i\epsilon G_\xi f(z) \tag{7.73}$$

where the generator is given by (the factor $-i$ being conventional)

$$G_\xi = -i\xi(z)\frac{\partial}{\partial z} = -i\sum_{n=-\infty}^{\infty} \xi_n z^{n+1}\frac{\partial}{\partial z} \equiv i\sum_{n=-\infty}^{\infty} \xi_n \ell_n. \tag{7.74}$$

This defines operators

$$\ell_n = -z^{n+1}\frac{\partial}{\partial z}, \tag{7.75}$$

which can be considered as a basis of the infinitesimal transformation (7.71). These operators satisfy the algebra

$$[\ell_m, \ell_n] = (m - n)\ell_{m+n}, \tag{7.76}$$

which can be easily verified by a direct calculation.

In the same way we can obtain operators $\bar{\ell}_n$ for an infinitesimal conformal transformation that is anti-holomorphic, $\bar{z} \to \bar{w}(\bar{z}) = \bar{z} + \epsilon\bar{\xi}(\bar{z})$. These operators obey the corresponding algebra, $\left[\bar{\ell}_m, \bar{\ell}_n\right] = (m-n)\bar{\ell}_{m+n}$. Moreover, the mixed commutator vanishes, $\left[\ell_m, \bar{\ell}_n\right] = 0$. This algebra is often called loop, or Witt, algebra.

We see from the consideration that leads to (7.75) that the number of independent infinitesimal conformal transformations is infinite. This observation, which is special to two dimensions, will have far-reaching consequences.

We note that in general the generators ℓ_n and $\bar{\ell}_n$ are not globally defined but exhibit singularities either at $z = 0$ or $w = 1/z = 0$. Only the generators $\ell_{-1}, \ell_0, \ell_1$, and $\bar{\ell}_{-1}, \bar{\ell}_0, \bar{\ell}_1$ are globally defined conformal transformations. It will be interesting to connect them to the global conformal transformations: translation, rotation, dilatation, and special conformal transformation.

Global translations are generated by ℓ_{-1} and $\bar{\ell}_{-1}$

$$\ell_{-1} = -\frac{\partial}{\partial z}, \qquad \bar{\ell}_{-1} = -\frac{\partial}{\partial \bar{z}} \tag{7.77}$$

as can be seen from considering, for instance,

$$\delta f = f(z-\epsilon) - f(z) = -\epsilon\frac{\partial}{\partial z}f(z) \equiv \epsilon\ell_{-1}f(z). \tag{7.78}$$

We may also work with the combinations

$$\ell_{-1} + \bar{\ell}_{-1} = -\frac{\partial}{\partial x}, \qquad i\left(\ell_{-1} - \bar{\ell}_{-1}\right) = -\frac{\partial}{\partial y}. \tag{7.79}$$

EXERCISE 7.14 Generators of rotation, dilatation, and special conformal transformation Show that ℓ_0 and $\bar{\ell}_0$ describe dilatations and rotations, while ℓ_1 and $\bar{\ell}_1$ describe special conformal transformations.

Hint: For the first part of the exercise it may be useful to consider the complex variable in polar representation, $z = \rho e^{i\theta}$.

We now direct our attention to the behaviour of order parameter fields and their correlation functions under conformal transformations.

7.5 Order parameter fields and correlation functions

Section 7.1 showed how correlation functions of order parameter fieldsare assumed to behave under conformal transformations. Accordingly, we introduce a terminology for order parameter fields that indeed transform under a conformal transformation in a way described in section 7.1. We call them *quasi-primary* order parameter fields, if, under a global conformal transformation in two dimensions,[1]

$$z \to w(z), \tag{7.80}$$

$$\bar{z} \to \bar{w}(\bar{z}), \tag{7.81}$$

they transform as

$$\phi_j(z,\bar{z}) \to \phi_j(w,\bar{w}) = \left(\frac{\partial w}{\partial z}\right)^{-h_j} \left(\frac{\partial \bar{w}}{\partial \bar{z}}\right)^{-\bar{h}_j} \phi_j(z,\bar{z}). \tag{7.82}$$

The exponents h_j and $\bar{h}_j$ are real numbers, called the conformal weights or conformal dimensions of the field ϕ_j. If this transformation behaviour is valid also for local conformal transformations, the corresponding fields are called *primary* fields. Fields that do not show this simple behaviour under local conformal transformations are called *secondary* fields. Secondary fields may or may not be quasi-primary.

Take again the example discussed in section 7.1: a pure rescaling $w(z) = \lambda z$ and $\bar{w}(\bar{z}) = \lambda \bar{z}$. This yields for a primary field ϕ_j

$$\phi_j(w,\bar{w}) = \lambda^{-(h_j+\bar{h}_j)} \phi_j(z,\bar{z}). \tag{7.83}$$

A rotation by a fixed angle θ, i.e. $w(z) = e^{i\theta} z$ and $\bar{w}(\bar{z}) = e^{-i\theta}\bar{z}$, leads to

$$\phi_j(w,\bar{w}) = e^{-i(h_j-\bar{h}_j)\theta} \phi_j(z,\bar{z}). \tag{7.84}$$

We can thus identify the conformal weights with the scaling dimensions x_j and a new quantity s_j, called, because of its connection with rotations, spin of the order parameter or scaling field ϕ_j, namely

$$x_j = h_j + \bar{h}_j, \qquad s_j = h_j - \bar{h}_j. \tag{7.85}$$

For an infinitesimal holomorphic conformal transformation (7.71) and its anti-holomorphic counterpart, we obtain from (7.82) for the variation of a quasi-primary or primary field

$$\begin{aligned} \delta_{\epsilon\bar{\epsilon}}\phi_j(z,\bar{z}) &= \phi_j(w,\bar{w}) - \phi_j(z,\bar{z}) = \epsilon \frac{d\phi_j(w,\bar{w})}{d\epsilon}|_{\epsilon=0} + \bar{\epsilon}\frac{d\phi_j(w,\bar{w})}{d\bar{\epsilon}}|_{\bar{\epsilon}=0} \\ &= -\phi_j(z,\bar{z})\left(\epsilon h_j \xi'(z) + \bar{\epsilon}\bar{h}_j\bar{\xi}'(\bar{z})\right) - \left(\epsilon\xi(z)\frac{\partial}{\partial z} + \bar{\epsilon}\bar{\xi}(z)\frac{\partial}{\partial \bar{z}}\right)\phi_j(z,\bar{z}). \end{aligned} \tag{7.86}$$

[1] We restrict our attention here to the two-dimensional case.

As a consequence of the definition (7.82) of quasi-primary fields, n-point correlation functions of quasi-primary fields satisfy

$$\langle \phi_1(z_1, \bar{z}_1)\phi_2(z_2, \bar{z}_2) \dots \phi_n(z_n, \bar{z}_n)\rangle = \left(\prod_i \left(\frac{\partial w_i}{\partial z_i}\right)^{h_i} \left(\frac{\partial \bar{w}_i}{\partial \bar{z}_i}\right)^{\bar{h}_i}\right) \times \langle \phi_1(w_1, \bar{w}_1)\phi_2(w_2, \bar{w}_2) \dots \phi_n(w_n, \bar{w}_n)\rangle, \quad (7.87)$$

where the expectation value is defined, as introduced in the last chapter, by the weight e^{-S} determined by the action $S\,[\{\phi_i\}]$ or Hamiltonian $\mathcal{H}$

$$\langle \dots \rangle = \frac{\int \prod_i \mathcal{D}[\{\phi_i\}]\,(\dots)e^{-S[\{\phi_i\}]}}{\int \prod_i \mathcal{D}[\{\phi_i\}]\,e^{-S[\{\phi_i\}]}} = \frac{1}{Z}\int \prod_i \mathcal{D}[\{\phi_i\}]\,(\dots)e^{-S[\{\phi_i\}]}. \quad (7.88)$$

Conformal invariance assumes that the action or Hamiltonian is invariant under a conformal transformation $z \to w(z)$ and $\bar{z} \to \bar{w}(\bar{z})$.

According to the assumption of conformal invariance (cp. section 7.1), the variation of an n-point correlation function of quasi-primary or primary fields vanishes

$$\delta_{\epsilon\bar{\epsilon}}\langle \phi_1(z_1, \bar{z}_1)\phi_2(z_2, \bar{z}_2) \dots \phi_n(z_n, \bar{z}_n)\rangle = 0. \quad (7.89)$$

This assumption, using (7.86), leads to a differential equation for the n-point correlation function

$$\sum_{i=1}^{n} \left(\epsilon\xi(z_i)\frac{\partial}{\partial z_i} + \epsilon h_i \frac{\partial \xi(z_i)}{\partial z_i} + \bar{\epsilon}\bar{\xi}(\bar{z}_i)\frac{\partial}{\partial \bar{z}_i} + \bar{\epsilon}\bar{h}_i \frac{\partial \bar{\xi}(\bar{z}_i)}{\partial \bar{z}_i}\right) \times \langle \phi_1(z_1, \bar{z}_1)\phi_2(z_2, \bar{z}_2) \dots \phi_n(z_n, \bar{z}_n)\rangle = 0. \quad (7.90)$$

This differential equation can now be used to determine various correlation functions by exploiting them for different forms of the functions $\xi(z)$ and $\bar{\xi}(\bar{z})$ in the infinitesimal holomorphic conformal transformations (7.71) and its anti-holomorphic counterpart.

Let us start by considering the two-point correlation function $\langle \phi_1(z_1, \bar{z}_1)\phi_2(z_2, \bar{z}_2)\rangle$. As discussed, e.g. a translation is generated by the elements ℓ_{-1} and $\bar{\ell}_{-1}$ and corresponds to $\xi(z) = \bar{\xi}(\bar{z}) = 1$. The corresponding differential equation (7.90) separates into the two equations

$$\left(\frac{\partial}{\partial z_1} + \frac{\partial}{\partial z_2}\right)\langle \phi_1(z_1, \bar{z}_1)\phi_2(z_2, \bar{z}_2)\rangle = 0, \quad (7.91)$$

and

$$\left(\frac{\partial}{\partial \bar{z}_1} + \frac{\partial}{\partial \bar{z}_2}\right)\langle \phi_1(z_1, \bar{z}_1)\phi_2(z_2, \bar{z}_2)\rangle = 0, \quad (7.92)$$

which imply that $\langle\phi_1(z_1,\bar{z}_1)\phi_2(z_2,\bar{z}_2)\rangle$ does not depend on the centre-of-mass coordinates z_1+z_2 and $\bar{z}_1+\bar{z}_2$, but rather only on the relative coordinates z_1-z_2 and $\bar{z}_1-\bar{z}_2$, which is nothing more than a consequence of translation invariance.

Next, for rotations and dilatations, we have $\xi(z)=z$ and $\bar{\xi}(\bar{z})=\bar{z}$, and hence (7.90) separates into the two equations

$$\left(z_1\frac{\partial}{\partial z_1}+h_1+z_2\frac{\partial}{\partial z_2}+h_2\right)\langle\phi_1(z_1,\bar{z}_1)\phi_2(z_2,\bar{z}_2)\rangle=0, \tag{7.93}$$

$$\left(\bar{z}_1\frac{\partial}{\partial \bar{z}_1}+\bar{h}_1+\bar{z}_2\frac{\partial}{\partial \bar{z}_2}+\bar{h}_2\right)\langle\phi_1(z_1,\bar{z}_1)\phi_2(z_2,\bar{z}_2)\rangle=0. \tag{7.94}$$

These differential equations can be used to determine the two-point correlation functions to be

$$\langle\phi_1(z_1,\bar{z}_1)\phi_2(z_2,\bar{z}_2)\rangle=(z_1-z_2)^{-h_1-h_2}(\bar{z}_1-\bar{z}_2)^{-\bar{h}_1-\bar{h}_2}, \tag{7.95}$$

which can be verified by insertion into the differential equations. However, it can also be explicitly demonstrated in a more constructive way only using translation and scale invariance directly.

Translation invariance (and this also follows from (7.91) and (7.92)) combined with scale invariance means that

$$\begin{aligned}\langle\phi_1(z_1,\bar{z}_1)\phi_2(z_2,\bar{z}_2)\rangle &= G(z_1-z_2,\bar{z}_1-\bar{z}_2)\\ &= \lambda^{h_1+h_2}\bar{\lambda}^{\bar{h}_1+\bar{h}_2}\langle\phi_1(\lambda z_1,\bar{\lambda}\bar{z}_1)\phi_2(\lambda z_2,\bar{\lambda}\bar{z}_2)\rangle\\ &= \lambda^{h_1+h_2}\bar{\lambda}^{\bar{h}_1+\bar{h}_2}G\big(\lambda(z_1-z_2),\bar{\lambda}(\bar{z}_1-\bar{z}_2)\big).\end{aligned} \tag{7.96}$$

Choosing $\lambda=(z_1-z_2)^{-1}$ and $\bar{\lambda}=(\bar{z}_1-\bar{z}_2)^{-1}$, this produces (7.95) where the constant $G(1,1)$ has been chosen to be unity.

EXERCISE 7.15 Conformal weights Starting from (7.95) and using the special conformal transformations

$$z\to w(z)=\frac{z}{1-\epsilon z},\qquad \bar{z}\to\bar{w}(\bar{z})=\frac{\bar{z}}{1-\bar{\epsilon}\bar{z}} \tag{7.97}$$

show that the conformal weights must be equal $h_1=h_2$ and $\bar{h}_1=\bar{h}_2$ for any non-vanishing conformally invariant two-point correlation function.

The result of this exercise can also be confirmed by inserting (7.95) into (7.90) for the functions $\xi(z)=z^2$ and $\bar{\xi}(\bar{z})=\bar{z}^2$ of the infinitesimal special conformal transformation (cp. exercise 7.14).

7.6 Energy-momentum tensor

One of the most useful objects in the discussion of conformal transformations and conformally invariant correlation functions of order parameter fields is the energy-momentum tensor, which we introduce in this section and employ extensively in what is to follow. This quantity, the energy-momentum tensor, incorporates many of the essential features of conformally invariant theories.

Let us go back for a moment to the general case of a Minkowskian space-time $\mathbb{R}^{p,q}$ $(p + q = d)$. The dynamics of a scalar field $\phi(\mathbf{r})$ is determined by the action

$$S[\phi] = \int d^d x \, \mathcal{L}\left(\phi, \frac{\partial \phi}{\partial r^\mu}\right) \tag{7.98}$$

with the Lagrangian density $\mathcal{L}(\phi, \partial\phi/\partial r^\mu)$ which satisfies the Euler–Lagrange equations

$$\frac{\partial \mathcal{L}}{\partial \phi} - \sum_\mu \frac{\partial}{\partial r^\mu}\left(\frac{\partial \mathcal{L}}{\partial\left(\frac{\partial \phi}{\partial r^\mu}\right)}\right) = 0. \tag{7.99}$$

Now consider a conformal transformation of the coordinates $\mathbf{r} \to \tilde{\mathbf{r}} = \mathbf{f}(\mathbf{r})$ and assume for simplicity that the conformal weights of the scalar field ϕ vanish, i.e.

$$\phi(\mathbf{r}) \to \psi(\tilde{\mathbf{r}}) = \phi(\mathbf{r}). \tag{7.100}$$

The field theory will then be invariant under this conformal transformation if

$$\begin{aligned}\mathcal{L}\left(\phi(\mathbf{r}), \frac{\partial \phi(\mathbf{r})}{\partial r^\mu}\right) d^d r &= \mathcal{L}\left(\psi(\tilde{\mathbf{r}}), \frac{\partial \psi(\tilde{\mathbf{r}})}{\partial \tilde{r}^\mu}\right) d^d \tilde{r} \\ &= \mathcal{L}\left(\phi(\mathbf{r}), \sum_\nu \frac{\partial \phi(\mathbf{r})}{\partial r^\nu}\frac{\partial r^\nu}{\partial \tilde{r}^\mu}\right) \det\left|\frac{\partial \tilde{r}^\rho}{\partial r^\sigma}\right| d^d r,\end{aligned} \tag{7.101}$$

which implies $S[\phi] = S[\psi]$. For an infinitesimal transformation

$$r^\mu \to \tilde{r}^\mu = f^\mu(\mathbf{r}) = r^\mu + \epsilon \xi^\mu(\mathbf{r}) \tag{7.102}$$

we obtain

$$\frac{\partial \tilde{r}^\rho}{\partial r^\sigma} = \delta^\rho_\sigma + \epsilon \frac{\partial \xi^\rho(\mathbf{r})}{\partial r^\sigma} \tag{7.103}$$

and

$$\det \left| \frac{\partial \tilde{r}^\rho}{\partial r^\sigma} \right| = 1 + \epsilon \sum_\rho \frac{\partial \xi^\rho(\mathbf{r})}{\partial r^\rho}. \tag{7.104}$$

Finally,

$$\frac{\partial \phi(\mathbf{r})}{\partial r^\nu} \frac{\partial r^\nu}{\partial \tilde{r}^\mu} = \frac{\partial \phi(\mathbf{r})}{\partial r^\mu} - \epsilon \frac{\partial \phi(\mathbf{r})}{\partial r^\nu} \frac{\partial \xi^\nu}{\partial r^\mu}. \tag{7.105}$$

The vanishing difference between the left-hand side and the right-hand side of (7.101) can then be expressed as

$$\delta \mathcal{L} = \epsilon \sum_{\mu\nu} \frac{\partial}{\partial r^\mu} \left\{ \left(\eta^{\mu\nu} \mathcal{L} - \frac{\partial \phi}{\partial r_\nu} \frac{\partial \mathcal{L}}{\partial \left(\frac{\partial \phi}{\partial r^\mu} \right)} \right) \xi_\nu \right\} = 0. \tag{7.106}$$

EXERCISE 7.16 Variation of the Lagrangian density Derive the expression for the variation of the Lagrangian density (7.106).

The variation of the Lagrangian density (7.106) suggests the introduction of a tensor, called the energy-momentum or stress tensor $T^{\mu\nu}(\mathbf{r})$

$$T^{\mu\nu}(\mathbf{r}) = \frac{\partial \phi}{\partial r_\nu} \frac{\partial \mathcal{L}}{\partial \left(\frac{\partial \phi}{\partial r^\mu} \right)} - \eta^{\mu\nu} \mathcal{L}, \tag{7.107}$$

which, according to (7.106), satisfies a continuity equation

$$\sum_\mu \frac{\partial \mathcal{J}^\mu}{\partial r^\mu} = 0, \quad \text{with} \quad \mathcal{J}^\mu = \sum_\nu T^{\mu\nu} \xi_\nu. \tag{7.108}$$

This is the Noether theorem, which states that every continuous symmetry of the Langrangian density results in a conserved current $\mathcal{J}^\mu(\mathbf{r})$.

A conserved current is always associated with a conserved charge. In order to appreciate this, let us allow the case $p = 1$ and $q = 3$, where now $r^1 = x^0 = t$ and $(r^2, r^3, r^4)^T = (x^1, x^2, x^3)^T$ The conserved quantity is

$$Q = \int d^3x \, \mathcal{J}^0 \tag{7.109}$$

where the integral is over a volume-like 'surface' at constant time t. This can be seen from (7.108), which implies

$$\int d^3r \sum_\mu \frac{\partial \mathcal{J}^\mu}{\partial r^\mu} = \int d^3x \left(\frac{\partial \mathcal{J}^0}{\partial t} + \sum_{i=1}^{3} \frac{\partial \mathcal{J}^i}{\partial r^i} \right) = 0. \tag{7.110}$$

By the divergence theorem, the second part of the integral vanishes and we obtain

$$\int d^3r \frac{\partial \mathcal{J}^0}{\partial t} = \frac{d}{dt} \int d^3r \, \mathcal{J}^0 = \frac{dQ}{dt} = 0. \tag{7.111}$$

Before we focus the discussion of the energy-momentum tensor to its significance for conformally invariant systems in two dimensions, let us discuss a number of its salient features in general, which can be derived by considering special transformations.

For space-time translations, where $\xi_\nu(\mathbf{r}) = a_\nu$ is a constant vector, we obtain one conserved current for every component a_ν

$$\sum_\mu \frac{\partial T^{\mu\nu}}{\partial r^\mu} = 0, \qquad \mathcal{J}^\mu = T^{\mu\nu}, \qquad \nu = 1, 2, \ldots, d, \tag{7.112}$$

i.e. the energy-momentum tensor of a translationally invariant Lagrangian density is itself the conserved current.

Again, in four-dimensional space-time, a conserved charge is associated with each conserved current. Explicitly, these conserved charges are the total energy

$$H = \int d^3x \, T^{00} \tag{7.113}$$

and the total momentum

$$P^i = \int d^3x \, T^{0i}. \tag{7.114}$$

EXERCISE 7.17 Further properties of the energy-momentum tensor Show that the energy-momentum tensor of a conformally invariant Lagrangian density is symmetric

$$T^{\mu\nu} = T^{\nu\mu} \tag{7.115}$$

and its trace vanishes

$$\sum_\mu T^\mu{}_\mu = 0 \tag{7.116}$$

The former property follows from rotation invariance, while the latter follows from general conformal invariance using (7.39).

7.7 Energy-momentum tensor in two dimensions

The results of the previous section for the energy-momentum tensor of an arbitrary Minkowskian space-time $\mathbb{R}^{p,q}$ translate directly to two Euclidian dimensions $\mathbb{R}^{0,2}$.

Translation invariance implies (cp. (7.112))

$$\frac{\partial T^{00}}{\partial r^0} + \frac{\partial T^{10}}{\partial r^1} = 0, \qquad \frac{\partial T^{01}}{\partial r^0} + \frac{\partial T^{11}}{\partial r^1} = 0, \tag{7.117}$$

while tracelessness and symmetry of the energy-momentum tensor become

$$T^{00} = -T^{11}, \qquad T^{01} = T^{10}, \tag{7.118}$$

i.e. we have only two independent components of the energy-momentum tensor. Taking the latter relations into account, the former become

$$\frac{\partial T^{00}}{\partial r^0} + \frac{\partial T^{10}}{\partial r^1} = 0, \qquad \frac{\partial T^{00}}{\partial r^1} - \frac{\partial T^{10}}{\partial r^0} = 0. \tag{7.119}$$

The energy-momentum tensor in complex coordinates becomes, according to the general transformation relation

$$T^{\mu\nu} = \sum_{\alpha\beta} \frac{\partial r_\alpha}{\partial r_\mu} \frac{\partial r_\beta}{\partial r_\nu} T^{\alpha\beta} \tag{7.120}$$

with $z = r^0 + ir^1$ and $\bar{z} = r^0 - ir^1$

$$T_{zz} = \frac{1}{4}\left(T^{00} - iT^{01} - iT^{10} - T^{11}\right) = \frac{1}{2}\left(T^{00} - iT^{10}\right), \tag{7.121}$$

$$T_{\bar{z}\bar{z}} = \frac{1}{4}\left(T^{00} + iT^{01} + iT^{10} - T^{11}\right) = \frac{1}{2}\left(T^{00} + iT^{10}\right), \tag{7.122}$$

$$T_{z\bar{z}} = T_{\bar{z}z} = \frac{1}{4}\left(T^{00} + T^{11}\right) = 0, \tag{7.123}$$

where we have used (7.118).

The two relations (7.119) can be interpreted as the Cauchy–Riemann equations of the two different non-vanishing combinations (7.121) and (7.122) of components of the energy-momentum tensor

$$T_{zz} = T(z, \bar{z}) = \frac{1}{2}\left(T^{00} - iT^{10}\right), \qquad T_{\bar{z}\bar{z}} = \bar{T}(z, \bar{z}) = \frac{1}{2}\left(T^{00} + iT^{10}\right), \tag{7.124}$$

the first of which depends only on $z = r^0 + ir^1$, the second only on $\bar{z} = r^0 - ir^1$, i.e.

$$\frac{\partial T}{\partial \bar{z}} = 0 \quad \Rightarrow \quad T = T(z), \qquad \frac{\partial \bar{T}}{\partial z} = 0 \quad \Rightarrow \quad \bar{T} = \bar{T}(\bar{z}). \tag{7.125}$$

Fields, like the energy-momentum tensor, which depend only on one variable, either z or $\bar{z}$, are called chiral fields. The energy-momentum tensor will play a decisive role in the understanding of the critical properties of two-dimensional statistical systems based on their conformal symmetry.

For use in the next section, we note the components of the metric in complex coordinates. Using the metric (7.59), we obtain $ds^2 = \sum_{\mu\nu} g_{\mu\nu} dx^\mu dx^\nu = (dx)^2 + \left(dy^2\right) = dz d\bar{z}$ in the complex plane and thus

$$g_{zz} = g_{\bar{z}\bar{z}} = 0, \qquad g_{z\bar{z}} = g_{\bar{z}z} = \frac{1}{2}. \tag{7.126}$$

Furthermore, the total energy and momentum, (7.113) and (7.114), become in two dimensions

$$H = \int dr^1 \, T^{00} \tag{7.127}$$

and

$$P = \int dr^1 \, T^{10}. \tag{7.128}$$

7.8 Conformal Ward identities

In section 7.6, when we derived the energy-momentum tensor, we assumed that the fields ϕ_j do not change under a conformal transformation, which was tantamount to the assumption that the conformal weights vanish. Let us now relinquish this assumption but immediately switch our focus to the two-dimensional case. We write a general conformal transformation as

$$(z, \bar{z}) \to (\tilde{z}, \bar{\tilde{z}}) = (w(z), \bar{w}(\bar{z})) \tag{7.129}$$

such that a primary field becomes

$$\phi_j(z, \bar{z}) \to \tilde{\phi}_j(\tilde{z}, \bar{\tilde{z}}) = \phi_j(z, \bar{z}) + \delta\phi_j(z, \bar{z}). \tag{7.130}$$

An n-point correlation function, defined in (7.88), can then be written as

$$\langle\tilde{\phi}_1(\tilde{z}_1,\bar{\tilde{z}}_1)\ldots\tilde{\phi}_n(\tilde{z}_n,\bar{\tilde{z}}_n)\rangle = \frac{1}{Z}\int\tilde{\mathcal{D}}\left[\{\tilde{\phi}_i\}\right]\tilde{\phi}_1(\tilde{z}_1,\bar{\tilde{z}}_1)\ldots\tilde{\phi}_n(\tilde{z}_n,\bar{\tilde{z}}_n)e^{-\tilde{S}\left[\{\tilde{\phi}_i\}\right]}. \quad (7.131)$$

We now assume that the measure of this functional integral is invariant under conformal transformations, $\tilde{\mathcal{D}}\left[\{\tilde{\phi}_i\}\right] = \mathcal{D}\left[\{\phi_i\}\right]$. This assumption has to be included in the assumptions made for conformally invariant field theories. It can only be corroborated for specific examples. However, the action itself does change

$$\tilde{S}\left[\{\tilde{\phi}_i\}\right] = S\left[\{\phi_i\}\right] + \delta S\left[\{\phi_i\}\right]. \quad (7.132)$$

Now, we use the central tenet of conformal field theory that the correlation functions are invariant under conformal transformations. Using (7.130), the n-point correlation function becomes formally

$$\begin{aligned}\langle\tilde{\phi}_1(\tilde{z}_1,\bar{\tilde{z}}_1)\ldots\tilde{\phi}_n(\tilde{z}_n,\bar{\tilde{z}}_n)\rangle_{\left(\tilde{S}\left[\{\tilde{\phi}_i\}\right]\right)} &= \\ \langle(\phi_1(z_1,\bar{z}_1)+\delta\phi_1(z_1,\bar{z}_1))\ldots(\phi_n(z_n,\bar{z}_n)+\delta\phi_n(z_n,\bar{z}_n))\rangle_{(S[\{\phi_i\}]+\delta S[\{\phi_i\}])}&. \end{aligned} \quad (7.133)$$

Using the abbreviation $\Phi = \phi_1(z_1,\bar{z}_1)\ldots\phi_n(z_n,\bar{z}_n)$, the right–hand side of this equation can be evaluated to lowest order

$$\begin{aligned}\langle(\phi_1(z_1,\bar{z}_1)+\delta\phi_1(z_1,\bar{z}_1))\ldots(\phi_n(z_n,\bar{z}_n)+\delta\phi_n(z_n,\bar{z}_n))\rangle_{(S[\{\phi_i\}]+\delta S[\{\phi_i\}])} &= \\ \frac{1}{Z}\int\mathcal{D}\left[\{\phi_i\}\right]\delta\Phi e^{-S[\{\phi_i\}]} - \frac{1}{Z}\int\mathcal{D}\left[\{\phi_i\}\right]\Phi\delta S\left[\{\phi_i\}\right]e^{-S[\{\phi_i\}]}&\end{aligned} \quad (7.134)$$

where

$$\delta\Phi = \sum_{i=1}^{n}\phi_1(z_1,\bar{z}_1)\ldots(\phi_i(z_i,\bar{z}_i)+\delta\phi_i(z_i,\bar{z}_i))\ldots\phi_n(z_n,\bar{z}_n). \quad (7.135)$$

Thus, we need to have

$$\delta\langle\Phi\rangle \equiv \langle\delta\Phi\rangle = \langle\delta S\Phi\rangle \quad (7.136)$$

in order for the n-point correlation function to remain invariant under the conformal transformation. The right–hand side of this equation is given by (7.90) together with (7.86), using the explicit form of the infinitesimal holomorphic conformal transformation (7.90) and its anti-holomorphic counterpart

$$\delta_{\epsilon\bar{\epsilon}}\langle\Phi\rangle = -\sum_{i=1}^{n}\left(\epsilon\xi(z_i)\frac{\partial}{\partial z_i} + \epsilon h_i\frac{\partial\xi}{\partial z_i} + \bar{\epsilon}\bar{\xi}(\bar{z}_i)\frac{\partial}{\partial\bar{z}_i} + \bar{\epsilon}\bar{h}_i\frac{\partial\bar{\xi}}{\partial\bar{z}_i}\right)\langle\Phi\rangle. \qquad (7.137)$$

The left-hand side can be related to the energy-momentum tensor using (7.98), (7.106), and (7.107)[2]

$$\delta S\left[\{\phi_i\}\right] = -\frac{\epsilon}{2\pi}\int d^2r\,\delta\mathcal{L} = -\frac{\epsilon}{2\pi}\int d^2r\sum_{\mu\nu}\frac{\partial}{\partial r^\mu}\left(T^{\mu\nu}\xi_\nu\right). \qquad (7.138)$$

This integral can be rewritten with the help of the divergence theorem in two dimensions

$$\int d^2r\sum_{\mu\nu}\frac{\partial}{\partial r^\mu}\left(T^{\mu\nu}\xi_\nu\right) = \sum_\nu\oint\left(dr^0\,\xi_\nu T^{1\nu} - dr^1\,\xi_\nu T^{0\nu}\right). \qquad (7.139)$$

Introducing the complex variables and functions

$$z = r^0 + ir^1, \quad \bar{z} = r^0 - ir^1, \quad \xi(z) = \xi_0 + i\xi_1, \quad \text{and } \bar{\xi}(\bar{z}) = \xi_0 - i\xi_1, \qquad (7.140)$$

the integrand can be rewritten as

$$\sum_\nu\left(dr^0\,\xi_\nu T^{1\nu} - dr^1\,\xi_\nu T^{0\nu}\right) = idz\,\xi(z)T(z) - id\bar{z}\,\bar{\xi}(\bar{z})\bar{T}(\bar{z}) \qquad (7.141)$$

where we used the symmetry and tracelessness of the components $T^{\mu\nu}$ (see (7.118)) as well as $T(z)$ and $\bar{T}(\bar{z})$ given in (7.124) and (7.125).

A general conformal transformation (7.129), if it is not a global conformal transformation represented by the Möbius transformations, and if it is not a trivial constant, will only be conformal locally, not globally. If we assume the infinitesimal conformal transformation employed here to be locally conformal in a region U containing all points $(z_i, \bar{z}_i)$ of the n-point correlation function, then δS in (7.138) will vanish in this region, but generally not in the complementary region $\bar{U}$ outside of it. It is in this region $\bar{U}$ where (7.138) has a non-vanishing contribution.

[2] The factor $-1/2\pi$ has been introduced here for later convenience.

With this interpretation of (7.138), we finally obtain

$$
\begin{aligned}
&\delta_{\epsilon\bar{\epsilon}} \langle \phi_1(z_1, \bar{z}_1) \ldots \phi_n(z_n, \bar{z}_n) \rangle = \\
-&\sum_{i=1}^{n} \left(\epsilon\xi(z_i) \frac{\partial}{\partial z_i} + \epsilon h_i \frac{\partial \xi}{\partial z_i} + \bar{\epsilon}\bar{\xi}(\bar{z}_i) \frac{\partial}{\partial \bar{z}_i} + \bar{\epsilon}\bar{h}_i \frac{\partial \bar{\xi}}{\partial \bar{z}_i} \right) \langle \phi_1(z_1, \bar{z}_1) \ldots \phi_n(z_n, \bar{z}_n) \rangle = \\
&\frac{1}{2\pi i} \int_C dz\, \epsilon\xi(z) \langle T(z) \phi_1(z_1, \bar{z}_1) \ldots \phi_n(z_n, \bar{z}_n) \rangle \\
-&\frac{1}{2\pi i} \int_C d\bar{z}\, \bar{\epsilon}\bar{\xi}(\bar{z}) \langle \bar{T}(\bar{z}) \phi_1(z_1, \bar{z}_1) \ldots \phi_n(z_n, \bar{z}_n) \rangle
\end{aligned}
\tag{7.142}
$$

where the curve C is the boundary curve between the regions U and $\bar{U}$, i.e. it encloses all the points $(z_i, \bar{z}_i)$. These relations are called the conformal Ward identities. In the case of global conformal transformations the right-hand side vanishes and (7.142) reduces to (7.90).

Comparing the conformal Ward identities (7.142) with (7.73) and (7.74) suggests the interpretation of the energy-momentum tensor as the generator of infinitesimal transformations on correlation functions of order parameter fields or, in an abstract notation, on order parameter fields directly

$$
\begin{aligned}
\delta_{\epsilon\bar{\epsilon}} \left(\phi_1(z_1, \bar{z}_1) \ldots \phi_n(z_n, \bar{z}_n) \right) = &\frac{1}{2\pi i} \int_C dz\, \epsilon\xi(z) T(z) \phi_1(z_1, \bar{z}_1) \ldots \phi_n(z_n, \bar{z}_n) \\
-&\frac{1}{2\pi i} \int_C d\bar{z}\, \bar{\epsilon}\bar{\xi}(\bar{z}) \bar{T}(\bar{z}) \phi_1(z_1, \bar{z}_1) \ldots \phi_n(z_n, \bar{z}_n).
\end{aligned}
\tag{7.143}
$$

We observe that the conformal Ward identities (7.142) separate into a holomorphic and an anti-holomorphic part. Restricting ourselves to the holomorphic part and using the residue theorem, we can write the left–hand side of the conformal Ward identities (7.142) as

$$
\begin{aligned}
&-\sum_{i=1}^{n} \left(\epsilon\xi(z_i) \frac{\partial}{\partial z_i} + \epsilon h_i \frac{\partial \xi}{\partial z_i} \right) \langle \phi_1(z_1, \bar{z}_1) \ldots \phi_n(z_n, \bar{z}_n) \rangle = \\
-&\frac{1}{2\pi i} \sum_{i=1}^{n} \oint_{C_i} dz\, \epsilon\xi(z) \left(\frac{h_i}{(z - z_i)^2} + \frac{1}{z - z_i} \frac{\partial}{\partial z_i} \right) \langle \phi_1(z_1, \bar{z}_1) \ldots \phi_n(z_n, \bar{z}_n) \rangle
\end{aligned}
\tag{7.144}
$$

where the integration contours C_i are around the points z_i wholly contained in the region U. However, $\xi(z)$ is holomorphic in all of the region U, so that we can finally write

$$
\oint_C dz\, \epsilon\xi(z) \left\langle \left[T(z) + \sum_{i=1}^{n} \left(\frac{h_i}{(z - z_i)^2} + \frac{1}{z - z_i} \frac{\partial}{\partial z_i} \right) \right] \phi_1(z_1, \bar{z}_1) \ldots \phi_n(z_n, \bar{z}_n) \right\rangle = 0
\tag{7.145}
$$

where now the contour C encloses the whole of U. Since this equation must hold for all holomorphic functions $\xi(z)$ bounded by C, we can conclude that the part in the expectation brackets $\langle \ldots \rangle$ is also a holomorphic function inside of C. This can only be the case if $\langle T(z)\phi_1(z_1,\bar{z}_1)\ldots\phi_n(z_n,\bar{z}_n)\rangle$ in (7.145) is a meromorphic function, i.e. a holomorphic function except for isolated singularities. Moreover, these singularities must cancel the singularities in the other part of the expression (7.145). Thus, we can identify

$$\langle T(z)\phi_1(z_1,\bar{z}_1)\ldots\phi_n(z_n,\bar{z}_n)\rangle =$$
$$-\sum_{i=1}^{n}\left(\frac{h_i}{(z-z_i)^2}+\frac{1}{z-z_i}\frac{\partial}{\partial z_i}\right)\langle\phi_1(z_1,\bar{z}_1)\ldots\phi_n(z_n,\bar{z}_n)\rangle. \tag{7.146}$$

This expression also bears the name conformal Ward identities. It is valid for an arbitrary product of primary fields in the n-point correlation function. This statement is also true for the conformal Ward identities in the form (7.142), the latter holding also for arbitrary $\epsilon\xi$ and $\bar{\epsilon}\bar{\xi}$. Thus, we can write the conformal Ward identities as the formal expression

$$T(w)\phi_j(z,\bar{z}) = \left(\frac{h_i}{(w-z)^2}+\frac{1}{w-z}\frac{\partial}{\partial z}\right)\phi_j(z,\bar{z}) + R(w), \tag{7.147}$$

where $R(w)$ represents a holomorphic function defined inside C, which is, in particular, regular at the point $w \to z$ and does therefore not contribute to the integral in (7.142). This formal expression is an example of a so-called operator product expansion (OPE). Its meaning is given by (7.146), i.e. by inserting the operator expression into an appropriate correlation function.

7.8.1 Operator product expansion

In general, an operator product expansion describes the behaviour of a product of local operators $\mathcal{O}_i(z,\bar{z})$ and $\mathcal{O}_j(w,\bar{w})$ if the arguments of the operators come close $w \to z$ and $\bar{w} \to \bar{z}$. It expresses the product formally in terms of all other operators of the theory

$$\mathcal{O}_i(z,\bar{z})\mathcal{O}_j(w,\bar{w}) = \sum_k C_{ij}^k(z-w,\bar{z}-\bar{w})\mathcal{O}_k(w,\bar{w}). \tag{7.148}$$

Due to translational invariance, the coefficient functions $C_{ij}^k(z-w,\bar{z}-\bar{w})$ depend only on the distance between the operators. In dealing with operator product expansions, it is important to remember that they are always to be interpreted as relations *within* correlation functions

$$\langle\mathcal{O}_i(z,\bar{z})\mathcal{O}_j(w,\bar{w})\ldots\rangle = \sum_k C_{ij}^k(z-w,\bar{z}-\bar{w})\langle\mathcal{O}_k(w,\bar{w})\ldots\rangle \tag{7.149}$$

where the ... stand for an arbitrary string of operators.

We also need to emphasize that we have now slipped into the terminology of conformal quantum field theory. While the objects most relevant to us are order parameter fields ϕ_j, these objects are quantum field operators ϕ_j, their derivatives $\partial^n\phi$, or even more complicated composite operators, such as $e^{i\phi}$, in the conformal *quantum* field theory context.

Moreover, the string of operators in (7.149) is assumed to have some kind of ordering, e.g. time ordering, if this is appropriate for the problem at hand. Because of this assumed ordering, the OPE commutes: $\mathcal{O}_i(z,\bar{z})\mathcal{O}_j(w,\bar{w}) = \mathcal{O}_j(w,\bar{w})\mathcal{O}_i(z,\bar{z})$, except for possible signs in case of Fermi operators. In all expressions containing strings of operators we have encountered so far, such an ordering procedure is assumed.

The only requirement for the otherwise arbitrary operators in (7.149) is that they are far enough away from $(z,\bar{z})$ and $(w,\bar{w})$. This requirement also determines the radius of convergence within which the OPEs are exact. This radius is given by the distance from $(z,\bar{z})$ and $(w,\bar{w})$ of the closest other operator in (7.149).

The following explores the singular behaviour as $z \to w$ and $\bar{z} \to \bar{w}$ of OPEs, and how it determines how operators, that is in our context order parameter fields, and their correlation functions transform under conformal symmetry.

OPEs of the energy-momentum tensor of primary operators have the special form

$$T(z)\mathcal{O}(w,\bar{w}) = h\frac{\mathcal{O}(w,\bar{w})}{(z-w)^2} + \frac{\partial_w\mathcal{O}(w,\bar{w})}{(z-w)} + R, \tag{7.150}$$

$$\bar{T}(\bar{z})\mathcal{O}(w,\bar{w}) = \bar{h}\frac{\mathcal{O}(w,\bar{w})}{(\bar{z}-\bar{w})^2} + \frac{\partial_{\bar{w}}\mathcal{O}(w,\bar{w})}{(\bar{z}-\bar{w})} + \bar{R}, \tag{7.151}$$

i.e. there are no terms with stronger singularities. R and $\bar{R}$ represent non-singular terms.

EXERCISE 7.18 Primary operators Show that (7.150) and (7.151) can serve as definitions of primary operators equivalent to (7.82).

7.9 Energy-momentum tensor and the Virasoro algebra

This section explains the behaviour under infinitesimal conformal transformations of the energy-momentum tensor, for which we shall need the operator product expansion of the energy-momentum tensor with itself. We shall see that the energy-momentum tensor is not a primary (chiral) field. Its transformation properties are described by the Virasoro (Lie) algebra, which is the central extension of the Witt (or loop) algebra (7.76).

First, we give some heuristic arguments for the form of the operator product expansion of the energy-momentum tensor with itself, which are supported by direct calculations of simple model systems.

7.9.1 Heuristic argument

The scale invariance or, more generally, conformal invariance of the action S can be used in an argument for the scaling dimension of the energy-momentum tensor. From the relation of the action S and the Lagrangian density, expressed in (7.98), and the variation of the Lagrangian density (7.106) that defines the energy-momentum tensor via (7.107), we can infer that in two dimensions the holomorphic energy-momentum tensor should have conformal weights $(h, \bar{h}) = (2, 0)$ and, hence, scaling dimension $x = 2$ and spin $s = 2$. The anti-holomorphic energy-momentum tensor then has conformal weights $(h, \bar{h}) = (0, 2)$, scaling dimension $x = 2$, and spin $s = -2$. These heuristic considerations suggest that the operator product expansion of the energy-momentum tensor with itself, $T(z)T(w)$ is an object with scaling dimension $x = 4$. It therefore contains a piece (cp. (7.150))

$$T(z)T(w) = \ldots + \frac{T(w)}{(z-w)^2} + \frac{\partial_w T(w)}{z-w} + \ldots. \tag{7.152}$$

Therefore, according to this reasoning, every other term which can appear on the right–hand side must have scaling dimension $x = 4$, i.e. be of the form

$$\frac{\mathcal{O}_m}{(z-w)^n} \tag{7.153}$$

with $m + n = 4$. If we exclude the possibility of negative conformal weights, as shown for unitary conformal field theories (see section 7.9.9.5), then there are two possibilities left: $n = 4$, which implies a term multiplied by a constant, and $n = 3$, a term we can exclude as it violates the symmetry $z \leftrightarrow w$.

Thus, we expect the operator product expansion of the energy-momentum tensor with itself to take the form

$$T(z)T(w) = \frac{c/2}{(z-w)^4} + \frac{2T(z)}{(z-w)^2} + \frac{\partial_z T(z)}{(z-w)} + \ldots, \tag{7.154}$$

$$\bar{T}(\bar{z})\bar{T}(\bar{w}) = \frac{\bar{c}/2}{(\bar{z}-\bar{w})^4} + \frac{2\bar{T}(\bar{z})}{(\bar{z}-\bar{w})^2} + \frac{\partial_{\bar{z}} \bar{T}(\bar{z})}{(\bar{z}-\bar{w})} + \ldots. \tag{7.155}$$

The mixed product $T(z)\bar{T}(\bar{w})$ has a non-singular expansion. This form of the operator product expansion indicates that the energy-momentum tensor is not a primary field. In order to have the correct scaling dimension, the amplitude c, called the central charge or conformal anomaly, of the most singular term must be a number. Yielding particular values for c, this form is also obtained for particular free field models, e.g. for the free massless Boson (see §5.3 of Di Francesco *et al.*, 1997 for some detailed calculations of this and other field theories) one obtains $c = 1$ (to obtain this simple number for the massless Boson case is the reason for choosing $c/2$ as amplitude).

We remark here that the term $\partial_w T(w)/(z-w)$ does *not* violate the $z \leftrightarrow w$–symmetry, as can be easily demonstrated. Exchanging the arguments in (7.154)

$$T(w)T(z) = \ldots + \frac{T(z)}{(z-w)^2} + \frac{\partial_z T(z)}{w-z} + \ldots \tag{7.156}$$

and expanding $T(z)$ around $z = w$

$$T(z) = T(w) + \partial_w T(w)(z - w) + \ldots, \qquad \partial_z T(z) = \partial_w T(w) + \ldots \tag{7.157}$$

we obtain

$$T(w)T(z) = \ldots + \frac{2T(w) + 2\partial_w T(w)(z - w)}{(z - w)^2} - \frac{\partial_w T(w)}{z - w} + \ldots = T(z)T(w). \tag{7.158}$$

Section 7.9.8 demonstrates that this form for the operator product expansion of the energy-momentum tensor is consistent with the Virasoro algebra (see section 7.9.3) as the algebra of generators of conformal transformations for order parameter fields in analogy of the Witt algebra (see section 7.4.1) as the algebra of generators of conformal coordinate transformations.

First, however, we look at the behaviour of the energy-momentum tensor under conformal transformations.

7.9.2 Conformal transformation of the energy-momentum tensor

The energy-momentum tensor, at least as long as the conformal charge does not vanish, cannot be a primary field, which can be inferred by comparing (7.154) and (7.150) (together with exercise 7.18). Writing (7.142) for the holomorphic energy-momentum tensor itself

$$\delta_\epsilon T(w) = \oint \frac{dz}{2\pi i} \epsilon \xi(z) T(z) T(w) \tag{7.159}$$

with the infinitesimal conformal transformation $z \to w = z + \epsilon\xi(z)$ and inserting the operator product expansion (7.154), we obtain

$$\delta_\epsilon T(w) = \oint \frac{dz}{2\pi i} \epsilon \xi(z) \left(\frac{c/2}{(z - w)^4} + \frac{2T(w)}{(z - w)^2} + \frac{\partial_w T(w)}{(z - w)} + \ldots \right) \tag{7.160}$$

$$= \epsilon \left(\frac{c}{12} \partial_w^3 \xi(w) + 2T(w)\partial_w \xi(w) + \xi(w)\partial_w T(w) \right), \tag{7.161}$$

where we use a Taylor expansion of $\xi(z)$ around $z = w$ up to third order in $(z - w)$ to capture all relevant terms in the residue, which yields the final result.

We state the general result for an arbitrary conformal transformation $w \to f(w)$ without derivation

$$\tilde{T}(w) = \left(\frac{\partial f}{\partial w} \right)^2 T(f(w)) + \frac{c}{12} S(f(w); w) \tag{7.162}$$

where $S(f, w)$ is the Schwarzian derivative, defined as

$$S(f;w) \equiv \frac{\partial_w^3 f}{\partial_w f} - \frac{3}{2}\left[\frac{\partial_w^2 f}{\partial_w f}\right]^2. \tag{7.163}$$

This expression shows again that the energy-momentum tensor is not a primary field. However, it may still be a quasi-primary field. The following exercise investigates this possibility.

EXERCISE 7.19 Schwarzian derivative of a Möbius transformation Calculate the Schwarzian derivative $S(w;z)$ for a Möbius transformation (7.70).

EXERCISE 7.20 Infinitesimal conformal transformation of the energy-momentum tensor Show that for the infinitesimal conformal transformation $w \to f(w) = w + \epsilon\xi(w)$ the conformal transformation (7.162) of the energy-momentum tensor is given by (7.161).

In addition to the vanishing of the Schwarzian derivative for the Möbius transformation (see exercise 7.19), it can also be shown that Möbius transformations are the only transformations for which the Schwarzian derivative vanishes. The result of exercise 7.20 then shows that the energy-momentum tensor $T(z)$ is a quasi-primary field and confirms that the conformal weights are $(h, \bar{h}) = (2, 0)$.

7.9.3 Virasoro algebra

The Lie algebra (7.76), the Witt algebra, and its anti-holomorphic counterpart, allow for a generalization, the so-called central extension. This central extension consists in an additional element, a number $c \in \mathbb{C}$, which commutes with all other elements ℓ_n of the algebra. This extension of the Witt algebra (7.76) called Virasoro algebra, will prove useful in the following developments.

We denote the elements of the Virasoro algebra by L_n with $n \in \mathbb{Z}$ and write their commutation relations as

$$[L_m, L_n] = (m-n)L_{m+n} + cp(m,n). \tag{7.164}$$

The constant $c \in \mathbb{C}$, called central charge, commutes with all other elements, $[L_n, c] = 0$. In the context of conformal symmetry, c is also called conformal anomaly. It can be shown (see exercise 7.21, or, e.g. Blumenhagen and Plauschinn, 2009) that

$$p(n,-n) = \frac{c}{12}(n+1)n(n-1) = \frac{c}{12}n(n^2-1) \tag{7.165}$$

while all other $p(m, n) = 0$. We have chosen a normalization $p(2, -2) = 1/2$ such that the standard example, the model of free Bosons, has a simple value for c, namely $c = 1$.

EXERCISE 7.21 Virsoro algebra Show that (7.165) follows from (7.164).

In order to do so, use that $p(m,n) = -p(n,m)$. Furthermore, consider the redefinitions

$$\tilde{L}_n = L_n + c\frac{p(n,0)}{n}, \quad n \neq 0, \qquad \tilde{L}_0 = L_0 + c\frac{p(1,-1)}{2}. \tag{7.166}$$

What choices follow for $p(n,0)$ and $p(1,-1)$?

Finally, exploit the special Jacobi identities

$$[[L_m, L_n], L_0] + [[L_n, L_0], L_m] + [[L_0, L_m], L_n] = 0, \tag{7.167}$$

$$[[L_{-n+1}, L_n], L_{-1}] + [[L_n, L_{-1}], L_{-n+1}] + [[L_{-1}, L_{-n+1}], L_n] = 0. \tag{7.168}$$

The latter of these identities gives rise to a recursion relation for which we have chosen $p(2,-2) = 1/2$ as initial value or as normalization, as mentioned previously.

An analogous algebra $\bar{L}_n$ can be obtained for anti-holomorphic infinitesimal conformal transformations that is independent of the holomorphic algebra. In summary, we have

$$[L_m, L_n] = (m-n)L_{m+n} + \frac{c}{12}m(m^2-1)\delta_{m+n,0}, \tag{7.169}$$

$$\left[\bar{L}_m, \bar{L}_n\right] = (m-n)\bar{L}_{m+n} + \frac{\bar{c}}{12}m(m^2-1)\delta_{m+n,0}, \tag{7.170}$$

$$\left[L_m, \bar{L}_n\right] = 0. \tag{7.171}$$

The interpretation of the energy-momentum tensor pointed to by the conformal Ward identities (7.142) and (7.143) as the generator of infinitesimal transformations on correlation functions of order parameter fields suggests that we can introduce operators L_n and $\bar{L}_n$ in analogy to the operators ℓ_n and $\bar{\ell}_n$ of the Witt algebra (7.76) as the coefficients of a Laurent series

$$T(z) = \sum_{n=-\infty}^{\infty} \frac{L_n}{z^{n+2}}, \tag{7.172}$$

$$\bar{T}(\bar{z}) = \sum_{n=-\infty}^{\infty} \frac{\bar{L}_n}{\bar{z}^{n+2}}, \tag{7.173}$$

where L_n and $\bar{L}_n$ are thus given by

$$L_n = \frac{1}{2\pi i}\oint dz\, z^{n+1}\, T(z), \tag{7.174}$$

$$\bar{L}_n = \frac{1}{2\pi i}\oint d\bar{z}\, \bar{z}^{n+1}\, \bar{T}(\bar{z}). \tag{7.175}$$

We now want to show that the operator product expansion of the energy-momentum tensor with itself is given by

$$T(z)T(w) = \frac{c/2}{(z-w)^4} + \frac{2T(z)}{(z-w)^2} + \frac{\partial_z T(z)}{(z-w)} + \ldots, \tag{7.176}$$

$$\bar{T}(\bar{z})\bar{T}(\bar{w}) = \frac{\bar{c}/2}{(\bar{z}-\bar{w})^4} + \frac{2\bar{T}(\bar{z})}{(\bar{z}-\bar{w})^2} + \frac{\partial_{\bar{z}} \bar{T}(\bar{z})}{(\bar{z}-\bar{w})} + \ldots. \tag{7.177}$$

In order to demonstrate this result, we show that it is consistent with the Laurent expansion coefficients L_n and $\bar{L}_n$ in (7.172) and (7.173), forming a Virasoro algebra with central extensions c and $\bar{c}$, respectively.

For this demonstration, we need a few preparations, the first of which is a construction, radial quantization, which is quite natural when using conformal symmetry in string theory, while in our context it will be of help in the subsequent calculations. The second is an addition to our discussion in sections 7.6 and 7.7 of the the energy-momentum tensor associated with conserved currents. Thirdly, we need the notion of radial ordering of fields in a correlation function that is directly connected to radial quantization since the usual time ordering of correlation functions will be mapped onto a radial ordering.

7.9.4 Radial quantization

We compactify the Euclidean space (r^0, r^1) in the r^1 direction by introducing the complex notation

$$w = r^0 + ir^1 \tag{7.178}$$

where we identify w and $w + 2\pi i$. This choice is appropriate for a theory defined on a cylinder $-\infty < r^0 < \infty$ and $0 \leq r^1 \leq 2\pi$.

The points on the cylinder are now mapped onto the complex plane by

$$z = e^w = e^{r^0} e^{ir^1}. \tag{7.179}$$

In particular, $r^0 = -\infty$ is mapped onto the origin $z = 0$ while r^1 is the angular coordinate of concentric rings around the origin with radius e^{x^0}.

The translations $r^0 \to r^0 + a^0$, which may be interpreted as time translations, now become dilatations $z \to e^{a^0} z$ and the translations $r^1 \to r^1 + a^1$, which can be chosen to represent space translations, become rotations $z \to e^{ia^1} z$.

From relativistic quantum mechanics, we know that the generator of time translations is the Hamiltonian, which thus in radial quantization corresponds to the dilatation operator. Moreover, the generator for space translations is the momentum operator now corresponding to rotations. Recalling the results from exercise 7.14, we can thus identify

$$\mathcal{H} = L_0 + \bar{L}_0 = \ell_0 + \bar{\ell}_0, \tag{7.180}$$

$$\mathcal{P} = i\left(L_0 - \bar{L}_0\right) = i\left(\ell_0 - \bar{\ell}_0\right). \tag{7.181}$$

We also need to construct states of a Hilbert space out of the order parameter fields we have discussed thus far.

7.9.5 Asymptotic states

We now want to employ the analogy between two-dimensional statistical mechanics and quantum field theory in one space and one time dimension (discussed in chapter 6). This language will prove convenient to elucidate the representations of the Virasoro algebra.

A primary order parameter field $\phi(z, \bar{z})$ with conformal weights h and $\bar{h}$ can be expanded in a Laurent series at $z = \bar{z} = 0$

$$\phi(z, \bar{z}) = \sum_{m,\bar{m}=-\infty}^{\infty} z^{-m-h} \bar{z}^{-\bar{m}-\bar{h}} \phi_{m\bar{m}} \tag{7.182}$$

where we can now quantize the modes of the Laurent expansion $\phi_{m\bar{m}}$ and interpret them as operators which create states from a vacuum state $|0\rangle$.

In order to construct these states, we recall from (7.179) that $r^0 \to -\infty$, which we interpret as the infinite past, is transformed to $z = 0$ and $\bar{z} = 0$. Motivated by this observation, we define an asymptotic state $|\phi\rangle$ by

$$|\phi\rangle = \lim_{z,\bar{z}\to 0} \phi(z, \bar{z})|0\rangle = \lim_{z,\bar{z}\to 0} \sum_{m,\bar{m}=-\infty}^{\infty} z^{-m-h} \bar{z}^{-\bar{m}-\bar{h}} \phi_{m\bar{m}}|0\rangle. \tag{7.183}$$

This state is only non-singular at $z = 0$ or $\bar{z} = 0$, if we require that

$$\phi_{m\bar{m}}|0\rangle = 0 \qquad \text{for} \qquad m > -h, \quad \text{or} \quad \bar{m} > -\bar{h} \tag{7.184}$$

i.e. the asymptotic state corresponding to the infinite past is given by

$$|\phi\rangle = \lim_{z,\bar{z}\to 0} \phi(z, \bar{z})|0\rangle = \phi_{-h,-\bar{h}}|0\rangle. \tag{7.185}$$

Correspondingly, we need the asymptotic state corresponding to the infinite future, which according to (7.179) corresponds to $r^0 \to +\infty$ and therefore to $z \to \infty$ and $\bar{z} \to \infty$. This state will be the Hermitian conjugate state to (7.185). We obtain this state by performing the conformal transformations $z \to 1/\bar{z}$ and $\bar{z} \to 1/z$ which, according to (7.179), is the equivalent in radial quantization to a time reversal transformation. Therefore, it maps the asymptotic state at $r^0 \to -\infty$ to the asymptotic state at $r^0 \to +\infty$ or, on the complex plane $z = 0$ to $z \to \infty$. Taking into account also the transformation behaviour (7.82) of primary fields, we first arrive at

$$\phi^\dagger(z,\bar{z}) = \bar{z}^{-2h} z^{-2\bar{h}} \phi\left(\frac{1}{\bar{z}}, \frac{1}{z}\right) = \sum_{m\bar{m}} z^{\bar{m}-\bar{h}} \bar{z}^{m-h} \phi_{m,\bar{m}}. \tag{7.186}$$

This expression can now be compared with the direct Hermitian conjugation of (7.182)

$$\phi^\dagger(z,\bar{z}) = \sum_{m,\bar{m}=-\infty}^{\infty} \bar{z}^{-m-h} z^{-\bar{m}-\bar{h}} \phi^\dagger_{m\bar{m}} \tag{7.187}$$

from which we conclude that $\phi^\dagger_{m\bar{m}} = \phi_{-m,-\bar{m}}$. In analogy the asymptotic state (??), we obtain for the asymptotic state at infinity the condition

$$\langle\phi| = \lim_{z,\bar{z}\to 0} \langle 0|\phi^\dagger(z,\bar{z}) \lim_{w,\bar{w}\to\infty} w^{2h}\bar{w}^{2\bar{h}} \langle 0|\phi(w,\bar{w}) \tag{7.188}$$

where $w = 1/\bar{z}$ and $\bar{w} = 1/z$. This expression remains finite if we can satisfy the condition

$$\langle 0|\phi_{m\bar{m}} = 0 \qquad \text{for} \qquad m < h, \quad \text{or} \quad \bar{m} < \bar{h} \tag{7.189}$$

so that we can finally write

$$\langle\phi| = \lim_{w,\bar{w}\to\infty} \langle 0|\phi(w,\bar{w}) = \langle 0|\phi_{h,\bar{h}} \tag{7.190}$$

for the asymptotic state at infinity analogous to the asymptotic state at the origin (7.185).

Next, we use the observation that conserved currents imply conserved charges to derive an expression for the conserved charge associated to the conformal symmetry in terms of the energy-momentum tensor.

7.9.6 Conserved charge

Associated with the conserved current $\mathcal{J}_\mu = \sum_\nu T_{\mu\nu}\xi^\nu$ is a conserved charge

$$Q = \int dx^1\, \mathcal{J}_0 \tag{7.191}$$

at constant x^0. The proof uses the derivative of Q with respect to x^0 and the continuity equation for the conserved current $\sum_\mu \frac{\partial \mathcal{J}^\mu}{\partial r^\mu} = 0$. The change of coordinates (7.178) implies that a constant r^0 translates into a constant $|z|$ such that the integration in (7.191) yielding the conserved charge Q becomes a contour integral

$$Q = \frac{1}{2\pi i} \oint_C \left(dz\, \epsilon\xi(z) T(z) + d\bar{z}\bar{\epsilon}\bar{\xi}(\bar{z})\bar{T}(\bar{z})\right) \tag{7.192}$$

where, for convenience and as before, we have introduced a factor $1/2\pi$ in a redefinition of the energy-momentum tensor. The conserved charge can be used as an alternative way to derive the conformal Ward identities. This is done using the observation that the conserved charge Q is the generator of symmetry transformations for any operator ϕ, i.e.

$$\delta\phi = [Q, \phi]. \tag{7.193}$$

We are not dwelling on this possibility, except for the following demonstration. Choosing a particular infinitesimal conformal transformation

$$\epsilon\xi(z) = -\epsilon\xi_n z^{n+1}, \tag{7.194}$$

we obtain for, e.g. the holomorphic part of the conserved charge (7.192)

$$Q_n = \oint \frac{dz}{2\pi i} T(z) \left(-\epsilon\xi_n z^{n+1}\right) = -\epsilon\xi_n \sum_{m=-\infty}^{\infty} \oint \frac{dz}{2\pi i} L_m z^{n-m-1} = -\epsilon\xi_n L_n, \tag{7.195}$$

which reveals the Laurent modes L_n of the energy-momentum tensor as the generators of infinitesimal conformal transformations. An analogous statement holds for the anti-holomorphic part of the conserved charge (7.192).

7.9.7 Radial ordering

In order to demonstrate that the operator product expansions (7.176) and (7.177) are consistent with the Laurent expansion coefficients L_n and $\bar{L}_n$ in (7.172) and (7.173) forming a Virasoro algebra with central extensions c and $\bar{c}$, respectively, we need to perform contour integrations. In these contour integrations, the relative positions of the variables z and w are essential.

We recall that the operator product expansions are to be understood as residing inside correlation functions and are, hence, initially time ordered products where the product of two operators $A(x_a, t_a)$ and $B(x_b, t_b)$ can be written, with the help of the Hamiltonian $\mathcal{H}$ of the system, as

$$A(x_a, t_a)B(x_b, t_b) = e^{i\mathcal{H}t_a} A(x_a, 0) e^{-i\mathcal{H}t_a} e^{i\mathcal{H}t_b} B(x_b, 0) e^{-i\mathcal{H}t_b}. \tag{7.196}$$

The factor in the middle $e^{-i\mathcal{H}(t_a - t_b)}$ that occurs in this expression, though quite innocent looking, becomes $e^{-\mathcal{H}(\tau_a - \tau_b)}$ upon a Wick rotation $t \to -i\tau$. Since the Hamiltonian is bounded from below, but not from above, the exponential may become arbitrarily large if $\tau_a < \tau_b$ and the expectation values of the operator product will not be properly defined. Hence, time ordering has to be assumed in operator products

$$T\left(A(x_a, t_a)B(x_b, t_b)\right) = \begin{cases} A(x_a, t_a)B(x_b, t_b) & \text{for} \quad t_a > t_b, \\ \eta B(x_b, t_b)A(x_a, t_a) & \text{for} \quad t_a < t_b, \end{cases} \tag{7.197}$$

where $\eta = \pm 1$ takes care of the Bose and Fermi cases, respectively.

Using the change of coordinates (7.179) from the cylinder to the complex plane, time ordering now becomes radial ordering. This means that operator product expansions always are to be understood as radially ordered

$$R\,(A(z)B(w)) \equiv \begin{cases} A(z)B(w) & \text{for} \quad |z| > |w|, \\ \eta B(w)A(z) & \text{for} \quad |w| > |z|. \end{cases} \tag{7.198}$$

On the cylinder a correlation function has thus the time ordered form

$$\langle 0|\, T\,(A_1(x_1,t_1)A_2(x_2,t_2)\ldots A(x_n,t_n))\,|0\rangle, \tag{7.199}$$

where the states $|0\rangle$ and $\langle 0|$ are asymptotic states at $t \to -\infty$ and $t \to +\infty$, respectively. After the conformal transformation (7.179), these correlation functions become radially ordered

$$\langle 0|R\,(A_1(z_1,\bar{z}_1)A_2(z_2,\bar{z}_2)\ldots A(z_n,\bar{z}_n))\,|0\rangle. \tag{7.200}$$

The states $|0\rangle$ and $\langle 0|$ are asymptotic states at $z = 0$ and $z \to \infty$, respectively. Section 7.9.5 demonstrated how such states can be constructed from primary order parameter fields.

7.9.8 Operator product expansion of the energy-momentum tensor and Virasoro algebra

We now return to the demonstration that the proposed operator product expansion of the energy-momentum tensor is consistent with the Laurent expansion of this operator where the expansion coefficients form the Virasoro algebra. We restrict ourselves to the demonstration for the holomorphic operator product expansion, the demonstration for the anti-holomorphic operator product expansion being completely analogous. Using the result of the previous section, the core point of the demonstration rests on the observation that

$$\begin{aligned}\oint dz\,[A(z),B(w)] &= \oint_{|z|>|w|} dz\,A(z)B(w) - \oint_{|z|<|w|} dz\,B(w)A(z) \\ &= \oint_{C(w)} dz\,R\,(A(z)B(w))\,. \end{aligned} \tag{7.201}$$

Taking into account the radial ordering (7.198), the contour integrals have to be taken with the correct radial ordering, which is different for the two parts of the commutator. The difference of the two contours $|z| > |w|$ and $|z| < |w|$ is a contour $C(w)$ encircling the point w.

We now calculate the commutator (7.169) of the Virasoro algebra using the representation (7.174) of the operators L_n as contour integrals

$$[L_m, L_n] = \oint \frac{dw}{2\pi i} \oint \frac{dz}{2\pi i} z^{m+1} w^{n+1} \left[T(z), T(w)\right]. \tag{7.202}$$

Using the prescription (7.201) for the radial ordering of operators, we first obtain

$$[L_m, L_n] = \oint \frac{dw}{2\pi i} \oint_{C(w)} \frac{dz}{2\pi i} z^{m+1} w^{n+1} R\left(T(z)\, T(w)\right) \tag{7.203}$$

assuming a fixed value w in the contour integration with respect to z. Inserting the proposed operator product expansion (7.176) yields

$$[L_m, L_n] = \oint \frac{dw}{2\pi i} \oint_{C(w)} \frac{dz}{2\pi i} z^{m+1} w^{n+1} \left(\frac{c/2}{(z-w)^4} + \frac{2T(w)}{(z-w)^2} + \frac{\partial_w T(w)}{(z-w)} \right). \tag{7.204}$$

EXERCISE 7.22 Commutator of the Virasoro operators The integral along the contour $C(w)$ can be evaluated using the residue theorem. In order to capture all non-vanishing terms of the residue at $z = w$, we need to expand z^{m+1} in a Taylor series around the point $z_0 = w$ up to third order in $(z - w)$. The remaining contour integral with respect to w can again be partly evaluated by the residue theorem, while the other parts, after a partial integration, can be identified as proportional to the Virasoro operator L_{m+n}.

In summary, the operator product expansion between two energy-momentum tensors (7.176) is consistent with the Virasoro algebra (7.169). An analogous calculation shows the consistency between the operator product expansion (7.177) and the Virasoro algebra (7.170).

7.9.9 Representation of the Virasoro algebra

In order to work effectively with any given algebra, it is important to find its representations using an appropriate set of eigenstates from a Hilbert space. A well-known example of this procedure is the angular momentum algebra. We encountered the algebra of angular momentum in chapter 3, where the three operators $L^{\pm}$ and $L_3 \equiv L_z$ form the generators of the algebra

$$\left[L_3, L^{\pm}\right] = \pm L^{\pm}, \qquad \left[L^+, L^-\right] = 2L_3. \tag{7.205}$$

There we found, without actually using this terminology, that all unitary representations are labelled by an integer or half-odd-integer l, which is the largest eigenvalue of the operator L_3. This state is annihilated by L^+ while it is lowered by unity by L^-. As seen in chapter 3, we can thus generate the representation of the algebra from the state with maximal eigenvalue l of the operator L_3 by successive application of L^-. In the mathematical literature, such a state is called a *highest weight state*, the corresponding representation a highest weight representation.

We now emulate this procedure for the Virasoro algebra. As with angular momentum, we start by determining a set of commuting operators (in the case of angular momentum, these are L^2 and L_3). We shall find that a convenient choice for the Virasoro algebra will consist in L_0 and the central charge c. This choice is motivated by our previous identification of $L_0 + \bar{L}_0$ as the Hamiltonian 7.180 (note that for $n = 0$, we have $L_0 = \ell_0$).

There are usually many representations of a given algebra. The one we consider here is the so-called unitary highest weight representation, which we shall select, as we did for angular momentum, by requiring some physically motivated conditions.

7.9.9.1 *Unitary representation of the Virasoro algebra*

A representation of the Virasoro algebra has to satisfy two conditions in order to qualify as a unitary representation. Firstly, there must be a Hilbert space such that all generators L_n of the Virasoro algebra are operators acting on this Hilbert space. Secondly, the condition that $L_n^\dagger = L_{-n}$ (and $\bar{L}_n^\dagger = \bar{L}_{-n}$) must hold. The condition $L_n^\dagger = L_{-n}$ implies in particular that $T(z)$ is a Hermitean operator that can be seen on the cylinder, where we have in Minkowskian space-time coordinates $(r^0, r^1) = (r, -it)$, i.e. $w = r^0 + ir^1 = r + t$

$$T(w) + \bar{T}(\bar{w}) = \sum_{n=-\infty}^{\infty} \left(L_n e^{-in(r+t)} + \bar{L}_n e^{-in(r-t)} \right) \qquad (7.206)$$

where the two parts, $T(w)$ and $\bar{T}(\bar{w})$, are Hermitian if ${L_n}^\dagger = L_{-n}$ and $\bar{L}_n^\dagger = \bar{L}_{-n}$.

7.9.9.2 *Highest weight representation*

A highest weight representation is defined as a representation containing a state with a largest or smallest value of the eigenvalues of one of the generators. This state is then annihilated by other generators of the algebra while yet other generators will lower or raise, as the case may be, the eigenvalue when acting on the state with highest or lowest eigenvalue and thereby create new states. In the case of angular momentum (cp. chapter 3), the state with the largest eigenvalue is $|l, l\rangle$ which satisfies $L_3|l, l\rangle = l|l, l\rangle$ and $L^+|l, l\rangle = 0$ while there is also a state with a lowest eigenvalue, $L_3|l, -l\rangle = -l|, -ll\rangle$ with $L^-|l, -l\rangle = 0$. As demonstrated in chapter 3, by applying L^- on the state with highest eigenvalue or, respectively, by applying L^+ on the state with lowest eigenvalue, all other angular momentum states can be generated.

Although not all representations have this property, it is reasonable to require it in a physical theory, since we have identified $L_0 + \bar{L}_0$ as the Hamiltonian, which, as a Hamiltonian, is bounded from below. Even though this is rather a *lowest* weight representation, the name *highest weight* for a state with lowest energy and correspondingly the designation highest weight representation, although perhaps somewhat confusing, has established itself.

The next task is to determine the highest weight state of $L_0 + \bar{L}_0$.

7.9.9.3 Ground state of the Virasoro generator

The Hamiltonian (7.180), i.e. $\mathcal{H} = L_0 + \bar{L}_0$, is connected to the scale transformation or dilatation $z \to (1+\epsilon)z$, its eigenvalues must therefore be numbers characteristic of the scale invariance of the system. Such numbers are the scaling dimensions x. Therefore, we expect the eigenvalues to be related to the scaling dimensions x of the order parameter fields. Indeed, the eigenvalues are the scaling dimensions. Thus, there will be a state $|\psi\rangle$ for which

$$\left(L_0 + \bar{L}_0\right)|\psi\rangle = x|\psi\rangle = (h + \bar{h})|\psi\rangle \tag{7.207}$$

or separately

$$L_0|\psi\rangle = h|\psi\rangle, \tag{7.208}$$

$$\bar{L}_0|\psi\rangle = \bar{h}|\psi\rangle. \tag{7.209}$$

This state will also satisfy

$$\left(L_0 - \bar{L}_0\right)|\psi\rangle = s|\psi\rangle = (h - \bar{h})|\psi\rangle \tag{7.210}$$

for the momentum operator (7.181). From the Virasoro algebra, we can immediately infer that acting with the operators L_n decreases the eigenvalue of L_0 by n units

$$L_0 L_n|\psi\rangle = (L_n L_0 - nL_n)|\psi\rangle = (h - n)L_n|\psi\rangle \tag{7.211}$$

and thus creates further states $L_n|\psi\rangle$ with eigenvalues $h - n$. In other words, the operators L_n are raising and lowering operators depending on whether n is positive or negative. Assuming $n > 0$, then L_n lowers the eigenvalue by n units while L_{-n} raises the eigenvalue by n units. If the spectrum of $L_0 + \bar{L}_0$ is indeed bounded from below, there must exist some states which are annihilated by all operators L_n and $\bar{L}_n$ for $n > 0$. Such states, obeying

$$L_n|\psi\rangle = \bar{L}_n|\psi\rangle = 0 \qquad \text{for all} \quad n > 0 \tag{7.212}$$

are called primary states. They are indeed the highest weight states we have been discussing. We denote these states by $|h, \bar{h}\rangle$ and discuss them and their connection with the asymptotic states we introduced in section 7.9.5 in the subsequent section. The states which are generated by acting with L_{-n} ($n > 0$) on a highest weight state $|h, \bar{h}\rangle$ are called descendent states.

7.9.9.4 Vacuum state

We now want to characterize the highest weight state (7.207) in more detail as a vacuum state. The asymptotic state (7.185)

$$\lim_{z,\bar{z}\to 0} \phi(z,\bar{z})|0\rangle = \phi(0,0)|0\rangle = \phi_{-h,-\bar{h}}|0\rangle \equiv |h,\bar{h}\rangle \tag{7.213}$$

acts as a vacuum state for all L_n with $n > 0$, i.e.

$$L_n\phi_{-h,-\bar{h}}|0\rangle = 0 \qquad \text{for} \qquad n > 0. \tag{7.214}$$

This statement can be demonstrated by considering the expression (where we suppress the dependence on the barred variables)

$$\langle 0|\phi_m(z_m)\phi_{m-1}(z_{m-1})\dots\phi_2(z_2)L_n\phi_{-h,-\bar{h}}|0\rangle =$$

$$\oint_C \frac{dz}{2\pi i} z^{n+1} \langle 0|\phi_m(z_m)\phi_{m-1}(z_{m-1})\dots\phi_2(z_2)T(z)\phi_{-h,-\bar{h}}|0\rangle \tag{7.215}$$

where we used primary fields at positions $z_2, z_3, \dots, z_m$, which are radially ordered, i.e. $|z_m| > |z_{m-1}| > \dots > |z_2|$ and where the contour circles the origin while all points $z_2, z_3, \dots, z_m$ are outside the contour. The integrand of this expression is determined by the Ward identity (7.146)

$$\langle 0|\phi_m(z_m)\phi_{m-1}(z_{m-1})\dots\phi_2(z_2)T(z)\phi_{-h,-\bar{h}}|0\rangle =$$

$$\left(\frac{h}{z^2} + \frac{1}{z}\frac{\partial}{\partial z_1}\right)\langle 0|\phi_m(z_m)\phi_{m-1}(z_{m-1})\dots\phi_2(z_2)\phi(z_1)|0\rangle|_{z_1=0} + \dots. \tag{7.216}$$

For all $n > 0$, the contour integral (7.216) vanishes and we conclude that

$$L_n\phi_{-h,-\bar{h}}|0\rangle = L_n|h,\bar{h}\rangle = 0. \tag{7.217}$$

For $n = 0$, however, there remains a residue $h\langle 0|\phi_m(z_m)\phi_{m-1}(z_{m-1})\dots\phi_2(z_2)\phi(0)|0\rangle$. Therefore, we have

$$L_0\phi_{-h,-\bar{h}}|0\rangle = L_0|h,\bar{h}\rangle = h. \tag{7.218}$$

In summary, we have shown that the highest weight state $|h,\bar{h}\rangle$, an eigenstate of L_0 with eigenvalue h (and $\bar{L}_0$ with eigenvalue $\bar{h}$), is annihilated by all Virasoro generators L_n (and $\bar{L}_n$) with $n > 0$. Recalling (7.211), we can further conclude that L_{-n} with $n > 0$ creates further states with eigenvalues $h + n$. These states are the descendent states introduced in the previous section.

The vacuum state $|0\rangle$ itself is characterized by the requirement that it is annihilated by the maximally possible number of generators L_n. Because of the central extension of the algebra, the vacuum state $|0\rangle$ cannot be annihilated by all generators L_n. If L_2, L_{-2}, and L_0 annihilate $|0\rangle$, then also the commutator $[L_2, L_{-2}]$ annihilates $|0\rangle$. However, the Virasoro algebra gives

$$[L_2, L_{-2}] = 4L_0 + \frac{c}{2}. \tag{7.219}$$

The statement can thus only be true for $c = 0$.

From the commutator $[L_1, L_{-1}] = 2L_0$ we conclude that any highest, weight state annihilated by L_0 is also annihilated by L_1 and L_{-1}, and vice-versa. The maximal number of generators of a Virasoro algebra with non–vanishing c annihilating the vacuum $|0\rangle$ is therefore given by

$$L_n|0\rangle = 0, \qquad \text{for} \qquad n \geq -1. \tag{7.220}$$

Its Hermitean conjugate $\langle 0|$ satisfies $\langle 0|L_n = 0$ for $n \leq 1$. We assume that there is one state only with these properties.

7.9.9.5 Positivity of c and h

We can use the results of the previous sections on the representation of the Virasoro algebra to determine constraints on the possible values of the central charge c and the conformal weights h—the two numbers that characterize the unitary highest weight representation of the Virasoro algebra. We achieve this goal by expressing the norms of states in terms of the central charge c and the the scaling dimension h and demanding that these norms must be non-negative.

First, we conclude that the central charge is non-negative by considering the commutator

$$[L_n, L_{-n}] = 2nL_0 + \frac{c}{12}\left(n^3 - n\right) \tag{7.221}$$

whose expectation value in the vacuum state is

$$\langle 0|\,[L_n, L_{-n}]\,|0\rangle = \frac{c}{12}\left(n^3 - n\right). \tag{7.222}$$

This expectation value can be rewritten as

$$\langle 0|\,[L_n, L_{-n}]\,|0\rangle = \langle 0|L_n L_{-n}|0\rangle = \langle 0|L^{\dagger}_{-n} L_{-n}|0\rangle = ||L_{-n}|0\rangle|| \geq 0 \tag{7.223}$$

since $L_n|0\rangle = 0$ and $L_n = L_{-n}{}^{\dagger}$. This implies that $c \geq 0$ for $n \geq 2$. But since c is a number characterizing the whole Virasoro algebra, it must be independent of n and, hence, we can conclude that $c \geq 0$.

Doing a similar analysis for an arbitrary highest weight state $|h, \bar{h}\rangle$ (and suppressing $\bar{h}$ in the following formulas), we first obtain

$$\langle h| \left[L_n, L_{-n}\right] |h\rangle = \left(2nh + \frac{c}{12}\left(n^3 - n\right)\right) \langle h|h\rangle. \tag{7.224}$$

The expectation value of the commutator, this time with the highest weight state $|h\rangle$, can be treated in the same way as before for the vacuum state $|0\rangle$ in (7.223). This time the norm becomes therefore

$$0 \leq ||L_{-n}|h\rangle|| = \left(2nh + \frac{c}{12}\left(n^3 - n\right)\right) \langle h|h\rangle. \tag{7.225}$$

From this expression we can again conclude that $c \geq 0$ but now also, taking $n = 1$, that $h \geq 0$.

7.9.9.6 Summary of some further results

Many more restrictions on the central charge c and the conformal weights h (and $\bar{h}$) can be obtained by considering the norms of further states, especially also including descendent states, and requiring that they all be non-negative. Since these considerations quickly become very involved, we shall restrict ourselves in this section to a summary of the final result that can be derived with methods refining the elementary ones applied in the previous section.

These results, however, play no major role, neither indirectly nor directly, for the understanding of the uses of conformal symmetry we have in mind in later chapters of this book, in particular in part VI on the Bethe ansatz for finite-size systems. We therefore refer for these and further developments to the literature cited at the beginning of this chapter.

It can be shown that the central charge can only acquire a discrete set of points in the range $0 < c \leq 1$ given by

$$c = 1 - \frac{6}{m(m+1)}, \qquad m \geq 3. \tag{7.226}$$

Each value of c or, respectively, m, has a corresponding discrete set of allowed values for the conformal weights

$$h = h_{p,q} = \frac{[(m+1)p - mq]^2 - 1}{4m(m+1)}, \qquad p = 1, 2, \ldots, m-1, \quad q = 1, 2, \ldots, p. \tag{7.227}$$

These equations determine the famous discrete series of minimal unitary Virasoro models and show that the central charge or conformal anomaly is the hallmark quantity that classifies and labels the two-dimensional models of statistical mechanics into separate universality classes.

Moreover, these relations provide the fundamental explanation of the longstanding observation that critical exponents of two-dimensional systems are rational numbers. For instance, the critical two-dimensional Ising model corresponds to the smallest value of $m = 3$, i.e. to a value for the central charge of $c = 1/2$. The associated values of the conformal weights are $h = 1/2$ and $h = 1/16$ (we shall not discuss further the trivial value $h = 0$ corresponding to the identity). With $h = \bar{h}$, we then have for the correlation function exponent $\eta = 2x = 2(h + \bar{h})$ the two values $\eta = 2$ and $\eta = 1/4$. These are indeed the values of the two-dimensional Ising model of the energy density and the spin-spin correlation functions, respectively, which are independently well-known from Onsager's exact solution of this model. Many other values of m or, respectively, the corresponding central charge c, have been identified with critical exponents of further two-dimensional models of statistical mechanics. We again refer to the literature cited at the beginning of this chapter for a more detailed discussion.

We now return to the issue of finite-size corrections that allow direct access to the critical data c and h from calculations of model systems of finite size.

7.10 Finite-size corrections revisited

In section 7.1.2, we studied the finite-size behaviour of the two-point correlation function for which we obtained the result (7.14). We evaluated (7.14) for large distances $u_1 - u_2 \gg L$ along the strip and obtained the important result (7.16), which connected the critical exponent η of the infinite system to the correlation length $\xi(L)$ for a finite system. However, the expression (7.14) contains considerably more information and therefore warrants a more careful evaluation, which will then be connected with an evaluation of the correlation function employing the transfer matrix technique encountered in previous chapters (cp. especially sections 4.16 and 6.2), and which we shall use again as a major technique later in part II.

7.10.1 Finite-size corrections of the correlation function

Employing the generalized binomial theorem valid for real exponents α

$$(1 + x)^{\alpha} = \sum_{k=0}^{\infty} \binom{\alpha}{k} x^k, \tag{7.228}$$

where

$$\binom{\alpha}{k} = \frac{\alpha(\alpha - 1)\ldots(\alpha - (k - 1))}{k!} = \frac{\Gamma(\alpha + k)}{\Gamma(\alpha)k!} \quad \text{if} \quad k \geq 1, \quad \binom{\alpha}{0} = 1, \tag{7.229}$$

(7.14) can be exactly rewritten for $u_1 > u_2$ as

$$\langle\phi(w_1)\phi(w_2)\rangle_{\text{strip}} = \left(\frac{2\pi}{L}\right)^{2x}$$
$$\times \sum_{m,\bar{m}=1}^{\infty} A_m A_{\bar{m}} \exp\left(-\frac{2\pi}{L}((x+m+\bar{m})(u_1-u_2)+i(m-\bar{m})(v_1-v_2))\right) \quad (7.230)$$

where the scaling dimension $x = x_\phi$ relates to the primary operator $\phi(w)$ in the correlation function and where the coefficients are given by

$$A_m = \frac{\Gamma(x+m)}{\Gamma(x)m!}. \quad (7.231)$$

EXERCISE 7.23 Two-point correlation function in strip geometry Convince yourself that the expression (7.230) for the two-point correlation function in the strip geometry is equivalent to the expression (7.14).

We now rederive this result employing the transfer matrix technique introduced and used in previous chapters. We interpret the quantum field theory in one space and one time dimension on the cylinder now as a two-dimensional statistical mechanical system according to the equivalence discussed in chapter 6. The transfer matrix $\mathcal{T}$, the object in statistical mechanics equivalent to the time evolution operator in quantum mechanics or to the propagator in quantum field theory, describes the evolution of a statistical configuration in one direction. We assume the statistical mechanical system to remain situated on the cylinder and further assume the evolution to be along the cylinder with coordinate $r^0 \equiv u$. The transfer matrix associated with an infinitesimal step δu is given by

$$\mathcal{T} = e^{-\mathcal{H}\delta u}, \quad (7.232)$$

which defines in its turn a one-dimensional quantum Hamiltonian $\mathcal{H}$. For this Hamiltonian and the momentum operator $\mathcal{P}$ in the direction across the cylinder

$$\mathcal{P} = -i\frac{\partial}{\partial v} \quad (7.233)$$

we now introduce a complete set of simultaneous eigenstates

$$\mathcal{H}|n,k\rangle = E_n|n,k\rangle, \quad (7.234)$$

$$\mathcal{P}|n,k\rangle = k_n|n,k\rangle. \quad (7.235)$$

The two-point correlation function (7.230) can now be written with the help of the complete set of states $\{|n,k\rangle\}$ as

$$\langle\phi(w_1)\phi(w_2)\rangle = \sum_{n,k}\langle 0|\phi(w_1)|n,k\rangle\langle n,k|\phi(w_2)|0\rangle \quad (7.236)$$

where we interpret the order parameter field $\phi(w)$ as operators in the Hilbert space of the one-dimensional quantum Hamiltonian $\mathcal{H}$. The operator $\phi(w) = \phi(r^0, r^1)$, where r^0 is the coordinate along and r^1 the coordinate across the cylinder, can be written in the Heisenberg picture after a Wick rotation $r^0 \to -iu$ and renaming $r^1 \equiv v$

$$\phi(u, v) = e^{\mathcal{H}u}\phi(v)e^{-\mathcal{H}u}. \tag{7.237}$$

The two-point function then becomes

$$\begin{aligned}\langle\phi(w_1)\phi(w_2)\rangle &= \sum_{n,k}\langle 0|\phi(v_1)|n,k\rangle\langle n,k|\phi(v_2)|0\rangle e^{-(E_n-E_0)(u_1-u_2)} \\ &= \sum_{n,k}|\langle 0|\phi|n,k\rangle|^2 e^{-((E_n-E_0)(u_1-u_2)+ik_n(v_1-v_2))}\end{aligned} \tag{7.238}$$

where we have also used

$$\phi(v) = e^{iv\mathcal{P}}\phi e^{-iv\mathcal{P}} \tag{7.239}$$

and where E_0 is the ground state energy $\mathcal{H}|0\rangle = E_0|0\rangle$.

Now we can compare (7.238) and (7.230) and identify

$$E_n - E_0 = \frac{2\pi}{L}(x + m + \bar{m}), \tag{7.240}$$

$$k_n = \frac{2\pi}{L}(m - \bar{m}). \tag{7.241}$$

The ground state energy E_0 in the first of these expressions is as yet undetermined. This ground state energy is itself a finite-size quantity $E_0 = E_0(L)$. If it were known, these two equations could be used to determine the scaling dimensions x of the critical infinite system from calculations on finite systems. The following section shows how the finite-size behaviour of the ground state energy can be determined from conformal symmetry, in particular, that the central charge c is the decisive quantity that specifies the leading-order finite-size corrections. Lastly, a concrete physical interpretation of the meaning of the central charge will emerge through these calculations.

7.10.2 Ground state energy and a physical interpretation of the central charge

We remind ourselves that the free energy (in units of temperature T) is given by the partition function Z, which we can express via the Hamiltonian or via the functional integral over the action (cp. section 6.1)

$$e^{-F} = Z = \sum_{\{S_i\}} e^{-\mathcal{H}(S_i)} = \int \prod_i \mathcal{D}[\{\phi_i\}] \, e^{-S[\{\phi_i\}]}. \tag{7.242}$$

The free energy per unit length of an infinitely long strip of width L at criticality has, assuming periodic boundary conditions connecting the upper and lower edges of the strip, the finite-size scaling form

$$f(L) = \frac{F}{L} = f_\infty L + \frac{A}{L} + \dots \tag{7.243}$$

where numerical data and renormalization group arguments (Nightingale and Blöte, 1983; Privman and Fisher, 1984) suggest that the amplitude of the lowest finite-size correction proportional to L^{-1} is universal. The free energy per area f_∞ of the infinite system, the bulk energy, is in general not universal. In case of free or fixed boundary conditions, there is generally a surface energy term present in (7.243) that carries no dependence on the width L of the strip and which is also not universal.

Using the equivalence between two-dimensional systems of statistical mechanics and one-dimensional quantum systems, we can reinterpret (7.243) as the ground state energy of a quantum system of finite length L

$$E_0(L) = e_\infty L + \frac{A}{L} + \dots \tag{7.244}$$

where now ϵ_∞ is the ground state energy density of the infinite system.

We now demonstrate that the central charge c, the universal quantity separating statistical mechanical model systems into universality classes, is essentially characterizing the universal amplitude $A \propto c$. The central charge is connected to the introduction of a length scale in the system that locally breaks the conformal symmetry. Let us look at a specific example. The transformation used in section 7.1.2 from the whole complex plane to a cylinder with circumference L

$$z \to w = \frac{L}{2\pi} \ln z = u + iv \tag{7.245}$$

induces a transformation of the energy-momentum tensor on the plane $T(z) = T_{\text{plane}}(z)$ to the energy-momentum tensor on the cylinder $\tilde{T}(w) = T_{\text{cylinder}}(w)$ according to (7.162)

$$T_{\text{cylinder}}(w) = \left(\frac{2\pi}{L}\right)^2 \left[T_{\text{plane}}(z) z^2 - \frac{c}{24}\right], \tag{7.246}$$

where we used the Schwarzian derivative $S(w; z) = 1/(2z^2)$.

The Hamiltonian on the cylinder is given in terms of the energy-momentum operator as

$$\mathcal{H} = \frac{1}{2\pi}\int_0^L dv\,\left(T_{\text{cylinder}}(v) + \bar{T}_{\text{cylinder}}(v)\right) \tag{7.247}$$

for constant u, which we choose as $u = 0$. Using

$$z = e^{\frac{2\pi}{L}w} = e^{\frac{2\pi}{L}}e^{\frac{2\pi i}{L}v} = re^{i\theta} \tag{7.248}$$

we can evaluate this as a complex integral with the unit circle ($r = 1$) as integration contour.

Using $v = L\theta/2\pi$ and (7.246) we can write for, the holomorphic part of this integral

$$\mathcal{H} = \frac{1}{2\pi}\int_0^L dv\,T_{\text{cylinder}}(v) = \frac{L}{2\pi}\int_0^{2\pi}\frac{d\theta}{2\pi}\left(\frac{2\pi}{L}\right)^2\left(e^{2i\theta}\,T_{\text{plane}}\left(e^{i\theta}\right) - \frac{c}{24}\right). \tag{7.249}$$

The part of this integral proportional to c is trivial, and the second part can be written as

$$\int_0^{2\pi}\frac{d\theta}{2\pi}e^{2i\theta}\,T_{\text{plane}}\left(e^{i\theta}\right) = \oint\frac{dz}{2\pi i}\,zT_{\text{plane}}(z) = L_0 \tag{7.250}$$

where $z = e^{i\theta}$ on the unit circle. Hence, it can be identified with the Virasoro generator L_0.

In summary, we obtain

$$\mathcal{H} = \frac{2\pi}{L}\left(L_0 + \bar{L}_0 - \frac{c}{24} - \frac{\bar{c}}{24}\right). \tag{7.251}$$

Furthermore, we obtain for the momentum operator

$$k = \frac{1}{2\pi}\int_0^L dv\,\left(T_{\text{cylinder}}(v) - \bar{T}_{\text{cylinder}}(v)\right) = \frac{2\pi}{L}\left(L_0 - \bar{L}_0\right). \tag{7.252}$$

We conclude from (7.251) and (7.252) that the eigenstates of the operators $\mathcal{H}$ and k are in exact correspondence to the eigenstates of the operators $L_0 + \bar{L}_0$ and $L_0 - \bar{L}_0$ for which Belavin *et al.* (1984*b*) were able to show, *inter alia*, and which are described in section 7.9.9:

- The eigenvalues of $(L_0, \bar{L}_0)$ are in exact correspondence with the conformal weights $(h, \bar{h})$;
- The ground state of $(L_0, \bar{L}_0)$ corresponds to $(h = 0, \bar{h} = 0)$; and
- The eigenstates of $(L_0, \bar{L}_0)$ are among the highest-weight representations of the Virasoro algebra where the highest weight state in a given representation (by a

definitional accident with the lowest value of h and $\bar{h}$) corresponds to a primary operator. The eigenvalues of the other states differ by integers from h or $\bar{h}$.

These results yield the eigenvalues of (7.251) and (7.252)

$$E = E_0 + \frac{2\pi x}{L} \tag{7.253}$$

$$k = \frac{2\pi s}{L} \tag{7.254}$$

where $x = h+\bar{h}$ and $s = h-\bar{h}$ are the scaling dimension and the spin of the corresponding operator, respectively. The first result is in agreement with the more general result (7.240). The second result can be combined with (7.241) to yield the general expression for the momentum

$$k = \frac{2\pi}{L}(s + m - \bar{m})\,. \tag{7.255}$$

Moreover, the ground state energy of (7.251) is

$$E_0 = -\frac{\pi c}{6L} \tag{7.256}$$

assuming that $c = \bar{c}$, a famous result due simultaneously to Blöte *et al.* (1986) and Affleck (1986).

The derivation of (7.256) pertains to the finite strip geometry only and thus there will be no contribution from the ground state of the infinite system. Taking such a contribution into account, (7.256) becomes

$$E_0 = e_\infty L - \frac{\pi c}{6L} \tag{7.257}$$

in accordance with our expression (7.244).

Our derivation of (7.256) also shows that this energy term is not dependent on the state with which it is evaluated. It contributes a constant term or zero mode to the energy and, as such, it is the equivalent of the vacuum energy of a finite region in the Casimir effect (Casimir, 1948; for recent comprehensive treatments, see Bordag *et al.*, 2009 or Simpson and Leonhardt, 2015). The derivation of (7.256) essentially depends on (7.162), the behaviour of the energy-momentum tensor under conformal transformations where the extra term containing c does not depend on the energy-momentum tensor itself. This interpretation of the central charge c as the amplitude of the Casimir energy is one of the several physical interpretations possible for this central quantity of conformal symmetry. For other interpretations, we refer to the literature.

These results offer an important method of testing the conformal symmetry of critical statistical mechanical models by direct and independent calculations of the energies of a system of finite size. These direct and independent calculations of the value of the central charge c and the scaling dimensions can be numerical or analytical calculations, often even exact analytical calculations, if the finite-size model is amenable to an analytical extraction of its energies.

In part VI on the Bethe ansatz for finite systems, we shall encounter methods based on the Bethe ansatz that allow the detailed analytical determination of the energies of a finite Heisenberg quantum spin chain from which the scaling dimension x and the central charge c of this model can be extracted.

8

Models of Strongly Interacting Quantum Matter

The theory is not an ultimate one; its object is physical rather than metaphysical. From the point of view of the physicist, a theory of matter is a policy rather than a creed; its object is to connect or co-ordinate apparently diverse phenomena, and above all to suggest, stimulate, and direct experiment. It ought to furnish a compass which, if followed, will lead the observer further and further into previously unexplored regions.

Joseph John Thomson (1846–1940)

The theoretical description of strongly correlated quantum matter is arguably one of the most challenging areas of contemporary theoretical physics. Many fascinating phenomena have emerged in this area and still await complete understanding. Examples include Bose–Einstein condensation in atomic gases, high-temperture superconductivity in ceramic materials, the integer and fractional quantum Hall effects, heavy electron materials, the Kondo effect, and quantum critical phenomena. One of the most recent and quite exciting developments is connected to the advent of the technology of optical lattices, by which it is possible to create strongly interacting Bose and Fermi systems in the laboratory. Our focus, however, is on low-dimensional, especially one-dimensional, models of quantum matter (a topic to which, with different emphases, for instance, Voit (1994) and Giamarchi (2004) are devoted).

On the technical level, traditional perturbative many-particle techniques will not work satisfactorily in most cases. Novel non-perturbative approaches have to be employed and developed. To control these approaches, exactly solvable models are of paramount importance. In order to provide physical background for the methodologies for the mathematical treatment of exactly solvable and integrable models in this book, this chapter introduces a number of interacting or strongly interacting systems for which corresponding models exist and have been successfully solved.

From the plethora of possible strongly interacting quantum many-particle systems, we have selected a number of models representing Bosonic and Fermionic systems as well as systems where mostly magnetic moments are involved. Section 8.1 explores the Bose gas or fluid starting with the non-interacting Bose gas (see section 4.13.1), although here we include an external trapping potential. In order to improve the description of the condensate of Bose particles, we then treat the interacting Bose fluid with approximative schemes

Models of Quantum Matter: A First Course on Integrability and the Bethe Ansatz. Hans-Peter Eckle, Oxford University Press (2019). © Hans-Peter Eckle. DOI: 10.1093/oso/9780199678839.001.0001

of increasing accuracy, thereby deriving the Gross–Pitaevskii equations for the condensate wave function and applying the sophisticated microscopic Bogoliubov theory.

The following sections are devoted to Fermionic systems where, in most cases, we have electronic systems in mind. Section 8.2 presents an overview of electronic systems, which remains within an effective mean-field description. Section 8.3 then covers the full problem of electrons mutually interacting via their Coulomb potentials. The eminently fruitful Landau Fermi liquid theory is the focus of section 8.4. In low, especially one, dimensions, the Fermi liquid breaks down and has to be replaced by the Luttinger liquid concept, which is covered section 8.5. Eventually the theories and models of strongly interacting electrons lead also to an understanding of magnetism—a genuinely quantum phenomenon addressed in section 8.6. Sections 8.7 and 8.8 are then devoted to a closer look at two very prominent models of condensed matter theory that are also intimately related to magnetic phenomena: the Hubbard and Heisenberg models. Section 8.9 covers another interesting and challenging aspect of strong electronic interactions that consists in the interplay of delocalized electrons and the magnetic moments of localized electrons. Lastly, in section 8.10 we leave the realm of condensed matter theory and consider models for the strong interaction of light and matter and discuss the Rabi model, which has many descendent models.

The treatment here is necessarily sometimes cursory and interested readers are referred to the existing vast literature on the subject, e.g. Auerbach (1994), Douçot and Zinn-Justin (1995), Fazekas (1999), Fulde (2002), and the recent volume by Piers Coleman (2015). The latter work focuses on many of the advanced techniques in quantum many-body physics, which we cannot accommodate here. Our emphasis instead is on conceptual foundations rather than calculational (often approximative) techniques, examples of real systems, and applications.

8.1 Bose fluid

When matter is cooled to very low temperatures, its properties may change dramatically. Heike Kamerlingh Onnes succeeded in liquefying helium using an ingenious experimental set-up (for the story of the discovery, see van Delft (2008).

One of the most striking aspects of matter at very low temperatures is Bose–Einstein condensation (discussed in section 4.13.2), an intriguing macroscopic quantum phenomenon for a gas of non-interacting Bose particles. Bose–Einstein condensation is the phenomenon where, at sufficiently low temperatures, the dominant fraction of Bose particles condenses into the lowest, and hence macroscopically occupied, state, the ground state of the non-interacting Bose gas. Section 4.13.4 that this phenomenon has recently become the focus of active research investigating interactions between the Bose particles.

This section outlines the theoretical description of weakly interacting Bose gases, particularly the existence of an exactly solvable model of an interacting Bose gas, whose Bethe ansatz solution we discuss in detail in chapter 15. Moreover, the Gross–Pitaevskii equation, which describes an approximation to the wave function of the Bose–Einstein condensate and which we derive in section 8.1.2, is mathematically closely related to this exactly solvable model.

We restrict our attention to only a few basic discussions. We do not attempt to provide a complete overview of recent theoretical developments, let alone consider the richness of the experimental situation.

For a deeper penetration of the subject, interested readers are referred to Lewenstein *et al.* (2012), Pitaevskii and Stringari (2008), Bloch *et al.* (2008), Pethick and Smith (2008), Leggett (2006), and the Les Houches summer school volume Kaiser *et al.* (1999).

However, we first consider the experimentally highly relevant situation where a non-interacting Bose gas is confined by an external potential $V(\mathbf{r})$. We examine the simplest case of a so-called harmonic trap, where $V(\mathbf{r}) \propto |\mathbf{r}|^2$.

8.1.1 Non-interacting Bose gas in a harmonic trap

A non-interacting gas of Bosons is characterized by the Hamiltonian

$$\mathcal{H}_0 = \sum_n \left(-\frac{\nabla^2}{2m} + V(\mathbf{r}_n) \right). \tag{8.1}$$

The external potential $V(\mathbf{r})$ may have various forms that depend in practice on the experimental possibilities. For the present discussion, we shall choose a three-dimensional harmonic oscillator potential that may, however, be anisotropic with different frequencies ω_x, ω_y, and ω_z in the different Cartesian space directions

$$V(\mathbf{r}) = \frac{1}{2} m \left(\omega_x^2 x^2 + \omega_y^2 y^2 + \omega_z^2 z^2 \right). \tag{8.2}$$

In thermal equilibrium, an ensemble of N Bose particles is then described by the density operator of the grand canonical ensemble (cf. sections 4.12.3 and 4.13.1)

$$\rho = \frac{1}{\mathcal{Z}} \exp\left(-\beta \left[\mathcal{H} - \mu \mathcal{N}\right]\right), \tag{8.3}$$

where $\mathcal{N}$ is the number operator giving the total number of particles and μ is the chemical potential, which, in the following calculations, can again be conveniently combined into the fugacity

$$z = e^{\beta\mu}. \tag{8.4}$$

The harmonic potential $V(\mathbf{r})$ corresponds to three independent one-dimensional oscillators whose single particle eigenstates can thus be described by three numbers

$$\mathbf{m} = (m_x, m_y, m_z), \qquad m_\alpha = 0, 1, 2, \ldots, \qquad \alpha = x, y, z, \tag{8.5}$$

corresponding to the single-particle energy levels

$$\varepsilon_{\mathbf{m}} = \omega_x m_x + \omega_y m_y + \omega_z m_z. \tag{8.6}$$

For convenience, the zero-point energy $(\omega_x + \omega_y + \omega_z)/2$ may be absorbed into the chemical potential. In order to avoid complicating the equations unnecessarily, we now focus on the isotropic harmonic oscillator potential $\omega_x = \omega_y = \omega_z \equiv \omega$. Then, the energy levels become $\varepsilon_{\mathbf{m}} = \omega m$ with $m = m_x + m_y + m_z = 0, 1, 2, \ldots$ and a degeneracy of

$$(m+1)(m+2)/2. \tag{8.7}$$

The occupation number of the single particle eigenstates of the Bose particles in the trap is determined by the Bose–Einstein distribution

$$n_{\mathbf{m}} = \frac{1}{e^{\beta(\varepsilon_{\mathbf{m}}-\mu)} - 1} = \frac{1}{\frac{1}{z}e^{\beta\omega m} - 1}, \tag{8.8}$$

which requires $0 < z < 1$ in order to be always positive. The average number of particles is given by $N = \sum_{\mathbf{m}} \varepsilon_{\mathbf{m}}$. The occupation number of the ground state $m = 0$ is determined by the fugacity

$$N_0 = n_{\mathbf{0}} = \frac{z}{1-z}. \tag{8.9}$$

This occupation number diverges as $z \to 1$, which signals the presence of a Bose–Einstein condensate in the many-particle ground state of the non-interacting Bosons confined to a harmonic trap.

Since we are working with the grand canonical potential, the total number of particles is not fixed and instead will be determined in the thermodynamical equilibrium via the chemical potential. In particular, we want to see the Bose–Einstein condensate develop as the fugacity approaches unity, $z \to 1$. This limit determines the number of Bose particles in excited states

$$\delta N = N - N_0 = \sum_{\mathbf{m}\neq 0} n_{\mathbf{m}}, \tag{8.10}$$

which is bounded from above

$$\delta N = \sum_{\mathbf{m}\neq 0} \frac{1}{\frac{1}{z}e^{\beta\omega m} - 1} < \sum_{\mathbf{m}\neq 0} \frac{1}{e^{\beta\omega m} - 1} \equiv \delta N_{\max}. \tag{8.11}$$

The occupation number δN of excited states exhibits saturation, and $\delta N_{\max}$ is thus called the saturation number.

In order to calculate this saturation bound for the occupation number of the excited states, we need to take into account the degeneracy (8.7) of the single-particle energy levels (8.6) for the isotropic harmonic trap. The saturation number defined in (8.11) can then be written as

$$\delta N_{\text{max}} = \frac{1}{2}\sum_{m=1}^{\infty} \frac{(m+1)(m+2)}{e^{\beta\omega m} - 1}. \tag{8.12}$$

Provided $\beta\omega \ll 1$, we can replace this sum by an integral using the variable $x \equiv m\beta\omega$

$$\delta N_{\text{max}} \to \frac{1}{2(\beta\omega)^3}\int_{\beta\omega}^{\infty} dx\, \frac{(x+\beta\omega)(x+2\beta\omega)}{e^x - 1}. \tag{8.13}$$

The following exercise is useful to evaluate the integrals in this expression.

EXERCISE 8.1 Integrals involving the Bose–Einstein distribution function
Show that for $p > 1$, the following integrals are

$$I(p) = \int_0^{\infty} dx\, \frac{x^{p-1}}{e^x - 1} = \zeta(p)\Gamma(p) \tag{8.14}$$

given as a product of the gamma function

$$\Gamma(p) = \int_0^{\infty} d\rho\, e^{-\rho}\rho^{p-1} \tag{8.15}$$

and the Riemann ζ function

$$\zeta(p) = \sum_{n=1}^{\infty} \frac{1}{n^p}. \tag{8.16}$$

The saturation number (8.13) can be split into three integrals. The first two, with the numerator of the integrand proportional to x^2 and x, respectively, can be approximated by replacing the lower integration limit by zero since the integrand remains regular. Using the results of the exercise, these first two integrals contribute to the saturation number (8.13)

$$\delta N_{\text{max}} = \frac{\zeta(3)}{(\beta\omega)^3} + \frac{3\zeta(2)}{2(\beta\omega)^2} = \left(\frac{T}{\omega}\right)^3 \zeta(3) + \frac{3}{2}\left(\frac{T}{\omega}\right)^2 \zeta(2). \tag{8.17}$$

The third integral is more delicate. Setting the lower integration limit to zero would produce a logarithmic singularity. However, the integral can be evaluated also with the finite lower integration limit $\beta\omega$. Using the identity

$$\frac{1}{e^x - 1} = \frac{e^{-x}}{1 - e^{-x}} = \sum_{\nu=1}^{\infty} e^{-\nu x} \tag{8.18}$$

we obtain

$$\int_{\beta\omega}^{\infty} dx \, \frac{1}{e^x - 1} = \int_{\beta\omega}^{\infty} dx \sum_{\nu=1}^{\infty} e^{-\nu x} = \sum_{\nu=1}^{\infty} \frac{e^{-\nu\beta\omega}}{\nu} = -\ln\left(1 - e^{-\beta\omega}\right). \tag{8.19}$$

Recalling the requirement that $\beta\omega \ll 1$ and taking into account the powers of $\beta\omega$ that dominate the three contributions to (8.13), we conclude that the dominant contribution stems from the first integral and yields a leading T^3 behaviour. Correspondingly, the Bose–Einstein condensate behaves as

$$\frac{N_0}{N} \approx \frac{N - \delta N_{\max}}{N} \approx 1 - \left(\frac{T}{\omega}\right)^3 \zeta(3)\frac{1}{N} \equiv 1 - \left(\frac{T}{T_c^{\mathrm{ho}}}\right)^3. \tag{8.20}$$

At the critical temperature

$$T_c^{\mathrm{ho}} = \omega \left(\frac{N}{\zeta(3)}\right)^{1/3} \tag{8.21}$$

the Bosons start to accumulate in the lowest single-particle state with energy $\varepsilon_0 = 0$. Above this temperature all Bosons are in excited states. The corresponding value

$$\omega\beta_c^{\mathrm{ho}} = \left(\frac{\zeta(3)}{N}\right)^{1/3} \ll 1 \tag{8.22}$$

renders credibility to our assumption $\beta\omega \ll 1$ used in this calculation.

Note the difference in the temperature dependence for the formation of the condensate fraction in the Bose–Einstein condensate in the harmonic trap (8.20) and in the homogeneous case (4.113).

8.1.2 Condensate wave function: Gross–Pitaevskii equation

This section investigates how the wave function of the condensed state, the condensate, can be described in thermal equilibrium for low temperatures $T \ll T_c$. In this temperature regime, we assume that the condensate is almost pure, i.e. we consider the approximation that the cloud of non-condensed Bosons has only a negligible effect on the condensate. Perturbation theory fails to describe this situation because of the highly degenerate nature of the problem. However, this situation can be described by a non-linear wave equation of the Schrödinger type: the Gross–Pitaevskii equation, which describes the dynamics of the condensate at low temperatures well below the critical temperature and which can be considered as a mean-field approximation. This equation can be derived in several ways (see the literature mentioned). We choose a variational approach in the following.

We take as a variational wave function of the condensed Bose gas the state where all Bose particles are in the same single-particle ground state $\psi_0(\mathbf{r})$ or, in second quantized form, with the vacuum state $|0\rangle \equiv |vac\rangle = |0, 0, \ldots\rangle$

$$|1, 0, 0, \ldots\rangle = a_0^\dagger|0\rangle = \int d^3r\, \psi_0(\mathbf{r})\psi^\dagger(\mathbf{r})|0\rangle. \tag{8.23}$$

Hence, we shall use the many-particle wave function

$$\Psi(\mathbf{r}_1, \mathbf{r}_1, \ldots, \mathbf{r}_{N_0}) = \prod_{n=1}^{N_0} \psi_0(\mathbf{r}_n) \tag{8.24}$$

or, in second quantized form,

$$\begin{aligned} |\Psi\rangle \equiv |N_0, 0, 0, \ldots\rangle &= \frac{1}{\sqrt{N_0!}}\left(a_0^\dagger\right)^{N_0}|0\rangle \\ &= \frac{1}{\sqrt{N_0!}}\left(\int d^3r\, \psi_0(\mathbf{r})\psi^\dagger(\mathbf{r})\right)^{N_0}|0\rangle, \quad N_0 = 0, 1, 2, \ldots \end{aligned} \tag{8.25}$$

as trial wave function. The trial wave function can be assumed as normalized

$$\prod_{n=1}^{N_0} \psi_0^*(\mathbf{r}_n)\psi_0(\mathbf{r}_n) = 1 \qquad \text{or, equivalently} \qquad \psi_0^*(\mathbf{r})\psi_0(\mathbf{r}) = 1 \tag{8.26}$$

or

$$\langle\Psi|\Psi\rangle = \left(\int d^3r\, \psi_0^*(\mathbf{r})\psi_0(\mathbf{r})\right)^{N_0} = 1. \tag{8.27}$$

We thus seek an equation that determines the wave function $\psi_0(\mathbf{r})$ by extremizing a suitable energy functional of the wave function (8.24) or (8.25), respectively.

The energy functional we wish to make extremal

$$E(\psi_0) = \frac{\langle\Psi|\mathcal{H}|\Psi\rangle}{\langle\Psi|\Psi\rangle} \tag{8.28}$$

is formed using the interacting many-particle Hamiltonian

$$\mathcal{H} = \mathcal{H}_0 + \mathcal{H}_{\mathbf{int}} = \sum_{n=1}^{N}\left(-\frac{\nabla^2}{2m} + V(\mathbf{r}_n)\right) + \frac{1}{2}\sum_{n=1}^{N}\sum_{\substack{m=1\\ m\neq n}}^{N} U(\mathbf{r}_n - \mathbf{r}_m). \tag{8.29}$$

In second quantized form this Hamiltonian is

$$\begin{aligned}\mathcal{H} = \int d^3r\, \psi^\dagger(\mathbf{r})\left(-\frac{\nabla^2}{2m} + V(\mathbf{r}_n)\right)\psi(\mathbf{r}) \\ + \frac{1}{2}\int d^3r\, d^3r'\, U(\mathbf{r}-\mathbf{r}')\psi^\dagger(\mathbf{r})\psi^\dagger(\mathbf{r}')\psi(\mathbf{r}')\psi(\mathbf{r}).\end{aligned} \tag{8.30}$$

In the following, we use the first quantized formulation, which is the more transparent in this case.

Since we are interested in a situation where the number of particles will be variable, we do not seek the extremal value of (8.28) directly, but rather of an energy functional, which is similar to the grand canonical potential discussed in section 4.10

$$\Phi(\psi_0) = \langle\Psi|\mathcal{H}|\Psi\rangle - \mu\langle\Psi|\Psi\rangle. \tag{8.31}$$

Searching for the extremum of this expression can also be viewed as searching for the extremum under a constraint introduced via the Lagrange multiplier μ. Physically this Lagrange multiplier represents the chemical potential.

EXERCISE 8.2 Variation of the energy functional Show that the variation of the energy functional (8.31) is

$$\frac{\delta\Phi}{\delta\psi_0^*} = N\int d^3r\, F(\mathbf{r})\delta\psi_0^*(\mathbf{r}) \tag{8.32}$$

where

$$F(\mathbf{r}) = \left[-\frac{1}{2m}\nabla^2 + V(\mathbf{r}) + (N-1)\left(\int d^3r'\, U(\mathbf{r}-\mathbf{r}')|\psi_0(\mathbf{r}')|^2\right) - \mu\right]\psi_0(\mathbf{r}). \tag{8.33}$$

What would you get if you calculated the variation $\delta\Phi/\delta\psi_0$ instead?

In order to be an extremum, the variational expression (8.32) must vanish, which, for arbitrary variation $\delta\psi_0{}^*$, implies that (8.33) must be zero, which can be written as

$$\left[-\frac{1}{2m}\nabla^2 + V(\mathbf{r}) + (N-1)\left(\int d^3r'\, U(\mathbf{r}-\mathbf{r}')|\psi_0(\mathbf{r}')|^2\right)\right]\psi_0(\mathbf{r}) = \mu\psi_0(\mathbf{r}). \tag{8.34}$$

For large enough systems, we can approximate $N - 1$ by N. Detailed considerations from scattering theory (see the cited literature) imply that the two-particle potential can be approximated by a contact potential

$$U(\mathbf{r}) = U_0\delta(\mathbf{r}) \equiv \frac{4\pi}{m} a\delta(\mathbf{r}), \tag{8.35}$$

which introduces the scattering length a, a measure of the intensity of the interaction between the Bosons. Both attractive ($a < 0$) and repulsive ($a > 0$) interactions are possible. With this potential (8.34) becomes

$$\left(-\frac{1}{2m}\nabla^2 + V(\mathbf{r}) + U_0 N|\psi_0(\mathbf{r})|^2\right)\psi_0(\mathbf{r}) = \mu\psi_0(\mathbf{r}) \tag{8.36}$$

or, after the substitution $\sqrt{N}\psi_0 \to \phi_0$

$$\left(-\frac{1}{2m}\nabla^2 + V(\mathbf{r}) + U_0|\phi_0(\mathbf{r})|^2\right)\phi_0(\mathbf{r}) = \mu\phi_0(\mathbf{r}). \tag{8.37}$$

This differential equation is the (time-independent) Gross–Pitaevskii equation. It is a non-linear equation that is very similar to the non-linear Schrödinger equation. The integro-differential equation (8.34) may be considered the generalized Gross–Pitaevskii equation.

We interpret $\phi_0(\mathbf{r})$, the wave function of the condensate with $n_0(\mathbf{r}) = |\phi_0(\mathbf{r})|^2$ its local density, as a classical field in analogy to an order parameter in the theory of critical phenomena. The Gross–Pitaevskii equation thus provides a mean-field theory describing the interacting non-uniform Bose gas formally at zero temperature, i.e. at temperatures well below the critical temperature. The non-uniformity is a consequence of the trap potential V. Hence, the Gross–Pitaevskii equation provides a tool for investigating Bose–Einstein condensates in traps of various shapes and strengths. The interaction of the Bose particles is captured in the Gross–Pitaevskii equation in one parameter the scattering length a. Technically, as a mean-field equation the Gross–Pitaevskii equation allows the self-consistent calculation of the order parameter $\phi_0(\mathbf{r}, t)$. There is a time-dependent form of the Gross–Pitaevskii equation (interested readers are referred to the cited literature).

A solution $\phi_0(\mathbf{r})$ of the Gross–Pitaevskii equation allows the calculation of various observable quantities, the most elementary of which are: the local density of the condensate $n_0(\mathbf{r}) = |\phi_0(\mathbf{r})|^2$ and the local current density

$$\mathbf{j}(\mathbf{r}) = -\frac{i}{2m}\left(\phi_0^*(\mathbf{r})\nabla\phi_0(\mathbf{r}) - \phi_0(\mathbf{r})\nabla\phi_0^*(\mathbf{r})\right) = n_0(\mathbf{r})\mathbf{v}(\mathbf{r}). \tag{8.38}$$

The latter reveals an interesting property of the phase of the condensate wave function. If we decompose $\phi_0(\mathbf{r}) = \sqrt{n_0(\mathbf{r})}\exp(i\theta(\mathbf{r}))$ into amplitude $\sqrt{n_0}$ and phase θ, we can calculate the velocity

$$\mathbf{v}(\mathbf{r}) = \frac{1}{m}\nabla\theta(\mathbf{r}). \tag{8.39}$$

This velocity field is given by a gradient of a scalar function, which means that the motion of the condensate is described by a flow. Moreover, this result shows a further use of the Gross–Pitaevskii equation, beyond its usefulness in non-uniform situations imposed by an external trapping potential. Condensate configurations may also be stable for topological reasons. Since $\mathbf{v}$ in (8.39) is a gradient field, its rotation must vanish, as it is irrotational,

$$\nabla \times \mathbf{v} = \frac{1}{m}\nabla \times \nabla\theta(\mathbf{r}) = 0, \tag{8.40}$$

which, by Stokes's theorem, is equivalent to

$$\oint_C \mathbf{v} \cdot d\mathbf{s} = 0 \tag{8.41}$$

around a closed loop C. This argument breaks down, however, for a singular phase field, as is the case for a vortex in the condensate fluid. But, since the condensate wave function $\phi_0(\mathbf{r})$ must remain single-valued, the change in phase around a closed contour must be a multiple of 2π and hence we must have for the circulation

$$\Gamma \equiv \frac{1}{m}\Delta\theta = \oint_C \mathbf{v} \cdot d\mathbf{s} = \frac{2\pi}{m}\ell \tag{8.42}$$

with an integer ℓ, i.e. the circulation is quantized.

EXERCISE 8.3 Superfluidity of the Bose–Einstein condensate The time-dependence of the condensate wave function $\phi_0(\mathrm{r}, t)$ is governed by the time-dependent Gross–Pitaevskii equation

$$i\frac{\partial\phi_0}{\partial t} = \left(-\frac{1}{2m}\nabla^2 + V(\mathbf{r}) + U_0|\phi_0|^2\right)\phi_0. \tag{8.43}$$

Use the decomposition $\phi_0(\mathbf{r}) = \sqrt{n_0(\mathbf{r})}\exp\left(i\theta(\mathbf{r})\right)$ to derive equations for the density n_0 and phase θ of the condensate wave function.

You will obtain a continuity equation for the density n_0 and velocity $\mathbf{v} = \nabla\theta/m$ and a second equation that can be interpreted in analogy to a special case of the Navier–Stokes equation, called the Euler equation, as an equation of a fluid flow without the influence of viscosity.

The following section explores a more detailed microscopic many-particle theory of the weakly interacting Bose gas.

8.1.3 Microscopic description of the interacting Bose gas: Bogoliubov theory

The Bogoliubov theory of the weakly interacting Bose gas provides a strategy to improve on the Gross–Pitaevskii approximation discussed in the previous section. In its simplest form, it assumes a uniform condensate, i.e. one that is not under the influence of an external potential $V(\mathbf{r})$. The condensate wave function is then given by a constant density n as $\phi_0(\mathbf{r}) = \sqrt{n}$.

We shall draw on the second quantization formulation and work in a basis of momentum eigenstates $\{|\mathbf{k}\rangle\}$ (cf. chapter 2, especially sections 2.5–2.7). The many-particle Hamiltonian appropriate for a grand canonical treatment where the particle number is variable assumes the form

$$\mathcal{H} = \mathcal{H}_0 + \mathcal{H}_{\text{int}} = \sum_{\mathbf{k}} \varepsilon(\mathbf{k}) a^\dagger_{\mathbf{k}} a_{\mathbf{k}} + \frac{U_0}{2\Omega} \sum_{\substack{\mathbf{k}_1,\mathbf{k}_2,\mathbf{k}_3,\mathbf{k}_4 \\ \mathbf{k}_1+\mathbf{k}_2=\mathbf{k}_3+\mathbf{k}_4}} a^\dagger_{\mathbf{k}_1} a^\dagger_{\mathbf{k}_2} a_{\mathbf{k}_3} a_{\mathbf{k}_4} \tag{8.44}$$

where $\varepsilon(\mathbf{k}) = k^2/2m$, Ω is the volume, and we again use the contact potential (8.35).

When acting with the interaction part of this Hamiltonian, $\mathcal{H}_{\text{int}}$, on the approximation to the ground state (8.25) used in the Gross–Pitaevskii theory, only contributions survive from $\mathbf{k}_3 = \mathbf{k}_4 = 0$

$$\mathcal{H}_{\text{int}}|\Psi\rangle = \frac{U_0}{2\Omega} \sum_{\substack{\mathbf{k}_1,\mathbf{k}_2 \\ \mathbf{k}_1+\mathbf{k}_2=0}} a^\dagger_{\mathbf{k}_1} a^\dagger_{\mathbf{k}_2} a_0 a_0 |\Psi\rangle = \frac{U_0}{2\Omega} \sum_{\mathbf{k}} a^\dagger_{\mathbf{k}} a^\dagger_{-\mathbf{k}} a_0 a_0 |\Psi\rangle. \tag{8.45}$$

This suggests that an improvement of the Gross–Pitaevskii wave function will contain pairs of Bosons with opposite momenta, $\mathbf{k}$ and $-\mathbf{k}$.

We expect that the ground state of the weakly interacting Bose gas is close to the Gross–Pitaevskii ground state in the sense that there will only be a few excitations above this ground state. Quantitatively, the occupancy of the states with $\mathbf{k} = 0$ (zero-momentum states) will be much greater than of the states with momentum $\mathbf{k} \neq 0$. Thus, we break up the interaction part of the Hamiltonian into contributions according to the number of zero-momentum Bose creation or annihilation operators

$$\mathcal{H}_{\text{int}} = \mathcal{H}_{\text{int},4} + \mathcal{H}_{\text{int},2} + \mathcal{H}_{\text{int},1} + \mathcal{H}_{\text{int},0}. \tag{8.46}$$

There cannot be a contribution in this Hamiltonian with three creation or annihilation operators with zero-momentum $\mathbf{k} = \mathbf{0}$. Such a contribution would not conserve momentum, as we see immediately when writing down the other contributions explicitly where momentum conservation is possible. When all momenta are zero, we obtain the contribution

$$\mathcal{H}_{\text{int},4} = \frac{U_0}{2\Omega} a_0^\dagger a_0^\dagger a_0 a_0. \tag{8.47}$$

The Gross–Pitaevskii approximation consists in keeping only this contribution.

The next contribution contains two zero-momentum operators

$$\mathcal{H}_{\text{int},2} = \frac{U_0}{2\Omega} \sum_{\mathbf{k}\neq 0} \left(a_{\mathbf{k}}^\dagger a_{-\mathbf{k}}^\dagger a_0 a_0 + a_0^\dagger a_0^\dagger a_{\mathbf{k}} a_{-\mathbf{k}} + 4 a_{\mathbf{k}}^\dagger a_0^\dagger a_{\mathbf{k}} a_0 \right). \tag{8.48}$$

The factor four in this contribution with two zero-momentum operators stems from the fact that this contribution appears in four equivalent ways in the general expression. The contribution with only one zero-momentum operator is

$$\mathcal{H}_{\text{int},1} = \frac{U_0}{2\Omega} \sum_{\substack{\mathbf{k}_1,\mathbf{k}_2,\mathbf{k}_3\neq 0 \\ \mathbf{k}_1=\mathbf{k}_2+\mathbf{k}_3}} \Big(a_{\mathbf{k}_2}^\dagger a_{\mathbf{k}_3}^\dagger a_{\mathbf{k}_1} a_0 + a_0^\dagger a_{\mathbf{k}_1}^\dagger a_{\mathbf{k}_2} a_{\mathbf{k}_3} + a_{\mathbf{k}_1}^\dagger a_0^\dagger a_{\mathbf{k}_2} a_{\mathbf{k}_3} + a_{\mathbf{k}_2}^\dagger a_{\mathbf{k}_3}^\dagger a_0 a_{\mathbf{k}_1} \Big). \tag{8.49}$$

Finally, the contribution with no zero-momentum operators is

$$\mathcal{H}_{\text{int},0} = \frac{U_0}{2\Omega} \sum_{\substack{\mathbf{k}_1,\mathbf{k}_2,\mathbf{k}_3,\mathbf{k}_4\neq 0 \\ \mathbf{k}_1+\mathbf{k}_2=\mathbf{k}_3+\mathbf{k}_4}} a_{\mathbf{k}_1}^\dagger a_{\mathbf{k}_2}^\dagger a_{\mathbf{k}_3} a_{\mathbf{k}_4}. \tag{8.50}$$

In order to improve on the Gross–Pitaevskii approximation, we now include the contribution $\mathcal{H}_2$ containing all contributions with two zero-momentum operators. This is called the pair approximation. The Hamiltonian in this approximation we retain is

$$\begin{aligned} \mathcal{H}_{\text{pair}} = \sum_{\mathbf{k}} \varepsilon(\mathbf{k}) a_{\mathbf{k}}^\dagger a_{\mathbf{k}} + \frac{U_0}{2\Omega} a_0^\dagger a_0^\dagger a_0 a_0 \\ + \frac{U_0}{2\Omega} \sum_{\mathbf{k}\neq 0} \left(a_{\mathbf{k}}^\dagger a_{-\mathbf{k}}^\dagger a_0 a_0 + a_0^\dagger a_0^\dagger a_{\mathbf{k}} a_{-\mathbf{k}} + 4 a_{\mathbf{k}}^\dagger a_0^\dagger a_{\mathbf{k}} a_0 \right). \end{aligned} \tag{8.51}$$

This Hamiltonian is still quite complicated as it involves products of four creation and annihilation operators. Our goal is to reduce this complication further in order to finally obtain a Hamiltonian that contains only products of two creation and annihilation operators, i.e. a Hamiltonian of quadratic form.

First we must attend to the contribution that contains the four zero-momentum operators. Introducing

$$\delta N = \sum_{\mathbf{k}\neq 0} N_{\mathbf{k}} = \sum_{\mathbf{k}\neq 0} a_{\mathbf{k}}^\dagger a_{\mathbf{k}} \tag{8.52}$$

and

$$a_0^\dagger a_0 = N_0 = N - \delta N, \tag{8.53}$$

we can write this term as

$$a_0^\dagger a_0^\dagger a_0 a_0 = a_0^\dagger a_0 a_0^\dagger a_0 - a_0^\dagger a_0 = (N - \delta N)^2 - N + \delta N \tag{8.54}$$

$$= N(N-1) - 2\delta N N_0 + \mathcal{O}(\delta N) \tag{8.55}$$

$$= N(N-1) - 2\sum_{\mathbf{k}\neq 0} a_{\mathbf{k}}^\dagger a_{\mathbf{k}} a_0^\dagger a_0 \tag{8.56}$$

$$\approx N^2 - 2N_0 \sum_{\mathbf{k}\neq 0} a_{\mathbf{k}}^\dagger a_{\mathbf{k}} \approx N^2 - 2N \sum_{\mathbf{k}\neq 0} a_{\mathbf{k}}^\dagger a_{\mathbf{k}}, \tag{8.57}$$

observing that $N \approx N_0$ while $N, N_0 \gg \delta N$.

As a further and last simplification, we replace all zero-momentum operators still left in the Hamiltonian by $\sqrt{N_0} \approx \sqrt{N}$, so that we finally obtain the so-called Bogoliubov Hamiltonian

$$\mathcal{H}_{\text{pair}} = \sum_{\mathbf{k}} \varepsilon(\mathbf{k}) a_{\mathbf{k}}^\dagger a_{\mathbf{k}} + \frac{U_0}{2} Nn + \frac{U_0 n}{2} \sum_{\mathbf{k}\neq 0} \left(a_{\mathbf{k}}^\dagger a_{-\mathbf{k}}^\dagger + a_{\mathbf{k}} a_{-\mathbf{k}} + 2 a_{\mathbf{k}}^\dagger a_{\mathbf{k}} \right), \tag{8.58}$$

where $n = N/\Omega$ is the particle density.

The simplification regarding the condensate particle number N_0 can be justified by observing how typical products $a_{\mathbf{k}}^\dagger a_0$ and $a_{\mathbf{k}} a_0^\dagger$ appearing in the Hamiltonian (8.51) act on states of the Fock space. Factoring states from the Fock space into non-zero and zero-momentum parts $|\Psi_{\mathbf{k}\neq 0}\rangle |N_0\rangle$, we obtain

$$a_{\mathbf{k}}^\dagger a_0 \left(|\Psi_{\mathbf{k}\neq 0}\rangle |N_0\rangle \right) = \left(a_{\mathbf{k}}^\dagger |\Psi_{\mathbf{k}\neq 0}\rangle \right) \left(a_0 |N_0\rangle \right) = \left(a_{\mathbf{k}}^\dagger |\Psi_{\mathbf{k}\neq 0}\rangle \right) \sqrt{N_0} |N_0 - 1\rangle \tag{8.59}$$

and

$$a_{\mathbf{k}} a_0^\dagger \left(|\Psi_{\mathbf{k}\neq 0}\rangle |N_0\rangle \right) = \left(a_{\mathbf{k}} |\Psi_{\mathbf{k}\neq 0}\rangle \right) \left(a_0^\dagger |N_0\rangle \right) = \left(a_{\mathbf{k}} |\Psi_{\mathbf{k}\neq 0}\rangle \right) \sqrt{N_0 + 1} |N_0 + 1\rangle. \tag{8.60}$$

The number of Bose particles N_0 in the lowest quantum state is supposed to be very large, so we can approximately ignore the difference between N_0 and $N_0 + 1$. The approximation thus consists in the assumption that the particle number N_0 will not fluctuate appreciably and the matrix elements of the Hamiltonian will not change when we make this replacement.

EXERCISE 8.4 Bogoliubov transformation of the Bogoliubov Hamiltonian
Consider the canonical transformation

$$a_{\mathbf{k}}^{\dagger} = u_k b_{\mathbf{k}}^{\dagger} - v_k b_{-\mathbf{k}} \qquad \text{and} \qquad a_{\mathbf{k}} = u_k b_{\mathbf{k}} - v_k b_{-\mathbf{k}}^{\dagger} \tag{8.61}$$

with coefficients u_k and v_k assumed to be real and spherically symmetric. Show that the Bosonic commutation relations are preserved, i.e. that

$$\left[a_{\mathbf{k}}, a_{\mathbf{k}'}^{\dagger}\right] = \delta_{\mathbf{k}\mathbf{k}'}, \qquad \Leftrightarrow \qquad \left[b_{\mathbf{k}}, b_{\mathbf{k}'}^{\dagger}\right] = \delta_{\mathbf{k}\mathbf{k}'}, \tag{8.62}$$

$$\left[a_{\mathbf{k}}^{\dagger}, a_{\mathbf{k}'}^{\dagger}\right] = [a_{\mathbf{k}}, a_{\mathbf{k}'}] = 0, \qquad \Leftrightarrow \qquad \left[b_{\mathbf{k}}^{\dagger}, b_{\mathbf{k}'}^{\dagger}\right] = [b_{\mathbf{k}}, b_{\mathbf{k}'}] = 0. \tag{8.63}$$

Show that the transformed Bogoliubov Hamiltonian (8.58) becomes

$$\mathcal{H}_{\text{pair}} = \sum_{\mathbf{k}\neq 0} E(\mathbf{k}) b_{\mathbf{k}}^{\dagger} b_{\mathbf{k}} + E_0, \tag{8.64}$$

where the dispersion of the transformed Bosons is given by

$$E(\mathbf{k}) = \left((\varepsilon(\mathbf{k}) + U_0 n)^2 - (U_0 n)^2\right)^{1/2} = (\varepsilon(\mathbf{k})\,(\varepsilon(\mathbf{k}) + 2U_0 n))^{1/2} \tag{8.65}$$

and the ground state energy by

$$E_0 = \frac{U_0}{2} N n + \frac{1}{2} \sum_{\mathbf{k}\neq 0} (E(\mathbf{k}) - \varepsilon(\mathbf{k}) - U_0 n)\,. \tag{8.66}$$

Exercise 8.4 shows that, under the approximations made, the excitations of the weakly interacting Bose gas can be described as a non-interacting Bose gas with a spectrum given by (8.65), which is called the Bogoliubov excitation spectrum.

This spectrum has remarkable properties. For a small momentum $|\mathbf{k}| \to 0$, it behaves as

$$E(\mathbf{k}) \approx \sqrt{\frac{\mu}{m}} |\mathbf{k}| = \sqrt{\frac{U_0 n_0}{m}} |\mathbf{k}| \equiv v_0 |\mathbf{k}|, \tag{8.67}$$

which is the dispersion relation of a sound wave in a fluid with propagation speed v_0. For a large momentum $|\mathbf{k}| \to \infty$, we recover the dispersion of free particles

$$E(\mathbf{k}) \approx \frac{k^2}{2m} = \varepsilon(\mathbf{k}). \tag{8.68}$$

The ground state energy (8.66) diverges for large momentum $|\mathbf{k}| \to \infty$. This is an artefact of the approximation and can be remedied by a closer analysis of the approximation that leads to the use of a constant two-particle potential U_0. Interested readers are referred to a more detailed discussion found in section 35 of Fetter and Walecka (2003).

For a further discussion of the Bogoliubov approximation and more on the great variety of intriguing phenomena connected with Bose fluids, we refer again to the cited literature.

8.2 Electronic correlations

The classical Drude theory[1] of electrons in metals, which treats the electrons in analogy to the kinetic theory of gases, failed to predict the heat capacity and the magnetic susceptibility of metals correctly. These quantities could only be computed in semi-quantitative agreement with experiment after the advent of quantum mechanics in the mid 1920s through the contributions of Pauli, and especially of Sommerfeld[2]. The essential new insight due to Pauli is that electrons obey the exclusion principle. The semi-quantitatively correct results for the specific heat and magnetic susceptibility of metals constituted one of the early triumphs of quantum mechanics and greatly helped to establish its credibility.

We begin our discussion in the next section looking first at what we believe[3] to be the microscopic foundation for the electronic correlations in quantum matter. Then, we adopt a more heuristic viewpoint and build up the theory from the most drastic approximations: non-interacting and free electrons, respectively.

8.2.1 The 'Theory of Everything' Hamiltonian

We can easily write down a Hamiltonian that is the basis for many properties of condensed matter systems, i.e. much of condensed matter physics lies within the reach of this Hamiltonian. According to Robert Laughlin (1998) and Laughlin and David Pines (2000), this Hamiltonian constitutes the 'Theory of Everything' of condensed matter physics. In first quantization it is (Laughlin and Pines 2000; Laughlin 1998) for a condensed matter system consisting of N electrons of mass m and M ionic cores of masses M_α and charges Z_α

1 See, e.g. Ashcroft and Mermin (1976), chapter 1, which emphasizes that the theory, despite its shortcomings, can still be of use in condensed matter physics.

2 See the early handbook review on the electron theory of metals by Sommerfeld and Bethe (1933), which, according to Sommerfeld's footnote on page 1, has, except for the first chapter, been written by Bethe.

3 According to Xiao-Gang Wen (2004), 'As we cannot check the validity of the result obtained from the Schrödinger equation directly in full detail, our belief that the Schrödinger equation determines all the properties of a many-particle system is just a *faith*'.

$$\mathcal{H} = -\sum_{j=1}^{N} \frac{1}{2m}\nabla_j^2 - \sum_{\alpha=1}^{M} \frac{1}{2M_\alpha}\nabla_\alpha^2$$
$$-\sum_{j=1}^{N}\sum_{\alpha=1}^{M} \frac{Z_\alpha e^2}{|\mathbf{r}_j - \mathbf{R}_\alpha|} + \sum_{\substack{j,k=1\\ j<k}}^{N} \frac{e^2}{|\mathbf{r}_j - \mathbf{r}_k|} + \sum_{\substack{\alpha,\beta=1\\ \alpha<\beta}}^{M} \frac{Z_\alpha Z_\beta e^2}{|\mathbf{R}_\alpha - \mathbf{R}_\beta|} \tag{8.69}$$

where, to have charge neutrality, $N = \sum_{\alpha=1}^{M} Z_\alpha$. The first two terms in this Hamiltonian describe the kinetic energies of the electrons and ions, respectively, while the last two terms describe the Coulomb repulsion between electrons and ions, respectively. The third term describes the Coulomb attraction between electrons and ions.

For the time being, we are only interested in the electronic part of this Hamiltonian

$$\mathcal{H} = -\sum_{j=1}^{N}\left(\frac{1}{2m}\nabla_j^2 + U_{\text{ion}}(\mathbf{r}_j)\right) + \sum_{\substack{j,k=1\\ j<k}}^{N} \frac{e^2}{|\mathbf{r}_j - \mathbf{r}_k|} \tag{8.70}$$

In this (still very general) Hamiltonian, we have already made a number of further simplifications. It describes electrons interacting with the potential $U_{\text{ion}}(\mathbf{r})$ of a static lattice of background ions. That is, we are using the Born-Oppenheimer approximation (Born and Oppenheimer, 1927), which neglects the motion of the ion lattice. This approximation is appropriate as we are only interested in the interactions of electrons, and not in dynamical lattice effects like phonons or electron-phonon interactions. The Hamiltonian (8.70) describes adequately our interest: the behaviour of electrons in a solid.

We now use the Hamiltonian (8.70) as a reference theory for our description of electrons or, more generally, Fermions, which we, however, develop in a heuristic way starting with a gas of non-interacting electrons.

8.2.2 Non-interacting electron gas

Disregarding the electron-electron interaction in the Hamiltonian (8.70), we immediately observe that the many-particle Schrödinger equation

$$\mathcal{H}\Psi(\mathbf{r}_1,\ldots,\mathbf{r}_N) = \sum_{j=1}^{N}\left(-\frac{1}{2m}\nabla_j^2 + U_{\text{ion}}(\mathbf{r}_j)\right)\Psi(\mathbf{r}_1,\ldots,\mathbf{r}_N) = E\Psi(\mathbf{r}_1,\ldots,\mathbf{r}_N) \tag{8.71}$$

decomposes into single electron equations

$$\left(-\frac{1}{2m}\nabla^2 + U_{\text{ion}}(\mathbf{r})\right)\psi_j(\mathbf{r}) = E_j\psi_j(\mathbf{r}). \tag{8.72}$$

The eigenfunctions $\Psi(\mathbf{r}_1, \ldots, \mathbf{r}_N)$ of the many-particle problem (8.71) are products of the eigenfunctions $\psi_j(\mathbf{r})$ of the single electron problem (8.72)

$$\Psi(\mathbf{r}_1, \ldots, \mathbf{r}_N) = \frac{1}{N!} \sum_P (-1)^P \prod_{j=1}^{N} \psi_{p_j}(\mathbf{r}), \tag{8.73}$$

where the sum is over all permutations P of N integers with the sign $(-1)^P$ of the permutation, and p_j denotes the jth entry in the permuation P. The energy of the many-particle problem is given as the sum $E = \sum_{j=1}^{N} E_j$ over the single-electron energies.

While (8.71) already involved quite drastic approximations, this problem remains difficult enough. Therefore, it is worthwhile to first study a further simplified problem: the free electron or free Fermi gas

$$\sum_{j=1}^{N} \left(-\frac{1}{2m} \nabla_j^2 \right) \Psi(\mathbf{r}_1, \ldots, \mathbf{r}_N) = E\Psi(\mathbf{r}_1, \ldots, \mathbf{r}_N) \tag{8.74}$$

for which we now have neglected all interactions of the gas of free electrons, i.e. there is not even a background potential from the ions of the crystal lattice.

To solve (8.74), we need to specify boundary conditions. A convenient choice for calculations are periodic boundary conditions where we assume the system in a cubic box with side length L and hence volume $V = L^3$ is repeated indefinitely in all three spatial directions, for example

$$\Psi(x_1 + L, y_1, z_1, \ldots, x_N, y_N, z_N) = \Psi(x_1, y_1, z_1, \ldots, x_N, y_N, z_N), \tag{8.75}$$

and so on for all $3N$ coordinates. A physically more realistic choice would be to require the wave function $\Psi(\mathbf{r}_1, \ldots, \mathbf{r}_N)$ to vanish whenever one of its arguments reaches the boundaries of the system. Compared to the periodic boundary conditions, this more realistic choice would result in finite-size corrections of order $1/L$ for most physical quantities of interest—corrections that are negligible for macroscopic system sizes L.

Having now completely defined the free electron system, we state some important results (leaving out the details of the calculations which can be found in any text on solid state or condensed matter physics, e.g. Marder, 2010).

The one-electron solutions of (8.74) are plane waves

$$\psi_{\mathbf{k}}(\mathbf{r}) = \frac{1}{\sqrt{V}} e^{i\mathbf{k}\cdot\mathbf{r}} \quad \text{where the label } \mathbf{k} \text{ assumes the values} \quad \mathbf{k} = \frac{2\pi}{L}\mathbf{l} \tag{8.76}$$

where the triple of integers $\mathbf{l} = (l_x, l_y, l_z)^T$ range from $-\infty$ to ∞. The energy eigenvalue corresponding to this one-electron eigenfunction is

$$E^0(\mathbf{k}) = \frac{k^2}{2m} = \frac{1}{2m}\left(k_x^2 + k_y^2 + k_z^2\right). \tag{8.77}$$

The ground state of N electrons can be constructed from products of the plane wave solutions (8.76) respecting the Pauli exclusion principle in such a way that, starting with one electron per spin degree of freedom for $\mathbf{k} = 0$ up to a maximum value $|\mathbf{k}| = k_F$, the Fermi wave number, states with the lowest possible energy $E^0_{\mathbf{k}}$, (8.77), are filled. In other words, in reciprocal or $\mathbf{k}$ space, the free electron ground state occupies a sphere of radius k_F, the Fermi sphere, each state occupying a reciprocal volume, according to $\mathbf{k}$ as defined in (8.76), of $(2\pi/L)^3$. All states with $|\mathbf{k}| > k_F$ are unoccupied in the ground state and correspond, when occupied, to excitations of the free electron gas.

The *density of states* in $\mathbf{k}$ space is independent of $\mathbf{k}$

$$D(\mathbf{k}) = 2\frac{1}{(2\pi)^3} \tag{8.78}$$

where the factor 2 is due to the two possible spin directions of an electron. For a macroscopic system, the density of states in reciprocal space ($\mathbf{k}$ space) is useful to convert sums into integrals with an negligible error of order $1/L$. For a typical quantity $F(\mathbf{k})$, we can write

$$\sum_{\mathbf{k},\sigma} F(\mathbf{k}) = V\int d^3k\, D(\mathbf{k})F(\mathbf{k}) = \frac{2V}{(2\pi)^3}\int d^3k\, F(\mathbf{k}). \tag{8.79}$$

More interesting than $D(\mathbf{k})$ is the density of states as a function of energy, the energy density of states $D(E)$. For $F(\mathbf{k}) = F(E(\mathbf{k}))$, we can use (8.79) to write

$$\sum_{\mathbf{k},\sigma} F(E(\mathbf{k})) = V\int d^3k\, \frac{2}{(2\pi)^3} F(E(\mathbf{k})) \tag{8.80}$$

$$\equiv V\int dE\, F(E) \int d^3k\, \frac{2}{(2\pi)^3}\delta(E - E(\mathbf{k})) \tag{8.81}$$

$$\equiv V\int dE\, F(E)D(E). \tag{8.82}$$

This definition of the energy density of states $D(E)$ remains valid for more general forms of the energy dispersion than that for free electrons, (8.77).

For free electrons, we can now harvest a number of important results. The energy density of states itself can be derived from its definition and the free electron energy dispersion (8.77)

$$D(E) = \int d^3k\, \frac{2}{(2\pi)^3}\delta(E - E^0(\mathbf{k})) = \frac{m}{\pi^2}\sqrt{2mE}. \tag{8.83}$$

Introducing an occupation number $f(\mathbf{k})$, which is unity only if a state is occupied, i.e. has $|\mathbf{k}| < k_F$, and zero otherwise, we can calculate the number of electrons fitting into a sphere of radius k_F

$$N = \sum_{\mathbf{k},\sigma} f(\mathbf{k}) = V \int d^3k\, D(\mathbf{k}) f(\mathbf{k}) = V \frac{2}{(2\pi)^3} \int d^3k\, \theta(k_F - k) = V \frac{1}{3\pi^2} k_F^3. \quad (8.84)$$

For a given electron density $n = N/V$, this translates into equations for the Fermi wave number k_F, the Fermi energy $E_F = k_F^2/2m$, and the Fermi velocity $v_F = k_F/m$,

$$k_F = (3\pi^2 n)^{1/3}, \quad E_F = \frac{1}{2m}(3\pi^2 n)^{2/3}, \quad \text{and} \quad v_F = \frac{1}{m}(3\pi^2 n)^{1/3}. \quad (8.85)$$

As we shall see later, in many cases the energy density of states at the Fermi energy is of great significance. It can be expressed through the Fermi momentum k_F or through the Fermi energy E_F and the electron density n as

$$D(E_F) = \frac{mk_F}{\pi^2} = \frac{3}{2}\frac{n}{E_F}. \quad (8.86)$$

For later convenience, we state the density of states for arbitrary dimension D

$$D(\mathbf{k}) = 2\left(\frac{1}{2\pi}\right)^D. \quad (8.87)$$

The corresponding energy density of states in two dimensions is constant

$$D(E) = \frac{m}{\pi} \quad (8.88)$$

while in one dimension we obtain

$$D(E) = \frac{1}{\pi}\sqrt{\frac{m}{2E}}. \quad (8.89)$$

Much of the ground state structure discussed so far for the case of free electrons remains valid for non-interacting electrons in a potential $U_{\text{ion}}(\mathbf{r}_j)$ satisfying (8.71). The ground state energies of the non-interacting electrons E_j can be ordered by magnitude $E_0 \le E_1 \le E_2 \ldots$, the largest occupied energy being the Fermi energy. The energy density of states $D(E)dE$ continues to be defined as the number of single electron states per volume in an energy range between E and $E + dE$. For periodic potentials, as is typical for the ionic cores in crystals, the electron states will still be carrying a wave vector index $\mathbf{k}$. This leads to the important programme of electronic band structure calculations in metals and semiconductors.

In order to explain the thermodynamic properties of the non-interacting electron gas, we first compute the grand canonical partition function

$$Z_G = \sum_{\textbf{states}} e^{\beta(\mu N - E)} = \sum_{n_1=0}^{1} \sum_{n_2=0}^{1} \ldots e^{\beta \sum_j n_j(\mu - E_j)} \tag{8.90}$$

where the total number of electrons is $N = \sum_j n_j$ and the total energy $E = \sum_j n_j E_j$. Rearranging terms, we obtain

$$Z_G = \prod_j \left(\sum_{n_j=0}^{1} e^{\beta n_j(\mu - E_j)} \right) = \prod_j \left(1 + e^{\beta(\mu - E_j)} \right) \tag{8.91}$$

from which the grand canonical potential Φ follows

$$\Phi(T, \mu, V) = -T \ln Z_G = -T \sum_j \ln \left(1 + e^{\beta(\mu - E_j)} \right) \tag{8.92}$$

$$= -TV \int dE\, D(E) \ln \left(1 + e^{\beta(\mu - E)} \right), \tag{8.93}$$

where the last expression shows the usefulness of the definition of the density of states given in the preceding paragraph.

From the grand canonical partition function, or equivalently the grand canonical potential, all other thermodynamic quantities can be derived. For example, the average total number of electrons $\langle N \rangle = N$ follows from the thermodynamic relation

$$N = -\frac{\partial \Phi}{\partial \mu} = V \int dE\, D(E) \frac{e^{\beta(\mu - E)}}{1 + e^{\beta(\mu - E)}} \tag{8.94}$$

so that the electron density $n = N/V$ becomes

$$n = \int dE\, D(E) \frac{1}{e^{\beta(E - \mu)} + 1} = \int dE\, D(E) f(E) \tag{8.95}$$

where we have introduced the Fermi function

$$f(E) = \frac{1}{e^{\beta(E - \mu)} + 1}. \tag{8.96}$$

For $T \to 0$ or, equivalently, $\beta \to \infty$, the Fermi function becomes a step function

$$f(E) = \theta(\mu - E). \tag{8.97}$$

The total energy of the non-interacting electron gas can be calculated from

$$E - \mu N = \frac{\partial \beta \Phi}{\partial \beta} = V \int dE' \, D(E')(E' - \mu) f(E') \tag{8.98}$$

as

$$\frac{E}{V} = \int dE' \, D(E') E' f(E'). \tag{8.99}$$

8.2.3 Sommerfeld expansion

We can assign a formal temperature, the Fermi temperature, to the non-interacting electron gas by[4]

$$T_F = \frac{E_F}{k}. \tag{8.100}$$

Typical Fermi temperatures of metals are of the order of 10,000 K so that it makes quite a lot of sense to try to work out the low temperature properties of the non-interacting electron gas. For this purpose, the Sommerfeld expansion is applicable to expressions of the form

$$\int_{-\infty}^{\infty} dE \, g(E) f(E), \tag{8.101}$$

where $f(E)$ is the Fermi function and $g(E)$ some other function related to a quantity of physical interest. The initial step of the method consists in a partial integration

$$\int_{-\infty}^{\infty} dE \, g(E) f(E) = \int_{-\infty}^{\infty} dE \left(\int_{-\infty}^{E} dE' \, g(E') \right) \frac{\partial f}{\partial \mu}(E). \tag{8.102}$$

It is important to note that the function $\partial f / \partial \mu$ is essentially only non-zero in an interval around $E = \mu$, which is only a few T in width. Therefore a Taylor series of $h(E) = \int_{-\infty}^{E} dE' \, g(E')$ at $E = \mu$ can be used to evaluate the integral with which we started

$$h(E) = h(\mu) + \sum_{n=1}^{\infty} \left(\frac{(E-\mu)^n}{n!} \right) \left(\frac{\partial^n h(E)}{\partial E^n} \right)_{E=\mu}. \tag{8.103}$$

4 Since it serves clarity, we reintroduce Boltzmann's constant k for a moment before we set $k = 1$ again.

Using

$$\int_{-\infty}^{\infty} dE \, \frac{\partial f}{\partial \mu} = -\int_{-\infty}^{\infty} dE \, \frac{\partial f}{\partial E} = 1 \tag{8.104}$$

we obtain

$$\begin{aligned} &\int_{-\infty}^{\infty} dE \, g(E) f(E) = \\ &= \int_{-\infty}^{\mu} dE \, g(E) + \sum_{n=1}^{\infty} \int_{-\infty}^{\infty} dE \left(\frac{(E-\mu)^{2n}}{(2n)!} \right) \frac{\partial f}{\partial \mu} \left(\frac{\partial^{2n-1} g(E)}{\partial E^{2n-1}} \right)_{E=\mu} \end{aligned} \tag{8.105}$$

and finally, substituting $x = \beta(E-\mu) = (E-\mu)/T$

$$\int_{-\infty}^{\infty} dE \, g(E) f(E) = \int_{-\infty}^{\mu} dE \, g(E) + \sum_{n=1}^{\infty} a_n T^{2n} \left(\frac{\partial^{2n-1} g(E)}{\partial E^{2n-1}} \right)_{E=\mu} \tag{8.106}$$

where the numbers a_n are connected to the Riemann ζ function and the Bernoulli numbers B_n, respectively,

$$a_n = \int_{-\infty}^{\infty} dx \, \frac{x^{2n}}{(2n)!} \left(-\frac{d}{dx} \frac{1}{e^x + 1} \right) = \left(2 - \frac{1}{2^{2(n-1)}} \right) \zeta(2n) \tag{8.107}$$

$$= 2 \left(2^{2n-1} - 1 \right) \frac{\pi^{2n}}{(2n)!} B_n. \tag{8.108}$$

For most practical calculations, only few terms in the expansion are needed, predominantly only the first. The first two numbers a_n are: $a_1 = \pi^2/6$ and $a_2 = 7\pi^4/360$, so that the low temperature expansion of the integral becomes

$$\int_{-\infty}^{\infty} dE \, g(E) f(E) = \int_{-\infty}^{\mu} dE \, g(E) + \frac{\pi^2}{6} T^2 g'(\mu) + \frac{7\pi^4}{360} T^4 g'''(\mu). \tag{8.109}$$

8.2.4 Low temperature specific heat of the non-interacting electron gas

As an application of the Sommerfeld expansion we calculate the low temperature specific heat of the non-interacting electron gas, which is the important semi-quantitative result mentioned at the beginning of this chapter. The specific heat can be calculated from the thermodynamic relation

$$c_V = \frac{1}{V}\frac{\partial E}{\partial T}\bigg|_{NV} \tag{8.110}$$

involving the total energy E, (8.99), of the system of non-interacting electrons that we have to calculate first to order T^2 using the Sommerfeld expansion

$$\frac{E}{V} = \int dE' f(E')E'D(E') = \int_0^{\mu} dE'\, E'D(E') + \frac{\pi^2}{6}T^2\frac{d(\mu D(\mu))}{d\mu}. \tag{8.111}$$

In order to proceed further, we need the temperature dependence of the chemical potential $\mu = \mu(T)$, which we can obtain from the thermodynamic relation (see, e.g. Callen, 1985)

$$\frac{\partial\mu}{\partial T}\bigg|_{NV} = -\frac{\frac{\partial N}{\partial T}\Big|_{\mu V}}{\frac{\partial N}{\partial\mu}\Big|_{TV}}. \tag{8.112}$$

Hence, we can use the expression (8.94) for the total number of non-interacting electrons and apply the Sommerfeld expansion to order T^2

$$N = V\int dE f(E)D(E) = V\int_0^{\mu} dE\, D(E) + V\frac{\pi^2}{6}T^2D'(\mu). \tag{8.113}$$

Taking derivatives of N with respect to T and μ, respectively, and neglecting the term proportional to T^2 in the denominator $\partial N/\partial\mu$ of (8.112) gives

$$\frac{\partial\mu}{\partial T}\bigg|_{NV} = -\frac{\pi^2}{3}T\frac{D'(\mu)}{D(\mu)}. \tag{8.114}$$

Integrating this equation with respect to T and observing that $\mu(T=0) = E_F$, we obtain

$$\mu(T) = E_F - \frac{\pi^2}{6}T^2\frac{D'(E_F)}{D(E_F)} = E_F\left(1 - \frac{\pi^2}{12}\left(\frac{T}{T_F}\right)^2\right) \tag{8.115}$$

where, to order T^2, we could safely replace μ by E_F in D and D'. Likewise, we can replace μ by E_F in the second part of (8.111). Thus, in (8.111) $\mu(T)$ only remains in the upper integration limit, and we deal with this by writing $\mu = E_F + (\mu - E_F)$. To order T^2 we finally thus obtain for the total energy (taking advantage of a cancellation of certain terms due to (8.115))

$$\frac{E}{V} = \int_0^{E_F} dE'\, E'D(E') + \frac{\pi^2}{6}T^2D(E_F) \tag{8.116}$$

and for the specific heat

$$c_V = \frac{\pi^2}{3} T D(E_F) = \frac{\pi^2}{2} n \frac{T}{T_F}, \tag{8.117}$$

$$\propto T D(E_F) \tag{8.118}$$

which gives the experimentally observed linear temperature dependence of the specific heat at low temperatures and also shows the importance, as anticipated, of the density of states at the Fermi energy $D(E_F)$.

It is instructive to note that the coefficient of the linear temperature dependence of the specific heat $\gamma = \pi^2 D(E_F)/3$ is proportional to the mass of the electrons $\gamma \propto m$ and the cube root of the electron density $\gamma \propto n^{1/3}$. Therefore, we can deduce from specific heat measurements at low temperatures the accuracy of the description of the electrons as free or non-interacting. For some metals, e.g. silver, this description works very well, i.e. the ratio of experimental to theoretical value of γ is close to one. For other metals, e.g. the $3d$ transition metal nickel, the ratio is about 15, while for some metallic compounds involving $4f$ or $5f$ elements, e.g. UPt_3, the ratio can be of the order of 100 to 1000. The latter are called heavy Fermion or heavy electron compounds because it is necessary to introduce an effective mass to account for the discrepancy. A qualitative explanation for the high effective electron masses in these compounds and also for the effective electron masses in general are localization effects of the d or f electrons, although these electrons contribute to the electronic density of states at the Fermi level. Heavy Fermion compounds are at the focus of condensed matter research since their discovery as superconductors. It is suspected that the mechanisms of superconductivity in heavy Fermion and high-temperature ceramic superconductors are similarly related to the magnetic moments in these compounds.

8.2.5 The non-interacting electron picture and beyond

So far, we have not specified the potential of the ionic background $U_{\text{ion}}(\mathbf{r})$ in (8.71). In order to make further progress, even still within the framework of non-interacting electrons, we need to specify and treat specific potentials. This is a vast area of solid state and condensed matter physics that this section will cover only briefly, highlighting the most important results as we go along. The interested reader is referred to the literature on solid state and condensed matter physics, e.g. Marder (2010), and, for technical details, the literature on many-particle physics, e.g. Mahan (2000) and Coleman (2015).

8.2.5.1 Periodic potentials and Bloch's theorem

In a crystal lattice, the potential of the ionic background $U_{\text{ion}}(\mathbf{r})$ will be periodic

$$U_{\text{ion}}(\mathbf{r} + \mathbf{R}) = U_{\text{ion}}(\mathbf{r}) \tag{8.119}$$

with periodicity $\mathbf{R}$ given by the vectors of the Bravais lattice of the crystal. This periodicity has immediate consequences on the eigenstates and eigenvalues of (8.71). First, we note that the operator generating translations through a Bravais vector $\mathbf{R}$

$$T(\mathbf{R}) = e^{-i\mathbf{P}\cdot\mathbf{R}} \tag{8.120}$$

where $\mathbf{P}$ is the momentum operator commutes with the Hamiltonian of (8.71). The simultaneous eigenstates of $T(\mathbf{R})$ in (8.120) and the Hamiltonian in (8.71) are

$$\mathcal{H}|\psi_n(\mathbf{k})\rangle = E_n(\mathbf{k})|\psi_n(\mathbf{k})\rangle \quad \text{and} \quad T^\dagger(\mathbf{R})|\psi_n(\mathbf{k})\rangle = e^{i\mathbf{k}\cdot\mathbf{R}}|\psi_n(\mathbf{k})\rangle. \tag{8.121}$$

The eigenvalue equation for the translation operator, the second of the above equations, is Bloch's theorem. It can be re-expressed in position space as

$$\psi_{n\mathbf{k}}(\mathbf{r}+\mathbf{R}) = e^{i\mathbf{k}\cdot\mathbf{R}}\psi_{n\mathbf{k}}(\mathbf{r}) \quad \text{or} \quad u_{n\mathbf{k}}(\mathbf{r}+\mathbf{R}) = u_{n\mathbf{k}}(\mathbf{r}) \tag{8.122}$$

with the Bloch wave functions

$$\psi_{n\mathbf{k}}(\mathbf{r}) = e^{i\mathbf{k}\cdot\mathbf{r}}u_{n\mathbf{k}}(\mathbf{r}). \tag{8.123}$$

The wave functions of electrons in a periodic crystal lattice are plane waves undulating with the period of the lattice and modulated with another plane wave of a longer period. Moreover, there are generally many solutions for a given wave vector $\mathbf{k}$ that are labelled by the index n. For varying $\mathbf{k}$, the energies $E_n(\mathbf{k})$ form bands, and the index n is called the band index.

The band energies $E_n(\mathbf{k})$ contain a great wealth of information about the solid, e.g. whether the solid is a metal, a semiconductor, or an insulator; moreover, they allow the calculation of electron velocities and, hence, can be used to elucidate electrical transport properties. The usefulness of results from band structure calculations, which rely heavily on the exploitation of the crystal symmetries using group theoretical methods, cannot be overestimated.

Before we leave this well-established area of solid state physics, which lies outside the main scope of our considerations, we mention an important restatement of Bloch's theorem. The Fourier transformed single electron Schrödinger equation (8.72) is

$$\left(E - E^0_{\mathbf{q}}\right)\psi(\mathbf{q}) = \sum_{\mathbf{K}} U(\mathbf{K})\psi(\mathbf{q}-\mathbf{K}) \quad \text{with} \quad E^0_{\mathbf{q}} = \frac{q^2}{2m} \tag{8.124}$$

where $\mathbf{K}$ is a reciprocal lattice vector and we have dropped the subscript 'ion' on the Fourier transformed potential.

EXERCISE 8.5 Derivation of Bloch's theorem Derive (8.124). A pair of Fourier transformed functions is defined by

$$f(\mathbf{q}) = \int d^3r\, e^{-i\mathbf{q}\cdot\mathbf{r}} f(\mathbf{r}) \quad \text{and} \quad f(\mathbf{r}) = \frac{1}{V}\sum_{\mathbf{q}} f(\mathbf{q}) e^{i\mathbf{q}\cdot\mathbf{r}}. \tag{8.125}$$

Observe that $U(\mathbf{K})$ is not simply a Fourier transform but is related to

$$\bar{U}(\mathbf{q}) = \frac{1}{\mathcal{V}} \int_{\text{unit cell}} d^3r\, e^{-i\mathbf{q}\cdot\mathbf{r}} U(\mathbf{r}) \tag{8.126}$$

with $\mathcal{V}$ the volume of the unit cell.

The Fourier form of the single electron Schrödinger equation, (8.124), has many solutions for a fixed value of $\mathbf{q}$. These are the different bands discussed in connection with Bloch's theorem, (8.121). In other words, (8.124) is a restatement of Bloch's theorem, and is also a good starting point for realistic band structure calculations, which usually involve elaborate numerical implementations.

Nevertheless, it is helpful as background information to remind ourselves of a few approximative schemes that have proved successful and which also have (8.124) as their starting point.

8.2.5.2 Nearly free electron approximation

In the nearly free electron approximation, introduced by Rudolf Peierls after Bloch's more general insights, we go back one step and treat the strong Coulomb potential of the ionic cores as a weak periodic potential. This crude approximation, which makes perturbation theory applicable, delivers surprisingly accurate results for the Fermi surfaces of metals when compared to experiment.

Assuming the potential $U(\mathbf{K})$ to be weak, degenerate perturbation theory finds for the eigenvalue of (8.124)

$$E = U(0) + \frac{E^0(\mathbf{k}) + E^0(\mathbf{k}+\mathbf{K})}{2} \pm \sqrt{\frac{(E^0(\mathbf{k}) - E^0(\mathbf{k}+\mathbf{K}))^2}{4} + |U(\mathbf{K})|^2}. \tag{8.127}$$

For a derivation of this result, see e.g. Marder (2010). It is an important result because careful analyses of (8.127) explain, as anticipated, the experimentally observed gaps between bands on the Fermi surfaces of solids explaining whether the solid is a metal, a semiconductor, or an insulator.

8.2.5.3 Tightly bound electron approximation: Tight-binding model

A viewpoint different from the delocalized electrons considered thus far starts from atomic wave functions. Such a viewpoint is provided by the tight-binding method. It is complementary not in contradiction to the one developed so far. The fundamental ingredient of this atomic viewpoint are Wannier wave functions, which are centered at lattice sites $\mathbf{R}$ and build up from Bloch wave functions and are orthonormal

$$w_n(\mathbf{R},\mathbf{r}) = \frac{1}{\sqrt{N}}\sum_{\mathbf{k}} e^{-i\mathbf{k}\cdot\mathbf{R}}\psi_{n\mathbf{k}}(\mathbf{r}) \tag{8.128}$$

with

$$\int d^3r\, w_n(\mathbf{R},\mathbf{r})w_m^*(\mathbf{R}',\mathbf{r}) = \delta_{\mathbf{R}\mathbf{R}'}\delta_{nm}. \tag{8.129}$$

The Bloch functions can be recovered from the Wannier functions by

$$\psi_{n\mathbf{k}}(\mathbf{r}) = \frac{1}{\sqrt{N}}\sum_{\mathbf{R}} w_n(\mathbf{R},\mathbf{r})e^{i\mathbf{k}\cdot\mathbf{R}}. \tag{8.130}$$

Without going into details, let us state that it can be shown that the Wannier functions for insulators are localized around individual atoms, while for metals they may be delocalized.

The matrix elements of the Hamiltonian in (8.72) in the Wannier basis are

$$\mathcal{H}_{\mathbf{R}\mathbf{R}'} = \int d^3r\, w_n^*(\mathbf{R}',\mathbf{r})\left(-\frac{\nabla^2}{2m} + U(\mathbf{r})\right)w_n(\mathbf{R},\mathbf{r}). \tag{8.131}$$

This matrix element is particularly useful when it is only of appreciable magnitude for $\mathbf{R}$ and $\mathbf{R}'$ being nearest neighbours. Then, when appropriate symmetries obtain, $\mathcal{H}_{\mathbf{R}\mathbf{R}'}$ can be be replaced by two constants: t, called the hopping constant, if $\mathbf{R}$ and $\mathbf{R}'$ are nearest neighbours and ϵ_0, the on-site interaction, for $\mathbf{R} = \mathbf{R}'$. In second quantized form, the Hamiltonian, called the tight-binding Hamiltonian, can thus be written as

$$\mathcal{H}_{TB} = t\sum_{\mathbf{R},\delta} c_{\mathbf{R}}^\dagger c_{\mathbf{R}+\delta} + \epsilon_0\sum_{\mathbf{R}} c_{\mathbf{R}}^\dagger c_{\mathbf{R}}. \tag{8.132}$$

An extension of the tight-binding Hamiltonian will be used when we discuss the Hubbard model.

The tight-binding Hamiltonian (8.132) can be diagonalized exactly by introducing a pair of Fourier transformed operators

$$c_{\mathbf{R}} = \frac{1}{\sqrt{N}}\sum_{\mathbf{k}} e^{-i\mathbf{R}\cdot\mathbf{k}}c_{\mathbf{k}} \quad \text{and} \quad c_{\mathbf{k}} = \frac{1}{\sqrt{N}}\sum_{\mathbf{R}} e^{i\mathbf{R}\cdot\mathbf{k}}c_{\mathbf{R}} \tag{8.133}$$

for **k** in the first Brillouin zone. We obtain

$$\mathcal{H}_{TB} = \sum_{\mathbf{k}} \epsilon(\mathbf{k}) c_{\mathbf{k}}^{\dagger} c_{\mathbf{k}} \quad \text{where} \quad \epsilon(\mathbf{k}) = \epsilon_0 + t \sum_{\delta} e^{-i\mathbf{k}\delta}. \tag{8.134}$$

This is the simplest example of an energy band. The important notion of the band width $2W$, the difference between the maximal and minimal values of the energies or energy dispersion $\epsilon(\mathbf{k})$, gives in this simplest case $2W = 2z|t|$ with z the number of nearest neighbours.

So far, the strong electron-electron interaction has been completely neglected. This is, of course, a drastic approximation which could, in the last consequence, only be justified by the success of the approximations. This success is limited and the strong electron-electron interactions must be discussed eventually. The following sections remind us of a number of standard methods to deal with electron-electron interactions in an approximative way.

8.2.6 Electron-electron interaction

The electron-electron Coulomb interaction term in (8.70) and the Pauli exclusion principle induce collective motions and correlations of the electrons that render it more difficult to solve the Schödinger equation. The following two sections give a brief outline how the problem can be reduced again to an effective single electron problem.

8.2.6.1 Hartree approximation

In the Hartree approximation, a mean or effective field is introduced to account for the electron-electron Coulomb interaction in the Hamiltonian (8.70). The mean field acting on electron j from all other electrons $l \neq j$ is

$$U_{ee}(\mathbf{r}) = \int d^3r' \frac{e^2 n(\mathbf{r}')}{|\mathbf{r} - \mathbf{r}'|} \quad \text{with} \quad n(\mathbf{r}) = \sum_{l \neq j} |\psi_l(\mathbf{r})|^2 \tag{8.135}$$

the number density of electrons, the jth electron excluded. Inserting this mean-field potential into the Hamiltonian (8.70), we obtain the Hartree equations

$$\left(-\frac{1}{2m} \nabla^2 + U_{\text{ion}}(\mathbf{r}) + U_{ee}(\mathbf{r}) \right) \psi_j(\mathbf{r}) = E_j \psi_j(\mathbf{r}) \tag{8.136}$$

for the single electron wave functions $\psi_j(\mathbf{r})$.

This result can be derived from the variational principle that the eigenvalue problem (8.136) is equivalent to the variational problem to minimize

$$\langle \mathcal{H} \rangle = \frac{\langle \Psi | \mathcal{H} | \Psi \rangle}{\langle \Psi | \Psi \rangle} \tag{8.137}$$

with the variational many-particle wave function as a product of single electron wave functions: $|\Psi\rangle = \prod_j |\psi_j\rangle$.

Obviously, the main flaw of the Hartree approach is this assumption that the many-particle wave function is a simple product of single electron wave functions: $|\Psi\rangle = \prod_j |\psi_j\rangle$. Of course, such a many-particle wave function does not satisfy the Pauli exclusion principle.

8.2.6.2 Hartree–Fock approximation

The immediate way out of this flaw of the Hartree approximation is, of course, to construct an anti-symmetrized many-particle wave function out of the orthonormal single electron wave functions. In this prescription, the electron spin taken into account explicitly, the many-particle wave function is the Slater determinant of section 2.2

$$\Psi(\mathbf{r}_1\sigma_1, \mathbf{r}_2\sigma_2, \ldots, \mathbf{r}_N\sigma_N) = \frac{1}{\sqrt{N!}} \sum_P (-1)^P \psi_{p_1}(\mathbf{r}_1\sigma_1)\psi_{p_2}(\mathbf{r}_2\sigma_2)\ldots\psi_{p_N}(\mathbf{r}_N\sigma_N), \tag{8.138}$$

where the sum is over all permutations P of 1, 2, . . ., N and $\sigma_i = \pm 1$ is the spin index. As long as the Hamiltonian is not explicitly spin dependent, as it is in the case we are considering, the single electron wave functions can be written as $\psi_j(\mathbf{r}, \sigma) = \phi_j(\mathbf{r})\chi_j(\sigma)$ where $\chi_j(\sigma) = \delta_{1,\sigma}$ for spin up, or $\chi_j(\sigma) = \delta_{-1,\sigma}$ for spin down.

Using this many-particle wave function in the variational calculation yields

$$\left(-\frac{1}{2m}\nabla^2 + U_{\text{ion}}(\mathbf{r}) + e^2 \sum_{l\neq j} \int d^3r' \frac{|\phi_l(\mathbf{r}')|^2}{|\mathbf{r}-\mathbf{r}'|}\right)\phi_j(\mathbf{r})$$
$$-\left(\delta_{\chi_j\chi_l} e^2 \sum_{l\neq j} \int d^3r' \frac{\phi_l^*(\mathbf{r}')\phi_j(\mathbf{r}')}{|\mathbf{r}-\mathbf{r}'|}\right)\phi_l(\mathbf{r}) = E_j\phi_j(\mathbf{r}). \tag{8.139}$$

Comparing with the Hartree equations (8.136), we see that the Hartree–Fock equations contain an additional term. Physically this is a spin-dependent exchange integral term. Mathematically, due to the additional term, we are now faced with a set of coupled equations. For the further treatment of the Hartree–Fock equations we refer to the literature, e.g. Madelung (1978).

8.3 Coulomb gas

Now we turn from the treatment of the electron-electron interaction reduced to an effective mean-field single electron problem to the full problem of electrons mutually interacting via their Coulomb potentials.

The interaction between charged particles, i.e. the electrons and the ions of the lattice, the Coulomb interaction is both strong and of long range. For example, the Coulomb energy of two electrons in a metal separated by an average atomic distance is comparable to the kinetic energy or band energy of an electron at the Fermi surface. How does it then come about that electrons in a metal behave as if they were nearly independent particles?

We already know that, in order to answer this question, we have to investigate the low-lying excitations of a many-particle system. In general, there are two kinds of such low-lying excitations: quasiparticles, which are single particle excitations modified by the interactions between particles, and collective modes, which arise from the correlations generated in the particle motions by the interactions between particles. This section on the Coulomb gas picture examining the correlation effects and section (8.4) on Landau Fermi liquid theory investigating the quasiparticle excitations are devoted to providing answers to this question.

The picture of the Coulomb gas of electrons assumes that the ions of the lattice provide a featureless static background of positive charge that guarantees overall charge neutrality. This model is also called the Jellium model of a metal. It has two important features. The first is charge oscillations, which are called plasma oscillations or, upon quantization, plasmons. The other, more interesting one, is the screening of charges. We restrict ourselves to a macroscopic discussion. A more complete discussion, especially also of the microscopic aspects of the Jellium model, can be found in Rickayzen (1980).

A macroscopic discussion of the Coulomb gas considers the behaviour of the gas when it is exposed to perturbations that vary slowly and with long wavelength. A small fluctuation in the charge density of the gas, i.e. a small local violation of charge neutrality, results in an electric field that will, in its attempt to restore charge neutrality, accelerate charges and thereby generate a charge density oscillation, i.e. the plasma oscillation.

Macroscopically, small fluctuations in the charge density can be described by the linearized hydrodynamic equations of motion. The local drift velocity $\mathbf{v}(\mathbf{r}, t)$ and the number density $n(\mathbf{r}, t)$ of the electrons obey

$$m\frac{\partial(n\mathbf{v})}{\partial t} = -\nabla p + en\mathbf{E} \quad \text{and} \quad \frac{\partial n}{\partial t} + \nabla \cdot (n\mathbf{v}) = 0 \tag{8.140}$$

where e and m are charge and mass of the electron, respectively, $p(\mathbf{r}, t)$ is the local pressure and $\mathbf{E}(\mathbf{r}, t)$ the local electric field. The second equation is the continuity equation reflecting charge conservation. The number density fluctuations $\delta n = n - n_0$ from the neutral equilibrium number density n_0 generate the local electric field

$$\nabla \cdot \mathbf{E} = 4\pi e(n - n_0) = 4\pi e\delta n. \tag{8.141}$$

The number density fluctuations δn and the electron velocity $\mathbf{v}$ are assumed to be small quantities of first order. Moreover, we assume that the electron motion is small enough to preserve equilibrium locally, which implies that the local pressure $p(\mathbf{r}, t)$ is a function of the local number density $p(\mathbf{r}, t) = p(n(\mathbf{r}, t))$. To first order in the small quantities, we therefore get

$$m\frac{\partial \mathbf{v}}{\partial t} = -\frac{1}{n_0}\frac{\partial p}{\partial n}\nabla\delta n + e\mathbf{E}. \tag{8.142}$$

The derivative of p with respect to n has to be taken at constant entropy.

Taking the divergence of this equation yields a wave equation for describing the density oscillations

$$m\frac{\partial^2 \delta n}{\partial t^2} - \frac{\partial p}{\partial n}\nabla^2\delta n + 4\pi e^2 n_0 \delta n = 0 \tag{8.143}$$

which allows plane wave solutions

$$\delta n \propto e^{i\mathbf{q}\cdot\mathbf{r}-i\omega t} \tag{8.144}$$

provided that the dispersion relation

$$\omega^2 = \frac{4\pi e^2 n_0}{m} + \frac{1}{m}\frac{\partial p}{\partial n}q^2 \tag{8.145}$$

then holds.

These oscillations are the plasma oscillations referred to earlier. In the limit of long wavelengths, their frequencies correspond for typical good metals to energies of the order of 5 to 10 electron volts.

The screening effect on the electrons can be studied by adding an additional external electric field $\mathbf{E}_e$ to the electric field $\mathbf{E}$ in the first equation of (8.140). The total electric field is now $\mathbf{E}_t = \mathbf{E} + \mathbf{E}_e$. Repeating the steps leading to the dispersion relation (8.145) determines the additional external field

$$\frac{4\pi e^2 n_0 \nabla\cdot\mathbf{E}_e}{m\omega^2 - 4\pi e^2 n_0 - \frac{\partial p}{\partial n}q^2} = \nabla\cdot\mathbf{E} = \nabla\cdot(\mathbf{E}_t - \mathbf{E}_e) = 4\pi e\delta n. \tag{8.146}$$

The total electric field can now be expressed in terms of the external electric field

$$\mathbf{E}_t = \frac{m\omega^2 - \frac{\partial p}{\partial n}q^2}{m\omega^2 - 4\pi e^2 n_0 - \frac{\partial p}{\partial n}q^2}\mathbf{E}_e = \frac{1}{\epsilon(q,\omega)}\mathbf{E}_e \tag{8.147}$$

where we have introduced a frequency- and wave number dependent dielectric constant

$$\epsilon(q,\omega) = 1 - \frac{4\pi e^2 n_0}{m\omega^2 - \frac{\partial p}{\partial n}q^2}. \tag{8.148}$$

This total electric field acts on any charge in the Coulomb gas.

For static charges, i.e. $\omega = 0$, the dielectric constant $\epsilon(q, 0)$ diverges for long wavelengths, $q \to 0$.

To demonstrate the screening of charges explicitly, we now assume that the external field $\mathbf{E}_e$ stems from an additional stationary charge of an electron corresponding to the potential $V(r) = e/r$ whose Fourier transform is

$$\frac{e}{r} = \frac{e}{2\pi^2} \int d^3q \, \frac{1}{q^2} e^{i\mathbf{q}\cdot\mathbf{r}}. \tag{8.149}$$

The Fourier transform of the total electric field $\mathbf{E}_t$ is then, according to (8.147),

$$\frac{e}{2\pi^2} \int d^3q \, \frac{1}{\epsilon(q, 0) q^2} e^{i\mathbf{q}\cdot\mathbf{r}} = e \frac{e^{-\kappa r}}{r} \tag{8.150}$$

where the screening length κ is given by

$$\kappa^{-2} = \frac{1}{4\pi e^2 n_0} \frac{\partial p}{\partial n}. \tag{8.151}$$

We have found that an additional charge is screened by all other charges of the Coulomb gas. This screening effect is also successful in reducing the potential acting between the original charges. For a good metal, the screening length can be estimated to be of the order of 1, nanometre, which is of the same order as the average distance between electrons in the metal. Due to screening, the effective screened interaction between electrons is hence of short range.

Both plasma oscillations and screening of the interaction between electrons are also derivable from a microscopic theory. Such a microscopic theory involves several approximative schemes, e.g. the Hartree approximation discussed earlier. Moreover, Green's function techniques may be applied usefully. For more information, interested readers are referred to the literature, e.g. Rickayzen (1980).

Nevertheless, in the following two sections, we derive two important results using simple microscopic considerations.

8.3.1 Lifetime of excitations

For later reference and because of its conceptual importance, the first of these results of the microscopic theory (which goes beyond the hydrodynamic approach outlined above), we review is the lifetime τ of an electron excited by an energy ϵ above the Fermi energy E_F for small $\epsilon \ll E_F$, which tends to infinity as $\tau \propto \epsilon^{-2}$ for small ϵ. Correspondingly, the level width $\propto 1/\tau$ of such a state of energy ϵ is proportional to ϵ^2 and, hence, much smaller than the energy ϵ itself for $\epsilon \to 0$. The state of the excited electron, the quasiparticle state, is therefore well defined. This result of the microscopic approach to the Coulomb gas forms a major ingredient of Landau's ingenious treatment of Fermi systems (see section 8.4).

We consider a filled Fermi sea whose state is given by

$$|\Psi_0\rangle = |F\rangle = \prod_{|\mathbf{k}|<k_F} \prod_{\sigma} c^{\dagger}_{\mathbf{k},\sigma} |0\rangle \tag{8.152}$$

and add one electron with momentum $\mathbf{k}$ and energy $\epsilon(\mathbf{k}) = \mathbf{k}^2/2m$ satisfying $\mathbf{k} > k_F$ and $\epsilon(\mathbf{k}) > \epsilon_F$. This state will interact with the electrons of the filled Fermi sea and decay through many-particle scattering processes, in which electron-hole pairs are created with momenta $\mathbf{k}' + \mathbf{q}$ (for the electrons lifted out of the Fermi sea) and $\mathbf{k}'$ (for the holes left behind in the Fermi sea). The additional electron gets scattered into a state with momentum $\mathbf{k} - \mathbf{q}$, $\mathbf{q}$ being the momentum transferred in the scattering process. The second quantized interaction part of the Hamiltonian for such a scattering process takes the form in momentum space

$$\mathcal{H}_{ee} = \sum_{\mathbf{k},\mathbf{k}',\mathbf{q}} \sum_{\sigma,\sigma'} U(\mathbf{q}) c^{\dagger}_{\mathbf{k}-\mathbf{q},\sigma} c^{\dagger}_{\mathbf{k}'+\mathbf{q},\sigma'} c_{\mathbf{k}',\sigma'} c_{\mathbf{k},\sigma}. \tag{8.153}$$

For the interaction potential $U(\mathbf{q})$ we use the short range screened potential (Yukawa potential) discussed in the last section

$$U(\mathbf{q}) = \frac{4\pi e^2}{q^2 \epsilon(\mathbf{q}, 0)} = \frac{4\pi e^2}{q^2 + \kappa^2}. \tag{8.154}$$

The transition described here is allowed whenever momentum and energy are conserved, in other words

$$\mathbf{k} = (\mathbf{k} - \mathbf{q}) - \mathbf{k}' + (\mathbf{k}' + \mathbf{q}), \tag{8.155}$$

$$\epsilon(\mathbf{k}) = \epsilon(\mathbf{k} - \mathbf{q}) - \epsilon(\mathbf{k}') + \epsilon(\mathbf{k}' + \mathbf{q}). \tag{8.156}$$

In order to calculate the lifetime $\tau_{\mathbf{k}}$ of the initial state of one additional electron above a filled Fermi sea, we employ Fermi's golden rule, i.e. first-order time-dependent perturbation theory, which states

$$\frac{1}{\tau_{\mathbf{k}}} = 2\pi \frac{1}{V^2} \sum_{\mathbf{k}',\mathbf{q}} \sum_{\sigma'} |V(\mathbf{q})|^2 n(\mathbf{k}', \sigma')(1 - n(\mathbf{k} - \mathbf{q}, \sigma))(1 - n(\mathbf{k}' + \mathbf{q}, \sigma'))$$
$$\times \delta(\epsilon(\mathbf{k} - \mathbf{q}) - \epsilon(\mathbf{k}) - (\epsilon(\mathbf{k}') - \epsilon(\mathbf{k}' + \mathbf{q}))). \tag{8.157}$$

The factor $n(\mathbf{k}', \sigma')(1 - n(\mathbf{k} - \mathbf{q}, \sigma))(1 - n(\mathbf{k}' + \mathbf{q}, \sigma'))$, where $n(\kappa, s) = \langle\Psi|c^{\dagger}_{\kappa,s} c_{\kappa,s}|\Psi\rangle$ is the momentum distribution in the ground state $|\Psi\rangle$ of the interacting electron system, ensures that the Pauli principle is satisfied. We relegate the details of the evaluation of this expression to exercise 8.6 and only state here the final result

$$\frac{1}{\tau_{\mathbf{k}}} = \frac{1}{8\pi}\frac{D(\epsilon_F)}{v_F^2}(\epsilon(\mathbf{k}) - \epsilon_F)^2 \int dq\,|V(q)|^2 \propto (\epsilon(\mathbf{k}) - \epsilon_F)^2, \tag{8.158}$$

which confirms the behaviour we anticipated for an electron with momentum $\mathbf{k}$ slightly above the Fermi surface.

EXERCISE 8.6 Lifetime In order to evaluate (8.157), we convert the summations over $\mathbf{k}'$ and $\mathbf{q}$ into integrations. The crucial step consists in determining the integration limits from the product of momentum distributions $n(\mathbf{k}', \sigma')(1 - n(\mathbf{k} - \mathbf{q}, \sigma))(1 - n(\mathbf{k}' + \mathbf{q}, \sigma'))$. Since the scattering process takes place very close to the Fermi surface (see figure 8.1), we can replace the momentum distributions by the momentum distribution of $n(\kappa, s)$ of the interacting electron system by the momentum distribution of the non-interacting electron system at $T = 0$:

$$n(\boldsymbol{\kappa}, s) \to n^{(0)}(\boldsymbol{\kappa}, s) = \Theta(k_F - |\boldsymbol{\kappa}|). \tag{8.159}$$

The error we incur in this way is comparable to the errors already entailed by the use of the perturbative approach via Fermi's golden rule and the consideration of only the lowest order effect, the creation of one electron–hole excitation. Therefore, the final result (8.158) to be calculated in this exercise will be accurate to lowest order in perturbation theory.

Perform the integrations that will lead to (8.158) in two steps.

First, integrate over $\mathbf{k}'$ by using the decomposition $\mathbf{k}' = \mathbf{k}'_{\parallel} + \mathbf{k}'_{\perp}$ into a component parallel to $\mathbf{q}$ and a component perpendicular to $\mathbf{q}$. This suggests using cylindrical coordinates in the $\mathbf{k}'$ integration.

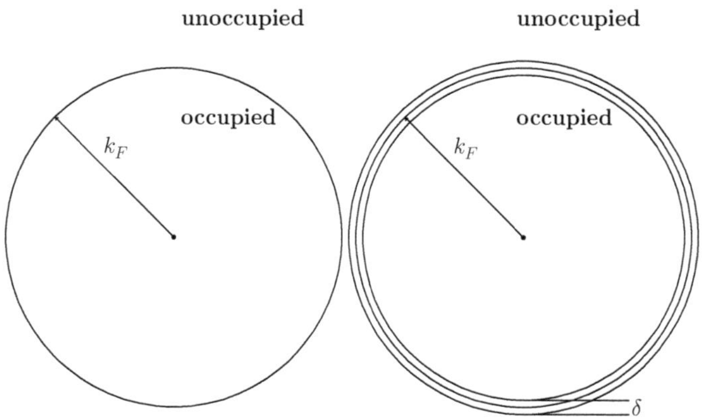

Figure 8.1 *Fermi surfaces of a non-interacting electron system (left) and an interacting electron system (right). Only states from a thin layer δ around the Fermi surface contribute in the interacting case.*

In the second step, integrate over the remaining integration variable $\mathbf{q}$ using spherical coordinates. Observe that the integration limits of this second integration are determined by the condition that $\mathbf{k}$ and $\mathbf{k} - \mathbf{q}$ are both outside the Fermi sphere.

The result for the lifetime (8.158) implies that the electron state $|\mathbf{k}, \sigma\rangle$ is a resonance of width $1/\tau_{\mathbf{k}}$. This resonance becomes arbitrarily sharp as the Fermi surface is approached

$$\frac{1/\tau_{\mathbf{k}}}{\epsilon_{\mathbf{k}} - \epsilon_F} \to 0 \quad \text{as} \quad k \to k_F, \tag{8.160}$$

which justifies the concept of a quasiparticle, which is a stable particle in this limit. Considered as an instance of time-energy uncertainty relation, this result signifies that the momentum of an electron close to the Fermi surface is a good quantum number. The Pauli exclusion principle and the short range of the screened Coulomb interaction are imperative for this result to hold. The former restricts the phase space available for decay processes of single particle states close to the Fermi surface to a thin layer of states in the vicinity of the Fermi surface.

8.3.2 Discontinuity in the momentum distribution function

The momentum distribution function of the electrons, which for non-interacting electrons at $T = 0$ is a simple step function due to the Pauli exclusion principle, plays an important role and therefore merits a closer examination also in the interacting case. Moreover, the assumption made in exercise 8.6 about the momentum distribution function of the interacting situation described in the previous section also needs some further justification. For this purpose, we need to specify the electron-electron interaction. The simplest choice is that of a contact potential in real space $U(\mathbf{r}) = U\delta(\mathbf{r}-\mathbf{r}')$, i.e. expressed in field operators

$$\mathcal{V} = \int d^3r\, d^3r'\, \psi_\uparrow^\dagger(\mathbf{r})\psi_\downarrow^\dagger(\mathbf{r}')U\delta(\mathbf{r} - \mathbf{r}')\psi_\downarrow(\mathbf{r}')\psi_\uparrow(\mathbf{r}) \tag{8.161}$$

$$= \frac{U}{V} \sum_{\mathbf{k},\mathbf{k}',\mathbf{q}} c_{\mathbf{k}+\mathbf{q}\uparrow}^\dagger c_{\mathbf{k}'+\mathbf{q}\downarrow}^\dagger c_{\mathbf{k}'\downarrow} c_{\mathbf{k}\uparrow}^\dagger \tag{8.162}$$

with V the system volume.

To calculate the momentum distribution function $n(\mathbf{k}, \sigma) = \langle c_{\mathbf{k}s}{}^\dagger c_{\mathbf{k}s}\rangle$, we employ Rayleigh–Schrödinger perturbation theory. The first order correction to the ground state, i.e. the filled Fermi sea of the non-interacting electrons is

$$|\Psi\rangle = |\Psi^{(0)}\rangle + |\Psi^{(1)}\rangle + \dots \tag{8.163}$$

where

$$|\Psi^{(0)}\rangle = |\Psi_0\rangle = |F\rangle \tag{8.164}$$

$$|\Psi^{(1)}\rangle = \frac{U}{V}\sum_{\mathbf{k}\mathbf{k}'\mathbf{q}}\sum_{\sigma\sigma'}\frac{c^{\dagger}_{\mathbf{k}-\mathbf{q}\sigma}c^{\dagger}_{\mathbf{k}'+\mathbf{q}\sigma'}c_{\mathbf{k}'\sigma'}c_{\mathbf{k}\sigma}}{\epsilon_{\mathbf{k}}+\epsilon_{\mathbf{k}'}-\epsilon_{\mathbf{k}-\mathbf{q}}-\epsilon_{\mathbf{k}'+\mathbf{q}}}|\Psi_0\rangle. \tag{8.165}$$

The correction to the ground state in lowest order of perturbation theory can be seen from this expression to involve electron-hole excitations. The momentum distribution function is then given by

$$n(\mathbf{k}\sigma) = \frac{\langle\Psi|c^{\dagger}_{\mathbf{k}\sigma}c_{\mathbf{k}\sigma}|\Psi\rangle}{\langle\Psi|\Psi\rangle} = n^{(0)}(\mathbf{k}\sigma) + \delta n^{(2)}(\mathbf{k}\sigma) + \ldots \tag{8.166}$$

where

$$n^{(0)}(\mathbf{k}\sigma) = \Theta(k_F - |\mathbf{k}|) \tag{8.167}$$

is the unperturbed momentum distribution of the non-interacting electron system at $T = 0$. The lowest order corrections to this below and above the Fermi surface, respectively, are given by

$$\delta n^{(2)}(\mathbf{k}\sigma) = 1 - \frac{U^2}{V^2}\sum_{\mathbf{k}_1\mathbf{k}_2}\frac{(1-n^{(0)}(\mathbf{k}_1))(1-n^{(0)}(\mathbf{k}_2))n^{(0)}(\mathbf{k}_1+\mathbf{k}_2-\mathbf{k})}{(\epsilon_{\mathbf{k}}+\epsilon_{\mathbf{k}_1+\mathbf{k}_2-\mathbf{k}}-\epsilon_{\mathbf{k}_1}-\epsilon_{\mathbf{k}_2})^2} \tag{8.168}$$

for $|\mathbf{k}| < k_F$ and

$$\delta n^{(2)}(\mathbf{k}\sigma) = \frac{U^2}{V^2}\sum_{\mathbf{k}_1\mathbf{k}_2}\frac{(n^{(0)}(\mathbf{k}_1)n^{(0)}(\mathbf{k}_2)(1-n^{(0)}(\mathbf{k}_1+\mathbf{k}_2-\mathbf{k}))}{(\epsilon_{\mathbf{k}_1}+\epsilon_{\mathbf{k}_2}-\epsilon_{\mathbf{k}}-\epsilon_{\mathbf{k}_1+\mathbf{k}_2-\mathbf{k}})^2} \tag{8.169}$$

for $|\mathbf{k}| > k_F$. For a more detailed discussion of these results, see Sólyom (2010), section 28.2.

We are mainly interested in the behaviour of these expressions when the Fermi surface is approached from below and above, respectively, where we find a finite jump, i.e. a discontinuity (cp. figure 8.2)

$$\delta n_{k_F-\varepsilon} - \delta n_{k_F+\varepsilon} = 1 - \left(\frac{UD(E_F)}{2}\right)^2 \ln 2 \quad \text{for} \quad \varepsilon \to 0. \tag{8.170}$$

The discontinuity at the Fermi surface depends on the interaction and is smaller than unity. The magnitude of the jump, usually denoted $Z = Z(k_F)$ and called the

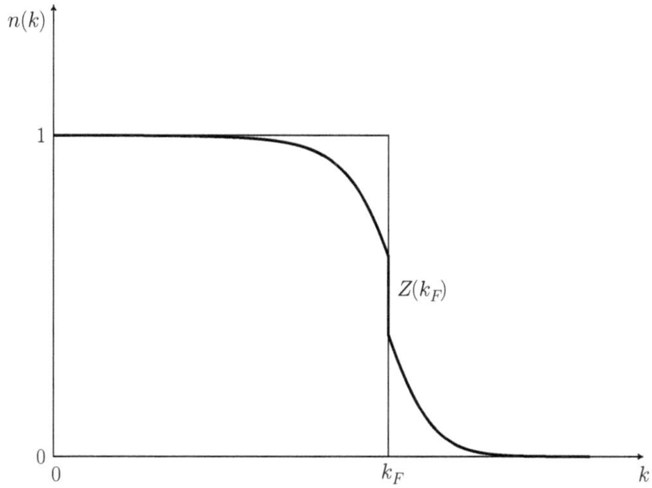

Figure 8.2 *Schematic representation of the momentum distribution function of interacting electrons. The step function represents the non-interacting case at* $T = 0$.

quasiparticle weight, reappears in other contexts as the weight $Z(\mathbf{k})$ of the spectral function of the spectral form of the Green function (see Sólyom, 2010, especially appendix K for the appropriate definitions). The spectral weight is a major feature distinguishing between electrons in three spatial dimensions, as in our calculations so far, and lower dimensional systems. In section 8.5.3 we find a vanishing Z as an indication for the breakdown of Fermi liquid theory for electrons in one dimension.

In higher orders of perturbation theory an increasing number of electron-hole pairs are created with the holes inside and the electrons outside of the Fermi sphere. Using more sophisticated methods of the many-particle theory, it can be shown that a sharp finite jump—the Fermi edge—survives in the momentum distribution function.

For interacting systems, the Fermi surface may still be defined in momentum space as the surface where the momentum distribution function is discontinuous. For isotropic systems, the discontinuity stays at the Fermi momentum k_F of the unperturbed, i.e. non-interacting, system. Hence, in isotropic systems the interaction does not affect the spherical Fermi surface. In more realistic cases the crystal periodicity of the potential of the ions leads to a non-spherical Fermi surface whose shape may be modified due to the the electron-electron interaction. However, the Luttinger theorem, which we shall not prove, states that the momentum space volume enclosed by the Fermi surface remains unchanged by the interaction.

Under quite general conditions, the existence of a sharp Fermi edge and the Luttinger theorem can be proved as long as perturbation theory is applicable and convergent. Such Fermi systems are called normal Fermi systems, which is the case, e.g. for ordinary metals. Their perturbed ground state evolves adiabatically from the non-interacting ground state.

There is, however, no way of deciding *a priori* whether or not a given system of Fermions can be described as a normal Fermi liquid. The only way of answering this question is to compare the predictions of Landau Fermi liquid theory with measurements of the system's properties.

8.4 Landau Fermi liquid theory

Despite the strong (compared to the electrons' kinetic energy) and long-range Coulomb interaction between electrons in a typical metal, the models describing them as free or non-interacting are remarkably successful in capturing qualitatively many properties of metals, e.g. the linear temperature dependence of the specific heat. Such successful descriptions are, moreover, not restricted to the *electron fluid* at low temperatures in metals, but also apply to such diverse systems as neutron stars or liquid ^{3}He in its normal state. Moreover, very recently, Fermi liquids have been designed in the laboratory by trapping atoms in atomic traps or optical lattices where the interactions are tunable by the experimenter.

We have learned three important lessons from the macroscopic hydrodynamic description of section 8.3:

- the plasma oscillations are high-energy collective excitations of the electron gas and, hence, play no significant role for the thermal and transport properties of metals that are due to low energy excitations;
- the strong long-range Coulomb potential of the electrons is screened in the Coulomb gas to an effective short-range interaction affecting electrons only over a distance comparable to the average electron-electron distance in a metal;
- low-lying excitations of the electron gas form nearly stable quasiparticles. This can be understood intuitively: due to the Pauli principle, the low-lying excitations can only decay into energetically available unoccupied states. However, since these excitations are just above the Fermi sea, which is almost completely filled at low temperatures, there is not much phase space available for them into which to decay.

Based on these results, how can we describe the quasiparticles as nearly free particles although their repulsive energy is comparable to their kinetic energy?

This question was resolved by Lev Landau using a series of brilliantly intuitive physical insights (Landau (1957a, 1957b, 1959). The major ingredient is the concept of *adiabatic continuity*, which helps to answer the general question about a many-particle system, i.e. how do we understand a many-particle system when, except for a few special cases, we cannot solve it exactly?

8.4.1 The concept of adiabatic continuity

The first ingredient of Landau's approach in answering this question is the concept of adiabatic continuity. In many cases, we can solve for its ground state $|g_0\rangle$ a simplified many-particle system with Hamiltonian $\mathcal{H}_0$, whereas we cannot calculate exactly the ground state $|g\rangle$ of the more realistic full many-particle Hamiltonian $\mathcal{H}$. The ground states in both cases, however, may have the same symmetry.

Let the Hamiltonian of the many-particle system slowly evolve from $t = -\infty$ to $t = 0$

$$\mathcal{H}(t) = \mathcal{H}_0 + \lambda(t)\mathcal{V} \quad \text{where} \quad \mathcal{V} = \mathcal{H} - \mathcal{H}_0 \quad \text{and} \quad \lambda(t) = e^{t\delta} \tag{8.171}$$

for some small $\delta > 0$.

Typically, as the system evolves adiabatically, the energy levels and eigenstates of the Hamiltonian will evolve from their ground state at $\lambda = 0$ to the fully interacting state at $\lambda = 1$. Distinct energy levels, if they come close to one another, will repel each other due to the influence of off-diagonal matrix elements of the interaction $\mathcal{V}$ between the unperturbed basis states (see Cohen-Tannoudji *et al.* 1977 for a discussion of an elementary example, the two-level system).

In some cases, however, an excited state with a different symmetry compared to the ground state can cross to a lower energy than the ground state energy at a critical value $\lambda = \lambda_c$ and, hence, become the new ground state. This scenario is possible when, because of the different symmetries, selection rules may lead to vanishing off-diagonal matrix elements. Such a situation, a *level crossing*, is connected with a spontaneous symmetry breaking, indicating a phase transition.

Section 8.5 discusses how the dimensionality of the space in which the electrons move also plays a central role. In one dimension, it can be demonstrated that the Fermi liquid concept breaks down (see section 8.5.3) and has to be replaced by the so-called Luttinger liquid concept. The situation in two dimensions (which is of the great theoretical interest) is, unfortunately, less clear at the moment.

If there is no symmetry breaking phase transition as we switch on the interaction $\mathcal{V}$, the adiabatic evolution indeed evolves the ground state from $|g_0\rangle$, the ground state of the non-interacting system to $|g\rangle$ the ground state of the fully interacting system. Then, the concept of adiabatic continuity establishes a one-to-one correspondence between ground states exhibiting the same qualitative physical behaviour and is one of the building blocks of Landau's theory of interacting Fermi liquids. However, it also plays a central part in other areas, e.g. in perturbation theory.

Landau developed his theory with the somewhat simpler case of liquid ^{3}He in mind where the complications of a long–range Coulomb interaction, as for electrons in a metal, are absent. The Fermions in liquid ^{3}He interact via short-range interactions. We first outline Landau's Fermi liquid theory below regarding the so-called *neutral* Fermi liquid and then comment on the *charged* Fermi liquid. The feature that distinguishes these cases is, however, not so much the electrical neutrality of the particles. After all, the electrons charges are balanced by the background of ions in a metal. The distinguishing feature is rather the range of the interactions: short-range interactions for ^{3}He versus long-range Coulomb interactions for electrons.

8.4.2 The quasiparticle concept

The concept of quasiparticles is a direct consequence of the assumption of adiabatic continuity. The quasiparticles of the interacting theory are assumed to have evolved adiabatically from the free particles of the non-interacting theory.[5]

8.4.2.1 The ideal Fermi gas

Let us briefly remind ourselves of the results obtained for the non-interacting, i.e. ideal Fermi gas in a language most suitable for the formulation of Fermi liquid theory.

For a three-dimensional isotropic system, the ground state momentum distribution function $n^{(0)}(\mathbf{k}) = \Theta(k_F - |\mathbf{k}|)$ defines the Fermi momentum k_F via the mean particle density n

$$n = \frac{1}{V}\sum_{\mathbf{k},\sigma} \Theta(k_F - |\mathbf{k}|) = 2\int \frac{d^3k}{(2\pi)^3}\,\Theta(k_F - |\mathbf{k}|) = \frac{k_F^3}{3\pi^2} \tag{8.172}$$

and leads to the ground state energy

$$E_0 = \sum_{\substack{\mathbf{k},\sigma \\ |\mathbf{k}|\le k_F}} \epsilon^{(0)}(\mathbf{k}) = \frac{3}{5} n\epsilon_F V \tag{8.173}$$

where $\epsilon^{(0)} = \mathbf{k}^2/2m$ is the free Fermion energy and $\epsilon_F = k_F^2/2m = \mu(T=0)$ is the Fermi energy, i.e. the chemical potential at zero temperature.

To prepare for the development of Fermi liquid theory in the interacting case, we now describe low-lying excited states via their momentum distribution function

$$n(\mathbf{k},\sigma) = n^{(0)}(\mathbf{k}) + \delta n(\mathbf{k},\sigma). \tag{8.174}$$

These low-lying excitations lie in a narrow shell around the Fermi surface: particles just above, holes just below the Fermi surface. The total energy due to the low-lying excitations changes by an amount

$$\delta E[n(\mathbf{k},\sigma)] = E[n(\mathbf{k},\sigma)] - E_0 = \sum_{\mathbf{k},\sigma} \epsilon^{(0)}(\mathbf{k})\delta n(\mathbf{k},\sigma). \tag{8.175}$$

[5] Elementary excitations are generally called quasiparticles, even if, like for phonons, there are no corresponding 'bare' particles. In the context of Fermi liquid theory, quasiparticles have the narrower meaning explained in this section.

The energy of the excitation is now given as the functional derivative

$$\epsilon^{(0)}(\mathbf{k}) = \frac{\delta(\delta E[n(\mathbf{k},\sigma)])}{\delta n(\mathbf{k},\sigma)} = \frac{\delta E[n(\mathbf{k},\sigma)]}{\delta n(\mathbf{k},\sigma)}, \tag{8.176}$$

while the group velocity of the excitation and the Fermi velocity are obtained as

$$\mathbf{v}(\mathbf{k}) = \nabla_{\mathbf{k}}\epsilon^{(0)}(\mathbf{k}), \quad v_F = |\mathbf{v}(\mathbf{k})|_{|\mathbf{k}|=k_F} = \left.\frac{\partial\epsilon^{(0)}(\mathbf{k})}{\partial|\mathbf{k}|}\right|_{|\mathbf{k}|=k_F} = \frac{k_F}{m}, \tag{8.177}$$

respectively. In these equations, the non-interacting character of the ideal Fermi gas is recovered.

Near the Fermi surface $|\mathbf{k}| = k_F$, the excitation energy can be linearized

$$\epsilon^{(0)}(\mathbf{k}) = \epsilon_F + v_F\,(|\mathbf{k}| - k_F) + O((\mathbf{k} - k_F)^2). \tag{8.178}$$

Hence, we have the equivalent expressions

$$n^{(0)}(\mathbf{k}) = \Theta(k_F - |\mathbf{k}|) = \Theta(\epsilon_F - \epsilon^{(0)}(\mathbf{k})), \tag{8.179}$$

which we note for later convenience.

In the ideal Fermi gas, all excited eigenstates consist of elementary excitations that are particles added to or particles removed from the ground state, the latter referred to as hole excitations. Since these particles and holes are non-interacting for the ideal Fermi gas, the total energy δE of an excited state is the sum of particle and hole excitation energies according to (8.175).

We now generalize this picture to the interacting case using the concept of adiabatic continuity outlined in section 8.4.1.

8.4.2.2 The interacting Fermi liquid

The crucial assumption of Landau Fermi liquid theory, the assumption of adiabatic continuity, assures that any state of the ideal Fermi gas characterized by a momentum distribution function $n(\mathbf{k},\sigma) = n^{(0)}(\mathbf{k}) + \delta n(\mathbf{k},\sigma)$ adiabatically corresponds to an eigenstate of the interacting Fermi system as the interactions are switched on. This eigenstate of the interacting Fermi system can therefore be described by the same momentum distribution function $n(\mathbf{k},\sigma)$ which, used in this way to describe the interacting system, is now called the quasiparticle distribution function of the interacting Fermi liquid (while it is really a property of the non-interacting Fermi gas). The quasiparticle distribution function of the interacting Fermi liquid is therefore different and should not be confused with the momentum distribution function $\langle c^\dagger_{\mathbf{k},\sigma} c_{\mathbf{k},\sigma}\rangle$ of the interacting Fermi liquid. For a review example of the latter for a particular model interaction and in lowest order perturbation theory see section 8.3.2.

The assumption of adiabatic continuity allows us also to repeat the steps for the ideal Fermi gas outlined in section 8.4.2.1. For the interacting system, which we still assume to be isotropic, the Fermi surface remains spherical and, according to the Luttinger theorem, bounds the same volume in momentum space as for the non-interacting system. Consequently, the Fermi momentum k_F remains unchanged when the interactions are switched on adiabatically. In the expressions for the interacting case corresponding to the energy (8.178) of a single particle linearized around $\mu(T=0)=\epsilon_F$,

$$\epsilon(\mathbf{k}) = \epsilon_F + v_F^* \left(|\mathbf{k}| - k_F\right), \quad v_F^* = \frac{k_F}{m^*}, \tag{8.180}$$

i.e. the energy of a single quasiparticle, and velocity, (8.177),

$$\mathbf{v}^*(\mathbf{k}) = \nabla_{\mathbf{k}}\epsilon(\mathbf{k}) = v_F^* \frac{\mathbf{k}}{|\mathbf{k}|} \tag{8.181}$$

of the ideal Fermi gas, only the mass has to be renormalized to the effective mass: $m \to m^*$. Moreover, the density of states on the Fermi surface, (8.86), only gets modified by replacing the mass by the effective mass

$$D^*(\epsilon_F) = \frac{m^* k_F}{\pi^2}. \tag{8.182}$$

It will now be the task of Fermi liquid theory to calculate the effective mass and to connect it to experimentally accessible quantities. To achieve this goal, an energy functional will be constructed in analogy to the energy functional (8.175) of the ideal Fermi gas.

8.4.2.3 Landau energy functional

The change in energy in the interacting Fermi liquid due to a change $\delta n(\mathbf{k}, \sigma)$ in the quasiparticle distribution function by the addition of a single quasiparticle with energy $\epsilon(\mathbf{k})$ is to first order in $\delta n(\mathbf{k}, \sigma)$

$$\delta E[n(\mathbf{k},\sigma)] = \sum_{\mathbf{k},\sigma} \epsilon(\mathbf{k}) \delta n(\mathbf{k},\sigma). \tag{8.183}$$

According to this expression, however, there would be no interaction between quasiparticles since the contributions of several quasiparticles to the total energy (8.183) would simply be additive. This led Landau to extend (8.183) to quadratic order

$$\delta E[n(\mathbf{k},\sigma)] = \sum_{\mathbf{k},\sigma} \epsilon(\mathbf{k}) \delta n(\mathbf{k},\sigma) + \frac{1}{2V} \sum_{\mathbf{k}\mathbf{k}'\sigma\sigma'} f(\mathbf{k},\mathbf{k}',\sigma,\sigma') \delta n(\mathbf{k},\sigma) \delta n(\mathbf{k}',\sigma'). \tag{8.184}$$

Both terms in (8.184) are of the same order λ^2 if we define $\lambda = ||\mathbf{k}| - k_F|$. It is important to observe that this expression does not require any knowledge of the ground state energy of the interacting Fermi liquid. Nor does the momentum distribution function of the interacting system play any role.

Moreover, although the quasiparticle distribution function $n(\mathbf{k}, \sigma)$ or $\delta n(\mathbf{k}, \sigma)$, respectively, and the unperturbed momentum distribution function $n^{(0)}(\mathbf{k})$ of the non-interacting Fermi system will be used in the intermediate steps of the calculations, neither of them will appear in the expressions for physical quantities that are experimentally accessible.

The quasiparticle energy $\hat{\epsilon}(\mathbf{k}, \sigma)$ is now given as the functional derivative of the energy functional (8.184)

$$\hat{\epsilon}(\mathbf{k}, \sigma) = \frac{\delta(\delta E)}{\delta(\delta n(\mathbf{k}, \sigma))} = \epsilon(\mathbf{k}) + \frac{1}{V} \sum_{\mathbf{k}', \sigma'} f(\mathbf{k}, \mathbf{k}', \sigma, \sigma') \delta n(\mathbf{k}', \sigma'). \tag{8.185}$$

The consistency requirement is fulfilled that the quasiparticle energy $\hat{\epsilon}(\mathbf{k}, \sigma)$ coincides with the quasiparticle energy of a single quasiparticle $\epsilon(\mathbf{k})$ (see (8.178)) when there are no other quasiparticles in the quasiparticle distribution function, i.e. when $\delta n(\mathbf{k}', \sigma') = 0$. Equation (8.185) reflects the mean-field character of the Fermi liquid approach: the change in the quasiparticle energy $\hat{\epsilon}(\mathbf{k}, \sigma)$ stems from the quasiparticle distribution function of all other quasiparticles.

8.4.2.4 Landau parameters

In order to make progress, we now specify the Landau function $f(\mathbf{k}, \mathbf{k}', \sigma, \sigma')$ for the physical situation with which we are dealing, i.e. an isotropic system that we also assume to be spin-rotation invariant. The latter invariance implies that we can split f in to two terms, a spin symmetric f^s and a spin antisymmetric f^a term

$$f(\mathbf{k}, \mathbf{k}', \sigma, \sigma') = f^s(\mathbf{k}, \mathbf{k}') + \sigma\sigma' f^a(\mathbf{k}, \mathbf{k}'). \tag{8.186}$$

As we are dealing with processes close to the Fermi surface, we can set $|\mathbf{k}| \approx k_F$ and $|\mathbf{k}'| \approx k_F$, and retain, for the isotropic system, as the relevant argument of f the angle α between $\mathbf{k}'$ and $\mathbf{k}'$ (or more precisely $\cos\alpha = \mathbf{k} \cdot \mathbf{k}'/|\mathbf{k}||\mathbf{k}'|$)

$$f(\mathbf{k}, \mathbf{k}', \sigma, \sigma') = f(\mathbf{k}_F, \mathbf{k}_F', \sigma, \sigma') = f^s(\alpha) + \sigma\sigma' f^a(\alpha). \tag{8.187}$$

The functions $f^s(\alpha)$ and $f^a(\alpha)$ can be expanded in Legendre polynomials with the transformation equations ($d\Omega = d\varphi d\alpha \sin\alpha$)

$$f^{s,a}(\alpha) \sum_{l=0}^{\infty} f_l^{s,a} P_l(\cos\alpha), \quad f_l^{s,a} = (2l+1) \int_0^{2\pi} \int_0^{\pi} \frac{d\Omega}{4\pi} f^{s,a}(\alpha) P_l(\cos\alpha). \tag{8.188}$$

Finally, to make them dimensionless, we define the so-called Landau parameters by multiplying the expansion coefficients $f_l^{s,a}$ by the density of states at the Fermi level, (8.182),

$$F_l^{s,a} = D^*(\epsilon_F) f_l^{s,a}. \tag{8.189}$$

So far, all our arguments have been for a system at zero temperature, $T = 0$. In order to generalize them to finite temperature, we introduce an expression for the entropy of the Fermi system. In keeping with the general spirit of Landau Fermi liquid theory, this entropy will depend on the quasiparticle distribution function.

8.4.2.5 Entropy and thermodynamic potential

The states of the interacting Fermi liquid are counted as in the non-interacting Fermi gas. Therefore, the entropy corresponding to the quasiparticle distribution function $n(\mathbf{k}, \sigma)$ is given by Fermi–Dirac statistics (see section 4.13.3)

$$S[n(\mathbf{k},\sigma)] = -\sum_{\mathbf{k},\sigma} [n(\mathbf{k},\sigma)\ln n(\mathbf{k},\sigma) + (1 - n(\mathbf{k},\sigma))\ln(1 - \ln n(\mathbf{k},\sigma))] \tag{8.190}$$

and the corresponding thermodynamic potential (grand canonical potential, cp. section 4.10)

$$\Phi[n(\mathbf{k},\sigma)] = E[n(\mathbf{k},\sigma)] - \mu N[n(\mathbf{k},\sigma)] - TS[n(\mathbf{k},\sigma)] \tag{8.191}$$

where $N[n(\mathbf{k},\sigma)] = \sum_{\mathbf{k},\sigma} n(\mathbf{k},\sigma)$ is the total number of quasiparticles.

The equilibrium quasiparticle distribution $\bar{n}(\mathbf{k},\sigma)$ is determined by the condition that the grand canonical potential $\Phi[n(\mathbf{k},\sigma)]$ is stationary: $\delta\Phi[n(\mathbf{k},\sigma)]/\delta n(\mathbf{k},\sigma) = 0$

$$\bar{n}(\mathbf{k},\sigma) = n_F(\hat{\epsilon}(\mathbf{k},\sigma) - \mu, T) = \frac{1}{e^{\beta(\hat{\epsilon}(\mathbf{k},\sigma)-\mu)} + 1}. \tag{8.192}$$

The quasiparticle distribution function $n(\mathbf{k},\sigma)$ for finite temperatures is replaced by the Fermi function $n_F(\hat{\epsilon}(\mathbf{k},\sigma) - \mu, T)$ where now

$$\hat{\epsilon}(\mathbf{k},\sigma) = \frac{\delta E[n(\mathbf{k},\sigma)]}{\delta n(\mathbf{k},\sigma)} = \epsilon(\mathbf{k}) + \frac{1}{V}\sum_{\mathbf{k}'\sigma'} f(\mathbf{k},\mathbf{k}',\sigma,\sigma')\left(\bar{n}(\mathbf{k}',\sigma') - n^{(0)}(\mathbf{k}')\right) \tag{8.193}$$

is the quasiparticle energy of the equilibrium quasiparticle distribution $\bar{n}(\mathbf{k},\sigma)$. In other words, (8.192) is an implicit equation for the finite-temperature quasiparticle distribution function, the quasiparticle energies (8.193) depend on the quasiparticle distribution function itself, as indicated by the notation $n_F(\hat{\epsilon}(\mathbf{k},\sigma) - \mu, T)$.

EXERCISE 8.7 Landau function Show that, to lowest order in $\delta n(\mathbf{k},\sigma)$, the thermodynamic potential Φ, (8.191), has the expansion

$$\Phi[\bar{n}+\delta n] - \Phi[\bar{n}] = \frac{1}{2}\sum_{\mathbf{k}\mathbf{k}'\sigma\sigma'}\left[\frac{\delta_{\sigma\sigma'}\delta_{\mathbf{k}\mathbf{k}'}}{n_F'(\hat{\epsilon}(\mathbf{k},\sigma)-\mu,T)} + \frac{1}{V}f(\mathbf{k},\mathbf{k}',\sigma,\sigma')\right]\delta n(\mathbf{k},\sigma)\delta n(\mathbf{k}',\sigma'). \tag{8.194}$$

Use this result to show that the Landau function f can be defined as

$$\frac{1}{V}f(\mathbf{k},\mathbf{k}',\sigma,\sigma') = \frac{\delta_{\sigma\sigma'}\delta_{\mathbf{k}\mathbf{k}'}}{n_F'(\hat{\epsilon}(\mathbf{k},\sigma)-\mu)} + \left.\frac{\delta^2\Phi[n(\mathbf{k},\sigma)]}{\delta n(\mathbf{k},\sigma)\delta n(\mathbf{k}',\sigma')}\right|_{\bar{n}}. \tag{8.195}$$

In both expressions

$$n_F'(x,T) = \left.\frac{dn_F(x,T)}{dx}\right|_{x=\hat{\epsilon}(\mathbf{k},\sigma)-\mu}. \tag{8.196}$$

We are now in a position to connect the parameters of Fermi liquid theory, the Landau parameters $F_l^{s,a}$, and, especially, the effective mass m^* to experimentally accessible physical quantities like, e.g., the specific heat.

8.4.2.6 Low-temperature properties of Fermi liquids

For the low-temperature properties of the Fermi liquid, we need to estimate the temperature dependences of (8.192) and (8.193). If the Landau function f itself is temperature independent, then it obviously suffices to estimate the temperature dependence of $\hat{\epsilon}(\mathbf{k},\sigma) - \epsilon(\mathbf{k})$. Assuming that we can replace $\hat{\epsilon}(\mathbf{k},\sigma)$ by $\epsilon(\mathbf{k})$ in the Fermi function (8.192), we can indeed show that $\hat{\epsilon}(\mathbf{k},\sigma) - \epsilon(\mathbf{k})$ is of order T^2, which justifies our initial assumption. The number of thermally excited quasiparticles is thus also of order T^2.

EXERCISE 8.8 Estimate the temperature dependence of the quasiparticle energy Under the assumption that $\hat{\epsilon}(\mathbf{k},\sigma)$ can be replaced by $\epsilon(\mathbf{k})$ in the Fermi function (8.192), approximate $\hat{\epsilon}(\mathbf{k},\sigma) - \epsilon(\mathbf{k})$ by replacing the sum over $\mathbf{k}'$ by an integral over $d\Omega = \sin\alpha' d\varphi' d\alpha'$ (angular average) and an integral over energy ϵ using the density of states $D^*(\epsilon)$. Using furthermore the orthogonality properties of the Legendre polynomials

$$\int_{-1}^{1} P_n(x)P_m(x) = \frac{2}{2n+1}\delta_{nm} \tag{8.197}$$

and the Sommerfeld expansion, show that

$$\hat{\epsilon}(\mathbf{k},\sigma) - \epsilon(\mathbf{k}) \approx f_0^{(s)}\frac{\pi^2}{6}T^2\left.\frac{dD^*(\epsilon)}{d\epsilon}\right|_{\epsilon=\epsilon_F}. \tag{8.198}$$

8.4.2.7 *Specific heat*

As in section 8.2.4, we calculate the specific heat from its thermodynamic definition

$$c_V(T) = \frac{1}{V}\frac{\partial E}{\partial T}\bigg|_{NV}, \tag{8.199}$$

where in

$$E[n(\mathbf{k},\sigma)] = E_0 + \sum_{\mathbf{k}\sigma} \epsilon(\mathbf{k})\delta n(\mathbf{k},\sigma) + \frac{1}{2V}\sum_{\mathbf{k}\mathbf{k}'\sigma\sigma'} f(\mathbf{k},\mathbf{k}',\sigma,\sigma')\delta n(\mathbf{k},\sigma)\delta n(\mathbf{k}',\sigma') \tag{8.200}$$

the quasiparticle interaction term involving $\delta n(\mathbf{k},\sigma)\delta n(\mathbf{k}',\sigma')$ can be neglected according to the results of the previous section. This implies that the only change compared to the calculation of the low-temperature specific heat for the non-interacting Fermi gas in section 8.2.4 consists in using the renormalized density of states at the Fermi surface $D^*(\epsilon_F)$, (8.182). We obtain

$$c_V(T) = \frac{\pi^2}{3}D^*(\epsilon_F)T = \frac{m^* k_F}{3}T \tag{8.201}$$

for low temperatures. Hence, the phenomenological parameter m^*, the effective mass, can be determined from measurements of the low temperature specific heat of the Fermi system.

8.4.2.8 *Effective mass*

The result for the specific heat of the previous section only contains limited information about the interacting Fermi system via the effective mass m^*. Further information about the interacting system is contained in the Landau parameters and it will therefore be desirable to connect the effective mass to these parameters.

For this purpose, we consider the expectation value of the current density operator of a many-particle system

$$\mathbf{j} = \frac{1}{V}\left\langle \sum_i \frac{\mathbf{p_i}}{m}\right\rangle. \tag{8.202}$$

Adding a further particle of mass m, spin σ, and wave vector $\mathbf{k}$ to the many-particle system changes the current (8.202) by

$$\mathbf{j}(\mathbf{k},\sigma) = \frac{1}{V}\frac{\mathbf{k}}{m}. \tag{8.203}$$

This is the current contribution in the laboratory frame. To find the influence of the interaction with the other particles we now apply a Galilean transformation to the whole system, which includes the added particle, to a frame moving with a uniform relative velocity $\mathbf{v}$. Only the kinetic part of the total energy is affected by the Galilean transformation. The interaction part remains unchanged. Therefore, total energy and total momentum in the moving frame are given by

$$E' = E - \mathbf{P} \cdot \mathbf{v} + \frac{M}{2} v^2, \tag{8.204}$$

$$\mathbf{P}' = \mathbf{P} - M\mathbf{v}, \tag{8.205}$$

where M is the total mass of the system. This moving frame is equivalent to shifting the wave vector of all particles in the system by the same amount $\mathbf{q}$ if the velocity is chosen to be $\mathbf{v} = -\mathbf{q}/m$. The transformed Hamiltonian of the many-particle system is then

$$\mathcal{H}' = \sum_i \frac{(\mathbf{p}_i + \mathbf{q})^2}{2m} + \mathcal{H}_{\text{int}} = \mathcal{H} + \mathbf{q} \cdot \sum_i \frac{\mathbf{p}_i}{m} + N \frac{(\mathbf{q})^2}{2m}, \tag{8.206}$$

where N is the total number of particles.

This Hamiltonian can be used to describe the single added particle of wave vector $\mathbf{k}$ and spin σ in the moving frame. It must be treated as a quasiparticle whose quasiparticle energy has two components due to the overall shift $\mathbf{q}$

$$\hat{\epsilon}(\mathbf{k} + \mathbf{q}, \sigma) = \epsilon(\mathbf{k}, \sigma) + \mathbf{q} \cdot \mathbf{v}(\mathbf{k}) + \frac{1}{V} \sum_{\mathbf{k}'\sigma'} f(\mathbf{k}, \mathbf{k}', \sigma, \sigma') \delta n_{\mathbf{q}}(\mathbf{k}', \sigma'), \quad \mathbf{v}(\mathbf{k}) = \frac{\mathbf{k}}{m^*}. \tag{8.207}$$

The overall shift of wave vectors by $\mathbf{q}$ is responsible for the change in the quasiparticle distribution function $\delta n_{\mathbf{q}}(\mathbf{k}', \sigma')$ in (8.207) by creating quasiparticles on one side of the Fermi surface and quasiholes on the other side, hence

$$\delta n_{\mathbf{q}}(\mathbf{k}', \sigma') = -\mathbf{q} \cdot \nabla_{\mathbf{k}} n^{(0)}(\mathbf{k}) = -\mathbf{q} \cdot \mathbf{v}(\mathbf{k}) \frac{\partial n^{(0)}(\mathbf{k})}{\partial \epsilon(\mathbf{k})}. \tag{8.208}$$

Neglecting terms proportional to q^2, we can use (8.206) together with (8.207) and (8.208) to write

$$\mathbf{j}(\mathbf{k}, \sigma) = \frac{1}{V} \nabla_{\mathbf{q}} \hat{\epsilon}(\mathbf{k} + \mathbf{q}, \sigma) = \frac{1}{V} \left[\frac{\mathbf{k}}{m^*} - \frac{1}{V} \sum_{\mathbf{k}'\sigma'} f(\mathbf{k}, \mathbf{k}', \sigma, \sigma') \frac{\mathbf{k}'}{m^*} \frac{\partial n^{(0)}(\mathbf{k}')}{\partial \epsilon(\mathbf{k}')} \right]. \tag{8.209}$$

The propagating quasiparticle interacts with the rest of the Fermi system while the other quasiparticles generate an additional contribution to the current, the so-called ‘backflow’ that is described by the second term in (8.209). The only contribution stems from quasiparticles in a narrow shell around the Fermi surface because

$$\frac{\partial n^{(0)}(\mathbf{k})}{\partial \epsilon(\mathbf{k})} = -\delta(\epsilon(\mathbf{k}) - \epsilon_F). \tag{8.210}$$

Comparing the current in the moving frame (8.209) with the current in the laboratory frame (8.203), we obtain

$$\frac{\mathbf{k}}{m} = \frac{\mathbf{k}}{m^*} + \frac{1}{V}\sum_{\mathbf{k}'\sigma'} f(\mathbf{k}, \mathbf{k}', \sigma, \sigma')\frac{\mathbf{k}'}{m^*}\delta(\epsilon(\mathbf{k}') - \epsilon_F). \tag{8.211}$$

Evaluating this expression leads to

$$\frac{m^*}{m} = 1 + \frac{1}{3}F_1^s \tag{8.212}$$

relating the effective mass m^* to the Landau parameter F_1^s. It is important to note that this relation restricts the permissible values of the Landau parameter to $F_1^s > -3$.

The calculation of other thermodynamic quantities lead to similar restrictions on the permissible values of the Landau parameters. For example, $F_0^s > -1$ must be satisfied for the isothermal compressibility to remain positive for the interacting Fermi liquid.

A general stability analysis shows that $F_l^s > -(2l+1)$ and $F_l^a > -(2l+1)$. If some of these conditions are not satisfied the spherical Fermi surface of the ground state becomes unstable. These are the so-called Pomeranchuk instabilities. See e.g. Sólyom (2010) or Baym and Pethick (2004) for more details.

EXERCISE 8.9 In (8.211) contributions from $|\mathbf{k}| = |\mathbf{k}'| = k_F$ dominate. Use this observation to evaluate (8.211), replacing the sum over $\mathbf{k}'$ by an appropriate integration.

For an alternative derivation of (8.212), which does not use the particle current, see Baym and Pethick (2004).

To gain access to the antisymmetric Landau parameters F_l^a we need to study Fermi systems subject to a magnetic field that breaks spin rotation invariance. The magnetization and magnetic susceptibility of the Fermi liquid are then related to the Landau parameter F_0^a that has to satisfy $F_1^a > -1$ if the Fermi liquid is to remain paramagnetic. Again, we refer to the cited literature for detailed results.

Leaving the discussion of Fermi liquids in three dimensions where the properties of strongly interacting electrons could be successfully approximated within Landau's Fermi liquid theory based on the concept of emergent non-interacting quasiparticles, we turn our attention to the special case of one-dimensional Fermi systems. These systems cannot be understood based on a picture of non-interacting quasiparticles and, in particular, Fermi liquid theory will turn out to be inadequate for Fermi systems in one dimension. One-dimensional Fermi systems turn out to be markedly different from their counterparts in three dimensions. A rich variety of unexpected new physical phenomena such as spin-charge separation unfold especially due to the way electron interactions operate when they are confined to thin quasi-one-dimensional channels.

The example of a one-dimensional Fermi system, the interacting electron gas in one dimension, is amenable to an exact or almost exact treatment after only a few, physically well-motivated, simplifying assumptions. This is one of its major attractions.

For a recent comparison of experiments on one-dimensional carbon nanotubes and nanowires with the theoretical concepts introduced for the description of of one-dimensional Fermi systems, see Deshpande *et al.* (2010).

Fermi systems in one dimension are today mostly discussed under the name of Luttinger liquids. The following section introduces the theory of Luttinger liquids.

8.5 Luttinger liquid theory

The investigation of one-dimensional systems is not a mere academic theoretical exercise. An ever-increasing number of new materials have been synthesized over the past decades that are quasi-one-dimensional in the sense that the electrons interact predominantly along one particular direction of the material, whereas the interactions perpendicular to this preferred direction are much—sometimes orders of magnitude weaker due to steep and high potential barriers. An example are the Bechgaard salts. Moreover, structures have been created experimentally, so-called quantum wires, in which the motion of the electrons is confined, to a very good approximation, in the one dimension prescribed by the wire. A prominent example of such structures are carbon nanotubes. In such quasi-one-dimensional materials, while Landau Fermi liquid theory fails qualitatively, the strong effect of interactions between Fermions confined to one dimension leads to new states of matter with fundamentally different properties from corresponding properties in three dimensions.

While many quasi-one-dimensional materials satisfy the requirements for one-dimensional behaviour by their structure on the molecular level, in other cases it may put considerable demands on the skills of the experimenter to produce artificial structures that are sufficiently one-dimensional. The following section estimates, these requirements for the case of a quantum wire, i.e. asks how thin such a wire would have to be to qualify as a good one-dimensional system.

8.5.1 Experimental dimensional requirements

The creation of one-dimensional structures, such as quantum wires, is still a great challenge in the laboratory. In order to estimate the requirements, let us consider a long quantum wire of length L in the z-direction and transverse diameter d in the x- and y-direction where $L \gg d$. Assuming a infinite square well potential to confine the electrons in the transverse direction while they are free in the z-direction, the electrons can be described by a wave function

$$\psi(\mathbf{r}) = \phi_{n_x}(x)\phi_{n_y}(y)\frac{1}{\sqrt{L}}e^{ik_z z} \tag{8.213}$$

where the square well part consists of standing waves

$$\phi_{n_x}(x) = \sqrt{\frac{2}{d}}\sin\left(\frac{\pi x n_x}{d}\right), \tag{8.214}$$

$$\phi_{n_y}(y) = \sqrt{\frac{2}{d}}\sin\left(\frac{\pi y n_y}{d}\right) \tag{8.215}$$

with spectrum given by

$$E_n = \frac{\pi^2}{2md^2}n^2, \qquad n = 1, 2, 3, \ldots, \qquad n = n_x \quad \text{or} \quad n = n_y. \tag{8.216}$$

On top of each energy level E_n sits an energy band

$$\epsilon(k_z) = \frac{k_z^2}{2m} \tag{8.217}$$

with, assuming periodic boundary conditions, quantized momentum

$$k_z = \frac{2\pi}{L}n_z, \qquad n_z = 0, \pm 1, \pm 2, \ldots. \tag{8.218}$$

Now we can formulate conditions under which the physics of the wire will be one-dimensional. First, we fill the Fermi sea densely with electrons, i.e. the spacing of the momentum is required to satisfy

$$\Delta k_z \equiv \frac{2\pi}{L} \ll k_F. \tag{8.219}$$

The corresponding filled Fermi sea with Fermi energy

$$\epsilon_F \equiv \epsilon(k_z = k_F) = \frac{1}{2m}k_F^2 \tag{8.220}$$

is required to be small compared to the level spacing $\Delta E = E_2 - E_1$ in the square well potential, i.e.

$$\epsilon_F \ll \Delta E \tag{8.221}$$

such that excitations to the level E_2 can be neglected. This condition implies

$$\frac{d}{\lambda_F} \ll \frac{3}{4} \qquad \text{or, simply} \qquad \frac{d}{\lambda_F} \ll 1 \tag{8.222}$$

with the Fermi wave length $\lambda_F = 2\pi/k_F$. Taken together, these conditions postulate for a truly one-dimensional wire

$$\frac{L}{d} \gg 1 \gg \frac{d}{\lambda_F}. \tag{8.223}$$

A useful estimate of the thickness of the quantum wires attempted can also be culled from the above. Let us assume that the Fermi energy of the quadratic energy band of the free electrons is $\epsilon_F \approx 1\text{eV}$. Then, in order to have

$$\Delta E = \frac{3}{2m}\left(\frac{\pi}{d}\right)^2 > \epsilon_F \tag{8.224}$$

requires

$$d < \sqrt{3}\pi\sqrt{\frac{a_0^2}{2ma_0^2\epsilon_F}} \approx 1\text{nm} \tag{8.225}$$

where we have introduced the Bohr radius $a_0 \approx 0.05\text{nm}$ and with it the Rydberg energy $1/(2ma_0^2) \approx 13.6\text{eV}$.

These estimates show that creating truly one-dimensional electron systems in the laboratory is certainly an ambitious undertaking. This is even more true since any imprecision along the wire may lead to unwanted localization effects in the one direction the wire is supposed to support propagation. Full control of the structure down to the atomic scale is thus required, particularly in the transverse direction.

Promising experimental procedures include the following techniques. The first is an extension of molecular beam epitaxy called cleaved-edge-overgrowth. Molecular beam epitaxy has become a standard technique for the growth of atomically perfect layered two-dimensional electron gas systems. By cleaving this two-dimensional system and overgrowing an edge can be produced along one of the crystal axes, which is also perfect on an atomic scale (Yacoby *et al.*, 1996). This experimental setup has been used to measure the collective excitation spectrum of interacting electrons in one dimension (Auslaender *et al.*, 2002).

A second possibility to obtain perfect structures on an atomic scale consists of using macromolecules that form nanotubes, the famous carbon nanotubes. Here it is the chemical composition that guarantees the tubes form perfect structures on the nanoscale. Experiments on carbon nanotubes demonstrated the signatures of a one-dimensional electronic many-particle system see for example (Bockrath *et al.*, 1999).

8.5.2 One-dimensional Fermi systems

As a first step toward the description of one-dimensional Fermi systems section 8.5.3 demonstrates the breakdown of the Fermi liquid description in one dimension.

It is worth highlighting a few of the remarkable features of one-dimensional Fermi systems, which include

- the absence of quasi-particles carrying quantum numbers of free Fermions in the vicinity of the Fermi surface, see section 8.5.3;
- that low-energy collective excitations consisting of density fluctuations are of paramount importance;
- that the charge and spin of an electron may be torn apart and then move independently of each other: spin-charge separation;
- that the momentum distribution function $n(k)$ exhibits no discontinuity but is given by a power law whose exponent depends on the interaction, see section 8.5.3; and
- power-law decay of correlation functions with non-universal, i.e. interaction dependent exponents.

Furthermore, it is instructive to outline briefly the theoretical development. After early work by Felix Bloch (1934) in the 1930s, Sin-Itiro Tomonaga (1950) early in the 1950s and Joaquin Luttinger (1963) in the 1960s proposed models for the one-dimensional interacting electron gas whose exact solutions were further investigated by Mattis and Lieb (1965).

Duncan Haldane 1980; 1981*b* demonstrated that the low-energy effective behaviour of many important one-dimensional models of quantum many-particle physics, e.g. the Hubbard model (see section 8.7 which provides a detailed explanation of the Hubbard model), can be described in the general framework of the Tomonaga–Luttinger models. Haldane developed the concept of the Luttinger liquid that has since become an active field of theoretical and experimental activity with numerous fruitful applications.

The theory of Luttinger liquid combines sophisticated mathematical and conceptual tools to produce well-controlled, in some cases even exact, analytical results for a fully interacting many-particle system. Moreover, recent advances in experimental physics at the mesoscopic and nanoscale, as well as in ultracold atomic gases in optical lattices and atom traps, furnish highly precise tests of these results.

Quite recently, for instance, experimental evidence for Luttinger liquid behaviour has been found in measurements involving two parallel one-dimensional quantum wires (Laroche *et al.*, 2014).

There is a close connection of Luttinger liquid theory and Bethe ansatz solvable models, the latter, since they admit an exact solution, providing particularly well-understood examples of the former. For a review on the connection between Luttiger liquid theory and integrable models, see Sirker (2012*a*) (also Sirker, 2012*b* for an errratum).

8.5.2.1 Key qualitative picture: particle density fluctuations

The key to an understanding of the peculiar properties of one-dimensional Fermi systems is provided by a simple qualitative picture. Interacting Fermions confined to one dimension can only attempt to minimize their energy by rearranging their positions. This creates fluctuations in the particle density $\rho(x)$ (see section 2.6, especially example 5 in 2.6.1). In higher-dimensional systems, however, Fermions are free to move around each other.

Therefore, it is suggestive to assume that the particle density operator $\rho(x)$ will play a central role for one-dimensional Fermi systems. For Fermions, this operator, being the product of Fermi operators, has a Bosonic character, a fact which will also be of paramount importance for one-dimensional Fermi systems. The Fourier transformed particle density operator will indeed play a central role in the discussion of the quantitative tools for one-dimensional Fermi systems outlined in section 8.5.4.

After demonstrating that the Fermi liquid concept fails in one dimension, section 8.5.4 introduces the models of Tomonaga and Luttinger.

More exhaustive presentations of Luttinger liquid theory than we can provide here may be found in Voit (1994), Giamarchi (2004), Sólyom (2010), and, with the emphasis on the important theoretical method of Bosonization, Gogolin *et al.* (1998).

8.5.3 Breakdown of the Fermi liquid description in one dimension

In section 8.3.2 we calculated the discontinuity of the momentum distribution function for interacting electrons in three dimensions. The interaction of the electrons can, however, not always be treated perturbatively. For example, this is the case for an electron system in one dimension where an attempt to calculate the first non-vanishing correction to the momentum distribution function demonstrates the breakdown of perturbation theory.

As a specific demonstration of this breakdown, we evaluate (8.169) for a one-dimensional system. To obtain non-vanishing contributions, the momenta k_1 and k_2 have to be chosen in the range $(-k_F, k_F)$ whereas $k_1 + k_2 - k$ must lie outside this range. These requirements can be satisfied by choosing k_1 in the vicinity of k_F and k_2 in the vicinity of $-k_F$, or vice versa. We therefore obtain an overall factor of 2. Moreover, since we are only concerned with the vicinity of the Fermi points $\pm k_F$, the energy dispersion may be linearized around these points within a momentum bandwidth of $2K$

$$\epsilon_k = \begin{cases} v_F(k - k_F) & \text{for} \quad k_F - K < k < k_F + K, \\ -v_F(k + k_F) & \text{for} \quad -k_F - K < k < -k_F + K, \end{cases} \tag{8.226}$$

where

$$v_F = \frac{\partial \epsilon_k}{\partial k}\bigg|_{k_F} = -\frac{\partial \epsilon_k}{\partial k}\bigg|_{-k_F} \tag{8.227}$$

is the Fermi velocity and the Fermi energies vanish $\epsilon_{k_F} = \epsilon_{-k_F} = 0$.

Replacing the sums in (8.169) by integrals

$$\sum_k \to L \int \frac{dk}{2\pi} \tag{8.228}$$

where L is the length of the one-dimensional system, we obtain

$$\delta n^{(2)}(k,\sigma) = 2U^2 \int_{k_F-K}^{k_F} \frac{dk_1}{2\pi} \int_{-k_F}^{-k_F+K} \frac{dk_2}{2\pi} \frac{\Theta(-k_F-(k_1+k_2-k))}{[2v_F(k-k_1)]^2} \tag{8.229}$$

$$= 2\left(\frac{U}{4\pi v_F}\right)^2 \int_{k_F-K}^{k_F} dk_1 \frac{1}{k-k_1} \tag{8.230}$$

$$= 2\left(\frac{U}{4\pi v_F}\right)^2 \ln\left(\frac{K}{k-k_F}\right) \quad \text{as} \quad k \to k_F. \tag{8.231}$$

This logarithmic divergence of the second order correction as k approaches k_F indicates that a perturbative approach is not feasible for one-dimensional systems.

Moreover, it can be shown that the distribution function is indeed continuous at the Fermi edge $k = k_F$ in one dimension, i.e. there is no sharp Fermi surface any more, with drastic consequences: the quasiparticle weight having vanished, there are no quasiparticles as elementary excitations. The whole Fermi liquid approach becomes inapplicable. One-dimensional electron or Fermi systems, hence, present essentially new physical properties that also need novel theoretical tools for their description.

8.5.4 Tomonaga and Luttinger models

The discussion in section 8.5.3 showed that different dimensionalities of the Fermi systems can produce qualitatively different behaviours. Materials synthesized in the last decades that are quasi-one dimensional concerning at least their electronic properties provide the investigation of such one-dimensional systems with more than just an academic interest.

Bloch (1934) realized that the electron-hole excitations in a non-interacting one-dimensional electron gas can be described as sound waves, i.e. as Bosonic excitations. Tomonaga (1950) obtained phonon-like Bosonic elementary excitations with a linear dispersion relation for a one-dimensional electron gas with a special electron-electron interaction. An exactly solvable modification of Tomonaga's model by Luttinger (1963; see also Mattis and Lieb, 1965, who corrected a subtle mistake in Luttinger's solution) exhibits surprising properties: low-energy Fermionic quasiparticles and a Fermi surface, as usually defined, are absent; moreover, correlation functions decay asymptotically at large distances and long times with power laws exhibiting non-universal exponents.

The models of Tomonaga and Luttinger, somewhat incorrectly often summarized under the rubric Tomonaga–Luttinger model, led Haldane 1980; 1981*b* to introduce the

Luttinger liquid concept that describes the low-energy properties of one-dimensional interacting Fermi systems that have no energy gap between the ground state energy and excitation energies. The systems are described by this concept in a universal way without regard for the special nature of the microscopic Hamiltonians governing the Fermions.

A special interest in Luttinger liquid theory in the context of this book consists in the fact that one-dimensional quantum models exactly solvable by Bethe ansatz, like the Heisenberg quantum spin chain and the one-dimensional Hubbard model, have played essential roles in the understanding and development of Luttinger liquid theory (Haldane, 1980, 1981*a*). These one-dimensional Bethe ansatz solvable models can be recast in the Luttinger liquid framework. Thus, the parameters describing them as Luttinger liquids can be calculated exactly from their Bethe ansatz solutions.

We present the models of Tomonaga and Luttinger in the following sections. We start by collecting some of the expressions obtained in chapter 2, which introduced the language of second quantization. These expressions are the one-particle operator for the kinetic energy (2.153) and the two-particle operator (2.167) together with (2.168) evaluated for the Coulomb energy between two electrons in momentum representation. For a one-dimensional system these expressions become

$$\mathcal{H} = \sum_k (\epsilon(k) - \epsilon_F)\, c^\dagger_{k\sigma} c_{k\sigma} + \frac{1}{2L} \sum_{kk'q\neq 0\sigma\sigma'} V_{\sigma\sigma'}(q) c^\dagger_{k+q\sigma} c^\dagger_{k'-q\sigma'} c_{k'\sigma'} c_{k\sigma} \tag{8.232}$$

where $V_{\sigma\sigma'}(q)$ is the Fourier transform of the interaction potential (cp. (2.168)), which may, in general, be spin-dependent (cp. Sólyom (2010), chapter 28 for a discussion of this point) and we have included the Fermi energy ϵ_F into the Hamiltonian.

In order to avoid too much complication and also because this approach was used in the early literature cited above, we shall mostly disregard spin in the following. Moreover, we shall develop most of the formalism of Luttinger liquid theory considering only the non-interacting, or free part of the Hamiltonian (8.232).

8.5.4.1 Non-interacting Fermions in one dimension

The basic features of quasi-one-dimensional materials can be derived from a simple electronic band structure with a single band near the Fermi energy (see figure 8.3) whose energy dispersion relation or spectrum

$$\xi(k) = \epsilon(k) - \epsilon_F \tag{8.233}$$

for the non-interacting Hamiltonian

$$\mathcal{H}_0 = \sum_k \xi(k) c^\dagger_{k\sigma} c_{k\sigma} \tag{8.234}$$

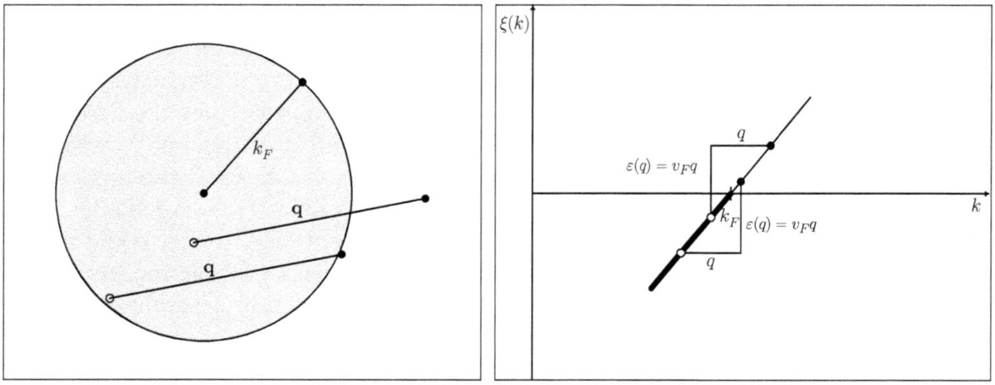

Figure 8.3 *Fermi disk in two dimensions or cross-section of Fermi sphere in three dimensions with radius k_F (left panel). For a fixed momentum transfer $\mathbf{q}$, particle-hole pair excitations with a continuum of different energies can be created. In one dimension the spectrum is approximately linear around the Fermi points $\pm k_F$ (right panel). As a consequence, for a given momentum transfer q, there are only the excitation energies $\varepsilon(q) = \pm v_F q$ possible.*

can be assumed to be parabolic for a weak periodic potential

$$\xi(k) = \frac{1}{2m}\left(k^2 - k_F^2\right), \tag{8.235}$$

see the left panel of figure 8.4, or

$$\xi(k) = -2t\cos ka \qquad \text{and} \qquad k_F = \frac{\pi}{2a} \tag{8.236}$$

(with a the lattice spacing) in the opposite regime, the tight-binding approximation (cp. section 8.2.5.3); see the right panel of figure 8.4.

Assuming the length of the system to be L and assuming furthermore periodic boundary conditions, successive values of the wave number are separated by $2\pi/L$.

In one dimension the 'Fermi surface' consists of only two points $\pm k_F$. The interesting physical properties of the system are determined within a thermal energy range of T by exciting single electrons to states just above the Fermi surface or by creating electron-hole pairs in the vicinity of the Fermi surface. States well below this range are practically always completely filled, while states well above this range are always empty. It suffices therefore to consider only a finite bandwidth of $2D$ about the Fermi energy.

Therefore, as indicated in figure 8.4 for the free electron and the tight-binding model, a linear approximation can be used for the physically relevant part of the energy-momentum dispersion relation. We therefore assume in all further calculations that the band energy measured from the Fermi energy, $\xi(k) = \epsilon_k - \epsilon_F$, has the linear form

$$\xi(k) = \begin{cases} v_F(k - k_F) & \text{for} \quad k_F - K < k < k_F + K, \\ -v_F(k + k_F) & \text{for} \quad -k_F - K < k < -k_F + K. \end{cases} \tag{8.237}$$

The bandwidth D and the momentum cut-off K are related by $D = v_F K$.

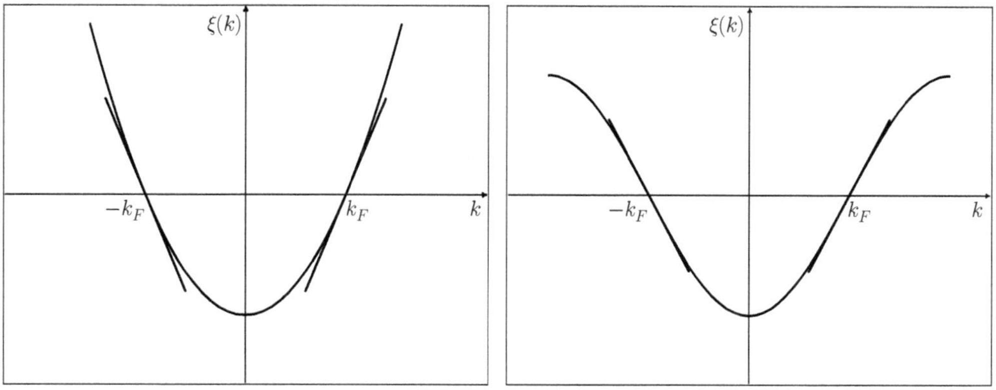

Figure 8.4 *Energy dispersion of a one–dimensional electron gas in the non-interacting electron model (left panel) and in the tight-binding approximation (right panel). The linear approximations around the Fermi points* $\pm k_F$ *are also indicated.*

Tomonaga now chose the linearized spectrum to hold for any momentum

$$\xi(k) = \begin{cases} v_F(k - k_F) & \text{for} \quad k > 0, \\ -v_F(k + k_F) & \text{for} \quad k < 0, \end{cases} \tag{8.238}$$

such that the spectrum is continuous for $k = 0$ with $\xi_{k=0} = -v_F k_F$ but has a tip at $k = 0$ where the left and right branches of the spectrum terminate. This procedure defines the Tomonaga model, whose Hamiltonian can now be written as

$$\mathcal{H}_0 = \sum_{k>0,\sigma} v_F(k - k_F) r^{\dagger}_{k\sigma} r_{k\sigma} - \sum_{k<0,\sigma} v_F(k + k_F) \ell^{\dagger}_{k\sigma} \ell_{k\sigma} \tag{8.239}$$

where we introduced creation and annihilation operators $r_{k\sigma}{}^{\dagger}$ and $r_{k\sigma}$, which create and annihilate Fermions on the right branch of the spectrum, so-called right-moving Fermions, or simply right movers with velocity v_F, and corresponding creation and annihilation operators $\ell_{k\sigma}{}^{\dagger}$ and $\ell_{k\sigma}$ for the left branch of the spectrum, so-called left-moving Fermions, or left movers with velocity $-v_F$. Note that the summation is over $k > 0$ in the first part and $k < 0$ in the second part of the Tomonaga Hamiltonian.

Luttinger (1963) chose a different procedure to handle the linearized spectrum. He extended both branches of the spectrum to infinity

$$\xi(k) = \begin{cases} v_F(k - k_F) & \text{for} \quad -\infty < k < \infty, \\ -v_F(k + k_F) & \text{for} \quad -\infty < k < \infty. \end{cases} \tag{8.240}$$

This choice is more subtle than Tomonaga's but renders the model exactly solvable.

In order to avoid infinite energy contributions, the two species of Fermions, the left movers and the right movers, have to be normal ordered in Luttinger's scheme

$$: r^\dagger_{k\sigma} r_{k\sigma} : = r^\dagger_{k\sigma} r_{k\sigma} - \langle r^\dagger_{k\sigma} r_{k\sigma} \rangle_0, \tag{8.241}$$

$$: \ell^\dagger_{k\sigma} \ell_{k\sigma} : = \ell^\dagger_{k\sigma} \ell_{k\sigma} - \langle \ell^\dagger_{k\sigma} \ell_{k\sigma} \rangle_0 \tag{8.242}$$

where

$$\langle r^\dagger_{k\sigma} r_{k\sigma} \rangle_0 = \Theta(k_F - k), \quad \langle \ell^\dagger_{k\sigma} \ell_{k\sigma} \rangle_0 = \Theta(k_F + k). \tag{8.243}$$

The Hamiltonian for the Luttinger model is formally the same, except for the normal ordering of the creation and annihilation operators and the summation, which is now over $-\infty < k < \infty$ in both parts of the Luttinger Hamiltonian

$$\mathcal{H}_0 = \sum_{k,\sigma} v_F(k - k_F) : r^\dagger_{k\sigma} r_{k\sigma} : - \sum_{k,\sigma} v_F(k + k_F) : \ell^\dagger_{k\sigma} \ell_{k\sigma} : . \tag{8.244}$$

The ground state energy of the Luttinger model vanishes due to the normal ordering of the operatots, $E_0 = 0$.

We shall not go into a detailed discussion of the differences between the three schemes outlined above: the model with finite-bandwidth, the Tomanaga and the Luttinger models. All three models describe essentially the same physics. In explicit calculations, we shall use the model with finite momentum cut-off $-K < k < K$.

Measuring the wave number k from the respective Fermi points and restricting the summation to the region $-K < k < K$, we finally arrive at the Hamiltonian we shall be using in the further discussion

$$\mathcal{H}_0 = \sum_{k,\sigma} v_F k \left\{ r^\dagger_{k_F + k\sigma} r_{k_F + k\sigma} - \ell^\dagger_{-k_F + k\sigma} \ell_{-k_F + k\sigma} \right\}. \tag{8.245}$$

We now turn our attention to the low-energy excitations in the non-interacting Fermi system where we make explicit the observation discussed in section 8.5.2.1.

8.5.4.2 Bosonic character of electron-hole excitations

The Bosonic character of electron-hole pair excitations of a one-dimensional Fermi system can be qualitatively understood by the following consideration in momentum space.

The Fermi 'sphere' of the one-dimensional system is given by the filled momentum states of the ground state in the interval $-k_F \leq k \leq k_F$ from the left to the right Fermi points. The Fermi 'surface', consists of two isolated points at $-k_F$ and k_F (see the two sets of figures 8.4 and 8.5). Generic higher-dimensional Fermi systems possess continuous Fermi surfaces that are simply connected. These extended Fermi surfaces provide considerable (momentum) space for low-energy scattering processes. The two isolated Fermi energy sectors available to low-energy processes of the one-dimensional system are, as we shall corroborate later, much more restrictive. (cp. figure 8.3)

We now encounter one of the particularities of Fermions in one dimension that can be discussed with the help of the linearized finite-bandwidth model. Only low-energy excitations around the Fermi energy and within the bandwidth are relevant. In the ground state all states below the Fermi level are occupied. Fermions from below the Fermi level leaving behind an empty state or a hole can be elevated above the Fermi level. In this way a pair consisting of a particle and a hole is created. Due to the peculiar Fermi 'surface' in one dimension, there are only two possible processes that can achieve this. For small momentum transfer $\Delta k \approx 0$, the particle-hole pairs are created on the same branch, while for large momentum transfer $\Delta k \approx 2k_F$ the particle-hole pairs are created on different branches. These two possibilities are depicted in figure 8.5.

As discussed qualitatively in section 8.5.2.1, the particle density fluctuations $\rho(x)$ are the dominant excitations in one-dimensional Fermi systems (see example 2.5 in section 2.6.1, especially the Fourier transformed particle density (2.158) for the the particle density operator and its Fourier transform). The Fourier components of the particle density fluctuations measured from the Fermi wave number k_F

$$c^\dagger_{\pm k_F+k\sigma}\, c_{\pm k_F+k+q\sigma} \tag{8.246}$$

create particle-hole pairs either close to one Fermi point or across from one Fermi point to the other with momentum transfer $\Delta k = q$ or $\Delta k = 2k_F + q$. The energy of the state

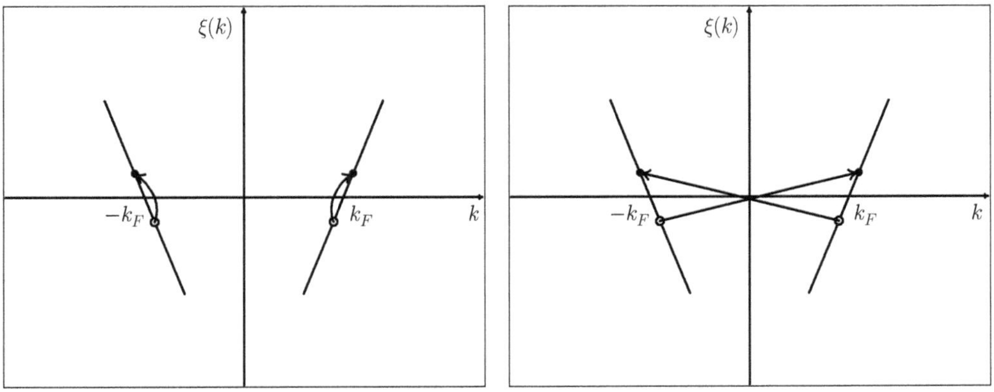

Figure 8.5 *Four types of low-energy particle-hole excitations in the one-dimensional Fermi system with a finite bandwidth. Left panel: creation of a particle-hole pair involving a small momentum transfer $q \approx 0$ on the left and right branches, respectively. Right panel: creation of a particle-hole pair with momentum transfer $q \approx 2k_F$ from the right to the left branch or vice versa. The interactions of the model will be described by processes of these kinds and two other kinds of processes to be introduced later. The coupling constants corresponding to the processes depicted in the left panel are denoted $\mathsf{g}_{2\sigma\sigma'}$, the coupling constants corresponding to the processes shown in the right panel by $\mathsf{g}_{1\sigma\sigma'}$.*

created by this operator is $\pm v_F q$ independent of the value of the wave number k and is therefore highly degenerate. It is therefore equivalent to employ the full particle density operator

$$\rho(q) = \sum_{k\sigma} c^{\dagger}_{\pm k_F + k\sigma} c_{\pm k_F + k + q\sigma}. \tag{8.247}$$

Accordingly, this operator can be termed particle-hole pair creation operator. The $q = 0$ component of this operator can be interpreted as counting states above the Fermi level

$$\rho(q = 0) = \sum_{k\sigma} c^{\dagger}_{\pm k_F + k\sigma} c_{\pm k_F + k\sigma} \tag{8.248}$$

which correspond to single-particle excitations.

We now restrict our further discussion for the moment to the case depicted in the left panel of figure 8.5, i.e. to processes with small momentum transfer $\Delta k = q$ but we will return to the other case later in section 8.5.4.4.

In the following, we shall frequently use the particle-hole pair creation operators for left and right movers

$$\rho_{r\sigma}(q) \equiv \rho_{+\sigma}(q) = \sum_{k\sigma} r^{\dagger}_{k_F + k\sigma} r_{k_F + k + q\sigma}, \tag{8.249}$$

$$\rho_{\ell\sigma}(q) \equiv \rho_{-\sigma}(q) = \sum_{k\sigma} \ell^{\dagger}_{-k_F + k\sigma} \ell_{-k_F + k + q\sigma} \tag{8.250}$$

where we have introduced an index, called the chirality index, $\lambda = \pm 1$, i.e. $\rho_{\lambda\sigma}(q)$, with $\lambda = 1$ for right movers and $\lambda = -1$ for left movers.

These operators are supposed to have Bosonic character. The different schemes of linearization by Tomonaga, Luttinger, and for the case of a finite bandwidth require due care in calculating the commutators. For a careful treatment of this point we refer to Sólyom (2010) and give here only the final result

$$\left[\rho_{\lambda\sigma}(q), \rho^{\dagger}_{\lambda'\sigma'}(q')\right] = \delta_{\lambda\lambda'}\delta_{\sigma\sigma'}\delta_{qq'}\frac{qL}{2\pi} \tag{8.251}$$

where

$$\rho_{\lambda\sigma}(-q) = \rho^{\dagger}_{\lambda\sigma}(q). \tag{8.252}$$

Note that, in order to avoid infinite contributions, the operators have to be normal ordered in Luttinger's scheme where the linearized spectrum is extended to infinitely large momenta $k \to \pm\infty$. Furthermore, these expressions should be taken for $q \neq 0$, or, more precisely, for $q > 0$ in the $\lambda = 1$ right-moving chirality case and for $q < 0$ in the $\lambda = -1$ left-moving chirality case.

The factor $qL/2\pi$ on the right-hand side of the commutator stems from the remaining Fermi operators, which are summed over a range of momentum values of q. The explicit calculation reveals that these operators correspond to states that are completely filled because they are deep below the Fermi energy. In a sample of length L, the possible wave numbers are separated by $2\pi/L$ and, hence, the number of states in a range of size q is $qL/2\pi$.

These statements concerning the commutators of the particle density operators in momentum space, the particle-hole pair creation operators, are only approximate for the Tomonaga model where, due to the choice of the linearized spectrum deep lying excited states may occur and produce corresponding particle-hole pairs. The smaller q, i.e. for long-wavelength fluctuations, the better the quality of the approximation. The same qualification must be made in the case of a finite bandwidth. However, for the choice made by Luttinger, the possibility that not all deep lying states are filled is excluded and the above statements become exact.

It is straightforward to construct proper Bosonic operators from the operators $\rho_{\lambda\sigma}(q)$. We obtain for the Bosonic annihilation operators

$$b_{q\sigma} = \sum_k \left(\frac{2\pi}{Lq}\right)^{1/2} r^\dagger_{k_F+k\sigma} r_{k_F+k+q\sigma} \qquad q > 0 \tag{8.253}$$

$$= \left(\frac{2\pi}{Lq}\right)^{1/2} \rho_{+\sigma}(q) \tag{8.254}$$

$$b_{q\sigma} = \sum_k \left(\frac{2\pi}{L|q|}\right)^{1/2} \ell^\dagger_{-k_F+k\sigma} \ell_{-k_F+k+q\sigma} \qquad q < 0 \tag{8.255}$$

$$= \left(\frac{2\pi}{L|q|}\right)^{1/2} \rho_{-\sigma}(q) \tag{8.256}$$

and for the creation operators

$$b^\dagger_{q\sigma} = \sum_k \left(\frac{2\pi}{Lq}\right)^{1/2} r^\dagger_{k_F+k+q\sigma} r_{k_F+k\sigma} \qquad q > 0 \tag{8.257}$$

$$= \left(\frac{2\pi}{Lq}\right)^{1/2} \rho_{+\sigma}(-q) \tag{8.258}$$

$$b^\dagger_{q\sigma} = \sum_k \left(\frac{2\pi}{L|q|}\right)^{1/2} \ell^\dagger_{-k_F+k+q\sigma} \ell_{-k_F+k\sigma} \qquad q < 0 \tag{8.259}$$

$$= \left(\frac{2\pi}{L|q|}\right)^{1/2} \rho_{-\sigma}(-q). \tag{8.260}$$

These choices are, however, not unique and there are differing ones to be found in the literature (see Sólyom, 2010) for a discussion of possible other choices). The Bosonic operators now satisfy proper Bose commutation relations

$$\left[b_{q\sigma}, b^{\dagger}_{q'\sigma'}\right] = \delta_{\sigma\sigma'}\delta_{qq'}. \tag{8.261}$$

The operators $b^{\dagger}_{q\sigma}$and $b_{q\sigma}$ are creation and annihilation operators of a superposition of many electron-hole pair excitations. As can be seen from the preceding expression, the annihilation operator annihilates the ground state of the non-interacting Fermi sea

$$b_{q\sigma}|FS\rangle = 0. \tag{8.262}$$

The filled Fermi sea $|FS\rangle$ is thus the vacuum state of the Bosons just constructed.

8.5.4.3 Non-interacting Hamiltonian in Bosonic form

In order to rewrite the non-interacting Hamiltonian of the Fermi system in terms of the Bosonic operators $b^{\dagger}_{q\sigma}$and $b_{q\sigma}$, we recall that these operators create and annihilate, respectively, a collective excitation of particle-hole pairs around the right and left Fermi points, respectively. These collective excitations have a kinetic energy of $v_F q$ with $q > 0$ for the excitation around the right Fermi point and $-v_F q$ with $q < 0$ for the excitation around the left Fermi point. This kinetic energy is thus expected to be produced by the non-interacting Hamiltonian

$$\mathcal{H}_0 = \sum_{q\neq 0\sigma} v_F|q| b^{\dagger}_{q\sigma} b_{q\sigma} \tag{8.263}$$

In order to confirm this expectation, we calculate the commutators

$$\left[\mathcal{H}_0, b_{q\sigma}\right] = -v_F b_{q\sigma} \tag{8.264}$$

$$\left[\mathcal{H}_0, b^{\dagger}_{q\sigma}\right] = v_F b^{\dagger}_{q\sigma}. \tag{8.265}$$

using, on the one hand, the Fermionic form of the non-interacting Hamiltonian (8.245) and the explicit forms of the Bosonic operators $b_{q\sigma}$ and $b_{q\sigma}{}^{\dagger}$ in (8.253–8.255) and (8.257–8.259) and, on the other hand, the expected form of the Hamiltonian (8.263) together with the Bosonic commutation relation (8.261). The results, obtained by straightforward calculation, are identical in both cases, although the calculations in the first case are a lot more tedious.

We can thus conclude from these considerations that for the collective excitations of small-momentum particle-hole pairs in the low-energy sector of the Hilbert space the one-dimensional non-interacting Fermi gas, described by the Fermionic Hamiltonian (8.245), is equivalent to a gas of non-interacting Bosons, described by the Bosonic Hamiltonian (8.263).

EXERCISE 8.10 Non-interacting Bosonic Hamiltonian Perform the two calculations mentioned above explicitly to convince yourself that the equivalence between the Fermionic and the Bosonic forms of the Hamiltonian for the one-dimensional non-interacting Fermi gas is valid.

8.5.4.4 Processes with chirality or spin change

The particle and the hole in the small-momentum excitations considered in the previous section are created with the same spin σ and chirality λ. These are the scattering processes represented in the left panel of figure 8.5. The other possible low-energy excitation processes, depicted in the right panel of figure 8.5, have not yet been included. They cannot be described by the Bosonic Hamiltonian (8.263).

For a full Bosonic description of the low-energy excitations of the non-interacting Fermi system, we need to include also these processes into the Hamiltonian, which can be achieved in the following way.

According to the right panel of figure 8.5 an electron-hole pair can also be created where the electron and the hole have opposite chiralities. In that case, the wave number of the pair is large but close to $\pm\, 2k_F$. In such a process, the spin of the electron can also be flipped to create a particle-hole pair with opposite spin and chirality.

Excitations of these kinds can be produced in two steps. There is an equal number of Fermions in each branch in the ground state. In the first step, by creating or annihilating states in either branch, we can, always observing the Pauli exclusion principle, create low-energy single particle or single hole states in the left and right branch of the spectrum. The number of these Fermions is denoted by $N_{\lambda\sigma}$. The creation of these $N_{\lambda\sigma}$ Fermions is accompanied by a shift in the wave number of $\Delta k_{\lambda\sigma} = 2\pi\lambda N_{\lambda\sigma}/L$ which add up to an average contribution of the $N_{\lambda\sigma}$ particles to the excitation energy of

$$\Delta E_{\lambda\sigma} = \frac{1}{2} v_F \Delta k_{\lambda\sigma} \left(\lambda N_{\lambda\sigma}\right) = v_F \frac{\pi}{L} \left(N_{\lambda\sigma}\right)^2 . \tag{8.266}$$

In the second step, we then create low-energy collective excitations of electron-hole pairs in the two branches that are labelled by the momentum transfer q. The excitation energy contributions of the particle-hole pairs are included into the Bosonic Hamiltonian (8.263). As a result, the Fermion excitations thus created, i.e. the added particles and the particle-hole pairs, are characterized by the quantum numbers $N_{\lambda\sigma}$ and q and the non-interacting Hamiltonian takes on the form

$$\mathcal{H}_0 = \sum_{q \neq 0 \sigma} v_F |q| b^\dagger_{q\sigma} b_{q\sigma} + \frac{\pi v_F}{L} \left(N_{\lambda\sigma}\right)^2 . \tag{8.267}$$

Note that all $q = 0$ contributions are contained in the second term of this Hamiltonian.

The excitations described by the quantum numbers $N_{\lambda\sigma}$ are often called topological excitations because their energy cannot be made to vanish by letting $q \to 0$.

8.5.4.5 Spin-charge separation

We are now in a position to demonstrate, for the non-interacting case, one of the most astonishing properties of Fermi systems in one dimension: spin-charge separation, where the spin and charge degrees of freedom of the Fermi system have been decoupled.

We introduce spin-symmetric

$$b_{qc} = \frac{1}{\sqrt{2}} \left(b_{q\uparrow} + b_{q\downarrow}\right), \tag{8.268}$$

$$b^\dagger_{qc} = \frac{1}{\sqrt{2}} \left(b^\dagger_{q\uparrow} + b^\dagger_{q\downarrow}\right), \tag{8.269}$$

and spin-antisymmetric Bosonic operators

$$b_{qs} = \frac{1}{\sqrt{2}} \left(b_{q\uparrow} - b_{q\downarrow}\right), \tag{8.270}$$

$$b^\dagger_{qs} = \frac{1}{\sqrt{2}} \left(b^\dagger_{q\uparrow} - b^\dagger_{q\downarrow}\right). \tag{8.271}$$

The spin and charge Bosons appear now separated in the Hamiltonian

$$\mathcal{H}_0 = \sum_{q\neq 0} \left(v_F|q| b^\dagger_{qc} b_{qc} + v_F|q| b^\dagger_{qs} b_{qs}\right) \tag{8.272}$$

The topological excitations can can also be written in a form that makes the separation of spin and charge manifest by introducing

$$N_c = N_{+\uparrow} + N_{+\downarrow} + N_{-\uparrow} + N_{-\downarrow}, \tag{8.273}$$

$$N_s = N_{+\uparrow} - N_{+\downarrow} + N_{-\uparrow} - N_{-\downarrow}, \tag{8.274}$$

$$\mathcal{J}_c = N_{+\uparrow} + N_{+\downarrow} - N_{-\uparrow} - N_{-\downarrow}, \tag{8.275}$$

$$\mathcal{J}_s = N_{+\uparrow} - N_{+\downarrow} - N_{-\uparrow} + N_{-\downarrow}, \tag{8.276}$$

for which

$$\sum_{\lambda\sigma} (N_{\lambda\sigma})^2 = \frac{1}{4} \left(N_c^2 + N_s^2 + \mathcal{J}_c^2 + \mathcal{J}_s^2\right). \tag{8.277}$$

These numbers are the total number of particles (N_c) and the total spin (N_s), and the charge imbalance ($\mathcal{J}_c$) and the spin imbalance ($\mathcal{J}_s$) between the branches of the spectrum. The imbalances are also conventionally called currents.

The full non-interacting Hamiltonian is thus a sum of two parts

$$\mathcal{H}_0 = \mathcal{H}_{0c} + \mathcal{H}_{0s} \tag{8.278}$$

with the spin and charge Hamiltonians

$$\mathcal{H}_{0c} = \sum_{q\neq 0} v_F b^\dagger_{qc} b_{qc} + \frac{\pi v_F}{4L}\left(N_c^2 + \mathcal{J}_c^2\right), \tag{8.279}$$

$$\mathcal{H}_{0s} = \sum_{q\neq 0} v_F b^\dagger_{qs} b_{qs} + \frac{\pi v_F}{4L}\left(N_s^2 + \mathcal{J}_s^2\right). \tag{8.280}$$

The second terms in each case have the following interpretation. To each branch of the spectrum $N_c/2$ particles are added, half with spin up and half with. This gives a contribution proportional to $(N_c)^2$ to the energy of the topological charge excitations. Moreover, creating the corresponding imbalance mentioned, $\mathcal{J}_c/2$ particles from the highest occupied states of one branch are transferred to the lowest unoccupied states of the other branch, again half this number per spin orientation, which remains the same in these processes. The term proportional to $\mathcal{J}_c^2$ represents the energy of a corresponding topological charge-current excitation. Analogous interpretations apply for the terms in the spin sector with index s.

8.5.4.6 *Interacting Fermions in one dimension*

We now turn to the interaction part of the Hamiltonian (8.232) for the one-dimensional Fermi system

$$\mathcal{H}_{\text{int}} = \frac{1}{2L} \sum_{kk'q\neq 0\sigma\sigma'} V_{\sigma\sigma'}(q) c^\dagger_{k+q\sigma} c^\dagger_{k'-q\sigma'} c_{k'\sigma'} c_{k\sigma}, \tag{8.281}$$

where $V_{\sigma\sigma'}(q)$ is the Fourier transform of the interaction energy which may, in general, be spin-dependent. The interaction Hamiltonian involves the creation of two pairs of particle-hole excitations.

The following considers the finite-bandwidth model (i.e. with $K < k_F$) and assumes that the Fourier transformed interaction $V_{\sigma\sigma'}(q)$ varies only very slowly with q. We can then classify the possible scattering processes according to the momentum transfer involved, i.e. according to whether the chirality remains unchanged in small momentum transfer processes or whether a chirality exchange is involved in processes with momentum transfer $2k_F$, and also to the spin orientation of the scattered particles. The two scattering processes depicted in figure 8.5 represented all the distinction needed to describe the processes involved in the non-interacting Hamiltonian of the one-dimensional Fermi system. However, to describe the interaction term (8.281) of the full Hamiltonian, we need to re-evaluate the possible scattering processes more carefully since we are now dealing with two pairs of particle-hole excitations. Indeed, we find that in addition to small momentum transfer processes on different branches as displayed in the left panel of figure 8.5, there are also small momentum transfer processes on the same branch as displayed in the left panel of figure 8.6. Both these types of processes are forward-scattering processes. Moreover, the particle-hole pairs created with accompanying momentum transfer of $q \approx 2k_F$ can occur in two ways. The first consists in a right mover scattered to become a left mover, and vice versa as shown in the right panel of figure 8.5. This process is a backward-scattering process. The second possibility are two right movers scattered to become two left movers, or vice versa, as shown in the right panel of figure 8.6. This type of process is an Umklapp process, which becomes only important in a half-filled band where $N_e = N/2$ and thus for a system of length $L = Na$ we have $k_F = \pi N_e/L = \pi/2a$ with a the lattice constant. In this situation, crystal momentum is conserved because the total momentum transfer of an Umklapp process of $\pm 4k_F = \pm 2\pi/a$ corresponds to a reciprocal lattice vector of the one-dimensional model. In all other processes, i.e. except the Umklapp process, momentum is conserved.

Since we have assumed that the potential energy $V_{\sigma\sigma'}(q)$ of the generic Hamiltonian (8.281) depends only very slowly on q, we can describe the above four possibilities for the creation of two particle-hole pairs by four spin-dependent but q-independent phenomenological sets of quantities $g_{n\sigma\sigma'}$ ($n = 1, 2, 3, 4$). These constants could also depend on the chirality, i.e. they could be asymmetric with respect to the branch or the direction in which particles are transfered from one branch to the other. However, we shall not introduce this complication but assume a left-right symmetry between branches.

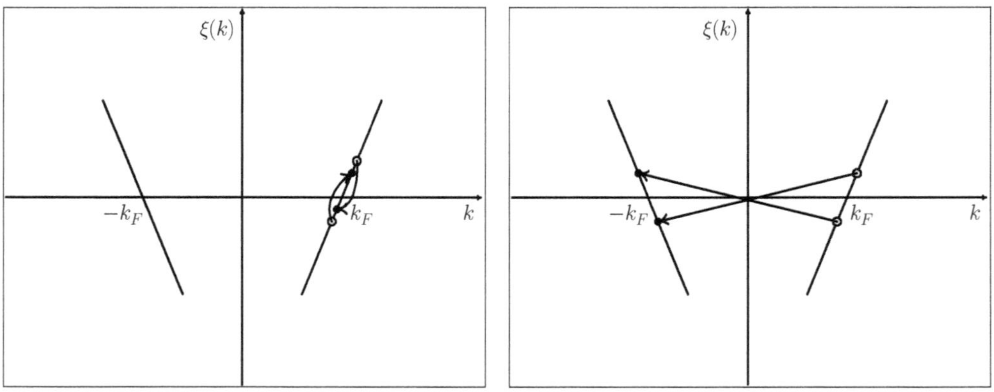

Figure 8.6 *Low-energy particle-hole excitations involving the creation of two particle-hole pairs in the one-dimensional Fermi system with a finite bandwidth. The processes in the left panel are described by the coupling constants $g_{4\sigma\sigma'}$, and the processes in the right panel by $g_{3\sigma\sigma'}$. The other two possibilities are depicted in figure 8.5.*

The model described so far and whose Hamiltonian we write now is often called the g-ology model.

The part of the interaction Hamiltonian describing the creation of two pairs of small-momentum particle-hole excitations in the g-ology model consists of the four terms

$$\begin{aligned}\mathcal{H}_{\text{int}}^{(1)} = \frac{1}{2L} \sum_{kk'q\neq 0\sigma\sigma'} g_{1\sigma\sigma'} \Big\{ & r^{\dagger}_{k_F+k+q\sigma} \ell^{\dagger}_{-k_F+k'-q\sigma'} r_{k_F+k'\sigma'} \ell_{-k_F+k\sigma} \\ & + \ell^{\dagger}_{-k_F+k+q\sigma} r^{\dagger}_{k_F+k'-q\sigma'} \ell_{-k_F+k'\sigma'} r_{k_F+k\sigma} \Big\} \\ + \frac{1}{2L} \sum_{kk'q\neq 0\sigma\sigma'} g_{2\sigma\sigma'} \Big\{ & r^{\dagger}_{k_F+k+q\sigma} \ell^{\dagger}_{-k_F+k'-q\sigma'} \ell_{-k_F+k'\sigma'} r_{k_F+k\sigma} \\ & + \ell^{\dagger}_{-k_F+k+q\sigma} r^{\dagger}_{k_F+k'-q\sigma'} r_{k_F+k'\sigma'} \ell_{-k_F+k\sigma} \Big\} \\ + \frac{1}{2L} \sum_{kk'q\neq 0\sigma\sigma'} g_{3\sigma\sigma'} \Big\{ & r^{\dagger}_{k_F+k+q\sigma} r^{\dagger}_{k_F+k'-q\sigma'} \ell_{-k_F+k'\sigma'} \ell_{-k_F+k\sigma} \\ & + \ell^{\dagger}_{-k_F+k+q\sigma} \ell^{\dagger}_{-k_F+k'-q\sigma'} r_{k_F+k'\sigma'} r_{k_F+k\sigma} \Big\} \\ + \frac{1}{2L} \sum_{kk'q\neq 0\sigma\sigma'} g_{4\sigma\sigma'} \Big\{ & r^{\dagger}_{k_F+k+q\sigma} r^{\dagger}_{k_F+k'-q\sigma'} r_{k_F+k'\sigma'} r_{k_F+k\sigma} \\ & + \ell^{\dagger}_{-k_F+k+q\sigma} \ell^{\dagger}_{-k_F+k'-q\sigma'} \ell_{-k_F+k'\sigma'} \ell_{-k_F+k\sigma} \Big\}. \end{aligned} \tag{8.282}$$

In the g-ology notation the coefficients $g_{n\sigma\sigma'}$ correspond to the four two particle-hole excitation processes depicted in the right panel of figure 8.5: $g_{1\sigma\sigma'}$, the left panel of figure 8.5: $g_{2\sigma\sigma'}$, the right panel of figure 8.6: $g_{3\sigma\sigma'}$, and the left panel of figure 8.6: $g_{4\sigma\sigma'}$.

As discussed, the Umklapp processes with coupling constants $g_{3\sigma\sigma'}$ are only allowed in particular situations. Moreover, the other processes with large momentum transfer $\pm\, 2k_F$, the back-scattering processes with coupling constants $g_{1\sigma\sigma'}$, can only be distinguished from the forward-scattering processes with coupling constants $g_{2\sigma\sigma'}$ if the electrons involved in the scattering process have spins with opposite orientation.

In their original work, both Tomonaga and Luttinger considered the case where processes with large momentum transfer are neglected. They then showed that the model can be solved in a very good approximation in Tomonaga's choice of linearized spectrum or even exactly in Luttinger's choice of the linearized spectrum.

We too shall neglect the large momentum transfer processes in following discussion. The interaction Hamiltonian expressed in terms of the particle-hole pair creation operators then takes the form

$$\mathcal{H}_{\text{int}}^{(1)} = \frac{1}{2L} \sum_{q\neq 0\sigma\sigma'\lambda} g_{2\sigma\sigma'} \rho_{\lambda\sigma}(-q) \rho_{-\lambda\sigma'}(q) + \frac{1}{2L} \sum_{q\neq 0\sigma\sigma'\lambda} g_{4\sigma\sigma'} \rho_{\lambda\sigma}(-q) \rho_{\lambda\sigma'}(q) \tag{8.283}$$

or, expressed in terms of the Bosonic operators

$$\mathcal{H}_{\text{int}}^{(1)} = \frac{1}{4\pi} \sum_{q\neq 0\sigma\sigma'} |q| \left[g_{2\sigma\sigma'} \left\{ b_{q\sigma}^{\dagger} b_{-q\sigma'}^{\dagger} + b_{-q\sigma} b_{q\sigma'} \right\} + g_{4\sigma\sigma'} \left\{ b_{q\sigma}^{\dagger} b_{q\sigma'} + b_{-q\sigma} b_{-q\sigma'}^{\dagger} \right\} \right]. \tag{8.284}$$

The components of the particle-hole pair creation operators, or chiral densities, with $q = 0$ corresponding to particles added to the branches of the spectrum, which we have excluded so far, give rise to a second part of the interaction Hamiltonian

$$\mathcal{H}_{\text{int}}^{(2)} = \frac{1}{2L} \sum_{\sigma\sigma'\lambda} g_{2\sigma\sigma'} N_{\lambda\sigma} N_{-\lambda\sigma'} + \frac{1}{2L} \sum_{\sigma\sigma'\lambda} g_{4\sigma\sigma'} N_{\lambda\sigma} N_{\lambda\sigma'}. \tag{8.285}$$

The full Hamiltonian, composed of the non-interacting part (8.267) and the interacting parts (8.284) and (8.285), will, in the absence of external fields, be spin-reversal symmetric. In this case, the coupling constants depend only on the relative spin orientation and we can introduce the notation ($n = 2, 4$) for parallel and antiparallel spin orientation

$$g_{n\text{p}} = g_{i\uparrow\uparrow} = g_{n\downarrow\downarrow}, \tag{8.286}$$

$$g_{n\text{a}} = g_{i\uparrow\downarrow} = g_{n\downarrow\uparrow}. \tag{8.287}$$

Furthermore, with the help of the spin-symmetric and spin-antisymmetric Bosonic operators (8.268–8.271) and the corresponding combinations of the coupling constants ($n = 2, 4$)

$$g_{nc} = g_{n\text{a}} + g_{n\text{a}}, \tag{8.288}$$

$$g_{ns} = g_{n\text{a}} - g_{n\text{a}}, \tag{8.289}$$

we can rewrite the first part of the interaction Hamiltonian as

$$\mathcal{H}_{\text{int}}^{(1)} = \frac{1}{4\pi} \sum_{q\neq 0} g_{2c}|q| \left(b_{qc}^{\dagger} b_{-qc}^{\dagger} + b_{-qc} b_{qc} + b_{qs}^{\dagger} b_{-qs}^{\dagger} + b_{-qs} b_{qs} \right) + \frac{1}{4\pi} \sum_{q\neq 0} g_{4c}|q| \left(b_{qc}^{\dagger} b_{qc} + b_{-qc} b_{-qc}^{\dagger} + b_{qs}^{\dagger} b_{qs} + b_{-qs} b_{-qs}^{\dagger} \right). \tag{8.290}$$

Using (8.273–8.276), the second part becomes

$$\mathcal{H}_{\text{int}}^{(2)} = \frac{g_{2c}}{8L} \left(N_c^2 - \mathcal{J}_c^2 \right) + \frac{g_{2s}}{8L} \left(N_s^2 - \mathcal{J}_s^2 \right) + \frac{g_{4c}}{8L} \left(N_c^2 + \mathcal{J}_c^2 \right) + \frac{g_{4s}}{8L} \left(N_s^2 + \mathcal{J}_s^2 \right). \tag{8.291}$$

Thus, we see that the full Hamiltonian of the model we have introduced in this section displays spin-charge separation, i.e. one of the distinguishing features of the one-dimensional Fermi system that has no counterpart in three dimensions.

For easy reference, let us write down the charge

$$\mathcal{H}_c = v_F \sum_{q\neq 0} \left(1 + g_{4c}^R\right) |q| b_{qc}^\dagger b_{qc} + \frac{1}{2} v_F \sum_{q\neq 0} g_{2c}^R |q| \left(b_{qc}^\dagger b_{-qc}^\dagger + b_{-qc} b_{qc}\right)$$
$$+ \frac{\pi v_F}{4L} \left(\left(1 + g_{2c}^R + g_{4c}^R\right) N_c^2 + \left(1 - g_{2c}^R + g_{4c}^R\right) \mathcal{J}_c^2\right) \tag{8.292}$$

and spin parts of the full Hamiltonian explicitly

$$\mathcal{H}_s = v_F \sum_{q\neq 0} \left(1 + g_{4s}^R\right) |q| b_{qs}^\dagger b_{qs} + \frac{1}{2} v_F \sum_{q\neq 0} g_{2s}^R |q| \left(b_{qs}^\dagger b_{-qs}^\dagger + b_{-qs} b_{qs}\right)$$
$$+ \frac{\pi v_F}{4L} \left(\left(1 + g_{2s}^R + g_{4s}^R\right) N_s^2 + \left(1 - g_{2s}^R + g_{4s}^R\right) \mathcal{J}_s^2\right) \tag{8.293}$$

where we introduced renormalized coupling constants

$$g^R = \frac{g}{2\pi v_F}. \tag{8.294}$$

It is remarkable that if there is no spin dependence at all of the coupling constants, we have $g_{ns} = 0$ and the spin part of the Hamiltonian reduces to a non-interacting Hamiltonian.

Looking back at the part of the interacting Hamiltonian (8.283) that involves the chiral particle-hole densities $\rho_{\lambda\sigma}(q)$, we realize that it is the g_2 processes that mix chiral densities, i.e. couple the Bosonic density fluctuations of right- and left moving Fermions. These mixing terms can also be identified by the terms proportional to the coupling constant g_2 in the charge and spin Hamiltonians (8.292) and (8.293).

Bilinear Hamiltonians of the the form of (8.292) and (8.293) can be diagonailzed by a Bogoliubov transformation, as demonstrated in exercise 2.4. Obviously, (8.292) and (8.293) can be diagonalized completely independently of each other. Here, we write the transformation for the charge Bose operators

$$B_{qc} = u_{qc} b_{qc} + v_{qc} b_{-qc}^\dagger, \tag{8.295}$$

$$B_{qc}^\dagger = u_{qc} b_{qc}^\dagger + v_{qc} b_{-qc}. \tag{8.296}$$

In order to satisfy Bosonic commutation relations, i.e. $[B_{qc}, B_{qc}^\dagger] = 1$, we require

$$u_{qc}^2 - v_{qc}^2 = 1, \tag{8.297}$$

which can be parametrized by an variable θ_{qc}

$$u_{qc} = \cosh\theta_{qc}, \tag{8.298}$$

$$v_{qc} = \sinh\theta_{qc}. \tag{8.299}$$

The inverse transformation becomes

$$b^\dagger_{qc} = B^\dagger_{qc}\cosh\theta_{qc} - B_{-qc}\sinh\theta_{qc}, \tag{8.300}$$

$$b_{-qc} = B_{-qc}\cosh\theta_{qc} - B^\dagger_{qc}\sinh\theta_{qc}. \tag{8.301}$$

For a detailed derivation of the Bogoliubov transformation from a canonical transformation of Schrieffer–Wolff type, see appendix I in Sólyom (2009). In particular, it explains that there is only one set of parameters u_{qc} and v_{qc} for operators with positive and negative momentum q.

The Hamiltonian $\mathcal{H}_c$ becomes diagonal in the transformed operators if the condition

$$-\left(1+g^R_{4c}\right)\cosh\theta_{qc}\sinh\theta_{qc} + \frac{1}{2}g^R_{2c}\left(\cosh^2\theta_{qc} + \sinh^2\theta_{qc}\right) = 0 \tag{8.302}$$

holds which is equivalent to

$$\tanh 2\theta_{qc} = \frac{g^R_{2c}}{1+g^R_{4c}} \tag{8.303}$$

and

$$\cosh^2\theta_{qc} = \frac{1}{2}\left(\frac{1+g^R_{4c}}{\sqrt{(1+g^R_{4c})^2 - g^2_{2c}}} + 1\right), \tag{8.304}$$

$$\sinh^2\theta_{qc} = \frac{1}{2}\left(\frac{1+g^R_{4c}}{\sqrt{(1+g^R_{4c})^2 - g^2_{2c}}} - 1\right). \tag{8.305}$$

Finally, the diagonal Hamiltonian becomes

$$\mathcal{H}_c = \sum_{q\neq 0} u_c|q|B^\dagger_{qc}B_{qc} + \frac{\pi}{4L}\left(v_{N_c}N_c^2 + v_{\mathcal{J}_c}\mathcal{J}_c^2\right) \tag{8.306}$$

where the velocities

$$u_c = v_F\sqrt{\left(1+g_{4c}^R\right)^2 - \left(g_{2c}^R\right)^2} \tag{8.307}$$

and

$$v_{N_c} = v_F\left(1+g_{2c}^R+g_{4c}^R\right), \tag{8.308}$$

$$v_{\mathcal{J}_c} = v_F\left(1-g_{2c}^R+g_{4c}^R\right) \tag{8.309}$$

have been introduced. These velocities are not independent of each other, but satisfy

$$u_c^2 = v_{N_c}v_{\mathcal{J}_c}. \tag{8.310}$$

It is customary to define the interaction parameter

$$K_c = \frac{u_c}{v_{N_c}} = \frac{v_{\mathcal{J}_c}}{u_c} = \sqrt{\frac{1-g_{2c}^R+g_{4c}^R}{1+g_{2c}^R+g_{4c}^R}} \tag{8.311}$$

with the help of which the charge Hamiltonian finally becomes

$$\mathcal{H}_c = \sum_{q\neq 0} u_c|q|B_{qc}^\dagger B_{qc} + \frac{\pi}{4L}\left(\frac{u_c}{K_c}N_c^2 + u_cK_c\mathcal{J}_c^2\right). \tag{8.312}$$

We note that there are two parameters that describe the charge part of the Luttinger liquid, K_c and u_c.

In the very same way and with the analogous definitions we obtain for the spin part

$$\mathcal{H}_s = \sum_{q\neq 0} u_c|q|B_{qc}^\dagger B_{qs} + \frac{\pi}{4L}\left(\frac{u_s}{K_s}N_s^2 + u_sK_s\mathcal{J}_s^2\right). \tag{8.313}$$

We stop our exploration of the Luttinger liquid here and refer to the cited literature for further developments. In conclusion, we mention an important result for the momentum distribution function

$$n_\sigma(k) = \langle F|c_{k\sigma}^\dagger c_{k\sigma}|F\rangle \tag{8.314}$$

of the one-dimensional Fermi system where $|F\rangle$ is the state of the filled Fermi sea. In contrast to the three-dimensional case, the momentum distribution function in one dimension does not exhibit a finite jump at the Fermi momentum k_F (cp. section 8.3.2). Instead, there is a power law behaviour of the Fermi momentum distribution in one dimension in the vicinity of k_F with an exponent that depends on the coupling constants via the interaction parameters K_c and K_s

$$n_\sigma(k) \approx \frac{1}{2} - C|k - k_F|^{4(\Delta_c+\Delta_s)}\text{sgn}(k - k_F) \tag{8.315}$$

where

$$\Delta_c = \frac{1}{16}\left(\frac{1}{K_c} + K_c - 2\right), \tag{8.316}$$

$$\Delta_s = \frac{1}{16}\left(\frac{1}{K_s} + K_s - 2\right). \tag{8.317}$$

Figure 8.7 shows this behaviour of the momentum distribution function schematically.

Further insight into the low-energy properties of Fermi systems in one dimension is provided by the powerful analytical technique of Bosonization. This technique provides a unifying and very flexible framework that encompasses what we have discussed so far.

We shall not engage with the Bosonization technique but instead refer to a few selected works from the vast literature on Bosonization. In addition to the comprehensive book by Gogolin *et al.* (1998), for a first encounter, we recommend Eggert (2007). Further recommendations include the reprint volume edited and expertly introducing the subject by Stone (1994) and the long article by von Delft and Schoeller (1998).

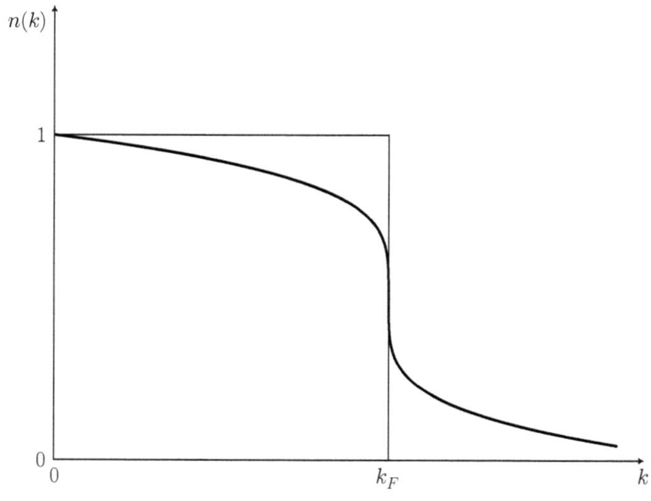

Figure 8.7 *Schematic representation of the power law behaviour of the momentum distribution function of The function is continuous, but has an infinite slope at* $k = k_F$*. The exponent of the power law behaviour of n(k) near* k_F *depends on the interactions of the model.*

8.6 Magnetism

This section provides a brief overview on how a quantum mechanical description of electrons in a solid can account for the solid's magnetic properties. Two models, the Hubbard and the Heisenberg model, play a prominent role in our discussions. For mean-field ideas, we refer back to section 4.15 where interacting magnetic moments are treated in mean-field theory, and to section 5.3 on Landau–Ginzburg theory.

Magnetism is a vast subject with a vast literature. Besides the chapters in most textbooks on solid state and condensed matter physics, we recommend Auerbach (1994), Fazekas (1999), Blundell (2001), Mattis (2006, which starts with a nice chapter on the history of magnetism), Nolting and Ramakanth (2009), and Parkinson and Farnell (2010); this list is certainly very incomplete.

We begin with a broad classification of magnetic behaviour in matter.

8.6.1 Magnetic matter

An external magnetic field **H** induces a magnetization in all matter

$$\mathbf{M} = \chi \mathbf{H} \tag{8.318}$$

where the magnetic susceptibility χ can be used for a first classification of the possible behaviours of magnetic matter.

8.6.1.1 Diamagnetism

A negative constant magnetic susceptibility $\chi < 0$ charcterizes diamagnetic matter where the external magnetic field induces magnetic dipoles which are oriented in the direction opposite to the field according to Lenz's law. All matter exhibits diamagnetism. However, it is usually a small that which may be overwhelmed by stronger magnetic responses of the material.

An interesting example are superconducting materials below T_c, which are perfect diamagnets with $\chi = -1$. This so-called Meissner–Ochsenfeld effect, the complete expulsion of the external magnetic field from the material sample, is indeed a hallmark of superconductivity.

8.6.1.2 Paramagnetism

A positive and temperature-dependent $\chi > 0$ is the hallmark of paramagnetic matter when there exist permanent magnetic moments or dipoles. These moments can be orientated by the external magnetic field, but the orientation is opposed by the thermal fluctuations of the moments. Depending on whether the permanent magnetic moments are localized or itinerant there are two subcategories of paramagnetic behaviour:

- **Localized moments**
  The localized moments originate from partially filled inner electron shells, e.g. the $3d$ shells of transition metals or the $4f$ and $5f$ shells of, respectively, rare earth and actinide metals.

This kind of paramagnetism is called Langevin paramagnetism and is characterized by a susceptibility with a Curie behaviour (cf. exercises 4.3 and 4.4)

$$\chi(T) \propto 1/T. \tag{8.319}$$

- **Itinerant moments**
  Conduction band electrons, which are to a first approximation free electrons, carry a magnetic moment of one Bohr magneton $\mu_B = \frac{e}{2m}$ each. These magnetic moments contribute a small paramagnetic susceptibility that is, again to a first approximation, temperature-independent. This contribution to the paramagnetic behaviour of materials bears the name Pauli paramagnetism.

8.6.1.3 Collective magnetism

This form of magnetism arises from an interaction between localized (e.g. the rare earth metal gadolinium) or itinerant (e.g. magnetic metals like iron and nickel) permanent magnetic moments, the so-called exchange interaction, which is a genuinely quantum mechanical mechanism and leads to a strong coupling between the magnetic moments (or dipoles). Direct dipole-dipole interaction can be shown to be far too weak to create the effects observed in collective magnetism.

In section 4.15 we investigated the statistical mechanics of the Ising model, a phenomenological model not yet rooted in quantum mechanics but that captures essential characteristics of the proper quantum mechanical model described in the following sections. We found that a distinguishing feature of collective magnetism is the existence of a critical temperature T^* below which spontaneous ordering of the magnetic moments occurs. Above T^* the magnetic moments are not ordered the material is paramagnetic.

The susceptibility of quantum matter exhibiting collective magnetism in general depends on temperature and external magnetic field $\chi = \chi(T, H)$, but can also depend on the history of how the sample has been treated so far (hysteresis). For a detailed description of the variety of possible collective magnetic behaviours, see Hurd (1982). We restrict ourselves to the two most important cases.

- **Ferromagnetism**
  The exchange interaction in ferromagnets, e.g. iron, favours parallel alignment of the magnetic moments below a critical temperature, which in this case is called Curie temperature $T^* = T_c$. The typical behaviour of the spontaneous magnetization and the Curie–Weiss behaviour of the susceptibility were obtained in section 4.15.1 from a mean-field analysis of the Ising model.
- **Antiferromagnetism**
  In antiferromagnets the exchange interaction favours antiparallel orientation of the magnetic moments on two sub-lattices A and B. Antiferromagnets therefore have a vanishing net magnetization. However, the susceptibility reveals two characteristic temperatures of an antiferromagnet. For temperatures above the critical temperature $T_N = T^*$, which is called Néel temperature in this context, the susceptibility behaves similar to a ferromagnet

$$\chi \propto \frac{1}{T + |\theta|} \tag{8.320}$$

but with a paramagnetic Curie temperature θ, which is negative. Below T_N there is a spontaneous sub-lattice magnetization. In general, the temperatures T_N and $|\theta|$ are different.

In this broad scheme, we have only included the most important manifestations of magnetism. There is a great variety of further possibilities, e.g. Landau diamagnetism, van Vleck paramagnetism, or, for collective magnetism, the various kinds of ferri- and antiferrimagnetism and much else. Again, we refer interested readers to the literature for a more complete exposition of the fascinating wealth of magnetic phenomena.

The next section explains that magnetism is a phenomenon that cannot be explained within classical physics; it is a genuine quantum phenomenon.

The major ingredient from quantum mechanics to understand magnetic phenomena will be presented further on. This is the exchange mechanism, which is fully quantum mechanical, with the Pauli exclusion principle for electrons as its crucial constituent together with the purely electrostatic Coulomb repulsion of the electrons.

8.6.2 Magnetism—a genuine quantum phenomenon

Before the advent of quantum mechanics, attempts were made to explain magnetism, i.e. relate an applied magnetic field to an induced magnetization via the magnetic susceptibility. Niels Bohr (1911) in his doctoral dissertation and, apparently independently, Hendrika Johanna van Leeuwen (1921), proved what is now called the Bohr–van Leeuwen theorem. The theorem precludes a classical explanation of magnetism based on the motion of electrons. More precisely it states that the magnetic susceptibility of any collection of moving electrons vanishes classically.

The argument can be summarized in the following observations. In classical electrodynamics a magnetic moment is associated with a circulating current: in the simplest case a charge q circling a given point at a distance $\mathbf{r}$ with velocity $\mathbf{v}$

$$\boldsymbol{\mu} = q\mathbf{r} \times \mathbf{v}. \tag{8.321}$$

A collection of N charges might give rise to an average magnetization

$$\langle \mathbf{M} \rangle = \langle \sum_i \boldsymbol{\mu}_i \rangle = \langle \sum_i q_i \mathbf{r}_i \times \mathbf{v}_i \rangle. \tag{8.322}$$

The average is to be taken according to classical statistical mechanics

$$\langle \mathbf{M} \rangle = \frac{1}{\mathscr{Z}} \int d^3p\, d^3r\, \mathbf{M} \exp\left(-\frac{H(\mathbf{p}_i, \mathbf{r}_i)}{T}\right), \tag{8.323}$$

with the classical Hamiltonian for particles in a magnetic field

$$H(\mathbf{p}_i, \mathbf{r}_i) = \sum_i \left(\frac{1}{2m_i} \left[\mathbf{p}_i + q_i \mathbf{A}(\mathbf{r}_i)\right]^2 + q_i \phi(\mathbf{r}_i) \right) + U(\mathbf{r}_1, \ldots, \mathbf{r}_N) \tag{8.324}$$

where the effect of the magnetic field $\mathbf{B}$ is a shift in the canonical momentum $\mathbf{p} \to \mathbf{p} - q\mathbf{A}(\mathbf{r})$ effected by the vector potential $\mathbf{A}$ corresponding to the magnetic field $\mathbf{B} = \nabla \times \mathbf{A}$. Now a change of variables in the momentum integrals of (8.323) from $\mathbf{p}_i$ to $\mathbf{v}_i = \left(\mathbf{p}_i - q_i \mathbf{A}(\mathbf{r}_i)\right) / m_i$ shows that all momentum integrals vanish by symmetry since the integrand is an odd function of $\mathbf{v}_i$. Hence, the magnetization vanishes in a classical system of circulating charges.

There is a simple picture that corroborates the conclusion of the Bohr–van Leeuwen theorem, which, from a classical perspective, seems quite counterintuitive at first. Imagine a finite sample of electrons in a magnetic field. Classically, the charges in the sample, if they are free to move, will create circular currents whose net orbital current should add up to a macroscopic magnetization of the sample. However, at the boundary of the sample, the current loops cannot be completed. Instead, the electrons scatter at the boundary and circle the sample boundary in half-loops moving in an orientation opposite to the current loops in the bulk of the sample. The two net orbital currents, the bulk orbital current and the boundary orbital current, cancel each other. No net current and, hence, no macroscopic magnetization results.

We thus conclude that magnetism is a macroscopic manifestation of the quantum nature of reality which could not exist if we lived in a classical world. Of course, the intrinsic magnetic moment of the electron, its spin, a genuine quantum phenomenon, was not discovered then.

However, does the internal magnetic moment, the spin, of the electron suffice to explain the strong macroscopic forces magnetic matter can exert? We briefly explore this possibility in the following section.

8.6.3 Magnetic dipole-dipole interactions

In fact, we shall see that it does not suffice to consider dipole-dipole interactions between magnetic moments because these turn out to be much too weak to account for collective magnetism, which can be estimated as follows.

The interaction energy between two magnetic dipoles μ_1 and μ_2 that are a distance $\mathbf{r}$ apart is (in SI units)

$$E = \frac{\mu_0}{4\pi r^3} \left(\boldsymbol{\mu}_1 \cdot \boldsymbol{\mu}_2 - \frac{3}{r^2} \left(\boldsymbol{\mu}_1 \cdot \mathbf{r}\right) \left(\boldsymbol{\mu}_2 \cdot \mathbf{r}\right) \right). \tag{8.325}$$

In order to obtain a rough estimate of typical values of this energy we may substitute typical values for an electron on a Bohr orbit: $|\boldsymbol{\mu}_1| = |\boldsymbol{\mu}_2| = \mu_B = \frac{e}{2m}$ and $r = 10^{-10}$m yields an energy corresponding to a temperature of about $1K$. This value is way too small to explain that, e.g. ferromagnetism persists at room temperature rather than being destroyed by thermal fluctuations.

In section 4.15 we considered the magnetic properties of interacting magnetic moments using the Ising model with classical variables $\sigma = \pm 1$. However, this cannot be the full story as these classical variables have historically been introduced *ad hoc*. They lack a proper quantum mechanical justification. Nonetheless, it must be emphasized that the ideas encapsulated in the Ising model have been an extremely far-reaching and clever anticipation that finds its expression even today in a vast research literature devoted to or using the Ising model and its many variants.

Finally, we turn to the quantum mechanical mechanism responsible for the observable magnetic phenomena.

8.6.4 Exchange interaction

As we have seen, a classical explanation of magnetism using classical current loops is impossible. Moreover, the dipole-dipole interaction of magnetic moments appears to be too weak to account for collective magnetism.

The mechanism indeed responsible for the strong effects of collective magnetism consists first and foremost of an interplay between electronic energy scales, the kinetic energy of the electrons and the Coulomb interaction between them, and the symmetry connection between the spatial and spin wave functions enforced by the Pauli exclusion principle. This interplay, the exchange interaction, is a genuine quantum effect. The energy scales involved in the exchange interaction are related to the different spatial configurations associated with a given spin configuration. These different spatial configurations can have significantly different energies large enough to account for the energy scales observed in collective magnetism.

However, as a consequence of the symmetry connection between spatial and spin wave functions, the effective Hamiltonians to describe collective magnetism are expressed in terms of the relative orientation of the spins or spin densities. However, the interaction between the spins or spin densities is mediated by the exchange interaction.

There are two types of exchange mechanism: direct or Coulomb potential exchange, and kinetic exchange.

- **Direct or Coulomb potential exchange**
  This type of exchange interaction considers the Coulomb interaction between electrons. In order to identify the ground state, we need to minimize this interaction which is achieved through antisymmetric spatial wave functions of the electrons because these have smaller overlap than symmetric spatial wave functions. This, on the other hand, favours symmetric spin wave functions and, consequently, ferromagnetic ordering of the spins.
- **Kinetic exchange**
  In order to find the ground state of electrons with respect to the kinetic energy, spatial wave functions are required to be extended or delocalized because the gradients of such wave functions and, hence, their contribution to the kinetic energy, are small. This description applies to symmetric wave functions and their superpositions.

This favours corresponding spin wave functions that are antisymmetric and, thus, antiferromagnetic ordering of the spins.

The mechanism of direct exchange interaction is most instructively demonstrated by considering the hydrogen molecule. The solution of the problem of the hydrogen molecule, a two-electron problem like the helium atom, indeed strongly influenced Heisenberg's (1928) quantum mechanical study of the ferromagnetism problem. Therefore, the next section presents a many-particle generalization of the observations gained from studying the hydrogen molecule, followed by a discussion of kinetic exchange.

8.6.4.1 Hydrogen molecule

The Hamiltonian of the hydrogen molecule can be written approximately as

$$\mathcal{H} = -\frac{1}{2m}\left(\nabla_1^2 + \nabla_2^2\right) - e^2\left(\frac{1}{r_{1A}} + \frac{1}{r_{1B}} + \frac{1}{r_{2A}} + \frac{1}{r_{2B}} - \frac{1}{r_{12}} - \frac{1}{r_{AB}}\right). \tag{8.326}$$

The two electron positions with respect to the protons are $r_{iA} = |\mathbf{r}_i - \mathbf{r}_A|$ and $r_{iB} = |\mathbf{r}_i - \mathbf{r}_B|$ $(i = 1, 2)$, their mutual distance is $r_{12} = |\mathbf{r}_1 - \mathbf{r}_2|$. We assume that the two protons, labelled A and B, have fixed positions a distance $r_{AB} = |\mathbf{r}_A - \mathbf{r}_B|$ apart. Their movement, and hence kinetic energy can be neglected compared to the electron movement because of the huge mass difference between protons and electrons. This is an approximation, going back to Born and Oppenheimer 1927, which is very commonly used in both molecular, and condensed matter physics.

In the corresponding Schrödinger equation

$$\mathcal{H}\psi = E\psi \tag{8.327}$$

the relative coordinate of the protons $R = r_{AB}$ acts as a parameter only.

Hence, the wave function of the hydrogen molecule depends in this approximation only on the spatial and spin coordinates of the electrons

$$\psi(\mathbf{r}_1, \mathbf{r}_2, \sigma_1, \sigma_2) = \varphi(\mathbf{r}_1, \mathbf{r}_2)\chi(\sigma_1, \sigma_2), \tag{8.328}$$

which can, furthermore, be written as a product wave function of spatial and spin wave function as the Hamiltonian does not depend explicitly on spin.

However, we must consider the spin part of the wave function in order to build in properly the Pauli exclusion principle, which demands a total wave function, spatial and spin part together, which is antisymmetric (cp. chapter 2). We therefore consider the spin part of the wave function first. Section 3.2 discussed the angular momentum states for two spin-1/2 particles and found first a singlet state with total spin $S = 0$ and thus corresponding azimuthal quantum number $m_S = 0$, which we can write as

$$\chi_s(\sigma_1, \sigma_2) = \frac{1}{\sqrt{2}}\left(\chi_\uparrow(1)\chi_\downarrow(2) - \chi_\uparrow(2)\chi_\downarrow(1)\right) \tag{8.329}$$

or, using Dirac's notation, as

$$|S = 0, m_S = 0\rangle \equiv |0, 0\rangle = \frac{1}{\sqrt{2}} (| \uparrow\downarrow\rangle - | \downarrow\uparrow\rangle). \tag{8.330}$$

Secondly, we found a triplet state with total spin $S = 1$ and corresponding azimuthal quantum number $m_S = 1, 0, -1$, which is

$$\chi_t(\sigma_1, \sigma_2) = \left\{ \begin{array}{c} \chi_\uparrow(1)\chi_\uparrow(2) \\ \frac{1}{\sqrt{2}} \left(\chi_\uparrow(1)\chi_\downarrow(2) + \chi_\uparrow(2)\chi_\downarrow(1)\right) \\ \chi_\downarrow(1)\chi_\downarrow(2) \end{array} \right. \tag{8.331}$$

or, again, in the intuitive Dirac notation used in chapter 3

$$|S = 1, m_S = 1\rangle \equiv |1, 1\rangle = | \uparrow\uparrow\rangle \tag{8.332}$$

$$|S = 1, m_S = 0\rangle \equiv |1, 0\rangle = \frac{1}{\sqrt{2}} (| \uparrow\downarrow\rangle + | \downarrow\uparrow\rangle) \tag{8.333}$$

$$|S = 1, m_S = -1\rangle \equiv |1, -1\rangle = | \downarrow\downarrow\rangle. \tag{8.334}$$

Before exploring the spatial wave functions corresponding to the singlet and triplet states, we must determine the strategy we want to apply to solve the Schrödinger equation (8.327) for the Hamiltonian (8.326). This is the variational approach used in the Heitler–London approximation . The energy eigenvalue E in the Schrödinger equation satisfies

$$E \leq \frac{\langle \psi | \mathcal{H} | \psi \rangle}{\langle \psi | \psi \rangle} \tag{8.335}$$

where the equal sign holds for the actual solution of the Schrödinger equation, i.e. the eigenfunction of the eigenvalue E, while for any other function the unequal sign holds. The practical task of the variational method as an approximation now consists in finding functions, sometimes called trial wave functions, which attempt to make the expression on the right hand side of (8.335) as small as possible, although they are not solutions of the Schrödinger equation. As the name trial wave function suggests, these functions are often educated guesses.

Heitler and London used the wave functions $\varphi_I(\mathbf{r}_i)$ ($i = 1, 2$ and $I = A, B$) as trial wave functions that satisfy the Schrödinger equation for $R \to \infty$, i.e. when the two protons are far from each other, and the problem reduces to four equations describing electron i forming a hydrogen atom with proton I

$$\left(-\frac{1}{2m}\nabla_i^2 - \frac{e^2}{r_{iI}}\right) \varphi_I(\mathbf{r}_i) = E_0 \varphi_I(\mathbf{r}_i) \qquad i = 1, 2; \quad I = A, B. \tag{8.336}$$

These equations, the hydrogen problem, are exactly solvable and the wave functions, e.g. the ground state wave function, are exactly known.

In order to find an approximation to the ground state of the hydrogen molecule via (8.335) we need to build a symmetric spatial wave function from the ground state solutions of (8.336) for the spin singlet to obtain a total wave function, which is antisymmetric

$$\varphi_s(\mathbf{r}_1,\mathbf{r}_2) = \frac{1}{\sqrt{2}}(\varphi_A(\mathbf{r}_1)\varphi_B(\mathbf{r}_2) + \varphi_A(\mathbf{r}_2)\varphi_B(\mathbf{r}_1)) \equiv |\varphi_s\rangle \tag{8.337}$$

and an antisymmetric wave function for the spin triplet

$$\varphi_a(\mathbf{r}_1,\mathbf{r}_2) = \frac{1}{\sqrt{2}}(\varphi_A(\mathbf{r}_1)\varphi_B(\mathbf{r}_2) - \varphi_A(\mathbf{r}_2)\varphi_B(\mathbf{r}_1)) \equiv |\varphi_a\rangle. \tag{8.338}$$

These spatial wave functions, together with the spin parts for singlet and triplet

$$|s\rangle = |\varphi_s\rangle|0,0\rangle \tag{8.339}$$

$$|a\rangle = |\varphi_a\rangle|1,m\rangle \quad m=-1,0,1 \tag{8.340}$$

can now be used to calculate approximate energies from (8.335) for the hydrogen molecule according to the variational method

$$E_{s/a} = \frac{\langle s/a|\mathcal{H}|s/a\rangle}{\langle s/a|s/a\rangle}. \tag{8.341}$$

The results of calculating these energies can be obtained to a large extent analytically, see e.g. Flügge (1979), where the problem is formulated and an extensive solution is given. The spin parts of the wave function play no role for the calculation of the energies itself.

For our purposes, the details of the calculation and the explicit expressions of the integrals involved in (8.341) are not important. Therefore, we give the results directly and discuss their consequences. The energies can be expressed in terms of integrals as (see, e.g. Haken and Wolf, 1984 for a detailed derivation)

$$E_s = 2E_0 + \frac{e^2}{R} + \frac{1}{1+|S|^2}[2A + U + 2DS + V] \tag{8.342}$$

$$= 2E_0 + \frac{e^2}{R} + \frac{1}{1+|S|^2}[E_C + E_E] \tag{8.343}$$

for the spatially symmetric wave function 8.337 and

$$E_a = 2E_0 + \frac{e^2}{R} + \frac{1}{1-|S|^2}[2A + U - 2DS - V] \tag{8.344}$$

$$= 2E_0 + \frac{e^2}{R} + \frac{1}{1 - |S|^2} [E_C - E_E] \tag{8.345}$$

for the spatially antisymmetric wave function 8.338.

EXERCISE 8.11 Derivation of the Heitler–London energies Derive the expressions (8.342–8.345).

Moreover, evaluate the integrals using the ground state wave functions

$$\varphi_I(\mathbf{r}_i) = \frac{1}{\sqrt{\pi a_0^3}} \exp\left[-\frac{|\mathbf{r}_i - \mathbf{r}_I|}{a_0}\right] \qquad I = A, B, \quad i = 1, 2 \tag{8.346}$$

of the hydrogen atom where $a_0 = \frac{1}{me^2}$ is the Bohr radius.

In the expressions (8.342–8.345) the integral

$$S = \int d^3r\, \varphi_A^*(\mathbf{r})\varphi_B(\mathbf{r}) \tag{8.347}$$

is called the overlap integral, which stems from the fact that the trial wave functions (8.337) and (8.338) are not orthogonal. However, the ground state wave functions $\varphi_I(\mathbf{r}_i)$ of the hydrogen Schrödinger equations (8.336) are orthonormal. Thus, in the denominator of (8.341) one summand is unity while the other summand is $\pm|S|^2$, the latter reflecting the exchange of the electrons between proton A and proton B, a first consequence of the Pauli exclusion principle.

The integrals A and U describe Coulomb interactions.

$$A = \int d^3r_1\, |\varphi_A(\mathbf{r}_1)|^2 \left(\frac{-e^2}{r_{B1}}\right) = \int d^3r_2\, |\varphi_B(\mathbf{r}_2)|^2 \left(\frac{-e^2}{r_{A2}}\right) < 0 \tag{8.348}$$

is the interaction of electron 1 in the state φ_A with the Coulomb field of proton B and of electron 2 in the state φ_B with the Coulomb field of proton A, respectively, which by symmetry are equal.

$$U = \int d^3r_1\, d^3r_2\, |\varphi_A(\mathbf{r}_1)|^2 |\varphi_B(\mathbf{r}_2)|^2 \frac{e^2}{r_{12}} \tag{8.349}$$

is the Coulomb repulsion between the electrons.

The integrals D and V describe exchange interactions.

$$D = \int d^3r\, \varphi_B^*(\mathbf{r}) \left(-\frac{e^2}{|\mathbf{r}_B - \mathbf{r}|}\right) \varphi_A(\mathbf{r}) = \int d^3r\, \varphi_A^*(\mathbf{r}) \left(-\frac{e^2}{|\mathbf{r}_A - \mathbf{r}|}\right) \varphi_B(\mathbf{r}) \tag{8.350}$$

is the single electron exchange integral that first appears in the solution of the hydrogen molecule ion H_2^+. Finally,

$$V = \int d^3r_1\, d^3r_2\, \varphi_B^*(\mathbf{r}_1)\varphi_A^*(\mathbf{r}_2)\left(\frac{e^2}{r_{12}}\right)\varphi_A(\mathbf{r}_1)\varphi_B(\mathbf{r}_2) \tag{8.351}$$

is the full (Coulomb) exchange interaction involving both electrons of the hydrogen molecule.

The integrals D and V are consequences of the Pauli exclusion principle. Therefore, it is justified to group the integrals together as

$$E_C = 2A + U \tag{8.352}$$

$$E_E = 2DS + V \tag{8.353}$$

as already done in (8.343) and (8.345).

Evaluation of the integrals and plotting E_s and E_a, (8.343) and (8.345), against the proton-proton separation $R = r_{AB}$ reveals that the symmetric solution E_s is the ground state and leads to a binding of the hydrogen atoms to form a molecule. This corresponds thus to antiparallel orientation of the spins of the electrons. The antisymmetric solution E_a with parallel spin orientation does not lead to molecule formation, since $E_a(R)$ is always positive and tends to zero as $R \to \infty$.

Finally, how can the results obtained so far be used to establish an effective Hamiltonian that acts on the spin states $|S, m_S\rangle$ directly and produces the same energies as eigenvalues? In other words, we seek

$$\mathcal{H}_{\text{eff}}|0,0\rangle = E_s|0,0\rangle \tag{8.354}$$

$$\mathcal{H}_{\text{eff}}|1,m\rangle = E_a|1,m\rangle \qquad m = -1, 0, 1. \tag{8.355}$$

To answer this question we recall the addition $\mathbf{S} = \mathbf{S}_1 + \mathbf{S}_2$ of two spins 1/2 from section 3.2. From

$$\mathbf{S}^2 = S(S+1) = \mathbf{S}_1^2 + \mathbf{S}_2^2 + 2\mathbf{S}_1 \cdot \mathbf{S}_2 = \frac{3}{2} + \frac{1}{2}\boldsymbol{\sigma}_1 \cdot \boldsymbol{\sigma}_2 \tag{8.356}$$

it follows for the spin singlet $S = 0$

$$\boldsymbol{\sigma}_1 \cdot \boldsymbol{\sigma}_2 = -3 \tag{8.357}$$

and for the spin triplet $S = 1$

$$\boldsymbol{\sigma}_1 \cdot \boldsymbol{\sigma}_2 = 1. \tag{8.358}$$

Writing

$$2\mathcal{J} = E_s - E_a < 0, \tag{8.359}$$

which we know from the evaluation of the integrals (see, e.g. Flügge, 1979) we find that the effective Hamiltonian

$$\mathcal{H}_{\text{eff}} = \frac{1}{4}\left(E_s + 3E_a - 2\mathcal{J}\boldsymbol{\sigma}_1 \cdot \boldsymbol{\sigma}_2\right) \tag{8.360}$$

exactly fits the bill. Acting with $\mathcal{H}_{\text{eff}}$ on a symmetric spin wave function produces the energy E_a of the corresponding antisymmetric spatial wave function, while acting on the antisymmetric spin wave function produces the energy E_s of the corresponding symmetric spatial wave function.

This result is the connection to collective magnetism that we seek in this section. The effective Hamiltonian can be rewritten, neglecting a constant contribution, as

$$\mathcal{H}_{\text{eff}} = -\frac{1}{2}\mathcal{J}\boldsymbol{\sigma}_1 \cdot \boldsymbol{\sigma}_2. \tag{8.361}$$

Heisenberg (1928) used the results of Heitler and London (1927) as a major motivation and ingredient in his attempt to build a theory of ferromagnetism.

The next section demonstrates how a Heisenberg exchange Hamiltonian, of which $\mathcal{H}_{\text{eff}}$ in (8.361) is a prototype, can be obtained in the many-particle second quantization formalism.

8.6.4.2 Direct exchange interaction in quantum matter: ferromagnetism

Section 2.7 introduced the two-particle operator in the second quantization formalism. Informed by the results of the hydrogen molecule, we now formulate the two-particle operator for the Coulomb interaction of electrons in condensed matter, e.g. in a crystal lattice.

The field operators creating an electron with spin σ at position $\mathbf{r}$ may be chosen as

$$\Psi_\sigma^\dagger(\mathbf{r}) = \sum_{\mathbf{R}} \phi_{\mathbf{R}}^*(\mathbf{r})\chi_\sigma c_{\mathbf{R}\sigma}^\dagger \tag{8.362}$$

where, e.g. for a crystal lattice $\phi_{\mathbf{R}}(\mathbf{r}) \equiv w(\mathbf{R}, \mathbf{r})$ are the orthonormal Wannier wave functions at lattice position $\mathbf{R}$ introduced in section 8.2.5.3. Here, we neglect a possible band or orbital index n. The spin wave functions for an electron are $\chi_\uparrow = (1, 0)$ and $\chi_\downarrow = (0, 1)$, which are also orthonormal $\chi_\sigma\chi_\sigma = 1$ and $\chi_\sigma\chi_{-\sigma} = 0$ such that, e.g. the number operator $\mathcal{N}_\sigma$ becomes

$$\int d^3r\, \Psi_\sigma^\dagger(\mathbf{r})\Psi_\sigma(\mathbf{r}) = \int d^3r \sum_{\mathbf{R}\mathbf{R}'} \phi_{\mathbf{R}}^*(\mathbf{r})\phi_{\mathbf{R}'}(\mathbf{r})\chi_\sigma\chi_\sigma c_{\mathbf{R}\sigma}^\dagger c_{\mathbf{R}'\sigma} = \sum_{\mathbf{R}} c_{\mathbf{R}\sigma}^\dagger c_{\mathbf{R}\sigma}. \tag{8.363}$$

Lastly, the Fermion creation and annihilation operators at lattice site $\mathbf{R}$ with spin σ satisfy the usual anti-commutation relations, e.g. $\{c_{\mathbf{R}\sigma}, c^{\dagger}_{\mathbf{R}'\sigma'}\} = \delta_{\mathbf{R}\mathbf{R}'}\delta_{\sigma\sigma'}$.

The two-particle operator of the Coulomb interaction using this representation becomes

$$\mathcal{H}_C = \frac{1}{2} \sum_{\mathbf{R}_1\mathbf{R}_2\mathbf{R}_3\mathbf{R}_4} \left\langle \mathbf{R}_1\mathbf{R}_2 \left| \frac{e^2}{|\mathbf{r}_1 - \mathbf{r}_2|} \right| \mathbf{R}_3\mathbf{R}_4 \right\rangle \sum_{\sigma\sigma'} c^{\dagger}_{\mathbf{R}_1\sigma} c^{\dagger}_{\mathbf{R}_2\sigma'} c_{\mathbf{R}_3\sigma'} c_{\mathbf{R}_4\sigma} \tag{8.364}$$

where

$$\left\langle \mathbf{R}_1\mathbf{R}_2 \left| \frac{e^2}{|\mathbf{r}_1 - \mathbf{r}_2|} \right| \mathbf{R}_3\mathbf{R}_4 \right\rangle = \int d^3r_1\, d^3r_2\, \phi^*_{\mathbf{R}_1}(\mathbf{r}_1)\phi^*_{\mathbf{R}_2}(\mathbf{r}_2)\frac{e^2}{|\mathbf{r}_1 - \mathbf{r}_2|}\phi_{\mathbf{R}_3}(\mathbf{r}_2)\phi_{\mathbf{R}_4}(\mathbf{r}_1). \tag{8.365}$$

The dominant contributions to the Coulomb Hamiltonian $\mathcal{H}_C$ stem from the summands with large overlap integrals, i.e. the direct Coulomb integral

$$K_{\mathbf{R}_1\mathbf{R}_2} = \left\langle \mathbf{R}_1\mathbf{R}_2 \left| \frac{e^2}{|\mathbf{r}_1 - \mathbf{r}_2|} \right| \mathbf{R}_2\mathbf{R}_1 \right\rangle \tag{8.366}$$

$$= \int d^3r_1\, d^3r_2\, |\phi_{\mathbf{R}_1}(\mathbf{r}_1)|^2 \frac{e^2}{|\mathbf{r}_1 - \mathbf{r}_2|} |\phi_{\mathbf{R}_2}(\mathbf{r}_2)|^2 \geq 0 \tag{8.367}$$

and the exchange integral

$$\mathcal{J}_{\mathbf{R}_1\mathbf{R}_2} = \left\langle \mathbf{R}_1\mathbf{R}_2 \left| \frac{e^2}{|\mathbf{r}_1 - \mathbf{r}_2|} \right| \mathbf{R}_1\mathbf{R}_2 \right\rangle \tag{8.368}$$

$$= \int d^3r_1\, d^3r_2\, \phi^*_{\mathbf{R}_1}(\mathbf{r}_1)\phi^*_{\mathbf{R}_2}(\mathbf{r}_2)\frac{e^2}{|\mathbf{r}_1 - \mathbf{r}_2|}\phi_{\mathbf{R}_1}(\mathbf{r}_2)\phi_{\mathbf{R}_2}(\mathbf{r}_1). \tag{8.369}$$

Thus, after rearranging the creation and annihilation operators, the two-particle Hamiltonian of the Coulomb interaction becomes

$$\mathcal{H}_C = \frac{1}{2} \sum_{\mathbf{R}_1\mathbf{R}_2\sigma\sigma'} \left(K_{\mathbf{R}_1\mathbf{R}_2} c^{\dagger}_{\mathbf{R}_1\sigma} c_{\mathbf{R}_1\sigma} c^{\dagger}_{\mathbf{R}_2\sigma'} c_{\mathbf{R}_2\sigma'} - \mathcal{J}_{\mathbf{R}_1\mathbf{R}_2} c^{\dagger}_{\mathbf{R}_1\sigma} c_{\mathbf{R}_1\sigma'} c^{\dagger}_{\mathbf{R}_2\sigma'} c_{\mathbf{R}_2\sigma} \right) \tag{8.370}$$

where we have neglected terms containing $c_{\mathbf{R}_1\sigma}{}^{\dagger} c_{\mathbf{R}_1\sigma}$ that correspond to one-particle potentials and are irrelevant for our purposes.

We can show two important inequalities of the direct and exchange Coulomb integrals, namely

$$\mathcal{J}_{\mathbf{R}_1\mathbf{R}_2} \geq 0, \tag{8.371}$$

$$K_{\mathbf{R}_1\mathbf{R}_2} \geq \mathcal{J}_{\mathbf{R}_1\mathbf{R}_2}. \tag{8.372}$$

The first inequality uses the Fourier transform of the Coulomb potential

$$\frac{e^2}{|\mathbf{r}_1 - \mathbf{r}_2|} = \int \frac{d^3q}{(2\pi)^3} e^{i\mathbf{q}\cdot(\mathbf{r}_1-\mathbf{r}_2)} \frac{4\pi e^2}{q^2} \tag{8.373}$$

such that

$$\mathcal{J}_{\mathbf{R}_1\mathbf{R}_2} = \int \frac{d^3q}{(2\pi)^3} \frac{4\pi e^2}{q^2} \int d^3r_1\, \phi^*_{\mathbf{R}_1}(\mathbf{r}_1)\phi_{\mathbf{R}_2}(\mathbf{r}_1)e^{i\mathbf{q}\cdot\mathbf{r}_1} \int d^3r_2\, \phi_{\mathbf{R}_1}(\mathbf{r}_2)\phi^*_{\mathbf{R}_2}(\mathbf{r}_2)e^{-i\mathbf{q}\cdot\mathbf{r}_2} \tag{8.374}$$

$$= \int \frac{d^3q}{(2\pi)^3} \frac{4\pi e^2}{q^2} \left| \int d^3r_1\, \phi^*_{\mathbf{R}_1}(\mathbf{r}_1)\phi_{\mathbf{R}_2}(\mathbf{r}_1)e^{i\mathbf{q}\cdot\mathbf{r}_1} \right|^2 \geq 0. \tag{8.375}$$

EXERCISE 8.12 Inequality between Coulomb and exchange integral Prove the second inequality $K_{\mathbf{R}_1\mathbf{R}_2} - \mathcal{J}_{\mathbf{R}_1\mathbf{R}_2} \geq 0$.

In order to interpret $\mathcal{H}_C$ further, let us rewrite the string of creation and annihilation operators in the second part of the sum in (8.370). First, we introduce number and spin operators

$$n_{\mathbf{R}} = \sum_{\sigma} c^{\dagger}_{\mathbf{R}\sigma} c_{\mathbf{R}\sigma} \tag{8.376}$$

$$\mathbf{s}^{\alpha}_{\mathbf{R}} = \frac{1}{2} \sum_{\sigma\sigma'} c^{\dagger}_{\mathbf{R}\sigma} \left(\boldsymbol{\sigma}_{\sigma\sigma'}\right) c_{\mathbf{R}\sigma'} \qquad \alpha = x, y, z \tag{8.377}$$

$$\mathbf{s}^{\pm}_{\mathbf{R}} = \mathbf{s}^{x}_{\mathbf{R}} \pm i\mathbf{s}^{y}_{\mathbf{R}}, \tag{8.378}$$

where σ is the vector of Pauli spin matrices, and observe that the following identity holds for the string of creation and annihilation operators in the second part of the sum in (8.370)

$$\begin{aligned}\sum_{\sigma\sigma'} c^{\dagger}_{\mathbf{R}_1\sigma} c_{\mathbf{R}_1\sigma'} c^{\dagger}_{\mathbf{R}_2\sigma'} c_{\mathbf{R}_2\sigma} &= \frac{1}{2}\left(c^{\dagger}_{\mathbf{R}_1\uparrow}c_{\mathbf{R}_1\uparrow} + c^{\dagger}_{\mathbf{R}_1\downarrow}c_{\mathbf{R}_1\downarrow}\right)\left(c^{\dagger}_{\mathbf{R}_2\uparrow}c_{\mathbf{R}_2\uparrow} + c^{\dagger}_{\mathbf{R}_2\downarrow}c_{\mathbf{R}_2\downarrow}\right) \\ &+ \frac{1}{2}\left(c^{\dagger}_{\mathbf{R}_1\uparrow}c_{\mathbf{R}_1\uparrow} - c^{\dagger}_{\mathbf{R}_1\downarrow}c_{\mathbf{R}_1\downarrow}\right)\left(c^{\dagger}_{\mathbf{R}_2\uparrow}c_{\mathbf{R}_2\uparrow} - c^{\dagger}_{\mathbf{R}_2\downarrow}c_{\mathbf{R}_2\downarrow}\right) \\ &+ c^{\dagger}_{\mathbf{R}_1\uparrow}c_{\mathbf{R}_1\downarrow}c^{\dagger}_{\mathbf{R}_2\downarrow}c_{\mathbf{R}_2\uparrow} + c^{\dagger}_{\mathbf{R}_1\downarrow}c_{\mathbf{R}_1\uparrow}c^{\dagger}_{\mathbf{R}_2\uparrow}c_{\mathbf{R}_2\downarrow}.\end{aligned} \tag{8.379}$$

From this identity we can read off

$$\sum_{\sigma\sigma'} c^{\dagger}_{\mathbf{R}_1\sigma} c_{\mathbf{R}_1\sigma'} c^{\dagger}_{\mathbf{R}_2\sigma'} c_{\mathbf{R}_2\sigma} = \frac{n_{\mathbf{R}_1} n_{\mathbf{R}_2}}{2} + s^{+}_{\mathbf{R}_1} s^{-}_{\mathbf{R}_2} + s^{-}_{\mathbf{R}_1} s^{+}_{\mathbf{R}_2} + 2 s^{z}_{\mathbf{R}_1} s^{z}_{\mathbf{R}_2} \tag{8.380}$$

$$= \frac{n_{\mathbf{R}_1} n_{\mathbf{R}_2}}{2} + 2\mathbf{s}_{\mathbf{R}_1} \cdot \mathbf{s}_{\mathbf{R}_2} \tag{8.381}$$

and obtain the final form of the Hamiltonian of the Coulomb interaction

$$\mathcal{H}_C = \frac{1}{2} \sum_{\mathbf{R}_1\mathbf{R}_2} \left(\left\{ K_{\mathbf{R}_1\mathbf{R}_2} - \frac{1}{2}\mathcal{J}_{\mathbf{R}_1\mathbf{R}_2} \right\} n_{\mathbf{R}_1} n_{\mathbf{R}_2} - 2\mathcal{J}_{\mathbf{R}_1\mathbf{R}_2} \mathbf{s}_{\mathbf{R}_1} \mathbf{s}_{\mathbf{R}_2} \right). \tag{8.382}$$

The result of the previous exercise implies that $K_{\mathbf{R}_1\mathbf{R}_2} - \frac{1}{2}\mathcal{J}_{\mathbf{R}_1\mathbf{R}_2} > 0$. Thus, the Coulomb term remains a repulsive interaction, although it has been reduced by an exchange contribution. Furthermore, the exchange interaction term in the Coulomb Hamiltonian can now be identified as an effective spin-spin interaction

$$\mathcal{H}_C^{\text{eff}} = -\sum_{\mathbf{R}_1\mathbf{R}_2} \mathcal{J}_{\mathbf{R}_1\mathbf{R}_2}\, \mathbf{s}_{\mathbf{R}_1} \cdot \mathbf{s}_{\mathbf{R}_2}. \tag{8.383}$$

Since $\mathcal{J}_{\mathbf{R}_1\mathbf{R}_2} \geq 0$, this interaction prefers a parallel orientation of the spins, and hence, it describes ferromagnetic order. The Hamiltonian (8.383) is the celebrated Heisenberg (1928) Hamiltonian.

EXERCISE 8.13 Hubbard U-term Assuming a Wannier wave function $\phi(\mathbf{r})$, which is the same for all lattice sites $\mathbf{R}$, show that (8.364) can be written in the form

$$\mathcal{H}_C = U \sum_{\mathbf{R}} n_{\mathbf{R}\uparrow} n_{\mathbf{R}\downarrow} \tag{8.384}$$

where U is the on-site interaction that plays a central role in the Hubbard model we explore in section 8.7.

In a particular choice of basis states, the second quantized Hamiltonian for the Coulomb interaction between electrons or, more generally Fermions, is

$$\mathcal{H}_C = \frac{1}{2} \sum_{\sigma\sigma'} \sum_{ijkl} \left\langle ij \left| \frac{e^2}{r} \right| kl \right\rangle c^{\dagger}_{i\sigma} c^{\dagger}_{j\sigma'} c_{k\sigma} c_{l\sigma'}. \tag{8.385}$$

The dominant contributions to this Hamiltonian stem from the summands with large overlap integrals, which, from our experience with the hydrogen molecule, we expect to be the direct and the exchange Coulomb terms, i.e. $|ij\rangle \to |ij\rangle$ and $|ij\rangle \to |ji\rangle$ such that

$$\mathcal{H}_C = \frac{1}{2}\sum_{\sigma\sigma'}\sum_{ij}\left(-\left\langle ij\left|\frac{e^2}{r}\right|ij\right\rangle c^\dagger_{i\sigma}c_{i\sigma}c^\dagger_{j\sigma'}c_{j\sigma'} + \left\langle ij\left|\frac{e^2}{r}\right|ji\right\rangle c^\dagger_{i\sigma}c_{i\sigma'}c^\dagger_{j\sigma'}c_{j\sigma}\right). \tag{8.386}$$

8.6.4.3 Kinetic exchange interaction in quantum matter: antiferromagnetism

The treatment in the previous section emphasized the local interaction of electrons via the Coulomb interaction. This section examines the case where, due to the hybridization between lattice sites, the kinetic energy of the electrons becomes important. The electrons can tunnel between lattice sites and become delocalized.

The very versatile Hubbard model (discussed in detail in section 8.7) exactly describes this situation. Its Hamiltonian has the form

$$\mathcal{H} = -t\sum_{\mathbf{R}\mathbf{R}'\sigma} c^\dagger_{\mathbf{R}'\sigma}c_{\mathbf{R}\sigma} + U\sum_{\mathbf{R}} c^\dagger_{\mathbf{R}\uparrow}c^\dagger_{\mathbf{R}\downarrow}c_{\mathbf{R}\downarrow}c_{\mathbf{R}\uparrow} \tag{8.387}$$

where the first term describes the hopping motion of electrons between lattice sites and the second term is an energy cost for double occupancy of a lattice site. This Hubbard U-term encountered in exercise 8.13, and it can be rewritten in the form

$$U\sum_{\mathbf{R}} c^\dagger_{\mathbf{R}\uparrow}c^\dagger_{\mathbf{R}\downarrow}c_{\mathbf{R}\downarrow}c_{\mathbf{R}\uparrow} = U\sum_{\mathbf{R}} c^\dagger_{\mathbf{R}\uparrow}c_{\mathbf{R}\uparrow}c^\dagger_{\mathbf{R}\downarrow}c_{\mathbf{R}\downarrow} = U\sum_{\mathbf{R}} n_{\mathbf{R}\uparrow}n_{\mathbf{R}\downarrow}. \tag{8.388}$$

We discuss the Hubbard model (and focus on particular aspects of it) in other parts of this book. Here, we are mostly interested in the magnetic properties of the model, which can be discussed with the help of a simple two-site toy model whose Hamiltonian is

$$\mathcal{H}_2 = -t\sum_{\sigma}\left(c^\dagger_{1\sigma}c_{2\sigma} + c^\dagger_{2\sigma}c_{1\sigma}\right) + U\sum_{i=1}^{2} n_{i\uparrow}n_{i\downarrow}. \tag{8.389}$$

This Hamiltonian describes the hopping of electrons between just two sites $i = 1, 2$ where double occupancy of a site costs an energy U. Each site can be in one of four states: empty, occupied by a single electron with spin up or down, and occupied by two electrons with opposite spin direction. These possibilities span a Fock space of dimension 4^2. Depending on the number of electrons in this two-site system, this Fock space splits into sectors. For the sector with two electrons the Fock space reduces to six dimensions and can be spanned by the six basis vectors

$$\{|m\rangle\} = \{|\uparrow\downarrow, 0\rangle, |0, \uparrow\downarrow\rangle, |\uparrow, \downarrow\rangle, |\downarrow, \uparrow\rangle, |\uparrow, \uparrow\rangle, |\downarrow, \downarrow\rangle\} \qquad m = 1, \ldots, 6, \tag{8.390}$$

where the first entry refers to site $i = 1$, and the second to site $i = 2$.

The following exercise is an invitation to investigate the two-site Hubbard model in detail for the case when the on-site repulsion U is much stronger than the intersite hopping t, i.e. when $U \gg t$.

EXERCISE 8.14 Two-site Hubbard model Diagonalize the two-site Hubbard model Hamiltonian, (8.389), in the basis (8.390) for the limiting case $U \gg t$. In order to do this, it is advantageous to replace the two basis vectors $|3\rangle = |\uparrow, \downarrow\rangle$ and $|4\rangle = |\downarrow, \uparrow\rangle$ in the basis (8.390) by the symmetric and antisymmetric linear combinations

$$|3'\rangle = \frac{1}{\sqrt{2}}(|\uparrow, \downarrow\rangle - |\downarrow, \uparrow\rangle) = \frac{1}{\sqrt{2}}(|3\rangle - |4\rangle) \tag{8.391}$$

and

$$|4'\rangle = \frac{1}{\sqrt{2}}(|\uparrow, \downarrow\rangle + |\downarrow, \uparrow\rangle) = \frac{1}{\sqrt{2}}(|3\rangle + |4\rangle), \tag{8.392}$$

respectively.

Especially, identify the ground state energy in this basis. Which basis vector describes approximately the ground state? What kind of magnetic ordering do your results imply for this two-site model?

We expect the energy levels of the doubly occupied states $|1\rangle = |\uparrow\downarrow, 0\rangle$ and $|2\rangle = |0, \uparrow\downarrow\rangle$ to correspond to high energies of the order of the Coulomb repulsion U. The detailed calculation performed in exercise 8.14 reveals that the two highest lying energies

$$E_1 = \frac{U}{2}\left(1 + \sqrt{1 + \frac{16t^2}{U^2}}\right) \approx U + \frac{4t^2}{U} \quad \text{for} \quad U \gg t \tag{8.393}$$

$$E_2 = U \tag{8.394}$$

correspond to the states $|1\rangle + |2\rangle$ and $|1\rangle - |2\rangle$. However, while the latter is an exact eigenstate of the two-site Hubbard model for the eigenvalue U, the former is only an approximate eigenstate. Before discussing this approximate eigenstate further, let us briefly look at the other eigenvalues of the Hamiltonian (8.389). These are

$$E_{4'} = E_5 = E_6 = 0, \tag{8.395}$$

the threefold degenerate eigenvalue corresponding exactly to the eigenstates $|4'\rangle$, $|5\rangle$, and $|6\rangle$, the three basis states forming a spin triplet. The last eigenvalue is

$$E_{3'} = \frac{U}{2}\left(1 - \sqrt{1 + \frac{16t^2}{U^2}}\right) \approx -\frac{4t^2}{U} < 0 \quad \text{for} \quad U \gg t. \tag{8.396}$$

This last eigenvalue, being the smallest, corresponds to the ground state. Its approximate eigenstate is $|3'\rangle$, the spin singlet state that corresponds to antiparallel orientation of the spins and, hence, antiferromagnetic order.

The two eigenvalues $E_1 \equiv E_+$ and $E_{3'} \equiv E_-$, i.e. the lowest and the highest energy eigenvalues, have exact eigenstates

$$|1\rangle + |2\rangle + \alpha_+|3'\rangle \quad \text{for} \quad E_+, \tag{8.397}$$

$$|1\rangle + |2\rangle + \alpha_-|3'\rangle \quad \text{for} \quad E_- \tag{8.398}$$

where

$$\alpha_\pm = -\frac{U}{2\sqrt{2}t}\left(1 \mp \sqrt{1 + \frac{16t^2}{U^2}}\right). \tag{8.399}$$

While

$$\alpha_+ \approx \frac{2\sqrt{2}t}{U} \ll 1 \quad \text{for} \quad U \gg t \tag{8.400}$$

justifies the state $|1\rangle + |2\rangle$ as an approximate eigenstate for the eigenvalue E_+,

$$\alpha_- \approx -\frac{U}{\sqrt{2}t}, \quad \text{with} \quad |\alpha_-| \gg 1 \quad \text{for} \quad U \gg t \tag{8.401}$$

justifies the singlet state $|3'\rangle$ as an approximate ground state with ground state energy $E_- < 0$.

The singlet state ($S = 0$) with antiparallel spins is thus lower in energy than the triplet states ($S = 1$) with parallel spins, i.e. the singlet state is the ground state and the interaction is antiferromagnetic. While for parallel spins hopping is blocked by the Pauli principle and there is no t-dependence of the eigenenergies $E_{4'} = E_5 = E_6$, antiparallel spins can lower the energy by hopping. This mechanism is therefore called kinetic exchange. The situation is reminiscent of the H_2 molecule where the ground state is also a singlet.

These results can be condensed in an effective Heisenberg Hamiltonian for two interacting spins $s_1 = s_2 = 1/2$

$$\mathcal{H}_{\text{eff}} = -\mathcal{J}\mathbf{s}_1 \cdot \mathbf{s}_2. \tag{8.402}$$

Rewritten in terms of the total spin $\mathbf{S} = \mathbf{s}_1 + \mathbf{s}_2$ this becomes

$$\mathcal{H}_{\text{eff}} = -\mathcal{J}\mathbf{s}_1 \cdot \mathbf{s}_2 = -\frac{\mathcal{J}}{2}\left[\mathbf{S} \cdot \mathbf{S} - \mathbf{s}_1 \cdot \mathbf{s}_1 - \mathbf{s}_2 \cdot \mathbf{s}_2\right] = -\frac{\mathcal{J}}{2}\left[S(S+1) - \frac{3}{2}\right], \tag{8.403}$$

which yields for difference between the singlet ($S = 0$) and the triplet ($S = 1$) energies

$$E_s - E_t = \mathcal{J}. \tag{8.404}$$

The comparison with the singlet energy (8.396) and the triplet energies (8.395) determines the exchange coupling of the effective Heisenberg Hamiltonian to be

$$\mathcal{J} = -\frac{4t^2}{U} \quad \text{for} \quad \frac{U}{t} \to \infty. \tag{8.405}$$

In summary, we could identify the spin singlet state$|3'\rangle = (|\uparrow,\downarrow\rangle - |\downarrow,\uparrow\rangle)/\sqrt{2}$ as the ground state in the limit $U/t \to \infty$ of the two-site Hubbard model, which signifies antiferromagnetic order.

We have thus identified two models for collective magnetism, the Heisenberg model of localized spins interacting via exchange interaction and the Hubbard model of itinerant electrons. The order of the spins in the former model favours ferromagnetism. In its simplest form, the two-site model, the latter exhibits antiferromagnetic order for large on-site repulsion.

8.6.5 Ground state of the Heisenberg model

The previous sections have introduced the Heisenberg and the Hubbard models, which are devised to understand the magnetic properties of matter. While we explored the Heisenberg model quite thoroughly, the introduction of the Hubbard model was rather more ad hoc. Section 8.7 provides a more detailed exploration of the Hubbard model from first principles.

Both models describe the different situations of localized (Heisenberg), and itinerant (Hubbard) , magnetic moments. These are, of course, limiting cases that are not always realized in pure form. However, the two models are also connected, which allow us to understand the transition between itinerant and localized magnetism depending on the filling of the lattice with electrons.

Before exploring these questions in section 8.8, we first devote time to the important problems of the ground state and the low-lying the Heisenberg and the Hubbard model excitations above the ground state in the Heisenberg model.

The basis for our discussion will be the Heisenberg Hamiltonian (8.383) on a lattice of N spins of magnitude s, which we write in the standard form

$$\mathcal{H} = -\frac{1}{2}\sum_{ij} \mathcal{J}_{ij}\, \mathbf{s}_i \cdot \mathbf{s}_j = -\frac{1}{4}\sum_{ij} \mathcal{J}_{ij} \left(s_i^+ s_j^- + s_i^- s_j^+\right) - \frac{1}{2}\sum_{ij} \mathcal{J}_{ij}\, s_i^z s_j^z. \tag{8.406}$$

The exchange couplings $\mathcal{J}_{ij} = \mathcal{J}_{ji}$ possess the translational symmetry of the lattice. They, especially their signs, will determine the magnetic properties of the Heisenberg model.

First we must discover which operators commute with the Heisenberg Hamiltonian (8.406). Exercise 8.15 is devoted to this question.

EXERCISE 8.15 Set of commuting operators for the Heisenberg Hamiltonian
Consider the Heisenberg Hamiltonian (8.406) and show that $\{\mathcal{H}, \mathbf{S}_{\text{tot}}, S^z_{\text{tot}}\}$ form a set of commuting operators, where $\mathbf{S}_{\text{tot}} = \sum_i \mathbf{s}_i$ is the total spin and $S^z_{\text{tot}} = \sum_i s^z_i$ is the total z-component of the spin.

Using the results of exercise 8.15 we can classify the eigenstates of the Heisenberg Hamiltonian (8.406) according to

$$|\Psi\rangle = |S_{\text{tot}}, M\rangle \tag{8.407}$$

$$M = -S_{\text{tot}}, -S_{\text{tot}} + 1, \ldots, S_{\text{tot}} \tag{8.408}$$

$$S_{\text{tot}} \leq Ns, \tag{8.409}$$

where M is the eigenvalue of the total z-component of spin S^z_{tot} and the total spin satisfies

$$\mathbf{S}^2_{\text{tot}} = S_{\text{tot}}\,(S_{\text{tot}} + 1). \tag{8.410}$$

These preparations allow us to address the problems of the ground state and the low-lying excitations above the ground state in the Heisenberg model.

8.6.5.1 Ferromagnetic case

To investigate the ground state of the Heisenberg model in the case when all exchange couplings in (8.406) are non-negative, $\mathcal{J}_{ij} \geq 0$, we start with a specific spin state for a system of N spin-s particles interacting according to the Heisenberg Hamiltonian (8.406)

$$|\Psi_{\text{fp}}\rangle = |s, m_1 = s\rangle|s, m_2 = s\rangle \ldots |s, m_N = s\rangle. \tag{8.411}$$

This is a state with maximal total spin $S = S_{\text{tot}} = Ns$ and an eigenstate of the Hamiltonian (8.406). We want to compare the energy of this fully polarized state with the energy of the basis states (occasionally called Ising states)

$$|\Phi\rangle = |s, m_1\rangle|s, m_2\rangle \ldots |s, m_N\rangle. \tag{8.412}$$

EXERCISE 8.16 Ground state energy of the ferromagnetic Heisenberg model
Calculate and compare the expressions

$$\langle\Psi_{\text{fp}}|\mathcal{H}|\Psi_{\text{fp}}\rangle \tag{8.413}$$

and

$$\langle\Phi|\mathcal{H}|\Phi\rangle \tag{8.414}$$

where $\mathcal{H}$ is the Heisenberg Hamiltonian (8.406) and $|\Psi_{\text{fp}}\rangle$ and $|\Phi\rangle$ are the states (8.411) and (8.412), respectively.

The result of exercise 8.16 is

$$\langle\Phi|\mathcal{H}|\Phi\rangle = -\frac{1}{2}\sum_{ij}\mathcal{J}_{ij}m_i m_j \geq -\frac{1}{2}s^2\sum_{ij}\mathcal{J}_{ij}. \tag{8.415}$$

The states $|\Phi\rangle$ form a basis of the Hilbert space associated with the system of N spins of magnitude s. However, they are not generally eigenstates of the Heisenberg Hamiltonian (8.406). This is precluded for most states $|\Phi\rangle$ by the terms $s_i^+ s_j^-$ or $s_i^- s_j^+$ in the Heisenberg Hamiltonian, and that, in general, connect states with different quantum numbers m_i or m_j. Corresponding terms in the expectation value $\langle\Phi|\mathcal{H}|\Phi\rangle$ vanish. According to (8.415) all eigenstates formed from basis states $|\Phi\rangle$ will have an energy larger than or equal to $-\frac{1}{2}s^2\sum_{ij}\mathcal{J}_{ij}$, which is the eigenvalue of the eigenstate $|\Psi_{\text{fp}}\rangle$. Therefore, the fully polarized state $|\Psi_{\text{fp}}\rangle$ must be a ground state of the Heisenberg model when all exchange couplings $\mathcal{J}_{ij}$ are non-negative, which therefore describes ferromagnetic order.

8.6.5.2 Antiferromagnetic case

The situation is much more involved if some of the exchange couplings $\mathcal{J}_{ij}$ in the Heisenberg Hamiltonian (8.406) are negative. In fact, there are no general exact results for this case and we must consider particular examples of Heisenberg spin systems on concrete lattices one at a time. The antiferromagnetic ground states of particular manifestations of the Heisenberg model are very much a research field still in constant flux. Even an exposition attempting an overview is therefore precluded. Our discussion focuses on the bipartite lattices which is a class of lattices for which some exact results are known.

In our context of magnetism, a bipartite lattice is defined as a lattice that can be subdivided into two lattices A and B such that each lattice site belongs to either sub-lattice A or sub-lattice B. Moreover, the exchange couplings $\mathcal{J}_{ij} = 0$ if the lattice sites i and j both belong to either sub-lattice A or sub-lattice B. There are only interactions $\mathcal{J}_{ij} \neq 0$ between sites belonging to different sub-lattices.

Simple examples of bipartite lattices are the linear chain in one dimension, the square lattice in two dimensions, and the simple cubic lattice in three dimensions with only nearest-neighbour couplings in all cases. An obvious example of a lattice which is not bipartite is the triangular lattice in two dimensions.

We restrict our discussion in the following to bipartite lattices in the above-defined sense. A simple guess for the case of a bipartite lattice when all couplings are negative $\mathcal{J}_{ij} \leq 0$, informed from our discussion of the ferromagnetic case, would be the Néel state, where each sub–lattice is fully polarized like the state (8.411) but in opposite directions on the two sub–lattices, i.e.

$$|\Psi_N\rangle = \prod_{i\in A} |s, m_i = s\rangle \prod_{j\in B} |s, m_j = -s\rangle. \tag{8.416}$$

However, this state is not an eigenstate of the Heisenberg Hamiltonian.

EXERCISE 8.17 Néel state not an eigenstate of the antiferromagnetic Heisenberg model Convince yourself that the Néel state (8.416) is not an eigenstate of the Heisenberg Hamiltonian (8.406) and, hence, nor can it be the ground state of the antiferromagnetic Heisenberg model .

Nevertheless, the Néel state can be used in a variational sense to obtain an upper bound for the ground state energy E_{gs} of the Heisenberg model with negative exchange couplings $\mathcal{J}_{ij} \leq 0$. We restrict our attention to the special case of a Heisenberg model on a bipartite lattice whose exchange coupling $\mathcal{J} > 0$ does not depend on the lattice sites and extends only to nearest-neighbour lattice sites $\langle ij\rangle$

$$\mathcal{H} = \mathcal{J}\sum_{\langle ij\rangle} \mathbf{s}_i \cdot \mathbf{s}_j = \mathcal{J}\sum_{\langle ij\rangle} \left(\frac{1}{2}\left(s_i^+ s_j^- + s_i^- s_j^+\right) + s_i^z s_j^z\right). \tag{8.417}$$

The expectation of this Hamiltonian in the Néel state (8.416) provides an upper bound for the ground state energy E_{gs}, which is

$$\langle\Psi_N|\mathcal{H}|\Psi_N\rangle = \langle\Psi_N|\mathcal{J}\sum_{\langle ij\rangle} s_i^z s_j^z|\Psi_N\rangle = -\frac{\mathcal{J}s^2 Nz}{2} > E_{gs} \tag{8.418}$$

where z is the number of nearest neighbours, the coordination number, and N the number of lattice sites. The number of nearest-neighbour pairs of spins is thus $Nz/2$.

The magnetization $s^z = \sum_i s_i^z$ of an antiferromagnet in the Néel state vanishes. A more useful quantity for antiferromagnets that is often used to replace the magnetization is the staggered magnetization

$$s^z_{\text{staggered}} = \sum_{i\in A} s_i^z - \sum_{i\in B} s_i^z. \tag{8.419}$$

We shall return to the staggered magnetization in exercise 8.24 when we discuss the spin wave theory of antiferromagnetic matter in section 8.6.6.2.

8.6.5.3 Marshall's theorems

There are are number of exact theorems. Two of them are due to Marshall (1955) and have been extended and sharpened by Lieb and Mattis (1962). For the proofs, we refer to the original literature (see also, e.g. Auerbach (1994)).

For the formulation of the Marshall theorems it is convenient to rotate the spins on the B sub-lattice about the z-axis, i.e. for $i \in B$

$$s_i^+ \rightarrow -\widetilde{s}_i^+, \tag{8.420}$$

$$s_i^- \rightarrow -\widetilde{s}_i^-, \tag{8.421}$$

$$s_i^z \rightarrow \widetilde{s}_i^z. \tag{8.422}$$

A factor in a basis state (8.412) then transforms into

$$|s, m_i\rangle \rightarrow |s, \widetilde{m}_i\rangle = \begin{cases} |s, m_i\rangle & i \in A \\ (-1)^{(s+m_i)}|s, m_i\rangle & i \in B \end{cases}. \tag{8.423}$$

These single-spin functions define a new set of basis states $|\widetilde{\Phi}\rangle$ in which we now use wave functions in sectors with fixed

$$M = \sum_i \widetilde{m}_i. \tag{8.424}$$

These wave functions have the form

$$|\Psi^M\rangle = \sum_\alpha g_\alpha^M |\widetilde{\Phi}_\alpha^M\rangle = \sum_\alpha g_\alpha^M \prod_{\sum_i \widetilde{m}_i^\alpha = M} |S, \widetilde{m}_i^\alpha\rangle. \tag{8.425}$$

With these preparations we can state the first theorem.

Theorem 8.1 Marshall Consider the bipartite Heisenberg Hamiltonian with exchange interactions $\mathcal{J}_{ij} \leq 0$ where any two sites are connected by a finite sequence of exchange interactions $\mathcal{J}_{nm}$ between intermediary sites. For any sector M, the lowest energy state $|\Psi_0^M\rangle$ can be chosen to be

$$|\Psi_0^M\rangle = \sum_\alpha g_\alpha^M |\Phi_\alpha^M\rangle \tag{8.426}$$

with all expansion coefficients g_α^M positive.

The absolute ground state $|\Psi_{gs}\rangle$ for a system where the sub-lattices A and B are of equal size is a singlet of total spin

$$\mathbf{S_{tot}}|\Psi_{gs}\rangle = 0. \tag{8.427}$$

In particular, the second part of Marshall's theorem will prove very useful in later chapters.

An important corollary of Marshall's theorem is that for a fixed quantum number M the ground state $|\Psi_{gs}\rangle$ is non-degenerate.

The next two sections introduce two useful concepts that help clarify the problem of the ground states of Heisenberg models with antiferromagnetic couplings, the transformation of spin degrees of freedom to Schwinger Bosons, and the valence bond states.

8.6.5.4 Schwinger Bosons

Section 5.7, especially sections 5.7.1 and 5.7.1.3 and the exercises 5.17 and 5.18, showed that it can be useful to transform spin degrees of freedominto equivalent Fermion degrees of freedom via the Jordan–Wigner transformation.

On different lattice sites spin operators for general spin s have vanishing commutators and thus exhibit Bosonic properties. It is therefore tempting to look for a way to fully transform the spin operators into Bosonic operators as Bosonic operators are generally easier to handle.

This section introduces a mapping of the Heisenberg model of interacting spins onto a different but equivalent model of Bosons and that will lead to new insights. In particular, this mapping works in any dimension and for arbitrary spin magnitudes. The simpler commutation relations of the creation and annihilation operators of the Bosons as compared to those of the spin operators facilitate the analysis of the models. Moreover, the Bosonic model allows for an elementary physical interpretation in terms of a system of in general coupled harmonic oscillators.

The Schwinger Boson operators at site i are related to the spin operators at site i by the transformation

$$s_i^+ = a_i^\dagger b_i, \tag{8.428}$$

$$s_i^- = b_i^\dagger a_i, \tag{8.429}$$

$$s_i^z = \frac{1}{2}\left(a_i^\dagger a_i - b_i^\dagger b_i\right). \tag{8.430}$$

In contrast to the Jordan–Wigner transformation the transformation to Schwinger Bosons has the advantage of being local. However, we need two operators, a_i and b_i which mutually commute

$$\left[a_i, a_j^\dagger\right] = \delta_{ij}, \quad \left[b_i, b_j^\dagger\right] = \delta_{ij}, \quad \left[a_i, b_j^\dagger\right] = \left[b_i, a_j^\dagger\right] = 0. \tag{8.431}$$

As a first consequence of the transformation to Schwinger Bosons we find

$$\mathbf{s}_i \cdot \mathbf{s}_i = \frac{1}{2}\left(a_i^\dagger a_i + b_i^\dagger b_i\right)\left(\frac{1}{2}\left(a_i^\dagger a_i + b_i^\dagger b_i\right) + 1\right) = s(s+1). \tag{8.432}$$

Since the quantum number s for the spins $\mathbf{s}_i$ is fixed, this puts a local constraint on the Schwinger Bosons, i.e. the number of Bosons of type a and type b

$$a_i^\dagger a_i + b_i^\dagger b_i = n_i^{(a)} + n_i^{(b)} = 2s \tag{8.433}$$

is locally conserved keeping the Bosons in the physically relevant subspace of fixed spin magnitude s. This constraint also 'conserves' the difficulty of the model.

The spin basis states $|s, m_i\rangle$ at site i labelled by the eigenvalues of $\mathbf{s}^2 = s(s+1)$ and s_i^z $(m_i = -s, -s+1, \ldots, s)$ can be expressed in terms of the Schwinger Boson creation operators

$$|s, m_i\rangle = \frac{\left(a_i^\dagger\right)^{s+m_i}}{\sqrt{(s+m_i)!}} \frac{\left(b_i^\dagger\right)^{s-m_i}}{\sqrt{(s-m_i)!}} |0\rangle \tag{8.434}$$

with $|0\rangle$ the Schwinger Boson vacuum state. For example, we obtain for spin magnitude $s = \frac{1}{2}$ (suppressing a site index for the moment)

$$\left|\frac{1}{2}, +\frac{1}{2}\right\rangle = |\uparrow\rangle = a^\dagger|0\rangle, \tag{8.435}$$

$$\left|\frac{1}{2}, -\frac{1}{2}\right\rangle = |\downarrow\rangle = b^\dagger|0\rangle. \tag{8.436}$$

The nearest-neighbour Heisenberg Hamiltonian with exchange coupling $\mathcal{J}$ expressed in Schwinger Boson operators becomes

$$\mathcal{H} = -\mathcal{J}\sum_{\langle ij\rangle} \mathbf{s}_i \cdot \mathbf{s}_j = -\frac{\mathcal{J}}{2}\sum_{\langle ij\rangle}\left(a_i^\dagger a_i a_j^\dagger a_j + b_i^\dagger b_i b_j^\dagger b_j + a_i^\dagger b_i b_j^\dagger a_j + b_i^\dagger a_i a_j^\dagger b_j\right), \tag{8.437}$$

where we have suppressed a constant term.

The biquadratic form of (8.437) explicitly reveals that the Bosonic Hamiltonian describes an interacting model. The first two terms in (8.437) describe a Boson density-density interaction, and the second two terms describe the hopping of an a or b Boson from site i to site j accompanied by the hopping of another a or b Boson from site j to site i, thus modelling correlated hopping. Furthermore, the Schwinger Boson Hamiltonian (8.437) conserves the total numbers of a and b Bosons $N_a = \sum_i a_i^\dagger a_i$ and $N_b = \sum_i b_i^\dagger b_i$ separately and the local Boson number ${a_i}^\dagger a_i + {b_i}^\dagger b_i = 2S$.

The following section demonstrates how to apply the Schwinger Boson representation to introduce the concept of the valence bond states, thereby gaining further insight into the problem of the ground state of the antiferromagnetic Heisenberg model.

8.6.5.5 *Valence bond states*

Section 8.6.5.2 discussed how finding the exact quantum ground state of the antiferromagnetic Heisenberg model is a difficult task. However, we have been able to use variational ansatz wave functions as good approximations. The valence bond states fall into this class of variational states.

The idea of the valence bond states consists of the following three points:

1) A lattice of N sites is divided into bond configurations (ij) (*not necessarily* nearest-neighbour bonds) in such a way that exactly $2s$ bonds emanate from each lattice site. Let us call such configurations $\mathscr{C}_\alpha$. For example, if $s = 1/2$, one bond connects exactly two sites. For $s = 1$, already many more configurations are possible as long as there are exactly two bonds emanating from a single lattice point. In this case, the simplest bond configuration consists of two sites that are, however, connected by two bonds.Therefore, this definition leads to disjoint clusters of sites connected by bonds. For $s = 1/2$, these clusters contain two sites, whereas, for $s = 1$, the clusters may already be of arbitrary size.

2) For each configuration $\mathscr{C}_\alpha$, we define a valence bond state with the help of the Schwinger Boson operators

$$|\mathscr{C}_\alpha\rangle = \prod_{(ij)\in\mathscr{C}_\alpha} \left(a_i^\dagger b_j^\dagger - b_i^\dagger a_j^\dagger\right)|0\rangle, \tag{8.438}$$

where $|0\rangle$ is the vacuum state with no Bosons on the lattice.Every term in the product of the valence bond state (8.438) containing a particular site creates exactly $2s$ Schwinger Bosons at that site, thus satisfying the constraint (8.433) for Schwinger Bosons.For example, in the case $s = 1/2$, if there is a bond between site i and j, the corresponding bond operator $\left(a_i^\dagger b_j^\dagger - b_i^\dagger a_j^\dagger\right)$ gives (using (8.434) or, more directly, (8.435) and (8.436))

$$\left(a_i^\dagger b_j^\dagger - b_i^\dagger a_j^\dagger\right)|0\rangle = |\uparrow_i, \downarrow_j\rangle - |\downarrow_i, \uparrow_j\rangle = |\uparrow_i\rangle|\downarrow_j\rangle - |\downarrow_i\rangle|\uparrow_j\rangle, \tag{8.439}$$

i.e. the (unnormalized) spin singlet state. Since, moreover, for $s = 1/2$, all contributions to $|\mathscr{C}_\alpha\rangle$ are spin singlets, and thus $|\mathscr{C}_\alpha\rangle$ itself is a spin singlet state. This observation is also true for higher spin $s > 1/2$, i.e. $|\mathscr{C}_\alpha\rangle$ is a spin singlet state for all magnitudes s of spin (see exercise 8.18).

3) A general valence bond state is a superposition of valence bond states with complex coefficients c_α of the form

$$|\{\mathscr{C}_\alpha\}, s\rangle = \sum_\alpha c_\alpha |\mathscr{C}_\alpha\rangle. \tag{8.440}$$

As a superposition of spin singlet states a general valence bond state is also a spin singlet state.

EXERCISE 8.18 Valence bond state for $s = 1$ For $s = 1$, using (8.434) write down the valence bond stateas a spin state for two sites connected by two bonds and convince yourself that it is a spin singlet state.

The valence bond states came to prominence when Anderson (1987) suggested his *resonating* valence bond theory as an explanatory model for high-temperature superconductors.

In order to demonstrate its usefulness, the following sections apply the concept of the valence bond state to the Heisenberg model for spin magnitude 1/2.

8.6.5.6 *Valence bond states for spin 1/2*

As shown, the states (8.435) and (8.436) form a basis for a single spin of magnitude 1/2

$$|\uparrow_i\rangle = a_i^\dagger|0\rangle, \tag{8.441}$$

$$|\downarrow_i\rangle = b_i^\dagger|0\rangle. \tag{8.442}$$

Furthermore, a valence bond state can thus be written as

$$|\mathscr{C}_\alpha\rangle = \prod_{(ij)\in\mathscr{C}_\alpha} \frac{1}{\sqrt{2}}\left(|\uparrow_i\rangle|\downarrow_j\rangle - |\downarrow_i\rangle|\uparrow_j\rangle\right). \tag{8.443}$$

This state is a spin singlet state that, in contrast to (8.439), we now write with the proper normalization.

We are interested in correlations between spins $\mathbf{s}_n$ and $\mathbf{s}_m$ in a valence bond state $|\mathscr{C}_\alpha\rangle$ defined by

$$g(n,m) = \langle\mathscr{C}_\alpha|\mathbf{s}_n\cdot\mathbf{s}_m|\mathscr{C}_\alpha\rangle - \langle\mathscr{C}_\alpha|\mathbf{s}_n|\mathscr{C}_\alpha\rangle\cdot\langle\mathscr{C}_\alpha|\mathbf{s}_m|\mathscr{C}_\alpha\rangle. \tag{8.444}$$

The spin expectation values $\langle\mathscr{C}_\alpha|\mathbf{s}_n|\mathscr{C}_\alpha\rangle$ in this expression vanish and can therefore be included in the definition with indemnity.

For $s = 1/2$, each site of the lattice belongs to only one single bond of the bond configuration $\mathscr{C}_\alpha$. Due to the construction of the valence bond states, there can thus only be correlations between sites that lie on a common bond, as sites on different bonds are not correlated. For the case $m = n$ we obtain

$$g(n,n) = s(s+1) = \frac{3}{4}. \tag{8.445}$$

In order to obtain a correlation for $n \neq m$ we need to have a bond operator $a_i^\dagger b_j^\dagger - b_i^\dagger a_j^\dagger$ that creates the corresponding state $\left(|\uparrow_i\rangle|\downarrow_j\rangle - |\downarrow_i\rangle|\uparrow_j\rangle\right)/\sqrt{2}$ in (8.443), where either $n = i$ and $m = j$, or vice versa. In both cases, an explicit calculation, using

$$\mathbf{s}_n \cdot \mathbf{s}_m = \frac{1}{2}\left(s_n^+ s_n^- + s_n^- s_m^+\right) + s_n^z s_m^z \tag{8.446}$$

yields

$$g(n, m) = -\frac{3}{4}. \tag{8.447}$$

In summary, we thus have for the spin-spin correlation function

$$g(n, m) = \begin{cases} 3/4 & \text{if} \quad n = m \\ -3/4 & \text{if} \quad (mn) \in \mathscr{C}_\alpha \quad \text{or} \quad (nm) \in \mathscr{C}_\alpha \\ 0 & \text{otherwise.} \end{cases} \tag{8.448}$$

If all bonds in a configuration $\mathscr{C}_\alpha$ are short-ranged, i.e. if all bonds are shorter than a finite length ξ, the valence bond state by construction, and hence also the superposition state (8.440), $|\{\mathscr{C}_\alpha\}, s = 1/2\rangle = \sum_\alpha c_\alpha |\mathscr{C}_\alpha\rangle$, contains only short-ranged correlations. Such a situation describes a so-called quantum *spin liquid* (see, e.g. Imai and Lee (2016) for a recent overview).

8.6.5.7 *Heisenberg quantum spin-1/2 chain*

As a further application of the concept of valence bond states, we next consider a chain of spins described by the Heisenberg Hamiltonian with antiferromagnetic exchange interactions between nearest neighbours, which we can write as

$$\mathcal{H} = \mathcal{J} \sum_i \mathbf{s}_i \cdot \mathbf{s}_{i+1} \tag{8.449}$$

with $\mathcal{J} > 0$. We further assume periodic boundary conditions $\mathbf{s}_{N+1} = \mathbf{s}_1$ and an even number N of sites.

Since the interactions are only between nearest neighbours, we can presume only valence bond states with nearest-neighbour bonds $(ij) = (ii + 1)$ in the configuration $\mathscr{C}_\alpha$ of bonds. There are then only two distinct sets of bond configurations, $\mathscr{C}_+$ connecting odd sites $2n - 1$ to even sites $2n$, and $\mathscr{C}_-$ connecting even sites $2n$ to odd sites $2n + 1$, where $n = 1, 2, \ldots, N/2$. The corresponding valence bond states

$$|\pm\rangle = \prod_{i=1}^{N/2} \frac{1}{\sqrt{2}} \left(|\uparrow_{2n\mp1}\rangle|\downarrow_{2n}\rangle - |\downarrow_{2n\mp1}\rangle|\uparrow_{2n}\rangle\right) \tag{8.450}$$

are called *dimer* states.

By symmetry, the expectation values of the Heisenberg Hamiltonian in the $|+\rangle$ state and in the $|-\rangle$ state are equal: $\langle+|\mathcal{H}|+\rangle = \langle-|\mathcal{H}|-\rangle$. The states $|+\rangle$ and $|-\rangle$ are thus degenerate, which disqualifies either of them from being candidates of the true ground state due to Marshall's theorem, which states that the true ground state is non-degenerate.

Although $|\pm\rangle$ are not eigenstates of the Heisenberg Hamiltonian (8.449), the expectation value is a useful quantity in the sense of a variational bound for the ground state energy. Explicit calculation using the results of the previous section for the spin-spin correlation functions yields

$$\langle+|\mathcal{H}|+\rangle = -\frac{3N\mathcal{J}}{8} < 0. \tag{8.451}$$

In order to assess how the valence bond state $|+\rangle$ fares in comparison to other variational states, let us compare (8.451) with the Néel state, which also is neither an eigenstate nor, again due to Marshall's theorem, a ground state

$$|\text{Néel}\rangle = |\uparrow_1\rangle|\downarrow_2\rangle|\uparrow_3\rangle|\downarrow_4\rangle\ldots|\uparrow_{N-1}\rangle|\downarrow_N\rangle. \tag{8.452}$$

For the Néel state we obtain

$$\langle\text{Néel}|\mathcal{H}|\text{Néel}\rangle = -\frac{N\mathcal{J}}{4} < 0. \tag{8.453}$$

Hence, we can infer that

$$\langle\pm|\mathcal{H}|\pm\rangle < \langle\text{Néel}|\mathcal{H}|\text{Néel}\rangle. \tag{8.454}$$

The valence bond states are thus much better approximations to the true ground state than the Néel state.

8.6.5.8 Majumdar–Ghosh spin chain

We now can ask if there is a Hamiltonian for which a given state or class of states provides the ground state? Hamiltonians with this property are called parent Hamiltonians.

This question cannot be answered in full generality. We focus on the valence bond states $|\pm\rangle$ introduced in the previous section, and hence on a one-dimensional spin problem. For these states, Majumdar and Ghosh (1969*a*; 1969*b*) and Majumdar (1970) have constructed a spin Hamiltonian for N spins of magnitude $s = 1/2$ with nearest- and next-nearest-neighbour interaction

$$\mathcal{H}_{\text{MG}} = \sum_i (\mathcal{J}_1 \mathbf{s}_i \cdot \mathbf{s}_{i+1} + \mathcal{J}_2 \mathbf{s}_i \cdot \mathbf{s}_{i+2}), \tag{8.455}$$

which we shall investigate for the special point $\mathcal{J}_2 = \mathcal{J}_1/2 \equiv \mathcal{J}/2$, the so-called Majumdar–Ghosh point, i.e.

$$\mathcal{H}_{\mathrm{MG}} = \mathcal{J} \sum_i \left(\mathbf{s}_i \cdot \mathbf{s}_{i+1} + \frac{1}{2} \mathbf{s}_i \cdot \mathbf{s}_{i+2} \right) \tag{8.456}$$

with $\mathcal{J} > 0$ and periodic boundary conditions $\mathbf{s}_{N+1} = \mathbf{s}_1$. The spins of both terms in the Hamiltonian, the nearest-neighbour and next-nearest neighbour terms, are thus coupled antiferromagnetically but the latter cannot be accommodated on a bipartite lattice. This can lead to a situation where the antiferromagnetic bonds, the nearest-neighbour and the next–nearest-neighbour bond, cannot both be satisfied. This phenomenon is called *frustration* and is rather common in spin systems.

In order to demonstrate how the construction works, let us rewrite the Majumdar–Ghosh Hamiltonian (8.456) in a way such that we sum only over odd sites but make it manifest why it is tailor-made for the valence bond states $|\pm\rangle$

$$\mathcal{H}_{\mathrm{MG}} = \mathcal{J} \sum_{i\,\mathrm{odd}} \left(\mathbf{s}_i \cdot \mathbf{s}_{i+1} + \frac{1}{2} \left(\mathbf{s}_i + \mathbf{s}_{i+1} \right) \mathbf{s}_{i-1} + \frac{1}{2} \left(\mathbf{s}_i + \mathbf{s}_{i+1} \right) \mathbf{s}_{i+2} \right). \tag{8.457}$$

In a singlet state the spin vector $\mathbf{s}_i + \mathbf{s}_{i+1}$ vanishes

$$\left(\mathbf{s}_i + \mathbf{s}_{i+1} \right) |+\rangle = 0 \quad \Leftrightarrow \quad \left(s_i^\alpha + s_{i+1}^\alpha \right) |+\rangle = 0, \quad \alpha = x, y, z. \tag{8.458}$$

The remaining part of the Hamiltonian (8.457) gives an energy $E_{\mathrm{odd}} = -3N\mathcal{J}/8$ because $\mathrm{s}_{i+1}|+\rangle = -\mathrm{s}_i|+\rangle$ and thus shows that $|+\rangle$ is an eigenstate of the Majumdar–Ghosh Hamiltonian. The same argument can be carried through for the other valence bond singlet state $|-\rangle$.

We must now demonstrate that the valence bond states $|\pm\rangle$ are ground states of the Majumdar–Ghosh Hamiltonian $\mathcal{H}_{\mathrm{MG}}$.

In order to do so, it is convenient that we first define a new spin vector encompassing three sites

$$\mathbf{g}_i = \mathbf{s}_{i-1} + \mathbf{s}_i + \mathbf{s}_{i+1} \tag{8.459}$$

for which the usual quantization condition holds

$$\mathbf{g}_i \cdot \mathbf{g}_i = g_i (g_i + 1). \tag{8.460}$$

The quantum numbers g_i can have the two values $g_i = 1/2$ or $g_i = 3/2$. Next, we define a projection operator

$$P_i = \frac{1}{3} \left(\mathbf{g}_i \cdot \mathbf{g}_i - \frac{3}{4} \right) \tag{8.461}$$

whose eigenvalues follow from those of $\mathbf{g}_i$

$$\frac{1}{3}\left(g_i(g_i+1)-\frac{3}{4}\right)=\begin{cases}0 & \text{for} \quad g_i=1/2\\ 1 & \text{for} \quad g_i=3/2.\end{cases} \tag{8.462}$$

From this equation, we see that $P_iP_i = P_i$ and P_i is indeed a projection operator. Moreover, it projects onto the subspace where the three successive spins combined together in the spin operator $\mathbf{g}_i$ have spin $g_i = 3/2$.

Finally, we connect the projector P_i with the Majumdar–Ghosh spin Hamiltonian. Observing that

$$P_i=\frac{1}{3}\left\{(\mathbf{s}_{i-1}+\mathbf{s}_i+\mathbf{s}_{i+1})\cdot(\mathbf{s}_{i-1}+\mathbf{s}_i+\mathbf{s}_{i+1})-\frac{3}{4}\right\} \tag{8.463}$$

$$=\frac{1}{2}+\frac{2}{3}\left(\mathbf{s}_{i-1}\cdot\mathbf{s}_i+\mathbf{s}_{i-1}\cdot\mathbf{s}_{i+1}+\mathbf{s}_i\cdot\mathbf{s}_{i+1}\right), \tag{8.464}$$

we can rewrite the Majumdar–Gosh spin Hamiltonian in terms of the projection operators (8.461)

$$\mathcal{H}_{\mathrm{MG}}=\frac{3}{4}\mathcal{J}\sum_i P_i-\frac{3\mathcal{J}N}{8}. \tag{8.465}$$

The advantage of this form of the Hamiltonian is that we can immediately evaluate it for the valence bond states $|\pm\rangle$. The projection operator P_i projects onto states with spin 3/2. However, the valence bond states $|\pm\rangle$ contain only singlet states with spin 0. Therefore, we obtain the simple result

$$P_i|\pm\rangle=0, \tag{8.466}$$

which makes the introduction of the projection operators worthwhile. Only the constant term of the Majumdar–Ghosh Hamiltonian gives a contribution

$$\mathcal{H}_{\mathrm{MG}}|\pm\rangle=-\frac{3\mathcal{J}N}{8}|\pm\rangle, \tag{8.467}$$

which again shows that the valence bond states $|\pm\rangle$ are indeed eigenstates of this Hamiltonian with eigenvalue $-\,3\mathcal{J}N/8<0$.

What remains for us is to demonstrate that these states are indeed ground states of the Hamiltonian. For this, we observe from (8.462) that the possible eigenvalues of $\frac{3}{4}\mathcal{J}\sum_i P_i$ are non-negative: $0, \frac{3}{4}\mathcal{J}, 2\frac{3}{4}\mathcal{J}, \ldots$. Thus, no other state can produce an expectation value that is smaller than the one we obtain for the valence bond states $|\pm\rangle$. This conclusion does not, however, exclude the possibility that there could be another eigenstate different from $|\pm\rangle$ that has the same energy and would thus be degenerate with the ground states $|\pm\rangle$.

8.6.6 Spin wave theory

This section introduces a general method widely applicable for different spin systems describing collective magnetism. It provides results for the energies of the elementary excitations or spin waves above the ground states of these spin systems. Knowledge of these energetically low-lying excitations is indispensable for an understanding of the properties of magnetic matter at low temperatures. The method, referred to as spin wave theory, can be applied to ferromagnets and antiferromagnets and is applicable in any dimension and for any value of the spin magnitude s, not just for spin $s = 1/2$. We have encountered spin wave theory in a simple situation already in exercise 3.10.

Here, we look at the ferromagnetic and antiferromagnetic cases in turn. The study of the ferromagnetic spin waves is less involved and yields exact results in most cases because we start from an exact ground state (cp. section 8.6.5.1). The antiferromagnetic spin waves require a more complicated analysis and the results are only approximate.

Despite not being a generally exact method, spin wave theory has a number of attractive features, including that

- it can be applied to many different spin Hamiltonians, in any number of spatial dimensions, and for arbitrary spin;
- nevertheless, it is relatively straightforward to implement;
- approximate results for non-zero temperature can be derived from it;
- the picture that emerges for the excitations can be easily interpreted physically.

8.6.6.1 Spin waves in ferromagnetic matter

The spin wave theory for ferromagnets presented here goes back to Bloch (1930). Bloch's work has been a major motivation for Bethe's later work and it is therefore worthwhile to demonstrate Bloch's treatment of the spin wave problem. The next section presents the spin wave theory for the antiferromagnetic case using a different method from Holstein and Primakoff (1940). We could have used the latter method here for the ferromagnet (cf. the exercise at the end of the next section).

Section 8.6.5.1 discussed how the ground state of a Heisenberg model (8.406), whose exchange couplings are all non-negative, $\mathcal{J}_{ij} \geq 0$, is the completely polarized state (8.411) where all spins are aligned with their s^z-component maximal, i.e. $m_i = s$ on all sites i

$$|\Psi_{\mathrm{gs}}\rangle = |s, m_1 = s\rangle |s, m_2 = s\rangle \dots |s, m_N = s\rangle. \tag{8.468}$$

The excitations above this ground state are of paramount interest. From our discussion in section 8.6.5.1 it is tempting to investigate the excited states obtained by lowering in the ground state (8.468) the state of one spin by unity, e.g.

$$|i\rangle = s_i^- |\Psi_{\mathrm{gs}}\rangle = |s, m_1 = s\rangle |s, m_2 = s\rangle \dots, |s, m_i = s-1\rangle, \dots |s, m_N = s\rangle. \tag{8.469}$$

Due to the spin-flip terms $s_i^+ s_j^- + s_i^- s_j^+$ in the Heisenberg Hamiltonian (8.406), this cannot in general be an eigenstate. Moreover, the excitation energy of the state

$$\Delta E = \langle i|\mathcal{H}|i\rangle - \langle \Psi_{\rm gs}|\mathcal{H}|\Psi_{\rm gs}\rangle \tag{8.470}$$

turns out to be very high of the order of the critical temperature T_c where the ferromagnet loses its magnetization due to thermal fluctuations.

EXERCISE 8.19 Energy of spin-flip excitation Calculate the spin-flip excitation energy (8.470) for the spin-flip state (8.469).

This result raises the question whether there are other excitations whose energy above the ground state energy is lower, possibly even connecting to the ground state energy without an energy gap.

In response, we write the spin operator for a single spin in the Heisenberg picture

$$\mathbf{s}_i(t) = e^{i\mathcal{H}t}\mathbf{s}_i e^{-i\mathcal{H}t} \tag{8.471}$$

from which we obtain the Heisenberg equation of motion for the spin operator

$$\frac{d}{dt}\mathbf{s}_l = i\,[\mathcal{H}, \mathbf{s}_l]. \tag{8.472}$$

If we extend the Heisenberg Hamiltonian (8.406) to include the effect of a homogeneous external magnetic field $\mathbf{h}$ (in suitable units)

$$\mathcal{H} = -\frac{1}{2}\sum_{ij} \mathcal{J}_{ij}\, \mathbf{s}_i \cdot \mathbf{s}_j - \mathbf{h}\cdot\sum_i \mathbf{s}_i = -\frac{1}{2}\sum_{ij} \mathcal{J}_{ij}\, \mathbf{s}_i \cdot \mathbf{s}_j - h\sum_i s_i^z \tag{8.473}$$

where we have chosen the magnetic field to be along the z-direction. Using this Hamiltonian to calculate the commutator in the Heisenberg equation of motion (8.472), we obtain

$$\frac{d}{dt}\mathbf{s}_l = i\,[\mathcal{H}, \mathbf{s}_l] = -\mathbf{h}_l \times \mathbf{s}_l \tag{8.474}$$

where

$$\mathbf{h}_l = \sum_j \mathcal{J}_{lj}\mathbf{s}_j + h\mathbf{e}_z. \tag{8.475}$$

In order to derive this result it is helpful to observe that

$$\left[\mathbf{s}_i \cdot \mathbf{s}_j, \mathbf{s}_l\right] = 0 \quad \text{if} \quad l \neq i \quad \text{and} \quad l \neq j \tag{8.476}$$

and also

$$[\mathbf{s}_l \cdot \mathbf{s}_l, \mathbf{s}_l] = 0 \tag{8.477}$$

such that the only remaining contributions come from

$$[\mathbf{s}_i \cdot \mathbf{s}_l, \mathbf{s}_l] \quad \text{for} \quad i \neq l \quad \text{and} \quad \left[\mathbf{s}_l \cdot \mathbf{s}_j, \mathbf{s}_l\right] \quad \text{for} \quad j \neq l. \tag{8.478}$$

The equations of motion for the spin at site l in components is

$$\frac{d}{dt} s_l^x = -\sum_j \mathcal{J}_{lj} \left(s_j^y s_l^z - s_j^z s_l^y \right) + h s_l^y \tag{8.479}$$

$$\frac{d}{dt} s_l^y = -\sum_j \mathcal{J}_{lj} \left(s_j^z s_l^x - s_j^x s_l^z \right) - h s_l^x \tag{8.480}$$

$$\frac{d}{dt} s_l^z = -\sum_j \mathcal{J}_{lj} \left(s_j^x s_l^y - s_j^y s_l^x \right). \tag{8.481}$$

For the low-energy excitations we are interested in, we expect the state of the system to deviate only little from the fully polarized ground state (8.468). We can therefore replace the z-component of the spin at site l everywhere in the equations of motion by its maximal eigenvalue s

$$s_i^z \to s \quad \Rightarrow \quad \frac{d}{dt} s_i^z \approx 0. \tag{8.482}$$

The equations of motion then become

$$\frac{d}{dt} s_l^x = -s \sum_j \mathcal{J}_{lj} \left(s_j^y - s_l^y \right) + h s_l^y \tag{8.483}$$

$$\frac{d}{dt} s_l^y = -s \sum_j \mathcal{J}_{lj} \left(s_l^x - s_j^x \right) - h s_l^x \tag{8.484}$$

or, using spin raising and lowering operators $s_i^{\pm} = s_i^x \pm i s_i^y$

$$\frac{d}{dt} s_l^{\pm} = \mp i \left\{ s \sum_j \mathcal{J}_{lj} \left(s_l^{\pm} - s_j^{\pm} \right) + h s_l^{\pm} \right\}. \tag{8.485}$$

These two equations can be solved by Fourier transform

$$s_{\mathbf{q}}^{\pm} = \frac{1}{\sqrt{N}} \sum_i s_i^{\pm} e^{-i\mathbf{q}\cdot\mathbf{R}_i} \tag{8.486}$$

$$s_i^{\pm} = \frac{1}{\sqrt{N}} \sum_{\mathbf{q}} s_{\mathbf{q}}^{\pm} e^{i\mathbf{q}\cdot\mathbf{R}_i} \tag{8.487}$$

where, for simplicity, we assume the sites i to lie on a Bravais lattice $\mathbf{R}_i$.

EXERCISE 8.20 Spin wave dispersion Solve the equations of motion (8.485) by Fourier transform and derive the spin wave dispersion relation $\omega(\mathbf{q})$.

The solution of the exercise are plane waves

$$\mathbf{s}_i^{\pm}(t) = \frac{1}{\sqrt{N}} \sum_{\mathbf{q}} \mathbf{s}_{\mathbf{q}}^{\pm}(0) \exp\left(i\mathbf{q}\cdot\mathbf{R}_i \mp i\omega(\mathbf{q})t + i\phi_{\mathbf{q}}\right) \tag{8.488}$$

with an arbitrary phase $\phi_{\mathbf{q}}$ and the spin wave dispersion

$$\omega(\mathbf{q}) = s\left(\mathcal{J}(0) - \mathcal{J}(\mathbf{q})\right) + h \tag{8.489}$$

and the Fourier transformed exchange couplings

$$\mathcal{J}(\mathbf{q}) = \sum_{\mathbf{R}_i-\mathbf{R}_j} \mathcal{J}_{\mathbf{R}_i-\mathbf{R}_j} e^{-i\mathbf{q}(\mathbf{R}_i-\mathbf{R}_j)} \tag{8.490}$$

where we have assumed that the exchange couplings $\mathcal{J}_{ij}$ depend only on the distance between sites $\mathbf{R}_i - \mathbf{R}_j$: $\mathcal{J}_{ij} = \mathcal{J}_{\mathbf{R}_i-\mathbf{R}_j}$.

The spin wave dispersion relation can be further evaluated for a nearest-neighbour exchange on a simple cubic lattice where

$$\mathcal{J}(\mathbf{q}) = \sum_{\langle ij\rangle} \mathcal{J} e^{-i\mathbf{q}\cdot(\mathbf{R}_i-\mathbf{R}_j)} = 2\mathcal{J}\left(\cos(q_x a) + \cos(q_y a) + \cos(q_z a)\right), \tag{8.491}$$

and hence in the absence of a magnetic field,

$$\omega(\mathbf{q}) = 2\mathcal{J}s\left(3 - \cos(q_x a) - \cos(q_y a) - \cos(q_z a)\right). \tag{8.492}$$

For small $\mathbf{q}$ we thus obtain the typical ferromagnetic q^2-behaviour

$$\omega(\mathbf{q}) = \mathcal{J}sa^2q^2. \tag{8.493}$$

The Heisenberg equation of motion (8.472) for the spin operator at site i can be readily rewritten in Fourier transformed form

$$\frac{d}{dt}s_{\mathbf{q}}^{-} = i\omega(\mathbf{q})s_{\mathbf{q}}^{-} = i\left[\mathcal{H}, s_{\mathbf{q}}^{-}\right], \tag{8.494}$$

which yields when applied to the fully polarized ferromagnetic ground state (8.411)

$$\mathcal{H}\left(s_{\mathbf{q}}^{-}|\Psi_{\text{fp}}\rangle\right) = \left(E_{\text{gs}} + \omega(\mathbf{q})\right)\left(s_{\mathbf{q}}^{-}|\Psi_{\text{fp}}\rangle\right). \tag{8.495}$$

In the approximation we have used, essentially replacing the z-component of spin s_i^z by its maximal value s, the interpretation of this equation is that $s_{\mathbf{q}}^{-}$ has created one spin wave quantum, called a magnon, with energy $\omega(\mathbf{q})$ and that $s_{\mathbf{q}}^{-}|\Psi_{\text{fp}}\rangle$ is an spin wave eigenstate of the Heisenberg Hamiltonian $\mathcal{H}$ with eigenvalue $E_{\text{gs}} + \omega(\mathbf{q})$. Spin wave states with more than one magnon, e.g. the two-magnon state $s_{\mathbf{q}}^{-}s_{\mathbf{q}'}^{-}|\Psi_{\text{fp}}\rangle$, are also eigenstates within the approximation we employed in this section and which therefore is equivalent to an approximation of non-interacting magnons. Beyond this approximation magnons will interact with each other. For more details on these issues and other developments of ferromagnetic spin wave theory, e.g. the low-energy thermodynamics of magnons (see exercise 8.21) or magnon-magnon interactions, we refer to the literature.

EXERCISE 8.21 Average magnetization of the ferromagnet in three dimensions at low temperature At low temperatures, we expect the approximation of non-interacting magnons to be reasonably well justified. We can write for the energy of n magnons assuming this approximation, i.e. for n non-interacting magnons

$$E_n = E_0 + n\omega(\mathbf{q}). \tag{8.496}$$

Calculate the average number $\langle n_{\mathbf{q}}\rangle$ of magnons with wave vector $\mathbf{q}$ and the total number $\langle n\rangle$ of magnons, the latter in the thermodynamic limit by replacing the discrete summation over $\mathbf{q}$ by an appropriate integral over the Brillouin zone.

Each magnon reduces the total spin by one. Thus, the magnetization of $\langle n\rangle$ magnons is

$$\langle m\rangle = \frac{1}{V}\left(Ns - \langle n\rangle\right) \equiv m_{\text{sat}} - \frac{1}{V}\langle n\rangle \tag{8.497}$$

with the saturation magnetization m_{sat} in the absence of any magnons and V the volume of the system.

Calculate the magnetization $\langle m\rangle$ using the fact that the integral determining the average number of magnons $\langle n\rangle$ is dominated by the contributions with small momenta q. Extract from your result the low temperature behaviour of the magnetization $\langle m\rangle$. This result is Bloch's law, named after Felix Bloch.

A useful integral:

$$\zeta\left(\frac{3}{2}\right) = \frac{1}{\Gamma\left(\frac{3}{2}\right)} \int_0^\infty dx \frac{\sqrt{x}}{e^x - 1} \tag{8.498}$$

where $\zeta(s)$ is the Riemann ζ function and $\gamma(s)$ is the Gamma function with, especially

$$\Gamma\left(\frac{3}{2}\right) = \frac{\sqrt{\pi}}{2}. \tag{8.499}$$

We now explore spin waves in Heisenberg models with antiferromagnetic exchange interactions.

8.6.6.2 Spin waves in antiferromagnetic matter

From our discussion in section 8.6.5.2 we expect that describing excitations in the antiferromagnetic case will not be as easy as in the ferromagnetic case. Indeed, we restrict ourselves to bipartite models without frustration and more specifically mostly to bipartite nearest-neighbour models. Under these restrictions, the fully spin polarized Néel state, i.e. the state where the two sub-lattices are fully polarized in opposite directions, provides the ground state in the mean-field approximation, which is not the true ground state as we know.

Similar to the transformation to Schwinger Bosons introduced in section 8.6.5.4, the basic idea of the spin wave theory for antiferromagnets (and ferromagnets, too, see exercise 8.25) is to replace spin operators by Bosonic operators. For spin operators on different lattice sites, this is easy to achieve; the problem is how to express the spin operators in terms of Bosons on the same lattice site.

Despite starting from an only approximately correct ground state to describe the low-lying excitations, we observe that the transformation (which we shall be using) to Holstein–Primakoff Bosons based on the Néel state not only yields good results but also furnishes an improved ground state.

We introduce another transformation of spin operators to Bosons, the Holstein–Primakoff transformation, which is similar to the Schwinger Boson transformation in section 8.6.5.4. For sub-lattice A and site index $i \in A$ we define Bose creation and annihilation operators $a_i{}^\dagger$ and a_i

$$s_i^+ = \sqrt{2s - a_i^\dagger a_i}\, a_i \tag{8.500}$$

$$s_i^- = a_i^\dagger \sqrt{2s - a_i^\dagger a_i} \tag{8.501}$$

$$s_i^z = s - a_i^\dagger a_i \tag{8.502}$$

while for sub-lattice B and site index $j \in B$ we define Bose creation and annihilation operators $b_j{}^\dagger$ and b_j

$$s_j^+ = b_j^\dagger \sqrt{2s - b_j^\dagger b_j} \tag{8.503}$$

$$s_j^- = \sqrt{2s - b_j^\dagger b_j}\, b_j \tag{8.504}$$

$$s_j^z = b_j^\dagger b_j - s. \tag{8.505}$$

The Holstein–Primakoff representation can also be employed for other purposes, e.g. the discussion of magnon-magnon interactions in ferromagnets.

In order to apply the Holstein–Primakoff transformation, we first focus on the transformation (8.500-8.502) for the spin operators on the lattice sites i of the sub-lattice A. For the Holstein–Primakoff transformation (8.503-8.505) for the spin operators on the lattice sites j of the sub-lattice B the argument will require only slight modification. We recall that we have for the spin operators

$$s_i^\pm |s, m_i\rangle = \sqrt{s(s+1) - m_i(m_i \pm 1)}|s, m_i \pm 1\rangle \tag{8.506}$$

$$s_i^z |s, m_i\rangle = m_i |s, m_i\rangle. \tag{8.507}$$

We now introduce Boson creation and annihilation operators $a_i{}^\dagger$ and a_i for each lattice site i, which reproduce most of these properties. Of the various possibilities to do this we select arguably the most useful and most widely used Holstein–Primakoff transformation.

As the vacuum state $|0\rangle$ of the Bosons at lattice site i that satisfies $a_i{}^\dagger a_i|0\rangle = 0$ we interpret the state with maximal z-component of spin $|s, m_i = s\rangle$. Furthermore the number of Bosons in an arbitrary state $|s, m_i\rangle$ is taken to be the number of deviations $s - m_i$ from the state $|s, m_i = s\rangle$, i.e.

$$a_i^\dagger a_i |s, m_i\rangle = (s - m_i)\, |s, m_i\rangle. \tag{8.508}$$

Thus, we obtain (8.502)

$$s_i^z = \left(s - a_i^\dagger a_i\right) |s, m_i\rangle = m_i |s, m_i\rangle. \tag{8.509}$$

The state $a_i|s, m_i\rangle$ is the state where the deviation from maximal spin z-component has been decreased by unity, i.e. the s_i^z quantum number m_i is increased by one

$$a_i |s, m_i\rangle = A_{m_i} |s, m_i + 1\rangle, \tag{8.510}$$

which gives the normalization in the usual way

$$\langle s, m_i | a_i^\dagger a_i | s, m_i\rangle = (s - m_i)\, \langle s, m_i | s, m_i\rangle = |A_{m_i}|^2 \langle s, m_i - 1 | s, m_i - 1\rangle \tag{8.511}$$

and hence, choosing A_{m_i} to be real,

$$A_{m_i} = \sqrt{S - m_i}. \tag{8.512}$$

Similarly we get for

$$a_i^\dagger |s, m_i\rangle = B_{m_i} |s, m_i - 1\rangle \tag{8.513}$$

that

$$B_{m_i} = \sqrt{S - m_i}. \tag{8.514}$$

But, how do we rewrite the spin ladder operators $s_i^\pm$ in terms of the Boson operators in order to reproduce (8.506)? The spin ladder operators $s_i^\pm$ must be proportional to the Boson annihilation and creation operators a_i and $a_i{}^\dagger$, respectively. The choices (8.500) and (8.501) satisfy these criteria and moreover disconnect the physical from the unphysical subspaces (see exercise 8.22).

EXERCISE 8.22 Holstein–Primakoff representation Consider only the Holstein–Primakoff transformation as defined in (8.500–8.502), which can be used for the spin wave theory of ferromagnets (cp. also exercise 8.25).

Show that, in order to be physical, we have to require of the Holstein–Primakoff Bosons the constraint

$$n_i = a_i^\dagger a_i \leq 2s. \tag{8.515}$$

Show that the definition of the Holstein–Primakoff Bosons implies that

$$s_i^- |s, m_i = -s\rangle = 0, \tag{8.516}$$

$$s_i^+ |s, m_i = -s - 1\rangle = 0, \tag{8.517}$$

and that, thus the physical, $n_i \leq 2s$, and the unphysical, $n_i \geq 2s + 1$, subspaces are not connected.

The vacuum state $|0\rangle$ on the A and B sub-lattices satisfies

$$a_i^\dagger a_i |0\rangle = b_j^\dagger b_j |0\rangle = 0, \tag{8.518}$$

which corresponds, using (8.502) and (8.505), to the Néel state

$$|0\rangle = \prod_{i \in A} |s, m_i = s\rangle \prod_{j \in B} |s, m_j = -s\rangle. \tag{8.519}$$

For the spins on the sub-lattice B, the above arguments have to be modified. On sub-lattice B, we identify as vacuum state the state $|s, m_j = -s\rangle$. The number of deviations from this state, which we again identify with the number of Bosons, this time on sub-lattice B, for a general state $|s, m_j\rangle$ is $s + m_j$. Then, the same arguments as above lead to the Holstein–Primakoff transformation (8.503–8.505) on sub-lattice B.

For later use, we note that the energy in the Néel state of the nearest-neighbour antiferromagnetic ($\mathcal{J} > 0$) Heisenberg Hamiltonian

$$\mathcal{H} = \mathcal{J} \sum_{\langle ij \rangle} \mathbf{s}_i \cdot \mathbf{s}_j, \tag{8.520}$$

to whose investigation we now turn, is given by

$$E_0^{\mathrm{N}} = \langle 0|\mathcal{H}|0\rangle = \mathcal{J} \sum_{\langle ij \rangle} \langle 0|\mathbf{s}_i \cdot \mathbf{s}_j|0\rangle = -\mathcal{J}s^2 z \frac{N}{2} = -dN\mathcal{J}s^2 \tag{8.521}$$

where the last equality holds for a simple cubic lattice in d dimensions for which the number of nearest neighbours is $z = 2d$.

Rewritten in terms of Holstein–Primakoff Bosons, the Heisenberg Hamiltonian (8.520), expanded in terms of $1/s$ up to $(1/s)^{-1}$, becomes

$$\mathcal{H} = -\frac{Nz\mathcal{J}s^2}{2} + \mathcal{J}s \sum_{\langle ij \rangle} \left(a_i^\dagger a_i + b_i^\dagger b_i + a_i b_j + a_i^\dagger b_j^\dagger \right) + \mathcal{O}(1). \tag{8.522}$$

The terms we dropped in this $1/s$-expansion of the Hamiltonian are interacting terms involving four Bose operators and higher interacting terms.

EXERCISE 8.23 Higher order terms in 1/*s* Give the expansion of $\mathcal{H} = \mathcal{J} \sum_{\langle ij \rangle} \mathbf{s}_i \cdot \mathbf{s}_j$ up to order $\mathcal{O}(1)$ in $1/s$.

The Hamiltonian (8.522) can be diagonalized in two steps: first, a Fourier transformation

$$a_i = \sqrt{\frac{2}{N}} \sum_{\mathbf{k}} a_{\mathbf{k}} e^{i\mathbf{k}\cdot\mathbf{R}_i}, \tag{8.523}$$

$$b_i = \sqrt{\frac{2}{N}} \sum_{\mathbf{k}} b_{\mathbf{k}} e^{-i\mathbf{k}\cdot\mathbf{R}_j}. \tag{8.524}$$

Note that we have introduced opposite signs in the two exponentials of the Fourier transforms and that the N sites of the lattice are divided into $N/2$ sites for each of the sub-lattices A and B. The Fourier-transformed Hamiltonian is

$$\mathcal{H} = -\frac{Nz\mathcal{J}s^2}{2} + \mathcal{J}s \sum_{\mathbf{k}} \sum_{j\in \mathrm{nn}(i)} \left(a_{\mathbf{k}}^{\dagger} a_{\mathbf{k}} + b_{\mathbf{k}}^{\dagger} b_{\mathbf{k}} + a_{\mathbf{k}} b_{\mathbf{k}} e^{-i\mathbf{k}\cdot(\mathbf{R}_j - \mathbf{R}_i)} \right.$$
$$\left. + \, a_{\mathbf{k}}^{\dagger} b_{\mathbf{k}}^{\dagger} e^{i\mathbf{k}\cdot(\mathbf{R}_j - \mathbf{R}_i)} \right), \tag{8.525}$$

where the inner sum $\sum_{j\in \mathrm{nn}(i)}$ is over the nearest-neighbour sites j of a given site i. This sum is independent of the site i.

For a simple cubic lattice in d dimensions, i.e. a linear chain in $d = 1$, a square lattice in $d = 2$, or a simple cubic lattice in the narrow sense in $d = 3$, we have $z = 2d$ nearest neighbours. The Hamiltonian thus becomes

$$\mathcal{H} = -\frac{Nz\mathcal{J}s^2}{2} + \mathcal{J}sz \sum_{\mathbf{k}} \left(a_{\mathbf{k}}^{\dagger} a_{\mathbf{k}} + b_{\mathbf{k}}^{\dagger} b_{\mathbf{k}} + \rho_{\mathbf{k}} a_{\mathbf{k}} b_{\mathbf{k}} + \rho_{\mathbf{k}} a_{\mathbf{k}}^{\dagger} b_{\mathbf{k}}^{\dagger} \right) \tag{8.526}$$

where

$$\rho_{\mathbf{k}} = \frac{1}{d} \sum_{n=1}^{d} \cos(k_n a). \tag{8.527}$$

In the second step, this bilinear Hamiltonian can finally be diagonalized employing a Bogoliubov transformation

$$a_{\mathbf{k}} = \cosh\theta_{\mathbf{k}} A_{\mathbf{k}} + \sinh\theta_{\mathbf{k}} B_{\mathbf{k}}^{\dagger}, \tag{8.528}$$

$$b_{\mathbf{k}} = \sinh\theta_{\mathbf{k}} A_{\mathbf{k}}^{\dagger} + \cosh\theta_{\mathbf{k}} B_{\mathbf{k}}, \tag{8.529}$$

where, in order to have mixed terms containing $A_{\mathbf{k}} B_{\mathbf{k}}$ and $A_{\mathbf{k}}{}^{\dagger} B_{\mathbf{k}}{}^{\dagger}$ vanish, the condition

$$\rho_{\mathbf{k}} = -\tanh 2\theta_{\mathbf{k}} \tag{8.530}$$

must be imposed. The Hamiltonian, which is an approximate effective Hamiltonian neglecting magnon-magnon interactions, thus takes the diagonalized form

$$\mathcal{H} = -\frac{Nz\mathcal{J}s(s+1)}{2} + \sum_{\mathbf{k}} \omega_{\mathbf{k}} \left(A_{\mathbf{k}}^{\dagger} A_{\mathbf{k}} + \frac{1}{2} + B_{\mathbf{k}}^{\dagger} B_{\mathbf{k}} + \frac{1}{2} \right) \tag{8.531}$$

where the dispersion is given by

$$\omega_{\mathbf{k}} = \mathcal{J}sz\sqrt{1 - \rho_{\mathbf{k}}^2}. \tag{8.532}$$

The Bogoliubov transformed Bosons define a new ground state $|0_{\rm eff}\rangle$, the vacuum state of the approximate Hamiltonian (8.531), by

$$A^\dagger_{\mathbf{k}} A_{\mathbf{k}} |0_{\rm eff}\rangle = B^\dagger_{\mathbf{k}} B_{\mathbf{k}} |0_{\rm eff}\rangle = 0 \tag{8.533}$$

with ground state energy

$$E_{\rm eff} = -\frac{Nz\mathcal{J}s(s+1)}{2} + \sum_{\mathbf{k}} \omega_{\mathbf{k}} = -dN\mathcal{J}s^2 \left(1 + \frac{1}{s}\left(1 - \frac{2}{N}\sum_{\mathbf{k}} \omega_{\mathbf{k}}\right)\right). \tag{8.534}$$

This energy is lower than the energy (8.521) of the Néel state with which we started. However, the correction terms proportional to $1/s$ become smaller for higher dimensions. They have to be calculated numerically except for the case $d = 1$, where an analytical evaluation is possible.

EXERCISE 8.24 Staggered magnetization Derive the staggered magnetization

$$m = \langle s^z_i \rangle|_{i\in A} = -\langle s^z_j \rangle|_{j\in B} \tag{8.535}$$

in the state $|0_{\rm eff}\rangle$.

Show especially that there is a logarithmic divergence for small momenta k in one spatial dimension. This signifies the absence of long-range order even in the ground state of the antiferromagnetic Heisenberg quantum spin chain.

Lastly, we are interested in the dispersion relation for small momenta $\mathbf{k}$ in the non-interacting magnon approximation derived here. We obtain in this limit

$$\omega_{\mathbf{k}} = 2\sqrt{d}\mathcal{J}ska, \tag{8.536}$$

i.e. a linear dispersion for the antiferromagnetic spin waves in contrast to the quadratic dispersion (8.493) obtained for the ferromagnetic spin waves.

EXERCISE 8.25 Spin wave theory of the ferromagnet using the Holstein–Primakoff transformation In the case of the ferromagnetic Heisenberg model, we can also apply the method of Holstein–Primakoff Bosons to obtain the spin wave spectrum or dispersion relation. The starting point, the fully polarized state (8.468), is the same. Assume a nearest-neighbour Heisenberg Hamiltonian. Instead of the Heisenberg equation of motion for the spin operators, employ the Holstein–Primakoff transformation but only for one species of Bosons, e.g. a_i. Furthermore use the approximation

$$\sqrt{1 - \frac{a^\dagger_i a_i}{2s}} \approx 1, \tag{8.537}$$

which can be regarded as 'simple' spin wave theory. The Heisenberg Hamiltonian bilinear in the Holstein–Primakoff Bosons can be diagonalized by Fourier transform. Will a Bogoliubov transformation be needed?

Once you have the general dispersion relation for the nearest-neighbour Heisenberg Hamiltonian, evaluate it for the linear spin chain and furthermore for $s = 1/2$.

Remark: a more sophisticated 'interacting' spin wave theory would use the approximation

$$\sqrt{1 - \frac{a_i^\dagger a_i}{2s}} \approx 1 - \frac{a_i^\dagger a_i}{2s}. \tag{8.538}$$

8.6.6.3 Antiferromagnetic Heisenberg quantum spin chain

In view of the related discussions of its Bethe ansatz solution, it is necessary to understand some general results obtained for the antiferromagnetic Heisenberg quantum spin chain.

Theorem 7 Lieb, Schultz, and Mattis Consider the Heisenberg model with nearest-neighbour antiferromagnetic interaction $\mathcal{J} > 0$ on a one-dimensional chain with periodic boundary conditions, i.e.

$$\mathcal{H} = \mathcal{J} \sum_{i=1}^{N} \mathbf{s}_i \cdot \mathbf{s}_{i+1} \tag{8.539}$$

where $\mathbf{s}_{N+1} = \mathbf{s}_1$ and the number of sites N even.

For half-odd integer spin $s = 1/2, 3/2, \ldots$, there is an excited state whose energy E_1 approaches the ground state energy E_{gs} in the thermodynamic limit $N \to \infty$.

For a proof we recommend Lieb *et al.*, 1961 paper. The proof only works for half-odd integer spin magnitudes and leaves open what would happen for integer spin.

1983*a*; 1983*b* addressed the question of integer spin and conjectured that there is a qualitative difference between antiferromagnetic Heisenberg quantum spin chains with half-odd integer and integer spin magnitude s. Haldane predicted the existence of an energy gap between the ground state and the first excited state of linear chains with an isotropic Heisenberg interaction. From the Lieb–Schultz–Mattis theorem 7 this gap is known to vanish in the thermodynamic limit $N \to \infty$ for $s = 1/2$. By contrast, Haldane predicted that for $s = 1$ and any other integer value of s there would be a non-zero gap. Haldane's method used a continuum limit of the chain and the application of field theoretical methods, although these methods were not appropriate to furnish a rigorous proof and left room for scepticism about the conclusions drawn.

However, numerical results by Botet and Jullien (1983), Parkinson and Bonner (1985), and others, and eventually by White and Huse (1993) using the density-matrix renormalization-group (DMRG) technique (which is very accurate especially in one dimension) kept corroborating Haldane's conjecture.

A non-zero gap indicates that the antiferromagnetic Heisenberg spin chain with integer spin exhibits a completely different kind of ground state. While for spin chains with half-odd integer spin the evidence points to an ordered ground state with algebraically decaying spin-spin correlations, for the spin chains with integer spin the ground state is disordered and spin-spin correlations decay exponentially.

8.6.6.4 *Spinons*

While magnons are excitations of the antiferromagnetic nearest-neighbour Heisenberg chain and our discussion in section 8.6.6.2 applies in this case, too, there is another class of fundamental excitations.

Flipping a single spin $s = 1/2$ from the Néel state with a change in the total z-component of spin of $|\Delta s^z| = 1$ suggests itself as a fundamental excitation. However, the energy cost of such a spin flip is high of the order of the exchange interaction $|\mathcal{J}|$. Distributing spin flips over the whole lattice and forming superpositions leads to magnon excitations with much lower energy whose dispersion $\omega(\mathbf{q})$ approaches zero in the limit of small q, i.e. in the long-wavelength limit.

However, in one dimension there is another mechanism. A single spin flipped in the Néel state and represented by

$$\uparrow\downarrow\uparrow\downarrow\uparrow\downarrow\uparrow\downarrow \boxed{\uparrow\uparrow\uparrow} \downarrow\uparrow\downarrow\uparrow\downarrow\uparrow\downarrow\uparrow \tag{8.540}$$

can be split into two kinks or domain walls in the Néel order, i.e.

$$\uparrow\downarrow\uparrow\downarrow\uparrow\downarrow \boxed{\uparrow\uparrow} \downarrow\uparrow\downarrow \boxed{\uparrow\uparrow} \downarrow\uparrow\downarrow\uparrow\downarrow\uparrow \tag{8.541}$$

which can freely move through the lattice

$$\uparrow\downarrow\uparrow\downarrow \boxed{\uparrow\uparrow} \downarrow\uparrow\downarrow\uparrow\downarrow\uparrow\downarrow \boxed{\uparrow\uparrow} \downarrow\uparrow\downarrow\uparrow \tag{8.542}$$

without an additional cost of energy or change of the total spin s^z. Again, like for a spin flip, a single kink costs an energy of the order of $|\mathcal{J}|/2$. Superpositions of kinks over the whole chain lead to excitations whose energy dispersion approaches zero in the long-wavelength limit $q \to 0$ and which are called spinons. They are the counterpart of magnons in spin wave theory, carry spin $s = 1/2$, as can be seen in the spin configurations (8.541) and (8.542), and can thus be viewed as 'half a magnon'. A single spin-flip, i.e. the amount of spin change transferred, for example in neutron scattering experiments, generates a pair of spinons, that then freely propagate as deconfined spin-1/2 'particles'.

We return to the topic of spinons after discussing the Bethe ansatz solution of the Heisenberg quantum spin chain in parts II and III, where we derive the spinon dispersion relation

$$\omega_{\text{spinon}}(\mathbf{q}) = \frac{\pi}{2}\mathcal{J}|\sin q|, \tag{8.543}$$

which is the lower edge of a continuum of excitations whose upper edge is bounded by

$$\omega(\mathbf{q}) = \pi \mathcal{J} \cos(q/2). \tag{8.544}$$

The continuum of excitations develops because spinons are always created in pairs, and therefore the momentum of the two spinons can be distributed in a continuum of different ways. Neutron scattering experiments on quasi-one-dimensional materials like $KCuF_3$ have corroborated the picture outlined here (see, e.g. Tennant *et al.*, 1995).

In dimensions higher than one, separating a flipped spin into a pair of kinks, or a magnon into a pair of spinons, costs energy, which thus confines spinons in dimensions $d \geq 2$.

The next section we constructs the Hubbard model from first principles and then show how the Heisenberg model can be obtained from the Hubbard model for half-filling and in the limit of strong on-site repulsion.

8.7 Hubbard model

The Hubbard model presents one of the simplest ways to obtain an understanding of the mechanisms through which interactions between electrons in a solid can give rise to insulating versus conducting, magnetic, and even novel superconducting behaviour. The preceding sections of this chapter more or less neglected these interaction or correlation effects between the electrons in a solid, or treated them summarily in a mean-field or quasiparticle approach (cf. sections 8.2 to 8.5). While the Hubbard model was first discussed in quantum chemistry in the early 1950s (Pariser and Parr, 1953; Pople, 1953), it was introduced in its modern form and used to investigate condensed matter problems in the 1960s independently by Gutzwiller (1963), Hubbard (1963), and Kanamori (1963). Their proposals of the model were motivated by different physical problem situations. Gutzwiller used the model to study the transition between metallic and insulating phases of solids, Hubbard's research focused on electron correlations in narrow energy bands of transition metals, and Kanamori's objective was the study of itinerant ferromagnetism.

Despite its simplicity, the Hubbard model has proven to be a versatile model with many applications to condensed matter systems. For a careful exposition of the materials and phenomena discussed on the basis of the Hubbard model, especially the high-temperature superconductors, see Fazekas (1999). However, in a recent editorial 2013 on the occasion of the fiftieth anniversary of the work of Gutzwiller, Hubbard, and Kanamori, a second upsurge of interest in the Hubbard model emerged occurred as a result of the experimental possibilities made available by the novel experimental techniques to trap ultracold atoms in optical lattices (see, e.g. Bloch (2005) for a review). These techniques allow the experimental investigation of the original Fermionic Hubbard (Joerdens *et al.*, 2008) model discussed here, but were also used to explore its Bosonic version (Greiner *et al.*, 2002) a number of years earlier.

This section an understanding of how the Hubbard model can be justified from a microscopic 'first principles' perspective. As in section 8.2.1, we begin with the electronic part of the 'Theory of Everything' Hamiltonian Laughlin and Pines (2000); Laughlin (1998) of condensed matter theory

$$\mathcal{H} = \mathcal{H}_{\text{kin}} + \mathcal{H}_{\text{int}} \tag{8.545}$$

$$= \sum_{\sigma} \int d^3 r \psi_{\sigma}^{\dagger}(\mathbf{r}) \left[-\frac{}{2m} \nabla^2 + U_{\text{ion}}(\mathbf{r}) \right] \psi_{\sigma}(\mathbf{r})$$
$$+ \sum_{\sigma,\sigma'} \int d^3 r \int d^3 r' \psi_{\sigma}^{\dagger}(\mathbf{r}) \psi_{\sigma'}^{\dagger}(\mathbf{r}') V_{\text{ee}}(\mathbf{r} - \mathbf{r}') \psi_{\sigma'}(\mathbf{r}') \psi_{\sigma}(\mathbf{r}), \tag{8.546}$$

which, unlike Laughlin in his Nobel presentation 1998, we again have written in second quantized form with field operators $\psi_{\sigma}{}^{\dagger}(\mathbf{r})$.

This very general Hamiltonian describes electrons interacting with the potential $U_{\text{ion}}(\mathbf{r})$ of a static lattice of ions, i.e. employs the Born–Oppenheimer approximation. Thus, we neglect the motion of the ion lattice since we are only interested in the interactions of electrons and not in dynamical lattice effects, such as phonons. Moreover, the electrons interact via Coulomb repulsion

$$V_{\text{ee}}(\mathbf{r} - \mathbf{r}') \propto \frac{1}{|\mathbf{r} - \mathbf{r}'|}. \tag{8.547}$$

In second quantization language, the Hamiltonian (8.546) describes the situation adequately.

The Hamiltonian (8.546) is quite complicated still, and in the following we make a number of approximations to reach a more tractable Hamiltonian. We attempt to make the approximations in such a way that still captures the essential physics of strongly interacting electrons. The method of second quantization introduced and discussed in chapter 2 is very well adapted to this task.

The first ingredient in our analysis is *Bloch's theorem*. The lattice potential $U_{\text{ion}}(\mathbf{r})$ is periodic

$$U_{\text{ion}}(\mathbf{r} + \mathbf{R_i}) = U_{\text{ion}}(\mathbf{r}) \tag{8.548}$$

with periodicity given by a lattice vector $\mathbf{R_i}$ (assuming a primitive lattice with one atom per unit cell)

$$\mathbf{R_i} = i_1 \mathbf{a}_1 + i_2 \mathbf{a}_2 + i_3 \mathbf{a}_{\jmath} = \sum_{\jmath=1}^{3} i_{\jmath} \, \mathbf{a}_{\jmath} \tag{8.549}$$

where $\{\mathbf{a}_{\mathcal{J}}\}$ are a set of basis vectors of the lattice and the numbers $i_{\mathcal{J}} \in \mathbb{Z}$ ($\mathcal{J} = 1, 2, 3$) describe an infinite lattice or a lattice with some appropriately chosen boundary conditions. For the triple of numbers $i_{\mathcal{J}}$ we have already used a vector notation: **i**.

Bloch's theorem states that the solution of the Schrödinger equation in a periodic lattice potential $U_{\text{ion}}(\mathbf{r})$ is given by wave functions that have the form of so-called Bloch functions $u_{\mathbf{k},\alpha}(\mathbf{r})$ and energy eigenvalues in the form of electronic bands $\epsilon_{\mathbf{k},\alpha}$.

The Bloch functions are functions that have been Fourier transformed with respect to the periodic lattice $\mathbf{R_i}$. Their inverse Fourier transformed counterparts are the Wannier functions

$$\phi_{\mathbf{i}\alpha}(\mathbf{r}) \equiv \phi_\alpha(\mathbf{r} - \mathbf{R_i}) = \frac{1}{\sqrt{N}} \sum_{\mathbf{k}} e^{-i\mathbf{k}\cdot\mathbf{R_i}} u_{\mathbf{k},\alpha}(\mathbf{r}), \tag{8.550}$$

which are localized at site $\mathbf{R_i}$ of the ionic lattice. Bloch and Wannier functions are connected by a unitary transformation and provide equivalent descriptions of the physical situation.

We now define creation operators for electrons in a Wannier state

$$c^\dagger_{\mathbf{i}\alpha\sigma} = \int d^3r\, \phi_{\mathbf{i}\alpha}(\mathbf{r}) \psi^\dagger_\sigma(\mathbf{r}), \tag{8.551}$$

which have the inverse relation

$$\psi^\dagger_\sigma(\mathbf{r}) = \sum_{\mathbf{i},\alpha} \phi^*_{\mathbf{i}\alpha}(\mathbf{r}) c^\dagger_{\mathbf{i}\alpha\sigma}. \tag{8.552}$$

These operators, being Fermionic, obey anti-commutation relations

$$\{c_{\mathbf{j}\alpha\sigma}, c^\dagger_{\mathbf{l}\beta\sigma'}\} = \delta_{\mathbf{jl}}\delta_{\alpha\beta}\delta_{\sigma\sigma'}, \quad \{c_{\mathbf{j}\alpha\sigma}, c_{\mathbf{l}\beta\sigma'}\} = \{c^\dagger_{\mathbf{j}\alpha\sigma}, c^\dagger_{\mathbf{l}\beta\sigma'}\} = 0. \tag{8.553}$$

The Hamiltonian 8.546 becomes in the Wannier basis

$$\mathcal{H} = \sum_{\mathbf{ij}\alpha\sigma} t^\alpha_{\mathbf{ij}} c^\dagger_{\mathbf{i}\alpha\sigma} c_{\mathbf{j}\alpha\sigma} + \sum_{\mathbf{ijmn}} \sum_{\alpha\beta\mu\nu} \sum_{\sigma\sigma'} v^{\alpha\beta\mu\nu}_{\mathbf{ijmn}} c^\dagger_{\mathbf{i}\alpha\sigma} c^\dagger_{\mathbf{j}\beta\sigma'} c_{\mathbf{n}\nu\sigma'} c_{\mathbf{m}\mu\sigma} \tag{8.554}$$

with the hopping

$$t^\alpha_{\mathbf{ij}} = \left\langle \mathbf{i}\alpha \left| \left[-\frac{1}{2m}\nabla^2 + U_{\text{ion}}(\mathbf{r}) \right] \right| \mathbf{j}\alpha \right\rangle \tag{8.555}$$

$$= \int d^3r\, \phi^*_{\mathbf{i}\alpha}(\mathbf{r}) \left[-\frac{1}{2m}\nabla^2 + U_{\text{ion}}(\mathbf{r}) \right] \phi_{\mathbf{j}\alpha}(\mathbf{r}) \tag{8.556}$$

and the interaction matrix elements

$$v_{\mathbf{ijmn}}^{\alpha\beta\mu\nu} = \langle \mathbf{i}\alpha, \mathbf{j}\beta | V_{\text{ee}}(\mathbf{r} - \mathbf{r}') | \mathbf{m}\mu, \mathbf{n}\nu \rangle \tag{8.557}$$

$$= \int d^3r \int d^3r' \; \phi_{\mathbf{i}\alpha}^*(\mathbf{r}) \phi_{\mathbf{j}\beta}^*(\mathbf{r}') V_{\text{ee}}(\mathbf{r} - \mathbf{r}') \phi_{\mathbf{m}\mu}(\mathbf{r}) \phi_{\mathbf{n}\nu}(\mathbf{r}'). \tag{8.558}$$

Apart from the Born–Oppenheimer approximation of a static ionic lattice, we have not yet made any further approximation. We just have rewritten the Hamiltonian (8.546) in a form that is more suitable for our purpose of describing interacting localized magnetic moments.

Following Hubbard 1963, Gutzwiller (1963), and Kanamori (1963), we now make a number of simplifying assumptions. Firstly, we assume that all except the lowest band have very high energies and are, thus, energetically unavailable. Therefore, we can drop all band indices α, β, μ, ν, and the corresponding summations. Secondly, we assume that the remaining band has rotational symmetry, i.e. is an s-band. This implies that the hopping matrix elements depend only on the distance between the sites **i** and **j**, i.e.

$$t_{\mathbf{ij}} = t(|\mathbf{R_i} - \mathbf{R_j}|). \tag{8.559}$$

We arrive, thus, at the simplified Hamiltonian

$$\mathcal{H} = \sum_{\mathbf{ij}\sigma} t_{\mathbf{ij}} c_{\mathbf{i}\sigma}^\dagger c_{\mathbf{j}\sigma} + \sum_{\mathbf{ijmn}} \sum_{\sigma\sigma'} v_{\mathbf{ijmn}} c_{\mathbf{i}\sigma}^\dagger c_{\mathbf{j}\sigma'}^\dagger c_{\mathbf{n}\sigma'} c_{\mathbf{m}\sigma}. \tag{8.560}$$

The matrix elements decrease fast with increasing distance $|\mathbf{R_i} - \mathbf{R_j}|$, so that we can restrict the summation over sites to nearest-neighbour sites $\langle \mathbf{ij} \rangle$ and arrive at the *generalized Hubbard model*

$$\begin{aligned}
\mathcal{H} = &-t \sum_{\langle \mathbf{ij} \rangle} \sum_{\sigma} \left(c_{\mathbf{i}\sigma}^\dagger c_{\mathbf{j}\sigma} + c_{\mathbf{j}\sigma}^\dagger c_{\mathbf{i}\sigma} \right) + U \sum_{\mathbf{i}} n_{\mathbf{i}\uparrow} n_{\mathbf{i}\downarrow} + V \sum_{\langle \mathbf{ij} \rangle} n_{\mathbf{j}} n_{\mathbf{j}} \\
&+ X \sum_{\langle \mathbf{ij} \rangle} \sum_{\sigma} \left(c_{\mathbf{i}\sigma}^\dagger c_{\mathbf{j}\sigma} + c_{\mathbf{j}\sigma}^\dagger c_{\mathbf{i}\sigma} \right) \left(n_{\mathbf{i},-\sigma} + n_{\mathbf{j},-\sigma} \right) + \mathcal{J} \sum_{\langle \mathbf{ij} \rangle} \mathbf{S_i} \cdot \mathbf{S_j} \\
&+ Y \sum_{\langle \mathbf{ij} \rangle} \left(c_{\mathbf{i}\uparrow}^\dagger c_{\mathbf{i}\downarrow}^\dagger c_{\mathbf{j}\downarrow} c_{\mathbf{j}\uparrow} + c_{\mathbf{j}\downarrow}^\dagger c_{\mathbf{j}\uparrow}^\dagger c_{\mathbf{i}\uparrow} c_{\mathbf{i}\downarrow} \right),
\end{aligned} \tag{8.561}$$

where the only surviving matrix elements are the hopping matrix element

$$t = -t_{\mathbf{ij}}, \tag{8.562}$$

describing single-electron hopping between nearest-neighbour sites $\mathbf{i}$ and $\mathbf{j}$, the on-site or Hubbard repulsion[6]

$$U = v_{\mathbf{iiii}}, \tag{8.563}$$

the Coulomb interaction between electrons on neighbouring sites

$$V = v_{\mathbf{ijij}}, \tag{8.564}$$

the so-called bond-charge interaction

$$X = v_{\mathbf{iiij}}, \tag{8.565}$$

the spin-spin, or Heisenberg exchange interaction between neighbouring sites

$$\mathcal{J} = -2v_{\mathbf{ijji}}, \tag{8.566}$$

and, finally, the term describing the hopping of pairs of electrons

$$Y = v_{\mathbf{iijj}}. \tag{8.567}$$

The bond-charge interaction X describes the hopping of single electrons where the hopping depends on the occupation of the sites involved. It is proportional to the charge, i.e. the number of electrons, located on the sites of the bond $\langle\mathbf{ij}\rangle$ between sites $\mathbf{i}$ and $\mathbf{j}$.

In the generalized Hubbard Hamiltonian (8.561), we introduced the following operators: the electron number operators at site $\mathbf{i}$

$$n_{\mathbf{i}\sigma} = c^{\dagger}_{\mathbf{i}\sigma} c_{\mathbf{i}\sigma}, \tag{8.568}$$

and

$$n_{\mathbf{i}} = n_{\mathbf{i}\uparrow} + n_{\mathbf{i}\downarrow}, \tag{8.569}$$

and the spin-$\frac{1}{2}$ operators at site $\mathbf{i}$

$$\mathbf{S_i} = \frac{1}{2}\sum_{\alpha,\beta} c^{\dagger}_{\mathbf{i}\sigma} \boldsymbol{\sigma}_{\alpha\beta} c_{\mathbf{i}\sigma}, \tag{8.570}$$

where $\sigma_{\alpha\beta}$ are the components of the vector σ of Pauli matrices (cf. chapter 3).

Some of the symmetries of the generalized Hubbard Hamiltonian (8.561), i.e. some of the operators with which (8.561) commutes, are:

[6] The attractive case is also sometimes considered.

- the number of electrons

$$\mathcal{N} = \sum_{\mathbf{i}} n_{\mathbf{i}}. \tag{8.571}$$

On a lattice of N sites we can at most place $2N$ electrons. Note the important special case of a so-called half-filled band, where in the ground state one has $N_e = \frac{N}{2}$ electrons in the lattice, one electron per site.

- the magnetization

$$\mathcal{M} = \mathcal{N}_{\uparrow} - \mathcal{N}_{\downarrow} = \sum_{\mathbf{i}} \left(n_{\mathbf{i}\uparrow} - n_{\mathbf{i}\downarrow} \right), \tag{8.572}$$

and the

- total spin of the chain

$$\mathbf{S} = \sum_{\mathbf{i}} \mathbf{S}_{\mathbf{i}}. \tag{8.573}$$

In other words, the total number of electrons $\mathcal{N}$, the magnetization $\mathcal{M}$, and the total spin $\mathbf{S}$ are conserved, the latter because the Heisenberg interaction $\mathcal{J}$ is isotropic. This latter property of the generalized Hubbard model can, in principle, be relaxed to allow anisotropic exchange interactions with $\mathcal{J}^x \neq \mathcal{J}^y \neq \mathcal{J}^z$ in the most general case. In fact, the anisotropic Heisenberg model retains all but the spin-spin interaction of the generalized Hubbard model (8.561) . Obviously, in such a model no electron motion is possible that would require at least a hopping term t. Therefore, it cannot describe itinerant magnetism. However, it is a perfectly reasonable model to describe the interaction of localized magnetic moments in magnetic insulators where the magnetism is indeed due to the interaction between local magnetic moments. The one-dimensional version of this model, the Heisenberg quantum spin chain, is one of our prime examples of a model solvable by Bethe ansatz.

EXERCISE 8.26 Generalized Hubbard model Convince yourself explicitly that the generalized Hubbard Hamiltonian (8.561) is obtained from the Hamiltonian (8.560) if we only retain nearest-neighbour interactions.

Hubbard (1963) provided a simple estimate for the order of magnitude of the various matrix elements in the generalized Hubbard model (8.561) in the case of transition metals like iron, cobalt, or nickel: $t \approx 1eV$, $U \approx 10eV$, $V \approx 2 - 3eV$, $X \approx 0.5eV$, and $\mathcal{J}$, $Y \ll 1eV$. Motivated by these orders of magnitude, most attention was henceforth focused on the case where all terms in the generalized Hubbard model (8.561) are neglected except the on-site Coulomb repulsion U, and the single-electron hopping t, the latter being necessary to still describe itinerant magnetism using this model.

The remaining model Hamiltonian

$$\mathcal{H} = -t \sum_{\langle \mathbf{ij} \rangle \sigma} \left(c_{\mathbf{i}\sigma}^{\dagger} c_{\mathbf{j}\sigma} + c_{\mathbf{j}\sigma}^{\dagger} c_{\mathbf{i}\sigma} \right) + U \sum_{\mathbf{i}} n_{\mathbf{i}\uparrow} n_{\mathbf{i}\downarrow} \equiv \mathcal{H}_t + \mathcal{H}_U \tag{8.574}$$

today bears the name Hubbard model, and constitutes arguably one of the models, alongside the Ising model and the Heisenberg model, to which a great deal of research effort in condensed matter physics is devoted. For calculations involving the grand canonical ensemble, a chemical potential μ and a corresponding term in the Hamiltonian

$$\mathcal{H}_{\mu} = -\mu \sum_{\mathbf{i}\sigma} n_{\mathbf{i}\sigma} \tag{8.575}$$

are introduced. Furthermore, an external magnetic field $\mathbf{h} = h\mathbf{e}_z$ will couple to the z-component of the total spin adding a further term

$$\mathcal{H}_h = h \sum_{\mathbf{i}} S_{\mathbf{i}}^z = \frac{h}{2} \sum_{\mathbf{i}} \left(n_{\mathbf{i}\uparrow} - n_{\mathbf{i}\downarrow} \right) \tag{8.576}$$

to the Hubbard Hamiltonian.

An alternative form for the on-site part of the Hubbard Hamiltonian is

$$\mathcal{H}_U' = U \sum_{\mathbf{i}} \left(n_{\mathbf{i}\uparrow} - \frac{1}{2} \right) \left(n_{\mathbf{i}\downarrow} - \frac{1}{2} \right) \tag{8.577}$$

such that

$$\mathcal{H}_U' = \mathcal{H}_U - \frac{U}{2} \sum_{\mathbf{i}} \left(n_{\mathbf{i}\uparrow} + n_{\mathbf{i}\downarrow} \right) + \frac{NU}{4}. \tag{8.578}$$

This form shifts the chemical potential $\mu \to \mu + U/2$ and introduces an overall additive constant $NU/4$ to the energy. However, the particle-hole symmetry discussed in section 8.7.1 is manifest in this form of the Hubbard Hamiltonian.

In summary, it is fair to say that

> the Hubbard model describes—in the simplest possible fashion—a system of interacting electrons. It can be viewed in just this way, as a toy model, or it can be viewed, a bit more realistically in the repulsive case, as a serious model of π-electrons hopping between localized Wannier orbitals in some molecule such as benzene (with $N = 6$) Heilmann and Lieb (1970); the half-filled band (one electron per site) is then especially important because it corresponds to

neutrality. The ultra-short range interaction is supposed to mimic a highly screened Coulomb potential. From the latter viewpoint it was known first in the chemistry literature as the Pariser–Parr–Pople Pariser and Parr (1953); Pople (1953) model; molecules having a bipartite structure were called 'alternant molecules'. It was a decade later that Hubbard (1963), Gutzwiller (1963), and Kanamori (1963) realized its importance for bulk matter—paraphrased from Lieb (1993)

8.7.1 Particle-hole symmetry of the Hubbard model

We now consider the bipartite lattice introduced in section 8.6.6.2.

On bipartite lattices we may consider new Fermionic creation and annihilation operators according to

$$C^{\dagger}_{\mathbf{i}\sigma} = (-1)^{\mathbf{i}} c_{\mathbf{i}\sigma} \qquad \text{and} \qquad C_{\mathbf{i}\sigma} = (-1)^{\mathbf{i}} c^{\dagger}_{\mathbf{i}\sigma} \tag{8.579}$$

where the factor $(-1)^{\mathbf{i}}$ is defined to be $(+1)$ on sub-lattice A and (-1) on the sub-lattice B. These new operators exchange the roles of creation and annihilation Fermions. The transformation to the new operators is called a particle-hole transformation, which is justified by the observation

$$C^{\dagger}_{\mathbf{i}\sigma} C_{\mathbf{i}\sigma} = 1 - c^{\dagger}_{\mathbf{i}\sigma} c_{\mathbf{i}\sigma}. \tag{8.580}$$

Among the symmetries exhibited by the Hubbard model (8.574), the particle-hole symmetry is of particular practical importance, e.g. in quantum Monte Carlo simulations.

This symmetry is manifest for the Hubbard Hamiltonian with the on-site interaction written in the form (8.577)

$$\mathcal{H}_t + \mathcal{H}'_U \to \mathcal{H}_t + \mathcal{H}'_U. \tag{8.581}$$

However, using (8.574) directly produces upon a particle-hole transformation (8.579) extra terms

$$\mathcal{H}_t + \mathcal{H}_U \to \mathcal{H}_t + \mathcal{H}_U + UN - U \sum_{\mathbf{i}} \left(n_{\mathbf{i}\uparrow} + n_{\mathbf{i}\downarrow}\right), \tag{8.582}$$

the first of which is an irrelevant constant. However, the second term acts as a shift in the chemical potential (8.575), which itself behaves as

$$\mathcal{H}_{\mu} \to \mathcal{H}_{-\mu} - 2\mu N \tag{8.583}$$

under a particle-hole transformation.

A simple application of the particle-hole transformation (8.579) of the Hubbard Hamiltonian (8.574) finds for the site occupation number or particle density

$$n(\mu, T) = \frac{1}{N}\langle n_\uparrow + n_\downarrow \rangle = \frac{1}{NZ}\mathrm{Tr}\left[(n_\uparrow + n_\downarrow)e^{-\beta\mathcal{H}_t-\beta\mathcal{H}_U-\beta\mathcal{H}_\mu}\right] = \frac{1}{N\beta}\frac{\partial \ln Z}{\partial \mu} \tag{8.584}$$

with

$$Z = \mathrm{Tr}\left[e^{-\beta\mathcal{H}_t-\beta\mathcal{H}_U-\beta-\beta\mathcal{H}_\mu}\right]. \tag{8.585}$$

Using the transformation behaviour of the various parts of the Hubbard Hamiltonian discussed above, under the particle-hole transformation, the site occupation number becomes

$$\frac{1}{N\beta}\frac{\partial}{\partial\mu}\ln\left[e^{-\beta\mathcal{H}_t-\beta\mathcal{H}_U-\beta\mathcal{H}_{(U-\mu)}-\beta(U-2\mu)N}\right] = -n(U-\mu, T) + 2. \tag{8.586}$$

Hence, we have the relation for the site occupation number

$$n(\mu, T) = 2 - n(U-\mu, T), \tag{8.587}$$

which implies for half-filling, i.e. $n = 1$, that we need $\mu = U/2$ for the particle-hole symmetric Hubbard model independent of temperature.

Exercise 8.27 concerns a similar application of the particle-hole symmetry. See exercise 8.28 for a discussion of further symmetries of the Hubbard Hamiltonian in the special case of one dimension.

....

EXERCISE 8.27 The single-site Hubbard model Consider the Hubbard Hamiltonian (8.574) together with the chemical potential term (8.575) and a magnetic field term (8.576), i.e. $\mathcal{H} = \mathcal{H}_t + \mathcal{H}_U + \mathcal{H}_\mu + \mathcal{H}_h$. Putting the hopping matrix element to zero, $t = 0$, in this Hamiltonian is tantamount to considering a collection of identical single-site Hubbard Hamiltonians, i.e. $\mathcal{H} = \sum_\mathbf{j} \mathcal{H}_\mathbf{j}$.

Calculate the partition function of the single-site Hubbard Hamiltonian

$$Z = \mathrm{Tr}\left[\exp(-\beta\mathcal{H}_\mathbf{j})\right] \tag{8.588}$$

and the site occupation number

$$\rho(\mu, h, T) \equiv \langle n_\uparrow + n_\downarrow \rangle \equiv Z^{-1}\mathrm{Tr}\left[\left(n_\uparrow + n_\downarrow\right)\exp(-\beta\mathcal{H}_\mathbf{j})\right], \tag{8.589}$$

which are, of course, independent of the site index **j**, as a function of chemical potential μ, temperature $T = 1/\beta$, and magnetic field h.

Convince yourself that, in the limit $T \to 0$ and for $h = 0$, the site occupation number $\rho(\mu, T)$ changes in steps: $\rho(\mu) = 0$ for $\mu < -U/2$, $\rho(\mu) = 1$ for $-U/2 < \mu < U/2$,

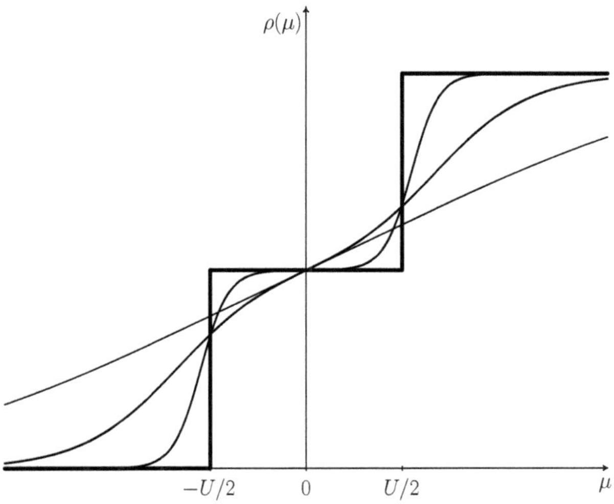

Figure 8.8 *Site occupation number Hubbard model. The so-called Mott plateau between $\mu = -U/2$ and $\mu = U/2$ with occupation number $\rho = 1$, i.e. half-filling, develops for decreasing temperature (thicker lines correspond to lower temperatures). In the limit $\mathsf{T} \to \mathsf{0}$ the occupation is $\rho = 0$ for $\mu < -U/2$. It jumps to $\rho = 1$ at $\mu = -U/2$. A cost in energy of U has to be expended to add a second electron.*

and $\rho(\mu) = 2$ for $\mu > U/2$. For finite temperature T, these steps are washed out but still clearly discernible for low temperatures (see figure 8.8).

....

In one dimension, the Hubbard model can be solved by Bethe ansatz. Investigating the properties of the one-dimensional Hubbard model is therefore a very useful exercise.

....

EXERCISE 8.28 The one-dimensional Hubbard model The Hubbard model in one dimension describes N_e electrons on a lattice of N lattice sites that are allowed to hop to nearest-neighbour sites. Double occupancy of a site with two electrons (which, of course, must have opposite spin orientation) costs an energy U. The second quantized Hamiltonian for this model is

$$\mathcal{H} = -\,t \sum_{\substack{j=1,\\ \sigma=\pm 1}}^{N} \left\{ \psi_\sigma^\dagger(x_j)\psi_\sigma(x_j + a) + \psi_\sigma^\dagger(x_j)\psi_\sigma(x_j - a) \right\}$$
$$+ \frac{U}{2} \sum_{\substack{j=1,\\ \sigma=\pm 1}}^{N} \left\{ \psi_\sigma^\dagger(x_j)\psi_\sigma(x_j)\psi_{-\sigma}^\dagger(x_j)\psi_{-\sigma}(x_j) \right\}, \tag{8.590}$$

where $x_j = ja, j = 1, 2, \ldots N$, and a is the lattice spacing.

The Fermi operators obey anti-commutation relations

$$\left\{\psi_\sigma(x_j), \psi_{\sigma'}^\dagger(x_l)\right\} = \delta_{\sigma,\sigma'}\delta_{jl}. \tag{8.591}$$

a) Diagonalize the hopping part of this Hamiltonian by Fourier transformation

$$\psi_\sigma(x) = \frac{1}{\sqrt{N}}\sum_k \exp(ikx)c_{k,\sigma} \tag{8.592}$$

and determine the allowed k-values assuming periodic boundary conditions $x_{j+N} = x_j$.

b) Show that

$$\mathcal{M}_\uparrow = \sum_j \psi_\uparrow^\dagger(x_j)\psi_\uparrow(x_j) \tag{8.593}$$

and $\mathcal{M}_\downarrow$ defined analogously are conserved quantities, i.e. commute with $\mathcal{H}$. Therefore $\mathcal{M}_\uparrow$ and $\mathcal{M}_\downarrow$ provide quantum numbers $M_\uparrow$ and $M_\downarrow$ of the spectrum of $\mathcal{H}$

$$E = E(M_\uparrow, M_\downarrow; t, U). \tag{8.594}$$

c) Determine the symmetries of $E = E(M_\uparrow, M_\downarrow; t, U)$ under the transformations

$$\psi_\sigma(x_j) = (-1)^j c_{j,\sigma} \tag{8.595}$$

and

$$\psi_\uparrow(x_j) = (-1)^j c_{j,\uparrow}^\dagger, \tag{8.596}$$

$$\psi_\downarrow(x_j) = c_{j,\downarrow}. \tag{8.597}$$

Are the new operators in both cases again Fermi operators?

8.8 Heisenberg model

Section 8.6 on magnetism, especially section 8.6.4 on the electrostatic origin of the exchange interaction in the hydrogen molecule, discussed the Heisenberg model as a model for localized magnetism. There is also another route to this model as an approximative limit starting from the Hubbard model, arguably one of the central models in the physics of strongly interacting quantum matter, and that is connected in the limit of large on-site interaction U to two other important models, the t–$\mathcal{J}$ model and, eventually, the Heisenberg model.

While the Hubbard model describes itinerant magnetic behaviour of the conduction band electrons for arbitrary filling of the band, the Heisenberg model is a model of fully localized strongly interacting spins that arises for an exactly half-filled conduction band. The t–$\mathcal{J}$ model occupies an intermediate position where the conduction band is away from half-filling and electrons can still hop between lattice sites because the penalty for double occupancy of a site is not too large.

In the following two sections we derive these two models. Our starting point is the Hamiltonian $\mathcal{H} = \mathcal{H}_t + \mathcal{H}_U$, (8.574), of the Hubbard model introduced in section 8.7. We focus on the Hubbard model on a bipartite lattice, which simplifies the discussion considerably. There are various techniques available for these derivations, see, e.g. Fazekas (1999), chapters 4 and 5, or Auerbach (1994), chapter 3. The method we choose for the derivation, the Schrieffer–Wolff transformation, was introduced by Schrieffer and Wolff (1966) in connection with quantum impurity models to derive the Kondo model from the Anderson model. However, in order to derive the Kondo Hamiltonian from the Anderson Hamiltonian in section 8.9, we employ a different method.

The Schrieffer–Wolff transformation is a unitary transformation used in many instances to extract the effective low-energy physics of strongly interacting quantum systems. It can be used if the energy scales in a Hamiltonian can be chosen such that only a subset of the states available in principle can be reached for low energies.

Under the condition of strong on-site repulsion $U \gg t$, double occupancy in the Hubbard model is suppressed. Depending then on the filling N_e/N of the N lattice sites with a total number of $N_e = N_\uparrow + N_\downarrow$ electrons we obtain either the t–$\mathcal{J}$ model for filling $N_e/N \neq 1$, i.e. away from half-filling, or the Heisenberg model for filling $N_e/N = 1$, i.e. at half-filling. We concentrate on the case of less than half-filling, $N_e/N < 1$. Close to half-filling most of the N sites of the lattice sites are filled with only one electron in the ground state configuration. At exactly half-filling, single occupancy of all N lattice sites defines the ground state. However, these ground states are highly degenerate.

8.8.1 From the Hubbard to the t–J model: non-half filled band case

We begin by looking at two sites only. By doing this, we obtain an overview of the most important processes possible in the Hubbard model and how they contribute as the parameters U, t, and the filling factor $\nu = N_e/N$ vary.

There are $4^2 = 16$ possible states to consider for two sites with $N_e = 0, 1, 2, 3, 4$ electrons. As discussed in the previous section, the number of electrons can be adjusted by a variation of the chemical potential μ. Here, we assume appropriate values of μ such that the lattice exhibits the desired regimes of electron filling.

For the strong-coupling limit $U \gg t$ we are interested in here, it is natural to start our discussion in the extreme limit, where $t = 0$, i.e. the atomic limit. In this limit, there are no hopping processes. The ground state consists of a lattice of singly occupied lattice sites and is highly degenerate. If we now switch on hopping, i.e. $t > 0$, the ground state degeneracy gets lifted and the four hopping processes

$$\text{(a)}\qquad \underline{\ \uparrow\ }\ \underline{\ \ \ }\ \Longrightarrow\ \underline{\ \ \ }\ \underline{\ \uparrow\ } \tag{8.598}$$

$$\text{(b)}\qquad \underline{\uparrow\downarrow}\ \underline{\downarrow}\ \Longrightarrow\ \underline{\downarrow}\ \underline{\uparrow\downarrow} \tag{8.599}$$

$$\text{(c)}\qquad \underline{\ \uparrow\ }\ \underline{\downarrow}\ \Longrightarrow\ \underline{\ \ \ }\ \underline{\uparrow\downarrow} \tag{8.600}$$

$$\text{(d)}\qquad \underline{\uparrow\downarrow}\ \underline{\ \ \ }\ \Longrightarrow\ \underline{\downarrow}\ \underline{\ \uparrow\ } \tag{8.601}$$

and their spin reversed versions must be considered. While in the hopping processes (a) and (b) the number of doubly occupied states remain unchanged, process (c) creates a new doubly occupied site, and process (d) lowers this number by one. Therefore, only the hopping processes (c) and (d) contribute an on-site energy U.

These considerations can now be used as input into the Schrieffer–Wolff transformation where we need a separation of energy scales in the Hamiltonian. Our goal with this unitary transformation is to eliminate the high-energy contributions from the Hubbard Hamiltonian and thus obtain an effective low-energy description.

We thus split up the hopping part $\mathcal{H}_t$ of the Hubbard Hamiltonian (8.574) into pieces that describe the different hopping processes

$$\mathcal{H}_t = \mathcal{H}_{t,0} + \mathcal{H}_{t,2} + \mathcal{H}_{t,d+} + \mathcal{H}_{t,d-} \equiv \mathcal{H}_t^{(0)} + \mathcal{H}_t^{(\pm)} \tag{8.602}$$

where

$$\mathcal{H}_{t,0} = -t \sum_{\langle \mathbf{ij} \rangle \sigma} \left\{ (1 - n_{\mathbf{i}-\sigma})\, c_{\mathbf{i}\sigma}^{\dagger} c_{\mathbf{j}\sigma} \left(1 - n_{\mathbf{j}-\sigma}\right) + \left(1 - n_{\mathbf{j}-\sigma}\right) c_{\mathbf{j}\sigma}^{\dagger} c_{\mathbf{i}\sigma} \left(1 - n_{\mathbf{i}-\sigma}\right) \right\} \tag{8.603}$$

represents processes where unoccupied sites hop by one lattice site, while

$$\mathcal{H}_{t,2} = -t \sum_{\langle \mathbf{ij} \rangle \sigma} \left\{ n_{\mathbf{i}-\sigma} c_{\mathbf{i}\sigma}^{\dagger} c_{\mathbf{j}\sigma} n_{\mathbf{j}-\sigma} + n_{\mathbf{j}-\sigma} c_{\mathbf{j}\sigma}^{\dagger} c_{\mathbf{i}\sigma} n_{\mathbf{i}-\sigma} \right\} \tag{8.604}$$

characterizes processes where doubly occupied sites hop by one lattice site. The Hamilton operator

$$\mathcal{H}_{t,d+} = -t \sum_{\langle \mathbf{ij} \rangle \sigma} \left\{ n_{\mathbf{i}-\sigma} c_{\mathbf{i}\sigma}^{\dagger} c_{\mathbf{j}\sigma} \left(1 - n_{\mathbf{j}-\sigma}\right) + n_{\mathbf{j}-\sigma} c_{\mathbf{j}\sigma}^{\dagger} c_{\mathbf{i}\sigma} \left(1 - n_{\mathbf{i}-\sigma}\right) \right\} \tag{8.605}$$

indicates hopping processes that *increase* the number of doubly occupied sites by one, while

$$\mathcal{H}_{t,d-} = -t \sum_{\langle \mathbf{ij} \rangle \sigma} \left\{ (1 - n_{\mathbf{i}-\sigma})\, c^{\dagger}_{\mathbf{i}\sigma} c_{\mathbf{j}\sigma} n_{\mathbf{j}-\sigma} + (1 - n_{\mathbf{j}-\sigma})\, c^{\dagger}_{\mathbf{j}\sigma} c_{\mathbf{i}\sigma} n_{\mathbf{i}-\sigma} \right\} \tag{8.606}$$

corresponds to hopping processes that *decrease* the number of doubly occupied sites by one.

Thus, the first two parts, $\mathcal{H}_{t,0}$ and $\mathcal{H}_{t,2}$, describe the hopping processes that leave the double occupancy unchanged. The last two parts, $\mathcal{H}_{t,d+}$ and $\mathcal{H}_{t,d-}$, change the double occupancy by $+1$ and -1, respectively, and thus change the energy by U. Moreover, we can also write $\mathcal{H}_t$ as two parts: a part $\mathcal{H}_t^{(0)}$ that leaves double occupancy unchanged, and a part $\mathcal{H}_t^{(\pm)}$ that changes double occupancy by one.

We now apply the Schrieffer–Wolff transformation to the present situation where we want to eliminate the high-energy contributions of the term $\mathcal{H}_t^{(\pm)}$ in the Hubbard Hamiltonian in order to generate an effective Hamiltonian.

The Schrieffer–Wolff transformation is a unitary canonical transformation of the form

$$\tilde{\mathcal{A}} = e^{\mathcal{S}} \mathcal{A} e^{-\mathcal{S}} \tag{8.607}$$

where $\mathcal{S}$, called the action operator, is required to be an anti-Hermitian operator, i.e.

$$\mathcal{S}^{\dagger} = -\mathcal{S} \tag{8.608}$$

such that, if $\mathcal{A}$ is Hermitian, then so is $\tilde{\mathcal{A}}$. Expectation values remain unchanged if the state is also transformed as

$$|\tilde{\psi}\rangle = e^{\mathcal{S}} |\psi\rangle. \tag{8.609}$$

In particular, eigenvalues of $\mathcal{A}$ and $\tilde{\mathcal{A}}$ are identical.

EXERCISE 8.29 Proof of the Schrieffer–Wolff transformation formula The Schrieffer–Wolff transformation gains its usefulness from the expansion

$$\begin{aligned} \tilde{\mathcal{A}} &= e^{\mathcal{S}} \mathcal{A} e^{-\mathcal{S}} \\ &= \mathcal{A} + [\mathcal{S}, \mathcal{A}] + [\mathcal{S}, [\mathcal{S}, \mathcal{A}]] + \ldots = \mathcal{A} + \sum_{n=1}^{\infty} [\mathcal{S}, [\mathcal{S}, \ldots [\mathcal{S}, \mathcal{A}] \ldots]]_n \end{aligned} \tag{8.610}$$

where the index $n = 1, 2, \ldots$ signifies an n-fold commutator.

Prove the validity of this expansion.

Here, we thus are looking for an operator $\mathcal{S}$ such that the part of the Hubbard Hamiltonian $\mathcal{H}_t^{(\pm)}$ that changes the number of doubly occupied sites is eliminated from the Hubbard Hamiltonian in lowest order

$$\tilde{\mathcal{H}} = e^{\mathcal{S}} \left(\mathcal{H}_t^{(0)} + \mathcal{H}_t^{(\pm)} + \mathcal{H}_U \right) e^{-\mathcal{S}} = \mathcal{H}_t^{(0)} + \mathcal{H}_U + \mathcal{O}\left(\frac{t^2}{U}\right). \tag{8.611}$$

Using the expansion derived in exercise 8.29, we can write a transformed Hamiltonian

$$\tilde{\mathcal{H}} = \mathcal{H} + [\mathcal{S}, \mathcal{H}] + \frac{1}{2} [\mathcal{S}, [\mathcal{S}, \mathcal{H}]] + \ldots \tag{8.612}$$

$$= \mathcal{H}_U + \mathcal{H}_t^{(0)} + \mathcal{H}_t^{(\pm)} + [\mathcal{S}, \mathcal{H}_U] + \left[\mathcal{S}, \mathcal{H}_t^{(0)}\right] + \left[\mathcal{S}, \mathcal{H}_t^{(\pm)}\right] + \frac{1}{2} [\mathcal{S}, [\mathcal{S}, \mathcal{H}]] + \ldots . \tag{8.613}$$

The first task now consists of finding an anti-Hermitian operator $\mathcal{S}$ such that

$$\mathcal{H}_t^{(\pm)} + [\mathcal{S}, \mathcal{H}_U] = \mathcal{H}_{t,d+} + \mathcal{H}_{t,d-} + [\mathcal{S}, \mathcal{H}_U] = 0 \tag{8.614}$$

and, hence, to lowest order, the Schrieffer–Wolff transformed operator does not change the number of doubly occupied sites.

In order to find this operator, we observe that

$$\left[\mathcal{H}_{t,d\pm}, \mathcal{H}_U\right] = \mp U \mathcal{H}_{t,d\pm}, \tag{8.615}$$

which suggests that the explicit form

$$\mathcal{S} = \frac{1}{U} \left(\mathcal{H}_{t,d+} - \mathcal{H}_{t,d-}\right) \tag{8.616}$$

satisfies the requirement (8.614).

EXERCISE 8.30 Commutators in the Schrieffer–Wolff expansion Convince yourself of the commutator relations (8.615).

However, with this choice of $\mathcal{S}$, the commutator

$$\left[\mathcal{S}, \mathcal{H}_t^{(0)}\right] \tag{8.617}$$

again introduces hopping processes that change the double occupancy of sites. There are two ways out of this predicament. The first is to consider the orders of magnitude of the terms. The eliminated terms are $\mathcal{O}(t)$, while the terms of (8.617) are of order $\mathcal{O}(t^2/U)$, i.e. of one order of magnitude smaller than the eliminated ones.

A more formal way to deal with this problem is to consider our choice of $\mathcal{S}$ as the first term in an expansion in the small parameter t/U and introduce a second order correction

$$\mathcal{S} \to \mathcal{S} + \mathcal{S}^{(2)}, \tag{8.618}$$

which is of order $(t/U)^2$ and which thus eliminates the unwanted term

$$\left[\mathcal{S}^{(2)}, \mathcal{H}_U\right] = -\left[\mathcal{S}, \mathcal{H}_{t,0}\right] = -\frac{1}{U}\left[\mathcal{H}_{t,d+} - \mathcal{H}_{t,d-}, \mathcal{H}_{t,0}\right]. \tag{8.619}$$

The remaining first-order commutators $\left[\mathcal{S}, \mathcal{H}_t^{(\pm)}\right]$ of the expansion (8.613) are of order of magnitude

$$\left[\mathcal{S}, \mathcal{H}_{t,d\pm}\right] = \frac{1}{U}\left[\mathcal{H}_{t,d+}, \mathcal{H}_{t,d-}\right] = \mathcal{O}(t^2/U), \tag{8.620}$$

which is the same order of magnitude as we obtain for

$$\frac{1}{2}\left[\mathcal{S}, \left[\mathcal{S}, \mathcal{H}_U\right]\right] = -\frac{1}{U}\left[\mathcal{H}_{t,d+}, \mathcal{H}_{t,d-}\right] = \mathcal{O}(t^2/U), \tag{8.621}$$

while we have for the remaining parts of the second-order in the expansion (8.613) as order of magnitude

$$\left[\mathcal{S}, \left[\mathcal{S}, \mathcal{H}\right]\right] = \left[\mathcal{S}, \left[\mathcal{S}, \mathcal{H}_U\right]\right] + \mathcal{O}(t^3/U^2). \tag{8.622}$$

Thus, collecting terms with order lower than $\mathcal{O}(t^3/U^2)$, the Schrieffer–Wolff transformed Hamiltonian of the Hubbard model becomes the effective Hamiltonian

$$\tilde{\mathcal{H}} = \mathcal{H}_{\mathbf{eff}} = \mathcal{H}_t^{(0)} + \mathcal{H}_U + \frac{1}{U}\left[\mathcal{H}_{t,d+}, \mathcal{H}_{t,d-}\right] + \mathcal{O}(t^3/U^2). \tag{8.623}$$

The main remaining task now is to work out the term $\frac{1}{U}\left[\mathcal{H}_{t,d+}, \mathcal{H}_{t,d-}\right]$ in the effective Hamiltonian. We are interested in the low-energy regime near half-filling where most sites are singly occupied except for a small number of unoccupied sites or holes. Therefore, the product $\mathcal{H}_{t,d+}\mathcal{H}_{t,d-}$ does not contribute as it would require doubly occupied sites. For the same reason, $\mathcal{H}_U$ does not give a contribution. We are left with $\mathcal{H}_{t,d-}\mathcal{H}_{t,d+}$, which creates a doubly occupied site and destroys one. This can be done in two ways, where either the product of operators acts only on two neighbouring sites $(\mathbf{i}, \mathbf{j})$ creating and annihilating a doubly occupied site, or involving three adjacent sites $(\mathbf{i}, \mathbf{j}, \mathbf{k})$ where the doubly occupied site is created and annihilated on the middle site $\mathbf{j}$.

Exercise 8.31 focuses is on the two-site process. The three-site process is discussed in Fazekas (1999), chapter 5, where the useful tool of Hubbard (projection) operators is introduced and used to discuss the t–$\mathcal{J}$ model. In Fazekas (1999), the discussion is framed in the notions of lower and upper Hubbard band, which we have not introduced. The t–$\mathcal{J}$ model is then the effective description of the low-energy physics or the physics of the lower Hubbard band.

EXERCISE 8.31 ***t–J* model** Considering only the two-site processes, show that the effective Hamiltonian (8.623) can be written as

$$\mathcal{H}_{t\mathcal{J}} = -\,t\sum_{\langle \mathbf{ij}\sigma\rangle}\Big(\left(1-n_{\mathbf{i}-\sigma}\right)c^{\dagger}_{\mathbf{i}\sigma}c_{\mathbf{j}\sigma}\left(1-n_{\mathbf{j}-\sigma}\right)+\left(1-n_{\mathbf{j}-\sigma}\right)c^{\dagger}_{\mathbf{j}\sigma}c_{\mathbf{i}\sigma}\left(1-n_{\mathbf{i}-\sigma}\right)\Big)$$
$$+\frac{4t^2}{U}\sum_{\langle\mathbf{ij}\rangle}\left(\mathbf{S_i}\cdot\mathbf{S_j}-\frac{1}{4}n_{\mathbf{i}}n_{\mathbf{j}}\right),\qquad n_{\mathbf{i}}=n_{\mathbf{i}\uparrow}+n_{\mathbf{i}\downarrow},\tag{8.624}$$

which is the Hamiltonian of the t–$\mathcal{J}$ model. The first term reflects the contributions of $\mathcal{H}_t^{(0)}$ in (8.623). Recall the local spin operators for electrons calculated in exercise 3.11.

The t–$\mathcal{J}$ Hamiltonian (8.624) will have to be supplemented by three-site terms. However, these are often neglected in applications of the model.

8.8.2 From the Hubbard to Heisenberg model: half-filled band case

As previously mentioned, the exactly half-filled band case is characterized by a ground state with one electron per site, i.e. there are no empty sites or holes. In this situation the first part of (8.623), stemming from $\mathcal{H}_{t,0}$, which corresponds to hopping of empty sites, ceases to contribute to the effective Hamiltonian and we are left with a pure antiferromagnetic Heisenberg Hamiltonian

$$\mathcal{H}=\frac{4t^2}{U}\sum_{\langle\mathbf{ij}\rangle}\mathbf{S_i}\cdot\mathbf{S_j}\equiv\mathcal{J}\sum_{\langle\mathbf{ij}\rangle}\mathbf{S_i}\cdot\mathbf{S_j}\tag{8.625}$$

with $\mathcal{J}>0$. This is the case of the t–$\mathcal{J}$ model with $n_{\mathbf{i}}=n_{\mathbf{i}\uparrow}+n_{\mathbf{i}\downarrow}=1$. We omitted the constant term arising from the density-density interaction $n_{\mathbf{i}}n_{\mathbf{j}}$ in (8.623). For further discussion of the consequences of this model's provenance from the Hubbard model, interested readers are referred to chapter 5 of Fazekas (1999).

8.9 Magnetic quantum impurity models

An important avenue to probe strongly interacting quantum matter consists in the study of the effects of quantum impurities. Quantum impurity problems present theoretical and experimental settings in which to analyse various aspects of strong electron correlations. In particular, magnetic quantum impurities induce an effective electron-electron interaction that increases as the energy scale is lowered, the so-called Kondo effect. The range of these effective electron-electron correlations is known as the Kondo screening cloud because the spin of the magnetic impurities becomes effectively screened by the spins of the correlated electrons.

Progress in experiments on mesoscopic and nanoscale systems now enables *controlled* investigations of a single quantum impurity interacting with conduction electrons. In groundbreaking experiments in the late 1990s, semiconductor quantum dots, connected capacitively to a gate and via tunnel junctions to electrodes, were shown to exhibit a tunable Kondo effect. Specifically, below a characteristic temperature T_K, a single electron occupying the highest spin-degenerate level of the dot forms a spin singlet with electrons in the leads, producing a Kondo resonance at the Fermi level. This and subsequent developments have turned quantum impurity physics into an essential piece of modern nanoscience (cp. section 8.9.4 for more details on these developments).

This section investigates the formation of and the interaction between a single localized magnetic moment of an atom or ion in an non-magnetic host metal, i.e. its spin, and the host metal's delocalized conduction electrons. This scenario is realized in non-magnetic metals with isolated impurity atoms carrying a magnetic moment, such as for unpaired d or f electrons in transition or rare earth metals, but also in compounds where a lattice of localized ions with magnetic moments due to unpaired d or f orbitals interacts with a band of delocalized s electrons. However, restrict our attention to isolated impurities that can be described by single impurity models and where, thus, we need not account for any interaction between impurities.

The class of models describing this scenario form an important class of quantum impurity models. We concentrate our attention on these models also because many among them can be reduced to a one-dimensional form and are amenable to an exact solution by Bethe ansatz.

The fundamental model we use to begin our discussion is the Anderson model (Anderson, 1961) whose atomic limit allows us to classify the possible scenarios of interaction between localized electrons and conduction band electrons. We are especially interested in the formation of a localized magnetic moment on the impurity atom and derive an effective Hamiltonian, the Kondo or *s-d* exchange Hamiltonian to describe the low-energy interactions of this localized moment with the conduction band electrons.

Magnetic quantum impurity problems continue to provide challenges for theoretical and experimental physics. In particular, the mathematical description of these systems constitutes a difficult problem for the arsenal of many-particle tools where perturbative approaches do not suffice in many cases. Instead, non-perturbative techniques, e.g. Bosonization, conformal field theory, or Bethe ansatz methods, must be invoked.

Experimental techniques arising from advances in nanotechnology have enabled the quantum engineering of systems of magnetic quantum impurities with unprecedented and pioneering ways of controlling them and have thus continued to provide new challenges that cross–fertilize theory and experiment in this field of research.

Historically, an effect of magnetic quantum impurities has been observed for the first time experimentally in electrical resistivity measurements of non-magnetic metals at low temperatures. However, at the time it was not known that these observations were connected with magnetic impurities.

Two comprehensive reviews of the physics of magnetic quantum impurities are Hewson (1993) and Kondo (2012).

8.9.1 Resistivity minimum

Several mechanisms determine the electrical resistivity, ρ, of non-magnetic metals at low temperatures. The scattering of the conduction electrons off the atomic cores of the underlying vibrating lattice contributes a phonon part, ρ_{Phonon}, which increases with temperature since more and more lattice vibrations are excited. Conversely, at lower and lower temperatures, this contribution becomes less and less important and a residual temperature-independent resistivity, ρ_0, remains due to the scattering of the electrons with defects, impurities, and vacancies that are always present in real materials. Unless a metal becomes superconducting at low temperatures, quantitatively the temperature dependence of the resistivity is therefore expected to be

$$\rho = \rho_0 + \rho_{\text{Phonon}} = \rho_0 + cT^5 \tag{8.626}$$

with a constant c (see, e.g. Marder, 2010, chapter 18, for a detailed discussion of resistivity calculations).

It was therefore surprising when de Haas and colleagues (1934) observed a resistivity minimum at low temperatures in the resistivity of gold in measurements. Instead of approaching a finite temperature-independent value, the resistivity, after passing through a minimal value, increased again as the temperature was further lowered. More examples of metals exhibiting a resistivity minimum at low temperatures were subsequently observed corroborating these findings, indicating an unknown additional scattering mechanism whose strength increases as the temperature is lowered. This additional mechanism remained a mystery for almost 30 years.

It was only in the early 1960s that it became apparent that the resistivity minima were connected to tiny amounts of impurity atoms that possess localized magnetic moments due to the spin of unpaired electrons in their atomic d or f shells. This insight opened the way to a theoretical understanding of the phenomenon of the resistivity minimum in non-magnetic metals due to magnetic impurities.

The following section introduces Anderson's (1961) proposed model to describe the simplest such situation: a single impurity atom with an unpaired d (or f) electron in a continuum of delocalized conduction band electrons.

8.9.2 Anderson model of localized magnetic moment formation

Anderson (1961), combining two previous insights, proposed a model for the formation of localized magnetic moments in metals with small amounts of transition or rare earth impurity atoms (for a review, see Anderson, 1978).

The first insight goes back to Mott and Peierls (1937) and Mott (1949). It argues that a strong repulsive Coulomb interaction between the electrons of an atomic state, like in a localized impurity atom, reduces and eventually blocks the path of electrons. This mechanism of a transition from metal to insulator is called a *Mott* transition.

The other insight envisages the formation of a resonance due to the tunnelling of electrons between the localized atoms and the conduction band. This process has been first proposed by Friedel (1956) (1958*a*) (1958*b*).

Anderson's (1961) model combines these insights. It describes the localized electron states, created by the operator $d^\dagger$, of energy ε_d of a spatially confined impurity atom where the addition of a second electron with opposite spin costs a Coulomb energy U

$$\mathcal{H}_{\mathrm{Imp}} = \mathcal{H}^{(0)}_{\mathrm{Imp}} + \mathcal{H}^{\mathrm{U}}_{\mathrm{Imp}} = \sum_\sigma \varepsilon_d d^\dagger_\sigma d_\sigma + U n_\uparrow n_\downarrow, \tag{8.627}$$

where $n_\sigma = d^\dagger_\sigma d_\sigma$ and the Coulomb interaction U of a doubly occupied d-state is given by the electron density $\rho_d(\mathbf{r}) = |\psi_d(\mathbf{r})|^2$ of the atomic orbital $\psi_d(\mathbf{r})$ as

$$U = \int d^3r\, d^3r' \frac{e^2}{|\mathbf{r} - \mathbf{r}'|} \rho_d(\mathbf{r}) \rho_d(\mathbf{r}'). \tag{8.628}$$

The Coulomb interaction U is the crucial quantity determining the magnetic properties of the Anderson model.

EXERCISE 8.32 Creation operator of localized electron state Express the operator $d^\dagger_\sigma$ that creates a localized d-electron state in terms of the corresponding electron quantum field operator $\psi^\dagger_\sigma(\mathbf{r})$.

The localized electron state interacts with delocalized conduction band electrons. The conduction band electrons that created by operators $c^\dagger_{\mathbf{k},\sigma}$ are described by the Hamiltonian

$$\mathcal{H}_{\mathrm{CB}} = \sum_{\mathbf{k},\sigma} \varepsilon(\mathbf{k}) c^\dagger_{\mathbf{k},\sigma} c_{\mathbf{k},\sigma}, \tag{8.629}$$

where the energies of the band electrons $\varepsilon(\mathbf{k}) = E(\mathbf{k}) - \mu$ with respect to the chemical potential μ are confined to a band with bandwidth D: $-D < \varepsilon(\mathbf{k}) < D$.

When the atom is placed in the metal, the electron states of the localized electrons and of the conduction band electrons are hybridized due to tunnelling processes in and out of the localized atomic states. This situation is described by the mixing or hybridization Hamiltonian

$$\mathcal{H}_{\mathrm{Mix}} \equiv \mathcal{H}_V = \sum_{\mathbf{k},\sigma} \left\{ V(\mathbf{k}) d^\dagger_\sigma c_{\mathbf{k},\sigma} + V^*(\mathbf{k}) c^\dagger_{\mathbf{k},\sigma} d_\sigma \right\} \tag{8.630}$$

with the hybridization matrix element of the atomic potential V_{atomic} from first-order perturbation theory

$$V(\mathbf{k}) = \langle \mathbf{k} | V_{\text{atomic}} | d \rangle = \int d^3 r\, e^{-i\mathbf{k}\cdot\mathbf{r}} V_{\text{atomic}}(\mathbf{r}) \psi_d(\mathbf{r}) \tag{8.631}$$

assuming plane waves as the Bloch waves of the conduction band electrons.

The Hamiltonian of the single-impurity Anderson model can now be written as the sum of the contributions discussed

$$\mathcal{H} = \mathcal{H}_{\text{CB}} + \mathcal{H}_{\text{Imp}} + \mathcal{H}_V \tag{8.632}$$

$$= \sum_{\mathbf{k},\sigma} \varepsilon(\mathbf{k}) c^\dagger_{\mathbf{k},\sigma} c_{\mathbf{k},\sigma} + \sum_\sigma \varepsilon_d d^\dagger_\sigma d_\sigma + U n_\uparrow n_\downarrow + \sum_{\mathbf{k},\sigma} \left\{ V(\mathbf{k}) d^\dagger_\sigma c_{\mathbf{k},\sigma} + V^*(\mathbf{k}) c^\dagger_{\mathbf{k},\sigma} d_\sigma \right\}. \tag{8.633}$$

The interaction term $\mathcal{H}^{\text{U}}_{\text{Imp}}$ of the impurity part of this Hamiltonian can be expressed in terms of fluctuations of the number operator $n_\sigma = d^\dagger_\sigma d_\sigma$ of d-electrons with spin component σ

$$\Delta n_\sigma \equiv d^\dagger_\sigma d_\sigma - \langle d^\dagger_\sigma d_\sigma \rangle \tag{8.634}$$

around its average value $\langle d^\dagger_\sigma d_\sigma \rangle$.

Anderson (1961) used this mean-field approximation of the model to discuss the conditions for the formation of a localized magnetic moment or spin, i.e. the conditions that the localized level is occupied on average by one, rather than two or zero, electrons. The result is that magnetic moment formation occurs when the impurity level ε_d is (well) below the Fermi level ε_F of the conduction band electrons, i.e. $\varepsilon_d < \varepsilon_F$ (or even $\varepsilon_d \ll \varepsilon_F$), while the Coulomb repulsion is large enough such that $\varepsilon_d + U > \varepsilon_F$ (or even $\varepsilon_d + U \gg \varepsilon_F$).

The same qualitative result can also be obtained by a simple consideration of the Anderson model in the atomic limit where the hybridization is completely neglected.

8.9.2.1 *Anderson model in the atomic limit*

The atomic physics of the isolated impurity atom is described by the simple Hamiltonian

$$\mathcal{H}_{\text{atomic}} = \varepsilon_d n_d + U n_{d\uparrow} n_{d\downarrow} \tag{8.635}$$

where $n_d = n_{d\uparrow} + n_{d\downarrow} = d_\uparrow{}^\dagger d_\uparrow + d_\downarrow{}^\dagger d_\downarrow$. This Hamiltonian is diagonal with eigenvalues

$$E_0 = 0, \tag{8.636}$$

$$E_2 = 2\varepsilon_d + U \tag{8.637}$$

for the non-magnetic unoccupied, $n_d = 0$, and doubly occupied, $n_d = 2$, states, and has eigenvalue

$$E_1 = \varepsilon_d \tag{8.638}$$

for the twofold degenerate magnetic so-called Kramers doublet state with $n_d = 1$. Therefore, removing or adding an electron from or to the singly occupied magnetic state involves the energies

$$E_2 - E_1 = \varepsilon_d + U \tag{8.639}$$

$$E_0 - E_1 = -\varepsilon_d. \tag{8.640}$$

If these energy differences are both positive, the singly occupied magnetic state is the ground state, i.e. it is the ground state if

$$\varepsilon_d < 0 \quad \text{and} \quad \varepsilon_d + U > 0, \tag{8.641}$$

which corroborates the mean-field theory result claimed earlier if we assume the Fermi energy $\varepsilon_F = 0$.

Another important question is whether the localized spin will become screened by interactions with the fluctuating spins of the conduction band electrons. In order to investigate this question, the following section considers the effective low-energy behaviour of the Anderson model.

8.9.3 Kondo model

The effective low-energy behaviour of the Anderson model leads to the Kondo model, which describes the interaction of conduction band electrons with a localized magnetic impurity, i.e. a localized spin. Provided the Coulomb energy is $\varepsilon_d + U > \varepsilon_F$ (and $\varepsilon_d < \varepsilon_F$), the Kondo model describes the low-energy physics of a single magnetic impurity in a sea of conduction electrons. Here, we shall assume $\varepsilon_F = 0$ for simplicity.

The Hamiltonian of the Kondo model can be derived in several ways. For its insight into the underlying physics, we choose to exploit what we learned from considering the atomic limit. The Hilbert space of the Anderson Hamiltonian can be split into subspaces where the impurity level is unoccupied, singly, or doubly occupied. The Kondo Hamiltonian can be derived by projecting the Anderson Hamiltonian on these subspaces and effectively eliminating the unoccupied and doubly occupied states from the Schrödinger equation, which is then treated perturbatively to obtain an effective Hamiltonian describing the low-energy physics of the excitations above the singly occupied ground state. For this procedure to work, the hybridization $V(k)$ is assumed to be small. The resulting Kondo Hamiltonian describes the interaction between the localized magnetic moment or spin of the impurity and the conduction band electrons.

A more generally applicable method that can be used to derive the Kondo Hamiltonian consists in using a canonical transformation, the Schrieffer–Wolff transformation (Schrieffer and Wolff, 1966), which also projects onto the subspace of singly occupied impurity states. The assumptions for the Schrieffer–Wolff transformation are the same as for the first method. We have used this method before in section 8.8 for the derivation of the t–$\mathcal{J}$ and Heisenberg models as the limit of the Hubbard model for large on-site interaction U.

8.9.3.1 Derivation of the Kondo Hamiltonian using projectors

Instead of carrying out a perturbation expansion in $V(k)$ directly, we first project onto the impurity states with occupancy $n = 0, 1, 2$. This can be achieved through the following projection operators. The projector on an unoccupied impurity state is

$$P_0 = \left(1 - n_\uparrow\right)\left(1 - n_\downarrow\right) \tag{8.642}$$

while the projectors on singly and doubly occupied impurity states are

$$P_\uparrow = n_\uparrow \left(1 - n_\downarrow\right) \tag{8.643}$$

$$P_\downarrow = n_\downarrow \left(1 - n_\uparrow\right) \tag{8.644}$$

$$P_1 = P_\uparrow + P_\downarrow = \left(n_\uparrow - n_\downarrow\right)^2 = n_\uparrow + n_\downarrow - 2n_\uparrow n_\downarrow \tag{8.645}$$

$$P_2 = n_\uparrow n_\downarrow . \tag{8.646}$$

As projectors these operators satisfy

$$P_n^2 = P_n \quad \text{for} \quad n = 0, 1, 2 \quad \text{or} \quad n = \uparrow, \downarrow, \tag{8.647}$$

and

$$P_n P_m = 0 \quad \text{for} \quad n \neq m, \tag{8.648}$$

$$\sum_{n=0}^{2} P_n = \mathbb{I}. \tag{8.649}$$

Formally, the Anderson Hamiltonian (8.633) can therefore be written as

$$\mathcal{H}_{nm} = P_n \mathcal{H} P_m \quad n, m = 0, 1, 2. \tag{8.650}$$

Because the Anderson Hamiltonian is Hermitian, the projected Hamiltonians satisfy

$$\mathcal{H}^*_{nm} = \mathcal{H}_{mn}. \tag{8.651}$$

The projected states are defined by

$$|0\rangle = P_0|\psi\rangle, \tag{8.652}$$

$$|1\rangle = P_1|\psi\rangle, \tag{8.653}$$

$$|2\rangle = P_2|\psi\rangle, \tag{8.654}$$

such that an arbitrary state can be decomposed as

$$|\psi\rangle = |0\rangle + |1\rangle + |2\rangle = P_0|\psi\rangle + P_1|\psi\rangle + P_2|\psi\rangle. \tag{8.655}$$

The Schrödinger equation

$$\mathcal{H}|\psi\rangle = E|\psi\rangle \tag{8.656}$$

can now be formally manipulated using the projector property $P^2 = P$

$$\mathcal{H}|0\rangle + \mathcal{H}|1\rangle + \mathcal{H}|2\rangle = \mathcal{H}P_0|\psi\rangle + \mathcal{H}P_1|\psi\rangle + \mathcal{H}P_2|\psi\rangle \tag{8.657}$$

$$= \mathcal{H}P_0P_0|\psi\rangle + \mathcal{H}P_1P_1|\psi\rangle + \mathcal{H}P_2P_2|\psi\rangle \tag{8.658}$$

$$= \mathcal{H}P_0|0\rangle + \mathcal{H}P_1|1\rangle + \mathcal{H}P_2|2\rangle \tag{8.659}$$

$$= E|\psi\rangle. \tag{8.660}$$

Multiplication of the last equation from the left with P_0 gives

$$P_0\mathcal{H}P_0|0\rangle + P_0\mathcal{H}P_1|1\rangle + P_0\mathcal{H}P_2|2\rangle = \mathcal{H}_{00}|0\rangle + \mathcal{H}_{01}|1\rangle + \mathcal{H}_{02}|2\rangle \tag{8.661}$$

$$= EP_0|\psi\rangle = E|0\rangle. \tag{8.662}$$

This and similar expressions obtained by multiplication with P_1 and P_2 from the left can be collected and written in matrix form as

$$\begin{pmatrix} \mathcal{H}_{00} & \mathcal{H}_{01} & \mathcal{H}_{02} \\ \mathcal{H}_{10} & \mathcal{H}_{11} & \mathcal{H}_{12} \\ \mathcal{H}_{20} & \mathcal{H}_{21} & \mathcal{H}_{22} \end{pmatrix} \begin{pmatrix} |0\rangle \\ |1\rangle \\ |2\rangle \end{pmatrix} = E \begin{pmatrix} |0\rangle \\ |1\rangle \\ |2\rangle \end{pmatrix}. \tag{8.663}$$

There are no terms in the Anderson Hamiltonian that add or remove two electrons to or from the impurity, thus the matrix elements

$$\mathcal{H}_{02} = \mathcal{H}_{20} = 0 \tag{8.664}$$

vanish. This allows us to eliminate the states $|0\rangle$ and $|2\rangle$ between the equations of (8.663). The first equation gives

$$|0\rangle = (E - \mathcal{H}_{00})^{-1}\,\mathcal{H}_{01}|1\rangle \tag{8.665}$$

while the third equation gives

$$|2\rangle = (E - \mathcal{H}_{22})^{-1}\,\mathcal{H}_{21}|1\rangle. \tag{8.666}$$

Hence, the second equation becomes

$$\Big(\mathcal{H}_{10}\,(E - \mathcal{H}_{00})^{-1}\,\mathcal{H}_{01} + (\mathcal{H}_{11} - E) + \mathcal{H}_{12}\,(E - \mathcal{H}_{22})^{-1}\,\mathcal{H}_{21}\Big)\,|1\rangle = 0. \tag{8.667}$$

The original Schrödinger equation for the Anderson Hamiltonian is thus recast entirely as an eigenvalue equation involving only states where the impurity is singly occupied. This approach is appropriate for the physical situation we envisage where we tune the energies ε_d and U such that single occupation of the impurity is expected. The terms in this eigenvalue equation involve virtual occupation of impurity states with $n = 0$ and $n = 2$. We have not yet made any approximation and diagonalizing (8.667) would give all energies and states of the full problem.

The next step consists of identifying explicit expressions for the projected Hamiltonians $\mathcal{H}_{nm}$ from the Anderson Hamiltonian, for which we shall be forced to consider a perturbative approximation where the hybridization $V(\mathbf{k})$ sets the scale of the perturbation.

We begin with the projected Hamiltonian $\mathcal{H}_{10}$. From its definition it acts on a state where the impurity is unoccupied and produces a state where the impurity is singly occupied. There is a part of the Anderson Hamiltonian that partly fits this bill. It is

$$\sum_{\mathbf{k}\sigma} V(\mathbf{k}) d_\sigma^\dagger c_{\mathbf{k}\sigma}. \tag{8.668}$$

This operator raises the occupation of the impurity by one, i.e. from zero occupancy to single occupancy, or from single occupancy to double occupancy. However, we need an operator that guarantees an unoccupied impurity state. To achieve this, we need to multiply with an appropriate projector from the right. Thus,

$$\mathcal{H}_{10} = P_1 \sum_{\mathbf{k}\sigma} V(\mathbf{k}) d_\sigma^\dagger c_{\mathbf{k}\sigma} P_0 \tag{8.669}$$

$$= \sum_{\mathbf{k}\sigma} V(\mathbf{k}) d_\sigma^\dagger c_{\mathbf{k}\sigma} P_0 \tag{8.670}$$

$$= \sum_{\mathbf{k}\sigma} V(\mathbf{k}) d_\sigma^\dagger c_{\mathbf{k}\sigma} \left(1 - n_\uparrow\right)\left(1 - n_\downarrow\right) \tag{8.671}$$

$$= \sum_{\mathbf{k}\sigma} V(\mathbf{k}) d_\sigma^\dagger \left(1 - n_{-\sigma}\right) c_{\mathbf{k}\sigma} \tag{8.672}$$

where in the first expression the projector P_1 is superfluous because $d_\sigma^\dagger$ can only create a singly occupied state and in the final expression we could safely remove the superfluous second factor $(1 - n_\sigma)$, because the factor $d_\sigma^\dagger$ will already annihilate a possible occupied spin σ impurity state.

EXERCISE 8.33 Further projected Hamiltonians Show that using considerations similar to those that led to the projected Hamiltonian $\mathcal{H}_{10}$ in (8.672) the other projected Hamiltonians are

$$\mathcal{H}_{00} = P_0 \sum_{\mathbf{k},\sigma} \varepsilon(\mathbf{k}) c_{\mathbf{k}\sigma}^\dagger c_{\mathbf{k}\sigma} P_0 = P_0 \sum_{\mathbf{k},\sigma} \varepsilon(\mathbf{k}) c_{\mathbf{k}\sigma}^\dagger c_{\mathbf{k}\sigma}, \tag{8.673}$$

$$\mathcal{H}_{01} = P_0 \sum_{\mathbf{k}\sigma} V^*(\mathbf{k}) c_{\mathbf{k}\sigma}^\dagger d_\sigma P_1 = \sum_{\mathbf{k}\sigma} V^*(\mathbf{k}) c_{\mathbf{k}\sigma}^\dagger d_\sigma \left(1 - n_{-\sigma}\right), \tag{8.674}$$

$$\mathcal{H}_{10} = \mathcal{H}_{01}^*, \tag{8.675}$$

$$\mathcal{H}_{11} = P_1 \left\{ \varepsilon_d + \sum_{\mathbf{k},\sigma} \varepsilon(\mathbf{k}) c_{\mathbf{k}\sigma}^\dagger c_{\mathbf{k}\sigma} \right\} P_1 = P_1 \left\{ \varepsilon_d + \sum_{\mathbf{k},\sigma} \varepsilon(\mathbf{k}) c_{\mathbf{k}\sigma}^\dagger c_{\mathbf{k}\sigma} \right\}, \tag{8.676}$$

$$\mathcal{H}_{21} = P_2 \sum_{\mathbf{k},\sigma} V(\mathbf{k}) d_\sigma^\dagger c_{\mathbf{k}\sigma} P_1 = \sum_{\mathbf{k},\sigma} V(\mathbf{k}) d_\sigma^\dagger n_{-\sigma} c_{\mathbf{k}\sigma}, \tag{8.677}$$

$$\mathcal{H}_{12} = \mathcal{H}_{21}^*, \tag{8.678}$$

$$\mathcal{H}_{22} = P_2 \left\{ 2\varepsilon_d + U + \sum_{\mathbf{k},\sigma} \varepsilon(\mathbf{k}) c_{\mathbf{k}\sigma}^\dagger c_{\mathbf{k}\sigma} \right\} P_2 = P_2 \left\{ 2\varepsilon_d + U + \sum_{\mathbf{k},\sigma} \varepsilon(\mathbf{k}) c_{\mathbf{k}\sigma}^\dagger c_{\mathbf{k}\sigma} \right\}. \tag{8.679}$$

With the explicit expressions for the projected Hamiltonians $\mathcal{H}_{nm}$ we now find an approximate solution of (8.667). In the first term of (8.667)

$$\mathcal{H}_{10} \left(E - \mathcal{H}_{00}\right)^{-1} \mathcal{H}_{01} \tag{8.680}$$

we need to shift the operator $\mathcal{H}_{01}$ to the left of $(E - \mathcal{H}_{00})^{-1}$. This amounts to calculating an operator $\mathcal{A}_{11}$ satisfying

$$(E - \mathcal{H}_{00})^{-1}\, \mathcal{H}_{01} = \mathcal{H}_{01}\mathcal{A}_{11}^{-1}, \tag{8.681}$$

which is equivalent to

$$\mathcal{H}_{01}\mathcal{A}_{11} = (E - \mathcal{H}_{00})\, \mathcal{H}_{01} = \mathcal{H}_{01}\left(E - \bar{\mathcal{H}}_{11}\right) - \tilde{\mathcal{H}}_{01} \tag{8.682}$$

where

$$\bar{\mathcal{H}}_{11} = P_1 \sum_{\mathbf{k}\sigma} \varepsilon(\mathbf{k}) c^{\dagger}_{\mathbf{k}\sigma} c_{\mathbf{k}\sigma} P_1 = P_1 \sum_{\mathbf{k}\sigma} \varepsilon(\mathbf{k}) c^{\dagger}_{\mathbf{k}\sigma} c_{\mathbf{k}\sigma} = \mathcal{H}_{11} - P_1 \varepsilon_d \tag{8.683}$$

and

$$\tilde{\mathcal{H}}_{01} = P_0 \sum_{\mathbf{k}\sigma} \varepsilon(\mathbf{k}) V^*(\mathbf{k}) c^{\dagger}_{\mathbf{k}\sigma} d_\sigma P_1 = \sum_{\mathbf{k}\sigma} \varepsilon(\mathbf{k}) V^*(\mathbf{k}) \left(1 - n_{-\sigma}\right) c^{\dagger}_{\mathbf{k}\sigma} d_\sigma . \tag{8.684}$$

The latter term stems from the anti-commutation of the conduction band electrons. In the region around the Fermi level $\mu = \varepsilon_F$, to which we from now on restrict our attention, we have $\varepsilon(\mathbf{k}) = E(\mathbf{k}) - \mu \approx 0$ so that this term can be neglected.

We finally find for the operator $\mathcal{A}_{11}$

$$\mathcal{A}_{11} = E - \mathcal{H}_{11} + P_1 \varepsilon_d \tag{8.685}$$

and the first term of (8.667) becomes

$$\begin{aligned}\mathcal{H}_{10}\mathcal{H}_{01}\mathcal{A}_{11}^{-1} = \\ &\sum_{\mathbf{k}\mathbf{k}'\sigma\sigma'} V(\mathbf{k}') V^*(\mathbf{k}) d^{\dagger}_{\sigma'} c_{\mathbf{k}'\sigma'} \left(1 - n_{-\sigma'}\right) \left(1 - n_{-\sigma}\right) c^{\dagger}_{\mathbf{k}\sigma} d_\sigma \cdot \\ &\cdot (E - \mathcal{H}_{11} + \varepsilon_d)^{-1}.\end{aligned} \tag{8.686}$$

This term is of order $|V(\mathbf{k})|^2$ in the hybridization energy. The inverse operator can be rewritten as

$$(E - \mathcal{H}_{11} + \varepsilon_d)^{-1} = \varepsilon_d^{-1} \left(1 - \frac{\mathcal{H}_{11} - E}{\varepsilon_d}\right)^{-1} \approx \varepsilon_d^{-1}. \tag{8.687}$$

In order to justify this approximation, we need to look at the second term of (8.667). The expression $(\mathcal{H}_{11} - E)$ contains no terms of the order of $V(\mathbf{k})$ and is thus the leading, i.e. zeroth, order on the energy scale set by the hybridization energy $V(\mathbf{k})$. As seen above the first term, and as we convince ourselves in exercise 8.34 also the third term, of (8.667) are of second order in $V(\mathbf{k})$. Thus, to second order in $V(\mathbf{k})$, (8.667) can be replaced by

$$(\mathcal{H}_{11} - E)\,|1\rangle = 0, \tag{8.688}$$

and, hence, $(\mathcal{H}_{11} - E)$ be neglected in (8.687).

We finally obtain

$$\mathcal{H}_{10}\mathcal{H}_{01}\mathcal{A}_{11}^{-1} = \sum_{\mathbf{k}\mathbf{k}'\sigma\sigma'} \frac{V(\mathbf{k}')V^*(\mathbf{k})}{\varepsilon_d} d^\dagger_{\sigma'} c_{\mathbf{k}'\sigma'} \left(1 - n_{-\sigma'}\right)(1 - n_{-\sigma})\, c^\dagger_{\mathbf{k}\sigma} d_\sigma \tag{8.689}$$

$$= \sum_{\mathbf{k}\mathbf{k}'\sigma\sigma'} \frac{V(\mathbf{k}')V^*(\mathbf{k})}{-\varepsilon_d} c^\dagger_{\mathbf{k}\sigma} c_{\mathbf{k}'\sigma'} d^\dagger_{\sigma'} d_\sigma + 2\sum_{\mathbf{k}} \frac{|V(\mathbf{k})|^2}{\varepsilon_d} P_1 \tag{8.690}$$

where the factors $\left(1 - n_{-\sigma'}\right)(1 - n_{-\sigma})$ can be safely omitted since they have no influence on the result as long as we only act with this operator on singly occupied states. The change in sign of the first term in (8.690) stems from the anti-commutation of the conduction band electrons that also produces the second term in (8.690), a spin-independent constant, which we shall discard in the following.

EXERCISE 8.34 Virtual double occupancy Repeat the above considerations for the third term in (8.667)

$$\mathcal{H}_{12}\,(E - \mathcal{H}_{22})^{-1}\,\mathcal{H}_{21} \tag{8.691}$$

of virtual double occupancy.

The result of the exercise is

$$\begin{aligned}\mathcal{H}_{12}\,(E - \mathcal{H}_{22})^{-1}\,\mathcal{H}_{21} = &\sum_{\mathbf{k}\mathbf{k}'\sigma\sigma'} \frac{V(\mathbf{k}')V^*(\mathbf{k})}{(U + \varepsilon_d)} c^\dagger_{\mathbf{k}\sigma} c_{\mathbf{k}'\sigma'} d^\dagger_{\sigma'} d_\sigma \\ &- \sum_{\mathbf{k}\mathbf{k}'\sigma} \frac{V(\mathbf{k}')V^*(\mathbf{k})}{(U + \varepsilon_d)} c^\dagger_{\mathbf{k}\sigma} c_{\mathbf{k}'\sigma}.\end{aligned} \tag{8.692}$$

In both expressions, (8.690) and (8.692), the essential product of creation and annihilation operators for impurity states and conduction band states is

$$c^\dagger_{\mathbf{k}\sigma} c_{\mathbf{k}'\sigma'} d^\dagger_{\sigma'} d_\sigma. \tag{8.693}$$

For $\sigma' = -\sigma$, the action of $d^{\dagger}_{-\sigma} d_{\sigma}$ is to flip the spin of the singly occupied impurity state if it is a σ–spin and to annihilate it otherwise. By inspection, we can thus identify

$$d^{\dagger}_{\uparrow} d_{\downarrow} = S^{+} \quad \text{and} \quad d^{\dagger}_{\downarrow} d_{\uparrow} = S^{-} \tag{8.694}$$

the spin raising and lowering operators S^{+} and S^{-}.

For $\sigma' = \sigma$ the identity

$$\begin{aligned} n_{\uparrow} c^{\dagger}_{\mathbf{k}\uparrow} c_{\mathbf{k}'\uparrow} + n_{\downarrow} c^{\dagger}_{\mathbf{k}\downarrow} c_{\mathbf{k}'\downarrow} =& \frac{1}{2} \left(n_{\uparrow} - n_{\downarrow}\right) \left(c^{\dagger}_{\mathbf{k}\uparrow} c_{\mathbf{k}'\uparrow} - c^{\dagger}_{\mathbf{k}\downarrow} c_{\mathbf{k}'\downarrow}\right) \\ &+ \frac{1}{2} \left(n_{\uparrow} + n_{\downarrow}\right) \left(c^{\dagger}_{\mathbf{k}\uparrow} c_{\mathbf{k}'\uparrow} + c^{\dagger}_{\mathbf{k}\downarrow} c_{\mathbf{k}'\downarrow}\right) \end{aligned} \tag{8.695}$$

holds. In the subspace of single occupancy

$$S^{z} = \frac{1}{2} \left(n_{\uparrow} - n_{\downarrow}\right) \tag{8.696}$$

measures the spin on the impurity. Recalling that $\left(n_{\uparrow} + n_{\downarrow}\right) = 1$ in this subspace, we obtain

$$n_{\uparrow} c^{\dagger}_{\mathbf{k}\uparrow} c_{\mathbf{k}'\uparrow} + n_{\downarrow} c^{\dagger}_{\mathbf{k}\downarrow} c_{\mathbf{k}'\downarrow} = S^{z} \left(c^{\dagger}_{\mathbf{k}\uparrow} c_{\mathbf{k}'\uparrow} - c^{\dagger}_{\mathbf{k}\downarrow} c_{\mathbf{k}'\downarrow}\right) + \frac{1}{2} \sum_{\sigma} c^{\dagger}_{\mathbf{k}\sigma} c_{\mathbf{k}'\sigma}. \tag{8.697}$$

Using these identifications, we can rewrite (8.667) as an effective Hamiltonian

$$\begin{aligned} \mathcal{H}_{\text{eff}} = \mathcal{H}_{11} &+ \sum_{\mathbf{kk}'} \mathcal{J}_{\mathbf{kk}'} \left\{ S^{+} c^{\dagger}_{\mathbf{k}\downarrow} c_{\mathbf{k}'\uparrow} + S^{-} c^{\dagger}_{\mathbf{k}\uparrow} c_{\mathbf{k}'\downarrow} + S^{z} \left(c^{\dagger}_{\mathbf{k}\uparrow} c_{\mathbf{k}'\uparrow} - c^{\dagger}_{\mathbf{k}\downarrow} c_{\mathbf{k}'\downarrow}\right) \right\} \\ &+ \sum_{\mathbf{kk}'\sigma} K_{\mathbf{kk}'} c^{\dagger}_{\mathbf{k}\sigma} c_{\mathbf{k}'\sigma} \end{aligned} \tag{8.698}$$

with the effective exchange coupling

$$\mathcal{J}_{\mathbf{kk}'} = V(\mathbf{k}') V^{*}(\mathbf{k}) \left[\frac{1}{-\varepsilon_d} + \frac{1}{U + \varepsilon_d} \right]. \tag{8.699}$$

The effective exchange coupling is positive, i.e. antiferromagnetic, $\mathcal{J}_{\mathbf{kk}'} > 0$, for the situation we envisage where $\varepsilon_d < 0$ and $U + \varepsilon_d > 0$, the so-called local moment regime. Furthermore, there is a potential scattering term with coefficients

$$K_{\mathbf{kk}'} = \frac{V(\mathbf{k}') V^{*}(\mathbf{k})}{2} \left[\frac{1}{-\varepsilon_d} - \frac{1}{U + \varepsilon_d} \right]. \tag{8.700}$$

The potential scattering term does not involve the internal spin degree of freedom of the impurity. Therefore we shall not further discuss this term in our context where we are mainly interested in the dynamics of the internal degree of freedom, the localized moment, of the impurity. Furthermore, this term vanishes in the symmetric case where $\varepsilon_d = -(U + \varepsilon_d)$, i.e. where the impurity levels are symmetric around zero.

Our main result is the term in the effective Hamiltonian (8.698) containing the effective exchange coupling (8.699). The conduction band creation and annihilation operators in this term can be written as spin density operators

$$\mathbf{s}_{\mathbf{k}\mathbf{k}'} = \frac{1}{2}\sum_{\sigma\sigma'} c^{\dagger}_{\mathbf{k}\sigma}\boldsymbol{\sigma}_{\sigma\sigma'}c_{\mathbf{k}'\sigma'} \quad \text{and} \quad s^{\pm}_{\mathbf{k}\mathbf{k}'} = s^{x}_{\mathbf{k}\mathbf{k}'} \pm i s^{y}_{\mathbf{k}\mathbf{k}'} \tag{8.701}$$

or, to cancel factors of 2 in the final expressions for the Hamiltonian

$$\boldsymbol{\sigma}_{\mathbf{k}\mathbf{k}'} = \sum_{\sigma\sigma'} c^{\dagger}_{\mathbf{k}\sigma}\boldsymbol{\sigma}_{\sigma\sigma'}c_{\mathbf{k}'\sigma'} \tag{8.702}$$

where $\boldsymbol{\sigma}_{\sigma\sigma'}$ are the matrix elements of the vector of Pauli spin matrices. The spin exchange part of the effective Hamiltonian (8.698) thus becomes

$$\mathcal{H}_{\text{spin}} = \sum_{\mathbf{k}\mathbf{k}'} \mathcal{J}_{\mathbf{k}\mathbf{k}'}\left\{S^{+}s^{-}_{\mathbf{k}\mathbf{k}'} + S^{-}s^{+}_{\mathbf{k}\mathbf{k}'} + 2S^{z}s^{z}_{\mathbf{k}\mathbf{k}'}\right\} \tag{8.703}$$

$$= 2\sum_{\mathbf{k}\mathbf{k}'} \mathcal{J}_{\mathbf{k}\mathbf{k}'}\left\{S^{x}s^{x}_{\mathbf{k}\mathbf{k}'} + S^{y}s^{y}_{\mathbf{k}\mathbf{k}'} + S^{z}s^{z}_{\mathbf{k}\mathbf{k}'}\right\} = 2\sum_{\mathbf{k}\mathbf{k}'} \mathcal{J}_{\mathbf{k}\mathbf{k}'}\,\mathbf{S}\cdot\mathbf{s}_{\mathbf{k}\mathbf{k}'} \tag{8.704}$$

$$= \sum_{\mathbf{k}\mathbf{k}'} \mathcal{J}_{\mathbf{k}\mathbf{k}'}\,\mathbf{S}\cdot\boldsymbol{\sigma}_{\mathbf{k}\mathbf{k}'}. \tag{8.705}$$

This Hamiltonian describes the interaction of the localized magnetic moment, represented by spin $\mathbf{S}$ and the spin density $\mathbf{s}_{\mathbf{k}\mathbf{k}'}$ of the conduction band electrons. The first two terms of (8.703) represent spin-flips on the impurity via virtual exchanges of spins with the conduction band electrons. The last term of (8.703) represents the exchange interaction of the localized spin on the impurity and the z-component of the spin density of the conduction band electrons.

Taking also the Hamiltonian of the conduction band electrons into account, i.e. writing the full effective Hamiltonian (8.698), we arrive at the Kondo Hamiltonian, which is the paradigmatic Hamiltonian describing the interaction of conduction band electrons with a localized magnetic moment

$$\mathcal{H}_{\text{Kondo}} = \sum_{\mathbf{k}\sigma} \varepsilon(\mathbf{k}) c^{\dagger}_{\mathbf{k}\sigma} c_{\mathbf{k}\sigma} + \sum_{\mathbf{k}\mathbf{k}'} \mathcal{J}_{\mathbf{k}\mathbf{k}'} \mathbf{S} \cdot \boldsymbol{\sigma}_{\mathbf{k}\mathbf{k}'}. \tag{8.706}$$

In this Hamiltonian, we subtracted a constant energy contribution $\varepsilon_d(n_\uparrow + n_\downarrow) = \varepsilon_d$ stemming from the occupation of the impurity state with a single electron.

It is important to keep in mind that these results are only valid in the subspace of single occupancy of the impurity. Moreover, they are only valid for conduction band energies in the ranges

$$|\varepsilon_{\mathbf{k}}| \ll |\varepsilon_d| \quad \text{and} \quad |\varepsilon_{\mathbf{k}}| \ll |U + \varepsilon_d|. \tag{8.707}$$

Often, the $\mathbf{k}$-dependence of the hybridization coupling strength $V(\mathbf{k}) \to V$ between the conduction band and impurity electrons can be discarded. Under this condition, the exchange coupling becomes

$$\mathcal{J}_{\mathbf{k}\mathbf{k}'} \to \mathcal{J} = V^2 \left[\frac{1}{-\varepsilon_d} + \frac{1}{U + \varepsilon_d} \right] \tag{8.708}$$

and the Kondo Hamiltonian takes the simpler form

$$\mathcal{H}_{\text{Kondo}} = \sum_{\mathbf{k}\sigma} \varepsilon(\mathbf{k}) c^{\dagger}_{\mathbf{k}\sigma} c_{\mathbf{k}\sigma} + \mathcal{J} \mathbf{S} \cdot \boldsymbol{\sigma}(0) \tag{8.709}$$

where at spin density of the conduction band at the position, $\mathbf{r} = 0$, of the impurity is

$$\boldsymbol{\sigma}(0) = \sum_{\sigma\sigma'} \psi^{\dagger}_{\sigma}(0) \boldsymbol{\sigma} \psi_{\sigma'}(0) \tag{8.710}$$

with the Fourier transformed electron field operator

$$\psi^{\dagger}_{\sigma}(\mathbf{r}) = \frac{1}{V} \sum_{\mathbf{k}} e^{-i\mathbf{k}\cdot\mathbf{r}} c^{\dagger}_{\mathbf{k}\sigma} \tag{8.711}$$

at $\mathbf{r} = 0$

$$\psi^{\dagger}_{\sigma}(0) = \frac{1}{V} \sum_{\mathbf{k}} c^{\dagger}_{\mathbf{k}\sigma}. \tag{8.712}$$

More generally, we can write the Kondo Hamiltonian (8.709) wholly in position space

$$\mathcal{H}_{\text{Kondo}} = \int d^3 r \left(\left(\sum_{\sigma} \psi^{\dagger}_{\sigma}(\mathbf{r}) \left[-\frac{\nabla^2}{2m} \right] \psi_{\sigma}(\mathbf{r}) \right) + \delta(\mathbf{r}) \mathcal{J} \mathbf{S} \cdot \boldsymbol{\sigma}(0) \right) = \mathcal{H}_0 + \mathcal{H}_K. \tag{8.713}$$

This representation of the Kondo Hamiltonian will be most useful later on, especially when we discuss the exact solution of the Kondo problem in part IV.

As mentioned in section 8.8, Schrieffer and Wolff (1966) developed their eponymous canonical transformation to map the low-energy physics of the Anderson model onto the Kondo model. The Schrieffer–Wolff transformation, thus, devises a way of implementing a perturbation theory. Since we already used the Schrieffer–Wolff transformation to extract the low-energy effective models, the t–$\mathcal{J}$ model and the Heisenberg model, of the Hubbard model in sections 8.8.1 and 8.8.2, respectively, we shall not demonstrate its usefulness again here. Detailed expositions of the Schrieffer–Wolff transformation applied to the Anderson Hamiltonian may be found in Coleman (2015) or Phillips (2012).

We emphasize, however, that both methods, the projector technique and the Schrieffer–Wolff transformation, consistently produce the same low-energy local moment physics described by the Kondo Hamiltonian $\mathcal{H}_{\text{Kondo}}$.

8.9.3.2 One-dimensional nature of the Kondo problem

Due to the impurity, the Kondo model is not translationally invariant. However, it is rotationally invariant with the impurity at the centre if we assume a spherically symmetric energy dispersion relation as we have, for instance, for free electrons

$$\varepsilon(\mathbf{k}) = \frac{k^2}{2m} - \frac{k_F^2}{2m} = \frac{k^2}{2m} - \varepsilon_F. \tag{8.714}$$

Thus, we can use a basis of spherical harmonics for the operators $c^\dagger_{\mathbf{k},\sigma}$, i.e. $c^\dagger_{klm,\sigma}$

$$c^\dagger_{\mathbf{k},\sigma} = \sum_l \sum_{m=-l}^{l} Y_{lm}\left(\frac{\mathbf{k}}{|\mathbf{k}|}\right) c^\dagger_{klm,\sigma} \tag{8.715}$$

where Y_{lm} are spherical harmonics depending on the direction of the $\mathbf{k}$-vector. We shall assume that only s-waves with $l = m = 0$ couple to the impurity, which is warranted by the δ-function Kondo interaction (cp. (8.713)).

However, putting straightforwardly $l = m = 0$ into (8.715), we obtain operators projected on the s-wave spherical harmonic $Y_{00}(\theta,\phi) = 1/\sqrt{4\pi}$

$$c^\dagger_{k00\sigma} = \int d\theta d\phi \, \sin\theta \; Y_{00} c^\dagger_{\mathbf{k}\sigma} = \frac{1}{\sqrt{4\pi}} \int d\Omega \, c^\dagger_{\mathbf{k}\sigma} \tag{8.716}$$

that do not obey the correct canonical anti-commutation relations

$$\left\{c^\dagger_{k00\sigma}, c_{k'00\sigma'}\right\} = \delta_{\sigma\sigma'}\delta(k-k'). \tag{8.717}$$

In order to remedy this situation, we follow a prescription proposed by Affleck and Ludwig (1991). We first use the creation operator in spherical coordinates

$c^{\dagger}_{\mathbf{k}\sigma} = c^{\dagger}_{\sigma}(k,\theta,\phi) = c^{\dagger}_{\sigma}\ (k,\Omega)$ and then define the projection onto the s-wave spherical harmonic $Y_{00}(\theta,\phi) = 1/\sqrt{4\pi}$

$$c^{\dagger}_{k00,\sigma} \equiv c^{\dagger}_{\sigma}(k) = \frac{k}{\sqrt{4\pi}} \int d\Omega\, c^{\dagger}_{\sigma}(k,\Omega). \tag{8.718}$$

The so-defined operators do obey the correct canonical anti-commutation relations (8.717), the demonstration of which is the object of exercise.

EXERCISE 8.35 Anti-commutation relations Convince yourself that (8.718) leads to the correct anti-commutation relations (8.717) while (8.716) does not. Hint: Recall the representation of the three-dimensional Dirac δ-function in spherical coordinates.

The three-dimensional Fermion operator consistent with (8.718) has thus the expansion to lowest harmonic order

$$c^{\dagger}_{\sigma}(\mathbf{k}) = \frac{1}{\sqrt{4\pi}\,k} c^{\dagger}_{\sigma}(k) + \ldots \tag{8.719}$$

where here and in the following expressions the dots signify higher harmonic contributions. Upon Fourier transformation, which is defined by

$$\psi^{\dagger}_{\sigma}(\mathbf{r}) = \frac{1}{V}\sum_{\mathbf{k}} e^{-i\mathbf{k}\cdot\mathbf{r}} c^{\dagger}_{\mathbf{k}\sigma} \to \int \frac{d^3k}{(2\pi)^{3/2}}\, e^{-i\mathbf{k}\cdot\mathbf{r}} c^{\dagger}_{\mathbf{k}\sigma}, \tag{8.720}$$

the position space operator becomes to lowest harmonic order

$$\psi^{\dagger}_{\sigma}(\mathbf{r}) = \frac{1}{\sqrt{4\pi}}\frac{1}{(2\pi)^{3/2}} \int_0^{\infty} \frac{k^2}{k} dk\, c^{\dagger}_{\sigma}(k) \int_0^{2\pi} d\phi \int_0^{\pi} \sin\theta\, d\theta\, e^{-ikr\cos\theta} + \ldots \tag{8.721}$$

$$= \frac{1}{\sqrt{2}\pi r}\int_0^{\infty} dk\, \sin(kr) c^{\dagger}_{\sigma}(k) + \ldots. \tag{8.722}$$

From now on, we focus on wave vectors

$$k = k_F + q, \tag{8.723}$$

which lie in a sufficiently small band with width 2Λ around the Fermi wave vector k_F

$$-\Lambda < q < \Lambda, \qquad \Lambda \ll k_F, \tag{8.724}$$

and introduce a linearized spectrum

$$\varepsilon(k) \approx v_F(k - k_F) = v_F q \tag{8.725}$$

where, in the following, we eventually choose the Fermi velocity $v_F = k_F/m = 1$. The creation operators $c_\sigma^\dagger(k) = c_\sigma^\dagger(k_F + q)$ (and the annihilation operators $c_\sigma(k_F + q)$) act symmetrically around the Fermi momentum k_F, i.e.

$$c_\sigma^\dagger(k_F + q) = c_\sigma^\dagger(k_F - q). \tag{8.726}$$

The expression (8.721) for the Fermi creation operator then reduces, again to lowest harmonic order, to

$$\psi_\sigma^\dagger(\mathbf{r}) = \frac{1}{2\sqrt{2\pi}\, ir}\left(e^{ik_F r}\psi_{\sigma,L}^\dagger(r) - e^{-ik_F r}\psi_{\sigma,R}^\dagger(r)\right) + \ldots \tag{8.727}$$

where we have introduced incoming (left-moving) and outgoing (right-moving) fields

$$\psi_{\sigma,L/R}^\dagger(r) = \int_{-\Lambda}^{\Lambda} dq\, e^{\pm iqr} c_\sigma^\dagger(k_F + q). \tag{8.728}$$

Because of the finite integration limits $\pm\Lambda$, these fields obey anti-commutation relations only approximately (and with an unusual normalization factor of 2π)

$$\left\{\psi_{\sigma,L/R}(r), \psi_{\sigma',L/R}^\dagger(r')\right\} = \delta_{\sigma\sigma'}\frac{2\sin\Lambda(r - r')}{r - r'} \approx 2\pi\,\delta_{\sigma\sigma'}\delta(r - r'), \tag{8.729}$$

but are spread out over a region of order $1/\Lambda$.

Evaluating (8.727) and (8.728) at position $r = 0$, we obtain

$$\psi_{\sigma,L}^\dagger(0) = \psi_{\sigma,R}^\dagger(0). \tag{8.730}$$

Applying the foregoing results, we are now in a position to extract from the Hamiltonian (8.713) its effective low-energy one-dimensional approximation. Let us first consider the non-interacting part $\mathcal{H}_0$ of (8.713). Using (8.727) together with (8.725), the non-interacting part of (8.713) becomes

$$\mathcal{H}_0 = -i\frac{k_F}{2\pi m}\int_0^\infty dr \sum_\sigma \left(\psi_{\sigma,L}^\dagger(r)\frac{\partial}{\partial r}\psi_{\sigma,L}(r) - \psi_{\sigma,R}^\dagger(r)\frac{\partial}{\partial r}\psi_{\sigma,R}(r)\right) + \ldots, \tag{8.731}$$

a Hamiltonian describing non-interacting relativistic (Dirac) Fermions in one dimension. The Fermi velocity $v_F = k_F/m$ is explicitly shown, but will later be set to unity again.

EXERCISE 8.36 Non-interacting part of the one-dimensional Kondo Hamiltonian Convince yourself of the expression (8.731) for the non-interacting part of the one-dimensional Kondo Hamiltonian. Hint: Neglect terms of order $e^{\pm 2k_F r}$.

The Kondo interaction $\mathcal{H}_K$ in (8.713) is written in terms of the three-dimensional field at $\mathbf{r} = \mathbf{0}$ which, using (8.727), (8.728), and (8.730) becomes

$$\psi_\sigma^\dagger(\mathbf{0}) = \frac{k_F}{2\sqrt{2}\pi}\left(\psi_{\sigma,L}^\dagger(0) + \psi_{\sigma,R}^\dagger(0)\right) = \frac{k_F}{\sqrt{2}\pi}\psi_{\sigma,L}^\dagger(0) = \frac{k_F}{\sqrt{2}\pi}\psi_{\sigma,R}^\dagger(0). \tag{8.732}$$

Thus, $\mathcal{H}_K$ may be written as

$$\mathcal{H}_K = \mathcal{J}\mathbf{S}\cdot\boldsymbol{\sigma}(\mathbf{0}) = \mathcal{J}\frac{k_F^2}{8\pi^2}\mathbf{S}\cdot\sum_{\sigma\sigma'}\left[\psi_{\sigma,L}^\dagger(0) + \psi_{\sigma,R}^\dagger(0)\right]\boldsymbol{\sigma}_{\sigma\sigma'}\left[\psi_{\sigma',L}(0) + \psi_{\sigma',R}(0)\right] \tag{8.733}$$

with the renormalized Kondo exchange interaction $\mathcal{J}_K/4 \equiv \mathcal{J}k_F^2/8\pi^2 = \mathcal{J}\rho v_F/8\pi$ where $\rho = k_F m/\pi$.

The Kondo Hamiltonian $\mathcal{H} = \mathcal{H}_0 + \mathcal{H}_K$ of (8.731) and (8.733), which describes one-dimensional relativistic (Dirac) Fermions defined on a half-line interacting with the impurity spin at the origin, can be further manipulated. The Dirac Fermi fields $\psi_{\sigma,L}(r)$ and $\psi_{\sigma,R}(r)$ ($r \geq 0$) can be interpreted as incoming (left-going) and outgoing (right-going) fields with time-dependences $v_F t + r$ and $v_F t - r$, respectively. The boundary condition (8.730) then suggests that we consider $\psi_{\sigma,R}^\dagger$ as the 'analytic' continuation of $\psi_{\sigma,L}^\dagger$ to the negative r-axis

$$\psi_{\sigma,R}^\dagger(r) = \psi_{\sigma,L}^\dagger(-r), \qquad r \geq 0. \tag{8.734}$$

Dropping the index L from now on altogether and putting again $v_F = 1$, we thus arrive at the one-dimensional Kondo Hamiltonian

$$\mathcal{H} = \frac{-i}{2\pi}\int_{-\infty}^{\infty} dr \sum_\sigma \psi_\sigma^\dagger(r)\frac{\partial}{\partial r}\psi_\sigma(r) + \mathcal{J}_K\mathbf{S}\cdot\sum_{\sigma\sigma'}\psi_\sigma^\dagger(0)\boldsymbol{\sigma}_{\sigma\sigma'}\psi_{\sigma'}(0). \tag{8.735}$$

This Hamiltonian is the starting point of the Bethe ansatz solution of the Kondo problem, which will form the core of our considerations in part IV on the generalization of the Bethe ansatz to electronic problems: the nested Bethe ansatz technique.

8.9.4 Magnetic quantum impurities in nanostructures

The preceding sections have explored magnetic impurity models, especially the Kondo model, from what could be termed a 'classical' condensed matter point of view. In such a condensed matter setting, for a given impurity and host material, the parameters that enter the model Hamiltonian are fixed quantities.

Over the previous couple of decades, however, this approach to condensed matter problems has been very successfully supplemented by the possibility to design and manipulate artificial systems. Decisive progress has been made in creating semiconductor structures, so-called semiconductor hetero-junctions, which opened up versatile possibilities for the controlled design and study of devices at the meso- and nanoscale, especially of their quantum transport properties. Recent monographs on this subject include Nazarov and Blanter (2009) and Ihn (2010).

Of particular interest in our context is the seminal experimental demonstration, in 1998, by Goldhaber-Gordon *et al.* 1998 and Cronenwett *et al.* 1998 that the Kondo effect can be observed in quantum dots. The existence of a Kondo regime in quantum dots had already been predicted a decade earlier by Ng and Lee (1988) and Glazman and Raikh (1988). This research area continues to be very active, see, e.g. Le Hur (2015).

The basic mechanism that may lead to the Kondo effect in quantum dots consists of electrons being transmitted by tunnel processes from an electron reservoir to the dot, or vice versa. During this transmission the spin of the electron on the dot and the spins of the electrons in the electron reservoirs may become correlated in a way similar to the correlation of the spin of the impurity electron and the spins of the conduction electrons in a metal. This formation of a correlated state leads to an increase in the probability of electron scattering off the impurity. For a magnetic impurity imbedded in bulk material, this higher scattering probability is connected with an increasing resistivity of the sample. For an impurity located in a tunnelling barrier separating two leads, however, an increased scattering probability facilitates electron tunnelling through the barrier. Instead of the resistivity, as in the Kondo effect in bulk matter, it is the conductance that is enhanced in quantum dot devices due to the Kondo effect.

This section provides a short introductory overview of this basic mechanism of the physics in the Kondo regime of quantum dots. Interested readers are referred to Pustilnik (2006) and the references therein.

A quantum dot is an artificially fabricated device made of semiconductor materials that have been used to study a wide range of physical phenomena. It consists of a small patch or droplet of semiconductor material charged with a well-defined number of electrons. The experimental set-up and colleagues, called a single-electron transistor, consists of two leads and a gate that are close to the quantum dot and weakly coupled to it. The dot is thus almost isolated. The physical properties of the quantum dots can, thus, be studied by manipulating the leads and gates close to the dot.

The leads, suggestively called source and drain, are coupled capacitively to the dot and through tunnel barriers. They thus act as electron reservoirs for changing the number of electrons on the dot and are often modelled as one-dimensional band electrons

$$\mathcal{H}_{\text{leads}} = \sum_{k\alpha\sigma} \varepsilon(k) c^{\dagger}_{k\alpha\sigma} c_{k\alpha\sigma} \tag{8.736}$$

where the leads are labelled by $\alpha = R, L$ (cp. the corresponding part (8.629) of the Anderson Hamiltonian (8.633)). The coupling to the gate is purely capacitive to regulate the electrical potential on the dot so that electrons can move on and off the dot by tunnelling from and onto leads. This coupling between the dot and the leads can be modelled by the Hamiltonian

$$\mathcal{H}_{\text{tunnel}} = \sum_{kn\alpha\sigma} \left(t_{n\alpha} c^{\dagger}_{k\alpha\sigma} d_{n\sigma} + t^{*}_{n\alpha} d^{\dagger}_{n\sigma} c_{k\alpha\sigma} \right) \tag{8.737}$$

with the tunnelling amplitudes $t_{n\alpha}$, where n enumerates the discrete energy levels of the quantum dot. For simplicity, we mostly focus on only one level $n = i$ in the following. The currents and/or voltages applied through these leads are used to probe the current response of the dot and, hence, the quantum states of the electrons on the dot.

In a semiconductor quantum dot, although consisting of roughly a million atoms and their corresponding number of electrons, the number of electrons not bound in the atomic cores and free to participate in electronic transport is typically very small. With today's experimental techniques, quantum dots can be prepared that are occupied by only one or two electrons. These electrons can be added and removed from the quantum dot in a controlled way. Similar to electrons in atomic orbitals, the electrons occupy discrete quantum levels ϵ_i. Another characteristic of the quantum dot is its charging energy, $E_C = e^2/2C$, the energy required to add or remove a single electron from the dot. Here, $C = C_{\text{source}} + C_{\text{drain}} + C_{\text{gate}}$ is the capacitance of the dot. The electrostatic energy of one electron on the dot is thus

$$E = \frac{e^2}{2C} - \frac{eC_{\text{gate}} V_{\text{gate}}}{C} = E_C \left(1 - N_0\right)^2 - \frac{(eN_0)^2}{2C} \tag{8.738}$$

where $N_0 = C_{\text{gate}} V_{\text{gate}}/e$ and V_{gate} is the applied gate voltage. Guided by this expression and focusing on the case of one energy level ϵ_i with one or two electrons on the dot, respectively, the quantum dot can be described by the Hamiltonian

$$\mathcal{H}_i = \sum_{\sigma} \epsilon_i d^{\dagger}_{i\sigma} d_{i\sigma} + E_C \left(\sum_{\sigma} d^{\dagger}_{i\sigma} d_{i\sigma} - N_0 \right)^2 \tag{8.739}$$

where we absorbed the constant $(eN_0)^2/2C$ in the Hamiltonian. The operator $d^{\dagger}_{i\sigma}$ creates an electron with spin σ in state i and energy ϵ_i on the dot. The terms of the second part of this expression can be absorbed in a redefinition of the energy level ϵ_i, except for the term $2E_C d^{\dagger}_{\uparrow} d_{\uparrow} d^{\dagger}_{\downarrow} d_{\downarrow}$.

The challenge mastered by Goldhaber-Gordon *et al.* (1998) and Cronenwett *et al.* (1998) has been to fine-tune the experimental parameters of the system consisting of the quantum dot and the leads and gate such that it could operate in the Kondo regime. For a comprehensive review, see Grobis *et al.* (2007).

We now briefly describe the quantum dot set-up to identify these parameters and compare them with the Anderson model Hamiltonian (8.633) and, eventually, with the Kondo Hamiltonian (8.735) since, as it will turn out, this system is very well described by a one-dimensional Kondo model.

If the quantum dot is occupied by one electron in the single electron state i with energy ϵ_i, then, to put a second electron of opposite spin in this state, an energy $\epsilon_i + E_C$ is required where the charging energy $E_C = e^2/2C$ is, to a good approximation, assumed to be independent of the state i. This charging energy is responsible for the so-called Coulomb blockade effect.

8.9.4.1 Coulomb blockade

At low temperatures, $T \ll E_C = e^2/2C$, and small voltage difference between source and drain (bias voltage), $eV_{\text{bias}} \ll E_C$, the energy to add an extra electron onto the dot exceeds the thermal energy, and there will be no current through the dot. This is the so-called Coulomb blockade effect.

We can discuss the condition for charge transport in some more detail by considering the generalization of (8.738) to the case of N electrons on the dot, neglecting an unimportant N–independent constant,

$$E_N = E_C\,(N - N_0)^2. \tag{8.740}$$

Comparing this expression for $N + 1$ and N electrons gives the chemical potential for adding one more electron to the N electrons on the dot

$$\mu_{N+1} \equiv E_{N+1} - E_N = 2E_C\left(N + \frac{1}{2} - N_0\right). \tag{8.741}$$

Thus, whenever the gate voltage $V_{\text{gate}} = eN_0/C_{\text{gate}}$ is tuned such that $N_0 = N + 1/2$, i.e. when N_0 assumes a half-integer value, an electron can tunnel onto and off the dot and there is a finite conductance. Away from these values, charge transport is suppressed at low temperatures, i.e. we have Coulomb blockade. For a non-zero bias voltage, there will be a staircase-like increase in the occupancy of the dot with jumps at $N_0 = N + 1/2$. Correspondingly, at these values of the gate voltage, there will be periodic conductance peaks. Both the staircase steps in the dot occupation and the peaks in the conductance are broadened by effects not discussed in detail here.

However, besides these single-electron processes, there are other processes that can lead to a finite conductance through the quantum dot even if the condition (8.741) is not met.

8.9.4.2 Kondo resonance

As previously argued, the quantum dot connected weakly to leads is a system analogous to the system of an impurity atom embedded in a metal. The leads, taking the role of the metal, are filled up to the Fermi level μ with electrons. However, in contrast to the condensed matter system of impurity atom and host metal, the dot can be artificially tuned by leads and gate. For example, the electrical potential of the dot can be modified by the electrical potential of the gate.

We now assume that the Fermi level $\mu_{\text{source}} = \mu_L$ of the source lead is slightly higher than the Fermi level $\mu_{\text{drain}} = \mu_R$ of the drain lead and that the single electron level of the dot ϵ_i is below, and the double occupancy level $\epsilon_i + E_C$ is above, these Fermi levels. If the dot is occupied by a single electron, a virtual or cotunnelling process can take place where the electron on the dot is transmitted to the drain while simultaneously an electron from the source is transmitted to the dot. This process is called the Kondo resonance. It can be described by the Anderson Hamiltonian (8.633) if we make the following identifications. The energy ε_d of the localized level becomes the single electron level ϵ_i, the Coulomb interaction U becomes the charging energy $2E_C$, and the the role of the hybridization energy V is played by the tunnelling potential barriers t_α $(\alpha = L, R)$ between the leads and the quantum dot. The Hamiltonians (8.736), (8.739), and (8.737) put together therefore constitute the single-impurity Anderson model Hamiltonian for the single-electron transistor that can thus be written as

$$\mathcal{H} = \sum_{k\alpha\sigma} \varepsilon(k) c^\dagger_{k\alpha\sigma} c_{k\alpha\sigma} + \sum_{\sigma} \varepsilon_i d^\dagger_{i\sigma} d_{i\sigma} + 2E_C n_{i\uparrow} n_{i\downarrow} + \sum_{k\alpha\sigma} \left\{ t_\alpha c^\dagger_{k\alpha\sigma} d_\sigma + t^*_\alpha d^\dagger_\sigma c_{k\alpha\sigma} \right\}. \tag{8.742}$$

With these identifications, the analysis of the Anderson model presented in section 8.9.2 and the derivation of the Kondo Hamiltonian from the Anderson Hamiltonian presented in section 8.9.3 can be taken over to the present situation. Eventually, we arrive again at the one-dimensional Kondo Hamiltonian (8.735).

EXERCISE 8.37 Canonical transformation of the tunnelling Hamiltonian Show that the last, tunnelling term of the Hamiltonian (8.742) can be simplified by introducing the electron operators

$$\begin{pmatrix} C_1 \\ C_2 \end{pmatrix} = \frac{1}{\sqrt{t_R^2 + t_L^2}} \begin{pmatrix} t_R & t_L \\ -t_L & t_R \end{pmatrix} \begin{pmatrix} c_R \\ c_L \end{pmatrix}. \tag{8.743}$$

8.9.4.3 Conductance

For the Kondo model of the single-electron transistor, we state the expression for the conductance at zero temperature without going into the details of the derivation (interested readers are referred to Pustilnik (2006) and references cited therein). The Landauer formula (see, e.g. Nazarov and Blanter, 2009) which relates the conductance to the transmission probabilities t_α through the dot yields for the present situation

$$G = G_0 \sin^2\left(\pi \frac{N_i}{2}\right) \tag{8.744}$$

where

$$N_i = \left\langle \sum_\sigma d^\dagger_{i\sigma} d_{i\sigma} \right\rangle \tag{8.745}$$

and

$$G_0 = \frac{2e^2}{h}\left[\frac{2t_R t_L}{t_R^2 + t_L^2}\right]^2. \tag{8.746}$$

The maximum conductance $G = 2e^2/h$ is thus reached for a singly occupied dot $N_i = 1$ and symmetrical leads $t_R = t_L$.

The last section of this chapter looks at strongly interacting quantum models that have their origin in quantum optics.

8.10 Quantum Rabi model

The quantum Rabi model describes the interaction between light and matter in the simplest possible way and is used as a basic model in many fields of physics (Haroche and Raimond, 2006). It was originally introduced, in semiclassical form, as the basis to understand nuclear magnetic resonance (Rabi, 1936; Rabi, 1937) and has since been applied to physical systems ranging from quantum optics to condensed matter physics, e.g. cavity and circuit quantum electrodynamics, quantum dots, trapped ions, and superconducting qubits. Moreover, it is used to describe nanoelectromechanical devices where the role of the photons is taken by phonons (see, e.g. Geller and Cleland, 2005 and Tian, 2011). All of these physical systems are under investigation as candidates for the physical realization of quantum information processing, i.e. quantum computers (see Nielsen and Chuang, 2010, chapter 7, for a review). The quantum Rabi model is, hence, of considerable importance in quantum information theory. In the language of quantum information theory, the quantum Rabi model describes a qubit interacting with a single Bosonic degree of freedom. After a further approximation, the rotating or resonant wave approximation, which is often justified in quantum optics, the model, then called the quantum Jaynes–Cummings model (Jaynes and Cummings, 1963), becomes exactly solvable by elementary means.

This chapter's main objective is to introduce the quantum Rabi model and its decendents in the context of cavity quantum electrodynamics. Especially, we shall derive from the quantum Rabi model, by applying the rotating wave approximation, the quantum Jaynes–Cummings model and its N qubit generalization, the quantum Tavis–Cummings model (Tavis and Cummings, 1968; Tavis and Cummings, 1969).

8.10.1 Quantum optics

Quantum optics is the study of the quantum mechanical nature of light, i.e. of photons, and of the interaction of light and matter. More precisely, quantum optics studies the interaction, described by *non-relativistic quantum mechanics*, of individual photons with quantum matter, e.g. atoms or molecules, or the conduction electrons in a solid. Consequently, the basis of quantum optics is non-relativistic quantum electrodynamics where matter is described by the Schrödinger equation rather than the Dirac equation, as would be the case in relativistic quantum electrodynamics. Furthermore light is described by second quantizing the Maxwell equations that are already relativistic, i.e. Lorentz invariant. However, sometimes a semiclassical approach suffices where the electromagnetic radiation field is treated according to the wave equations of classical electrodynamics and only matter is treated quantum mechanically. However, we shall not further investigate the semiclassical theory of quantum optics.

These final sections provide a brief and necessarily incomplete overview of those aspects of the vast subject of quantum optics that lead us most directly to the formulation of simple models, which, however, permit exact solutions. Detailed expositions of quantum optics may be found, e.g. in Fox (2006), Garrison and Chiao (2008), or Gerry and Knight (2005). A nice exposition, especially of the experimental situation, can also be found in Haroche and Raimond (2006).

We start with the quantization of the electromagnetic field, derive the Hamiltonian of an electron in an electromagnetic field, and conclude by deriving the interaction between an atom and the electromagnetic field.

8.10.1.1 Quantization of the electromagnetic field: photons

The dynamics of the electromagnetic field in vacuum with space- and time, dependent distributions of charges and currents, given by the charge density $\rho(\mathbf{r}, t)$ and the current density $\mathbf{j}(\mathbf{r}, t)$, is determined by Maxwell's equations (given here in Gaussian units)

$$\nabla \cdot \mathbf{E} = 4\pi\rho \tag{8.747}$$

$$\nabla \cdot \mathbf{B} = 0 \tag{8.748}$$

$$\nabla \times \mathbf{E} + \frac{1}{c}\frac{\partial}{\partial t}\mathbf{B} = 0 \tag{8.749}$$

$$\nabla \times \mathbf{B} - \frac{1}{c}\frac{\partial}{\partial t}\mathbf{E} = \frac{4\pi}{c}\mathbf{j} \tag{8.750}$$

for the space and time dependent electric and magnetic fields, $\mathbf{E}(\mathbf{r}, t)$ and $\mathbf{B}(\mathbf{r}, t)$.

Equations (8.747) and (8.750) yield the continuity equation that charge and current densities have to obey (using the mathematical identity $\nabla\cdot(\nabla\times\mathbf{V}) \equiv 0$, which holds for arbitrary vector fields $\mathbf{V}(\mathbf{r})$):

$$\frac{\partial\rho}{\partial t} + \nabla \cdot \mathbf{j} = 0. \tag{8.751}$$

Equation (8.748), which means that there are no magnetic monopoles, is identically satisfied (using again the same identity as for the continuity equation (8.751)) if we introduce another vector field, the vector potential $\mathbf{A}(\mathbf{r}, t)$ by

$$\mathbf{B} = \nabla \times \mathbf{A}. \tag{8.752}$$

Now, (8.749) can be written as

$$\nabla \times \left(\mathbf{E} + \frac{1}{c}\frac{\partial \mathbf{A}}{\partial t} \right) = 0. \tag{8.753}$$

The identity $\nabla\times (\nabla f) \equiv 0$, which is valid for arbitrary scalar fields $f(\mathbf{r})$, can now be used. Introducing a scalar potential $\varphi(\mathbf{r}, t)$ through the choice

$$\mathbf{E} + \frac{1}{c}\frac{\partial \mathbf{A}}{\partial t} = -\nabla\phi, \tag{8.754}$$

satisfies (8.749) identically.

The other two Maxwell equations, (8.747) and (8.750), now become in terms of the potentials

$$\frac{1}{c}\frac{\partial}{\partial t}\nabla \cdot \mathbf{A} + \nabla^2\phi = -4\pi\rho \tag{8.755}$$

and, using the identity for vector fields, $\nabla\times (\nabla\times\mathbf{V}) \equiv \nabla(\nabla\cdot\mathbf{V}) - \nabla^2\mathbf{V}$, and rearranging terms

$$\nabla \left(\nabla \cdot \mathbf{A} + \frac{1}{c}\frac{\partial}{\partial t}\phi \right) - \nabla^2\mathbf{A} + \frac{1}{c^2}\frac{\partial^2}{\partial t^2}\mathbf{A} = \frac{4\pi}{c}\mathbf{j}. \tag{8.756}$$

The four coupled partial differential equations, (8.755) and (8.756), can be simplified and, more importantly, decoupled by using a symmetry of the potentials in (8.752) and (8.754), which are invariant under the gauge transformations

$$\mathbf{A}' = \mathbf{A} + \nabla\chi \tag{8.757}$$

$$\phi' = \phi - \frac{1}{c}\frac{\partial \chi}{\partial t}, \tag{8.758}$$

with the arbitrary gauge potential $\chi = \chi(\mathbf{r}, t)$. The primed and the unprimed potentials lead to the same fields, which implies that we can choose a convenient gauge potential χ to simplify and decouple (8.755) and (8.756).

Looking at (8.756) suggests a choice such that the first term on the left-hand side vanishes, i.e. we would like to have

$$\nabla \cdot \mathbf{A} + \frac{1}{c}\frac{\partial}{\partial t}\phi = 0. \tag{8.759}$$

Indeed, if this equation were true, (8.755) and (8.756) would immediately become decoupled. Using the gauge transformed (i.e. primed) potentials in (8.755) and (8.756) shows that if the gauge potential, χ, satisfies

$$\nabla \cdot \mathbf{A}' + \frac{1}{c}\frac{\partial}{\partial t}\phi' = -\left(\nabla^2 - \frac{1}{c^2}\frac{\partial^2}{\partial t^2}\right)\chi = 0 \tag{8.760}$$

we have the desired simplified and decoupled equations for the primed potentials

$$\left(\nabla^2 - \frac{1}{c^2}\frac{\partial^2}{\partial t^2}\right)\mathbf{A}' = -\frac{4\pi}{c}\mathbf{j}, \tag{8.761}$$

$$\left(\nabla^2 - \frac{1}{c^2}\frac{\partial^2}{\partial t^2}\right)\phi' = -4\pi\rho. \tag{8.762}$$

The condition, (8.760), is called the Lorentz condition or Lorentz gauge. Of course, because of the gauge symmetry, we can drop the primes again now.

In a next step, we restrict ourselves to the case where there are no charges and currents: $\rho = 0$ and $\mathbf{j} = 0$. Furthermore, we choose $\phi = 0$, which turns the Lorentz gauge into the Coulomb gauge

$$\nabla \cdot \mathbf{A} = 0. \tag{8.763}$$

Thus, we are left with the wave equation

$$\left(\nabla^2 - \frac{1}{c^2}\frac{\partial^2}{\partial t^2}\right)\mathbf{A} = 0 \tag{8.764}$$

for the vector potential, which has plane waves as solutions:

$$\mathbf{A}(\mathbf{r}, t) = 2\mathbf{e}|\mathbf{A}_0|\cos(\mathbf{k}\cdot\mathbf{r} - \omega t + \alpha) \tag{8.765}$$

$$= \mathbf{A}_0 \exp\left(i(\mathbf{k}\cdot\mathbf{r} - \omega t)\right) + \text{c.c.}, \tag{8.766}$$

where we have used the complex amplitude

$$\mathbf{A}_0 = |\mathbf{A}_0|\mathbf{e}\exp(i\alpha). \tag{8.767}$$

The direction of propagation of the plane waves is given by the wave vector $\mathbf{k}$. Upon insertion into the wave equation, (8.764), of the plane wave solution we obtain the dispersion relation for light $\omega^2 = c^2|\mathbf{k}|^2$. The Coulomb gauge condition tells us that the plane waves are polarized perpendicular to the direction of propagation $\mathbf{e} \perp \mathbf{k}$, i.e. $\mathbf{e} \cdot \mathbf{k} = 0$.

The electric and magnetic fields are now easily obtained from (8.765)

$$\mathbf{E}(\mathbf{r}, t) = -\frac{1}{c}\frac{\partial}{\partial t}\mathbf{A}(\mathbf{r}, t) = -2|\mathbf{k}|\mathbf{e}|\mathbf{A}_0| \sin(\mathbf{k}\cdot\mathbf{r} - \omega t + \alpha) \tag{8.768}$$

$$\mathbf{B}(\mathbf{r}, t) = \nabla \times \mathbf{A}(\mathbf{r}, t) = -2\mathbf{k} \times \mathbf{e}|\mathbf{A}_0| \sin(\mathbf{k}\cdot\mathbf{r} - \omega t + \alpha)\,. \tag{8.769}$$

The plane wave solution for the vector potential, (8.766), is not the most general solution of the wave equation (8.764). In order to obtain a more general solution, it is advisable to now specify a volume for the field of the vector potential (loosely speaking, the 'light' field) and boundary conditions the field has to obey at the boundaries of the chosen volume. This allows us to describe the field with a discrete set of variables rather than the continuous description we are dealing with so far. A simple choice is a cubic volume $V = L^3$ of side length L and periodicity of the vector potential on the walls of the cubic volume

$$\mathbf{A}(L, y, z, t) = \mathbf{A}(0, y, z, t) \tag{8.770}$$

$$\mathbf{A}(x, L, z, t) = \mathbf{A}(x, 0, z, t) \tag{8.771}$$

$$\mathbf{A}(x, y, L, t) = \mathbf{A}(x, y, 0, t). \tag{8.772}$$

This allows us to perform what is called a normal mode analysis.

The general solution is then given by a normal mode or Fourier decomposition of $\mathbf{A}(\mathbf{r}, t)$

$$\mathbf{A}(\mathbf{r}, t) = \sum_{\mathbf{k}, k_z > 0} \sum_{\sigma=1,2} \left\{ a_{\mathbf{k},\sigma}(t)\mathbf{A}_{\mathbf{k},\sigma}(\mathbf{r}) + a^*_{\mathbf{k},\sigma}(t)\mathbf{A}^*_{\mathbf{k},\sigma}(\mathbf{r}) \right\}, \tag{8.773}$$

where we have introduced normal modes

$$\mathbf{A}_{\mathbf{k},\sigma}(\mathbf{r}) = A_{\mathbf{k}}\mathbf{e}_{\mathbf{k},\sigma} \exp(i\mathbf{k}\cdot\mathbf{r}) \tag{8.774}$$

with a real normalization factor $A_{\mathbf{k}}$ that we shall determine later (however, we anticipate that $A_{\mathbf{k}} = A_k = A_{-k}$) and Fourier coefficients or components $a_{\mathbf{k},\sigma}(t)$ ($\sigma = 1, 2$) of the vector potential. These Fourier coefficients will later, upon quantization, be identified with creation and annihilation operators for harmonic oscillator modes. The restriction to wave vectors with z-component $k_z > 0$ avoids double counting of the modes. (Recall that for every $\mathbf{k}$ with $k_z > 0$, i.e. in the upper $\mathbf{k}$-half space, there is a $\mathbf{k}$ with $k_z < 0$ in the lower $\mathbf{k}$-half space.) The two unit polarization vectors $\mathbf{e}_{\mathbf{k},\sigma}$ ($\sigma = 1, 2$) can, in the plane perpendicular to the direction of propagation given by the wave vector $\mathbf{k}$, be chosen orthogonal

$$\mathbf{e}_{\mathbf{k},\sigma} \cdot \mathbf{e}_{\mathbf{k},\sigma'} = \delta_{\sigma\sigma'}. \tag{8.775}$$

Furthermore, the periodic boundary conditions discretize the allowed **k** values to the set

$$\mathbf{k} = \frac{2\pi}{L}\mathbf{n} \quad \text{with} \quad n_x, n_y, n_z = 0, \pm 1, \pm 2, \ldots . \tag{8.776}$$

Inserting the Fourier decomposition, (8.773), using (8.774), into the wave equation (8.764) yields for the Fourier coefficients

$$\frac{d^2}{dt^2} a_{\mathbf{k},\sigma}(t) + \omega_{\mathbf{k}}^2 a_{\mathbf{k},\sigma}(t) = 0 \tag{8.777}$$

with ${\omega_{\mathbf{k}}}^2 = c^2|\mathbf{k}|^2$. This is, of course, nothing else but the well-known equation of motion of the harmonic oscillator for the Fourier coefficients $a_{\mathbf{k},\sigma}(t)$, whose solution is

$$a_{\mathbf{k},\sigma}(t) = a^{(1)}_{\mathbf{k},\sigma}(0)e^{-i\omega_k t} + a^{(2)}_{\mathbf{k},\sigma}(0)e^{i\omega_k t}. \tag{8.778}$$

Inserting this expression in (8.773) and choosing the time-independent Fourier coefficients ${a_{\mathbf{k},\sigma}}^{(i)}(0)$ in (8.778) in a clever way, we can get rid of the restriction $k_z > 0$ in (8.773). For $k_z > 0$, we relabel

$$a^{(1)}_{\mathbf{k},\sigma}(0) = a_{\mathbf{k},\sigma}(0) \quad \text{and} \quad a^{(2)}_{\mathbf{k},\sigma}(0) = -(-1)^\sigma a^*_{-\mathbf{k},\sigma}(0). \tag{8.779}$$

Using

$$\mathbf{e}_{\mathbf{k},\sigma} = -(-1)^\sigma \mathbf{e}_{-\mathbf{k},\sigma} \tag{8.780}$$

we obtain the result for the vector potential field

$$\mathbf{A}(\mathbf{r}, t) = \sum_{\mathbf{k}} \sum_{\sigma} A_k \mathbf{e}_{\mathbf{k},\sigma} \left\{ a_{\mathbf{k},\sigma}(t)e^{i\mathbf{k}\cdot\mathbf{r}} + a^*_{\mathbf{k},\sigma}(t)e^{-i\mathbf{k}\cdot\mathbf{r}} \right\}. \tag{8.781}$$

EXERCISE 8.38 Vector potential Complete the steps that lead to (8.781) observing that to every **k** with $k_z > 0$ there corresponds one, and only one, $\mathbf{k}'$ with $k_z < 0$. Hence, $\mathbf{k} \to -\mathbf{k}$ corresponds uniquely to going from the upper half-space ($k_z > 0$) to the lower half-space ($k_z < 0$) in **k**-space.

The observation that the Fourier coefficients

$$a_{\mathbf{k},\sigma}(t) = a_{\mathbf{k},\sigma}(0)e^{-i\omega_k t} \tag{8.782}$$

can be considered as the solutions of the first order differential equation

$$\frac{d}{dt}a_{\mathbf{k},\sigma}(t) = -i\omega_k a_{\mathbf{k},\sigma}(t) \tag{8.783}$$

plays a role in quantum mechanics. So far, all of this has been within the realm of classical electrodynamics. The last step before we quantize the electromagnetic field is writing down the classical Hamiltonian (also called the Hamiltonian function) of the electromagnetic field utilizing the results for the vector potential accumulated so far. Any textbook on *classical electrodynamics* (e.g. Jackson, 1999) tells us how to calculate the energy stored in the electromagnetic field in vacuum. We take this expression as the (classical) Hamiltonian:

$$\mathcal{H} = \frac{1}{8\pi}\int_V d^3r \left\{\mathbf{E}^2 + \mathbf{B}^2\right\}. \tag{8.784}$$

In the Coulomb gauge, we have been using, i.e. $\nabla\cdot\mathbf{A} = 0$ and $\phi = 0$, the Hamiltonian becomes

$$\mathcal{H} = \frac{1}{8\pi}\int_V d^3r \left\{\frac{1}{c^2}\frac{\partial \mathbf{A}}{\partial t}\cdot\frac{\partial \mathbf{A}}{\partial t} + (\nabla\times\mathbf{A})\cdot(\nabla\times\mathbf{A})\right\}. \tag{8.785}$$

Using (8.781), a straightforward calculation gives

$$\mathcal{H} = \sum_{\mathbf{k},\sigma}\frac{\omega_k^2}{4\pi c^2}A_k^2 L^3\left(a_{\mathbf{k},\sigma}a_{\mathbf{k},\sigma}^* + a_{\mathbf{k},\sigma}^* a_{\mathbf{k},\sigma}\right) \tag{8.786}$$

where, anticipating the role the Fourier coefficients will soon assume as creation and annihilation operators, we have been careful not to interchange any coefficients. Furthermore, anticipating that this expression will become the Hamiltonian of a collection of harmonic oscillators, we can now choose the normalization

$$A_k = \sqrt{\frac{2\pi}{L^3\omega_k}}, \tag{8.787}$$

thereby justifying the assumptions for this quantity we made earlier. We obtain

$$\mathcal{H} = \frac{1}{2}\sum_{\mathbf{k},\sigma}\omega_k\left(a_{\mathbf{k},\sigma}a_{\mathbf{k},\sigma}^* + a_{\mathbf{k},\sigma}^* a_{\mathbf{k},\sigma}\right). \tag{8.788}$$

The somewhat lengthy calculations to derive this result make up exercise 8.39.

EXERCISE 8.39 Hamiltonian of the electromagnetic field Derive the classical expression (8.788) for the Hamiltonian of the electromagnetic field. Some useful relations to do this are

$$\int_V e^{i\mathbf{k}\cdot\mathbf{r}} e^{-i\mathbf{k}'\cdot\mathbf{r}} = V\delta_{\mathbf{k},\mathbf{k}'}, \tag{8.789}$$

$$\mathbf{e}_{\mathbf{k},\sigma} \cdot \mathbf{e}_{\mathbf{k},\sigma'} = \delta_{\sigma,\sigma'}, \tag{8.790}$$

$$\mathbf{e}_{\mathbf{k},\sigma} \cdot \mathbf{e}_{-\mathbf{k},\sigma'} = -(-1)^{\sigma}\delta_{\sigma,\sigma'}. \tag{8.791}$$

You may also want to use (8.783).

In order to quantize the classical Hamiltonian (8.788) of the electromagnetic field, it is instructive to go through the following elementary exercise.

EXERCISE 8.40 Heisenberg equations of motion for the harmonic oscillator Introduce (commuting) complex variables a and a^* to write the Hamiltonian of the one-dimensional harmonic oscillator

$$\mathcal{H} = \frac{p^2}{2m} + \frac{m\omega^2}{2}q^2 \tag{8.792}$$

in the form

$$\mathcal{H} = \frac{\omega}{2}\left(a^*a + aa^*\right). \tag{8.793}$$

Show that the canonical equations

$$\frac{\partial\mathcal{H}}{\partial q} = -\dot{p} \quad \text{and} \quad \frac{\partial\mathcal{H}}{\partial p} = \dot{q} \tag{8.794}$$

lead to the classical equations of motion

$$\dot{a} = -i\omega a \quad \text{and} \quad \dot{a}^* = i\omega a^*. \tag{8.795}$$

Use the canonical quantization relation $[p, q] = -i$ to make the complex variables a and a^* into non-commuting operators a and $a^\dagger$, the creation and annihilation operator, respectively, whose commutator is $[a, a^\dagger] = 1$.

These operators satisfy the Heisenberg equations of motion

$$\frac{da}{dt} = i[\mathcal{H}, a] \quad \text{and} \quad \frac{da^\dagger}{dt} = i[\mathcal{H}, a^\dagger]. \tag{8.796}$$

The important point here is that we can show that these Heisenberg equations of motion have the same form as the classical equations of motion obtained from the canonical formalism.

Generalizing the results of exercise 8.40 for the Hamiltonian (8.781) and, thus, quantizing it, is now just a matter of calculating the commutator $[\mathcal{H}, a_{\mathbf{k},\sigma}]$ such that

$$\frac{d}{dt} a_{\mathbf{k},\sigma} = i[\mathcal{H}, a_{\mathbf{k},\sigma}] = -i\omega_k a_{\mathbf{k},\sigma}. \tag{8.797}$$

In order for (8.797) to hold, the creation and annihilation operators have to satisfy the commutation relations

$$[a_{\mathbf{k},\sigma}, a_{\mathbf{k}',\sigma'}] = \delta_{\mathbf{k},\mathbf{k}'}\delta_{\sigma,\sigma'}. \tag{8.798}$$

Finally the Hamiltonian (operator) takes the form

$$\mathcal{H} = \sum_{\mathbf{k},\sigma} \omega_k \left(a^\dagger_{\mathbf{k},\sigma} a_{\mathbf{k},\sigma} + \frac{1}{2} \right). \tag{8.799}$$

For completeness sake, we note down the expression for the operator of the vector field that we obtain from (8.781) by replacing the Fourier coefficients by annihilation and creation operators and inserting the expression (8.787) for the normalization

$$\mathbf{A}(\mathbf{r}) = \sum_{\mathbf{k}} \sum_{\sigma} \sqrt{\frac{2\pi}{L^3 \omega_k}} \mathbf{e}_{\mathbf{k},\sigma} \left\{ a_{\mathbf{k},\sigma} e^{i\mathbf{k}\cdot\mathbf{r}} + a^\dagger_{\mathbf{k},\sigma} e^{-i\mathbf{k}\cdot\mathbf{r}} \right\}. \tag{8.800}$$

The quantum states of the collection of harmonic oscillators is given by

$$|\ldots, n_{\mathbf{k},\sigma}, \ldots, n_{\mathbf{k}',\sigma'}, \ldots\rangle = |\ldots\rangle \ldots |n_{\mathbf{k},\sigma}\rangle \ldots |n_{\mathbf{k}',\sigma'}\rangle \ldots |\ldots\rangle. \tag{8.801}$$

The action of the annihilation and creation operators is to lower and raise, respectively, the occupation number by one of the mode $\mathbf{k}$, σ

$$a_{\mathbf{k},\sigma} |\ldots, n_{\mathbf{k},\sigma}, \ldots\rangle = \sqrt{n_{\mathbf{k},\sigma}} |\ldots, n_{\mathbf{k},\sigma} - 1, \ldots\rangle, \tag{8.802}$$

$$a^\dagger_{\mathbf{k},\sigma} |\ldots, n_{\mathbf{k},\sigma}, \ldots\rangle = \sqrt{n_{\mathbf{k},\sigma} + 1} |\ldots, n_{\mathbf{k},\sigma} + 1, \ldots\rangle. \tag{8.803}$$

In summary, this section demonstrated that light can be described quantum mechanically as a collection of independent harmonic oscillators, or independent *modes*.

The next section investigates the interaction of light with charged matter, again starting from the classical viewpoint.

8.10.1.2 Electron in an electromagnetic field

In order to understand the interaction of the electron and an electromagnetic field, we first consider the classical case. Our goal is to derive the Hamiltonian function describing the motion of the electron in an electromagnetic field. In the non-relativistic case in which we are interested, the electron of mass m and charge e experiences the Lorentz force, which we write in the form

$$\mathbf{F}_L = \frac{d}{dt}(m\mathbf{v}) = e\mathbf{E} + \frac{e}{c}\mathbf{v} \times \mathbf{B}. \tag{8.804}$$

In a first step, we want to find a Lagrange function such that it yields (8.804) as Euler–Lagrange equations. In order to achieve this, we use again the scalar and vector potentials ϕ and $\mathbf{A}$ from which the electric and magnetic fields can be calculated

$$\mathbf{E} = -\nabla\phi - \frac{1}{c}\frac{\partial \mathbf{A}}{\partial t} \tag{8.805}$$

and

$$\mathbf{B} = \nabla \times \mathbf{A}. \tag{8.806}$$

We insert these expressions for $\mathbf{E}$ and $\mathbf{B}$ into the Lorentz force (8.804) and restrict ourselves to the x-component. Rearranging terms, we obtain

$$\frac{d}{dt}\left(m\dot{x} + \frac{e}{c}A_x\right) = \frac{\partial}{\partial x}\left\{-e\phi + \frac{e}{c}(\dot{x}A_x + \dot{y}A_y + \dot{z}A_z)\right\}. \tag{8.807}$$

This expression for the Lorentz force is the Euler–Lagrange equation for the Lagrange function

$$L = \frac{m}{2}\left(\dot{x}^2 + \dot{y}^2 + \dot{z}^2\right) - e\phi + \frac{e}{c}\left(\dot{x}A_x + \dot{y}A_y + \dot{z}A_z\right). \tag{8.808}$$

The x-component of the canonical momentum is now

$$p_x = \frac{\partial L}{\partial \dot{x}} = m\dot{x} + \frac{e}{c}A_x. \tag{8.809}$$

Note that the canonical momentum, $\mathbf{p} = m\mathbf{v} + \frac{e}{c}\mathbf{A}$, is now different from the kinetic momentum $m\mathbf{v}$.

The Hamiltonian function can now be derived in the usual way and we obtain

$$\mathcal{H} = \frac{1}{2m}\left(\mathbf{p} - \frac{e}{c}\mathbf{A}\right)^2 + e\phi. \tag{8.810}$$

This Hamiltonian function describes the interaction of an electron with an electromagnetic field. Together with the quantized electromagnetic field in the absence of matter, which we derived in section 8.10.1.1, this will constitute the starting point for a number of approximations that will lead to Hamiltonians of models that admit an exact solution.

8.10.1.3 Interaction of light and matter

For our present purpose, it suffices to characterize matter as an assembly of mass points m_i and electric charges e_i ($i = 1, 2, \ldots, N$) located at positions $\mathbf{r}_i$ whose interaction is given by a pair potential $V(1, \ldots, \mathbf{r}_i, \ldots, \mathbf{r}_j, \ldots, N)$, e.g. the Coulomb potential

$$V(\{\mathbf{r}_i\}) = \sum_{i \neq j} \frac{e_i e_j}{|\mathbf{r}_i - \mathbf{r}_j|}. \tag{8.811}$$

Hence, non-relativistic matter can be described in our context by the many-particle Hamiltonian

$$\mathcal{H}_{MB} = \sum_i \frac{\mathbf{p}_i^2}{2m_i} + V(\{\mathbf{r}_i\}). \tag{8.812}$$

Section 8.10.1.2 showed that a charged particle is coupled to an electromagnetic field through the minimal coupling prescription

$$\mathbf{p}_i \to \mathbf{p}_i - \frac{e_i}{c}\mathbf{A}(\mathbf{r}_i). \tag{8.813}$$

Hence, the full Hamiltonian describing the interaction of non-relativistic matter and the electromagnetic field of light is given by replacing $\mathbf{p}_i$ in (8.812) with (8.813) and adding the Hamiltonian of the free electromagnetic field (8.799), i.e.

$$\begin{aligned} \mathcal{H} &= \sum_i \frac{1}{2m_i}\left(\mathbf{p}_i - \frac{e_i}{c}\mathbf{A}(\mathbf{r}_i)\right)^2 + V(\{\mathbf{r}_i\}) + \int_V d^3r \frac{1}{8\pi}(\mathbf{E}^2 + \mathbf{B}^2) \\ &= \sum_i \frac{1}{2m_i}\left(\mathbf{p}_i - \frac{e_i}{c}\mathbf{A}(\mathbf{r}_i)\right)^2 + V(\{\mathbf{r}_i\}) + \sum_{\mathbf{k},\sigma} \omega_k a^\dagger_{\mathbf{k},\sigma} a_{\mathbf{k},\sigma}. \end{aligned} \tag{8.814}$$

In this expression, we have assumed that the zero-point energy $\sum_{\mathbf{k},\sigma} \omega_k/2$ can be subtracted. This can be justified by the procedure of *normal ordering*, whose justification is discussed in section 2.2 in Garrison and Chiao (2008).

In the next step, we rewrite the Hamiltonian (8.814) such that it is split into three parts

$$\mathcal{H} = \mathcal{H}_{MB} + \mathcal{H}_{EMF} + \mathcal{H}_I, \tag{8.815}$$

where the first two terms represent the pure many-particle part without field, (8.812), and the pure light field, respectively, whereas the third part contains the interaction between light and matter. This latter part can be written as

$$\mathcal{H}_I = \sum_i \left\{ -\frac{e_i}{m_i c}\mathbf{A}(\mathbf{r}_i)\cdot\mathbf{p}_i + \frac{e_i^2}{2m_i c^2}\mathbf{A}^2(\mathbf{r}_i) \right\} = \mathcal{H}_I^{(1)} + \mathcal{H}_I^{(2)}. \tag{8.816}$$

The first term in (8.816) stems from the observation

$$[\mathbf{p}\cdot\mathbf{A}(\mathbf{r})]f(\mathbf{r}) = -i(\nabla\cdot\mathbf{A}(\mathbf{r}))f(\mathbf{r}) - i\mathbf{A}(\mathbf{r})\cdot\nabla f(\mathbf{r}) = \mathbf{A}(\mathbf{r})\cdot\mathbf{p}, \tag{8.816}$$

where we have used the Coulomb gauge, $\nabla\cdot\mathbf{A}(\mathbf{r}) = 0$.

We now focus on the case of one charged particle in the electromagnetic field. Using (8.800), $\mathcal{H}_I^{(1)}$ becomes

$$\mathcal{H}_I^{(1)} = -\frac{e}{mc}\sum_{\mathbf{k},\sigma}\sqrt{\frac{2\pi}{L^3\omega_k}}\left(a_{\mathbf{k},\sigma}e^{i\mathbf{kr}} + a^\dagger_{\mathbf{k},\sigma}e^{-i\mathbf{kr}}\right)\mathbf{e}_{\mathbf{k},\sigma}\cdot\mathbf{p}, \tag{8.817}$$

which describes atomic transitions where one photon is either created or annihilated.

Using (8.800) again to express $H_I^{(2)}$ reveals that the corresponding processes involve two photons. We are not going to write the expression for $H_I^{(2)}$ down explicitly, but only note that it is proportional to the fine structure constant $\alpha = e^2$ and hence, the corresponding processes are of higher order in perturbation theory. We shall not further consider $H_I^{(2)}$.

The next section introduces the final approximations that allow us to describe a reasonably simple, yet physically relevant, model: the quantum Rabi model, which is the main goal of this chapter.

8.10.2 The quantum Rabi Hamiltonian

We first summarize the results of section 8.10.1 for the special case of a single non-relativistic electron.

The Hamiltonian of a non-relativistic electron in a Coulomb potential $V(\mathbf{r})$ and a quantized electromagnetic field has been found to be given by

$$\mathcal{H} = \frac{1}{2m}(\mathbf{p} - e\mathbf{A}(\mathbf{r}))^2 + V(\mathbf{r}) + \sum_k \omega_k a_k^\dagger a_k. \tag{8.818}$$

Inserting the vector potential

$$\mathbf{A}(\mathbf{r}) = \sum_k \sqrt{\frac{\pi}{\omega_k L^3}}\mathbf{e}_k\left\{a_k e^{i\mathbf{k}\cdot\mathbf{r}} + a_k^\dagger e^{-i\mathbf{k}\cdot\mathbf{r}}\right\} \tag{8.819}$$

leads to a three-term Hamiltonian

$$\mathcal{H} = \mathcal{H}_A + \mathcal{H}_F + \mathcal{H}_I, \tag{8.820}$$

with the atomic Hamiltonian

$$\mathcal{H}_A = \frac{1}{2m}\mathbf{p}^2 + V(\mathbf{r}), \tag{8.821}$$

the field Hamiltonian

$$\mathcal{H}_F = \sum_k \omega_k a_k^\dagger a_k, \tag{8.822}$$

and the interaction Hamiltonian between atom and field

$$\mathcal{H}_I = -\frac{e}{m}\mathbf{A}\cdot\mathbf{p}. \tag{8.823}$$

As discusses earlier, we have neglected the term proportional to $\mathbf{A}^2$.

The operators $a_k^\dagger$ and a_k are Boson creation and annihilation operators that satisfy the commutation relations

$$[a_k, a_{k'}] = [a_k^\dagger, a_{k'}^\dagger] = 0, \quad [a_k, a_{k'}^\dagger] = \delta_{k,k'}. \tag{8.824}$$

We now simplify these expressions by introducing a number of assumptions or approximations.

First, we consider the atom as a two-level system with ground state $|g\rangle = |-\rangle$ and excited state $|e\rangle = |+\rangle$, which satisfy the Schrödinger equation with the atomic Hamiltonian

$$\mathcal{H}_A|\pm\rangle = E_\pm|\pm\rangle. \tag{8.825}$$

This two-level approximation can often be excellently justified by using light sources, mostly lasers, tuned to a specific transition frequency of a multi-level atom. In our case, the transition frequency is $\omega_0 = E_+ - E_-$. This gives the matrix elements

$$(\mathcal{H}_A)_{++} = \langle +|\mathcal{H}_A|+\rangle = E_+ \tag{8.826}$$

$$(\mathcal{H}_A)_{--} = \langle -|\mathcal{H}_A|-\rangle = E_- \tag{8.827}$$

$$(\mathcal{H}_A)_{+-} = (\mathcal{H}_A)_{-+} = 0 \tag{8.828}$$

or representeded in a 2 × 2 matrix

$$\mathcal{H}_A = \begin{pmatrix} E_+ & 0 \\ 0 & E_- \end{pmatrix} = \frac{E_+ + E_-}{2}\mathbb{I} + \frac{\omega_0}{2}\sigma^z \tag{8.829}$$

where $\mathbb{I}$ is the 2 × 2 unit matrix

$$\mathbb{I} = \begin{pmatrix} 1 & 0 \\ 0 & 1 \end{pmatrix}, \tag{8.830}$$

and the Pauli matrices are

$$\sigma^x = \begin{pmatrix} 0 & 1 \\ 1 & 0 \end{pmatrix}, \quad \sigma^y = \begin{pmatrix} 0 & -i \\ i & 0 \end{pmatrix}, \quad \sigma^z = \begin{pmatrix} 1 & 0 \\ 0 & -1 \end{pmatrix}, \tag{8.831}$$

which satisfy the commutation relations $[\sigma^x, \sigma^y] = 2i\sigma^z$ (and cyclic permutations thereof). These matrices act in a two-dimensional Hilbert space, which we denote as $\mathscr{H} = \mathbb{C}^2$. In order to have simpler expressions, we choose the energy zero such that $E_+ + E_- = 0$. Then

$$\mathcal{H}_A = \frac{\omega_0}{2}\sigma^z, \tag{8.832}$$

i.e.

$$E_+ = \frac{1}{2}\omega_0. \tag{8.833}$$

The next approximation uses that optical wavelengths, $\lambda = 2\pi/k$, are much greater than the atomic dimensions r at which the light interacts with the electrons of the atom, hence

$$e^{\pm i\mathbf{k}\cdot\mathbf{r}} \approx 1 \tag{8.834}$$

in (8.819).

Furthermore, in order to make effective use of our two-level approximation, we assume that the light is monochromatic $\omega_k = \omega$ with polarization $\mathbf{e}$, i.e. we retain only a single mode. These assumptions lead to

$$\mathbf{A}(\mathbf{r}) = \sqrt{\frac{\pi}{\omega L^3}}\mathbf{e}\left\{a + a^\dagger\right\} \tag{8.835}$$

and

$$\mathcal{H}_F = \omega a^\dagger a. \tag{8.836}$$

Next, we calculate the matrix elements of the interaction Hamiltonian $\mathcal{H}_I$ in the two-level basis $|\pm\rangle$. To do this, the equation

$$[\mathbf{r}, \mathcal{H}_A] = \frac{i}{m}\mathbf{p} \tag{8.837}$$

will be useful, the proof of which we leave to exercise 8.41.

EXERCISE 8.41 Prove (8.837).

We obtain

$$(\mathcal{H}_I)_{++} = (\mathcal{H}_I)_{--} = 0, \tag{8.838}$$

$$(\mathcal{H}_I)_{+-} = -i\omega_0 \mathbf{g} \cdot \mathbf{A}, \quad (\mathcal{H}_I)_{-+} = i\omega_0 \mathbf{g} \cdot \mathbf{A}. \tag{8.839}$$

We have introduced the dipole moment operator $\mathbf{g} = e\langle +|\mathbf{r}|-\rangle = e\langle -|\mathbf{r}|+\rangle$.

Another simple exercise:

EXERCISE 8.42 Using (8.837), convince yourself that (8.838) and (8.839) are correct.

We can now combine the results (8.838) and (8.839) with (8.835) to obtain the interaction Hamiltonian in 2×2 matrix form

$$\mathcal{H}_I = g\sigma^y(a + a^\dagger), \tag{8.840}$$

where the coupling constant is

$$g = e\omega_0\sqrt{\frac{\pi}{\omega L^3}}\mathbf{g} \cdot \mathbf{e}. \tag{8.841}$$

It is more convenient to express H_I using spin raising and lowering operators

$$\sigma^+ = \frac{1}{2}(\sigma^x + i\sigma^y) = \begin{pmatrix} 0 & 1 \\ 0 & 0 \end{pmatrix}, \quad \sigma^- = \frac{1}{2}(\sigma^x - i\sigma^y) = \begin{pmatrix} 0 & 0 \\ 1 & 0 \end{pmatrix}, \tag{8.842}$$

in terms of which the interaction Hamiltonian becomes

$$\mathcal{H}_I = -ig(\sigma^+ - \sigma^-)(a + a^\dagger). \tag{8.843}$$

A canonical transformation of the complete Hamiltonian, i.e. a rotation about the z-axis in spin space

$$\mathcal{H} \to e^{i\frac{\pi}{4}\sigma^z}\mathcal{H}e^{-i\frac{\pi}{4}\sigma^z} \tag{8.844}$$

transforms

$$-i\sigma^+ \to \sigma^+ \tag{8.845}$$

$$i\sigma^- \to \sigma^- \tag{8.846}$$

$$\sigma^z \to \sigma^z \tag{8.847}$$

so that we arrive at the Hamiltonian

$$\mathcal{H} = \omega a^\dagger a + \frac{\omega_0}{2}\sigma^z + g(\sigma^+ + \sigma^-)(a + a^\dagger). \tag{8.848}$$

This Hamiltonian describes the interaction of a single two-level system with a single-mode photon field in the simplest way. It is called the quantum Rabi model but is also known, for obvious reasons, under the name spin-Boson model. A schematic experimental arrangement described by the quantum Rabi model is given in Figure 8.9. The quantum Rabi model describes the case when the interaction between the atom and the single photon (g) is larger than the dissipative loss of photons (at rate κ) or the emission of the atom into other modes at rate γ or the inverse of the transfer time of the atoms through the cavity $1/t_{\text{transfer}}$; i.e. when $g \gg \kappa, \gamma, 1/t_{\text{transfer}}$, cp. Schoelkopf and Girvin (2008).

Despite its simplicity, the quantum Rabi model has not been solved exactly until very recently. The recent solution by Braak (2011) is quite involved and requires mathematical

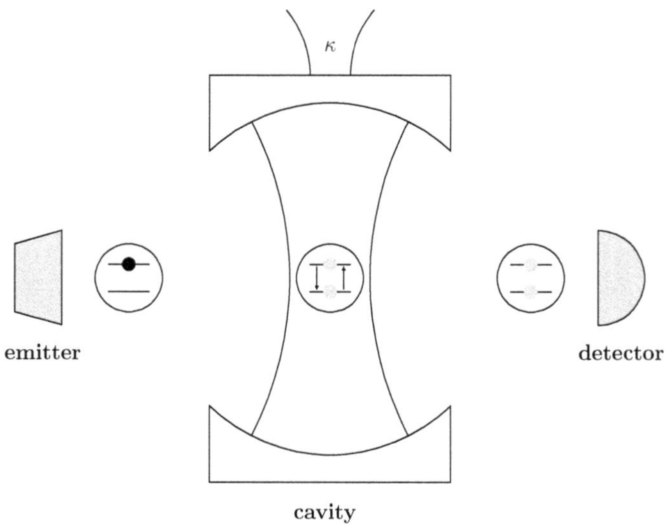

Figure 8.9 *Scheme of a cavity quantum electrodynamics system: a two-level atom trapped by mirrors forming the cavity.*

techniques not covered here. It has, however, kindled a great deal of new activity devoted to the quantum Rabi and related models.

EXERCISE 8.43 Quantum Rabi model including two-photon processes In order to derive the Hamiltonian (8.848) of the quantum Rabi model, we neglected the term proportional to the square of the vector potential in (8.816). However, this is not necessary.

Include the neglected term proportional to $\mathbf{A}^2(\mathbf{r})$ again in the case of one electron, i.e. in the Hamiltonian (8.818) and use the single mode approximation. The Hamiltonian now contains

$$\mathcal{H}_I^{(2)} = K(a + a^\dagger)^2. \tag{8.849}$$

Determine the constant K.

Show that the Bogoliubov transformation for the annihilation and creation operators

$$a = b\cosh\gamma - b^\dagger \sinh\gamma \tag{8.850}$$

$$a^\dagger = -b\sinh\gamma + b^\dagger \cosh\gamma \tag{8.851}$$

can be used to recover the form of the quantum Rabi Hamiltonian (8.848). Only the constants ω, ω_0, and g will be renormalized and a constant shift of energy will be acquired. As a consistency check, recover the original result (8.848) by choosing a suitable limiting value for K.

We now turn to the study of a simpler model derived from the the quantum Rabi Hamiltonian (8.848) using a further approximation, the so-called rotating or resonant wave approximation (RWA). This approximation renders the quantum Rabi model solvable by quite elementary means and is, thus, quite useful. The resulting model, the quantum Jaynes–Cummings model (Jaynes and Cummings, 1963), is, however, also a physically relevant model in its own right as it exhibits important physical effects.

8.10.3 The rotating wave approximation

The rotating wave approximation neglects the terms

$$g a^\dagger \sigma^+ \quad \text{and} \quad g a \sigma^- \tag{8.852}$$

in (8.848) and results in the quantum Jaynes–Cummings model Hamiltonian

$$\mathcal{H}_{\mathcal{JC}} = \omega a^\dagger a + \frac{\omega_0}{2}\sigma^z + g(\sigma^+ a + \sigma^- a^\dagger) = \mathcal{H}_F + \mathcal{H}_A + \mathcal{H}_{I,\mathcal{JC}}. \tag{8.853}$$

In order to justify the rotating wave approximation, we write the interaction part

$$\mathcal{H}_{R,I} = g(\sigma^+ + \sigma^-)(a + a^\dagger) = g(\sigma^+ a + \sigma^- a^\dagger + \sigma^+ a^\dagger + \sigma^- a) \tag{8.854}$$

of the Hamiltonian (8.848) of the quantum Rabi model in the interaction picture

$$e^{i\mathcal{H}_0 t}\mathcal{H}_{R,I}e^{-i\mathcal{H}_0 t} \tag{8.855}$$

where

$$\mathcal{H}_0 = \mathcal{H}_F + \mathcal{H}_A = \omega a^\dagger a + \frac{\omega_0}{2}\sigma^z. \tag{8.856}$$

We obtain

$$\begin{aligned}\mathcal{H}_{R,I}(t) &= e^{i\mathcal{H}_0 t}\mathcal{H}_{R,I}e^{-i\mathcal{H}_0 t} \\ &= g\left(\sigma^+ a e^{i(\omega_0-\omega)t} + \sigma^- a^\dagger e^{-i(\omega_0-\omega)t} + \sigma^+ a^\dagger e^{i(\omega_0+\omega)t} + \sigma^- a e^{-i(\omega_0+\omega)t}\right).\end{aligned} \tag{8.857}$$

The four terms in this expression have a markedly different time behaviour. While the first two terms vary slowly with time, $e^{\pm i(\omega_0-\omega)t}$, the last two terms vary rapidly with time, $e^{\pm i(\omega_0+\omega)t}$. The first two terms, corresponding to the absorption of a photon by the atom elevated to its excited state and the opposite process, an emission of a photon by the atom de-excited to the ground state, are therefore kept in the rotating wave approximation. This approximation is valid for $|\omega_0 - \omega| \ll \omega_0$, i.e. close to resonance of the photon with the splitting of the two-level atom.

The other two terms, corresponding to simultaneous emission of a photon and excitation of the atom and to simultaneous absorption of a photon and de-excitation of the atom, respectively, which are processes occurring at much higher energies, are discarded in this approximation. The latter processes can only occur in the framework of the quantum Rabi model when the coupling strength g is very strong.

More precisely, the following regimes are discussed in the literature (see Braak, 2013, and references cited therein). In the weak coupling regime, where the ratio of the coupling strength to the photon frequency $\frac{g}{\omega} \ll 1$, the rotating wave approximation may be applicable, whereas for $g/\omega \approx 1$, the so-called ultra-strong coupling regime, and for $g/\omega \leq 1$ or even $g/\omega > 1$, the so-called deep strong coupling regime, the full quantum Rabi model must be considered.

EXERCISE 8.44 Rotating wave approximation In order to derive the expression (8.857) for the interaction part of the quantum Rabi model Hamiltonian, derive the useful relations:

$$e^{\xi\frac{\sigma^z}{2}}\sigma^\pm e^{-\xi\frac{\sigma^z}{2}} = e^{\pm\xi}\sigma^\pm \tag{8.858}$$

$$e^{\xi a^\dagger a} a e^{-\xi a^\dagger a} = e^{-\xi} a \tag{8.859}$$

$$e^{\xi a^\dagger a} a^\dagger e^{-\xi a^\dagger a} = e^{\xi} a^\dagger \tag{8.860}$$

for an arbitrary complex number ξ.

The fastest way to derive these results is to establish a first order differential equation by differentiating the left-hand sides by ξ. The solution of these differential equations satisfying a suitable initial condition yields the right-hand sides.

For details of the rotating wave approximation, see Fox (2006), chapter 9.

EXERCISE 8.45 Paramagnetic resonance Consider an electron localized in a crystal lattice in such a way that the spin states of the electron are the only remaining degrees of freedom. A strong, time-independent magnetic field B_0 is applied in the z-direction and a weak oscillating field $B_1 \cos \omega t$ is applied in the x-direction.

1. Derive the Hamiltonian of this two-level system.
2. Introduce two more frequencies, besides ω, to express the two-level Hamiltonian in terms of these three frequencies. Approximately solve the time-dependent Schrödinger equation in the limiting case, where $|2\frac{\omega_0}{\omega} - 1| \ll 1$, i.e. use the rotating wave approximation. Describe the time dependence of the wave function if initially ($t = 0$) the spin of the electron points in the positive z-direction. Calculate the probability of the spin to point in the negative z-direction for $t > 0$.

8.10.4 The quantum Jaynes–Cummings model

This section follow, with some variations, the excellent exposition of Garrison and Chiao's (2008) chapter 12.

In order to solve the quantum Jaynes–Cummings model, we first consider the non-interacting case. The product states

$$|\pm, n\rangle^{(0)} = |\pm\rangle|n\rangle, \quad n = 0, 1, 2, \ldots, \tag{8.861}$$

where the states $|\pm\rangle$ are the atomic eigenstates of the two-level system (see (8.825)) and $|n\rangle$ are the number states of the single mode photon field

$$\mathcal{H}_F|n\rangle = \omega a^\dagger a|n\rangle = \omega n|n\rangle \tag{8.862}$$

solve the non-interacting Schrödinger equation

$$(\mathcal{H}_F + \mathcal{H}_A)|\pm, n\rangle^{(0)} = (E_\pm + \omega n)|\pm, n\rangle^{(0)}. \tag{8.863}$$

This justifies the name *bare states* for $|\pm, n\rangle^{(0)}$.

In order to further pursue our goal to solve the fully interacting case, we first observe that a straightforward calculation yields

$$\mathcal{H}_{I,\mathcal{JC}}|-, 0\rangle^{(0)} = 0. \tag{8.864}$$

EXERCISE 8.46 Show this result. What happens when $\mathcal{H}_I$ of the quantum Rabi model (8.848) acts on the state $|-,0\rangle^{(0)}$?

The result (8.864) indicates that there is no spontaneous absorption from the bare vacuum $|-,0\rangle^{(0)}$ in the rotating wave approximation. As a consequence, the ground state energy and the ground state vector of the combined system of two-level atom and electromagnetic field are

$$E_g = E_- = -\frac{\omega_0}{2} \quad \text{and} \quad |0\rangle = |-,0\rangle^{(0)}, \tag{8.865}$$

respectively.

For an arbitrarily given photon number, n, the pair of bare states $|+,n\rangle^{(0)}$ and $|-,n+1\rangle^{(0)}$ satisfy

$$\mathcal{H}_{I,\mathcal{JC}}|+,n\rangle^{(0)} = g\sqrt{n+1}|-,n+1\rangle^{(0)}, \tag{8.866}$$

$$\mathcal{H}_{I,\mathcal{JC}}|-,n+1\rangle^{(0)} = g\sqrt{n+1}|+,n\rangle^{(0)}. \tag{8.867}$$

EXERCISE 8.47 Derive these results.

The Hilbert space of the quantum Jaynes–Cummings model is therefore split into subspaces: a one-dimensional subspace $\mathscr{H}_g$ for the vacuum or ground state with photon number $n=0$ spanned by $|-,0\rangle^{(0)}$ and two-dimensional subspaces $\mathscr{H}_n$ with photon number $n=0,1,2,\ldots$ spanned by the pairs $|+,n\rangle^{(0)}$ and $|-,n+1\rangle^{(0)}$:

$$\mathscr{H}_{\mathcal{JC}} = \mathscr{H}_g \oplus \mathscr{H}_0 \oplus \mathscr{H}_1 \oplus \mathscr{H}_2 \oplus \ldots \tag{8.868}$$

In each subspace $\mathscr{H}_n$, the Hamiltonian of the quantum Jaynes–Cummings model can be written as a 2×2 matrix

$$\begin{aligned}\mathcal{H}_{\mathcal{JC},n} &= {}^{(0)}\langle n_\pm, \pm|\mathcal{H}_{\mathcal{JC}}|\pm, n_\pm\rangle^{(0)} \\ &= \left(n+\frac{1}{2}\right)\omega\begin{pmatrix}1 & 0\\ 0 & 1\end{pmatrix} + \frac{1}{2}\begin{pmatrix}\delta & -2g\sqrt{n+1}\\ 2g\sqrt{n+1} & -\delta\end{pmatrix},\end{aligned} \tag{8.869}$$

where $n_\pm = n$ for the excited atomic state $|+\rangle$ and $n_\pm = n+1$ for the atomic ground state $|-\rangle$. Furthermore, we introduced the detuning $\delta = \omega_0 - \omega$.

EXERCISE 8.48 Go through the steps to derive $\mathcal{H}_{\mathcal{JC},n}$ (8.869).

The solution of the full Schrödinger equation is, thus, reduced to the diagonalization of the 2×2 Hamiltonian matrix $\mathcal{H}_{\mathcal{J}C,n}$ in the subspace $\mathscr{H}_n$, i.e.

$$\mathcal{H}_{\mathcal{J}C,n}|\varphi_n\rangle = \epsilon_{\pm,n}|\varphi_n\rangle, \tag{8.870}$$

where

$$|\varphi_n\rangle = a|+,n\rangle^{(0)} + b|-,n+1\rangle^{(0)}. \tag{8.871}$$

A straightforward diagonalization of the 2×2 matrix Hamiltonian $H_{\mathcal{J}C,n}$ (8.869) yields

$$\epsilon_{+,n} = \left(n - \frac{1}{2}\right)\omega + \frac{\Omega_n}{2} \tag{8.872}$$

$$\epsilon_{-,n} = \left(n - \frac{1}{2}\right)\omega - \frac{\Omega_n}{2} \tag{8.873}$$

with

$$\Omega_n = \sqrt{\delta^2 + 4g^2(n+1)} \tag{8.874}$$

for the eigenenergies and

$$|+,n\rangle = \sin\theta_n|+,n\rangle^{(0)} + \cos\theta_n|-,n+1\rangle^{(0)} \tag{8.875}$$

$$|-,n\rangle = \cos\theta_n|+,n\rangle^{(0)} - \sin\theta_n|-,n+1\rangle^{(0)} \tag{8.876}$$

with

$$\cos\theta_n = \frac{\Omega_n - \delta}{\sqrt{(\Omega_n - \delta)^2 + 4g^2(n+1)}} \tag{8.877}$$

$$\sin\theta_n = \frac{2g\sqrt{n+1}}{\sqrt{(\Omega_n - \delta)^2 + 4g^2(n+1)}} \tag{8.878}$$

for the eigenstates of the quantum Jaynes–Cummings model in the subspace $\mathscr{H}_n$.

We recover the bare eigenenergies (8.863) for the non-interacting case, $g = 0$ in the form

$$\epsilon^{(0)}_{+,n} = \left(n + \frac{1}{2}\right)\omega + \frac{\delta}{2} \tag{8.879}$$

$$\epsilon^{(0)}_{-,n} = \left(n + \frac{1}{2}\right)\omega - \frac{\delta}{2}, \tag{8.880}$$

which makes their degeneracy obvious for vanishing detuning, $\delta = 0$, i.e. at resonance. However, for any finite interaction, $g > 0$, the exact eigenvalues satisfy

$$\epsilon_{+,n} - \epsilon_{-,n} = \Omega_n \geq 2g\sqrt{n+1}. \tag{8.881}$$

This is an example of the phenomenon of avoided level crossing or level repulsion that occurs when two quantum states are perturbatively coupled.

We have now solved exactly the time-independent Schrödinger equation of the quantum Jaynes–Cummings Hamiltonian. In particular, we obtained the exact eigenstates, the ground state $|0\rangle$, and the dressed states $|+, n\rangle$ and $|-, n\rangle$. These dressed states have become entangled states through the atom-field interaction of the bare atomic and field states.

Using the dressed eigenstates of the quantum Jaynes–Cummings model in the subspace $\mathscr{H}_n$, we can write down the general time dependent state

$$|\psi(t)\rangle = c_0 e^{-iE_g t}|0\rangle + \sum_{n=0}^{\infty} \sum_{\ell=+,-} c_{\ell,n} e^{-i\epsilon_{\ell,n} t}|\ell, n\rangle. \tag{8.882}$$

EXERCISE 8.49 Rabi oscillations Use the state $|\psi(t)\rangle$ (8.882) to calculate the probabilities for the states

$$P_{+,m} = \left| {}^{(0)}\langle +, m|\psi(t)\rangle \right|^2 \tag{8.883}$$

$$P_{-,m} = \left| {}^{(0)}\langle -, m+1|\psi(t)\rangle \right|^2 \tag{8.884}$$

at resonance, $\delta = 0$.

Interpret especially the case $m = 0$, i.e. $P_{+,0}$ and $P_{-,1}$, describing the so-called Rabi flopping or Rabi oscillations (consult, e.g. Garrison and Chiao (2008), chapter 12).

8.10.5 The quantum Tavis–Cummings model

The quantum Tavis–Cummings model (Tavis and Cummings, 1968, 1969) is a natural generalization of the quantum Jaynes–Cummings model. The model actually dates back to Dicke (1954). An ensemble of N two-level atoms is placed in a single-mode photon field. The N two-level atoms are assumed to be identical and not to interact with each other directly. They interact, however, with the photon field via the same mechanism as in the quantum Jaynes–Cummings model.

The exposition in this section partly follows Bogoliubov *et al.* (1996) as well as a set of lecture notes by Bogoliubov (1996).

In the following, in order to avoid factors of 1/2, we use the spin operators

$$s^z = \frac{1}{2}\sigma^z, \quad s^{\pm} = \sigma^{\pm}, \tag{8.885}$$

with commutation relations

$$[s^+, s^-] = 2s^z, \quad [s^z, s^\pm] = \pm s^\pm. \tag{8.886}$$

The Hamiltonian of the quantum Tavis–Cummings model is hence obtained by replacing the spin operators representing a single two-level system by the corresponding spin operators for an ensemble of two-level systems

$$\tilde{\mathcal{H}}_{TC} = \omega a^\dagger a + \omega_0 \sum_{j=1}^{N} s_j^z, +g \left(a \sum_{j=1}^{N} s_j^+ + a^\dagger \sum_{j=1}^{N} s_j^- \right). \tag{8.887}$$

The spin operators satisfy the commutation relations

$$\left[s_i^+, s_j^-\right] = 2\delta_{ij} s_j^z \quad \left[s_i^z, s_j^\pm\right] = \pm \delta_{ij} s_j^\pm. \tag{8.888}$$

They act on the N-fold tensor product Hilbert space $\mathscr{H} = (\mathbb{C}^2)^{\otimes N}$ that has dimension 2^N as the tensor product

$$s_n^\ell = I \otimes I \ldots \otimes \underbrace{s^\ell}_{n} \otimes \ldots \otimes I \tag{8.889}$$

where I is the 2×2 unit matrix and s^ℓ is one of the spin operators $\ell = x, y, z, +, -$ acting on the nth two-level atom.

8.10.6 Elementary properties of the quantum Tavis–Cummings model

Introducing the total spin operators

$$S^z = \sum_{j=1}^{N} s_j^z \quad S^\pm = \sum_{j=1}^{N} s_j^\pm \tag{8.890}$$

which satisfy the $su(2)$ commutation relations

$$\left[S^+, S^-\right] = 2S^z \quad \left[S^z, S^\pm\right] = \pm S^\pm \tag{8.891}$$

the quantum Tavis–Cummings Hamiltonian (8.887) can be written in the compact form

$$\tilde{\mathcal{H}}_{TC} = \omega a^\dagger a + \omega_0 S^z + g\left(aS^+ + a^\dagger S^-\right). \tag{8.892}$$

The generalized number operator

$$\mathcal{M} = S^z + a^\dagger a \tag{8.893}$$

and the Casimir operator

$$\mathbf{S}^2 = S^+ S^- + S^z(S^z - 1) = S^- S^+ + S^z(S^z + 1) \tag{8.894}$$

commute with the quantum Tavis–Cummings Hamiltonian and among themselves

$$[\tilde{\mathcal{H}}_{TC}, \mathcal{M}] = [\tilde{\mathcal{H}}_{TC}, \mathbf{S}^2] = [\mathcal{M}, \mathbf{S}^2] = 0. \tag{8.895}$$

Hence, $\mathcal{M}$ and $\mathbf{S}^2$ are constants of the motion, the eigenstates of $\tilde{\mathcal{H}}_{TC}$ can be chosen as simultaneous eigenstates of M, $\mathbf{S}^2$, and $\tilde{\mathcal{H}}_{TC}$.

These observations can also be used to obtain a further simplification. Instead of the quantum Tavis–Cummings Hamiltonian $\tilde{\mathcal{H}}_{TC}$ (8.892) we can equivalently treat the Hamiltonian

$$\mathcal{H}_{TC} = \frac{1}{g}\left(\tilde{\mathcal{H}}_{TC} - \omega \mathrm{M}\right) = \Delta S^z + aS^+ + a^\dagger S^-, \tag{8.896}$$

where

$$\Delta = \frac{\omega_0 - \omega}{g} = \frac{\delta}{g} \tag{8.897}$$

is the detuning δ renormalized by the coupling constant g.

In a first step, we consider the eigenstates of the spin operators $\mathbf{S}^2$ and S^z

$$\mathbf{S}^2|S, m\rangle = S(S+1)|S, m\rangle, \quad \text{and} \quad S^z|S, m\rangle = m|S, m\rangle, \tag{8.898}$$

where the numbers S take the values $0 \leq S \leq \frac{N}{2}$, i.e. S is integer or half-odd integer depending on whether N is even or odd. For fixed $S, m = -S, -S+1, \ldots, S-1, S$ and the states $m|S, m\rangle$ form the basis of the irreducible representation of dimension $2S+1$. Starting from the lowest state $|S, -S\rangle$ with

$$S^-|S, -S\rangle = 0, \quad \text{and} \quad S^z|S, -S\rangle = -S|S, -S\rangle, \tag{8.899}$$

we construct the basis states

$$|S, m\rangle = g_m \left(S^+\right)^{m+S} |S, -S\rangle, \tag{8.900}$$

where the the normalization $\langle m, S|S, m\rangle = 1$ for $-S \leq m \leq S$ determines the coefficients

$$g_m = \sqrt{\frac{(S-m)!}{(S+m!)(2S+1)!}}. \tag{8.901}$$

The vacuum state $|\Omega_S\rangle$ is the product state of the lowest spin state $|S, -S\rangle$ and the vacuum state of the photon field, i.e. the Fock vacuum state, $|0\rangle$ with $a|0\rangle = 0$

$$|\Omega_S\rangle = |0\rangle|S, -S\rangle. \tag{8.902}$$

The vacuum state satisfies

$$\mathbf{S}^2|\Omega_S\rangle = S(S+1)|\Omega_S\rangle, \quad \text{and} \quad \mathrm{M}|\Omega_S\rangle = -S|\Omega_S\rangle. \tag{8.903}$$

Because $\mathcal{H}_{TC}$ commutes with both $\mathcal{M}$ and $\mathbf{S}^2$, we can now search for the simultaneous eigenstates, which we denote $|\Phi_{S,M}{}^{\rho}\rangle$

$$\tilde{\mathcal{H}}_{TC}|\Phi^{\rho}_{S,M}\rangle = \tilde{E}^{\rho}_{S,M}|\Phi^{\rho}_{S,M}\rangle \tag{8.904}$$

$$\mathcal{M}|\Phi^{\rho}_{S,M}\rangle = (M-S)|\Phi^{\rho}_{S,M}\rangle \tag{8.905}$$

$$\mathbf{S}^2|\Phi^{\rho}_{S,M}\rangle = S(S+1)|\Phi^{\rho}_{S,M}\rangle \tag{8.906}$$

$$0 \leq M < \infty, \quad 0 \leq S \leq \frac{N}{2}. \tag{8.907}$$

The first two equations are combined to

$$\mathcal{H}_{TC}|\Phi^{\rho}_{S,M}\rangle = \frac{1}{g}\left(\tilde{H}_{TC} - \omega \mathrm{M}\right)|\Phi^{\rho}_{S,M}\rangle = \frac{1}{g}\left(\tilde{E}^{\rho}_{S,M} - \omega(M-S)\right)|\Phi^{\rho}_{S,M}\rangle \tag{8.908}$$

when we use the compact form of the quantum Tavis–Cummings Hamiltonian (8.896). The wave function will now be expressed in terms of the vacuum state $|\Omega_S\rangle$ as

$$|\Phi^{\rho}_{S,M}\rangle = \sum_{m=0}^{min(2S,M)} A^{\rho}_{S,M,m}\left(S^+\right)^m\left(a^\dagger\right)^{M-m}|\Omega_S\rangle. \tag{8.909}$$

The significance of the index ρ will become clear later.

With the help of the observation

$$\mathcal{M}S^+ = S^+(\mathcal{M}+1) \quad \text{and} \quad \mathcal{M}a^\dagger = a^\dagger(\mathcal{M}+1) \tag{8.910}$$

and

$$[\mathbf{S}^2, S^+] = 0 \tag{8.911}$$

the two equations of (8.905) and (8.906) can easily be shown to hold.

A traditional approach to solve the eigenvalue equation would be to derive a difference equation for the coefficients $A^{\rho}_{S,M,m}$ from the eigenvalue equation (8.904). In order to obtain these equations, we use the property of the annihilation operators

$$a(a^\dagger)^n = (a^\dagger)^n a + n(a^\dagger)^{n-1} \tag{8.912}$$

to move them from left to right until they can act on the vacuum state and gather terms with equal powers of S^+ and $a^\dagger$.

We are not pursuing this path here further, but only remark that we would also find that the matrix eigenvalue problem we would be able to create from the difference equation would be of dimension $K = \min(2S, M) + 1$ for fixed values of S and M. Hence, for later use, we can extract the information that the index $\rho = 1, 2, \ldots, K$.

In part II, we introduce the algebraic Bethe ansatz method and return to the solution of the eigenvalue equation (8.904) as an application of this method.

Part 2

Algebraic Bethe Ansatz

I would rather discover one cause than gain the kingdom of Persia.

Democritus (c. 460 BC—c. 370 BC)

Hans Bethe's 1931 seminal paper expanded on the one-dimensional version of the model of interacting magnetic moments, suggested by Heisenberg (1928) only a few years earlier. In today's terminology, the method Bethe developed is called the coordinate Bethe ansatz. We develop this approach in part III.

Another exact solution of a model for magnetism, the two-dimensional Ising model, was discovered by Onsager (1944). The Ising model, though conceived by Lenz (1920) to describe magnetism, and solved in its one-dimensional form by Ising (1925) as a contribution to the theory of ferromagnetism, is really a classical model as the 'spins' in this model are described by numbers rather than quantum mechanical operators. However, this fact neither diminishes the usefulness of the Ising model nor does it belittle the formidable mathematical feat of finding its exact solution in two dimensions.

Over the years a number of further exact solutions of two-dimensional classical statistical mechanical models and one-dimensional quantum mechanical models were found. It became apparent that these models and their solutions bore strong connections that were then formalized in the intimately connected quantum inverse scattering and algebraic Bethe ansatz methods, the former used to construct the models and demonstrate their integrability, the latter to find equations determining their spectra.

It must be noted that quantum integrability, as opposed to classical integrabilty, is far from being a uniquely defined concept. It remains an open question what relation there is between competing concepts of quantum integrability and whether and in what sense they may be equivalent. For discussions of the ongoing quest for a consistent definition of quantum integrability, see Weigert (1992), Caux and Mossel (2011), Larson (2013), and Batchelor and Zhou (2015). Here, however, we only develop the quantum inverse scattering method and the algebraic Bethe ansatz and the associated concept of quantum integrability, most recently referred to as Yang–Baxter integrability. We will not explore the relation of this quantum integrability concept with other proposals.

It is also for the sake of this concept of quantum integrability that we chose an ahistorical sequence in which to present the algebraic and the coordinate Bethe ansatz. The latter clearly demonstrates exact solutions of the interacting one-dimensional quantum models for which it works but leaves open a deeper understanding of the reasons why these models admit exact solutions, and does also not shed light on whether and in what sense these models might be quantum integrable. Starting our exposition with the algebraic Bethe ansatz thus provides a much deeper grounding also for our later exposition of the coordinate Bethe ansatz method, which remains a valuable tool in the arsenal of Bethe ansatz methods, especially since it may be considered to be more intuitive.

Our general strategy to introduce the algebraic Bethe ansatz is via the relation between two-dimensional classical statistical mechanical models whose prototypes are the Ising and ice-type models and one-dimensional quantum mechanical models whose prototype is the nearest–neighbour spin-1/2 Heisenberg quantum spin chain. In chapter 6, we reviewed the general connection based on the analogy between thermal and quantum fluctuations between statistical mechanics on the one side and quantum mechanics and quantum field theory on the other side. Here we discuss model systems in two and one dimensions, respectively, where this connection can be implemented in an exact way so that exact solutions of the two-dimensional model translate into exact solutions of the one-dimensional model.

The original physical motivation to study vertex models came from a peculiar observation of a thermodynamic property of (three-dimensional) ice at very low temperatures. Therefore, chapter 9 considers this property of ice and draw motivation from it to study a two-dimensional version of 'ice'.

Chapter 10, looks at general vertex models and introduce the machinery to compute the partition function and, hence, the thermodynamic properties of these models. As the key mathematical construct, we will again discuss the transfer matrix introduced in earlier chapters.

Chapter 11 develops the concept of quantum integrability for one-dimensional quantum systems will emerge. We develop this concept by specializing to the two-dimensional six-vertex model. The connection between the two- and one-dimensional models is brought about by the transfer matrix. On the one hand, its diagonalization is achieved in an algebraic manner and leads to a set of algebraic equations that encode the solution of the two-dimensional model. On the other hand, an infinite number of commuting quantum Hamiltonians can be constructed from the transfer matrix, among them the Hamiltonian of a Heisenberg quantum spin chain. Thus, with the transfer matrix, we have simultaneously diagonalized these quantum Hamiltonians and demonstrated that this Heisenberg quantum spin chain is a quantum integrable model.

Chapters 10 and 11 are partly inspired by Leon Takhtajan's (1985) review and, in particular, a special lecture series by Tuong Truong (1987)—Ferenc Woynarovich and the author were the only students.

Finally, chapter 12 demonstrates how the algebraic version of the Bethe ansatz can be used for models when there is no underlying two-dimensional model, or if we do not know of one. In such cases, it is sometimes possible to conjecture the building blocks of the transfer matrix. The model we consider for this purpose is from quantum optics describing strong light-matter interactions, the Tavis–Cummings model introduced in section 8.10 as one of the descendent models of the Rabi model.

For vertex models and their exact solutions, the classical text by Baxter (1982) is invaluable. A similarly classical text on the quantum inverse scattering method and the algebraic Bethe ansatz is Korepin *et al.* (1993), while Slavnov (2018) leads up to recent developments.

9

Ice Model

There must be some definite cause why, whenever snow begins to fall, its initial formation invariably displays the shape of a six-cornered starlet. For if it happens by chance, why do they not fall just as well with five corners or with seven?

Johannes Kepler (1571–1630)

The crystal structure of several of the phases of ice shows a peculiarity associated with a special type of disorder (Ziman, 1979). In the words of Linus Pauling (1960): 'there is one proton along each oxygen-oxygen axis, closer to one or the other of the two oxygen atoms.' This peculiarity, arising from a double well potential along the oxygen-oxygen axis, so that the much smaller proton can sit in one or the other potential minimum, depends only on the configuration and is, hence, independent of temperature. It gives rise to a finite entropy of ice, even at zero temperature: the *residual entropy*.

Ice is a particular example of a class of hydrogen-bonded crystalline materials. There are, for instance, compounds like potassium dihydrogen phosphate (KH_2PO_4), commonly known as KDP where the electric dipoles of the crystal exhibit parallel order and which is, thus, a ferroelectric (cp. section 11.2.1). In ammonium dihydrogen phosphate ($NH_4H_2PO_4$, ADP), the electric dipoles have a preferred antiparallel orientation.

We use the observation of a residual entropy in these systems in section 9.1 as a physical motivation to study, in chapters 10 and 11, a certain type of *two-dimensional* statistical mechanical models, the so-called *vertex* models. A precise definition of the exemplary vertex model, the ice model, appears in section 9.2.

9.1 Physical motivation for the square lattice ice model

Careful experiments by Giauque and Stout (1936) on the heat capacity of ice down to very low temperatures led to the extrapolation that there will remain a residual entropy of ice crystals at zero temperature. Experimentally they obtained the value[1]

$$\frac{S}{N} \approx 0.41 \approx \ln(1.5) \tag{9.1}$$

for the entropy per oxygen atom.

[1] Boltzmann's constant has been set to unity.

Models of Quantum Matter: A First Course on Integrability and the Bethe Ansatz. Hans-Peter Eckle, Oxford University Press (2019). © Hans-Peter Eckle. DOI: 10.1093/oso/9780199678839.001.0001

Concurrent with these experiments, Pauling (1935) suggested an explanation for the residual entropy of ice, based on the observation discussed above.

Many different microscopic configurations of hydrogen atoms with respect to the oxygen atoms are possible which lead to the same macroscopic state. Then already the simplest assumption leads to a finite value for the residual entropy of ice: Assume a lattice of N oxygen atoms as vertices with one proton per bond (i.e. oxygen-oxygen axis), hence $2N$ hydrogen atoms in total. The proton can be in one of two positions: near to or far from one of the two oxygen atoms, and, consequently, in the opposite position with respect to the other oxygen atom. Therefore we have, ignoring boundaries and, more importantly, possible restrictions on them, 2^{2N} possible configurations. The number of possible microscopic configurations is therefore

$$Z = 2^{2N} = 4^N, \tag{9.2}$$

which leads to the residual entropy per lattice site

$$\frac{S}{N} = \ln(4). \tag{9.3}$$

This value is much too large compared to the experimental result. A simple consideration, going back to Bernal and Fowler (1933) and applied to the problem of the residual entropy of ice by Pauling (1935, 1960), improves this result considerably.

Since the oxygen atoms in ice are fourfold coordinated (cf. figure 9.1), each oxygen atom can be surrounded by either zero, one, two, three, or four hydrogen atoms in the adjacent position. This accounts for a total of 16 configurations. Pauling argued that only six out of these 16 configurations are, in fact, allowed, namely, those where there are two hydrogen atoms surrounding the oxygen atom. Therefore, we get, instead of (9.2),

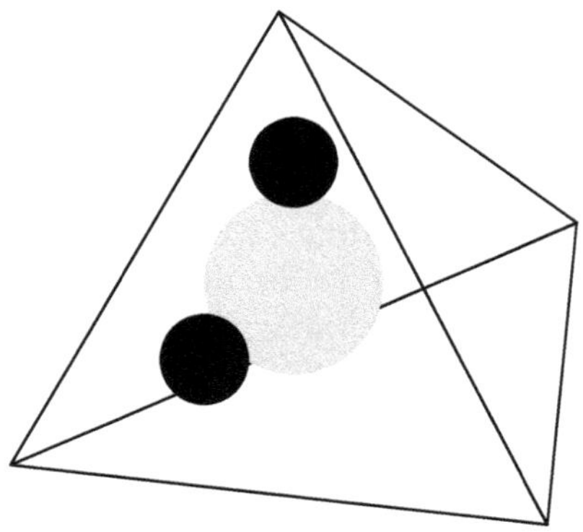

Figure 9.1 *Schematic representation of the coordination of water in an ice crystal.*

$$Z = 2^{2N}\left(\frac{6}{16}\right)^N = \left(\frac{3}{2}\right)^N, \tag{9.4}$$

and, thus,

$$\frac{S}{N} = \ln\left(\frac{3}{2}\right). \tag{9.5}$$

This value is in good agreement with the experimental result of Giauque and Stout (1936).

Introducing the number W such that $S = N\ln(W)$, numerical calculations by Nagle (1966) confirmed the value of $W \approx 1.5$.

Moreover, numerical calculations, also by Nagle (1966), for a caricature model of ice in two spatial dimensions yield

$$W \approx 1.540 \pm 0.001. \tag{9.6}$$

The latter result prompted mathematical physicist Elliott Lieb to investigate whether an exact solution of the problem of two-dimensional square ice was within reach (Lieb, 1967*a*, 1967*b*): 'It seemed worthwhile to try and find the value of W exactly for the following reasons: (a) It will serve as a check of Nagle's calculations. (b) It is an interesting graph theoretic problem. (c) It is the first step toward the solution of much more interesting problems having phase transitions . . .' Lieb (1967*b*).

Let us give a definition of the ice model before we return to Lieb's result for W.

Square ice, also called the *ice model*, is defined on a two-dimensional square lattice. Put N 'oxygen' atoms at the vertices of the two-dimensional square lattice and 'decorate' the edges with one of the $2N$ 'hydrogen' atoms. Each of the $2N$ 'hydrogen' atoms can be in one of the two positions, close or far with respect to a given oxygen atom, as depicted in figure 9.2.

Lieb (1967*a*, 1967*b*) solved the ice model using transfer matrices. He obtained the exact value

$$W = \left(\frac{4}{3}\right)^{\frac{3}{2}} = 1.5396007\ldots \tag{9.7}$$

Although we shall, in the following, present the transfer matrix method in some detail, our main interest is not in two-dimensional statistical mechanical models such as the ice model. These we rather use as a convenient entry point to study one-dimensional quantum mechanical models. In particular, the approach via two-dimensional statistical mechanical models will enable us to understand the integrability of the corresponding one-dimensional quantum models. For the same reason—our primary interest in one-dimensional quantum models—we shall also not present the details of Lieb's calculation of W for the ice model.

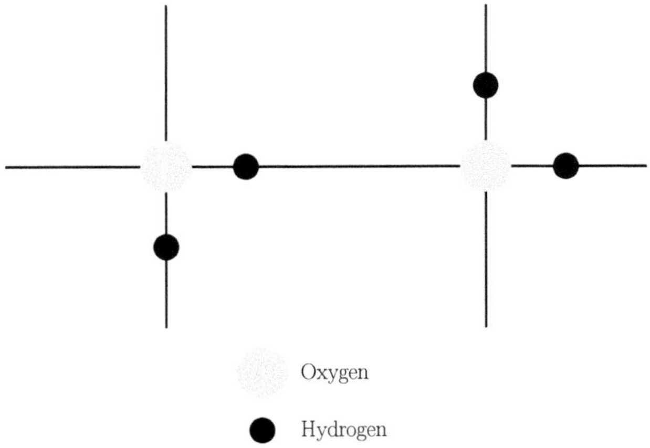

Figure 9.2 *Square ice.*

Let us reiterate the definition of the square lattice ice model and introduce some useful alternative graphical representation in section 9.2.

9.2 Definition of the ice model

The ice model is a two-dimensional model for square ice, i.e. it is a statistical mechanical model, as just described. However, we also need the (figure 9.3):

- *Ice rule*: there are exactly two hydrogen atoms near each oxygen atom.

We translate the 'ice picture' into the *vertex picture*: a hydrogen atom close to the oxygen atom, the vertex, is represented by an arrow pointing toward the oxygen atom, and a hydrogen atom far from the oxygen atom by an arrow pointing outward. Another way of decorating the lattice consists of using thick lines for arrows pointing downward or to the left.

While we will not use thick and thin lines to describe the ice model, we use the arrow picture, and call the ice model from now on mostly the six-vertex model, the oxygen atoms forming the vertices to or from which the arrows on the bonds or edges point.

The following exercise encourages playing with a small lattice to become familiar with the notions just introduced. Playing with small lattices is always a good idea to understand what is going on[2].

[2] Legend has it that Lars Onsager started his investigation of the two-dimensional Ising model by playing with small lattices.

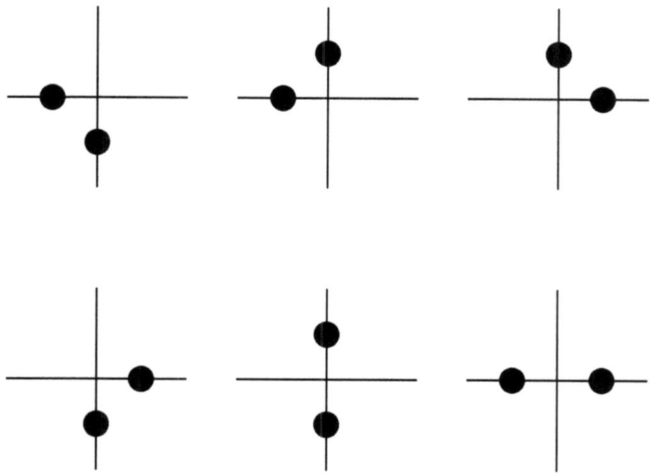

Figure 9.3 *The six local configurations for square ice allowed by the ice rule. The oxygen atoms at the vertices are not shown.*

EXERCISE 9.1 Various ways of decorating a small square lattice Draw a small square lattice, e.g. a 3×3 lattice. The vertices of the lattice correspond to oxygen atoms as described in the text. Decorate this lattice with hydrogen atoms such that the ice rule is satisfied. Now draw the lattice again and decorate it with the arrow configuration corresponding to the first lattice. Finally, draw the lattice a third time, and decorate it with thick and thin lines corresponding to the first two configurations.

- How did you treat the horizontal and vertical boundaries of the small finite lattice?
- What is the advantage of the last picture of thick/thin lines?

10

General Square Lattice Vertex Models

Whoever, in the pursuit of science, seeks after immediate practical utility, may generally rest assured that he will seek in vain.

Hermann von Helmholtz (1821–1894)

The class of models introduced in the previous chapter, the vertex models, is of physical interest in its own right. However, we study them for a particular reason: vertex models are a convenient path to introduce one-dimensional *integrable* quantum mechanical models and the Bethe ansatz. Quite apart from the questions of integrability, there is a correspondence between statistical mechanical lattice models, which are really classical models, and quantum mechanical models. As discussed in chapter 6, this correspondence relates the statistical mechanical models in $(d + 1)$ dimensions to quantum mechanical models in d dimensions (see, e.g. section 1 'Introduction' in Shenker (1982) for an elementary discussion).

Relaxing the ice rule, we can define vertex models on the square lattice with more than six possible vertices. Even more general vertex models are possible, as we shall briefly outline in section 10.1. The general two-dimensional vertex model on a square lattice has $n = 16$ different types of vertices, eight of which, shown in the upper two rows of figure 10.1, have an even number of outward and inward arrows. Especially vertex type 7 and 8 have all arrows outward and inward, respectively, and are, also respectively, sources and sinks of arrows. The other eight vertex types, in the lower two rows of figure 10.1, have either three inward and one outward arrow, or vice versa. Section 10.2 briefly summarizes the sixteen- and eight-vertex models before we examine them in the first step using concepts from chapter 4 on statistical mechanics, in particular, the partition function and the Boltzmann weight (section 10.3) for a vertex. These concepts are then tailored to the special case of the two-dimensional square lattice vertex models by the introduction of the R-matrix in section 10.4, which represents the possible vertex weights in matrix form. With the help of the R-matrix, another object which we encountered before, i.e. the transfer matrix, can be constructed in a systematic way. The transfer matrix also encodes the integrability of the two-dimensional vertex models, discussed in sections 10.5 and 10.6. In order to shed further light on the integrability concept, we introduce two more mathematical objects, the monodromy matrix and the L-matrix in section 10.7, the latter object being further investigated in section 10.8. Eventually, the concept of integrability can be subsumed in the Yang–Baxter relations discussed in section 10.9, expanded on in section 10.10, and then further exploited in section 10.11.

Models of Quantum Matter: A First Course on Integrability and the Bethe Ansatz. Hans-Peter Eckle, Oxford University Press (2019). © Hans-Peter Eckle. DOI: 10.1093/oso/9780199678839.001.0001

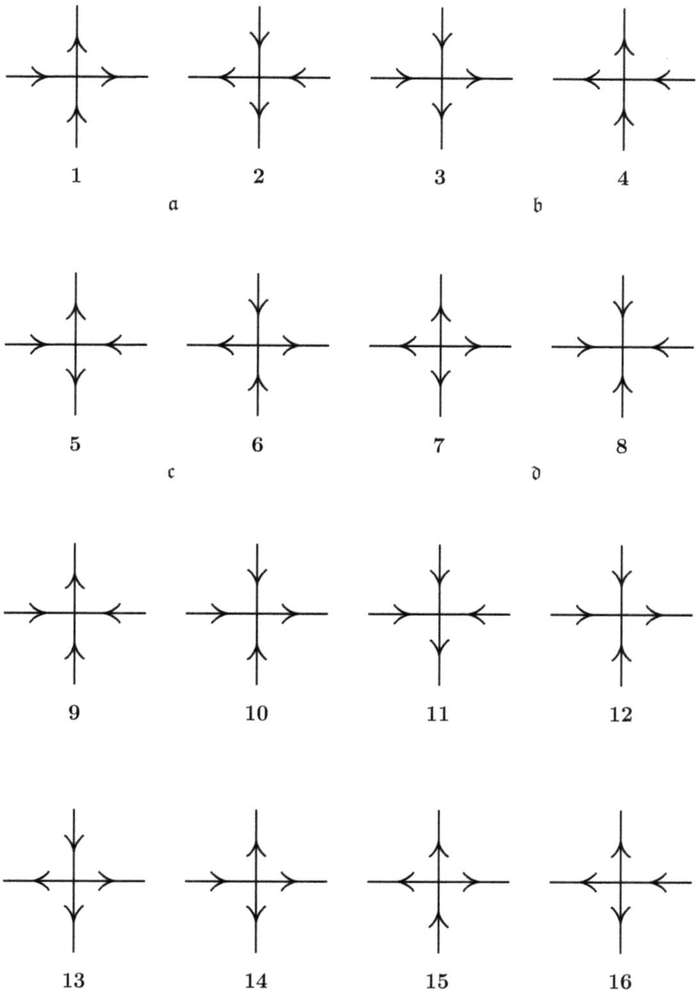

Figure 10.1 *The sixteen possible vertex configurations of the general square lattice vertex model (sixteen-vertex model). The energies and vertex weights (cf. main text) are* ϵ_j *and* $v_j, j = 1, \ldots, 16$*, respectively. The first eight vertex configurations are grouped into arrow reversal symmetric pairs labelled by* $\mathfrak{a}$, $\mathfrak{b}$, $\mathfrak{c}$, *and* $\mathfrak{d}$. *Furthermore, these eight vertex configurations exhibit an even number of arrows going in or out, respectively, two arrows going in or out for configurations* $\mathfrak{a}$, $\mathfrak{b}$, *and* $\mathfrak{c}$, *four arrows in or out for configuration* $\mathfrak{d}$. *These eight vertices constitute the aptly named eight-vertex model. Considering only the first six vertices, those which obey the ice rule, leads to the six-vertex model.*

We learn that there are no vertex weights and corresponding R-matrix known for the general sixteen-vertex model that satisfy the Yang–Baxter relations. However, there are solutions in this sense for the eight- and six-vertex models. Chapter 11 pursues the solution of the six-vertex model, the eight-vertex model being considerably more involved technically.

10.1 Vertex models in two dimensions

If we allow also for bonds with no arrows, and/or other lattices, e.g. triangular or honeycomb lattices, other numbers of allowed vertices (e.g. seven and nineteen) can be considered. Such models are of interest, e.g. to study surface and interface phenomena, since bonds with no arrows can be interpreted as bonds with no interaction between the sites connected.

For example, Batchelor *et al.* (1989) discuss a nineteen-vertex model on a square lattice, Batchelor and Blöte (1989) consider a seven-vertex model on a honeycomb lattice, and Blum and Shapir (1990) explore a seven-vertex model on a square lattice.

Here, we concentrate our attention on the sixteen-vertex model on the square lattice and, in particular, on its special cases, the eight- and six-vertex models.

10.2 Sixteen- and eight-vertex models

For the time being, the number of allowed vertices will make no difference. Therefore, we can keep the discussion general for some time. Our focus will, for the present, be on the sixteen-vertex model in figure 10.1, which we shall eventually abandon for the eight-vertex model. The reason for this focus is that, while the eight-vertex model (Sutherland, 1970) has been solved in zero external field by Baxter (1971; see also Baxter, 1982), there is no solution known for the sixteen-vertex model. However, the sixteen-vertex model has been of some interest in the proof of the equivalence of boundary conditions of the zero field six-vertex model (Brascamp *et al.*, 1973),[1] where the sixteen-vertex model was used at an intermediate step of the proof, after which the limiting case to the six-vertex model was taken, i.e. $v_j \to 0$ for $j = 9, 10, \ldots, 16$.

10.3 Vertex Boltzmann weights and the partition function

In order to have some flexibility, let us call s the number of allowed vertices. We assign an energy

$$\epsilon_j, \qquad j = 1, \ldots s \tag{10.1}$$

to each of the s vertices. Equivalently, a vertex of type j can be assigned a Boltzmann or statistical weight, called vertex weight in this connection, by

$$v_j = e^{-\beta\epsilon_j}. \tag{10.2}$$

Then, a given arrangement or configuration of vertices has an energy

[1] I thank Tony Dorlas for drawing my attention to this point and the relevant literature.

$$\mathcal{E} = n_1\epsilon_1 + n_2\epsilon_2 + \ldots + n_s\epsilon_s = \sum_{j=1}^{s} n_j\epsilon_j, \tag{10.3}$$

where n_j is the number of vertices of type j in this configuration of vertices. The partition function becomes

$$Z = \sum_{\text{allowed configurations}} \exp(-\beta\mathcal{E}), \tag{10.4}$$

i.e. a vertex appearing n_j times in the configuration contributes with a statistical weight of $\exp(-\beta n_j\epsilon_j) = v_j^{n_j}$ to the partition function. Arrows are a good way to draw and visualize configurations of vertices, but they are a bit tedious for algebraic purposes. Therefore we assign the numbers $\sigma = \pm 1$ to the arrows according to the following correspondences

$$\uparrow \text{ and } \rightarrow \quad \Leftrightarrow \quad \sigma = +1, \tag{10.5}$$

$$\downarrow \text{ and } \leftarrow \quad \Leftrightarrow \quad \sigma = -1. \tag{10.6}$$

The numbers σ are called spins or spin variables in the sense of the classical spin variables of, e.g., the Ising model. They are *not* quantum mechanical spins.

The ice rule for the six-vertex model, written in spin variables is simply (cf. figure 10.2)

$$\alpha' + \gamma' = \alpha + \gamma. \tag{10.7}$$

The s energies ϵ_j or, equivalently, vertex weights v_j characterize the vertex model under consideration. They must be chosen according to the physics we want to describe with the

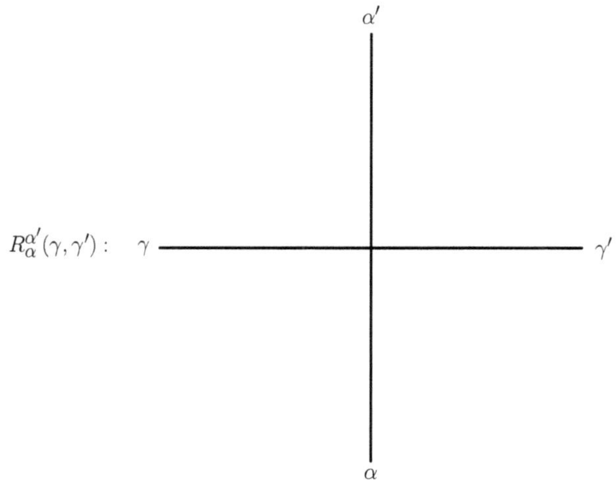

Figure 10.2 *The R-matrix element $R_\alpha^{\alpha'}(\gamma,\gamma')$.*

vertex model. However, these s values, energies or vertex weights, are fine to describe an individual vertex, but they are not yet very suitable to describe a whole two-dimensional lattice of vertices, or even just a single row of vertices. They have yet to be arranged in a proper way for that purpose.

This proper arrangement can be achieved by an object, called the R-matrix.

10.4 R-matrix: matrix of Boltzmann weights of a vertex

Each appearance of one of the $s = 16$ vertices contributes with a Boltzmann weight

$$v_j = \mathrm{e}^{-\beta\epsilon_j} \qquad j = 1, \ldots, 16 \tag{10.8}$$

to the partition function. It will soon prove important to arrange these 16 vertex weights in a 4×4 matrix, the so-called R-matrix. More precisely, a vertex configuration depends on four spin or arrow variables which we denote by $\alpha = \pm 1, \alpha' = \pm 1$ and $\gamma = \pm 1, \gamma' = \pm 1$, such that the Boltzmann weights of a single vertex can be represented by a 4×4 matrix, the R-matrix. The matrix elements

$$R_\alpha^{\alpha'}(\gamma, \gamma') \tag{10.9}$$

are defined by figure 10.2 and have entries

$$v_j = \exp(-\beta\epsilon_j) \qquad j = 1, \ldots, 16. \tag{10.10}$$

Figure 10.1 show, the sixteen possible arrow arrangements that may be translated into elements of the R-matrix with the help of the rules (10.5) and (10.6).

As the matrix elements in (10.9) initially may not look like proper matrix elements, let us spent a moment to clarify this point. Group together the unprimed and primed indices in (10.9) (cf. also figure 10.2) and make the following assignation

$$\{(\alpha, \gamma)\} = \{(+,+), (+,-), (-,+), (-,-)\} \quad \text{corresponding to} \quad \{1, 2, 3, 4\} \tag{10.11}$$

such that, with a similar assignation for $\{(\alpha', \gamma')\}$, the index pair $\{(\alpha, \gamma)\}$ corresponds to a row index and the index pair $\{(\alpha', \gamma')\}$ corresponds to a column index for the R-matrix elements.

We shall later often use a notation of the R-matrix where it is partitioned in blocks of 2×2 matrices according to

$$R = \left(R_\alpha^{\alpha'}(\gamma, \gamma')\right) = \begin{pmatrix} R_+^+(\gamma, \gamma') & R_+^-(\gamma, \gamma') \\ R_-^+(\gamma, \gamma') & R_-^-(\gamma, \gamma') \end{pmatrix} \equiv \begin{pmatrix} \hat{\alpha} & \hat{\beta} \\ \hat{\gamma} & \hat{\delta} \end{pmatrix}, \tag{10.12}$$

where $\{(\gamma, \gamma')\} = \{(+,+), (+,-), (-,+), (-,-)\}$.

From the vertices of figure 10.1 and the correspondences between arrows and spin variables, the arrangement of the weights v_j in the R-matrix, and hence the R-matrix itself, is now determined (once we have assigned the weights v_j).

The most important cases we discuss in some more detail further on and that are the integrable cases, i.e. the eight- and six-vertex models. Let us therefore give the R-matrices for these two cases explicitly introducing the convenient '$\mathfrak{abcd}$'-notation for the vertex weights (cf. the pairs of vertices labelled $\mathfrak{a}$, $\mathfrak{b}$, $\mathfrak{c}$, and $\mathfrak{d}$ in figure 10.1), where due to arrow reversal symmetry in the absence of external fields we have

$$R_+^+(+,+) = v_1 = R_-^-(-,-) = v_2 \equiv \mathfrak{a}, \tag{10.13}$$

$$R_-^-(+,+) = v_3 = R_+^+(-,-) = v_4 \equiv \mathfrak{b}, \tag{10.14}$$

$$R_-^+(+,-) = v_5 = R_+^-(-,+) = v_6 \equiv \mathfrak{c}, \tag{10.15}$$

$$R_-^+(-,+) = v_7 = R_+^-(+,-) = v_8 \equiv \mathfrak{d}, \tag{10.16}$$

or in matrix notation with the conventions (10.11) and (10.12), respectively,

$$R_{8\mathrm{v}} = \begin{pmatrix} \mathfrak{a} & 0 & 0 & \mathfrak{d} \\ 0 & \mathfrak{b} & \mathfrak{c} & 0 \\ 0 & \mathfrak{c} & \mathfrak{b} & 0 \\ \mathfrak{d} & 0 & 0 & \mathfrak{a} \end{pmatrix} = \begin{pmatrix} \hat{\alpha} & \hat{\beta} \\ \hat{\gamma} & \hat{\delta} \end{pmatrix}. \tag{10.17}$$

For the six-vertex model, the vertex weight $\mathfrak{d} = 0$. Later we shall evaluate the R-matrix in the '$\mathfrak{abc}$'-notation in order to find the exact solution of the six-vertex model.

For later use, it is also helpful to note that the R-matrix of the eight-vertex model can be written as a tensor product of Pauli matrices σ^j for $j = 1, 2, 3$ (or $j = x, y, z$) and the 2×2 unit matrix $\sigma^4 \equiv I_2$

$$R_{8\mathrm{v}} = \sum_{j=1}^{4} w_j \sigma^j \otimes \sigma^j. \tag{10.18}$$

For an introduction to and more on the tensor product, see section 10.7.1.

EXERCISE 10.1 Vertex weights of the eight-vertex model Determine the relations between the vertex weights $\{v_j\}$ and $\{w_j\}$ of the eight-vertex model.

But first, before we eventually focus on the integrable eight- and six-vertex cases, we further pursue the general properties of the R-matrix.

The R-matrix is the fundamental building block of the vertex model that we use to build up the two-dimensional lattice step by step, beginning with a single row, then attaching further rows.

If we put two vertices together, described by the R-matrix elements $R_{\alpha_1}^{\alpha'_1}(\gamma_1, \gamma'_1)$ and $R_{\alpha_2}^{\alpha'_2}(\gamma_2, \gamma'_2)$, respectively, the 'inner' spin variables γ'_1 and γ_2 have to be equal or the arrows have to point in the same direction (i.e. both left or both right), i.e. we shall have to equate the 'inner' spin indices $\gamma'_1 = \gamma_2$ (cf. figure 10.3).

Moreover, we can sum over this inner variable. We obtain the following object

$$\sum_{\gamma_2=\pm} R_{\alpha_1}^{\alpha'_1}(\gamma_1, \gamma_2) R_{\alpha_2}^{\alpha'_2}(\gamma_2, \gamma'_2), \tag{10.19}$$

which represents the statistical weight of two vertices. There are six unspecified spin variable $\alpha'_{1,2}$, $\alpha_{1,2}$, and γ_1 and γ'_2, and thus (10.19) represents $2^6 = 2^3 \cdot 2^3 = 64$ matrix elements. It is important to note that this construction, although it may superficially resemble one, is *not* an element of a matrix product. Below we describe how it can be expressed as a tensor product of the two R-matrices, each written as a 2×2 matrix of 2×2 blocks of matrix elements according to the convention (10.12).

In the same way, we can build a whole row of, say, N vertices and we obtain the matrix elements of a large $2^{N+1} \times 2^{N+1}$ matrix

$$T_{\{\alpha\},\{\alpha'\}}(\gamma_1, \gamma'_N) = \sum_{\gamma_2} \cdots \sum_{\gamma_N} R_{\alpha_1}^{\alpha'_1}(\gamma_1, \gamma_2) R_{\alpha_2}^{\alpha'_2}(\gamma_2, \gamma_3) \ldots R_{\alpha_N}^{\alpha'_N}(\gamma_N, \gamma'_N) \tag{10.20}$$

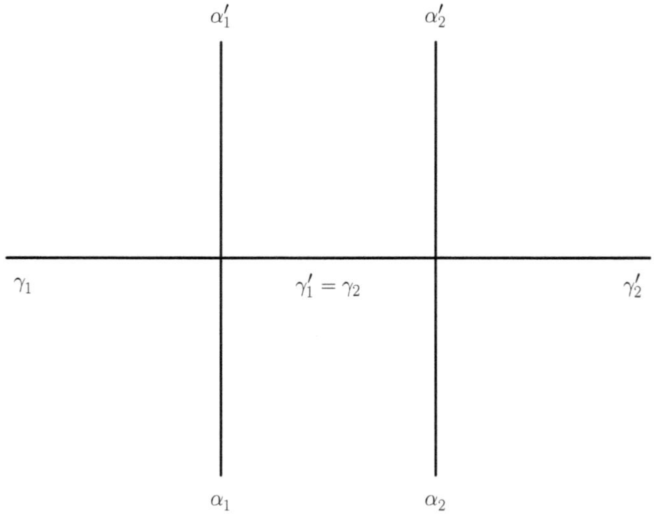

Figure 10.3 $\sum_{\gamma_2=\pm} R_{\alpha_1}^{\alpha'_1}(\gamma_1, \gamma_2) R_{\alpha_2}^{\alpha'_2}(\gamma_2, \gamma'_2)$: *The R-matrices of two vertices combined together to obtain a matrix element of their statistical weight.*

with N unspecified upper spin variables $\{\alpha'\}$ and N unspecified lower spin variables $\{\alpha\}$. Also, the first and last horizontal spin variable γ_1 and γ'_N are not yet specified. The latter we specify through boundary conditions, the simplest of which (and the ones we shall adopt) are periodic ones, i.e. $\gamma_1 = \gamma'_N$.[2] Therefore, we can also sum over $\gamma_1 = \pm 1$ in (10.20) to obtain the $2^N \times 2^N$ matrix

$$T_{\{\alpha\},\{\alpha'\}} = \sum_{\gamma_1} \dots \sum_{\gamma_N} R_{\alpha_1}^{\alpha'_1}(\gamma_1, \gamma_2) R_{\alpha_2}^{\alpha'_2}(\gamma_2, \gamma_3) \dots R_{\alpha_N}^{\alpha'_N}(\gamma_N, \gamma_1). \tag{10.21}$$

This matrix will prove to be another central object, called the transfer matrix. It has still many 'dangling', i.e. unspecified or free, indices: $\{\alpha\}, \{\alpha'\}$.

The transfer matrix, in a sense a one-dimensional object since it describes a row of vertices, can now be used to build up the two-dimensional square lattice row by row. For example, the statistical weight of a lattice consisting of only two rows can now be written as the matrix product of two transfer matrices

$$T^2_{\{\alpha\},\{\alpha''\}} = \sum_{\{\alpha'\}} T_{\{\alpha\},\{\alpha'\}} T_{\{\alpha'\},\{\alpha''\}}. \tag{10.22}$$

Note that (10.22), in contradistinction to (10.19), (10.20), and (10.21), is indeed a matrix product.

We repeat this procedure M times. The resulting matrix elements will depend on the lower and upper row of outer horizontal spin variables $\{\alpha\} = \alpha_1, \dots, \alpha_N$ and $\{\alpha^{(M)}\} = \alpha_1^{(M)}, \dots, \alpha_N^{(M)}$, and, hence, will still be the matrix elements of a $2^N \times 2^N$ matrix. This matrix is the Mth power T^M of the transfer matrix T.

By imposing periodic boundary conditions also in the vertical direction, hence using the topology of a torus (toroidal boundary conditions), we can finally equate and then sum over the last remaining free spin variables $\{\alpha\} = \{\alpha^{(M)}\}$. With this last summation we have performed a trace in the space of the $2^N \times 2^N$ matrices. On the other hand, in total we have summed the statistical weights of all possible configurations of the two-dimensional square lattice with toroidal boundary conditions, which is nothing else than the partition function of the lattice model. We are therefore allowed to write

$$Z = \mathrm{Tr}(T^M). \tag{10.23}$$

The computation of the partition function Z has now been reduced to the task of calculating the eigenvalues of the transfer matrix T.

The transfer matrix, by construction from the non-negative vertex weights, is a matrix with non-negative matrix elements. The Perron–Frobenius theorem of matrix theory then tells us that, for such a matrix, there is a unique positive real eigenvalue λ_1 such that for all other eigenvalues λ_j

[2] It should be emphasized at this point that other boundary conditions *are* possible, the use of which may lead to quite interesting physics.

$$\lambda_1 > |\lambda_j| \quad j \neq 1 \tag{10.24}$$

holds.

Supposing that we had achieved this task of diagonalizing the transfer matrix T, then, according to the Perron–Frobenius theorem, the partition function becomes

$$Z = \sum_{j=1}^{2^N} \lambda_j^M = \lambda_1^M \left(1 + \left(\frac{\lambda_2}{\lambda_1}\right)^M + \left(\frac{\lambda_3}{\lambda_1}\right)^M + \dots\right), \tag{10.25}$$

which, in the limit $M \to \infty$ reduces to just

$$Z = \lambda_1^M. \tag{10.26}$$

This result looks marvelously simple, but we are, of course, not there yet. We first have to find a way to achieve the diagonalization of the transfer matrix.

10.5 Integrability and the transfer matrix

The transfer matrix allows us to address the question of integrability. We pose the following question: can we diagonalize a whole family of transfer matrices simultaneously? In other words, can we find arbitrary sets of Boltzmann weights v_j and v'_j and the corresponding R-matrices such that the corresponding transfer matrices $T = T(v_j)$ and $T' = T(v'_j)$ commute

$$TT' - T'T \equiv [T, T'] = 0. \tag{10.27}$$

Before we search for an answer to this question, we shall pause for a moment and investigate the implications of (10.27).

10.6 Commuting transfer matrices

Recall the expression we found for the partition function of vertex models for the case $s = 8$, i.e. the eight-vertex model

$$Z = \sum_{\text{allowed configurations}} \exp\left(-\beta\mathcal{E}\right), \tag{10.28}$$

where

$$\mathcal{E} = n_1\epsilon_1 + n_2\epsilon_2 + \ldots + n_8\epsilon_8 = \sum_{j=1}^{8} n_j\epsilon_j. \tag{10.29}$$

We can also write the partition function in the following form

$$Z = \sum_{\text{allowed configurations}} \mathfrak{a}^{n_\mathfrak{a}}\mathfrak{b}^{n_\mathfrak{b}}\mathfrak{c}^{n_\mathfrak{c}}\mathfrak{d}^{n_\mathfrak{d}}, \tag{10.30}$$

using the $\mathfrak{abcd}$-notation for the vertex weights (cf. figure 10.1)

$$\mathfrak{a} = v_1 = v_2, \quad \mathfrak{b} = v_3 = v_4, \quad \mathfrak{c} = v_5 = v_6, \quad \mathfrak{d} = v_7 = v_8, \tag{10.31}$$

and the numbers of times the vertices appear in a configuration

$$n_\mathfrak{a} = n_1 + n_2, \quad n_\mathfrak{b} = n_3 + n_4, \quad n_\mathfrak{c} = n_5 + n_6, \quad n_\mathfrak{d} = n_7 + n_8. \tag{10.32}$$

In the rectangular lattice of $N \times M$ lattice sites we are considering, these numbers sum up such that $n_\mathfrak{a} + n_\mathfrak{b} + n_\mathfrak{c} + n_\mathfrak{d} = NM$. We assume that there is a spin or arrow inversion symmetry, i.e. flipping all spin variables, $\sigma \to -\sigma$ or, equivalently, reversing all arrows, of a vertex does not change the energy. This symmetry holds quite obviously as long as there are no external fields that might break the symmetry.

Writing the partition function in the form of (10.30) is quite instructive. First, we note that it can be written, e.g. as

$$Z = \mathfrak{d}^{NM} \sum_{\text{allowed configurations}} \left(\frac{\mathfrak{a}}{\mathfrak{d}}\right)^{n_\mathfrak{a}} \left(\frac{\mathfrak{b}}{\mathfrak{d}}\right)^{n_\mathfrak{b}} \left(\frac{\mathfrak{c}}{\mathfrak{d}}\right)^{n_\mathfrak{c}}. \tag{10.33}$$

The partition function thus depends only on the three ratios of vertex weights

$$\frac{\mathfrak{a}}{\mathfrak{d}}, \quad \frac{\mathfrak{b}}{\mathfrak{d}}, \quad \text{and} \quad \frac{\mathfrak{c}}{\mathfrak{d}}. \tag{10.34}$$

We can, thus, describe the eight-vertex model partition function in the three-dimensional space formed by these three vertex weight ratios.

The same conclusions hold also for the transfer matrix T. Generically, transfer matrices for different sets of vertex weights will not commute. However, we shall also see that we can find families of curves, parametrized by a parameter u, in the space of vertex weights $\mathbf{w} \equiv \left(\frac{\mathfrak{a}}{\mathfrak{d}}, \frac{\mathfrak{b}}{\mathfrak{d}}, \frac{\mathfrak{c}}{\mathfrak{d}}\right)^T$

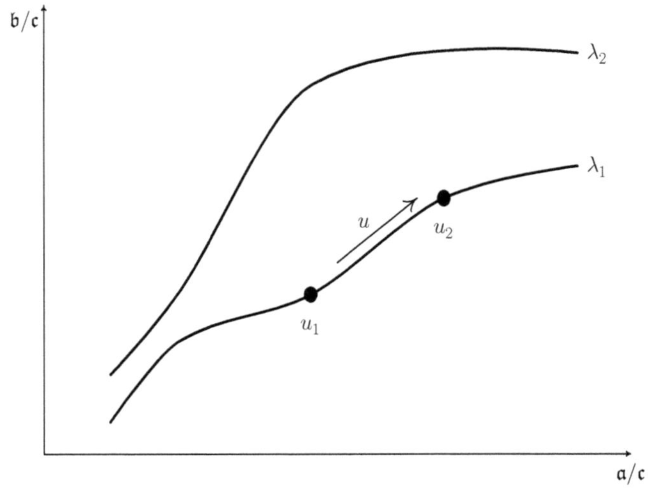

Figure 10.4 *Schematic representation of the families of curves $\mathfrak{w}(u)$ in the parameter space of the six-vertex model. The different families are specified by a parameter λ, the crossing parameter. For a fixed value of λ, pairs of transfer matrices for different positions along the curve, e.g. u_1 and u_2, commute.*

$$\mathbf{w} = \mathbf{w}(u) \equiv \begin{pmatrix} \frac{\mathfrak{a}}{\mathfrak{d}}(u) \\ \frac{\mathfrak{b}}{\mathfrak{d}}(u) \\ \frac{\mathfrak{c}}{\mathfrak{d}}(u) \end{pmatrix}, \tag{10.35}$$

such that the transfer matrices for different positions, i.e. different values of u, on those curves do commute. Different members of the families of curves are distinguished by a parameter λ, the crossing parameter. The parameter u is called the spectral parameter for reasons that will become clear a little later.

In figure 10.4, this is illustrated for the six-vertex model where there are only two parameters $\left(\frac{\mathfrak{a}}{\mathfrak{c}}, \frac{\mathfrak{b}}{\mathfrak{c}}\right)$ spanning the parameter space, since the $\mathfrak{d}$-vertex in the six-vertex model is not allowed by the ice rule.

Now we return to how we can actually perform this programme of finding commuting transfer matrices.

10.7 Monodromy matrix

In order to make progress toward an answer of the question posed in section 10.5, we recall an object encountered above in (10.20)

$$\mathcal{T} \equiv T_{\{\alpha\},\{\alpha'\}}(\gamma, \gamma') = \sum_{\gamma_2} \dots \sum_{\gamma_N} R_{\alpha_1}^{\alpha'_1}(\gamma, \gamma_2) R_{\alpha_2}^{\alpha'_2}(\gamma_2, \gamma_3) \dots R_{\alpha_N}^{\alpha'_N}(\gamma_N, \gamma'), \tag{10.36}$$

namely, the transfer matrix where no boundary conditions have been specified, i.e. where the first and last horizontal spin variables are free. This object bears the name *monodromy matrix*. If we focus on the dependence of the monodromy matrix of these first and last horizontal spin variables, it can be regarded as a 2×2 matrix

$$\mathcal{T} = \mathcal{T}(\gamma, \gamma') \tag{10.37}$$

whose four matrix elements are, however, themselves $2^N \times 2^N$ matrices. The transfer matrix can then be written as a trace in the two-dimensional space of these $2^N \times 2^N$ matrices

$$T = \mathrm{Tr}\, \mathcal{T} = \mathcal{T}(+,+) + \mathcal{T}(-,-). \tag{10.38}$$

The monodromy matrix $\mathcal{T}$, as well as the transfer matrix T, obviously are objects that describe an entire row of vertices. The derivation of $\mathcal{T}$ or T given above is built on a local object, the R-matrix. Next, we want to construct the monodromy matrix $\mathcal{T}$ from a different set of elementary matrices, which are, on the one hand also local objects in the sense that they describe a particular vertex. On the other hand, they are connected with a whole row as we shall see in a moment. The whole point is, similar to when we moved from vertex weights v_j to their neat arrangement as R-matrix, that we now write the R-matrix in such a way that the vertex to which it refers becomes easily localized in the row of vertices we consider. This is what the L-matrix does. The L-matrix carries a position label and is written more precisely as L_n, where n refers to the nth vertex of the row. The R-matrix does not carry such a position label.

First, let us write the R-matrix in a more explicit form as

$$R_\alpha^{\alpha'}(\gamma, \gamma')(w_{ij}) = \sum_{i,j=1}^{4} w_{ij} \sigma^i_{\gamma,\gamma'} \sigma^j_{\alpha,\alpha'} \tag{10.39}$$

where the matrices

$$\sigma^j = \begin{pmatrix} \sigma^j_{+,+} & \sigma^j_{+,-} \\ \sigma^j_{-,+} & \sigma^j_{-,-} \end{pmatrix} \qquad j = 1, 2, 3, 4 \tag{10.40}$$

are the Pauli matrices and the 2×2 unit matrix

$$\sigma^1 = \sigma^x = \begin{pmatrix} 0 & 1 \\ 1 & 0 \end{pmatrix}, \qquad \sigma^2 = \sigma^y = \begin{pmatrix} 0 & -i \\ i & 0 \end{pmatrix},$$
$$\sigma^3 = \sigma^z = \begin{pmatrix} 1 & 0 \\ 0 & -1 \end{pmatrix}, \qquad \sigma^4 = I_2 = \begin{pmatrix} 1 & 0 \\ 0 & 1 \end{pmatrix}. \tag{10.41}$$

Using the tensor products of the Pauli and unit matrices (10.41), the R-matrix can be written in the form

$$R = \sum_{i,j=1}^{4} w_{ij}\sigma^i \otimes \sigma^j \tag{10.42}$$

whose special case for the eight-vertex model we already encountered.

We won't yet discuss the detailed form of the Boltzmann weights w_{ij}, especially their connection to the previously introduced weights v_j, as they are not relevant yet for the still more formal development of our arguments. We just note now that any complex 4×4 matrix can be written in the form (10.42).

As we shall see, the drawback of the R-matrix is that, although it clearly is an object describing a vertex, it cannot be easily used to locate one particular vertex within the lattice of vertices, i.e. it does not carry a spatial label or coordinate. Motivated by the form (10.42) of the R-matrix, however, we can write down a matrix that does depend on the position within the lattice. More specifically, this matrix will depend on the position of a specific vertex in a row of vertices. This matrix based on the R-matrix in the form (10.42) will be a $2^{N+1} \times 2^{N+1}$ matrix, called the local L-matrix or, more commonly, the L-operator, which is given by

$$L_n(w_{ij}) = \sum_{i,j=1}^{4} w_{ij}\sigma^i \otimes \sigma_n^j \tag{10.43}$$

where

$$\sigma_n^j = \overbrace{I_2 \otimes I_2 \otimes \ldots \underbrace{\sigma^j}_{n} \otimes \ldots I_2}^{N} \tag{10.44}$$

is a $2^N \times 2^N$ matrix that acts non-trivially only on the position n along the row of vertices. In the notation I_2 for the 2×2 unit matrix we added the subscript 2 to distinguish from unit matrices of other dimensions, which we shall introduce shortly.

The L_n matrices are thus matrices located at site n in the row of vertices, as we anticipated above. They are a local version of the R-matrix.

Let us now pause for a brief introduction of the tensor product that has crept up in the foregoing considerations and, thus, merits some attention before we continue.

10.7.1 Interlude: direct or tensor product of matrices

In the following, the direct or tensor product of two square matrices of dimension N is important. Moreover we have just used it in writing down (10.43) and (10.44). Therefore, let us define the tensor product explicitly. However, we restrict ourselves to the case $N = 2$ mainly to keep the equations neat and manageable. The generalization to arbitrary N and even to matrices partitioned into blocks, themselves square matrices, is straightforward.

For the 2×2 (complex) matrices

$$A = \begin{pmatrix} a_{11} & a_{12} \\ a_{21} & a_{22} \end{pmatrix} \quad \text{and} \quad B = \begin{pmatrix} b_{11} & b_{12} \\ b_{21} & b_{22} \end{pmatrix}, \tag{10.45}$$

the tensor product $A \otimes B$ is defined by[3]

$$\begin{pmatrix} a_{11} & a_{12} \\ a_{21} & a_{22} \end{pmatrix} \otimes \begin{pmatrix} b_{11} & b_{12} \\ b_{21} & b_{22} \end{pmatrix} = \left(\begin{array}{cc|cc} a_{11}b_{11} & a_{11}b_{12} & a_{12}b_{11} & a_{12}b_{12} \\ a_{11}b_{21} & a_{11}b_{22} & a_{12}b_{21} & a_{12}b_{22} \\ \hline a_{21}b_{11} & a_{21}b_{12} & a_{22}b_{11} & a_{22}b_{12} \\ a_{21}b_{21} & a_{21}b_{22} & a_{22}b_{21} & a_{22}b_{22} \end{array}\right). \tag{10.46}$$

There is an easy way to remember this

$$A \otimes B = \begin{pmatrix} a_{11}B & a_{12}B \\ a_{21}B & a_{22}B \end{pmatrix}. \tag{10.47}$$

The 2×2 matrices act on a two-dimensional complex vector space

$$V = \mathbb{C}^2, \tag{10.48}$$

whose elements are two-component vectors

$$x = \begin{pmatrix} x_1 \\ x_2 \end{pmatrix}, \tag{10.49}$$

with $x_i \in \mathbb{C}$.

The tensor product $A \otimes B$ acts on the tensor product space $V \otimes V$ whose elements are of the form

$$x \otimes y = \begin{pmatrix} x_1 \\ x_2 \end{pmatrix} \otimes \begin{pmatrix} y_1 \\ y_2 \end{pmatrix} = \left(\begin{array}{c} x_1y_1 \\ x_1y_2 \\ \hline x_2y_1 \\ x_2y_2 \end{array}\right). \tag{10.50}$$

3 The horizontal and vertical lines help to organize the matrix elements; they have no other significance, and can be omitted. This practice follows Nepomechie (1999).

It is useful to recall the equation

$$(A \otimes B)(x \otimes y) = Ax \otimes By. \tag{10.51}$$

which generalizes to

$$(A_1 \otimes A_2 \otimes \ldots A_n)(x_1 \otimes x_2 \otimes \ldots x_n) = A_1 x_1 \otimes A_2 x_2 \otimes \ldots A_n x_n. \tag{10.52}$$

In other words, a 2×2 matrix in the ℓ's position $(1 \leq \ell \leq n)$ of an n-fold tensor product acts non-trivially only on the two-dimensional vector space in the ℓ's position of the n-fold tensor product space

$$\overset{\overset{1}{\downarrow}}{V} \otimes \cdots \otimes \overset{\overset{\ell}{\downarrow}}{V} \otimes \cdots \otimes \overset{\overset{n}{\downarrow}}{V}, \tag{10.53}$$

i.e. on the corresponding vector in the ℓ's position of the n-fold tensor product of vectors.

There is an important connection between the ordinary matrix product and the tensor product that we shall be using frequently in the following. For matrices A, B, C, D, or, generally, A_ℓ and B_ℓ $(\ell = 1, \ldots, n)$ whose matrix elements are complex numbers, we have

$$(A \otimes B)\,(C \otimes D) = AC \otimes BD \tag{10.54}$$

or

$$(A_1 \otimes A_2 \otimes \ldots \otimes A_n)\,(B_1 \otimes B_2 \otimes \ldots \otimes B_n) = A_1 B_1 \otimes A_2 B_2 \otimes \ldots \otimes A_n B_n. \tag{10.55}$$

What follows are a few simple observations about the tensor product we just defined.

10.7.1.1 Simple applications of the tensor product

Many of the tensor products we are interested in are related to the Pauli matrices and the corresponding spinors

$$|+\rangle = |\uparrow\rangle = \begin{pmatrix} 1 \\ 0 \end{pmatrix}, \quad |-\rangle = |\downarrow\rangle = \begin{pmatrix} 0 \\ 1 \end{pmatrix}. \tag{10.56}$$

Consider a lattice of two sites only to each of which we attach a quantum spin–$\frac{1}{2}$ object. Then the Pauli matrices acting on the first object only will be

$$\sigma_1^x = \sigma^x \otimes I_2 = \begin{pmatrix} 0 & 1 \\ 1 & 0 \end{pmatrix} \otimes \begin{pmatrix} 1 & 0 \\ 0 & 1 \end{pmatrix} = \left(\begin{array}{cc|cc} 0 & 0 & 1 & 0 \\ 0 & 0 & 0 & 1 \\ \hline 1 & 0 & 0 & 0 \\ 0 & 1 & 0 & 0 \end{array}\right), \tag{10.57}$$

while for the second object, we have

$$\sigma_2^x = I_2 \otimes \sigma^x = \begin{pmatrix} 1 & 0 \\ 0 & 1 \end{pmatrix} \otimes \begin{pmatrix} 0 & 1 \\ 1 & 0 \end{pmatrix} = \left(\begin{array}{cc|cc} 0 & 1 & 0 & 0 \\ 1 & 0 & 0 & 0 \\ \hline 0 & 0 & 0 & 1 \\ 0 & 0 & 1 & 0 \end{array}\right). \tag{10.58}$$

Similarly, we can obtain the 4×4 matrices σ_i^y and σ_i^z $(i = 1, 2)$.

The following little exercise is suggested to gain practice.

EXERCISE 10.2 Two spins

Show that for two spins we have

$$\boldsymbol{\sigma}_1 \cdot \boldsymbol{\sigma}_2 = \sigma_1^x \sigma_2^x + \sigma_1^y \sigma_2^y + \sigma_1^z \sigma_2^z \tag{10.59}$$

$$= \sigma^x \otimes \sigma^x + \sigma^y \otimes \sigma^y + \sigma^z \otimes \sigma^z \tag{10.60}$$

$$= \begin{pmatrix} 1 & 0 & 0 & 0 \\ 0 & -1 & 2 & 0 \\ 0 & 2 & -1 & 0 \\ 0 & 0 & 0 & 1 \end{pmatrix}. \tag{10.61}$$

The permutation operator (or permutation matrix)

$$\mathcal{P} = \begin{pmatrix} 1 & 0 & 0 & 0 \\ 0 & 0 & 1 & 0 \\ 0 & 1 & 0 & 0 \\ 0 & 0 & 0 & 1 \end{pmatrix} \tag{10.62}$$

acts on the tensor product of two spinors of the tensor product space $V \otimes V$ as

$$\mathcal{P} x \otimes y = \begin{pmatrix} x_1 y_1 \\ x_2 y_1 \\ x_1 y_2 \\ x_2 y_2 \end{pmatrix} = \begin{pmatrix} y_1 \\ y_2 \end{pmatrix} \otimes \begin{pmatrix} x_1 \\ x_2 \end{pmatrix} = y \otimes x. \tag{10.63}$$

The permutation operator squares to the 4×4 identity matrix which can be written as the tensor product of the 2×2 identity matrix with itself

$$\mathcal{P}^2 = I_2 \otimes I_2 = I_4. \tag{10.64}$$

Furthermore, combining this with the exercise 10.2 we immediately observe

$$\boldsymbol{\sigma}_1 \cdot \boldsymbol{\sigma}_2 + I_4 = 2\mathcal{P}. \tag{10.65}$$

This is an interesting result to which we will return. For the moment, we note that the two-spin operator

$$h_{i,i+1} = \frac{1}{2}\left(\boldsymbol{\sigma}_i \cdot \boldsymbol{\sigma}_{i+1} + I_4\right) = \mathcal{P}, \tag{10.66}$$

acting on spinors at neighbouring sites, is just the permutation operator $\mathcal{P}$.

We end this interlude section with an exercise which brings us back to R-matrices.

EXERCISE 10.3 Tensor product of two R-matrices Using the convention (10.12) for partitioning an R-matrix into 2×2 blocks, show that the vertex weight of two vertices (10.19) can be written as a tensor product of R-matrices in block form. Writing the elements of this composite object, we have

$$\sum_{\gamma_2=\pm} R_{\alpha_1}^{\alpha_1'}(\gamma_1,\gamma_2) R_{\alpha_2}^{\alpha_2'}(\gamma_2,\gamma_2') = (R_1 \otimes R_2)_{\alpha_1\alpha_2}^{\alpha_1'\alpha_2'}(\gamma_1,\gamma_1') \tag{10.67}$$

where

$$R_n = \left(R_{\alpha_n}^{\alpha_n'}(\gamma_n,\gamma_n)\right) = \begin{pmatrix} R_+^+(\gamma_n,\gamma_n') & R_+^-(\gamma_n,\gamma_n') \\ R_-^+(\gamma_n,\gamma_n') & R_-^-(\gamma_n,\gamma_n') \end{pmatrix} \tag{10.68}$$

are the R-matrices of the two vertices $n = 1, 2$.

Note that the vertex weights w_{ij} entering the R-matrices R_1 and R_2 are the same for both R-matrices.

10.8 Further to the L-operator

Let us now return to the operator L_n. Its definition (10.43), together with (10.44) and (10.55), implies that L_n-operators for different sites commute

$$[L_n, L_m] = 0 \qquad \text{for} \qquad n \neq m. \tag{10.69}$$

Furthermore, the L_n-operators can be formally decomposed and written as a 2×2 matrix with elements $\hat{\alpha}_n, \hat{\beta}_n, \hat{\gamma}_n$ and $\hat{\delta}_n$, which are themselves matrices or operators of dimension $2^N \times 2^N$

$$L_n = \begin{pmatrix} \hat{\alpha}_n & \hat{\beta}_n \\ \hat{\gamma}_n & \hat{\delta}_n \end{pmatrix}. \tag{10.70}$$

EXERCISE 10.4 L_n-matrix Convince yourself the the L_n-matrix (10.43) can indeed be written in the form (10.70).

The monodromy matrix may thus be expressed in terms of a matrix product of the the local L_n-operators

$$\mathcal{T} = L_1 L_2 \dots L_N = \prod_{n=1}^{N} L_n \tag{10.71}$$

or, because of (10.70)

$$\mathcal{T} = \prod_{n=1}^{N} \begin{pmatrix} \hat{\alpha}_n & \hat{\beta}_n \\ \hat{\gamma}_n & \hat{\delta}_n \end{pmatrix} \equiv \begin{pmatrix} A & B \\ C & D \end{pmatrix}. \tag{10.72}$$

Note that the matrix product (10.71) has been performed in the two-dimensional space of the matrices $\hat{\alpha}$, $\hat{\beta}$, $\hat{\gamma}$, and $\hat{\delta}$ as explicitly expressed in (10.72).

Now, we can formally perform the trace of (10.38) to obtain

$$T = A + D. \tag{10.73}$$

So far, of course, this was all purely formal and we still have not diagonalized the transfer matrix T. This will only be possible once we specify the model, i.e. once we define the energies ϵ_j and therewith the Boltzmann weights v_j. However, we can now give a *sufficient* condition (not a *necessary* one, though!) for the transfer matrices to commute.

10.9 Yang–Baxter relations

For the commutation relations (10.27) to hold for arbitrary mutually different sets of vertex Boltzmann weights $\{v_j\}$ and $\{v'_j\}$ (or, alternatively, $\{w_{ij}\}$ and $\{w'_{ij}\}$), requires a large number of conditions to be satisfied, which appears to be a remote possibility.

Historically, the question of whether and when these conditions may be met has, of course, initially not been answered in full generality, but rather by considering many special examples, starting with the ice model. For an early overview of these developments, see Lieb and Wu (1972).

The sufficient condition we now formulate is thus to be considered as the fruit of combining many research efforts across the years. It builds on concepts developed thus far, in particular, the monodromy matrix $\mathcal{T}$ in the form (10.72).

A vanishing commutator of the tensor product of two monodromy matrices $\mathcal{T}$ and $\mathcal{T}'$ for two different sets of vertex weights would imply commutation of the corresponding transfer matrices, i.e. $\left[T, T'\right] = 0$. However, the tensor product of $\mathcal{T}$ and $\mathcal{T}'$ will in general also not commute. What could be inferred from the examples, however, was that it may be possible to find a third set of vertex weights $\{v''_j\}$ or $\{w''_{ij}\}$ and a corresponding R-matrix such that

$$\tilde{R}''(\mathcal{T} \otimes \mathcal{T}') = (\mathcal{T}' \otimes \mathcal{T})\tilde{R}'' \tag{10.74}$$

where the tensor product, using (10.72), is given by

$$\mathcal{T} \otimes \mathcal{T}' = \begin{pmatrix} A\mathcal{T}' & B\mathcal{T}' \\ C\mathcal{T}' & D\mathcal{T}' \end{pmatrix}. \tag{10.75}$$

This means we are trying to find an R-matrix depending on a third set of vertex weights that interchanges the vertex weights on the two rows of vertices. These are the sufficient conditions for commuting transfer matrices, called the *Yang–Baxter relations*, that could be distilled from the vertex models for which an exact solution could be found. In our context, we will exploit their further algebraic consequences for vertex and quantum spin models, including the connection of a notion of quantum integrability and the exact solutions via the Bethe ansatz.

We emphasize that the R-matrix and the two monodromy matrices are to be taken for different sets of vertex weights v_j or w_{ij}, which is indicated by no prime, prime, and double prime, i.e.

$$\mathcal{T} = \mathcal{T}(v_j), \qquad \mathcal{T}' = \mathcal{T}(v'_j), \qquad \text{and} \qquad \tilde{R}'' = \tilde{R}(v''_j). \tag{10.76}$$

The R-matrix entering the Yang–Baxter relations (10.74) is obtained from the R-matrix discussed so far by multiplication with the permutation matrix (10.62),

$$\tilde{R} = \mathcal{P}R. \tag{10.77}$$

This modification of the R-matrix is needed to properly connect it to the two rows of vertices whose statistical weight is represented by the two monodromy matrices $\mathcal{T}$ and $\mathcal{T}'$. This reason for the modification (10.77) will, however, become especially clear when we discuss the graphical proof of the Yang–Baxter relations in section 10.10.1.

The Yang–Baxter relations (10.74) implies that the transfer matrices for different sets of vertex weights commute, i.e that (10.27) holds. In order to see this, we rewrite the Yang–Baxter relations (10.74) as

$$(\mathcal{T} \otimes \mathcal{T}') = \tilde{R}''^{-1}(\mathcal{T}' \otimes \mathcal{T})\tilde{R}'' \tag{10.78}$$

and take the appropriate trace in the four-dimensional space of matrices (10.75), recalling that matrices can be permuted cyclicly under the trace operation.

EXERCISE 10.5 Commutation relation of the transfer matrix Convince yourself that taking the trace in an appropriate four-dimensional space of the previous equation, (10.78), is sufficient for the transfer matrices for different sets of vertex weights to commute, i.e. for (10.27) to hold.

Section 10.10.1 presents a useful graphical way to prove this result, after we discuss the Yang–Baxter relations for the local L-operators and also for the R-matrices.

Using the local L-operators we can also write down local Yang–Baxter relations

$$\tilde{R}''(L_n \otimes L'_n) = (L'_n \otimes L_n)\tilde{R}'' \tag{10.79}$$

or, using the weights w_{ij} explicitly

$$\tilde{R}(w''_{ij})\left(L_n(w_{ij}) \otimes L_n(w'_{ij})\right) = \left(L_n(w'_{ij}) \otimes L_n(w_{ij})\right)\tilde{R}(w''_{ij}). \tag{10.80}$$

A corresponding Yang–Baxter relation can be given involving only R-matrices of different sets of vertex weights

$$\tilde{R}(w''_{ij})\left(R(w_{ij}) \otimes R(w'_{ij})\right) = \left(R(w'_{ij}) \otimes R(w_{ij})\right)\tilde{R}(w''_{ij}), \tag{10.81}$$

which we shall discuss in more detail in section 10.10. The tensor products in this expression are formed when the R-matrices are interpreted as built up from 2×2 matrices as in the convention given by (10.12). Then, the remaining product is a product of 4×4 matrices where one factor acts as an ordinary 4×4 matrix, and the other as a 4×4 matrix where each element is a product of two 2×2 matrices. Note that only one R-matrix on each side of (10.81), the one acting as a 4×4 matrix, is modified by the permutation matrix $\mathcal{P}$. We shall encounter this type of Yang–Baxter relation in section 11.1 when we use it to solve the six-vertex model explicitly.

We pause for a moment to clarify what our results so far mean. We found the Yang–Baxter relations as a sufficient condition for the integrability of the transfer matrix of the vertex model. Integrability here means that we can diagonalize the transfer matrix of a given vertex model and hence, calculate the partition function.

10.10 More on Yang–Baxter relations

The Yang–Baxter relations appear in many different versions. For an overview of the various forms and their physical and mathematical backgrounds, see Perk and Au-Yang (2006). We have already encountered two versions, the first (10.74) involving, besides the R-matrix, the whole monodromy matrix $\mathcal{T}$ expressing the statistical weight of a row of vertices. In the second version (10.80), we have replaced the monodromy matrix $\mathcal{T}$ by the *local* L_n matrices, which suggests that there is a version of the Yang–Baxter relations that employs only R-matrices. We have encountered this version already in (10.81) but now discuss it in more detail, mostly in the form of an exercise.

While the versions of the Yang–Baxter relations involving the monodromy matrix $\mathcal{T}$ or the L_n matrices are connected to the square lattice of the vertex model, the version involving only R-matrices consists of just three vertices attached to each other to form (the topology of) a triangle (cf. figure 10.5). This combination of vertices does, of course, not occur normally in a square lattice. We get from the left-hand side of figure 10.5 to the right-hand side, and vice versa, by a parallel shift of one of the sides of the triangle through the corner opposite that side. It is important to note that this shift implies that the vertex at the bottom of the right-hand side of the figure becomes the vertex at the top of the left-hand side, and vice versa, albeit with different spin indices on the bonds.

EXERCISE 10.6 Yang–Baxter relations for the R-matrices Convince yourself that the Yang–Baxter relations corresponding to figure 10.5 is explicitly given by

$$\sum_{\alpha''}\sum_{\beta''}\sum_{\gamma''} R^{\alpha'}_{\alpha''}(\gamma,\gamma'')(w)R^{\beta'}_{\beta''}(\gamma'',\gamma')(w')R^{\alpha''}_{\alpha}(\beta,\beta'')(w'') =$$
$$\sum_{\alpha''}\sum_{\beta''}\sum_{\gamma''} R^{\alpha'}_{\alpha''}(\beta'',\beta')(w'')R^{\beta''}_{\beta}(\gamma,\gamma'')(w')R^{\alpha''}_{\alpha}(\gamma'',\gamma')(w). \tag{10.82}$$

We have suppressed the indices ij of the vertex weights w, w', and w''. These equations represent $2^6 = 64$ linear equations between the matrix elements of the R-matrices.

Note that, since we are dealing with matrix *elements*, their sequence in the expression above does not matter but is conventionally written in the order we use. In many other forms of the Yang–Baxter relation we deal with matrices or operators where the corresponding order *does* matter.

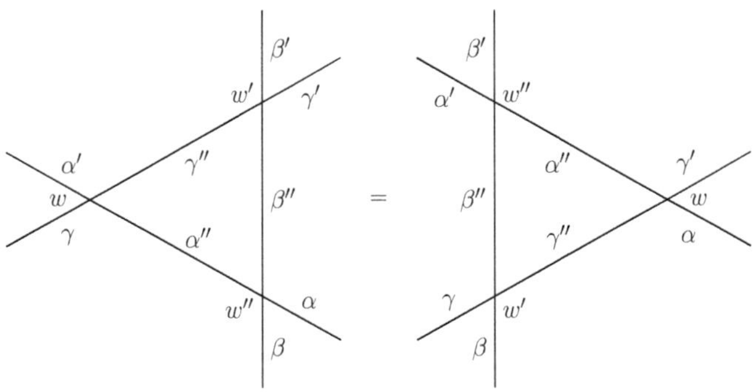

Figure 10.5 *Yang–Baxter relations for the R-matrices. The vertices are labelled by the weights w, w', and w'' (suppressing the indices ij) and four surrounding spin indices, cf. (10.39) or (10.84). Cf. section 10.10 and exercise 10.6.*

EXERCISE 10.7 Equivalence of Yang–Baxter relations Convince yourself that the Yang–Baxter relations in the forms (10.82) and (10.81) are indeed equivalent.

The local forms of the Yang–Baxter relations (10.82) and (10.81) for the R-matrices and their graphical representation in figure 10.5 are useful for a graphical demonstration of the commutation relation of the transfer matrix.

10.10.1 Graphical proof of the Yang–Baxter relation

First, we require a further relation to hold between the R-matrices. We assume that there are two sets of vertex weights w'''_{ij} and w''_{ij} such that the corresponding R-matrices satisfy

$$\sum_{v\nu} R^{\tau}_{v}(\sigma, \nu)(w'''_{ij}) R^{\nu}_{\beta}(v, \alpha)(w''_{ij}) = \delta_{\alpha\tau}\delta_{\beta\sigma}. \tag{10.83}$$

In other words, the two R-matrices in (10.83) are matrix inverses of each other.

EXERCISE 10.8 Inverse of the R-matrix Represent (10.83) graphically in a way similar to, e.g. figure 10.5.

Using (10.82) and (10.83) the graphical proof of the commutativity of the transfer matrix can be carried out; see figure 10.6 for the details.

10.11 Exploiting Yang–Baxter integrability

We now exploit the local Yang–Baxter relations (10.80), i.e. the integrability condition. In order to do so, let us finally establish an explicit relation between the vertex weights v_j and the elements of the R-matrix (cf. 10.39):

$$R^{\alpha'}_{\alpha}(\gamma, \gamma')(w_{ij}) = \sum_{i,j=1}^{4} w_{ij}\sigma^{i}_{\gamma,\gamma'}\sigma^{j}_{\alpha,\alpha'}. \tag{10.84}$$

The logic is the following: we are searching for vertex weights w_{ij} such that the local Yang–Baxter relations are satisfied.

So far, we have not imposed any restrictions on the model. Thus the R-matrix (10.84) still corresponds to the most general case of a sixteen-vertex model. Unfortunately, there are no vertex weights known for a sixteen-vertex model such that the corresponding R-matrix satisfies the local Yang–Baxter relations (10.80) or (10.81).

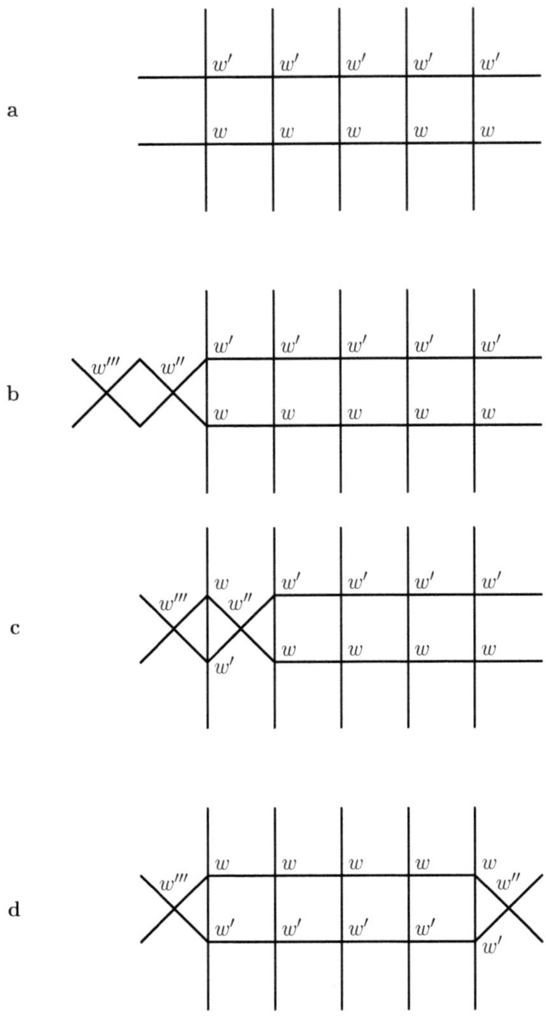

Figure 10.6 *Graphical proof of the Yang–Baxter relation (10.74). The figure in panel* **a** *represents the product of two transfer matrices as in (10.22) but with two different weights w and w' (the subscripts ij are again suppressed). All inner spin or arrow variables are identified and summed over. The rightmost and leftmost spin variables are also identified and summed over, i.e. we assume periodic boundary conditions in the horizontal direction. For the figure of panel* **b***, the matrix product (10.83) representing a unit matrix has been inserted at the left end, which does not cause any change. These weights w, w', and w'' are assumed to satisfy the Yang–Baxter relations. The Yang–Baxter relations in the form for the R-matrices is now used to move the vertex with weight w'' to the right (panel* **c***). Thereby, the upper vertex with weight w' and the lower vertex with weight w are exchanged. We can keep moving the vertex with weight w'' to the right thereby exchanging the upper and lower vertices until the end of the two rows is reached (panel* **d***). Due to the periodic boundary conditions, the vertices at the left and right ends (panel* **d***) can now be recombined to give a unit matrix. The net effect of this procedure is that the transfer matrices represented by the two rows of vertices are commuted. This constitutes a graphical proof the Yang–Baxter relations (10.74).*

We shall accordingly restrict our attention to the eight-vertex model and eventually even the six-vertex model where (10.84) simplifies to

$$R_{\alpha}^{\alpha'}(\gamma,\gamma')(w_j) = \sum_{j=1}^{4} w_j \sigma^j_{\gamma,\gamma'} \sigma^j_{\alpha,\alpha'} \tag{10.85}$$

and the L-operators are

$$L_n(w_j) = \sum_{j=1}^{4} w_j \sigma^j \otimes \sigma^j_n = \begin{pmatrix} w_4 I_n + w_3 \sigma^z_n & w_1 \sigma^x_n - i w_2 \sigma^y_n \\ w_1 \sigma^x_n + i w_2 \sigma^y_n & w_4 I_n - w_3 \sigma^z_n \end{pmatrix} = \begin{pmatrix} \hat{\alpha}_n & \hat{\beta}_n \\ \hat{\gamma}_n & \hat{\delta}_n \end{pmatrix}. \tag{10.86}$$

The R-matrix elements are given by (10.13) through (10.16) or, using (10.85), in terms of the weights w_j as

$$\mathfrak{a} = w_3 + w_4, \qquad \mathfrak{b} = w_4 - w_3, \qquad \mathfrak{c} = w_1 + w_2, \qquad \text{and} \qquad \mathfrak{d} = w_1 - w_2. \tag{10.87}$$

For the six-vertex model, the ice rule requires

$$\mathfrak{d} \equiv v_7 = v_8 = 0 \qquad \Leftrightarrow \qquad w_1 = w_2. \tag{10.88}$$

The local Yang–Baxter relations (10.80) relate the different sets of vertex weights w''_j, w'_j, and w_j. For the eight–vertex model, the Yang–Baxter relations describe an elliptic curve that can be parametrized by elliptic functions $\mathrm{sn}(u,k)$ (elliptic sine of modulus k). These remarks connect the present discussion with the remarks in section 10.6, especially figure 10.4.

Here, we finally focus on the six-vertex model (see chapter 11), and refer the interested reader to the literature (see Baxter (1982)) for the further detailed (and technically much more elaborate) treatment of the eight-vertex model.

11

Six-Vertex Model

Algebra is generous; she often gives more than is asked of her.

Jean-Baptiste le Rond d'Alembert (1717–1783)

We restrict our attention in this chapter to the six-vertex model, where the curves obtained from the local Yang–Baxter relations (10.80) or (10.81) and its equivalent relation (10.82) can be parametrized by trigonometric sine functions[1] instead of elliptic functions, as in the eight-vertex model. The trigonometric functions would also emerge from appropriate limits (i.e. $k \to 1$) of the elliptic functions.

In order to achieve this, we will exploit the Yang–Baxter relations for the example of the six-vertex model (see section 11.1) in the trigonometric parameterization (see section 11.2) and find Lieb's parameter Δ, which encapsulates integrability of the model, i.e. the commutation of the transfer matrices for different sets of vertex weights. We shall list the parameterizations and use these to interpret the six-vertex model as a model for two-dimensional ferroelectrics.

Section 11.3, given its parameterization, explores diagonalization of the transfer matrix of the six-vertex model, i.e. finding the Bethe ansatz equations as equations a set of parameters has to satisfy as conditions for this diagonalization to be valid. The algebraic procedure with which the diagonalization of the transfer matrix is achieved is called the algebraic Bethe ansatz.

Section 11.4 makes the correspondence explicit between a two-dimensional model of classical statistical mechanics and a corresponding one-dimensional quantum model, starting with the two-dimensional classical six-vertex model, and ending up with the one-dimensional anisotropic Heisenberg quantum spin chain, having established, along the way, integrability, and again in particular the role of the Yang–Baxter relations as the heart of integrability, for these models. From the transfer matrix, we can construct an infinite set of commuting operators; in particular, we find that the transfer matrix of the six-vertex model and the Hamiltonian of the quantum spin chain commute. Moreover, given the algebraic diagonalization of the transfer matrix, we have automatically also

[1] As we will see later, this is only one possible parameterization that corresponds to the disordered phase of the six-vertex model in 11.2.1, or to the anisotropic XXZ Heisenberg quantum spin chain. The Lieb parameter is $|\Delta| \leq 1$ and determines the position in the phase diagram in the six-vertex model while it describes the anisotropy of the quantum spin chain.

Models of Quantum Matter: A First Course on Integrability and the Bethe Ansatz. Hans-Peter Eckle, Oxford University Press (2019). © Hans-Peter Eckle. DOI: 10.1093/oso/9780199678839.001.0001

obtained the exact solution of the one-dimensional quantum spin chain Hamiltonian given by the same Bethe ansatz equations.

This is the constructive algebraic programme of the quantum inverse scattering method (QISM) and the *algebraic* Bethe ansatz. In the words of Evgeny Sklyanin (1992), 'The basic idea of QISM [. . .] is purely algebraic.'

Finally, section (11.5) demonstrates how the Yang–Baxter relations, which are still valid in spatially inhomogeneous situations, can be used to examine new classes of quantum integrable spin chains.

11.1 Yang–Baxter relation for the six-vertex model

Let us recall the R-matrix (10.17) for the six-vertex model for which the ice rule must be satisfied, which implies that we have to set $\mathfrak{d} = 0$

$$R_{6\text{v}} = \left(\begin{array}{cc|cc} \mathfrak{a} & 0 & 0 & 0 \\ 0 & \mathfrak{b} & \mathfrak{c} & 0 \\ \hline 0 & \mathfrak{c} & \mathfrak{b} & 0 \\ 0 & 0 & 0 & \mathfrak{a} \end{array}\right) \equiv \begin{pmatrix} \alpha & \beta \\ \gamma & \delta \end{pmatrix}. \tag{11.1}$$

EXERCISE 11.1 Two vertices Show that for a system consisting of only $N = 2$ vertices, the matrices $\hat{\alpha}_n$, $\hat{\beta}_n$, $\hat{\gamma}_n$, and $\hat{\delta}_n$ of the L_n-matrices of (10.70) with $n = 1, 2$ are related to the 2×2 matrices α, β, γ, and δ of the R-matrix (11.1) by

$$\hat{\alpha}_1 = \alpha \otimes I_2, \qquad \hat{\alpha}_2 = I_2 \otimes \alpha, \qquad \hat{\beta}_1 = \beta \otimes I_2, \qquad \hat{\beta}_2 = I_2 \otimes \beta, \tag{11.2}$$

$$\hat{\gamma}_1 = \gamma \otimes I_2, \qquad \hat{\gamma}_2 = I_2 \otimes \gamma, \qquad \hat{\delta}_1 = \delta \otimes I_2, \qquad \hat{\delta}_2 = I_2 \otimes \delta. \tag{11.3}$$

Show that for this system consisting of only two vertices the transfer matrix becomes

$$T_2 = \hat{\alpha}_1\hat{\alpha}_2 + \hat{\beta}_1\hat{\gamma}_2 + \hat{\gamma}_1\hat{\beta}_2 + \hat{\delta}_1\hat{\delta}_2 \tag{11.4}$$

in two ways:

- graphically by writing down the two vertices and summing over all horizontal spin or arrow variables; and
- by multiplying the two L_n-matrices and performing the trace.

Let us now evaluate the Yang–Baxter relations (10.81) in the $\mathfrak{abc}$-notation

$$\begin{pmatrix} \mathfrak{a}'' & 0 & 0 & 0 \\ 0 & \mathfrak{c}'' & \mathfrak{b}'' & 0 \\ 0 & \mathfrak{b}'' & \mathfrak{c}'' & 0 \\ 0 & 0 & 0 & \mathfrak{a}'' \end{pmatrix} \begin{pmatrix} \alpha\alpha' & \alpha\beta' & \beta\alpha' & \beta\beta' \\ \alpha\gamma' & \alpha\delta' & \beta\gamma' & \beta\delta' \\ \gamma\alpha' & \gamma\beta' & \delta\alpha' & \delta\beta' \\ \gamma\gamma' & \gamma\delta' & \delta\gamma' & \delta\delta' \end{pmatrix} = \begin{pmatrix} \alpha'\alpha & \alpha'\beta & \beta'\alpha & \beta'\beta \\ \alpha'\gamma & \alpha'\delta & \beta'\gamma & \beta'\delta \\ \gamma'\alpha & \gamma'\beta & \delta'\alpha & \delta'\beta \\ \gamma'\gamma & \gamma'\delta & \delta'\gamma & \delta'\delta \end{pmatrix} \begin{pmatrix} \mathfrak{a}'' & 0 & 0 & 0 \\ 0 & \mathfrak{c}'' & \mathfrak{b}'' & 0 \\ 0 & \mathfrak{b}'' & \mathfrak{c}'' & 0 \\ 0 & 0 & 0 & \mathfrak{a}'' \end{pmatrix}. \tag{11.5}$$

As in exercise 10.6, these are $16 \cdot 4 = 64$ linear equations for the three sets of three weight variables each: $(\mathfrak{a}, \mathfrak{b}, \mathfrak{c})$, $(\mathfrak{a}', \mathfrak{b}', \mathfrak{c}')$, and $(\mathfrak{a}'', \mathfrak{b}'', \mathfrak{c}'')$. A great number of these equations are, however, trivially satisfied. Of the remaining non-trivial equations, many turn out to be mutually identical. In the end, only three non-trivial independent equations survive.

EXERCISE 11.2 Linear equations of the six-vertex model Derive the three surviving linear equations of the six-vertex model. This can be done purely algebraically using (11.5) together with (11.1). The reduction in the number of independent equations can also be achieved by invoking the ice rule and by further symmetry arguments (see Baxter, 1982, section 9.6).

The remaining three linear equations are

$$\mathfrak{a}''\mathfrak{c}'\mathfrak{b} = \mathfrak{c}''\mathfrak{b}'\mathfrak{c} + \mathfrak{b}''\mathfrak{c}'\mathfrak{a}, \tag{11.6}$$

$$\mathfrak{a}''\mathfrak{a}'\mathfrak{c} = \mathfrak{c}''\mathfrak{c}'\mathfrak{a} + \mathfrak{b}''\mathfrak{b}'\mathfrak{c}, \tag{11.7}$$

$$\mathfrak{c}''\mathfrak{a}'\mathfrak{b} = \mathfrak{c}''\mathfrak{b}'\mathfrak{a} + \mathfrak{b}''\mathfrak{c}'\mathfrak{c}. \tag{11.8}$$

For given, but different, values of the weights $(\mathfrak{a}, \mathfrak{b}, \mathfrak{c})$ and $(\mathfrak{a}', \mathfrak{b}', \mathfrak{c}')$, we can view this set of equations as a homogeneous set of linear algebraic equations for the weights $(\mathfrak{a}'', \mathfrak{b}'', \mathfrak{c}'')$. The condition for the existence of such a solution is that the determinant of this set of equations vanishes, which is the case if $(\mathfrak{a}, \mathfrak{b}, \mathfrak{c})$ and $(\mathfrak{a}', \mathfrak{b}', \mathfrak{c}')$ satisfy

$$\frac{\mathfrak{a}^2 + \mathfrak{b}^2 - \mathfrak{c}^2}{2\mathfrak{a}\mathfrak{b}} = \frac{\mathfrak{a}'^2 + \mathfrak{b}'^2 - \mathfrak{c}'^2}{2\mathfrak{a}'\mathfrak{b}'} \equiv \Delta \tag{11.9}$$

where we have introduced a factor of 1/2 for later convenience. The quantity Δ is often called Lieb's parameter in honour of Elliott Lieb's seminal work on vertex models.

11.2 Parameterization of the six-vertex model

The condition (11.9), an algebraic curve, and, thus, the local Yang–Baxter relations (10.80) of the six-vertex model are satisfied, for instance, for the trigonometric parameterization

$$\mathfrak{a} = w_3 + w_4 = \sin(u + \eta), \tag{11.10}$$

$$\mathfrak{b} = w_4 - w_3 = \sin(u - \eta), \tag{11.11}$$

$$\mathfrak{c} = 2w_1 = \sin(2\eta), \tag{11.12}$$

i.e. we have two parameters, u and η, whose significance will become clear in the following.

In fact, we shall regard u, the so-called spectral parameter, as the main parameter, which will play a major role in diagonalizing the transfer matrix. The parameter η will turn out to describe different regimes of the model but can also be viewed as related to an interaction strength or an anisotropy parameter.

Other parameterizations are also possible, e.g. using hyperbolic functions

$$\mathfrak{a} = w_3 + w_4 = \sinh(u + \eta), \tag{11.13}$$

$$\mathfrak{b} = w_4 - w_3 = \sinh\left(\pm(u - \eta)\right), \tag{11.14}$$

$$\mathfrak{c} = 2w_1 = \sinh(2\eta). \tag{11.15}$$

For the trigonometric and the hyperbolic parameterizations, respectively, we obtain for Lieb's parameter Δ

$$\Delta = \cos(2\eta) \qquad \text{and} \qquad \Delta = \pm\cosh(2\eta). \tag{11.16}$$

Later we shall also encounter, in the limit $\eta \to 0$, a so-called rational parameterization, i.e. one in terms of rational functions of the parameters. After a short section on the interpretation of the six-vertex model to describe two-dimensional ferroelectrics, we thereafter concentrate on the trigonometric parameterization.

11.2.1 Two-dimensional ferroelectric models: phase diagram of the six-vertex model

The two-dimensional vertex models have a natural interpretation as models for two-dimensional ferroelectrics when the arrows on the bonds between vertices are interpreted now as electrical dipoles. In the two-dimensional lattice these electric dipoles can be ferroelectically or antiferroelectrically ordered, or they can be disordered according to the energies or, equivalently, vertex weights of the vertices.

The following gives a qualitative picture of the possible phases of the six-vertex model (cp. figure 11.1). Remember that we are always assuming *periodic* boundary conditions in both directions of the lattice, i.e. *toroidal* boundary conditions. More details going beyond a discussion of the possible phases of the six-vertex model can be found, in Lieb and Wu (1972) and in chapters 8 and 9 of Baxter (1982).

The phases of the six-vertex model depend on the three intervals of Lieb's parameter: $\Delta < -1$, $-1 < \Delta < 1$, and $\Delta > 1$. The free energy of the model has a different analytical form for each of these three intervals (see chapter 8 of Baxter, 1982, for the explicit calculation of the respective free energies). In our qualitative discussion, we distinguish the different phases by considering directly the vertices, their energies, and corresponding weights as shown in the first six panels of figure 10.1.

For all non-negative (and real) energies of the six vertices, the weights are restricted to the interval $0 < \mathfrak{a}, \mathfrak{b}, \mathfrak{c} < 1$ or, according to the discussion in section 10.6, to the interval $0 < \mathfrak{a}/\mathfrak{c}, \mathfrak{b}/\mathfrak{c} < 1$ of these two ratios.

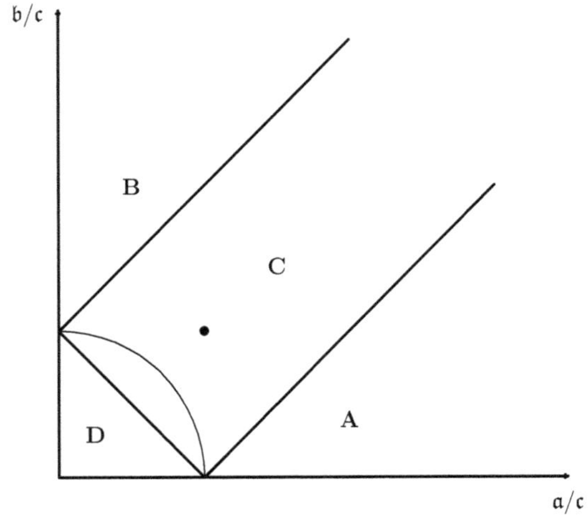

Figure 11.1 *Phase diagram of the six-vertex model. The dot represents the point in the phase diagram where the temperature becomes infinite:* $T = \infty$, *i.e.* $a = b = c = 1$. *The model is equivalent to a model of free Fermions for* $\Delta = 0$, *which corresponds to the quarter circle of the thin line, i.e.* $(a/c)^2 + (b/c)^2 = 1$.

Lieb's parameter is symmetric under the exchange of $a \leftrightarrow b$ so that the phase diagram will be symmetric with respect to the line $a/c = b/c$.

The phases described in the following can be discerned in figure 11.1.

A,B The *ferroelectric* phase is separated into two parts, depending on whether the ratios of the vertex weights satisfy either $a/c > b/c + 1$ or $b/c > a/c + 1$. This is the case if either $\epsilon_a < \epsilon_b$ and $\epsilon_a < \epsilon_c$ or $\epsilon_b < \epsilon_a$ and $\epsilon_b < \epsilon_c$ and $\Delta > 1$.

At very low temperatures, two ground states are possible. All vertices are either of type a or b. If the vertices are of type a, all dipoles point either up or to the right if the vertices are of subtype 1, or they point down or to the left if the vertices are of subtype 2. For the subtypes 3 and 4 of vertex type b, a similar statement holds.

We note in particular that there is a net polarization and thus the system exhibits a ferroelectric order. A detailed analysis reveals that excitations contribute only negligibly to the free energy, which is equal to ϵ_a or ϵ_b, respectively, and therefore the system remains frozen in one of the ground states (see Baxter, 1982, chapter 8).

D The antiferroelectric phase is characterized by $a/c + b/c < 1$, $\epsilon_c < \epsilon_a$ and $\epsilon_c < \epsilon_b$ such that $\Delta < -1$. The lowest energy is now obtained if there are only type c vertices where all neighbouring dipoles are antiparallel in either the vertical or horizontal direction. Hence, there are again two ground states that now exhibit antiferroelectric order.

C Arguably, the richest phase in the phase diagram is the one bounded by the three lines $a/c = b/c + 1$, $b/c = a/c + 1$, and $a/c + b/c = 1$. This region can be circumscribed in the symmetrical form $\mathfrak{a}, \mathfrak{b}, \mathfrak{c} < \frac{1}{2}(\mathfrak{a} + \mathfrak{b} + \mathfrak{c})$. In this region, Lieb's parameter is in the interval $-1 < \Delta < 1$.

In particular, the point $\mathfrak{a} = \mathfrak{b} = \mathfrak{c} = 1$ lies in this region. It corresponds to infinite temperature $T \to \infty$ where we thus expect a disordered phase. In fact, from the analysis of the free energy (see Baxter, 1982), it can be inferred that this whole region is a disordered phase.

Furthermore, the quarter circle $(\mathfrak{a}/\mathfrak{c})^2 + (\mathfrak{b}/\mathfrak{c})^2 = 1$ describes the locus of all points where $\Delta = 0$. In the detailed Bethe ansatz analysis, which can be found e.g. in Baxter (1982), this case is particularly simple and corresponds to a non-interacting spin chain or, equivalently, a chain of free Fermions.

Moreover, the Bethe ansatz solution of the eight-vertex model reveals that the entire region **C** corresponds to the critical eight-vertex model, i.e. the six-vertex model is critical in a whole region of the phase diagram, rather than just at a point or on a line.

We note that the region of $-1 < \Delta < 0$ is bounded by the quarter circle $(\mathfrak{a}/\mathfrak{c})^2 + (\mathfrak{b}/\mathfrak{c})^2 = 1$ from above and the line $\mathfrak{a}/\mathfrak{c} + \mathfrak{b}/\mathfrak{c} = 1$, whereas the region of $0 < \Delta < 1$ is above the quarter circle in the corridor limited by the lines $\mathfrak{b}/\mathfrak{c} = \mathfrak{a}/\mathfrak{c} \pm 1$.

Lastly, in our qualitative analysis, we remark that the point $\mathfrak{a}/\mathfrak{c} = \mathfrak{b}/\mathfrak{c} = 1$ is reached from all other points of the phase diagram as the temperature T increases from 0 to ∞. If the path traced out in this process starts in one of the regions **A**, **B**, or **D**, it will have to cross a phase boundary to eventually reach the point $\mathfrak{a}/\mathfrak{c} = \mathfrak{b}/\mathfrak{c} = 1$ in region **C**, i.e. one of the lines $\mathfrak{b}/\mathfrak{c} = \mathfrak{a}/\mathfrak{c} \pm 1$ and $\mathfrak{b}/\mathfrak{c} = -\mathfrak{a}/\mathfrak{c} + 1$. Again, a more detailed analysis, e.g. Baxter (1982), reveals that these transitions into the critical region possess only one critical transition temperature T_c where the free energy is singular and the correlation length is infinite.

For those statements in the classification of the phases of the six-vertex model that go beyond a qualitative description based on the vertex energies and weights, we again refer to Baxter (1982) and the references cited therein.

11.3 Algebraic Bethe ansatz solution of the six-vertex model

We now return to the Yang–Baxter relations and exploit them finally to solve a vertex model, the six-vertex model. In concrete terms, this means that we shall determine a set of parameters and derive the Bethe ansatz equations that determine these parameters.

The replacements in the parameterizations (11.10–11.12)

$$u \to v \qquad \text{and} \qquad u \to u - v + \eta \tag{11.17}$$

give sets w'_j and w''_j, respectively, such that the local Yang–Baxter relations (10.80) are satisfied. The local Yang–Baxter relations then take the form

$$\tilde{R}(u-v)(L_n(u)\otimes L_n(v)) = (L_n(v)\otimes L_n(u))\tilde{R}(u-v). \tag{11.18}$$

The L_n- and R-matrices can now be given explicitly as

$$L_n(u) = \begin{pmatrix} w_4(u)I_n + w_3(u)\sigma_n^z & \frac{\sin 2\eta}{2}\sigma_n^- \\ \frac{\sin 2\eta}{2}\sigma_n^+ & w_4(u)I_n - w_3(u)\sigma_n^z \end{pmatrix} \tag{11.19}$$

where

$$w_4(u) \pm w_3(u) = \sin(u \pm \eta) \tag{11.20}$$

and

$$\tilde{R}(u) = \begin{pmatrix} 1 & 0 & 0 & 0 \\ 0 & \mathfrak{c}(u) & \mathfrak{b}(u) & 0 \\ 0 & \mathfrak{b}(u) & \mathfrak{c}(u) & 0 \\ 0 & 0 & 0 & 1 \end{pmatrix} = \begin{pmatrix} 1 & 0 & 0 & 0 \\ 0 & b(u) & c(u) & 0 \\ 0 & c(u) & b(u) & 0 \\ 0 & 0 & 0 & 1 \end{pmatrix} \tag{11.21}$$

with[2]

$$b(u) = \frac{\sin 2\eta}{\sin(u+2\eta)} \quad \text{and} \quad c(u) = \frac{\sin u}{\sin(u+2\eta)}, \tag{11.22}$$

i.e. we have normalized the R-matrix by dividing by $\sin(u + 2\eta)$. Note that the ratio $b(u)/c(u)$ is an odd function of the spectral parameter u. This fact will be of some importance further on.

Now, we are in a position to start the diagonalization of the transfer matrix. The first step is to introduce the local vacuum state at site n of the row of vertices corresponding to the local L_n-matrix

$$\omega_n = \begin{pmatrix} 1 \\ 0 \end{pmatrix}_n, \tag{11.23}$$

i.e. the state defined as the eigenstate of the Pauli matrix σ_n^z with eigenvalue $+1$, i.e.

$$\sigma_n^z\omega_n = \sigma_n^z\begin{pmatrix} 1 \\ 0 \end{pmatrix}_n = +\begin{pmatrix} 1 \\ 0 \end{pmatrix}_n = +\omega_n. \tag{11.24}$$

[2] This change of notation, $\mathfrak{c} \to b$ and $\mathfrak{b} \to c$, will facilitate comparison with R-matrices of other models in later chapters and the conventional notation used there.

This state is annihilated by σ_n^+ while σ_n^- creates a new local state

$$\sigma_n^+\omega_n = \sigma_n^+ \begin{pmatrix} 1 \\ 0 \end{pmatrix}_n = 0 \qquad \text{and} \qquad \sigma_n^-\omega_n = \sigma_n^- \begin{pmatrix} 1 \\ 0 \end{pmatrix}_n = + \begin{pmatrix} 0 \\ 1 \end{pmatrix}_n, \tag{11.25}$$

the latter being an eigenstate of σ_n^z with eigenvalue (-1)

$$\sigma_n^z \begin{pmatrix} 0 \\ 1 \end{pmatrix}_n = - \begin{pmatrix} 0 \\ 1 \end{pmatrix}_n. \tag{11.26}$$

The L-operator accordingly acts on ω_n to produce

$$L_n(u)\omega_n = \begin{pmatrix} \alpha(u)\omega_n & * \\ 0 & \delta(u)\omega_n \end{pmatrix}. \tag{11.27}$$

which is an upper triangular matrix, where

$$\alpha(u) = \sin(u+\eta), \qquad \text{and} \qquad \delta(u) = \sin(u-\eta). \tag{11.28}$$

Because this is a triangular matrix, the matrix element indicated by an asterisk, *, will not play any further role in the analysis and we thus do not bother to write it down explicitly.

The reference state

$$\Omega = \prod_{n=1}^{N} \otimes\, \omega_n \equiv |0\rangle, \tag{11.29}$$

a global vacuum state, assumes the role of the local vacuum state for the monodromy matrix. Using the triangular form of (11.27), we immediately obtain for the monodromy matrix (10.71)

$$\mathcal{T}(u)|0\rangle = \prod_{n=1}^{N} L_n(u)|0\rangle = \begin{pmatrix} \alpha^N(u)|0\rangle & * \\ 0 & \delta^N(u)|0\rangle \end{pmatrix}. \tag{11.30}$$

Now, the formal developments introduced via a number of formal quantities are starting to pay off. The transfer matrix was

$$T(u) = \mathrm{Tr}\,(\mathcal{T}(u)) = A(u) + D(u). \tag{11.31}$$

We can now immediately read off an eigenvalue of the transfer matrix and also the eigenvalues of the matrices A and D

$$A(u)\Omega = \alpha^N(u)|0\rangle \qquad \text{and} \qquad D(u)\Omega = \delta^N(u)|0\rangle \tag{11.32}$$

and, finally, an eigenvalue of the transfer matrix itself

$$T(u)|0\rangle = \left(\alpha^N(u) + \delta^N(u)\right)|0\rangle. \tag{11.33}$$

Thus, we can note that the vacuum state $|0\rangle$ is an eigenstate of the transfer matrix with eigenvalue $\alpha^N(u) + \delta^N(u)$.

We now want to construct other eigenstates of the transfer matrix $T(u)$. The key ingredient in this construction is the fundamental Yang–Baxter relation using the R-matrix (11.21)

$$\tilde{R}(u-v)(\mathcal{T}(u) \otimes \mathcal{T}(v)) = (\mathcal{T}(v) \otimes \mathcal{T}(u))\tilde{R}(u-v). \tag{11.34}$$

It is this constructive algebraic approach that gives the whole approach its name: algebraic Bethe ansatz.

Writing the Yang–Baxter relations in components, we obtain 16 commutation relations between the operators A, B, C, and D, often called the fundamental commutation relations. Fortunately, many of them are rather trivial or redundant. In fact, we only need three of the 16 relations for our purpose. The operator B commutes for different values of the spectral parameter

$$[B(u), B(v)] = 0. \tag{11.35}$$

The other two commutation relations are more complicated and involve the special feature of two terms on the right-hand side,

$$A(u)B(v) = \frac{1}{c(v-u)}B(v)A(u) - \frac{b(v-u)}{c(v-u)}B(u)A(v) \tag{11.36}$$

and

$$D(u)B(v) = \frac{1}{c(u-v)}B(v)D(u) - \frac{b(u-v)}{c(u-v)}B(u)D(v). \tag{11.37}$$

They contain on the right-hand side a first term, where the arguments appearing in the operators on the left-hand side are preserved, and a second term where these arguments are exchanged. The two terms with the exchanged arguments in the second term are the major complication we encounter in the further analysis. We note that the operator $C(u)$ does not play a further role in our analysis. It can be shown that $C(u)$ acts as an annihilation operator in the same sense as $B(u)$ acts as a creation operator.

EXERCISE 11.3 Explicitly derive (11.35–11.37) from the Yang–Baxter relation (11.34) of the six-vertex model. You may want to derive further commutation relations.

The commutation relations (11.36) and (11.37) suggest an interpretation of $B(u)$ as a particular kind of creation operator. Acting with B-operators for M distinct values u_j

($j = 1, \ldots, M$) of the spectral parameter on the reference state $|0\rangle$, we may construct the product state

$$|u_1, u_2, \ldots, u_M\rangle \equiv |\{u_j\}\rangle = B(u_1)B(u_2)\ldots B(u_M)|0\rangle = \prod_{j=1}^{M} B(u_j)|0\rangle. \tag{11.38}$$

It is important to note that, due to their commutativity 11.35, we can permute the $B(u_j)$ operators in the product of this state arbitrarily.

A further observation that will play a key role in the following is that acting with the transfer matrix $T(u)$ (or, with the operators $A(u)$ and $D(u)$ separately) on the product state (11.38), only the $1 + M$ states

$$\prod_{j=1}^{M} B(u_j)|0\rangle \tag{11.39}$$

$$\prod_{j=1, j\neq n}^{M} B(u_j)|0\rangle, \qquad n = 1, \ldots, M \tag{11.40}$$

can be generated after commuting $T(u)$ through all operators $B(u_j)$ so that $T(u)$, acting on the reference state $|0\rangle$, creates the vacuum eigenvalue $\alpha^N(u) + \delta^N(u)$.

We expect that the state (11.38) will be an eigenstate of the transfer matrix. We shall see, however, that this is not immediately the case. Rather, we find that the commutation relations (11.36) and (11.37), due to their two terms with the exchanged arguments in the second term, create terms that are not proportional to the state $|\{u_j\}\rangle$. In fact, we obtain a large number of further terms

$$T(u)|\{u_j\}\rangle = (A(u) + D(u))\,|\{u_j\}\rangle = \Lambda(u)|\{u_j\}\rangle + \text{further terms}. \tag{11.41}$$

Provided we can make these further terms, called unwanted or sometimes even 'junk' terms, disappear, this equation would, however, indeed become an eigenvalue equation.

We shall find that the unwanted terms indeed do disappear, provided the distinct values of the spectral parameter $\{u_j\}$ satisfy

$$\frac{\alpha^N(u_j)}{\delta^N(u_j)} = \prod_{k=1, k\neq j}^{M} \frac{c(u_k - u_j)}{c(u_j - u_k)} \qquad j = 1, \ldots, M. \tag{11.42}$$

If this provision can be satisfied, we shall find the eigenvalue

$$\Lambda(u; u_1, \ldots, u_M) = \alpha^N(u)\prod_{j=1}^{M} \frac{1}{c(u_j - u)} + \delta^N(u)\prod_{j=1}^{M} \frac{1}{c(u - u_j)}. \tag{11.43}$$

In order to derive the statements related to (11.42) and (11.43), we use the commutation relations (11.35), (11.36), and (11.37). We may focus on the operator A and thus the commutation relations (11.35) and (11.36). The arguments will be analogous for the operator D and the commutation relations (11.35) and (11.37).

Since we already know how the operator A acts on the reference state $|0\rangle$, we are led to commute $A(u)$ through the string of operators $B(u_j)$ in the state (11.38) using (11.36) repeatedly until $A(u)$ directly acts on $|0\rangle$ to produce the eigenvalue $\alpha^N(u)$. Thus, if indeed there were only the first term on the right-hand side of the commutation relation (11.36), after M commutations, we would immediately reach the result

$$A(u)B(u_1)\dots B(u_M)|0\rangle = \alpha^N(u)\prod_{j=1}^{M}\frac{1}{c(u_j-u)}B(u_1)B\dots B(u_M)|0\rangle. \tag{11.44}$$

Thus, $B(u_1)B(u_2)\dots B(u_M)|0\rangle = |\{u_j\}\rangle$ would be the eigenvalue of $A(u)$ with eigenvalue $\alpha^N(u)\prod_{j=1}^{M} 1/c(u_j-u)$.

However, the second term on the right-hand side of (11.36) spoils this straightforward picture. Commuting $A(u)$ through the operators $B(u_j)$ one by one, two new terms are created each time. After M commutations, we thus would end up with 2^M terms. To handle this large number of terms seems a formidable task.

However, there are two observations that help and that we alluded to when we briefly discussed the $1+M$ states (11.39) and (11.40):

- Only in one of the terms have all arguments remained unchanged, i.e. the argument of the operator A is still u, the arguments of the B operators in the product state (11.38) are still u_j for $j=1,\dots,M$. This term is given by (11.44).
- Many terms can be grouped together to form M combined terms, each of which is characterized by one missing argument, say u_n, in the arguments of the B operators. The number of B operators in the product is, however, unchanged by applying the commutation relation (11.36). The argument u_n has instead become the argument of the operator A, while one of the B's now bears the argument u. We thus have M combined terms proportional to

$$B(u)\prod_{j\neq n}^{M}B(u_j)A(u_n)|0\rangle = \alpha^N(u_n)B(u)\prod_{j\neq n}^{M}B(u_j)|0\rangle \qquad n=1,\dots,M. \tag{11.45}$$

We emphasize again, that, because of the commutation relation (11.36), there can be no types of terms other than the two just described. What remains to be calculated are the proportionality functions for the second type of terms. These functions will depend on u and all $\{u_j\}$.

We can now state the intermediate result that the first part of (11.41), the equation yet to become an eigenvalue equation, pertaining only to the operator $A(u)$ of the transfer matrix $T(u)$ has the form

$$A(u)|\{u_j\}\rangle = \alpha^N(u) \prod_{j=1}^{M} \frac{1}{c(u_j - u)} |\{u_j\}\rangle$$
$$+ \sum_{n=1}^{M} \Lambda_n^{(A)} \left(u, \{u_j\}\right) B(u) \prod_{j=1,\, j\neq n}^{M} B(u_j)|0\rangle. \tag{11.46}$$

The functions $\Lambda_n^{(A)} \left(u, \{u_j\}\right)$ are yet to be determined. A similar expression will hold for the second part of (11.41) pertaining to $D(u)$ with as yet unknown functions $\Lambda_n^{(D)} \left(u, \{u_j\}\right)$.

If we can show that the sum of these two functions

$$\left(\Lambda_n^{(A)} \left(u, \{u_j\}\right) + \Lambda_n^{(D)} \left(u, \{u_j\}\right)\right) \tag{11.47}$$

vanishes and leads to an expression in the form of the conditions (11.42), we shall have attained our aim to demonstrate that (11.41) becomes an eigenvalue equation with the eigenvalue given by (11.43).

In order to achieve this, let us first consider the function with $n = 1$: $\Lambda_1^{(A)} \left(u, \{u_j\}\right)$. The part of the state (11.46) having this function as coefficient, i.e.

$$\Lambda_1^{(A)} \left(u, \{u_j\}\right) B(u)B(u_2)\ldots B(u_M)|0\rangle \tag{11.48}$$

is generated by applying the *second* part of the commutation relation (11.36) once to the product state (11.38). We thus obtain

$$-\frac{b(u_1 - u)}{c(u_1 - u)} B(u)A(u_1)B(u_2)\ldots B(u_M)|0\rangle. \tag{11.49}$$

Next, we pull $A(u_1)$ through all further $(M - 1)$ B-operators to its right, but apply only the *first* part of (11.36) this time. This procedure yields

$$A(u_1)B(u_2)\ldots B(u_M)|0\rangle = \frac{1}{c(u_2 - u_1)} B(u_2)A(u_1)B(u_3)\ldots B(u_M)|0\rangle, \tag{11.50}$$

$$= \prod_{j=2}^{M} \frac{1}{c(u_j - u_1)} B(u_2)\ldots B(u_M)A(u_1)|0\rangle, \tag{11.51}$$

$$= \alpha^N(u_1) \prod_{j=2}^{M} \frac{1}{c(u_j - u_1)} B(u_2)\ldots B(u_M)|0\rangle. \tag{11.52}$$

Only this procedure, applying the second part of the commutation relation once, the first part $(M-1)$ times, leads to a state proportional to

$$B(u)B(u_2)\dots B(u_M)|0\rangle = B(u)\prod_{j=2}^{M} B(u_j)|0\rangle. \tag{11.53}$$

Combining (11.49) and (11.52), we find the function

$$\Lambda_1^{(A)}\left(u,\{u_j\}\right) = -\alpha^N(u_1)\frac{b(u_1-u)}{c(u_1-u)}\prod_{j=2}^{M}\frac{1}{c(u_j-u_1)}. \tag{11.54}$$

Repeating the same arguments for the operator $D(u)$ applied to the reference state (11.38), we obtain the corresponding function

$$\Lambda_1^{(D)}\left(u,\{u_j\}\right) = -\delta^N(u_1)\frac{b(u-u_1)}{c(u-u_1)}\prod_{j=2}^{M}\frac{1}{c(u_1-u_j)}. \tag{11.55}$$

Finally, we recall the remarks after the definition of the state (11.38), which mean that we can write this state, for instance, in the form

$$|\{u_j\}\rangle = B(u_n)\prod_{j=1,\, j\neq n}^{M} B(u_j)|0\rangle. \tag{11.56}$$

Thus, we can go through the arguments above equally well for $B(u_n)$ instead of $B(u_1)$. In this general case, then, setting the sum of the functions $\Lambda_n^{(A)}$ and $\Lambda_n^{(D)}$ equal to zero implies the condition (11.42), which, in turn, renders (11.41) an eigenvalue equation.

The equations (11.42) are the celebrated Bethe ansatz equations, derived here within the algebraic framework.

The logic of what we have achieved now is the following. Suppose we find a solution, i.e. a set of M Bethe ansatz roots,[3] to the Bethe ansatz equations (11.42), then we succeed in diagonalizing the transfer matrix of the six-vertex model.

We are now only left with the task to show the connection of this approach with the Heisenberg quantum spin-1/2 chain. In order to do this, we have to consider the transfer matrix $T(u)$ in some more detail.

EXERCISE 11.4 Derive these results by directly applying the commutation relation (11.36), (11.35), and (11.37) on the state (11.38) respectively for the cases $M = 1$, which

[3] The solutions of the Bethe ansatz equations are sometimes also called Bethe ansatz roots. We shall be using both expressions.

is straightforward, and $M = 2$, which is already quite tedious. In the latter case, use the explicit forms of the functions $b(u)$ and $c(u)$. Larger values of M will obviously become unwieldy very rapidly.

11.4 Quantum Hamiltonians from the transfer matrix

We now construct quantum Hamiltonians from the transfer matrix $T(u)$. The latter can be viewed as the generating function of the former.

The trick is to find a special value, $u = u_0$, of the spectral parameter at which to evaluate the transfer matrix, or more precisely, its logarithm and logarithmic derivatives with respect to u, with the aim of writing these operators as sums of simpler operators.

The eigenvalue $\Lambda(u, \{u_j\})$ (11.43) suggests that we choose a value u_0 for the spectral parameter such that $\delta(u_0) = 0$, i.e. $u_0 = \eta$ in the six-vertex model.

Then, for this special value of the spectral parameter, $u = u_0$ with $\delta(u_0) = 0$, the operators

$$\log T(u_0) \qquad \text{and} \qquad \frac{d^k}{du^k} \log T(u)|_{u=u_0}, \qquad k = 1, 2, \ldots \tag{11.57}$$

all have an additive spectrum, i.e. a spectrum that corresponds to an operator that is a sum of simpler operators. This was the objective of our choice for the special value of the spectral parameter u_0.

We shall see that these operators are local quantum Hamiltonians, i.e. they can be written as sums of quantum operators on a one-dimensional lattice involving only a few lattice sites.

11.4.1 Isotropic case

We restrict our attention to the isotropic or XXX case (the reason for this terminology will become clear soon), which is obtained from the general six-vertex model after rescaling

$$u \to \kappa u \qquad \text{and} \qquad \eta \to \frac{i}{2}\kappa \tag{11.58}$$

in the limit $\kappa \to 0$.

Thus, with

$$\alpha(u) = u + \frac{i}{2} \qquad \text{and} \qquad \delta(u) = u - \frac{i}{2} \tag{11.59}$$

where also the functions $\alpha(u)$ and $\delta(u)$ have been rescaled, i.e.

$$\frac{1}{\kappa}\alpha(u) \to \alpha(u) \qquad \text{and} \qquad \frac{1}{\kappa}\delta(u) \to \delta(u), \tag{11.60}$$

the L_n-matrix becomes

$$L_n(u) = u + \frac{i}{2}\sum_{a=1}^{3} \sigma^a \otimes \sigma_n^a = \begin{pmatrix} u + \frac{i}{2}\sigma_n^z & \frac{i}{2}\sigma_n^- \\ \frac{i}{2}\sigma_n^+ & u - \frac{i}{2}\sigma_n^z \end{pmatrix}. \tag{11.61}$$

Moreover, with

$$b(u) = \frac{i}{u+i} \qquad \text{and} \qquad c(u) = \frac{u}{u+i} \tag{11.62}$$

we obtain the R-matrix of the isotropic six-vertex model explicitly

$$R(u) = \begin{pmatrix} 1 & 0 & 0 & 0 \\ 0 & \frac{i}{u+i} & \frac{u}{u+i} & 0 \\ 0 & \frac{u}{u+i} & \frac{i}{u+i} & 0 \\ 0 & 0 & 0 & 1 \end{pmatrix} = \frac{2iI_2 \otimes I_2 + u\left(I_2 \otimes I_2 + \sum_\alpha \sigma^\alpha \otimes \sigma^\alpha\right)}{2i + 2u}, \tag{11.63}$$

which is the simplest example of a rational R-matrix, i.e. an R-matrix where the matrix elements are rational functions of the spectral parameter u.

The special value of the spectral parameter is now

$$u = u_0 = \frac{i}{2}. \tag{11.64}$$

We have to evaluate $L_n(u)$, $T(u)$, etc, at this special value of $u = u_0 = \frac{i}{2}$.

At $u = u_0 = \frac{i}{2}$ we have

$$L_n\left(\frac{i}{2}\right) = \frac{i}{2}(\mathbb{I} + \boldsymbol{\sigma} \otimes \boldsymbol{\sigma}_n) = i\mathbb{P}_{0n} \tag{11.65}$$

where $\mathbb{I}$ is the 2×2 unit matrix where the matrix elements are themselves $2^N \times 2^N$ unit matrices and $\sigma \otimes \sigma_n$ stands for a 'scalar' product where the products of the components are, however, to be understood as tensor products

$$\boldsymbol{\sigma} \otimes \boldsymbol{\sigma}_n = \boldsymbol{\sigma} \otimes \left(\overbrace{I \otimes I \otimes \ldots \otimes \underbrace{\boldsymbol{\sigma}}_{n} \otimes \ldots \otimes I}^{N}\right). \tag{11.66}$$

The permutation operator $\mathbb{P}_{0n}$ (cp. (10.65)) can now be used to calculate the transfer matrix

$$T\left(\frac{i}{2}\right) = i^N \mathrm{Tr}\,(\mathbb{P}_{01} \ldots \mathbb{P}_{0N}) = i^N U_N^{-1}, \tag{11.67}$$

where U_N is the shift operator on the one-dimensional lattice

$$U_N \boldsymbol{\sigma}_n U_N^{-1} = \boldsymbol{\sigma}_{n+1} \tag{11.68}$$

with periodic boundary conditions understood.

EXERCISE 11.5 Consider the case $N = 2$ and show that

$$U_2 \sigma_1^x U_2^{-1} = \sigma_2^x. \tag{11.69}$$

A good starting point is to use the explicit form (11.61) for L_n at $u = \frac{i}{2}$.

The first logarithmic derivative of the transfer matrix is

$$\left(\frac{d}{du}\log T(u)\right)_{u=\frac{i}{2}} = \left(T^{-1}(u)\frac{d}{du}T(u)\right)_{u=\frac{i}{2}} = i^{-N} U_N \left(\frac{d}{du}T(u)\right)_{u=\frac{i}{2}} \tag{11.70}$$

$$= \frac{1}{i} U_N \sum_{n=1}^{N} \mathrm{Tr}\left(\mathbb{P}_{01}\ldots\underbrace{\mathbb{I}}_{n}\ldots\mathbb{P}_{0N}\right). \tag{11.71}$$

With

$$U_N \mathrm{Tr}\left(\mathbb{P}_{01}\ldots\underbrace{\mathbb{I}}_{n}\ldots\mathbb{P}_{0N}\right) = \mathbb{P}_{n,n+1} = \frac{1}{2}\left(\mathbb{I} + \boldsymbol{\sigma}_n \cdot \boldsymbol{\sigma}_{n+1}\right) \tag{11.72}$$

we finally get

$$i\left(\frac{d}{du}\log T(u)\right)_{u=\frac{i}{2}} = \sum_{n=1}^{N} \frac{1}{2}\left(\mathbb{I} + \boldsymbol{\sigma}_n \cdot \boldsymbol{\sigma}_{n+1}\right) \tag{11.73}$$

where again periodic boundary conditions $\sigma_{N+1} = \sigma_1$ are understood. The periodic boundary conditions are, of course, a reflection that we took the trace of the monodromy matrix $\mathcal{T}$ to obtain the transfer matrix T.

The right-hand side of (11.73) is the well-known Hamiltonian

$$\mathcal{H}_{\mathrm{XXX}} = \sum_{n=1}^{N} \frac{1}{2}\left(\mathbb{I} + \boldsymbol{\sigma}_n \cdot \boldsymbol{\sigma}_{n+1}\right) \tag{11.74}$$

of the isotropic Heisenberg quantum spin chain of spins $\frac{1}{2}$, the so-called XXX quantum spin chain. This explains the terminology we have been using throughout this section.

EXERCISE 11.6 Consider again the case $N = 2$ and show that

$$U_2 \mathrm{Tr}\,(\mathbb{P}_{01}\mathbb{I}) = \mathbb{P}_{1,2} = \frac{1}{2}(\mathbb{I} + \boldsymbol{\sigma}_1 \cdot \boldsymbol{\sigma}_2). \tag{11.75}$$

The eigenvalues of the transfer matrix determine the eigenvalues of the Heisenberg XXX quantum spin chain. Taking the logarithmic derivative of (11.43), recalling that we choose $\delta(u = \frac{i}{2}) = 0$, we obtain

$$h = -\sum_{j=1}^{M} \frac{1}{u_j^2 + \frac{1}{4}} \tag{11.76}$$

where the parameters u_j are to be determined by the Bethe ansatz equations (11.42).

EXERCISE 11.7 Derive the expression (11.76) for the eigenvalue of the XXX quantum spin chain. You may have to redefine the Heisenberg XXX Hamiltonian to subtract a constant term.

11.5 Inhomogeneous Yang–Baxter quantum integrable models

The previous sections of this chapter showed how, the local L_n matrices play a central oart in the algebraic form of the Bethe ansatz. Their decisive role in the analysis consists in their encoding the vertex weights for a whole row of vertices in such a way that the position at site n of the individual vertex in the row can be specified. However, so far we have only considered the homogeneous case where the vertex weights themselves are not site-dependent. The Yang–Baxter relations, e.g. in the form (10.80), remain valid if the vertex weights themselves that enter the L_n matrices become inhomogeneous, i.e. site-dependent. The inhomogeneities created in this way can be interpreted as local defects or impurities.

Integrable models with such site-dependent defects or impurities have been mentioned throughout the literature (Baxter, 1982; Takhtajan, 1985; Shastry, 1988*a*; Korepin *et al.*, 1993). This section only provides a brief overview of such models without detailed derivations.

11.5.1 Higher spin impurities

Andrei and Johannesson (1984) have considered an isotropic Heisenberg quantum spin chain with an impurity carrying a higher spin $S > 1/2$ for which they replaced one of the L_n matrices (11.61) by

$$L_n^{(S)}(u) = u + \frac{i}{2}\sum_{a=1}^{3}\sigma^a \otimes S_n^a \tag{11.77}$$

with the spin-S matrices S_n^a formed analogous to the spin-1/2 Pauli matrices in (10.44). The results of this procedure have been compared with those obtained for non-integrable spin chain impurity models by Sørensen *et al.* (1993) and generalized by Zvyagin and Schlottmann (1995) to spin chains with a finite concentration of spin-S impurities.

11.5.2 Impurities induced by local shifts in the spectral parameter

The spectral parameter u, the argument in the L_n matrices, offers another way to introduce inhomogeneities into the quantum spin chain when we replace it locally, i.e. $L_n(u) \rightarrow L_n(u_n)$ at one or several sites n. More specifically, the spectral parameter is shifted by site-dependent amounts $\{\nu_1, \ldots \nu_N\}$ such that, with $u_n = u + \nu_n$, the transfer matrix becomes

$$T(u, \{\nu_n\}) = \mathrm{Tr}\left(\prod_{i=1}^{N} L_n(u+\nu_n)\right) = A(u+\nu_n) + D(u+\nu_n). \tag{11.78}$$

The procedure was discussed in detail in section 11.4, and it can be applied to this transfer matrix to construct a quantum Hamiltonian of a Heisenberg spin chain with impurities (see Schmitteckert *et al.*, 1995; Eckle *et al.*, 1997; Eckle, 1998; and, for a similar treatment of the *t–J* model, see Bares, 1994; Foerster *et al.*, 1999).

The physical interest in these models is due to the general question of the effects of impurities and weak links in Fermi and, especially, Luttinger liquids. The interplay of (strong) interactions and impurities is here one of the focal points.

This class of quantum models with integrable impurities has continued to be of interest and most recently been extended to a Hubbard model with integrable impurity (Öz and Klümper, 2018). For further developments, see Guan *et al.* (2000), Yahagi *et al.* (2014), Yahagi *et al.* (2016), and, most recently, Essler *et al.* (2018) and the references in these works.

The Hamiltonians of the thus-constructed one-dimensional quantum models have a nearest-neighbour structure except in the vicinity of the local impurities that involve three-site interactions. The pertinent feature of these models is the absence of backscattering at the impurities: the impurities are transparent.

11.5.3 Isotropic Heisenberg quantum spin chain with a single impurity

A typical case is the isotropic Heisenberg quantum spin or XXX chain on a ring of size N with a single impurity at site m, i.e. all $\nu_n = 0$ except for $n = m$ with $\nu_m = \nu$. After a

Jordan–Wigner transformation, we obtain the Hamiltonian of a lattice gas of M spinless Fermions

$$\mathcal{H} = \mathcal{H}_{\mathrm{XXX}} + \mathcal{H}_{m,\mathrm{imp}} + 2M, \tag{11.79}$$

where

$$\mathcal{H}_{\mathrm{XXX}} = -\sum_{j=1}^{N}(c_{j+1}^{\dagger}c_j + h.c.) + 2\sum_{j=1}^{N}\rho_j\rho_{j+1} \tag{11.80}$$

is the Hamiltonian (11.74) of the clean XXX model. The operators $c_j^{\dagger}$ and c_j create and annihilate spinless Fermions on site j and $\rho_j = c_j^{\dagger}c_j$ is the density operator. The impurity interaction at site m consists of three terms

$$\mathcal{H}_{m,\mathrm{imp}} = \mathcal{H}_{m,t} + \mathcal{H}_{m,\rho} + \mathcal{H}_{m,\,j} \tag{11.81}$$

which are explicitly

$$\mathcal{H}_{m,t} = -(\xi + i\zeta)[c_m^{\dagger}c_{m-1} + c_{m+1}^{\dagger}c_m + c_{m-1}^{\dagger}c_{m+1}] + h.c., \tag{11.82}$$

$$\mathcal{H}_{m,\rho} = -2\xi[\rho_m - \rho_m(\rho_{m-1} + \rho_{m+1}) + \rho_{m-1}\rho_{m+1}], \tag{11.83}$$

$$\begin{aligned}\mathcal{H}_{m,c} = {} & 2\xi(c_{m+1}^{\dagger}c_{m-1} + c_{m-1}^{\dagger}c_{m+1})\rho_m + 2i\zeta(c_m^{\dagger}c_{m-1} - c_{m-1}^{\dagger}c_m)\rho_{m+1} \\ & + 2i\zeta(c_{m+1}^{\dagger}c_m - c_m^{\dagger}c_{m+1})\rho_{m-1},\end{aligned} \tag{11.84}$$

where the coupling constants as a function of ν are

$$\xi(\nu) = -\nu^2/(1+\nu^2) \qquad \text{and} \qquad \zeta(\nu) = -\nu/(1+\nu^2). \tag{11.85}$$

In the impurity Hamiltonian (11.81), we have three types of terms that can be interpreted in the following way:

1. $\mathcal{H}_{m,t}$: nearest and next-nearest neighbour hopping matrix elements are modified involving the sites m and $m \pm 1$;
2. $\mathcal{H}_{m,\rho}$: an on-site potential at site m and both nearest and next-nearest neighbour density-density interaction terms are induced;
3. $\mathcal{H}_{m,c}$: this term corresponds to a current-density interaction involving all three sites m and $m \pm 1$.

Thus, we see that a local translation of the spectral parameter $u \to u + v_m$ effectively creates interactions involving the site m and its two neighbors $m \pm 1$ and in addition includes next nearest-neighbour terms involving the sites $m - 1$ and $m + 1$.

For a first understanding of the properties of the integrable impurities, we look at the scattering of a single particle off a single impurity. The incoming and outgoing wave functions are, respectively,

$$\Psi_{\text{in}}(n) = e^{ikn} + R(k)e^{-ikn} \qquad \text{for} \qquad n < m, \tag{11.86}$$

$$\Psi_{\text{out}}(n) = T(k)e^{ikn} \qquad \text{for} \qquad n > m, \tag{11.87}$$

where $R(k)$ and $T(k)$ are the reflection and transmission amplitudes, respectively. It can be shown (Eckle *et al.*, 1997) that

1. the reflection amplitude $R(k) = 0$ for all momenta k, thus there is *no backscattering* of a particle due to the impurity;
2. unitarity of the scattering matrix requires $|T|^2 + |R|^2 = 1$ and consequently we have $|T(k)|^2 = 1$. Writing $T(k) = e^{i\delta_0(k,v)}$, the phase-shift δ_0 of the single-particle wave function due to the presence of a single impurity becomes

$$\delta_0(k, v) = -k + 2\tan^{-1}(\tan(k/2) - 2v). \tag{11.88}$$

Details of this calculation and its extension to many particles and many impurities can be found in Eckle *et al.* (1997). The literature cited in this section may be consulted for further information of these developments.

12 Quantum Tavis–Cummings Model

The man of true Physical instincts, endowed with the great faculty of scientific imagination, possessed for example by Lord Kelvin in a very remarkable degree, is for ever imagining models which shall enable him by their working to represent and depict the course of actual physical processes. The possibility and consistency of such models require Mathematical Analysis for their investigation.

Ernest William Hobson (1856–1933)

In this chapter, we want to apply what we learned about the algebraic Bethe ansatz for the XXZ Heisenberg quantum spin chain to the quantum Tavis–Cummings model, which was constructed to comprehend the strong interaction between light and quantum matter. However, unlike the in the case of the Heisenberg quantum spin chain, in the case of the quantum Tavis–Cummings model there is no underlying vertex or other classical statistical mechanical model to suggest the constituent building blocks of the algebraic Bethe ansatz approach, like the L-matrix or, ultimately, the transfer matrix.

Section 12.1 briefly recalls the concepts of the transfer and monodromy matrices in the algebraic Bethe ansatz. The algebraic Bethe ansatz is then first applied not to the quantum Tavis–Cummings Hamiltonian but to a modification of this Hamiltonian, which we obtain by adding a term called the Stark term (section 12.2). The model with the added Stark term is amenable to an algebraic Bethe ansatz, which we shall demonstrate in section 12.3 where we posit a transfer matrix and derive from this transfer matrix the conserved quantities, Hamiltonian and generalized number operator, of the model and in section 12.4 where we prove that transfer matrices commute for distinct values of its spectral parameter. Eventually, we shall come back to the unmodified quantum Tavis–Cummings model as a limit when the modifying Stark term vanishes. Section 12.5 shows how this limit can be performed but it is not trivial. Finally, section 12.6 derives the Bethe ansatz equations of the quantum Tavis–Cummings model.

12.1 Algebraic Bethe ansatz revisited

We recall from part II that central objects of the algebraic Bethe ansatz are the transfer matrix $T = T(\lambda)$ and the monodromy matrix $\mathcal{T}(\lambda)$, both of which depend on a complex parameter λ, the spectral parameter. They are related (cp. (10.72) and (10.73)) by

Models of Quantum Matter: A First Course on Integrability and the Bethe Ansatz. Hans-Peter Eckle, Oxford University Press (2019). © Hans-Peter Eckle. DOI: 10.1093/oso/9780199678839.001.0001

$$T(\lambda) = \mathrm{Tr}\,\mathcal{T}(\lambda) = A(\lambda) + D(\lambda). \tag{12.1}$$

The transfer matrix, $T(\lambda)$, plays the role of a generating function for the conserved quantities of the integrable model. As a consequence of the Yang–Baxter relations, the matrix elements of the monodromy matrix

$$\mathcal{T}(\lambda) = \begin{pmatrix} A(\lambda) & B(\lambda) \\ C(\lambda) & D(\lambda) \end{pmatrix} \tag{12.2}$$

satisfy an algebra, e.g. for the six-vertex model, the algebra of (11.35), (11.36), and (11.37).

As we lack an approach via a statistical mechanical model in two dimensions, our goal in this section is nevertheless to find a monodromy matrix $\mathcal{T}(\lambda)$ and a corresponding transfer matrix $T(\lambda)$, which can be used as starting points to derive the Hamiltonian and other conserved operators of the quantum Tavis–Cummings model for appropriate choices of the parameter λ. This is in analogy to the procedure for the six-vertex model, which yielded, *inter alia*, the Hamiltonian of the *XXZ* Heisenberg quantum spin chain.

12.2 Modified quantum Tavis–Cummings model

We shall see that the Hamiltonian $\mathcal{H}_{\mathrm{TC}}$ (8.896) is unsuitable to make contact with the algebraic treatment of the *XXZ* chain in a transparent way directly. It is more transparent to consider a modification of this Hamiltonian first

$$\mathcal{H}_{\mathrm{TCS}} = \Delta S^z + aS^+ + a^\dagger S^- + ca^\dagger aS^z \tag{12.3}$$

where the last term describes atomic transitions that depend on the photon number (Bogoliubov *et al.*, 1996).

Such a term proportional to $\sigma^z a^\dagger a$ for a two-level atom, also called a nonlinear Stark term (see Klimov and Chumakov, 2009, section 5.10), has recently attracted attention in relation to the quantum Rabi model with Hamiltonian

$$\mathcal{H}_{\mathrm{Rabi}} = \omega a^\dagger a + \Delta\sigma^z + g\sigma^x \left(a + a^\dagger\right) \tag{12.4}$$

in several respects.

On the one hand, the extended model, the quantum Rabi–Stark model with Hamiltonian

$$\mathcal{H} = \mathcal{H}_{\mathrm{Rabi}} + \gamma\sigma^z a^\dagger a, \tag{12.5}$$

has been shown to be an exactly solvable extension of the original quantum Rabi model by two different methods, one employed by Maciejewski *et al.* (2014*a*), Maciejewski *et al.* (2015), the other by Eckle and Johannesson (2017).

Both methods rely on the Bargmann space representation of the Hilbert space of the model. The first method pioneered by Braak (2011) for the original quantum Rabi model uses the wave functions of the Bargmann space directly to construct a transcendental function whose zeros provide the energy eigenvalue spectrum (Eckle and Johannesson, 2017). The other method devised by Maciejewski *et al.* (2014*b*) uses the wave functions to construct Wronskians and an analysis based on the Stokes phenomenon (Balser, 2000).

On the other hand, Braak (2011) proposed a criterion according to which the original quantum Rabi model is integrable. This criterion also applies to the quantum Rabi–Stark model (Eckle and Johannesson, 2017).

Let us now return to the quantum Tavis–Cummings model with Stark term. The generalized number operator

$$\mathcal{M} = S^z + a^\dagger a, \tag{12.6}$$

and the Casimir operator

$$\mathbf{S}^2 = S^+S^- + S^z(S^z - 1) \tag{12.7}$$

commute with the Hamiltonian (12.3)

$$[\mathcal{H}_{\text{TCS}}, \mathcal{M}] = [\mathcal{H}_{\text{TCS}}, \mathbf{S}^2] = 0, \tag{12.8}$$

which can be used to rewrite (12.3) in the form

$$\mathcal{H}^{(0)}_{\text{TCS}} = \mathcal{H}_{TCS} - \frac{c}{2}\mathcal{M}^2 = \Delta S^z + aS^+ + a^\dagger S^- - \frac{c}{2}a^\dagger a a^\dagger a - \frac{c}{2}\left(S^z\right)^2, \tag{12.9}$$

a form which we shall, however, not further use.

12.3 Transfer matrix of the modified quantum Tavis–Cummings model

We now postulate two 2×2 matrix operators, L–operators, one for Bosons $L_B(\lambda)$ and one for spins $L_S(\lambda)$, both depending on a complex parameter λ, in such a way that their product

$$\mathcal{T}(\lambda) = L_B(\lambda)L_S(\lambda) = \begin{pmatrix} A(\lambda) & B(\lambda) \\ C(\lambda) & D(\lambda) \end{pmatrix} \tag{12.10}$$

defines a transfer matrix via

$$T(\lambda) = \operatorname{Tr}\mathcal{T}(\lambda) = A(\lambda) + D(\lambda), \tag{12.11}$$

which we can use as a generating function to recover the conserved quantities $\mathcal{H}_{\text{TCS}}$ and $\mathcal{M}$.

Amico *et al.* (2007) carried out a study of more general cases of spin–Boson models that can also be constructed using the quantum inverse scattering method (Korepin *et al.*, 1993) and thus solved, by construction, via the algebraic Bethe ansatz.

In our case, the construction of the monodromy matrix $\mathcal{T}(\lambda)$ in the form (12.10) is indeed possible and the two L-operators may be chosen, respectively, as

$$L_B(\lambda) = \begin{pmatrix} c\lambda - c\Delta - 1 - c^2 a^\dagger a & c a^\dagger \\ ca & -1 \end{pmatrix}, \tag{12.12}$$

$$L_S(\lambda) = \begin{pmatrix} \lambda - cS^z & cS^+ \\ cS^- & \lambda + cS^z \end{pmatrix}. \tag{12.13}$$

EXERCISE 12.1 Show that (12.12) and (12.13) indeed lead to a transfer matrix $T(\lambda)$ from which the Hamiltonian (12.3) and the generalized number operator M (12.6) can be derived by suitable derivatives and, subsequently, a suitable choice of the parameter λ.

The results of exercise 12.1 are

$$A(\lambda) = (c(\lambda - \Delta) - 1 - c^2 a^\dagger a)(\lambda - cS^z) + c^2 a^\dagger S^-, \tag{12.14}$$

$$B(\lambda) = \lambda(c(a^\dagger + cS^+)) - c(1 + c\Delta)S^+ - c^2 a^\dagger S^z - c^3 a^\dagger a S^+ = \lambda X - Y, \tag{12.15}$$

$$C(\lambda) = -cS^- + ca(\lambda - cS^z), \tag{12.16}$$

$$D(\lambda) = c^2 a S^+ - (\lambda + cS^z). \tag{12.17}$$

Moreover,

$$T(\lambda) = \operatorname{Tr} \mathcal{T}(\lambda) = A(\lambda) + D(\lambda) = c\lambda^2 - \lambda(c^2\mathcal{M} + c\Delta + 2) + c^2 \mathcal{H}_{\text{TCS}} \tag{12.18}$$

with the Hamiltonian (12.3) and generalized number operator (12.6) so that

$$\mathcal{H}_{\text{TCS}} = \frac{1}{c^2} T(0) \tag{12.19}$$

and

$$\mathcal{M} = -\frac{1}{c^2}\left(\frac{\partial T(\lambda)}{\partial \lambda}\right)_{\lambda=0} - \frac{\Delta}{c} - \frac{2}{c^2}. \tag{12.20}$$

The crucial property of a transfer matrix is that the commutator

$$[T(\lambda), T(\mu)] = 0 \tag{12.21}$$

for arbitrary complex parameters λ and μ. Then, from (12.19) and (12.20), $T(\lambda)$ commutes with $\mathcal{H}$ and $\mathcal{M}$ and the eigenstates of $T(\lambda)$ are simultaneous eigenstates of $\mathcal{H}_{\mathrm{TCS}}$ and $\mathcal{M}$. Moreover, since we already know

$$[\mathcal{H}_{\mathrm{TCS}}, \mathbf{S}^2] = [\mathcal{M}, \mathbf{S}^2] = 0, \tag{12.22}$$

we see from (12.18) that also

$$[T, \mathbf{S}^2] = 0 \tag{12.23}$$

holds.

Before we prove (12.21) in Section 12.4, let us note in the following excercise some symmetry properties of the L-operators $L_B(\lambda)$ and $L_S(\lambda)$ given in (12.12) and (12.13) and the monodromy matrix $\mathcal{T}(\lambda)$ given in (12.10).

EXERCISE 12.2 Prove the symmetries of the monodromy matrices

$$e^{\alpha a^\dagger a} L_B(\lambda) e^{-\alpha a^\dagger a} = e^{\frac{\alpha}{2}\sigma^z} L_B(\lambda) e^{-\frac{\alpha}{2}\sigma^z}, \tag{12.24}$$

$$e^{\alpha S^z} L_S(\lambda) e^{-\alpha S^z} = e^{\frac{\alpha}{2}\sigma^z} L_S(\lambda) e^{-\frac{\alpha}{2}\sigma^z}, \tag{12.25}$$

$$e^{\alpha M} \mathcal{T}(\lambda) e^{-\alpha M} = e^{\frac{\alpha}{2}\sigma^z} \mathcal{T}(\lambda) e^{-\frac{\alpha}{2}\sigma^z} \tag{12.26}$$

for an arbitrary complex number α.

From the last result of the previous exercise, we obtain the important relations

$$\mathcal{M}B(\lambda) = B(\lambda)(\mathcal{M} + 1), \tag{12.27}$$

$$\mathcal{M}C(\lambda) = C(\lambda)(\mathcal{M} - 1) \tag{12.28}$$

which shows that $B(\lambda)$ acts as a creation operator while $C(\lambda)$ acts as an annihilation operator. Furthermore, we have

$$\mathbf{S}^2 B(\lambda) = B(\lambda)\mathbf{S}^2. \tag{12.29}$$

Now, we turn to the proof of the commutativity of the transfer matrix (12.21).

12.4 Commutativity of the transfer matrix

The commutativity of the transfer matrices for arbitrary complex parameters λ and μ

$$[T(\lambda), T(\mu)] = 0 \tag{12.30}$$

follows, of course, directly from the commutation relations (12.8) in conjunction with the explicit expression for the transfer matrix, (12.18).

However, it will be instructive and allow us to make contact with the algebraic treatment of the XXZ quantum spin chain, to derive this crucial property of the transfer matrix in a more general way which, moreover, can be applied also to other models.

In order to carry out the proof, we introduce the tensor product of monodromy matrices for arbitrary complex parameters λ and μ

$$\mathcal{T}(\lambda) \otimes \mathcal{T}(\mu). \tag{12.31}$$

Then

$$T(\lambda)T(\mu) = \mathrm{Tr}\,\mathcal{T}(\lambda) \otimes \mathcal{T}(\mu), \quad T(\mu)T(\lambda) = \mathrm{Tr}\,\mathcal{T}(\mu) \otimes \mathcal{T}(\lambda) \tag{12.32}$$

where we used that for the (respective) traces

$$\mathrm{Tr}\,A \otimes B = \mathrm{Tr}\,A \cdot \mathrm{Tr}\,B. \tag{12.33}$$

If we can find a 4×4 matrix with complex elements $R(\lambda, \mu)$ depending on the two complex parameters λ and μ such that

$$R(\lambda, \mu)T(\lambda) \otimes T(\mu) = T(\mu) \otimes T(\lambda)R(\lambda, \mu), \tag{12.34}$$

then, obviously, (12.30) holds, because

$$T(\lambda)T(\mu) = \mathrm{Tr}\,R(\lambda, \mu)\mathcal{T}(\lambda) \otimes \mathcal{T}(\mu)(R(\lambda, \mu))^{-1} \tag{12.35}$$

$$= \mathrm{Tr}\,\mathcal{T}(\mu) \otimes \mathcal{T}(\lambda)R(\lambda, \mu)R^{-1}(\lambda, \mu) \tag{12.36}$$

$$= \mathrm{Tr}\,\mathcal{T}(\mu) \otimes \mathcal{T}(\lambda) = T(\mu)T(\lambda). \tag{12.37}$$

We recognize (12.34) as the Yang–Baxter relations of the quantum Tavis–Cummings model.

Now, we need a result for tensor products that states that for the tensor product of the products of matrices AB and CD

$$AB \otimes CD = (A \otimes C)(B \otimes D) \tag{12.38}$$

only if the matrix elements of A and B mutually commute with the matrix elements of C and D.

Indeed, the matrix elements of $L_B(\lambda)$ (12.12) and $L_S(\lambda)$ (12.13) do commute mutually. Therefore,

$$\mathcal{T}(\lambda) \otimes \mathcal{T}(\mu) = L_B(\lambda)L_S(\lambda) \otimes L_B(\mu)L_S(\mu) \tag{12.39}$$

$$= (L_B(\lambda) \otimes L_B(\mu))(L_S(\lambda) \otimes L_S(\mu)). \tag{12.40}$$

Thus, if the R-matrix satisfies

$$R(\lambda, \mu)L_B(\lambda) \otimes L_B(\mu) = L_B(\mu) \otimes L_B(\lambda)R(\lambda, \mu), \tag{12.41}$$

and

$$R(\lambda, \mu)L_S(\lambda) \otimes L_S(\mu) = L_S(\mu) \otimes L_S(\lambda)R(\lambda, \mu), \tag{12.42}$$

then (12.34) is also satisfied.

This statement can, however, be easily verified and is left as the next exercise.

EXERCISE 12.3 Show that, if the R-matrix satisfies

$$R(\lambda, \mu)L_B(\lambda) \otimes L_B(\mu) = L_B(\mu) \otimes L_B(\lambda)R(\lambda, \mu), \tag{12.43}$$

and

$$R(\lambda, \mu)L_S(\lambda) \otimes L_S(\mu) = L_S(\mu) \otimes L_S(\lambda)R(\lambda, \mu), \tag{12.44}$$

then the Yang–Baxter relations (12.34) are also satisfied.

Thus, the task of finding the R-matrix is reduced from dealing with the Yang–Baxter relations for the monodromy matrix directly, i.e. with (12.34), to dealing with the simpler problem of the Yang–Baxter relations for the two L operators $L_B(\lambda)$ and $L_S(\lambda)$, i.e. with (12.41) and (12.42).

Since the commutation relations of the matrix elements of $L_B(\lambda)$ and $L_S(\lambda)$ are known, they can be used to determine $R(\lambda, \mu)$ from (12.41) and (12.42). We find

$$R(\lambda, \mu) = \begin{pmatrix} 1 + g(\lambda, \mu) & 0 & 0 & 0 \\ 0 & g(\lambda, \mu) & 1 & 0 \\ 0 & 1 & g(\lambda, \mu) & 0 \\ 0 & 0 & 0 & 1 + g(\lambda, \mu) \end{pmatrix} \tag{12.45}$$

$$= g(\lambda, \mu)I + \Pi, \tag{12.46}$$

$$g(\lambda, \mu) = \frac{c}{\lambda - \mu}, \tag{12.47}$$

where I is the unit matrix and Π is the permutation matrix in $\mathbb{C}^2 \otimes \mathbb{C}^2$, i.e.

$$\Pi = \begin{pmatrix} 1 & 0 & 0 & 0 \\ 0 & 0 & 1 & 0 \\ 0 & 1 & 0 & 0 \\ 0 & 0 & 0 & 1 \end{pmatrix}. \tag{12.48}$$

The Yang–Baxter relations (12.34) encode the 16 commutation relations of the four matrix elements $A(\lambda)$, $B(\lambda)$, $C(\lambda)$, and $D(\lambda)$, which form the monodromy matrix $\mathcal{T}(\lambda)$.

EXERCISE 12.4 Show that among the commutation relations there are

$$[A(\lambda), A(\mu)] = [B(\lambda), B(\mu)] = [C(\lambda), C(\mu)] = [D(\lambda), D(\mu)] = 0, \tag{12.49}$$

$$A(\mu)B(\lambda) = f(\lambda,\mu)B(\lambda)A(\mu) + g(\lambda,\mu)B(\mu)A(\lambda). \tag{12.50}$$

Show further that the same commutation relation holds when A is replaced by D. In addition to those six commutation relations, there are ten more.

At this point, we can start the algebraic programme to derive the Bethe ansatz equations: defining an appropriate vacuum state and letting the creation operators $B(\lambda)$ for different parameter values $\{\lambda_j\}$ act on this vacuum state to create states above the vacuum. The requirement that these states are eigenstates of the modified quantum Tavis–Cummings Hamiltonian (12.3) will lead to a set of consistency equations for the set of parameters $\{\lambda_j\}$, which are the Bethe ansatz equations. This programme can be carried out in complete analogy to the corresponding programme for the *XXZ* Heisenberg quantum spin chain.

We shall, however, not follow this programme here, but, instead, turn back to the unmodified quantum Tavis–Cummings model described by (8.896), the quantum Tavis–Cummings model without the Stark term $cS^z a^\dagger a$ describing photon number dependent atomic transitions of the two-level atom. The treatment of this model will prove less involved, although we shall encounter some modifications of the Bethe ansatz programme just outlined above and carried through in detail for the *XXZ* model in part II.

12.5 Simple quantum Tavis–Cummings model

The original quantum Tavis–Cummings model, described by the Hamiltonian (8.896), can be recovered formally from the modified quantum Tavis–Cummings model, described by the Hamiltonian (12.3), in the limit of small interaction parameter c. Obviously, on the level of the Hamiltonians, setting $c = 0$, the Hamiltonian (12.3) becomes the Hamiltonian of the original quantum Tavis–Cummings model, (8.896).

What, however, does $c \to 0$ imply for the L-operators, (12.12) and (12.13), and the corresponding monodromy matrices and the transfer matrix? By inspection, to order linear in the interaction c, we find

$$\mathbf{T}(\lambda) = -\lambda I + c\tilde{L}(\lambda) + O(c^2) \tag{12.51}$$

where

$$\tilde{L}(\lambda) = \begin{pmatrix} \lambda - \Delta + \frac{S^z}{\lambda} & a^\dagger - \frac{S^+}{\lambda} \\ a - \frac{S^-}{\lambda} & -\frac{S^z}{\lambda} \end{pmatrix}. \tag{12.52}$$

The monodromy matrix we obtain for $c = 0$ is obviously trivial, which indicates that we need to consider the order linear in the interaction c in the monodromy matrix (12.51). Substituting the expansion (12.51) into the Yang–Baxter relation (12.34) and keeping only terms up to linear order in the interaction parameter c, we arrive at the commutation relations

$$[X^+(\lambda), X^-(\mu)] = \frac{2}{\lambda - \mu}(X^z(\lambda) - X^z(\mu)), \tag{12.53}$$

$$[X^z(\lambda), X^\pm(\mu)] = \pm\frac{1}{\lambda - \mu}(X^\pm(\lambda) - X^\pm(\mu)), \tag{12.54}$$

$$[X^+(\lambda), X^+(\mu)] = [X^-(\lambda), X^-(\mu)] = [X^z(\lambda), X^z(\mu)] = 0. \tag{12.55}$$

Since a constant shift in the operator $\tilde{L}(\lambda)$

$$\tilde{L}(\lambda) \to \tilde{L}(\lambda) - \frac{1}{2}(\lambda - \Delta)\mathbb{I} \equiv L(\lambda) \tag{12.56}$$

leaves (12.53)–(12.55) unchanged, we arrive at

$$L(\lambda) = \begin{pmatrix} -\left(\frac{1}{2}(\Delta - \lambda) - \frac{S^z}{\lambda}\right) & a^\dagger - \frac{S^+}{\lambda} \\ a - \frac{S^-}{\lambda} & \left(\frac{1}{2}(\Delta - \lambda) - \frac{S^z}{\lambda}\right) \end{pmatrix} = \begin{pmatrix} -X^z & X^+ \\ X^- & X^z \end{pmatrix}. \tag{12.57}$$

EXERCISE 12.5 Expand the monodromy matrix (12.10) of the modified quantum Tavis–Cummings model to linear order in the interaction parameter c and derive the commutation relations (12.53–12.55) and the explicit form of the L operator, (12.57).

This L operator will now assume the role of the monodromy matrix, i.e. formally to linear order in c

$$\frac{1}{c}\left(\mathcal{T}(\lambda) + \lambda I\right) \to L(\lambda). \tag{12.58}$$

However, we observe that, nevertheless, the operator $L(\lambda)$ does not provide a transfer matrix, i.e. a generating function for conserved quantities, since

$$\mathrm{Tr}\, L(\lambda) = 0. \tag{12.59}$$

This, however, only means that we need to construct another invariant from $L(\lambda)$. The obvious next choice for such an invariant is

$$\tau(\lambda) = \frac{1}{2}\mathrm{Tr}\, L^2(\lambda) = \left(X^z(\lambda)\right)^2 + \frac{1}{2}X^+(\lambda)X^-(\lambda) + \frac{1}{2}X^-(\lambda)X^+(\lambda). \tag{12.60}$$

Explicitly, we have

$$\tau(\lambda) = \frac{1}{\lambda^2}\mathbf{S}^2 - \frac{1}{\lambda}\mathcal{H}_{TC} + \left(\frac{\Delta^2}{4} + \frac{1}{2} + \mathcal{M}\right) - \frac{\Delta}{2}\lambda + \frac{\lambda^2}{4}. \tag{12.61}$$

Moreover, from its definition and using (12.53)–(12.55), we obtain the following commutation relations

$$[\tau(\lambda), X^z(\mu)] = \frac{1}{\lambda - \mu}\left(X^-(\lambda)X^+(\mu) - X^-(\mu)X^+(\lambda)\right), \tag{12.62}$$

$$[\tau(\lambda), X^\pm(\mu)] = \frac{2}{\lambda - \mu}\left(X^\pm(\lambda)X^z(\mu) - X^\pm(\mu)X^z(\lambda)\right), \tag{12.63}$$

and, as most important fundamental property

$$[\tau(\lambda), \tau(\mu)] = 0, \tag{12.64}$$

which now encodes the mutual commutativity of the quantum Tavis–Cummings Hamiltonian $\mathcal{H}_{TC}$, the generalized number operator $\mathcal{M}$, and the Casimir operator $\mathbf{S}^2$, which can be explicitly calculated from $\tau(\lambda)$

$$\mathcal{H}_{TC} = -\left(\frac{d}{d\lambda}\lambda^2\tau(\lambda)\right)_{\lambda=0}, \tag{12.65}$$

$$\mathcal{M} = \frac{1}{2}\left(\frac{d^2}{d\lambda^2}\lambda^2\tau(\lambda)\right)_{\lambda=0} - \frac{1}{2} - \frac{\Delta^2}{4}, \tag{12.66}$$

$$\mathbf{S}^2 = \lambda^2\left(\tau(\lambda)\right)_{\lambda=0}. \tag{12.67}$$

Moreover,

$$[\tau(\lambda), \mathcal{H}_{TC}] = [\tau(\lambda), \mathcal{M}] = [\tau(\lambda), \mathbf{S}^2] = 0. \tag{12.68}$$

The eigenstates of $\tau(\lambda)$, the generating function of the conserved quantities, are simultaneous eigenstates of the conserved quantities $\mathcal{H}_{TC}$, $\mathcal{M}$, and $\mathbf{S}^2$, whose eigenvalues can be obtained from the eigenvalues of $\tau(\lambda)$ via (12.61).

12.6 Bethe ansatz solution of the quantum Tavis–Cummings model

We begin with the observation that the vacuum state

$$|\Omega_S\rangle = |0\rangle|S, -S\rangle \tag{12.69}$$

is an eigenstate of $\tau(\lambda)$

$$\tau(\lambda)|\Omega_S\rangle = k(\lambda)|\Omega_S\rangle \tag{12.70}$$

with

$$k(\lambda) = \frac{S(S+1)}{\lambda^2} + \frac{\Delta S}{\lambda} + \left(\frac{\Delta^2}{4} + \frac{1}{2} - S\right) - \frac{\Delta}{2}\lambda + \frac{\lambda^2}{4}. \tag{12.71}$$

Later it will prove useful to rewrite the eigenvalue of $\tau(\lambda)$ for the vacuum state $|\Omega_S\rangle$. In order to do this, we first note that in the limit $\mu \to \lambda$

$$[X^+(\lambda), X^-(\lambda)] = 2\frac{\partial}{\partial\lambda}X^z(\lambda) \tag{12.72}$$

and, hence, from (12.60)

$$\tau(\lambda)|\Omega_S\rangle = \left(\left(X^z(\lambda)\right)^2 - \frac{\partial}{\partial\lambda}X^z(\lambda)\right)|\Omega_S\rangle = \left(x^2(\lambda) - \frac{\partial}{\partial\lambda}x(\lambda)\right)|\Omega_S\rangle, \tag{12.73}$$

such that

$$k(\lambda) = x^2(\lambda) - \frac{\partial}{\partial\lambda}x(\lambda). \tag{12.74}$$

Moreover, the action of the operators $X^-(\lambda)$ and $X^z(\lambda)$ on the vacuum state is

$$X^-(\lambda)|\Omega_S\rangle = \left(a - \frac{S^-}{\lambda}\right)|\Omega_S\rangle = 0 \tag{12.75}$$

$$X^z(\lambda)|\Omega_S\rangle = \left(\frac{1}{2}(\Delta - \lambda) - \frac{S^z}{\lambda}\right)|\Omega_S\rangle = x(\lambda)|\Omega_S\rangle \tag{12.76}$$

with

$$x(\lambda) = \frac{1}{2}\left(\Delta - \lambda + 2\frac{S}{\lambda}\right). \tag{12.77}$$

The operator $X^+(\lambda)$ satisfies

$$\mathcal{M}X^+(\lambda) = X^+(\lambda)(\mathcal{M} + 1) \tag{12.78}$$

$$\mathbf{S}^2 X^+(\lambda) = X^+(\lambda)\mathbf{S}^2, \tag{12.79}$$

which makes it manifest that X^+ acts as a creation operator of the quasiparticles of the theory.

We can now formulate our task: we shall look for the M-particle eigenstate of $\tau(\lambda)$ in the usual form of the algebraic Bethe ansatz

$$|\Phi(\lambda_1, \lambda_2, \ldots, \lambda_M)\rangle = \left(\prod_{j=1}^{M} X^+(\lambda_j)\right)|\Omega_S\rangle \tag{12.80}$$

where it is the set of values λ_j that we need to determine.

Because $[X^+(\lambda), X^+(\mu)] = 0$, the state (12.80) is symmetric under pairwise exchange of its arguments. Moreover, similar to what we have found earlier

$$\mathcal{M}|\Phi(\lambda_1, \lambda_2, \ldots, \lambda_M)\rangle = (M - S)|\Phi(\lambda_1, \lambda_2, \ldots, \lambda_M)\rangle, \tag{12.81}$$

$$\mathbf{S}^2|\Phi(\lambda_1, \lambda_2, \ldots, \lambda_M)\rangle = S(S+1)|\Phi(\lambda_1, \lambda_2, \ldots, \lambda_M)\rangle. \tag{12.82}$$

A direct, but very lengthy, calculation, using the results obtained above, reveals that

$$\begin{aligned}\tau(\mu)|\Phi(\lambda_1, \lambda_2, \ldots, \lambda_M)\rangle &= \Theta_M(\mu)|\Phi(\lambda_1, \lambda_2, \ldots, \lambda_M)\rangle \\ &\quad + 2\sum_{j=1}^{M} \frac{f_j}{\mu - \lambda_j}|\Phi(\lambda_1, \ldots, \lambda_{j-1}, \mu, \lambda_{j+1}, \ldots, \lambda_M)\rangle\end{aligned} \tag{12.83}$$

where

$$\Theta_M(\mu) = x^2(\mu) - \frac{\partial}{\partial\mu}x(\mu) - 2\sum_{j=1}^{M} \frac{1}{\mu - \lambda_j}\left(x(\mu) - \sum_{k=1,k\neq j}^{M} \frac{1}{\lambda_j - \lambda_k}\right) \tag{12.84}$$

and

$$f_j = x(\lambda_j) - \sum_{k=1,k\neq j}^{M} \frac{1}{\lambda_j - \lambda_k}. \tag{12.85}$$

Obviously, $|\Phi(\lambda_1, \ldots, \lambda_M)\rangle$ is an eigenstate of $\tau(\mu)$ with eigenvalue $\Theta_M(\mu)$, if all f_j vanish, i.e.

$$x(\lambda_j) = \sum_{k=1,k\neq j}^{M} \frac{1}{\lambda_j - \lambda_k}. \tag{12.86}$$

These equations are the Bethe ansatz equations, which determine the Bethe ansatz roots λ_j. Note that the same requirement (12.86) follows from the observation that the left-hand side of (12.83) is an analytic function in μ, whereas the right-hand side has poles at $\mu = \lambda_j$. All of these poles are removed if the parameters λ_j satisfy the Bethe ansatz equations (12.86). The proof of (12.83) is quite lengthy, but straightforward and will be omitted here. It is, however, recommended as the following exercise.

EXERCISE 12.6 Eigenvalue equation and Bethe ansatz equations Derive the eigenvalue equation (12.83) and the Bethe ansatz equations (12.86). You will have to use the commutation relations (12.63) and (12.54), each for the operator X^+, i.e. the upper sign, to move the operators τ and X^z from left to right through the string of X^+ operators for distinct values λ_j until they act on the vacuum state Ω_S defined in (12.69).

The following relations will be useful:

$$\sum_{k\neq j} \frac{1}{u_k(u_k - u_j)} = -\sum_{k<j} \frac{1}{u_k u_j} \tag{12.87}$$

and

$$\frac{1}{(u-\lambda)(u-\mu)} = \frac{1}{\lambda-\mu}\left(\frac{1}{u-\lambda} - \frac{1}{u-\mu}\right). \tag{12.88}$$

If (12.86) holds, the eigenvalue of $\tau(\lambda)$ simplifies to

$$\Theta_M(\mu) = x^2(\mu) - \frac{\partial}{\partial\mu}x(\mu) - 2\sum_{j=1}^{M} \frac{x(\mu) - x(\lambda_j)}{\mu - \lambda_j}. \tag{12.89}$$

Now, we are only left with the task to express of the eigenvalues of $\mathcal{H}_{TC}$, $\mathcal{M}$, and $\mathbf{S}^2$ in terms of the Bethe roots λ_j.

First, let us write down the Bethe ansatz equations (12.86) using (12.77) for the quantum Tavis–Cummings model more explicitly as

$$\Delta - \lambda_k + 2\frac{S}{\lambda_k} - \sum_{j=1,\, j\neq k}^{M} \frac{2}{\lambda_k - \lambda_j} = 0, \quad 1 \leq k \leq M, \quad 0 \leq S \leq \frac{N}{2}. \tag{12.90}$$

Because of the structure of the vacuum state that becomes annihilated if the spin raising operator S^+ is applied more than $2S$ times, there are $K = \min(2S, M) + 1$ sets of solutions of these M Bethe ansatz equations that we label by ρ. The generating function $\tau(\lambda)$, thus, has K eigenvalues

$$\Theta_M^\rho(\mu) = \frac{S(S+1)}{\mu^2} - \frac{1}{\mu}\left(-\Delta S - 2S\sum_{j=1}^{M}\frac{1}{\lambda_j^\rho}\right)$$
$$+\left(\frac{\Delta^2}{4} + \frac{1}{2} + (M-S)\right) - \frac{\Delta}{2}\mu + \frac{\mu^2}{4}. \qquad (12.91)$$

Comparison with (12.61) shows that the eigenvalues of $\mathcal{H}_{TC}$, $\mathcal{M}$, and $\mathbf{S}^2$ are, respectively,

$$E_{S,M}^\rho = -S\left(2\sum_{j=1}^{M}\frac{1}{\lambda_j^\rho} + \Delta\right), \qquad (12.92)$$

$$M - S, \qquad (12.93)$$

$$S(S+1). \qquad (12.94)$$

Note that summation over the index k in the Bethe ansatz equations (12.90), because of

$$\sum_{k,j=1}^{M}\frac{1}{\lambda_k - \lambda_j} = 0, \qquad (12.95)$$

leads to an alternative and somewhat simpler expression for the eigenvalues of $\mathcal{H}_{TC}$

$$E_{S,M}^\rho = \Delta(M-S) - \sum_{j=1}^{M}\lambda_j^\rho. \qquad (12.96)$$

EXERCISE 12.7 Simple special cases Consider the cases

- $M = 1$ and $N \geq 1$.
- $M = 2$ at exact resonance $\Delta = 0$. Distinguish between the cases of a) even N and b) odd N.

We can now summarize and thereby generalize what we have learned in this section. Whenever we have a model whose operators are a representation of the algebra (12.53)–(12.55), we can construct the eigenstates of the generating function $\tau(\lambda)$. In order to do this, we first have establish a vacuum state $|\Omega\rangle$ that satisfies

$$X^{-}(\lambda)|\Omega\rangle = 0 \quad \text{and} \quad X^{z}(\lambda)|\Omega\rangle = x(\lambda)|\Omega\rangle. \tag{12.97}$$

The vacuum eigenvalue function $x(\lambda)$ of the operator $X^{z}(\lambda)$ relates to the vacuum eigenvalue $k(\lambda)$ of the generating function $\tau(\lambda)$, which is then given by

$$k(\lambda) = x^{2}(\lambda) - \frac{\partial}{\partial\lambda}x(\lambda). \tag{12.98}$$

The operator $X^{+}(\lambda)$ acts as creation operator and can be used to construct the Bethe ansatz eigenvalues of the model algebraically.

Part 3

Coordinate Bethe Ansatz

Have no fear of perfection; you will never reach it.
Marie Skłodowska Curie (1867–1934)

The perovskite compound $KCuF_3$ is a quasi-one-dimensional magnetic material which, in accordance with experimental findings is well described by the Heisenberg quantum spin chain with spin $S = 1/2$ (see, e.g. Steiner *et al.*, 1976), i.e. by the model that we have introduced via the six-vertex model in part II on the algebraic Bethe ansatz.

Of course, the Heisenberg quantum spin chain is simply and more naturally justified as the one-dimensional version of the Heisenberg model for localized magnetism we have encountered in chapter 8 on models of strongly interacting quantum matter and needs no justification via vertex models in two dimensions. However, there is a strong connection between the six-vertex model and the Heisenberg quantum spin chain. Both systems are exactly solvable and integrable by the same method.

The exact solution of the Heisenberg quantum spin chain, not using and indeed not knowing about the connection to vertex models, is the original form of Hans Bethe's eponymous ansatz. In 1931 he found a way to solve the (isotropic) Heisenberg model in one dimension, ingeniously guessing the wave function of the model. One of his major motivations was to correctly count the number of states of the spin chain which the spin-wave theory of Bloch (1930) had failed to produce. In pursuing this goal, Bethe not only was able to show that his trial wave function could generate the correct number of states, but he also furnished a highly versatile method to find exact solutions of many further quantum many-particle models.

Unfortunately he never kept the promise he gave at the end of his paper:

> 'In einer folgenden Arbeit soll die Methode auf räumliche Gitter ausgedehnt und die physikalischen Konsequenzen bezüglich Kohäsion, Ferromagnetismus und Leitfähigkeit gezogen werden.'
>
> 'In a future paper, this method will be extended to [three-dimensional] space lattices, and its physical implications for cohesion, ferromagnetism and electrical conductivity, will be derived.' Translation by Vince Frederick in Bethe (1997).

His promise was, namely, to generalize the method to higher dimensions.

Bethe realized the difficulties of a generalization of his method to higher dimensions two years later when he acknowledged, in a handbook article with Sommerfeld but predominantly his work, that 'the extension of the theory to higher dimensions appears to face unsurmountable difficulties' Sommerfeld and Bethe (1967).

In chapter 13, we expound the properties of the Heisenberg quantum spin chain and start out on the first steps to solve the Schrödinger eigenvalue equation for this model. Then, chapter 14 demonstrates the validity of Bethe's method, i.e. of his conjectured wave function, as solution of the eigenvalue problem. Bethe's method entails the Bethe ansatz equations, which encode the states of the spin chain but have to be examined carefully to extract definite information. We shall do this for the ground state and for low-lying excited states.

Chapter 15 then demonstrates the workings of the Bethe ansatz for a different one-dimensional model of strongly interacting quantum particles, a gas of Bosonic particles that interact on a line by short-ranged δ-function pair potentials. This latter model will prove to be less technically involved compared to the Heisenberg quantum spin chain. It is nevertheless worthwhile to treat this model in some detail by pursuing a programme similar to the Bethe ansatz for the quantum spin chain. We shall gain further familiarity with the Bethe ansatz method in a physical setting different from the quantum spin chains, but also for the intrinsic and in fact recent interest in this model.

Historically, this model was solved by Bethe ansatz many years after the Heisenberg quantum spin chain (Lieb and Liniger, 1963; Lieb, 1963) when interest in Bethe's work had picked up momentum again. It is also known as the Lieb–Liniger model.

In a series of articles, Karbach and Müller (1997), Karbach *et al.* (1998), and Karbach *et al.* (2000) presented a very accessible introduction to the coordinate Bethe ansatz for the isotropic Heisenberg quantum spin chain. Their discussion focuses on the low-lying part of the spectrum in both, the ferro- and antiferromagnetic regime. These articles are therefore recommended to supplement the exposition of this part.

13

The Anisotropic Heisenberg Quantum Spin Chain

Each problem that I solved became a rule, which afterwards served to solve other problems.

René Descartes (1596–1650)

The model we explain in this chapter is also often called the *Heisenberg–Ising* spin chain or model. A convenient shorthand notation, whose significance will soon become obvious, is the XXZ Heisenberg quantum spin chain.

Heisenberg 1928 introduced in 1928 a fully quantum mechanical model to describe magnetism, which is now known as the *Heisenberg model*. We discussed this model in sections 8.6 to 8.8 from several perspectives and concluded that it was a very reasonable model for magnetism caused by localized magnetic moments, typically insulating magnetic materials. The Heisenberg model couples localized spin variables, described by Pauli matrices, via exchange interactions. That it is therefore a good model to describe magnetic insulators where the magnetism is indeed due to the interaction between localized magnetic moments makes it, however, for the same reason, not a good model for itinerant magnetism as in, e.g. magnetic metals such as iron, nickel, or cobalt.

The one-dimensional version of the Heisenberg model, the Heisenberg quantum spin chain, provides a good picture of insulating magnetic materials such as $KCuF_3$, which is of particular interest in our context. This and similar materials belong to a class of insulating magnetic materials where the interaction of the magnetic moments in one particular direction is much larger than in the perpendicular directions, and which can, thus, be described with high accuracy as quasi-one-dimensional magnets.

The Heisenberg model fits into the picture of the the generalized Hubbard model of section 8.7, equation (8.561) if we retain only the spin-spin interaction term $\mathcal{J}$.

In section 13.1, we describe in some detail the one-dimensional Heisenberg model, the Heisenberg quantum spin chain and, in section 13.2, we discuss its various special cases. Finally, section 13.3 outlines more properties of the anisotropic Heisenberg quantum spin chain, the so-called XXZ spin chain. Here, we also prepare the solution of the Schrödinger eigenvalue equation for the energy eigenvalue spectrum of the anisotropic spin chain, which will immediately lead to the Bethe's conjecture for the ansatz solution of the wave function.

Models of Quantum Matter: A First Course on Integrability and the Bethe Ansatz. Hans-Peter Eckle, Oxford University Press (2019). © Hans-Peter Eckle. DOI: 10.1093/oso/9780199678839.001.0001

13.1 Description of the XXZ Heisenberg quantum spin chain

We consider a one-dimensional lattice, array, or chain of N spins of spin quantum number $s = \frac{1}{2}$, choosing, for simplicity, N to be an even number.[1] The spins interact via nearest-neighbour exchange interactions.

The general Hamiltonian of this spin chain is given by

$$\mathcal{H} = \sum_{n=1}^{N} \left(\mathcal{J}_x S_n^x S_{n+1}^x + \mathcal{J}_y S_n^y S_{n+1}^y + \mathcal{J}_z S_n^z S_{n+1}^z\right). \tag{13.1}$$

The spin operator at lattice site n is

$$\mathbf{S}_n = \frac{1}{2}\boldsymbol{\sigma}_n \tag{13.2}$$

where ${\sigma_n}^\alpha$, $\alpha = x, y, z$ is an N–fold tensor product

$$\sigma_n^\alpha = \overbrace{I \otimes I \otimes \ldots \underbrace{\sigma^\alpha}_{n} \otimes \ldots I}^{N} \tag{13.3}$$

with Pauli matrices σ^α in the nth position of the product. The product of two spin operators on the one-dimensional lattice of the spin chain is then given by

$$\mathbf{S}_n \cdot \mathbf{S}_m = \left(\overbrace{I \otimes \ldots \underbrace{\sigma^\alpha}_{n} \otimes \ldots I}^{N}\right) \cdot \left(\overbrace{I \otimes \ldots \underbrace{\sigma^\alpha}_{m} \otimes \ldots I}^{N}\right) \tag{13.4}$$

$$= \overbrace{I \otimes \ldots \underbrace{\sigma^\alpha}_{n} \otimes \ldots \underbrace{\sigma^\alpha}_{m} \otimes \ldots I}^{N}. \tag{13.5}$$

The model can easily be generalized to include the effect of an external magnetic field h pointing in the z-direction

$$\mathcal{H} = \sum_{n=1}^{N} \left(\mathcal{J}_x S_n^x S_{n+1}^x + \mathcal{J}_y S_n^y S_{n+1}^y + \mathcal{J}_z S_n^z S_{n+1}^z\right) - h \sum_{n=1}^{N} S_n^z. \tag{13.6}$$

For the most part, though, we shall consider the case without magnetic field $h = 0$.

[1] Occasionally, there are interesting effects depending on whether N is even or odd, see e.g. Stroganov (2001).

Spin raising and lowering operators are introduced via

$$S_n^{\pm} = S_n^x \pm i S_n^y \tag{13.7}$$

with commutation relations

$$\left[S_n^+, S_m^-\right] = 2\delta_{nm} S_n^z \tag{13.8}$$

and

$$\left[S_n^z, S_m^{\pm}\right] = \pm\delta_{nm} S_n^{\pm}. \tag{13.9}$$

Moreover, for spin $s = \frac{1}{2}$ operators we have

$$\left(S_n^+\right)^2 = \left(S_n^-\right)^2 = 0 \tag{13.10}$$

which reflects the fact that two spins cannot occupy the same lattice site, i.e. in this respect, the spins behave like Fermions.

Rather than treating the general case

$$\mathcal{J}_x \neq \mathcal{J}_y \neq \mathcal{J}_z, \tag{13.11}$$

the so-called XYZ chain, whose solution is rather involved (it corresponds to the eight-vertex model which was solved by Rodney Baxter (1982), we shall restrict ourselves to the case

$$\mathcal{J}_x = \mathcal{J}_y \equiv \mathcal{J} \neq \mathcal{J}_z, \tag{13.12}$$

the XXZ chain,[2] which corresponds to the six-vertex model.

The Hamiltonian can then be rewritten as

$$\mathcal{H}(\Delta) \equiv \frac{\mathcal{H}(\mathcal{J}, \mathcal{J}_z)}{\mathcal{J}} - \frac{N\Delta}{4} \tag{13.13}$$

$$= \sum_{n=1}^{N} \left[\frac{1}{2} \left(S_n^+ S_{n+1}^- + S_n^- S_{n+1}^+\right) + \Delta \left(S_n^z S_{n+1}^z - \frac{1}{4} \right) \right] \tag{13.14}$$

where

[2] There are a number of names in use for this model: anisotropic quantum spin-$\frac{1}{2}$ Heisenberg chain, Heisenberg–Ising spin chain, or some other combination to the same effect.

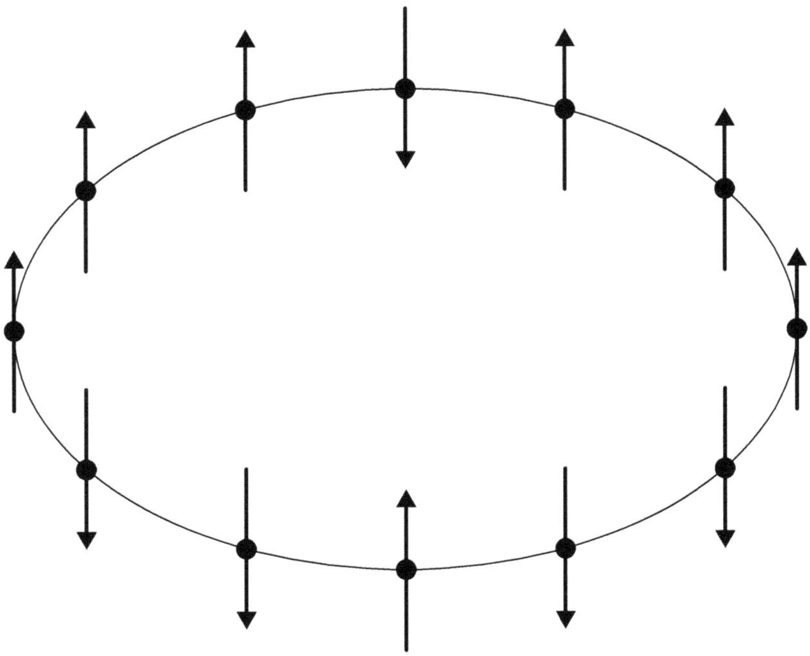

Figure 13.1 *Periodic the spin chain forms a ring.*

$$\Delta = \frac{\mathcal{J}_z}{\mathcal{J}} \tag{13.15}$$

is called the anisotropy parameter.

The additive constant, $-\frac{N\Delta}{4}$, has been chosen so as to have a convenient energy, namely $E = 0$, for the ferromagnetic reference state, which we shall introduce soon below.

The description or, if you like, definition of the model will be complete by specifying some boundary conditions. We choose, for technical simplicity, periodic boundary conditions (cf. figure 13.1)

$$\mathbf{S}_{n+N} = \mathbf{S}_n. \tag{13.16}$$

Other boundary conditions are, however, possible; see, e.g. Alcaraz *et al.* (1987), Hamer *et al.* (1987), Batchelor and Hamer (1990), and Eckle and Hamer (1991).

As a side remark, we note the Hamiltonian of an open quantum spin chain of N spins

$$\mathcal{H} = \sum_{n=1}^{N-1} \left(S_n^x S_{n+1}^x + S_n^y S_{n+1}^y + \frac{1}{2}(q + q^{-1}) S_n^z S_{n+1}^z \right) - \frac{1}{2}(q - q^{-1})(S_1^z - S_N^z), \tag{13.17}$$

where q is an arbitrary complex number. This model is not only integrable (Alcaraz *et al.*, 1987; Sklyanin, 1988), but the Hamiltonian also commutes with the generators of the quantum algebra $U_q[su(2)]$ (Pasquier and Saleur, 1990; Kulish and Sklyanin, 1991). The physical interpretation of this spin chain is as an open spin chain with 'surface' fields, i.e. fields acting only at he ends of the finite open chain (Alcaraz *et al.*, 1987).

13.2 Special cases of the XXZ Heisenberg quantum spin chain

This section presents the various special cases of the quantum spin chain Hamiltonian (13.1), starting with the most general one.

13.2.1 XYZ Heisenberg quantum spin spin chain

The fully anisotropic XYZ quantum spin chain where $\mathcal{J}_x \neq \mathcal{J}_y \neq \mathcal{J}_z$ has been the most challenging problem, which was only solved in the early 1970s by Rodney Baxter (1982 and the references therein). The solution is very involved mathematically, but the basic ideas are very similar to the strategy outlined in part II regarding the algebraic Bethe ansatz. The solution essentially depends on the relation between the eight-vertex model and the XYZ quantum spin chain. However, the XYZ model seems to have no realization as a magnetic system. It is, however, nonetheless of great theoretical interest, e.g. it exhibits continuously varying critical exponents; for details see Baxter (1982).

Here, we shall not delve any further into the technicalities of the XYZ model. The interested reader will find Baxter (1982) very enjoyable.

We note, however, as is obvious from the XYZ Hamiltonian and the discussion in the previous section, that there are two anisotropy parameters in the general model, e.g. with $\mathcal{J}_x \equiv \mathcal{J}$, we may choose $\gamma \equiv \mathcal{J}_y / \mathcal{J}$ and $\Delta \equiv \mathcal{J}_z / \mathcal{J}$.

13.2.2 XXZ Heisenberg quantum spin chain

The anisotropic spin chain $\mathcal{J} \equiv \mathcal{J}_x = \mathcal{J}_y \neq \mathcal{J}_z$ ($\Delta \equiv \frac{\mathcal{J}_z}{\mathcal{J}}$) is the model we focus on in this part of the book as regards the coordinate Bethe ansatz. We have discussed its relation to the six-vertex model. We shall soon continue with its full description and its solution with the help of the coordinate Bethe ansatz.

The parameter Δ distinguishes the regimes (see figure 13.2):

- $\Delta > 1$: easy-axis antiferromagnetic regime
- $-1 < \Delta < 1$: planar regime, where
  - $0 < \Delta < 1$: planar antiferromagnetic regime
  - $- < \Delta < 0$: planar ferromagnetic regime
- $\Delta < -1$: easy-axis ferromagnetic regime.

We shall find out more about the physical significance of these regimes as we go along (for a preview, see figure 13.2).

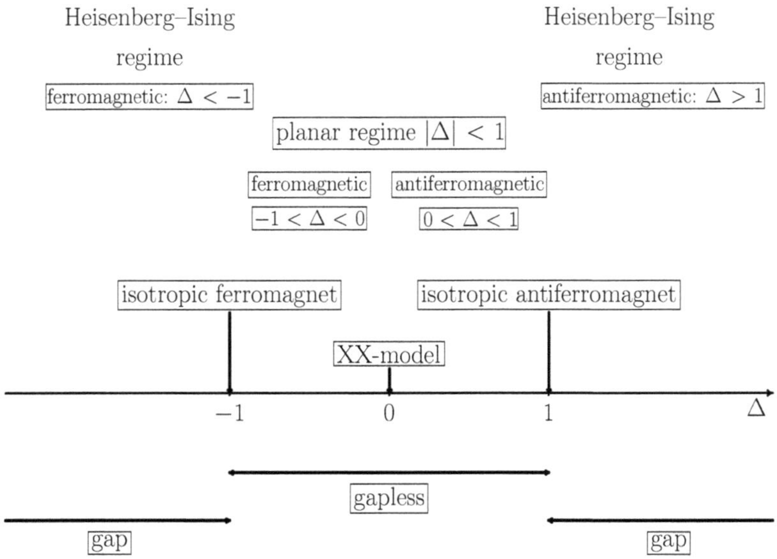

Figure 13.2 *Properties, special cases, and regimes of the quantum spin* $s = \frac{1}{2}$ *XXZ chain.*

There are a few special points, not included in the above inequalities, which merit special mention.

13.2.3 XXX Heisenberg quantum spin chain

Though perfectly possible, often the solution of the anisotropic XXZ spin chain in the coordinate Bethe ansatz framework will become somewhat intricate. Therefore, to not lose sight of the forest for the trees, we shall frequently carry through the full analysis of the Bethe ansatz to find some final results for physical quantities for the case of the isotropic spin chain $\mathcal{J}_x = \mathcal{J}_y = \mathcal{J}_z$, i.e. $\Delta = 1$, the XXX spin chain only. In doing so, we will keep the mathematics more manageable. We have done the same in part II on the algebraic Bethe ansatz.

13.2.4 XY Heisenberg quantum spin chain

A still simpler model is the XY spin chain, where $\Delta = 0$, which can be solved exactly (Lieb *et al.*, 1961) without Bethe ansatz. In our convention $\Delta = 0$ really corresponds to the isotropic model, the XX quantum spin chain. It can be generalized in an obvious way to an anisotropic model with different interaction strength in the x and y directions, the XY spin chain proper.

It is, in a certain sense, a non-interacting model. In fact, in its solution, the first step consists in a transformation of the spin operators to Fermionic operators, the Jordan–Wigner transformation (1928), which maps the XY spin chain into a one-dimensional free Fermion model. Nevertheless, the solution of the model is quite instructive, and, therefore recommended to the reader as an exercise.

EXERCISE 13.1 Solution of the XY spin chain

(a) Show that the Jordan–Wigner transformation (1928)

$$S_i^+ = \exp\left[-i\pi \sum_{l=1}^{i-1} c_l^\dagger c_l\right] c_i, \qquad S_i^- = c_i^\dagger \exp\left[i\pi \sum_{l=1}^{i-1} c_l^\dagger c_l\right], \qquad S_i^z = c_i^\dagger c_i - \frac{1}{2}, \tag{13.18}$$

where the operators $c_j^\dagger$ and c_i are Fermi creation and annihilation operators

$$\{c_i, c_j^\dagger\} = \delta_{ij}, \qquad \{c_i, c_j\} = \{c_i^\dagger, c_j^\dagger\} = 0, \tag{13.19}$$

can also be written as

$$S_i^+ = \prod_{l=1}^{i-1} \left(1 - 2c_l^\dagger c_l\right) c_i, \qquad S_i^- = c_i^\dagger \prod_{l=1}^{i-1} \left(1 - 2c_l^\dagger c_l\right), \qquad S_i^z = c_i^\dagger c_i - \frac{1}{2}. \tag{13.20}$$

Its inverse transformation is

$$c_i^\dagger = S_i^+ \exp\left[-i\pi \sum_{l<i} S_l^+ S_l^-\right], \qquad c_i = \exp\left[i\pi \sum_{l<i} S_l^+ S_l^-\right] S_i^-. \tag{13.21}$$

Use the Jordan–Wigner transformation to show that the one-dimensional anisotropic XY-model

$$\mathcal{H} = -\sum_{i=1}^{N-1} \left((1+\gamma)S_i^x S_{i+1}^x + (1-\gamma)S_i^y S_{i+1}^y\right) \tag{13.22}$$

can be written as a model of spinless Fermions

$$\mathcal{H} = -\sum_{i=1}^{N-1} \Big((c_{i+1}^\dagger c_i + \text{H.c.}) + \gamma(c_{i+1}^\dagger c_i^\dagger + \text{H.c.})\Big). \tag{13.23}$$

(b) Calculate the excitation spectrum for this model using, after a transformation to momentum space, an appropriate Bogoliubov transformation (cp. exercise 2.13) and sketch your results. Consider specifically the two cases

$$\gamma = 0 \tag{13.24}$$

and

$$\gamma = 1. \tag{13.25}$$

Hint: Do **not** consider boundary effects in (a) and (b), i.e. consider only the open chain where spins at the first and last site are not connected.

Reference: The original paper of Lieb *et al.* (1961) is quite detailed. The paper also appears in Lieb and Mattis (1966) and Mattis (1993).

13.2.5 Quantum Ising spin chain

If we formally put $\Delta \to \pm\infty$, we obtain the quantum Ising chain (Pfeuty, 1970) where only the $\sum_{n=1}^{N} S_i^z S_{i+1}^z$ part of the Hamiltonian remains. In our convention, $\Delta \to +\infty$ corresponds to the Ising antiferromagnet, while $\Delta \to -\infty$ corresponds to the Ising ferromagnet.

Section 5.7 on quantum phase transitions demonstrated an interesting application of a generalization of the quantum Ising spin chain, i.e. in a transverse magnetic field. This model is an exemplary model that exhibits a quantum phase transition.

13.3 Basic properties of the XXZ Heisenberg quantum spin chain

The Hamiltonian (13.14) has the remarkable symmetry

$$U\mathcal{H}(\Delta)U^{-1} = -\mathcal{H}(-\Delta), \tag{13.26}$$

under the unitary transformation

$$U = \exp\left(i\pi \sum_{n=1}^{N} n S_n^z\right), \tag{13.27}$$

which means that the spectra of eigenvalues for $\mathcal{H}(\Delta)$ and $-\mathcal{H}(-\Delta)$ are identical. Thus, it suffices to consider, e.g., $\Delta \geq 0$.

EXERCISE 13.2 Symmetry of the Heisenberg quantum spin chain Hamiltonian
The Hamiltonian of the anisotropic Heisenberg quantum spin chain

$$\mathcal{H} = \mathcal{H}(\Delta) = \sum_{n=1}^{N} \left\{S_n^x S_{n+1}^x + S_n^y S_{n+1}^y + \Delta S_n^z S_{n+1}^z\right\} \tag{13.28}$$

has the remarkable symmetry

$$U\mathcal{H}(\Delta)U^{-1} = -\mathcal{H}(-\Delta) \tag{13.29}$$

under the unitary transformation

$$U = \exp\left(i\pi \sum_{n=1}^{N} nS_n^z\right). \tag{13.30}$$

Prove this symmetry relation using an operator identity called the Baker–Campbell–Hausdorff formula (sometimes the second and sometimes even the first two names are omitted in the literature).

....

Next, the Hamiltonian (13.14) commutes with the z-component of the total spin

$$S^z = \sum_{n=1}^{N} S_n^z. \tag{13.31}$$

We note the Hamiltonian of the isotropic chain with $\mathcal{J}_x = \mathcal{J}_y = \mathcal{J}_z$ also commutes with the total spin

$$\mathbf{S} = \sum_{n=1}^{N} \mathbf{S}_n, \tag{13.32}$$

whereas the Hamiltonian of the general XYZ chain commutes with neither $\mathbf{S}$ nor S^z.

For N spins $s = \frac{1}{2}$, S^z can take the values

$$S^z = \frac{N}{2}, \frac{N}{2} - 1, \ldots, -\frac{N}{2} + 1, -\frac{N}{2} \tag{13.33}$$

and that we write in the more compact form

$$S^z = \frac{N}{2} - M \quad \text{for} \quad M = 0, 1, \ldots, N. \tag{13.34}$$

M is the number of spin deviations from the completely ferromagnetic state

$$|0\rangle = |\uparrow\rangle_1 |\uparrow\rangle_2 \ldots |\uparrow\rangle_N \equiv \bigotimes_{n=1}^{N} |\uparrow\rangle_n \tag{13.35}$$

$$|\uparrow\rangle_n = \left| s = \frac{1}{2}, S_n^z = +\frac{1}{2} \right\rangle, \tag{13.36}$$

which we shall use as the reference state mentioned previously.

A state with *one* spin deviation, e.g. at site m, will then be

$$| \uparrow\rangle_1 | \uparrow\rangle_2 \dots | \downarrow\rangle_m \dots | \uparrow\rangle_N = S_m^- |0\rangle \tag{13.37}$$

$$| \downarrow\rangle_m = \left| s = \frac{1}{2}, S_m^z = -\frac{1}{2} \right\rangle. \tag{13.38}$$

The reference state (13.35) is a simultaneous eigenstate of $\mathcal{H}(\Delta)$ and S^z with eigenvalues $E_0 = 0$ and $S^z = \frac{N}{2}$.

For a given number M of spin deviations at sites $1 \leq n_1, n_2, \dots, n_M \leq N$, the eigenstate $|M\rangle$ of the Hamiltonian $\mathcal{H}(\Delta)$ has the form

$$|M\rangle = \sum_{n_1, n_2, \dots, n_M} a(n_1, n_2, \dots, n_M) S_{n_1}^- S_{n_2}^- \dots S_{n_M}^- |0\rangle \tag{13.39}$$

$$\equiv \sum_{n_1, n_2, \dots, n_M} a(n_1, n_2, \dots, n_M) |n_1, n_2, \dots, n_M\rangle. \tag{13.40}$$

Because $\left(S^+\right)^2 = 0$, all positions n_j of the spin deviations have to be distinct. Otherwise, the state would vanish and not be an eigenstate of S^z. Moreover, since all S_n^- for different n commute with each other, we can adopt the convention

$$1 \leq n_1 < n_2 < \dots < n_m \leq N. \tag{13.41}$$

The number of amplitudes $a(n_1, n_2, \dots, n_M)$ in (13.40) is then given by the binomial coefficient

$$\binom{N}{M}. \tag{13.42}$$

The complete Hilbert space of the problem has the expected dimension

$$\sum_{M=0}^{N} \binom{N}{M} = 2^N. \tag{13.43}$$

In order to always respect the ordering of the positions n_j of the spin deviations, the periodic boundary conditions translated to the amplitudes must take the form

$$a(n_1, n_2, \dots, n_M) = a(n_2, n_3, \dots, n_M, n_1 + N). \tag{13.44}$$

Since we are looking for simultaneous eigenstates of S^z and $\mathcal{H}(\Delta)$, i.e. for states of the form (13.40) satisfying also

$$\mathcal{H}(\Delta)|M\rangle = E_M|M\rangle, \tag{13.45}$$

the remaining task of diagonalizing $\mathcal{H}(\Delta)$ consists of determining the amplitudes

$$a(n_1, n_2, \ldots, n_M) \tag{13.46}$$

from the Schrödinger equation

$$\mathcal{H}(\Delta)|M\rangle = E_M|M\rangle. \tag{13.47}$$

The Bethe ansatz[3] is an educated guess for these amplitudes. Chapter 14 introduces and justifies the ansatz.

[3] This is presumably the right place to give a translation of the German noun 'Ansatz'. As is generally the case, also *ansatz* has several meanings. We focus on the meaning relevant in our connection. A single-phrase translation would certainly be *educated guess*. So, we make an educated guess as to the solution to a physical or mathematical problem. The word *ansatz*, however, also indicates that there is a certain degree of variability that we can use to adapt our guess to the problem. An *ansatz* is, thus, often a procedure rather than just a formula. Instead of being a question of whether our *ansatz* does or does not solve the problem, our task is rather to make the *ansatz* work by adjusting it to the problem. Section 14.1 demonstrates how an *ansatz*, the Bethe ansatz, works.

14

Bethe Ansatz for the Anisotropic Heisenberg Quantum Spin Chain

The meaning of a proposition is the method of its verification.

Moritz Schlick (1882–1936)

Section 14.1 states and then verifies the conjecture for the wave function of the anisotropic Heisenberg quantum spin chain seen in section 14.3, namely, the Bethe ansatz wave function.

The Bethe ansatz can be expressed as the statement that the amplitudes in the state (13.40) are given by

$$a(n_1, n_2, \ldots, n_M) = \sum_P A_P \exp\left(i \sum_{j=1}^{M} k_{p_j} n_j\right). \tag{14.1}$$

The summation in (14.1) runs over all $M!$ permutations P of the M indices $1, 2, \ldots, M$.

Of course, this rather abrupt statement looks a bit daunting. In fact, equation (14.1) is rather the end result of a clever argument instead of its beginning. Therefore, Section 14.1 methodically examines first the case $M = 1$, then $M = 2$, and, finally, $M = 3$. In doing so, we discover how the ansatz (14.1) arises more naturally.

Moreover, we shall also find equations, the so-called Bethe ansatz equations, for the variables k_j in (14.1), the quasi-momenta, which, in turn, determine the energy eigenvalue

$$E_M = E_M(k_1, k_2, \ldots, k_M). \tag{14.2}$$

The Bethe ansatz equations are derived in section 14.2 as the consequence of, in our case, periodic boundary conditions. Thus, the solution to the Bethe ansatz equations will, finally, provide the diagonalization, i.e. the solution, of the XXZ spin chain.

However, solving the Bethe ansatz equations to determine the quasi-momenta is not an easy task. In order to make further progress, section 14.3 provides an appropriate parameterization of the quasi-momenta that will be very helpful to achieve a simplification of the Bethe ansatz equations. The new parameters, called rapidities, and the consideration of the thermodynamic limit will then allow us to extract information about

Models of Quantum Matter: A First Course on Integrability and the Bethe Ansatz. Hans-Peter Eckle, Oxford University Press (2019). © Hans-Peter Eckle. DOI: 10.1093/oso/9780199678839.001.0001

the ground state (section 14.4) and low-lying excitations (section 14.5) of the anisotropic quantum spin chain from the Bethe ansatz equations. Section 14.6 specifically examines the excitations of the isotropic quantum spin chain.

14.1 Verification of the Bethe ansatz

Despite the title of this section, it shows a systematic way to *guess* the Bethe ansatz wave function using a direct argumentation. In doing this, we shall also find conditions for the Bethe ansatz wave function necessary to satisfy the Schrödinger equation.

In order to start, we derive an equation for the amplitudes

$$a(n_1, n_2, \ldots, n_M) \tag{14.3}$$

from the Schrödinger equation (13.45)

$$\mathcal{H}(\Delta)|M\rangle = E_M|M\rangle, \tag{14.4}$$

which we shall solve explicitly for the quantum numbers $M = 1, 2,$ and 3 (the latter case will be given as a useful exercise). Here, we demonstrate the validity of the Bethe ansatz (14.1) and obtain expressions for the coefficients A_P and the quasi-momenta k.

In order to obtain the eigenvalue equation for $a(n_1, n_2, \ldots, n_M)$, we first calculate

$$\mathcal{H}_{n,n+1}|\sigma\rangle_n|\sigma'\rangle_{n+1} \tag{14.5}$$

where we have introduced the local Hamiltonian

$$\mathcal{H}_{n,n+1} = \frac{1}{2}\left(S_n^+ S_{n+1}^- + S_n^- S_{n+1}^+\right) + \Delta\left(S_n^z S_{n+1}^z - \frac{1}{4}\right). \tag{14.6}$$

The local Hamiltonian only acts on states at lattice sites n and $(n+1)$, i.e. on pairs of neighbouring sites.

For general angular momentum operators $\mathbf{L}$ (see also chapter 3), we recall that

$$L^{\pm}|l, m\rangle = [l(l+1) - m(m \pm 1)]^{\frac{1}{2}}\,|l, m \pm 1\rangle \tag{14.7}$$

$$L^z|l, m\rangle = m|l, m\rangle. \tag{14.8}$$

With these relations we obtain for the four possible states of spins on two neighbouring lattice sites

$$2\mathcal{H}_{n,n+1}|\uparrow\rangle_n|\downarrow\rangle_{n+1} = |\downarrow\rangle_n|\uparrow\rangle_{n+1} - \Delta|\uparrow\rangle_n|\downarrow\rangle_{n+1}, \tag{14.9}$$

$$2\mathcal{H}_{n,n+1}|\downarrow\rangle_n|\uparrow\rangle_{n+1} = |\uparrow\rangle_n|\downarrow\rangle_{n+1} - \Delta|\downarrow\rangle_n|\uparrow\rangle_{n+1}, \tag{14.10}$$

$$\mathcal{H}_{n,n+1}|\uparrow\rangle_n|\uparrow\rangle_{n+1} = 0, \qquad \mathcal{H}_{n,n+1}|\downarrow\rangle_n|\downarrow\rangle_{n+1} = 0. \tag{14.11}$$

Apparently pairs of neighbouring parallel spins do not contribute to

$$\mathcal{H}(\Delta)|n_1,\ldots,n_M\rangle, \tag{14.12}$$

whereas pairs of neighbouring antiparallel spins give two contributions each: one where the neighbouring antiparallel spins have exchanged their positions (the spins have been flipped), and one where they are unchanged.

Thus, we obtain for the state $|n_1,\ldots,n_M\rangle$

$$2\mathcal{H}(\Delta)|n_1,\ldots,n_M\rangle = \sum_{\{n'\}} |n'_1,\ldots,n'_M\rangle - N_a\Delta|n_1,\ldots,n_M\rangle \tag{14.13}$$

where N_a is the number of antiparallel spin pairs in the state $|n_1,\ldots,n_M\rangle$.

The state $|n'_1,\ldots,n'_M\rangle$ differs from the original state $|n_1,\ldots,n_M\rangle$ by one exchange of an antiparallel pair of spins, i.e.

$$n'_1 = n_1, n'_2 = n_2, \ldots, n'_\alpha = n_\alpha \pm 1, \ldots, n'_m = n_M \tag{14.14}$$

and the summation is over all N_a such configurations.

Equation (14.13) may also be written in the form

$$2\mathcal{H}(\Delta)|n_1,\ldots,n_M\rangle = \sum_{\{n'\}} \left\{|n'_1,\ldots,n'_M\rangle - \Delta|n_1,\ldots,n_M\rangle\right\}. \tag{14.15}$$

Inserting this into the Schrödinger equation yields

$$\sum_{\{n'\}} \left[a(n'_1,\ldots,n'_M) - \Delta a(n_1,\ldots,n_M)\right] = 2E_M a(n_1,\ldots,n_M). \tag{14.16}$$

Mathematically, this equation is a typical *difference* equation, a type of equation that often comes up in lattice problems and is therefore ubiquitous in statistical and condensed matter physics.

This is as far as we can easily work in the general case, i.e. with the general state $|M\rangle$. Let us maintain our focus and proceed by looking at the simplest cases $M = 1, 2,$ and 3.

Let us start with $M = 1$:

$$\uparrow\ \uparrow\ \uparrow\ \uparrow\ \downarrow\ \uparrow\ \uparrow\ \uparrow\ \uparrow \tag{14.17}$$

We have two antiparallel spin pairs (see (14.17)). Equation (14.16) thus yields

$$a(n-1)+a(n+1)-2\Delta a(n)=2E_1 a(n) \tag{14.18}$$

with the rather obvious plane wave solution

$$a(n)=A\exp(ikn) \tag{14.19}$$

and the corresponding eigenvalue

$$E_1=\cos k-\Delta. \tag{14.20}$$

This is not very exciting, but physically it suggests that one flipped spin behaves like a free particle in a vacuum consisting of the background of aligned spins, i.e. the completely ferromagnetically ordered state. This interpretation is important and we shall return to it.

Next, consider $M=2$, where we must distinguish between two cases.

- the two flipped spins are **not** on adjacent sites—they are 'far' apart, i.e. $n_2 > n_1+1$; there is at least one spin between the two flipped spins that is pointing in the original direction

$$\uparrow\ \uparrow\ \downarrow\ \uparrow\ \downarrow\ \uparrow\ \uparrow\ \uparrow\ \uparrow \tag{14.21}$$

  and there are four pairs of antiparallel spins in this case;
- the two flipped spins are on adjacent sites, i.e. $n_2=n_1+1$

$$\uparrow\ \uparrow\ \uparrow\ \downarrow\ \downarrow\ \uparrow\ \uparrow\ \uparrow\ \uparrow \tag{14.22}$$

  and there are only two pairs of antiparallel spins in this case.

The first of these two cases is easy to manage. We interpreted one flipped spin as a free particle, described by a plane wave, in the ferromagnetic background that we interpreted as a kind of vacuum. If the two flipped spins are 'far' apart, i.e. $n_2 > n_1+1$, they can be interpreted in a similar way as two free particles.

Equation (14.16) yields for our case of two flipped spins 'far' apart (see (14.21)).

$$\begin{aligned}a(n_1-1,n_2)+a(n_1+1,n_2)+a(n_1,n_2-1)+a(n_1,n_2+1)&\\ -4\Delta a(n_1,n_2)=2E_2 a(n_1,n_2).&\end{aligned} \tag{14.23}$$

This equation is solved by the linear combination of plane waves

$$a(n_1,n_2)=A_{12}\exp\left(ik_1n_1+ik_2n_2\right)+A_{21}\exp\left(ik_2n_1+ik_1n_2\right) \tag{14.24}$$

with *arbitrary* coefficients A_{12} and A_{21}. The corresponding energy eigenvalue is

$$E_2 = \cos k_1 - \Delta + \cos k_2 - \Delta. \tag{14.25}$$

Apparently this analysis of equation (14.16) could be generalized to an arbitrary number M of flipped spins, as long as no two flipped spins come close together.

Of course, there is no reason to exclude cases where two flipped spins come close together, i.e. are on adjacent lattice sites; see (14.22).

If $n_2 = n_1 + 1$, we have only two pairs of antiparallel spins and the eigenvalue equation (14.16) becomes

$$a(n_1 - 1, n_1 + 1) + a(n_1, n_1 + 2) - 2\Delta a(n_1, n_1 + 1) = 2E_2 a(n_1, n_1 + 1). \tag{14.26}$$

We do not want to abandon the ansatz in the form of a plane wave (equation (14.23)), and start over again. We want to keep this attractive and plausible form of wave function.[1] Thus, we have to adjust it to the case of two adjacent flipped spins. The first step to achieve this is to find a trick such that formally equation (14.26) looks the same as equation (14.23): we add *zero* on both sides of equation (14.26) in the form

$$a(n + 1, n + 1) + a(n, n) - 2\Delta a(n, n + 1) = 0 \tag{14.27}$$

Note that the coefficients $a(n, n)$ are only defined by equation (14.27); they have no physical significance yet.[2]

Now the plane wave solution (14.24) formally solves both cases, flipped spins far apart and flipped spins adjacent. However, equation (14.27) now has to be satisfied. As we shall see, this poses a condition on the coefficients A_{12} and A_{21}. Inserting the plane wave solution (14.24) into equation (14.27) yields

$$(A_{12} + A_{21}) \left[1 + \exp(ik_1 + ik_2)\right] = 2\Delta \left(A_{12}e^{ik_2} + A_{21}e^{ik_1}\right). \tag{14.28}$$

Since this equation only determines the ratio of the coefficients, we rewrite it as

$$\frac{A_{21}}{A_{12}} = -\frac{e^{i(k_1+k_2)} - 2\Delta e^{ik_2} + 1}{e^{i(k_1+k_2)} - 2\Delta e^{ik_1} + 1}. \tag{14.29}$$

Since

$$\left|\frac{A_{21}}{A_{12}}\right| = 1 \tag{14.30}$$

we write this as a phase factor

[1] Recall Bill Sutherlands (1985) dictum that the Bethe ansatz is '. . .a code name for a wave function . . .'.
[2] Here we see the flexibility of the ansatz and how it can be used to solve the eigenvalue problem.

$$\frac{A_{21}}{A_{12}} = e^{-i\Theta(k_1,k_2)} \tag{14.31}$$

where

$$\Theta(k_1,k_2) = -\Theta(k_2,k_1) = 2\cot^{-1}\left[\frac{\Delta\sin\left(\frac{k_1-k_2}{2}\right)}{\cos\left(\frac{k_1+k_2}{2}\right) - \Delta\cos\left(\frac{k_1-k_2}{2}\right)}\right] \tag{14.32}$$

$$= \pi - 2\tan^{-1}\left[\frac{\Delta\sin\left(\frac{k_1-k_2}{2}\right)}{\cos\left(\frac{k_1+k_2}{2}\right) - \Delta\cos\left(\frac{k_1-k_2}{2}\right)}\right]. \tag{14.33}$$

We shall occasionally use either form.

We are now in a position to add to the physical interpretation of our ansatz wave function (14.24). Flipped spins are interpreted as free particles in the vacuum of the ferromagnetic reference state where all spins point in the same direction. These particles, which live on discrete lattice sites, interact only when they come close, i.e. occupy adjacent lattice sites. The interaction or scattering process is described by the scattering phase (14.32).

In summary, if the coefficients A_{12} and A_{21} satisfy (14.31) and (14.32), then the plane wave solution (14.24) is the solution of the eigenvalue equation (14.23), whether or not the two flipped spins are nearest neighbours. Thus, (14.24) is the general solution of our problem in the case of $M = 2$ flipped spins.

The case $M = 3$ is left to the reader as exercise 14.1. In the particle scattering picture that has emerged, the case $M = 3$ will prove to factorize into two particle scattering processes. Therefore, the case $M = 3$ will need no further ingredients compared to the case $M = 2$, which we have just analysed.

EXERCISE 14.1 Eigenvalue equation: case $M = 3$ Consider the case of three flipped spins $M = 3$.

You will have to consider four cases for the eigenvalue equation

$$\sum_{\{n'\}}\left[a(n_1', n_2', n_3') - \Delta a(n_1, n_2, n_3)\right] = 2E_3 a(n_1, n_2, n_3), \tag{14.34}$$

namely

1. $n_2 > n_1 + 1, n_3 > n_2 + 1$, i.e. all three spins are 'far' apart;
2. $n_2 = n_1 + 1, n_3 > n_2 + 1$, i.e. spin 1 and 2 are adjacent, but spin 2 and 3 are 'far' apart;
3. $n_2 > n_1 + 1, n_3 = n_2 + 1$, now spin 2 and 3 are adjacent;
4. $n_2 = n_1 + 1, n_3 = n_2 + 1$, i.e. the three spins sit on three consecutive lattice sites.

Calculate the ratios of the $3! = 6$ coefficients A_P of the Bethe ansatz amplitudes $a(n_1, n_2, n_3)$ and express five of them through the sixth, which you may conveniently choose to be A_{123}.

Finally, express all six coefficients in a symmetric form.

Thus, having solved the cases $M = 1$ and $M = 2$, and having shown that $M = 3$ factorizes, we have found the general solution for arbitrary M.

The coefficients A_P for arbitrary M can be written in the symmetric form

$$A_P = N \exp\left[\frac{i}{2} \sum_{i<j} \Theta(k_{p_i}, k_{p_j})\right] \tag{14.35}$$

with corresponding energy eigenvalue

$$E_M = \sum_{i=1}^{M} (\cos k_i - \Delta). \tag{14.36}$$

This summarizes the general solution of the eigenvalue problem for the anisotropic Heisenberg quantum spin-$\frac{1}{2}$ chain. One ingredient, however, is still missing.

14.2 Periodic boundary conditions

The determination of the solution of the eigenvalue equation (14.16) for $M = 1, 2$ is not yet complete although this is not surprising. As with any differential or difference equation, we need to consider some conditions (initial conditions, boundary conditions) in order to finalize the solution and obtain a specific solution from the general one.

In our case, we employ *periodic boundary conditions*. Other boundary conditions are possible, e.g. free or fixed boundary conditions.

The boundary conditions determine the allowed values of the quasi-momenta k, providing, hence, a quantization condition.

Again, our strategy to consider the cases $M = 1, 2$, and leaving $M = 3$ as an exercise.

The periodic boundary conditions (13.44) for $M = 1$ are simply

$$a(n) = a(n + N), \tag{14.37}$$

which, inserted into the solution for $M = 1$

$$a(n) = Ae^{ikn} \tag{14.38}$$

leads to the simple quantization condition

$$e^{ikN} = 1 \Rightarrow kN = 2\pi\lambda, \quad \lambda = 0, 1, \ldots, N-1 \tag{14.39}$$

if we adopt the convention

$$0 \leq k < 2\pi \tag{14.40}$$

corresponding to a choice of the Brillouin zone. This is the familiar result for the quantization of free particles on a lattice with periodic boundary conditions.

Thus, there is nothing dramatic in the case $M = 1$, as we expected. More interesting is the case $M = 2$.

The periodic boundary conditions (13.44) are now for $M = 2$

$$a(n_1, n_2) = a(n_2, n_1 + N), \tag{14.41}$$

which leads, after inserting the plane wave solution (14.24), to

$$A_{12} = A_{21}e^{ik_1N} \tag{14.42}$$

and

$$A_{21} = A_{12}e^{ik_2N}. \tag{14.43}$$

Using the coefficient ratio (14.31), we obtain

$$k_1N = 2\pi\lambda_1 + \Theta(k_1, k_2) \tag{14.44}$$

$$k_2N = 2\pi\lambda_2 + \Theta(k_2, k_1) \tag{14.45}$$

$$\lambda_1, \lambda_2 = 0, 1, \ldots, N-1. \tag{14.46}$$

A rather simple, but unavoidable, exercise, is to repeat this for $M = 3$.

EXERCISE 14.2 Periodic boundary conditions: case $M = 3$ Derive the equations determining the quasi-momenta k_i, $i = 1, 2, 3$ for the case of $M = 3$ overturned spins from the periodic boundary conditions.

EXERCISE 14.3 Twisted boundary conditions Repeat the derivation of the Bethe ansatz equations for twisted bounday conditions

$$S^{\pm}_{N+1} = e^{i\Phi}S^{\pm}_1 \tag{14.47}$$

for some twist angle Φ.

Interpreting the twist angle Φ as a magnetic flux, the consideration of this kind of boundary conditions allows the calculation of so-called persistent currents in integrable models.

EXERCISE 14.4 Open boundary conditions Consider an open spin chain where the first and last spin of the finite chain are not connected. Instead of the plane wave solutions with which we started the analysis for periodic boundary conditions, start with 'standing wave' solutions of the form

$$a(n) = A(k)e^{ikn} - A(-k)e^{-ikn} \tag{14.48}$$

in this case of open boundary conditions and derive Bethe ansatz equations from there. Cp. Alcaraz *et al.* (1987).

....

The equations determining the quasi-momenta—the famous *Bethe ansatz equations*—are for arbitrary M

$$k_i N = 2\pi\lambda_i + \sum_{j=1,\, j\neq i}^{M} \Theta(k_i, k_j) \tag{14.49}$$

where the Bethe ansatz *quantum numbers* are

$$\lambda_i = 0, 1, \ldots, N-1; \quad i = 1, \ldots, M. \tag{14.50}$$

For a given set of quantum numbers λ_i, these equations determine the quasi-momenta and, thus, the Bethe ansatz diagonalization of the XXZ Heisenberg quantum spin chain Hamiltonian.

The equations (14.49) are a set of coupled, non-linear or, another way of saying the latter, transcendental equations. They are far from being easily solvable. Clearly, however, if we cannot solve them somehow, the Bethe ansatz would be basically useless,[3] apart from using them in a numerical attempt.[4] In a very concrete sense, dealing with the equations (14.49) is the real task to be performed when dealing with Bethe ansatz solvable models.

All Bethe ansatz problems lead to Bethe ansatz equations of the form (14.49).[5] In the case of several degrees of freedom per site, e.g. in problems of electrons with a charge and a spin degree of freedom, the Bethe ansatz equations (14.49) are generalized to a set of *coupled* equations, one for each degree of freedom. Important examples are the one-dimensional Fermi gas with δ-function interaction (Yang, 1967) and the Hubbard model (Lieb and Wu, 1968; 2003), the latter being, in a certain sense, a lattice version of the former.

The equations (14.49) are a coupled set of transcendental equations for the quasi-momenta k_i. The solution of (14.49) depend on the set of numbers λ_i, the Bethe

3 Another aspect of what 'ansatz' means: after you made the ansatz, in most cases the real work begins!

4 Even then, some, maybe even partial, analytical solution would be highly desirable as a check of the numerics.

5 Bill Sutherland (2004) calls (14.49) the *fundamental* equations.

ansatz quantum numbers. Determining the quasi-momenta k_i from (14.49) is a rather complicated problem and requires further simplification or additional assumptions. Section 14.4, provides a solution to (14.49) for the ground state of the isotropic antiferromagnetic chain in the thermodynamic limit $N \to \infty$. In this limit, the equations (14.49) simplify considerably. Further, section 14.5, discusses the simplest excitations for the ferromagnetic isotropic chain. Section 14.6 then discusses the isotropic antiferromagnetic spin chain.

Physically, equations (14.49) show that all quasi-momenta are coupled and how, moreover, they act back on one another.

Before proceeding, we remark that because of the antisymmetry of the phase shifts

$$\Theta(k_i, k_j) = -\Theta(k_j, k_i) \tag{14.51}$$

the total quasi-momentum is given by

$$K \equiv \sum_{i=1}^{M} k_i = \frac{2\pi}{N} \sum_{i=1}^{M} \lambda_i, \tag{14.52}$$

i.e. the sum of all quasi-momenta may be interpreted as the total crystal momentum of the chain that is determined by the quantum numbers λ_i only.

In order to discuss the solutions of (14.49), it is useful to introduce new variables, in terms of which the phases $\Theta(k_i, k_j)$ become functions of the difference of the new variables for two flipped spins at sites i and j.

Section 14.3 is devoted to this parameterization.

14.3 Parameterization of the quasi-momenta

Firstly, here we see that the quasi-momenta k_i and phases $\Theta(k_i, k_j)$ are complex quantities. We adopt the following convention for their real parts

$$0 \leq \Re(k_i) \leq 2\pi \tag{14.53}$$

and

$$-\pi \leq \Re\left(\Theta(k_i, k_j)\right) \leq \pi. \tag{14.54}$$

It will prove useful for the calculation of the ground state energy and the energy excitation spectrum of the XXZ quantum spin chain to introduce a variable transformation from the quasi-momenta k_i to new variables v_i called *rapidities*. The Bethe ansatz equations will become more tractable in the new variables mainly because we seek the new variables such that the phases $\Theta(k_i, k_j)$ become a function of the difference $v_i - v_j$ in the new rapidity variables.

The equation (14.32) for the phases can be rewritten

$$\cot\frac{\Theta(k_i,k_j)}{2}=\frac{\Delta\left(\cot\frac{k_i}{2}-\cot\frac{k_j}{2}\right)}{1+\Delta-(1-\Delta)\cot\frac{k_i}{2}\cot\frac{k_j}{2}}. \tag{14.55}$$

As sections 14.4 and 14.5 concentrate on the isotropic spin chain, $|\Delta|=1$, and to avoid too much cumbersome algebra, we shall give here only the parameterization for these two isotropic cases: antiferromagnetic $\Delta=1$ and ferromagnetic $\Delta=-1$, the latter as an exercise. For arbitrary anisotropy Δ, investing some more algebraic work, similar transformations can be found e.g. des Cloizeaux and Gaudin (1966); Izyumov and Skryabin (1988). For easy reference, we shall summarize the parameterizations for arbitrary anisotropy Δ at the end of this section.

14.3.1 Isotropic antiferromagnetic spin chain

For the isotropic antiferromagnetic chain $\Delta=1$, the equations for the phases in the form (14.55) reduce to

$$\cot\frac{\Theta(k_i,k_j)}{2}=\frac{1}{2}\left(\cot\frac{k_i}{2}-\cot\frac{k_j}{2}\right), \tag{14.56}$$

which readily suggests to introduce the new variables

$$v_i=\frac{1}{2}\cot\frac{k_i}{2}. \tag{14.57}$$

In the new variables, we have obviously

$$\Theta(k_i,k_j)\to\Theta(v_i,v_j)=\Theta(v_i-v_j), \tag{14.58}$$

which, of course, provides great technical simplifications.

For real k in the interval

$$0\le k_i\le 2\pi \tag{14.59}$$

the new variables sweep the whole real line

$$-\infty<v<\infty. \tag{14.60}$$

For later reference we note the inverse transformation

$$k_i=2\cot^{-1}2v_i=\frac{1}{i}\ln\frac{v_i+\frac{i}{2}}{v_i-\frac{i}{2}} \tag{14.61}$$

and, moreover, the phases are given explicitly by

$$\Theta(v_i - v_j) = 2\cot^{-1}(v_i - v_j) = \frac{1}{i}\ln\frac{v_i - v_j + i}{v_i - v_j - i}, \tag{14.62}$$

The equations (14.49) can be expressed in the equivalent form

$$e^{ik_iN} = \prod_{j=i,\, j\neq i}^{M} e^{i\Theta(k_i,k_j)}, \tag{14.63}$$

which, in the new variables, takes the form

$$\left(\frac{v_i + \frac{i}{2}}{v_i - \frac{i}{2}}\right)^N = \prod_{j=i,\, j\neq i}^{M} \frac{v_i - v_j + i}{v_i - v_j - i}. \tag{14.64}$$

We have used the mathematical identities

$$e^{2i\tan^{-1}x} = \frac{1+ix}{1-ix} \tag{14.65}$$

and

$$e^{2i\cot^{-1}x} = \frac{ix-1}{ix+1}. \tag{14.66}$$

The energy eigenvalue expressed in the new variables is

$$E_M = -\frac{1}{2}\sum_{i=1}^{M}\frac{1}{v_i^2 + \frac{1}{4}} = \frac{1}{2}\sum_{i=1}^{M}\frac{dk_i}{dv_i} \tag{14.67}$$

where we used

$$\sin^2\frac{k_i}{2} = \frac{1}{1+4v_i^2}. \tag{14.68}$$

EXERCISE 14.5 Isotropic ferromagnetic spin chain: $\Delta = -1$ Perform the analogous transformation to new variables v_i for the isotropic ferromagnetic case $\Delta = -1$. Also provide the inverse transform and the Bethe ansatz equations and energy eigenvalue in the transformed variables.

For later use, we note one of the results of the above exercise. For the isotropic ferromagnetic spin chain $\Delta = -1$, the transformation to rapidity variables is

$$v_i = \frac{1}{2} \tan \frac{k_i}{2}. \tag{14.69}$$

14.3.2 Anisotropic spin chain

For the sake of completeness, we briefly state the results for the anisotropic spin chain with $\Delta \neq \pm 1$ leaving the details of the calculation as an exercise to the reader.

Setting

$$\Delta = \cos 2\eta \tag{14.70}$$

with η not necessarily real, we parametrize the quasi-momenta k_i by

$$e^{ik_i} = \frac{\sin(v_i + \eta)}{\sin(v_i - \eta)}. \tag{14.71}$$

Using this parameterization in the equation for the phases $\Theta(k_i, k_j)$

$$e^{i\Theta(k_i,k_j)} = -\frac{e^{i(k_i+k_j)} - 2\Delta e^{ik_i} + 1}{e^{i(k_i+k_j)} - 2\Delta e^{ik_j} + 1} \tag{14.72}$$

we get, after some algebra,

$$e^{i\Theta(v_i-v_j)} = \frac{\sin(v_i - v_j + 2\eta)}{\sin(v_i - v_j - 2\eta)}. \tag{14.73}$$

This leads to the Bethe ansatz equations in the form

$$\left[\frac{\sin(v_i + \eta)}{\sin(v_i - \eta)}\right]^N = \prod_{j=1,\, j\neq i}^{M} \frac{\sin(v_i - v_j + 2\eta)}{\sin(v_i - v_j - 2\eta)}. \tag{14.74}$$

The case of the anisotropic spin chain is completed by the expression for the energy eigenvalue

$$E_M = \frac{1}{2} \sum_{i=1}^{M} \left[\frac{\sin(v_i + \eta)}{\sin(v_i - \eta)} \frac{\sin(v_i - \eta)}{\sin(v_i + \eta)} - 2\Delta\right]. \tag{14.75}$$

The anisotropic regime is further subdivided into the ranges

$$\Delta > 1, \tag{14.76}$$

the easy-axis anisotropic antiferromagnetic case

$$-1 < \Delta < 1, \tag{14.77}$$

the critical planar case, and

$$\Delta < -1, \tag{14.78}$$

the easy-axis anisotropic ferromagnetic case.

The cases $|\Delta| > 1$ have an energy gap between the ground state and the first excited state and are, hence, not quantum critical. For $|\Delta| < 1$, the energy spectrum is gapless and, hence, quantum critical.

14.3.2.1 Easy-axis antiferromagnetic spin chain

Setting

$$\eta = i\frac{\Phi}{2}, \quad 0 < \Phi < \infty \tag{14.79}$$

and

$$v_i = \frac{\varphi_i}{2}, \quad -\pi \leq \varphi_i \leq \pi \tag{14.80}$$

yields

$$\Delta = \cosh \Phi \tag{14.81}$$

and

$$e^{ik_i} = \frac{e^{i\varphi_i} - e^{\Phi}}{e^{i\varphi_i + \Phi} - 1} \tag{14.82}$$

or, equivalently,

$$\cot \frac{k_i}{2} = \coth \frac{\Phi}{2} \tan \frac{\varphi_i}{2}. \tag{14.83}$$

The equation for the phases thus becomes

$$\cot \frac{\Theta(\varphi_i - \varphi_j)}{2} = \coth \Phi \tan \frac{(\varphi_i - \varphi_j)}{2} \tag{14.84}$$

and the energy eigenvalue

$$E_M = -\sum_{i=1}^{M} \frac{\sinh^2 \Phi}{\cosh \phi - \cos \varphi_i}. \tag{14.85}$$

14.3.2.2 Critical planar spin chain

Setting

$$\eta = -\frac{\Psi}{2}, \quad 0 < \Psi < \pi \tag{14.86}$$

and

$$v_i = i\frac{\psi_i}{2} \tag{14.87}$$

so that

$$\Delta = \cos \Psi, \tag{14.88}$$

$$e^{ik_i} = \frac{e^{-\psi_i} - e^{\Psi}}{e^{i\psi - \psi_i} - 1} \tag{14.89}$$

or, equivalently,

$$\cot \frac{k_i}{2} = \cot \frac{\Psi}{2} \tanh \frac{\psi_i}{2}. \tag{14.90}$$

The phases thus are given implicitly by

$$\cot \frac{\Theta(\psi_i - \psi_j)}{2} = \cot \Psi \tanh \frac{(\psi_i - \psi_j)}{2}, \tag{14.91}$$

and the energy eigenvalue

$$E_M = -\sum_{i=1}^{M} \frac{\sin^2 \Psi}{\cosh \psi_i - \cos \Psi}. \tag{14.92}$$

14.3.2.3 Easy-axis ferromagnetic spin chain

Setting

$$\eta = \frac{\pi}{2} - i\frac{\tilde{\Phi}}{2}, \quad 0 < \tilde{\Phi} < \infty \tag{14.93}$$

and

$$v_i = -\frac{\varphi_i}{2}, \quad -\pi \leq \varphi_i \leq \pi \tag{14.94}$$

yields

$$\Delta = -\cosh \tilde{\Phi} \tag{14.95}$$

and

$$\cot \frac{k_i}{2} = \tanh \frac{\tilde{\Phi}}{2} \tanh \frac{\psi_i}{2}. \tag{14.96}$$

The equation for the phases thus becomes

$$\cot \frac{\Theta(\varphi_i - \varphi_j)}{2} = -\coth \tilde{\Phi} \tan \frac{(\varphi_i - \varphi_j)}{2} \tag{14.97}$$

and the energy eigenvalue

$$E_M = \sum_{i=1}^{M} \frac{\sinh^2 \tilde{\Phi}}{\cosh \tilde{\Phi} - \cos \varphi_i}. \tag{14.98}$$

Note, that, as we had announced as our intention for the introduction of rapidity variables, in all cases we have achieved that the two-particle phases Θ depend only on the difference of the rapidity variables. This will prove to be very important technically for the exact solvability of the spin chain problem.

14.4 Ground state of the XXZ Heisenberg quantum spin chain

The problem of solving the Schrödinger equation for the Heisenberg model, has been transformed into the problem of solving the Bethe ansatz equations (14.49) for a set of quantum numbers λ_i and in determining the corresponding eigenvalues and eigenstates.

14.4.1 Easy-axis ferromagnetic anisotropy regime

The simplest situation to consider is the ground state of the ferromagnetic chain, i.e. the range of anisotropy $\Delta \leq -1$. Flipping a spin costs energy if the exchange coupling is ferromagnetic, hence, the energetically favourable state is the ferromagnetic reference state (13.35)

$$|0\rangle = |\uparrow\rangle_1 |\uparrow\rangle_2 \ldots |\uparrow\rangle_N, \tag{14.99}$$

which has $M = 0$ and $E_0 = 0$. Section 14.5 demonstrates that flipping one spin from the state $|0\rangle$ leads to a higher energy $E_1 > E_0$.

14.4.2 Planar ferromagnetic and antiferromagnetic anisotropy regime

The anisotropy regime $\Delta \geq -1$ is more complicated. However, there are a number of theorems we can invoke, although we shall not prove these theorems in order to keep the treatment simple and focused. The theorems will enable us to calculate the ground state energy in the thermodynamic limit $N \to \infty$.

14.4.3 Isotropic antiferromagnetic case

In the isotropic antiferromagnetic case $\Delta = 1$, where we have the total spin $\mathbf{S}$ and the total z-component S^z as conserved quatities, Marshall (1955), Lieb *et al.* (1961), and Lieb and Mattis (1962) proved that the ground state has $S = 0$ and therefore $S^z = 0$, which is equivalent to $M = \frac{N}{2}$ (where, for simplicity, N has been assumed even).

Already Bethe (1931) had assumed that $S^z = 0$ or equivalently $M = \frac{N}{2}$ and showed then that the ground state is obtained by choosing

$$\lambda_i = 2i - 1, \quad i = 1, 2, \ldots, \frac{N}{2}. \tag{14.100}$$

Much later, Yang and Yang (1966*a*,*b*,*c*) proved Bethe's choice was the correct one to obtain the ground state energy in the whole region $\Delta \geq -1$.

We now want to calculate the ground state energy for the isotropic case $\Delta = 1$ using (14.100) in the thermodynamic limit $N \to \infty$. The first steps in this calculation will not depend on the value of the anisotropy Δ, so that we shall not specialize to the isotropic case $\Delta = 1$ until later. Of course, with a little more effort but in a very similar way, the ground state energy can also be calculated for arbitrary anisotropy Δ.

The first step is to replace the discrete variable i in (14.100) by

$$x = \frac{2i}{N} \tag{14.101}$$

and rewrite (14.100) for $N \to \infty$

$$\lambda = Nx, \quad 0 \leq x \leq 1. \tag{14.102}$$

The quasi-momenta k_i and the phases $\Theta(k_i, k_j)$ become functions of the variable x:

$$k_i \to k(x) \tag{14.103}$$

and

$$\Theta(k_i, k_j) \to \Theta(k(x), k(y)) \equiv \Theta(x, y). \tag{14.104}$$

The Bethe ansatz equations (14.49) transform thus into an integral equation

$$k(x) = 2\pi x + \frac{1}{2}\int_0^1 dy \Theta(x, y), \tag{14.105}$$

which we can parametrize with new variables ξ depending on the anisotropy Δ as in the previous section. Thus,

$$k(\xi(x)) = 2\pi x + \frac{1}{2}\int_0^1 dy \Theta(\xi(x), \xi(y)). \tag{14.106}$$

Using

$$\frac{dx}{d\xi} = -\rho(\xi), \tag{14.107}$$

which defines the function $\rho(\xi)$ as the density of values ξ in the interval ξ to $\xi + d\xi$, we differentiate (14.106) and obtain

$$\frac{dk(\xi)}{d\xi} = -2\pi\rho(\xi) - \frac{1}{2}\int_{\xi_0}^{\xi_1} d\eta \rho(\eta) \frac{\partial\Theta(\xi, \eta)}{\partial\xi} \tag{14.108}$$

with $\xi_1 = \xi(x = 1)$ and $\xi_0 = \xi(x = 0)$.

It can be shown that $\xi_1 = -\xi_0$. Thus, the integral equation takes the final form

$$-\frac{dk(\xi)}{d\xi} = 2\pi\rho(\xi) - \frac{1}{2}\int_{-\xi_0}^{\xi_0} d\eta \rho(\eta) \frac{\partial\Theta(\xi, \eta)}{\partial\xi}. \tag{14.109}$$

Importantly, note that $k(\xi)$ and $\Theta(\xi, \eta)$ are known functions. If the integral equation for the unknown function $\rho(\xi)$ can be solved, then we can calculate the ground state energy by a simple integration.

The integration limit ξ_0 is also unknown at this stage and has to be determined from the condition

$$\int_0^1 dx = \int_{-\xi_0}^{\xi_0} d\xi \rho(\xi) = 1. \tag{14.110}$$

Often, however, it is possible to guess the value of ξ_0 and then show the consistency of the solution of (14.109) using the guessed ξ_0 later.

In order to progress without too many technical complications, we now stop the general and formal discussion and focus on the isotropic case $\Delta = 1$.

For the isotropic case $\Delta = 1$, we have $\xi(x) = v(x)$ with $-\infty < v(x) < \infty$ and

$$k(v) = 2\cot^{-1} 2v, \tag{14.111}$$

$$\Theta(v, v') = 2\cot^{-1}(v - v'). \tag{14.112}$$

The integration boundary can be shown to be $v_0 = \infty$. Using

$$\frac{dk}{dv} = -\frac{1}{v^2 + \frac{1}{4}} \tag{14.113}$$

we obtain the integral equation (14.109) in the isotropic case $\Delta = 1$

$$\frac{1}{v^2 + \frac{1}{4}} = 2\pi\rho(v) - \frac{1}{2}\int_{-\infty}^{\infty} dv' \rho(v') \frac{\partial\Theta(v, v')}{\partial v}. \tag{14.114}$$

Because of our choice of the branch of the phase function $\Theta(v - v')$ (see the discussion in section 14.3 and figure 14.1), i.e. for the branch of the $\cot^{-1}$ function, care has to be taken when calculating the derivative

$$\frac{\partial\Theta(v - v')}{\partial v}. \tag{14.115}$$

The function $\Theta(v)$ has a jump at $v = 0$, and therefore, the derivative acquires a δ-function contribution.

Therefore,

$$\frac{\partial\Theta(v - v')}{\partial v} = 2\pi\delta(v - v') - \frac{2}{1 + (v - v')^2} \tag{14.116}$$

and we finally obtain the integral equation for the isotropic chain

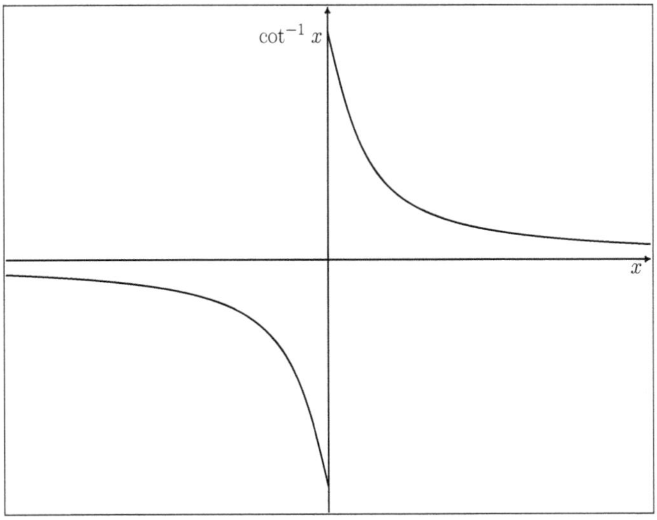

Figure 14.1 *The branch of the function* $\cot^{-1} x$ *such that* $-\frac{\pi}{2} \leq \cot^{-1} x \leq \frac{\pi}{2}$.

$$\frac{1}{v^2 + \frac{1}{4}} = \pi\rho(v) + \int_{-\infty}^{\infty} dv' \frac{\rho(v')}{1 + (v - v')^2}. \tag{14.117}$$

The solution of this integral equation is left to the reader as exercise 14.6.

EXERCISE 14.6 Integral equation for the ground state energy of the isotropic spin chain Solve the integral equation

$$\frac{1}{v^2 + \frac{1}{4}} = \pi\rho(v) + \int_{-\infty}^{\infty} dv' \frac{\rho(v')}{1 + (v - v')^2} \tag{14.118}$$

using Fourier transformation, i.e.

$$\rho(v) = \frac{1}{\sqrt{2\pi}} \int_{-\infty}^{\infty} du\, e^{iuv} r(u). \tag{14.119}$$

You may find the following integral useful

$$\int_{-\infty}^{\infty} dv \frac{e^{iuv}}{1 + v^2} = \pi e^{-|u|}. \tag{14.120}$$

The solution of the integral equation for the ground state of the isotropic chain is

$$\rho(v) = \frac{1}{\cosh \pi v}. \tag{14.121}$$

In section 14.3 we derived the energy eigenvalue

$$E_{\frac{N}{2}} = -\frac{1}{2}\sum_{i=1}^{\frac{N}{2}} \frac{1}{v_i + \frac{1}{4}}. \tag{14.122}$$

In the thermodynamic limit, this expression becomes an integral

$$E_{\frac{N}{2}} = -\frac{N}{4}\int_{-\infty}^{\infty} dv \frac{\rho(v)}{v^2 + \frac{1}{4}}, \tag{14.123}$$

which, using (14.121), yields the expression

$$E_{\frac{N}{2}} = -N\int_0^{\infty} \frac{dx}{(x^2+1)\cosh\frac{\pi x}{2}}. \tag{14.124}$$

This integral can be found in tables and has the value ln 2. The final result for the ground state energy of the isotropic antiferromagnetic chain.

$$E_{\frac{N}{2}} = -N \ln 2. \tag{14.125}$$

This is the famous result Lamek Hulthén obtained in his thesis work and then published in a long paper 1938; it was the first Bethe ansatz paper after Bethe's seminal work of 1931.

As discussed, the ground state energy for arbitrary $\Delta \geq -1$ can be calculated along similar lines using the parameterizations of section 14.3 to derive corresponding integral equations, which then can be solved with Fourier techniques.

14.5 Excitations of the XXZ Heisenberg quantum spin chain

For any given set of quantum numbers $\{\lambda_i\}$ the Bethe ansatz equations (14.49) describe a state. In general, we do not know beforehand which state this will be. This is one of the main problems to be solved within the framework of the coordinate Bethe ansatz. Generically, by prescribing a set of quantum numbers $\{\lambda_i\}$, we shall have hit on some excited state. At this stage, we would have to invest a lot of work to determine the low-lying excitations of the spin chain.

Instead, we restrict ourselves to a few simple examples that we examine in the thermodynamic limit $N \to \infty$, thereby simplifying the analysis considerably. The complexity of the Bethe ansatz equations will become apparent even for the simplest examples.

14.5.1 One magnon state in the ferromagnetic chain

The simplest case is the isotropic ferromagnetic spin chain, i.e. $\Delta = -1$, whose ground state is the reference state $|0\rangle$ introduced in section 13.3, (13.35). Section 14.3 demonstrated that the quasi-momenta for the isotropic ferromagnetic chain can be parametrized by

$$k_i = 2\tan^{-1} 2v_i \tag{14.126}$$

where, for real $0 \le k_i \le 2\pi$, the variables v_i are distributed over the whole real line. The phases, which, in the new variables, depend only on their difference, are given by

$$\Theta(k_i, k_j) = 2\cot^{-1}(v_i - v_j) \tag{14.127}$$

and the energy of the state by

$$E_M = \frac{1}{2}\sum_{j=1}^{M} \frac{1}{v_j + \frac{1}{4}} = \frac{1}{2}\sum_{j=1}^{M} \frac{dk_j}{dv_j}. \tag{14.128}$$

The Bethe ansatz equations in product form are then for even N

$$\left(\frac{v_i - \frac{i}{2}}{v_i + \frac{i}{2}}\right)^N = \prod_{j=1,\, j\neq i}^{M} \frac{v_i - v_j - i}{v_i - v_j + i}, \qquad i = 1, \ldots, M. \tag{14.129}$$

For $M = 1$ flipped spin, this reduces to

$$\left(\frac{v - \frac{i}{2}}{v + \frac{i}{2}}\right)^N = 1 \tag{14.130}$$

or, equivalently,

$$\frac{v - \frac{i}{2}}{v + \frac{i}{2}} = e^{ik} \tag{14.131}$$

with e^{ik} one of the Nth roots of unity, i.e.

$$k = \frac{2\pi\lambda}{N}, \qquad \lambda = 0, 1, 2, \ldots, N-1. \tag{14.132}$$

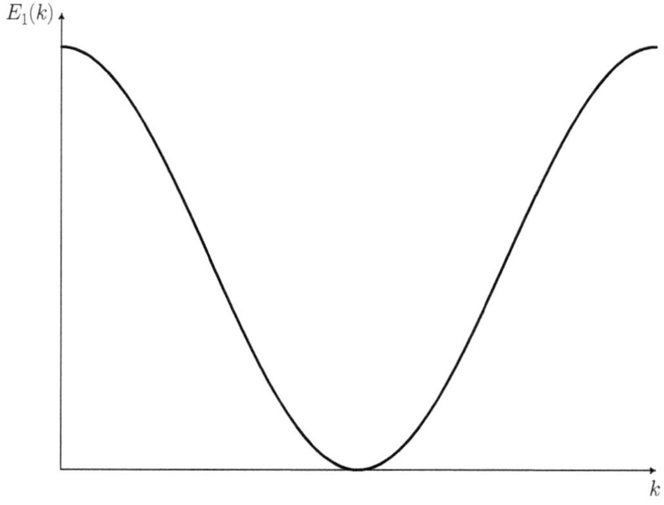

Figure 14.2 *Energy spectrum of one magnon excitations of the isotropic ferromagnetic Heisenberg model in the Brillouin zone $0 \leq k \leq 2\pi$.*

In the thermodynamic limit $N \to \infty$ the quasi-momentum k gets indeed distributed over the whole interval $0 \leq k \leq 2\pi$ and, hence, the rapidity variable v over the whole real line.

An excitation of this type is called a *magnon*, or spin wave. The energy of a magnon of the ferromagnetic spin chain is given by

$$E_1(k) = 1 + \cos k. \tag{14.133}$$

Since we choose the Hamiltonian in such a way that the ferromagnetic reference state $|0\rangle$ has energy $E_0 = 0$, we have the result that for the magnon energy for all quasi-momenta $0 \leq k \leq 2\pi$

$$E_1(k) \geq E_0. \tag{14.134}$$

This demonstrates that the energy of the reference state is indeed a good candidate for the absolute ground state of the spin chain.

Next, we discuss the more ambitious case of two flipped spins.

14.5.2 Bound states or spin complexes

In the case of two flipped spins, $M = 2$, we have the two coupled Bethe ansatz equations

$$\left(\frac{v_1 - \frac{i}{2}}{v_1 + \frac{i}{2}}\right)^N = \frac{v_1 - v_2 - i}{v_1 - v_2 + i}, \tag{14.135}$$

$$\left(\frac{v_2 - \frac{i}{2}}{v_2 + \frac{i}{2}}\right)^N = \frac{v_2 - v_1 - i}{v_2 - v_1 + i}. \tag{14.136}$$

14.5.2.1 Real solutions: two magnon state of the ferromagnetic spin chain

The first question to answer is whether these equations admit real solutions v_1 and v_2.

Again using the identity

$$\exp\left(2i \cot^{-1} x\right) = \frac{x + i}{x - i} \tag{14.137}$$

for real x, we can rewrite the two coupled equations

$$\frac{v_1 - \frac{i}{2}}{v_1 + \frac{i}{2}} = \left(\frac{v_1 - v_2 - i}{v_1 - v_2 + i}\right)^{\frac{1}{N}} = e^{-i\phi(v_1 - v_2)} = e^{ik_1} \tag{14.138}$$

$$\frac{v_2 - \frac{i}{2}}{v_2 + \frac{i}{2}} = \left(\frac{v_2 - v_1 - i}{v_2 - v_1 + i}\right)^{\frac{1}{N}} = e^{i\phi(v_1 - v_2)} = e^{ik_2} \tag{14.139}$$

where

$$k_i = \frac{2\pi\lambda_i}{N}, \quad \lambda_i = 0, 1, \ldots, N-1, \quad i = 1, 2 \tag{14.140}$$

and

$$\phi(v_1 - v_2) = \frac{1}{N}\Theta(v_1 - v_2) = \frac{2}{N}\cot^{-1}(v_1 - v_2) \tag{14.141}$$

using the principal value of the $\cot^{-1}$ function. Since the principal value of the $\cot^{-1}$ function is a bounded function, $\phi(v_1 - v_2)$ vanishes as $N \to \infty$. Therefore, in this limit,

$$\frac{v_i - \frac{i}{2}}{v_i + \frac{i}{2}} = e^{ik_i}, \quad i = 1, 2. \tag{14.142}$$

In other words, for real v_1 and v_2, the Bethe ansatz equations for two flipped spins decouple and reduce to two independent equations of the type we know from the discussion of the one-magnon state. Thus, we have the two-magnon state of the ferromagnetic spin chain with the quasi-momenta k_1 and k_2. The energy of this state is given by

$$E_2 = 1 + \cos k_1 + 1 + \cos k_2 = 2\left(1 + \cos\frac{k_1 + k_2}{2}\cos\frac{k_1 - k_2}{2}\right). \tag{14.143}$$

Again, we see, that $E_2 \geq E_0$.

However, in the case of two flipped spins, the two-magnon state does not exhaust the possible states of the spin chain. We have to take into account the possibility of complex solutions of the Bethe ansatz equations.[6] While the two-magnon solution discussed so far can be regarded as a state of two independent single magnon excitations, the solutions with complex rapidities will clearly show interaction effects.

14.5.2.2 *Complex solutions: bound state or spin complex*

We now investigate the Bethe ansatz equations (14.135) and (14.136) when the rapidities can take complex values

$$v_i = x_i + iy_i, \quad i = 1, 2. \tag{14.144}$$

Then we have for the moduli of the Bethe ansatz equations (14.135) and (14.136)

$$\left[\frac{x_1^2 + (y_1 - \frac{1}{2})^2}{x_1^2 + (y_1 + \frac{1}{2})^2}\right]^N = \frac{(x_1 - x_2)^2 + (y_1 - y_2 - 1)^2}{(x_1 - x_2)^2 + (y_1 - y_2 + 1)^2} \tag{14.145}$$

$$\left[\frac{x_2^2 + (y_2 - \frac{1}{2})^2}{x_2^2 + (y_2 + \frac{1}{2})^2}\right]^N = \frac{(x_1 - x_2)^2 + (y_1 - y_2 + 1)^2}{(x_1 - x_2)^2 + (y_1 - y_2 - 1)^2}, \tag{14.146}$$

which are symmetric with respect to exchange of the indices 1 and 2.

We examine (14.145) and (14.146) in the limit $N \to \infty$. Assuming that $y_1 > 0$, the left-hand side of (14.145) vanishes for $N \to \infty$, and moreover, it follows that

$$x_1 = x_2 \quad \text{and} \quad y_1 - y_2 = 1. \tag{14.147}$$

The same result follows from (14.146) assuming $y_2 < 0$. Using the exchange symmetry of the indices, in summary, we have either

$$x_1 = x_2 \quad \text{and} \quad y_1 - y_2 = 1 \quad \text{for} \quad y_1 > 0, \quad y_2 < 0 \tag{14.148}$$

or

$$x_1 = x_2 \quad \text{and} \quad y_2 - y_1 = 1 \quad \text{for} \quad y_1 < 0, \quad y_2 > 0. \tag{14.149}$$

Next, we multiply the Bethe ansatz equations (14.135) and (14.136) together, which yields using (14.148)

[6] This was where Bethe's analysis extended that of Bloch.

$$\left[\frac{x_1 + i(y_1 - \frac{3}{2})}{x_1 + i(y_1 + \frac{1}{2})}\right]^N = 1. \tag{14.150}$$

If we use (14.149) instead, we have to replace y_1 by y_2 in this expression. It follows that the complex number in the brackets of this expression must have unit modulus. For this to be true, we must either have $y_1 = \frac{1}{2}$ and, hence, $y_2 = -\frac{1}{2}$, or else $y_2 = \frac{1}{2}$ and $y_1 = -\frac{1}{2}$, depending on whether we use (14.148) or (14.149). In both cases, however, we end up with

$$\left[\frac{x_1 - i}{x_1 + i}\right]^N = 1 \tag{14.151}$$

or

$$\exp\left(2i \cot^{-1} x_1\right) = e^{ik} \tag{14.152}$$

where, again, we have to use the principal value of the $\cot^{-1}$ function and e^{ik} is one of the Nth roots of unity. In the limit $N \to \infty$, both k and $2\cot^{-1} x_1$ are defined on the same interval $[0, 2\pi]$. Hence, we do not find any restriction for the value of the real part x_1, which can thus assume any real value.

The complex solution of the Bethe ansatz equations (14.135) and (14.136) is thus found to be of the form (dropping the index 1 on the variable x)

$$v_1 = x \pm \frac{i}{2}, \quad v_2 = x \mp \frac{i}{2}, \quad -\infty < x < \infty, \tag{14.153}$$

i.e. they form a pair of conjugate complex values (of course, (14.153) only describes two, not four, rapidities).

The important question remaining is what kind of state is described by the complex solution (14.153). To answer this question, let us return from rapidities to quasi-momenta. Recalling the relation between the quasi-momenta k_i and the rapidities v_i for the isotropic ferromagnetic spin chain with $\Delta = -1$

$$k_i = 2\tan^{-1} 2v_i. \tag{14.154}$$

For the two flipped spins we are considering, we have

$$k_1 \pm k_2 = 2(\tan^{-1} 2v_1 \pm \tan^{-1} v_2), \tag{14.155}$$

which, with the help of the identity,

$$\tan^{-1} z = \frac{1}{2i} \ln \frac{1 + iz}{1 - iz} \tag{14.156}$$

after some algebra, yields

$$k_1 + k_2 = -2\cot^{-1} x \tag{14.157}$$

and

$$k_1 - k_2 = -i \ln\left[\frac{x^2}{1+x^2}\right]. \tag{14.158}$$

Written differently, this means that k_1 and k_2 are conjugate complex numbers

$$k_1 = \alpha + i\beta, \quad k_2 = \alpha - i\beta \tag{14.159}$$

where

$$\alpha = -\cot^{-1} x, \quad \beta = -\frac{1}{2}\ln\left[\frac{x^2}{1+x^2}\right]. \tag{14.160}$$

For pairs of conjugate complex quasi-momenta, Bethe (1931) showed that the probability $|a(n_1, n_2)|^2$ calculated from the Bethe ansatz wave function falls off roughly exponentially for increasing distance $n_2 - n_1$ of the two flipped spins. The two flipped spins are thus in a bound state, or what Bethe called a spin complex. We leave the details of this calculation as exercise 14.7.

EXERCISE 14.7 Bound state of the $M = 2$ sector of the isotropic ferromagnetic spin chain Calculate the probability $|a(n_1, n_2)|^2$ from the Bethe ansatz wave function for increasing distance $n_2 - n_1$ of the two flipped spins. This exercise is rather an enticement to read Bethe's original paper, either in the German original (Bethe, 1931) or in the English translation, available in Bethe (1997) or Mattis (1993).

It is also instructive to calculate the energy E_2^b of the bound state, which is what we shall do now. Recalling the results of section 14.3, especially of exercise 14.5, the energy for two flipped spins of the isotropic ferromagnetic chain is

$$E_2 = \frac{1}{2}\sum_{i=1}^{2} \frac{1}{v_i^2 + \frac{1}{4}} = \frac{1}{2}\sum_{i=1}^{2} \frac{dk_i}{dv_i}. \tag{14.161}$$

For the complex rapidities (14.153) this yields

$$E_2^b = \frac{1}{x^2+1} = \frac{1}{2}\left(1 - \cos(k_1 + k_2)\right), \tag{14.162}$$

the last expression following from equation (14.157).

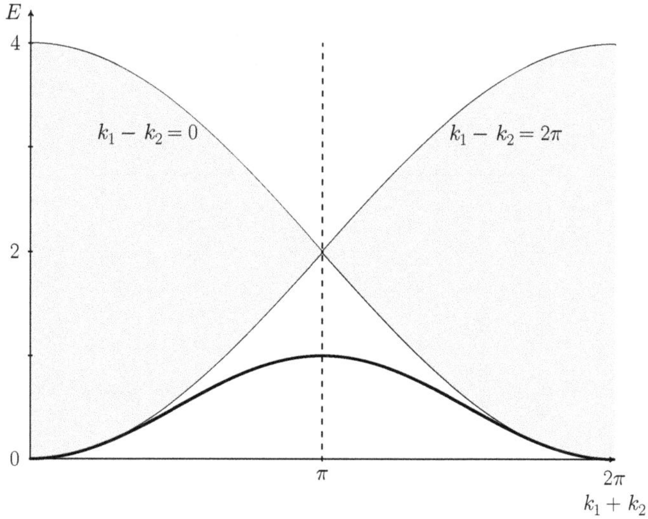

Figure 14.3 *Excitation spectrum of the isotropic ferromagnetic ($\Delta = -1$) spin chain for $M = 2$. The continuum of two-magnon excitations described by real solutions of the Bethe ansatz equations lies between the limiting curves labelled by $k_1 - k_2 = 0$ and $k_1 - k_2 = 2\pi$ (thin lines). Thick line: bound state energy of a branch of bound states corresponding to two-string solutions.*

The bound state energy and the two-magnon energy is plotted in figure 14.3.

It is apparent from the figure and the equations, too, that E_2^b is always above the ground state energy E_0: $E_2^b(k_1 + k_2) \geq E_0$, but also always below the two-magnon excitation energy: $E_2^b(k_1 + k_2) \leq E_2^S(k_1 + k_2)$.

It seems plausible to assume that these results can be generalized to three and more flipped spins. However, the algebra will also become more tedious. Furthermore, as we do not expect to gain any further physical insight, just more complicated excitation spectra, we shall not pursue this further. We close this section with a description of what we would find in the case $M = 3$ of three flipped spins from the totally ferromagnetic state.

In the case $M = 3$, the following types of solution occur:

1. All three v_i are real and independent; this solution characterizes the three-magnon state, three independent magnons with quasi-momenta k_1, k_2, and k_3;
2. One of the v_i, v_1 say, is real, and the other two form a conjugate complex pair; this corresponds to a one-magnon state with real quasi-momentum k_1 and a bound state or two-spin complex with real quasi-momentum $k = k_2 + k_3$;
3. The three rapidities v_i have the form $v_1 = x + i$, $v_2 = x$, and $v_3 = x - i$ with a common real part x; this is a three-spin complex, i.e. a bound state of the three flipped spins, with the real quasi-momentum $k = k_1 + k_2 + k_3$.

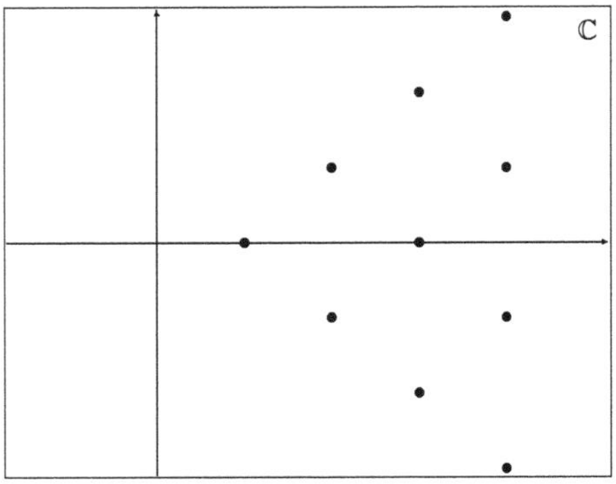

Figure 14.4 *Structure of strings in the complex plane ℂ: a one-, two-, three-, and four-string.*

Despite the slightly disparaging remark here about the physical interest in doing the corresponding calculations, it is nevertheless important to generalize the emerging picture of how the states are characterized by rapidities and quasi-momenta. One reason is that otherwise we cannot answer the question that started Bethe himself in his investigation,—the question of correctly enumerating *all* states of the spin chain. A second reason is connected with the extension of the Bethe ansatz to finite temperature: the thermodynamic Bethe ansatz,[7] where knowledge of the structure of the Bethe ansatz solutions is the starting point of the analysis; see part V.

We therefore continue with generalizing from what we learned in the cases $M = 1, 2, 3$ for the structure of the solutions of the Bethe ansatz equations.

14.5.3 String hypothesis

We now generalize from what we have seen previously in the chapter. The generalization of the picture of Bethe ansatz solutions that emerged for the case $M = 3$ suggests the following hypothesis. The solutions of the Bethe ansatz equations in product form

$$\left[\frac{v_i - \frac{i}{2}}{v_i + \frac{i}{2}}\right]^N = \prod_{j=1,\, j\neq i}^{M} \frac{v_i - v_j - i}{v_i - v_j + i}, \quad i = 1, \ldots, M, \quad M \leq \frac{N}{2} \tag{14.163}$$

for arbitrary M contains always spin complexes of various lengths n. An n-spin complex consists of an aggregate of n complex rapidities v_i, which all have a common real part and

[7] Not to be confused with the limit of large N (formally $N \to \infty$), the thermodynamic limit is not concerned with finite temperature.

whose imaginary parts differ by unity. The spin complex is thus described by n complex numbers

$$v^{n,j} = v^n + \frac{i}{2}(n+1-2j), \quad j = 1, 2, \ldots, n \tag{14.164}$$

where v^n is real. This aggregate is called a *string*[8] of length n. The hypothesis states that this is the general structure of the solutions of the Bethe ansatz equations.[9] We shall see below how the hypothesis (which basically goes back to Bethe himself) simplifies the analysis of the Bethe ansatz equations.

This hypothesis was extended to the anisotropic case $\Delta \neq \pm 1$. However, it is not correct as shown by Woynarovich (1982*a*, 1982*b*). There are indeed excitations of the Heisenberg quantum spin chain that do not obey the string hypothesis. However, the standard argument runs that these non-string excitations are not relevant in the thermodynamic limit because their number is small (i.e. remains microscopic of order unity) compared to the number of string excitations (which become macroscopic of order $\propto N$). Hence, they also do not contribute (see part V) to the thermodynamics of the spin chain. We are not going to discuss the issue of string and non-string solutions further here, but rather adopt the string hypothesis for the most part.

The examples discussed in detail here do indeed fit the string hypothesis. A one (see magnon state is described by a string of length $n = 1$, which is, of course, a rather fancy way of saying that it is described by a real rapidity. A two-spin complex, a bound state of two flipped spins, is described by a string of length $n = 2$, i.e. a pair of conjugate complex rapidities.

The string structure (14.164) is not yet the most general. We can have several, ν_n say, strings of the same length n that are distinguished by their real parts, i.e.

$$v_\alpha^{n,j} = v_\alpha^n + \frac{i}{2}(n+1-2j), \quad j = 1, 2, \ldots, n \quad \alpha = 1, 2, \ldots, \nu_n \tag{14.165}$$

where ν_n is the number of strings of length n. The index i for a rapidity in equation (14.163), thus, has been replaced by a combination of two indices $j = 1, \ldots, n$, enumerating the string, and $\alpha = 1, 2, \ldots, \nu_n$ of strings of length n with distinct real parts.

[8] No connection to string theory!

[9] As we have seen in the examples above, the strings take the form (14.164), strictly speaking, only in the thermodynamic limit $N \to \infty$. For finite N, there are small corrections that vanish rapidly as the chain length N increases; see e.g. Faddeev and Takhtajan (1984) and Gaudin (1983) and references therein for details. We ignore this fine detail here.

Therefore, the numbers ν_n are restricted by the condition

$$M = \sum_{n=1}^{M} n\nu_n. \tag{14.166}$$

In particular, this can easily be verified for the case $M = 3$ discussed in section 14.5.2.2. Indeed, for $M = 3$ we have the following possibilities

1. $\nu_1 = 3, \nu_2 = \nu_3 = 0$
2. $\nu_1 = \nu_2 = 1, \nu_3 = 0$
3. $\nu_1 = \nu_2 = 0, \nu_3 = 1$

which cover the cases we have discussed.

14.5.3.1 Bethe ansatz equations for the real parts of strings

We now examine the Bethe ansatz equations in product form

$$\left[\frac{v_i - \frac{i}{2}}{v_i + \frac{i}{2}}\right]^N = \prod_{j=1,\, j\neq i}^{M} \frac{v_i - v_j - i}{v_i - v_j + i} \tag{14.167}$$

inserting the string form for the rapidities v_i and v_j

$$v_i \to v_\alpha^{n,k} = v_\alpha^n + \frac{i}{2}(n+1-2k), \quad k = 1, 2, \ldots, n, \quad \alpha = 1, 2, \ldots, \nu_n, \tag{14.168}$$

$$v_j \to v_\beta^{m,l} = v_\beta^m + \frac{i}{2}(m+1-2l), \quad l = 1, 2, \ldots, m, \quad \beta = 1, 2, \ldots, \nu_m, \tag{14.169}$$

which yields the equations in the form

$$\left[V_1(v_\alpha^{n,k})\right]^N = \prod_{\substack{m,\beta \\ \{m,\beta\}\neq\{n,\alpha\}}} \prod_{l=1}^{m} \frac{v_\alpha^{n,k} - v_\beta^{m,l} - i}{v_\alpha^{n,k} - v_\beta^{m,l} + i} \tag{14.170}$$

$$= \prod_{\substack{m,\beta \\ \{m,\beta\}\neq\{n,\alpha\}}} \prod_{l=1}^{m} V_2(v_\alpha^{n,k} - v_\beta^{m,l}) \tag{14.171}$$

where

$$V_1(a) = \frac{a - \frac{i}{2}}{a + \frac{i}{2}}, \tag{14.172}$$

and

$$V_2(a) = \frac{a-i}{a+i}. \tag{14.173}$$

In the double product above, $m = 1, 2, \ldots, M$ and $\beta = 1, 2, \ldots, \nu_m$. The notation for the double product is used to indicate that, on the left-hand side, all terms are to be excluded where $\{m, \beta\} \neq \{n, \alpha\}$ would be fulfilled simultaneously. However, one must allow terms with $m = n$ as long as $\beta \neq \alpha$, and terms with $\beta = \alpha$ as long as $m \neq n$. Since $\{m, \beta\} \neq \{n, \alpha\}$ guarantees that we are either dealing with different real parts or strings of different length (i.e. different strings), the third product does not need a restriction.

For later use, we define also the functions for general index n

$$V_n(a) = \frac{a-\frac{n}{2}i}{a+\frac{n}{2}i}. \tag{14.174}$$

The inner product over l can be easily evaluated. First, using (14.169), we obtain

$$\prod_{l=1}^{m} \frac{v_\alpha^{n,k} - v_\beta^{m,l} - i}{v_\alpha^{n,k} - v_\beta^{m,l} + i} = \prod_{l=1}^{m} \frac{v + i(l-1)}{v + i(l+1)} \tag{14.175}$$

where we set

$$v \equiv v_\alpha^{n,k} - v_\beta^{m} - i\frac{m+1}{2} \tag{14.176}$$

as abbreviation. In this form the product is easily seen to be 'telescopic', i.e. most factors in the numerator and denominator of the product cancel and what remains is

$$\prod_{l=1}^{m} \frac{v + i(l-1)}{v + i(l+1)} = \frac{v}{v + i(m+1)} \frac{v+i}{v+im}. \tag{14.177}$$

The remaining two quotients on the right-hand side can be written with the help of the functions V_1 or V_n introduced above

$$\frac{v}{v + i(m+1)} \frac{v+i}{v+im} = V_1\left(\frac{v_\alpha^{n,k} - v_\beta^{m}}{m+1}\right) V_1\left(\frac{v_\alpha^{n,k} - v_\beta^{m}}{m-1}\right) \tag{14.178}$$

$$= V_{m+1}(v_\alpha^{n,k} - v_\beta^{m}) V_{m-1}(v_\alpha^{n,k} - v_\beta^{m}), \tag{14.179}$$

so that we get

$$\left[V_1(v_\alpha^{n,k})\right]^N = \prod_{m=1,m\neq n}^{M} \prod_{\beta=1,\beta\neq\alpha}^{\nu_n} V_1\left(\frac{v_\alpha^{n,k} - v_\beta^m}{m+1}\right) V_1\left(\frac{v_\alpha^{n,k} - v_\beta^m}{m-1}\right) \tag{14.180}$$

$$= \prod_{m=1,m\neq n}^{M} \prod_{\beta=1,\beta\neq\alpha}^{\nu_n} V_{m+1}(v_\alpha^{n,k} - v_\beta^m) V_{m-1}(v_\alpha^{n,k} - v_\beta^m). \tag{14.181}$$

Now we perform the product over $k (k = 1, \ldots, n)$ on both sides of this equation. The left-hand side again contains a telescopic product, so that the left-hand side reduces to

$$\prod_{k=1}^{n} \left[V_1(v_\alpha^{n,k})\right]^N = \left[V_1\left(\frac{v_\alpha^n}{n}\right)\right]^N = \left[V_n\left(v_\alpha^n\right)\right]^N, \tag{14.182}$$

and, thus, we are left with only the real parts of the string on the left-hand side. This yields the compact form of the Bethe ansatz equations

$$\left[V_n\left(v_\alpha^n\right)\right]^N = \prod_{m=1,m\neq n}^{M} \prod_{\beta=1,\beta\neq\alpha}^{\nu_n} V_{nm}(v_\alpha^n - v_\beta^m) \tag{14.183}$$

where

$$V_{nm}(v_\alpha^n - v_\beta^m) = \prod_{k=1}^{n} V_1\left(\frac{v_\alpha^{n,k} - v_\beta^m}{m+1}\right) V_1\left(\frac{v_\alpha^{n,k} - v_\beta^m}{m-1}\right) \tag{14.184}$$

$$= \prod_{k=1}^{n} V_{m+1}(v_\alpha^{n,k} - v_\beta^m) V_{m-1}(v_\alpha^{n,k} - v_\beta^m). \tag{14.185}$$

Still, we have not completely reduced the Bethe ansatz equations such that they contain only the real parts of the strings, which is, after all, our objective. The right-hand side still contains the full complex strings, equation (14.168). These can also be removed, with somewhat more effort, by inserting (14.168) into (14.185). It can then be shown that the Bethe ansatz equations can finally be recast in a form containing only the real parts of the (assumed) strings

$$\left[V_n(v_\alpha^n)\right]^N = \prod_{m=1,m\neq n}^{M} \prod_{\beta=1,\beta\neq\alpha}^{\nu_n} V_{nm}(v_\alpha^n - v_\beta^m), \tag{14.186}$$

where the functions V_{nm} now contain only the real centres of the strings, and are explicitly given by

$$V_{nm}(x) = V_{|n-m|}(x) V^2_{|n-m|+2}(x) \dots V^2_{n+m-2} V_{n+m}(x), \tag{14.187}$$

and

$$x = v^n_\alpha - v^m_\beta. \tag{14.188}$$

At this point we have arrived at a set of equations for the real parts of the strings only—which has been the goal of this (maybe slightly tedious) section.

However, we can obtain a further important result with almost no further work: the Bethe ansatz equations for the real parts of the strings are valid, not only for the isotropic ferromagnetic case $\Delta = -1$, but also for the isotropic antiferromagnetic case $\Delta = 1$. The Bethe ansatz equations (14.64) for the isotropic antiferromagnetic spin chain transform into the corresponding equations (14.129) for the ferromagnetic chain upon complex-conjugating, although for the rapidities v_i^* and v_j^* instead of v_i and v_j. This does not matter, though, since a look at the string structure equations (14.168) and (14.169) reveals that $\left(v_\alpha^{n,j}\right)^*$ and $\left(v_\beta^{m,l}\right)^*$ are the same set of complex numbers as (14.168) and (14.169) themselves. Thus, the Bethe ansatz equations (14.186) are generally valid for the isotropic spin chain $\Delta = \pm 1$.

In a first attempt to use the work invested, let us calculate physical quantities: crystal quasi-momentum and energy of a string solution.

In part V we will see that the tedious work to derive the Bethe ansatz equations for the real parts of the strings was necessary to correctly investigate the finite temperature thermodynamics of the spin chain.

14.5.3.2 Quasi-momentum and energy of a string solution

Starting with the ferromagnetic case $\Delta = -1$, we get for the crystal quasi-momentum from (14.52)

$$K = \sum_{j=1}^{M} k_j \tag{14.189}$$

and a result from exercise 14.5

$$k_j = 2\tan^{-1} 2v_j = \frac{1}{i} \ln \frac{\frac{i}{2} - v_j}{\frac{i}{2} + v_j} \tag{14.190}$$

the intermediate result

$$K = \sum_{j=1}^{M} k_j = \frac{1}{i} \prod_{j=1}^{M} \frac{\frac{i}{2} - v_j}{\frac{i}{2} + v_j}. \tag{14.191}$$

In order to make progress, we shall assume that all rapidities v_j ($j = 1, \dots, M$) belong to a single string of length M, i.e. to an M-string complex, thus

$$v_j = v + \frac{i}{2}(M + 1 - 2j), \quad j = 1, \ldots, M. \tag{14.192}$$

Again, the product is telescopic and we find

$$\prod_{j=1}^{M} \frac{\frac{i}{2} - v_j}{\frac{i}{2} + v_j} = (-1)^M V_1\left(\frac{v}{M}\right), \tag{14.193}$$

and thus

$$K = 2\tan^{-1}\frac{2v}{M} - \frac{\pi}{2}[1 + (-1)^{M+1}]. \tag{14.194}$$

The energy is then from

$$E_M(K) = \frac{1}{2}\sum_{i=1}^{M}\frac{dk_i}{dv_i} = \frac{1}{2}\frac{dK}{dv} \tag{14.195}$$

given by

$$E_M(K) = \frac{1}{M}[1 - (-1)^M \cos K], \tag{14.196}$$

which coincides with the special cases for $M = 1$ and $M = 2$, which we calculated above.

Using the corresponding equations for the isotropic antiferromagnetic spin chain $\Delta = 1$ (cf. 14.3.1), a similar calculation yields the results

$$K = 2\cot^{-1}\frac{2v}{M} \tag{14.197}$$

for the total crystal quasi-momentum, and

$$E_M(K) = -\frac{1}{M}(1 - \cos K) \tag{14.198}$$

for the energy.

Similar calculations can be performed for the case when there are several string solutions. We leave the example of a two-string solution as exercise 14.8.

EXERCISE 14.8 Quasi-momentum and energy of a two-string solution Repeat the calculations of section 14.5.3.2 leading to quasi-momentum and energy of a one-string solution of the Bethe ansatz equations for a two-string solution, i.e. for two strings of length $M - n$ and n, respectively. In this case, the total quasi-momentum splits into two parts

$$K = P + Q \tag{14.199}$$

one part, P say, corresponding to the string of length $M - n$, the other, Q, to the string of length n.

EXERCISE 14.9 Ferromagnetic versus antiferromagnetic spin chain Show that for fixed M and fixed K in the M-spin complex (one string of length M) the excitation energy of the ferromagnetic case is lowest, whereas the excitation energy of the antiferromagnetic case is highest.

14.6 Excitations of the isotropic antiferromagnetic spin chain

For the description of the antiferromagnetic spin chain it is much better to start from a different reference state, not from the totally ferromagnetic state $|0\rangle$. The discussion of the ground state of the spin chain (section 14.4) particularly of the isotropic antiferromagnetic chain in section 14.4.3, where we obtained for the ground state energy of the isotropic antiferromagnetic chain Hulthén's famous result

$$E_{\frac{N}{2}} = -N \ln 2 \tag{14.200}$$

suggests we start from a state that is completely antiferromagnetically ordered, i.e.

$$|AFM\rangle = |\uparrow\rangle_1 |\downarrow\rangle_2 |\uparrow\rangle_3 |\downarrow\rangle_4 \ldots |\downarrow\rangle_{N-1} \uparrow\rangle_N, \tag{14.201}$$

assuming, for simplicity, N to be even. This state is two-fold degenerate. This was the state whose Bethe ansatz solution we derived in section 14.4.3 and then evaluated in the limit $N \to \infty$ to obtain Hulthén's result for the ground state energy.

As noted, from the point of view of the Bethe ansatz equations, a state of the spin chain is characterized by a set of quantum numbers $\{\lambda_j\}$. However, we do not know offhand what state we shall obtain for a freely chosen set of $\{\lambda_j\}$. Thus, great care has to be taken in choosing such a set to guarantee we know which state we are talking about.

The ground state of the isotropic antiferromagnetic spin chain was derived from the choice Hulthén (1938)

$$\lambda_j = 2j - 1, \quad j = 1, 2, \ldots, \frac{N}{2} \tag{14.202}$$

or, written explicity,

$$\lambda_j = 1, 3, 5, \ldots, (N-3), (N-1), \tag{14.203}$$

which was proved by Yang and Yang (1966*a*, 1966*b*, 1966*c*) to be the correct choice. Note, that this set of λ_j's satisfies

$$\lambda_{j+1} = \lambda_j + 2. \tag{14.204}$$

The total spin quantum number S, or just spin S for short, of this state is

$$S = 0. \tag{14.205}$$

More generally, the lowest lying energy for spin S is obtained from the $M = \frac{N}{2} - S$ quantum numbers λ_j chosen, in analogy to (14.202), to be

$$\lambda_j = 2j - 1, \quad j = 1, 2, \ldots, \frac{N}{2} - S, \tag{14.206}$$

i.e. again satisfying

$$\lambda_{j+1} = \lambda_j + 2. \tag{14.207}$$

In general, we require that the $(\frac{N}{2} - S)$ quantum numbers λ_j satisfy

$$0 < \lambda_j < N, \quad \text{and} \quad \lambda_{j+1} \geq \lambda_j + 2. \tag{14.208}$$

for a spin S state (i.e. $M = \frac{N}{2} - S$) of the antiferromagnetic chain (recall that N was chosen to be even for convenience).

Now we ask how to describe low-lying excitations of the antiferromagnetic spin chain. This has not been an easy question to answer. In fact, it took approximately 20 years before the correct picture of certain low-lying excitations of the antiferromagnetic spin chain emerged. The problem as such seems to remain on the agenda; see, e.g., Woynarovich (1982*a*, 1982*b*), Alcaraz *et al.* (1988), and Razumov and Stroganov (2001).

Des Cloizeaux and Pearson 1962 were the first to tackle the problem but found only a partial solution: the spin wave spectrum or excitation spectrum above the antiferromagnetic ground state $\epsilon(q)$

$$E = E_{\frac{N}{2}} + \epsilon(q) = -N \ln 2 + \epsilon(q) \tag{14.209}$$

where

$$\epsilon(q) = \frac{\pi}{2} |\sin q| \,. \tag{14.210}$$

Nevertheless, their result was a definite improvement over the spectrum obtained from the approximate spin wave theory for the antiferromagnetic chain (Anderson, 1952), which yields

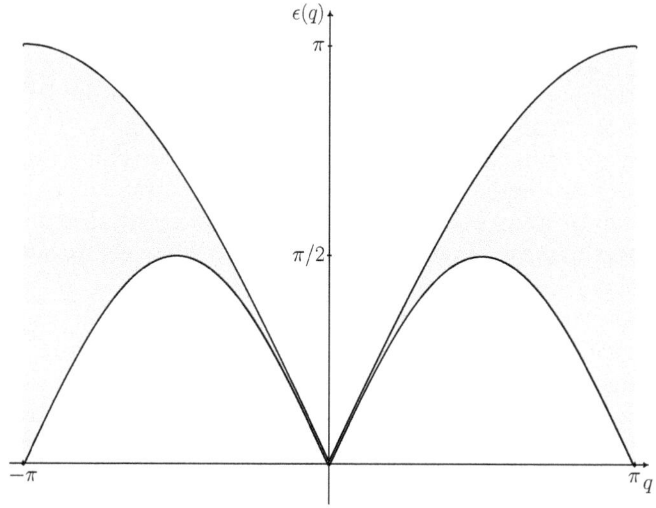

Figure 14.5 *The continuum of low-lying excited states. The lower and upper limits of the continuum are $\frac{\pi}{2}|\sin q|$ and $\pi|\sin\frac{q}{2}|$, respectively ($\mathcal{J}=1$). Shaded area: the continuum of lowest excited states of the isotropic antiferromagnetic chain.*

$$\epsilon_{\text{spinwave}} = |\sin q|. \tag{14.211}$$

It was observed that the spectrum found by des Cloizeaux and Pearson only forms the lower edge of a continuum of states (Yamada, 1969) whose upper edge is given by

$$\pi|\sin\frac{q}{2}|. \tag{14.212}$$

Figure 14.5 shows this continuum.

Next, Faddeev and Takhtajan (1981, 1984) observed that the spin of the excitations in the continuum is in fact neither a singlet ($S = 0$, $S^z = 0$) nor a triplet ($S = 1$, $S^z = -1, 0, 1$), but rather that the spin excitations form spin-$\frac{1}{2}$ objects which have since been named *spinons* in the different but related context (Haldane, 1988; Shastry, 1988*b*) of long-range models where the spinons are exact excitations also for finite chain lengths (we shall see later that this is not so for the short-range model we consider). Spinons are also discussed in two-dimensional physics and are believed to have to do with high-temperature superconductivity, especially the resonating valence-bond picture (Anderson, 1987; Kivelson *et al.*, 1987); see also Michael Gunn's article (1988).

Another closely related phenomenon is the separation of spin and charge in low-dimensional[10] electronic materials. A cartoon picture of the separation of spin and charge in a one-dimensional antiferromagnet would look like this:

10 Any number of dimensions below three is considered low, i.e. dimensions two, one, and, see quantum dots, also zero.

$$\uparrow\downarrow\uparrow\downarrow\uparrow \boxed{\uparrow} \downarrow\uparrow\downarrow\uparrow\downarrow\uparrow\downarrow \boxed{\uparrow\downarrow} \downarrow\uparrow\downarrow\uparrow \tag{14.213}$$

what the spin resides on the domain wall shown enclosed on the left and the charge resides on the doubly occupied site shown enclosed on the right.

After this overview of the history of low-lying excitations and the further announcement that the issues arising in this problem do have nice ramifications for condensed matter physics, let us do a real calculation for the isotropic antiferromagnetic Heisenberg quantum spin chain, based on what we have discussed about the Bethe ansatz quantum numbers λ_j for this model, equations (14.206) and (14.208).

In order to describe an excited state above this ground state, we now choose a distribution different from (14.202) or (14.206) but that still obeys (14.208). This can be done by shifting all quantum numbers to the right by one from a certain index m onwards until, at an index n we shift again by one, i.e. we consider the quantum numbers

$$\lambda_j = 1, 3, \ldots, (2m-1), (2m+2), (2m+4), \ldots, 2n, 2n+3, 2n+5, \ldots, (N-1). \tag{14.214}$$

In other words, we create two *holes* in the set of λ_js. These holes are, however, not fixed at a particular position (m, n), but can move freely.

We introduce a wave number to parametrize the two holes by

$$q = 2\pi\left(\frac{1}{2} - \frac{m+n+1}{N}\right), \tag{14.215}$$

such that

$$-\pi < q \leq \pi \tag{14.216}$$

if m and n run through the values allowed by (14.214).

Note at this point that this choice for the distribution of quantum numbers λ_j and corresponding wave number q will give rise to the continuum of excitations mentioned earlier. Des Cloizeaux and Pearson's 'mistake' was that they introduced only one hole, thereby unwittingly fixing the other hole at the edge of the λ_j distribution.

Now we are in a position to repeat, for the exited state described by (14.214), Hulthén's approach for the ground state in the thermodynamic limit $N \to \infty$. The ground state is proportional to the number of spins N, i.e. it is an extensive quantity. Thus, Hulthén could neglect all contributions proportional to $1/N$ when he transformed the Bethe ansatz equations into an integral equation in the limit $N \to \infty$. The excitation energy we are after now is of order unity. Thus, we must carefully incorporate contributions of order $\frac{1}{N}$ when transforming the Bethe ansatz equations into integral equations in the case of excitations.

With this proviso the analysis runs parallel to the one we discussed in section 14.4.3 for the ground state energy.

We prepare for the thermodynamic limit by introducing a variable x

$$x = \frac{2j-1}{N}, \quad 0 < x < 1 \quad \text{with} \quad j = 1, 2, \ldots, \frac{N}{2} - 1, \tag{14.217}$$

which will become a continuous variable in the limit $N \to \infty$, as in section 14.4.3. Note, though, that we keep the term proportional to $1/N$ in accordance with what we discussed earlier. For the ground state, such a $1/N$ contribution could be safely neglected.

Proper care has to be taken of the λ_j distribution, which can now be written in the variable x as

$$\lambda(x) = \frac{\lambda_j}{N} \tag{14.218}$$

and which, using (14.214), acquires three parts and becomes explicitly

$$\lambda(x) = \frac{2j-1}{N} = x, \quad \text{for} \quad j \leq m, \tag{14.219}$$

$$\lambda(x) = \frac{2j}{N} = x + \frac{1}{N}, \quad \text{for} \quad m < j \leq n, \quad \text{and}, \tag{14.220}$$

$$\lambda(x) = \frac{2j+1}{N} = x + \frac{2}{N}, \quad \text{for} \quad n < j < \frac{N}{2} - 1. \tag{14.221}$$

The quasi-momenta and phases also become functions of the variable x with $0 \leq x \leq 1$

$$k_i \to k(x), \tag{14.222}$$

and

$$\Theta(k_j, k_l) \to \Theta(k(x), k(y)) \equiv \Theta(x, y), \tag{14.223}$$

and the Bethe ansatz equations acquire the form

$$k(x) = 2\pi\lambda(x) + \frac{1}{2}\int_0^1 dy \Theta(x, y), \tag{14.224}$$

where

$$2\cot\frac{\theta(x, y)}{2} = \cot\frac{k(x)}{2} - \cot\frac{k(y)}{2}. \tag{14.225}$$

The energy becomes

$$E = -\frac{N}{2}\int_0^1 dx\,[1 - \cos k(x)]\,. \tag{14.226}$$

In the limit $N \to \infty$, these functions become functions of a continuous variable.

As discussed, the difference between Hulthén's calculation for the ground state and the present calculation of an excited state lies in the quantum numbers (14.214). Therefore, before we proceed as in section 14.4.3, we have to attend to the function $\lambda(x)$, defined in a rather cumbersome way in equations (14.219–14.221). Introducing the two variables

$$x_1 = \frac{2m+1}{N}, \tag{14.227}$$

and

$$x_2 = \frac{2n+1}{N}, \tag{14.228}$$

which describe the positions of the holes and which vary in the range

$$0 < x_1 + x_2 < 2. \tag{14.229}$$

We can write equations (14.219–14.221) in one compact form using the step function $\theta(x)$

$$\lambda(x) = x + \frac{1}{N}(1 + \theta(x - x_1) + \theta(x - x_2)), \tag{14.230}$$

and the quasi-momentum describing the holes

$$q = 2\pi\left(A - \frac{x_1 + x_2}{2}\right), \tag{14.231}$$

where

$$A = 0 \quad \text{for} \quad 0 < x_1 + x_2 < 1, \tag{14.232}$$

and

$$A = 1 \quad \text{for} \quad 1 < x_1 + x_2 < 2, \tag{14.233}$$

such that

$$-\pi < q \leq \pi\,. \tag{14.234}$$

Substituting the phrases (14.225) and (14.230) into (14.224), we obtain

$$k(x) = 2\pi \left[x - \frac{1}{N}(1 + \theta(x - x_1) + \theta(x - x_2)) \right] + \int_0^1 dy \cot^{-1}\left(\frac{1}{2}\cot\frac{k(x)}{2} - \frac{1}{2}\cot\frac{k(y)}{2} \right). \tag{14.235}$$

We have now highlighted what needs to be added to Hulthén's analysis of the ground state in order to calculate a particular class of excitations defined by (14.206). Of course, as we did in section 14.4.3, it will also be useful here to introduce rapidity variables, although this is not necessary. Hulthén himself did not use rapidity variables. However, as one would expect, the solution of the integral equation becomes more involved if one does not use rapidity variables.

However, rather than doing this, we leave the details to the reader as an exercise.

EXERCISE 14.10 Excitation spectrum of the isotropic antiferromagnetic chain Complete the derivation of the integral equations for the excitations of the isotropic antiferromagnetic spin chain along the lines of section 14.4.3.

Next, solve the integral equation and obtain the energy spectrum. Hulthén's original paper (1938) will be most helpful. Also useful are Yamada (1969) and des Cloizeaux and Pearson (1962), who performed this calculation originally. Be aware, however, of the 'mistake' in des Cloizeaux and Person's treatment, which prevented them from getting the continuous spectrum and only its lower edge.

For future reference, we need the final result, the expression for the excitation energy, of exercise 14.10

$$\epsilon = \frac{\pi}{2}\left[|\sin \pi x_1| + |\sin \pi x_2| \right]. \tag{14.236}$$

How can we understand this result?

First, we note that the excitation spectrum indeed describes a continuum of exited states as shown in figure 14.5.

The elementary excitations shown in figure 14.5 were, for a long time, interpreted as spin singlet and triplet excitations. However, in 1981 Faddeev and Takhtajan pointed out that this interpretation was wrong. In response to the question 'What is the spin of a spin wave?', they answered 'It is $\frac{1}{2}$!'

Moreover, they pointed out that these spin-$\frac{1}{2}$ excitations form domain walls in the antiferromagnetic chain and are, thus, kink-like excitations called *spinons*. A cartoon picture of the formation of these excitations has already been shown in (8.540–8.542) of section 8.6.6.4.

Let us repeat the qualitative picture of spinon formation in an antiferromagnetic spin chain. A single spin flipped in a section of the chain that represents a completely antiferromagnetic Néel state

$$\uparrow\ \downarrow\ \uparrow\ \downarrow\ \uparrow\ \downarrow\ \uparrow\ \downarrow\ \boxed{\uparrow\ \uparrow\ \uparrow}\ \downarrow\ \uparrow\ \downarrow\ \uparrow\ \downarrow\ \uparrow\ \downarrow\ \uparrow \tag{14.237}$$

can be split into two kinks or domain walls in the Néel order, i.e. the spin flip excitation becomes spatially separated into spinons carrying half a spin flip with each of the spinons.

$$\uparrow\ \downarrow\ \uparrow\ \downarrow\ \uparrow\ \downarrow\ \boxed{\uparrow\ \uparrow}\ \downarrow\ \uparrow\ \downarrow\ \boxed{\uparrow\ \uparrow}\ \downarrow\ \uparrow\ \downarrow\ \uparrow\ \downarrow\ \uparrow \tag{14.238}$$

The spinons form domain walls or kinks in the chain that can freely move through the lattice

$$\uparrow\ \downarrow\ \uparrow\ \downarrow\ \boxed{\uparrow\ \uparrow}\ \downarrow\ \uparrow\ \downarrow\ \uparrow\ \downarrow\ \uparrow\ \downarrow\ \boxed{\uparrow\ \uparrow}\ \downarrow\ \uparrow\ \downarrow\ \uparrow \tag{14.239}$$

where the separation of the domain walls costs no additional energy or changes the total spin S^z. This configuration has spin $S^z = 1$. From this picture we see that spinons can only be created in pairs.

Recent neutron scattering experiments confirming this picture are summarized in the National Institute of Standards annual report found at: http://www.ncnr.nist.gov/AnnualReport/FY2003_html/RH13/.

Spinons, like kinks and antikinks, can also be formed without flipping a spin from the antiferromagnetic ground state:

$$\uparrow\ \downarrow\ \uparrow\ \downarrow\ \boxed{\uparrow\ \uparrow}\ \downarrow\ \uparrow\ \downarrow\ \uparrow\ \downarrow\ \uparrow\ \boxed{\downarrow\ \downarrow}\ \uparrow\ \downarrow\ \uparrow\ \downarrow\ \uparrow \tag{14.240}$$

Such a two-spinon configuration has total spin 0.

The intuitive picture that has emerged can be supported by the analytical work of Ludwig Faddeev and Leon Takhtajan 1981 'What is the spin of a spin wave?', previously mentioned in section 14.6 (see also Faddeev and Takhtajan (1984)). Their work is based on their development of the quantum inverse scattering method and the algebraic Bethe ansatz introduced in chapters 10 and 11.

15

Bose Gas in One Dimension: Lieb–Liniger Model

All for one and one for all.
Alexandre Dumas (1802–1870)

In the early 1960s Lieb and Liniger (1963) and Lieb (1963) extended the coordinate Bethe ansatz technique to a model of a one-dimensional gas of Bosons. In this model, N Bosons mutually interact with a pair-wise repulsive δ-function potential of strength c on an interval or line of length L, which is repeated through periodic boundary conditions. This was the first demonstration that the coordinate Bethe ansatz technique can be used to treat a quantum many-particle system beyond the realm of the models of Heisenberg quantum spins chains treated so far.

Section 8.1 introduced some basic notions of the theory of weakly interacting Bose gases, specifically the Gross–Pitaevskii equation in 8.1.2 to describe the wave function of the condensate of Bose particles at very low temperatures. This equation has the form of a non-linear Schrödinger equation whose solution will in general be sought in a self-consistent manner.

In one spatial dimension and without an external potential, the Gross–Pitaevskii equation resembles the classical non-linear Schrödinger equation that belongs to the class of classical integrable systems.

The one-dimensional quantum model of which we gave a short rendering above and whose Bethe ansatz solution is arguably simpler than the one of the Heisenberg quantum spin chain, as mentioned has nevertheless only been treated at the beginning of the 1960s. It has regained much attention in recent years due to the experimental developments in the fields of ultracold gases and optical lattices. For recent reviews of these developments see Cazalilla *et al.* (2011) or Batchelor and Foerster (2016).

Section 15.1 begins with a short exposition of the classical non-linear Schrödinger equation, followed by section 15.2, where it is generalized to a quantum model. Section 15.3 explores an important limiting case of the δ-Bose gas, the Tonks–Girardeau gas. Returning to the general δ-Bose gas in section 15.4, we study the consequences of the special form of the δ-function potentials, which will turn out to supply a kind of boundary condition on the wave functions used in section 15.5 to further analyze the eigenfunctions of the Bethe ansatz. Section 15.6 constructs the Bethe ansatz wave function, which

Models of Quantum Matter: A First Course on Integrability and the Bethe Ansatz. Hans-Peter Eckle, Oxford University Press (2019). © Hans-Peter Eckle. DOI: 10.1093/oso/9780199678839.001.0001

we then examine in section 15.7 on the infinite line, and in section 15.8 for periodic boundary conditions. The latter will lead to the Bethe ansatz equations, which determine the momenta of the Bethe ansatz solution. Again, the solution of these equations is a formidable task in itself. We attempt solving them in the thermodynamic limit of an infinite system for the ground state in section 15.9, and for low-lying excited states in section 15.10.

The exposition of this chapter and chapter 17 on the thermodynamic Bethe ansatz for the Bose gas owes much to lecture notes by Anatoli Izergin (2000).

15.1 Classical non-linear Schrödinger equation

The classical non-linear Schrödinger equation is a partial differential equations for the non-relativistic classical field $\psi(x,t)$ in one space and one time dimension

$$i\frac{\partial\psi}{\partial t} = -\frac{\partial^2\psi}{\partial x^2} + 2c\psi^*\psi\psi. \tag{15.1}$$

In the classical case considered here, the sequence of the fields in this expression is inessential. The system derives from the Hamiltonian $\mathcal{H}$ given by

$$\mathcal{H} = \int dx \left(\frac{\partial\psi^*}{\partial x}\frac{\partial\psi}{\partial x} + c\psi^*\psi^*\psi\psi\right). \tag{15.2}$$

Momentum $\mathcal{P}$ and charge $\mathcal{Q}$ are given by

$$\mathcal{P} = \frac{i}{2}\int dx \left(\frac{\partial\psi^*}{\partial x}\psi - \psi^*\frac{\partial\psi}{\partial x}\right) = i\int dx \frac{\partial\psi^*}{\partial x}\psi = -i\int dx\, \psi^*\frac{\partial\psi}{\partial x}, \tag{15.3}$$

$$\mathcal{Q} = \int dx\, \psi^*\psi \tag{15.4}$$

where we have assumed that the fields decay fast enough at infinity so that the partial integration yields the stated result of the second and third equality in the first equation for $\mathcal{P}$. The equation of motion (15.1) can be derived by computing the Poisson brackets of the fields $\psi(x,t)$ and $\psi^*(x,t)$ with the Hamiltonian (15.2).

For two functionals $\mathcal{A}[\psi(x,t),\psi^*(x,t)]$ and $\mathcal{B}[\psi(x,t),\psi^*(x,t)]$ of the fields $\psi(x,t)$ and $\psi^*(x,t)$, the Poisson brackets are defined as

$$\{\mathcal{A},\mathcal{B}\} = -i\int dz \left(\frac{\delta\mathcal{A}}{\delta\psi(z)}\frac{\delta\mathcal{B}}{\delta\psi^*(z)} - \frac{\delta\mathcal{A}}{\delta\psi^*(z)}\frac{\delta\mathcal{B}}{\delta\psi(z)}\right), \tag{15.5}$$

where we assume the time t to be 'fixed', i.e. the Poisson brackets are equal time Poisson brackets. The definition of the functional derivative, i.e. the first variation of the functional, is given by

$$\delta\mathcal{A} = \mathcal{A}\left[\psi(x) + \epsilon\delta\psi(x), \psi^*(x) + \epsilon^*\delta\psi^*(x)\right] - \mathcal{A}\left[\psi(x), \psi^*(x)\right] \tag{15.6}$$

$$= \int dx \left(\frac{\delta\mathcal{A}}{\delta\psi(x)}\delta\psi(x) + \frac{\delta\mathcal{A}}{\delta\psi^*(x)}\delta\psi^*(x)\right). \tag{15.7}$$

where the functional derivatives we require for the Poisson brackets (15.5) are the expressions multiplying the variations of the fields $\delta\psi$ and $\delta\psi^*$. The canonical Poisson brackets for the fields $\psi(x,t)$ and $\psi^*(x,t)$ at equal time become

$$\{\psi(x,t), \psi^*(x',t)\} = -i\delta(x-x'), \tag{15.8}$$

$$\{\psi(x,t), \psi(x',t)\} = \{\psi^*(x,t), \psi^*(x',t)\} = 0. \tag{15.9}$$

With these definitions, the Poisson brackets of the fields ψ and ψ^* with the Hamiltonian (15.2)

$$\partial_t\psi = \{\psi, \mathcal{H}\}, \qquad \partial_t\psi^* = \{\psi^*, \mathcal{H}\} \tag{15.10}$$

produce the equation of motion (15.1) and its complex conjugate equation. Furthermore, we find that

$$-\frac{\partial\psi}{\partial x} = \{\psi, \mathcal{P}\}, \qquad -\frac{\partial\psi^*}{\partial x} = \{\psi^*, \mathcal{P}\}, \tag{15.11}$$

$$-i\psi = \{\psi, \mathcal{Q}\}, \qquad i\psi^* = \{\psi^*, \mathcal{Q}\}. \tag{15.12}$$

Charge $\mathcal{Q}$ and momentum $\mathcal{P}$ are 'commuting' integrals of motion, i.e.

$$\{\mathcal{H}, \mathcal{P}\} = \{\mathcal{H}, \mathcal{Q}\} = 0, \tag{15.13}$$

$$\{\mathcal{P}, \mathcal{Q}\} = 0. \tag{15.14}$$

In fact, there exist infinitely many commuting integrals of motion for the non-linear Schrödinger equation, which is, thus, a classical integrable system.

15.2 Quantum non-linear Schrödinger model

We can obtain a quantum system from the classical non-linear Schrödinger equation by replacing the classical fields ψ and ψ^* by quantum field operators ψ and $\psi^\dagger$. These quantum field operators are defined by a canonical quantization where the classical Poisson brackets (15.8) and (15.9) are replaced by equal time commutators

$$\left[\psi(x,t), \psi^\dagger(x',t)\right] = \delta(x-x'), \tag{15.15}$$

$$\left[\psi(x,t), \psi(x',t)\right] = \left[\psi^\dagger(x,t), \psi^\dagger(x',t)\right] = 0. \tag{15.16}$$

Note the absence of a factor $-i$. This is in correspondence with the transition in quantum mechanics from the classical Poisson bracket $\{p, q\} = 1$ to the corresponding quantum mechanical commutator $[p, q] = -i$.

The quantum non-linear Schrödinger equation describes a gas of Bose particles in one dimension where the interaction between the particles is given by a δ-function potential. Furthermore, we shall demonstrate that the quantum system, the so-called δ-Bose gas or Lieb–Liniger model, is exactly solvable by a Bethe ansatz.

The Hamiltonian and the momentum and charge operator are formally given by the classical expressions (15.2–15.4). In the quantum case, however, it is essential that the field operators are normal ordered, i.e. all annihilation operators ψ must be to the right of the creation operators $\psi^\dagger$.

The quantum field operators ψ and $\psi^\dagger$ act on the Fock space. The vacuum state $|0\rangle$ of the Fock space is defined by the requirement that it is annihilated by the field operator $\psi(x)$

$$\psi(x)|0\rangle = 0 \tag{15.17}$$

while the dual state $\langle 0| \equiv (|0\rangle)^*$ is annihilated by the field operator $\psi^\dagger(x)$

$$\langle 0|\psi^\dagger(x) = 0. \tag{15.18}$$

We can choose the vacuum state to be normalized $\langle 0|0\rangle = 1$. The vacuum state is an eigenstate of the Hamiltonian, the momentum, and the charge operators with eigenvalue zero

$$\mathcal{H}|0\rangle = \mathcal{P}|0\rangle = \mathcal{Q}|0\rangle = 0. \tag{15.19}$$

We now obtain all other states of the Fock space as linear combinations of states created by creation field operators $\psi^\dagger(x)$ acting on the vacuum state

$$|\Psi_N\rangle = \int dx_1 \dots dx_N \, \phi_N(\{x\}) \, \psi^\dagger(x_1)\psi^\dagger(x_2)\dots\psi^\dagger(x_N)|0\rangle, \quad N = 0, 1, 2, \dots \tag{15.20}$$

where $\phi_N(\{x\}) = \phi_N(x_1, x_2, \dots, x_N)$, the coefficient function of the linear combination, is the many-particle wave function of the state with N particles. Due to the commutativity 15.16 of the operators $\psi^\dagger(x_n)$, the wave function is symmetric under a permutation

$$\mathcal{Q}(x_1, x_2, \dots, x_n) = (x_{p_1}, x_{p_2}, \dots, x_{p_N}) \tag{15.21}$$

of the coordinates $\{x\} = x_1, x_2, \dots, x_N$, i.e.

$$\phi_N(x_{p_1}, x_{p_2}, \dots, x_{p_N}) = \phi_N(x_1, x_2, \dots, x_N). \tag{15.22}$$

This expresses the fact that we are dealing with particles obeying Bose–Einstein statistics. Using again the Bose commutation relations (15.15) and (15.16), we can derive the many-particle operators $\mathcal{H}_N$, $\mathcal{P}_N$, and $\mathcal{Q}_N$ from the action of the corresponding second quantized operators $\mathcal{H}$, $\mathcal{P}$, and $\mathcal{Q}$ on the many-particle state (15.20).

EXERCISE 15.1 Many-particle operators Show that the second quantized eigenvalue equations

$$\mathcal{H}|\Psi_N\rangle = E_N|\Psi_N\rangle, \tag{15.23}$$

$$\mathcal{P}|\Psi_N\rangle = P_N|\Psi_N\rangle, \tag{15.24}$$

$$\mathcal{Q}|\Psi_N\rangle = N|\Psi_N\rangle \tag{15.25}$$

imply the many-particle eigenvalue equations

$$\mathcal{H}_N\phi_N = E_N\phi_N, \tag{15.26}$$

$$\mathcal{P}_N\phi_N = P_N\phi_N, \tag{15.27}$$

$$\mathcal{Q}_N\phi_N = N\phi_N \tag{15.28}$$

where the many-particle Hamiltonian, the momentum, and particle number operators are given by

$$\mathcal{H}_N = -\sum_{n=1}^{N}\frac{\partial^2}{\partial x_n^2} + 2c\sum_{1\le n<m\le N}\delta(x_n - x_m), \tag{15.29}$$

$$\mathcal{P}_N = -i\sum_{n=1}^{N}\frac{\partial}{\partial x_n}, \tag{15.30}$$

$$\mathcal{Q}_N = N. \tag{15.31}$$

For $c > 0$, the interaction in the Hamiltonian (15.29) is repulsive, for $c < 0$ the interaction is attractive.

15.3 Lieb–Liniger model in the hard core limit: gas

The hard core limit $c \to \infty$ of the Lieb–Liniger model describes a gas of completely impenetrable Bose particles in one dimension. Tonks (1936) studied this limit for the classical gas, and it was generalized to the quantum regime by Girardeau (1960).

It is quite instructive to consider this limit to gain some insights that will prove useful for the general case of arbitrary finite coupling $c < \infty$. Moreover, the one-dimensional Tonks–Girardeau gas has recently been realized in experiments with optical lattices (Paredes *et al.*, 2004; Kinoshita *et al.*, 2004).

15.3.1 Eigenfunction and spectrum of the Tonks–Girardeau gas

In the limit $c \to \infty$, the eigenfunctions of the Hamiltonian (15.29) must vanish when two coordinates coincide

$$\phi_N(x_1, \ldots, x_i, \ldots, x_j, \ldots, x_N) \to 0, \quad \text{for} \quad x_i \to x_j, \quad i \neq j. \tag{15.32}$$

If all coordinates x_j are different, the eigenfunction must be a solution of the free Schrödinger equation. Section 2.2 discussed a wave function that fits this description, i.e. the Slater determinant (2.61), which we now write as a determinant of an $N \times N$ matrix whose elements are plane waves with parameters $(\{k\}) = (k_1, k_2, \ldots, k_N)$ and which play the role of momenta

$$\phi_N^{(0)}(x_1, \ldots, x_N | k_1, \ldots, k_N) = \frac{1}{\sqrt{N!}} \det \left[e^{ik_m x_n} \right]. \tag{15.33}$$

Note that the parameters $(\{k\})$ may in general be complex numbers. Since we intend to describe a Bosonic system, the Slater determinant, which is an antisymmetric wave function, has to be symmetrized. This can be achieved by multiplying (15.33) by a product of sign functions defined as

$$\text{sign}(x) = +1 \quad \text{for} \quad x > 0, \quad \text{sign}(x) = -1 \quad \text{for} \quad x < 0. \tag{15.34}$$

We have for a product of sign functions under a permutation $\mathcal{R} = (r_1, \ldots, r_N)$ of the positions $(x_1, \ldots, x_N)$, i.e. $(x_{r_1}, x_{r_2}, \ldots, x_{r_N})$

$$\prod_{m<n} \text{sign}(x_n - x_m) = (-1)^{\mathcal{R}} \prod_{m<n} \text{sign}(x_{r_n} - x_{r_m}). \tag{15.35}$$

The same sign change of $(-1)^{\mathcal{R}}$ occurs in the Slater determinant (15.33) under the same permutation of particle positions. Thus, the sign changes will be compensated. We have constructed a symmetric wave function for the Tonks–Girardeau gas

$$\phi_N(x_1, \ldots, x_N | k_1, \ldots, k_N) = \frac{C}{\sqrt{N!}} \det \left[e^{ik_m x_n} \right] \prod_{1 \le m < n \le N} \text{sign}(x_n - x_m) \tag{15.36}$$

where C is a normalization constant. This wave function is a common eigenfunction of the operators $\mathcal{H}_N$ and $\mathcal{P}_N$ with the respective eigenvalues

$$E_N = \sum_{n=1}^{N} k_n^2, \qquad P_N = \sum_{n=1}^{N} k_n. \tag{15.37}$$

The momenta ($\{k\}$) are as yet unrestricted. As usual, we can put restrictions on the momenta by imposing boundary conditions that the eigenfunctions (15.36) must satisfy. The conventional choice, which we also shall employ, are periodic boundary conditions that correspond to putting the system of Bosons on a ring of circumference L. The eigenfunction (15.36) is thus supposed to satisfy

$$\phi_N(x_1,\ldots,x_j=0,\ldots,x_N|\{k\}) = \phi_N(x_1,\ldots,x_j=L,\ldots,x_N|\{k\}) \tag{15.38}$$

for all coordinates $x_j, j=1,2,\ldots,N$. This condition on the wave function can be satisfied if all momenta $k_j, j=1,2,\ldots,N$, are subject to the condition

$$e^{ik_jL} = (-1)^{N-1}, \qquad j=1,\ldots,N, \tag{15.39}$$

which, in turn, is equivalent to

$$k_j = \frac{2\pi}{L}n_j, \qquad j=1,\ldots,N \tag{15.40}$$

where the numbers n_j are arbitrary integers (which have to be odd if the number of particles N is even, and even if N is odd). Again, the Slater determinant vanishes if two or more momenta k_j are equal. Thus, the one-dimensional gas of impenetrable Bosons shows Fermion-like behaviour.

15.3.2 Ground state of the Tonks–Girardeau gas

The quantum states of the Tonks–Girardeau gas are determined by the set of numbers $(\{n_j\}), j=1,\ldots,N$. For the ground state, these numbers are determined in such a way that the energy (15.37) becomes a minimum. This is achieved by choosing a distribution symmetric around zero, i.e.

$$k_j = \frac{2\pi}{L}\left(j-\frac{N+1}{2}\right), \qquad j=1,\ldots,N, \tag{15.41}$$

which yields a total momentum $\mathcal{P}_N = 0$. Doing the sum for the energy E_N and introducing the density of particles $\rho = N/L$, we find for the energy density

$$e_N \equiv \frac{E_N}{L} = \frac{\pi^2}{3}\rho^3 - \frac{\pi^2}{3L^2}\rho. \tag{15.42}$$

In the thermodynamic limit $N\to\infty$ and $L\to\infty$ with $\rho = N/L$ fixed, this becomes $e_N = \pi^2\rho^3/3$.

Let us now return to the general problem of an arbitrary coupling constant $c > 0$.

15.4 δ-potential as boundary condition

The solution of the many-particle problem with Hamiltonian (15.29), i.e. the many-particle wave function, must satisfy three conditions:

1. It must exhibit Bose symmetry;
2. It must be a solution of the Schrödinger equation, which, as we shall see, is equivalent to a boundary condition in configuration space; and
3. Further boundary conditions must be satisfied, for which we shall choose periodic boundary conditions.

The first condition, the symmetry (15.22) of the Bose wave function, implies that we need to construct a solution of (15.26) or (15.29), respectively, only for a certain sector of coordinates in configuration space determined by their ordering by magnitude. The wave function in all other sectors with a different ordering by magnitude of the coordinates can be constructed by permuting the ordering of the coordinates of this solution, i.e. by moving to a different sector.

It is the δ-potential in the Hamiltonian (15.29) that divides the configuration space into these sectors (or domains) along boundaries where two coordinates coincide, e.g. on hyperplanes $x_i = x_j$ for some pair of coordinates labelled i and j. These sectors are determined by an order relation between the coordinates, e.g.

$$x_{q_1} < x_{q_2} < \ldots < x_{q_N} \tag{15.43}$$

for some permutation $Q = (q_1, q_2, \ldots, q_N)$ of the labels $I = (1, 2, \ldots, N)$. We denote these sectors by $F_N^{(Q)}$. As the fundamental sector $F_N^{(I)}$ we choose the (natural) order of the coordinates by increasing magnitude

$$F_N^{(I)} = F_N^{(12\ldots N)} \quad : \quad x_1 < x_2 < \ldots < x_N. \tag{15.44}$$

In the fundamental sector $F_N^{(I)}$ or in any other sector reached by permuting coordinates, the N-particle wave function ϕ_N satisfies the free-particle Schrödinger equation

$$\mathcal{H}_N^{(0)} \phi_N = -\sum_{n=1}^{N} \frac{\partial^2}{\partial x_n^2} \phi_N = E_N \phi_N. \tag{15.45}$$

The wave function will thus be a continuous function of the coordinates $\{x\}$ except when two coordinates coincide. In this case, i.e. for some hyperplane $x_i = x_j$, because of the δ-potential, there will be a discontinuity of the first derivatives of the wave function while the wave function itself remains continuous. Thus, in order to solve the Schrödinger equation, the wave functions must meet these conditions at the boundaries of sectors in configuration space (the second condition indicated at the beginning of this section).

In order to derive these boundary conditions, consider two consecutive variables x_j and x_{j+1} in the fundamental sector $F_N^{(I)}$ with relative coordinate $x = x_{j+1} - x_j$ and centre of mass coordinate $X = (x_{j+1} + x_j)/2$. Then we have

$$\frac{\partial}{\partial x} = \frac{1}{2}\left(\frac{\partial}{\partial x_{j+1}} - \frac{\partial}{\partial x_j}\right), \tag{15.46}$$

$$\frac{\partial}{\partial X} = \frac{\partial}{\partial x_{j+1}} + \frac{\partial}{\partial x_j}, \tag{15.47}$$

$$\frac{\partial^2}{\partial x_j^2} + \frac{\partial^2}{\partial x_{j+1}^2} = \frac{1}{2}\frac{\partial^2}{\partial X^2} + 2\frac{\partial^2}{\partial x^2}, \tag{15.48}$$

$$dx_j\, dx_{j+1} = dx\, dX. \tag{15.49}$$

Using these relations, we now integrate the Schrödinger equation (15.26) over a small interval around $x_{j+1} = x_j$, i.e. over the interval $-\,\epsilon < x < \epsilon$, while all other coordinates $x_n \neq x_j, x_{j+1}$ are left unchanged and remain in the fundamental sector $F_N^{(I)}$. That is, we now look at the boundary of the fundamental sector $F_N^{(I)}$ with a sector $F_N^{(Q)}$ where only the two coordinates x_j and x_{j+1} have been permuted. In the limit $\epsilon \to 0$, we obtain the boundary condition from the δ-potential for the first derivative of the wave function using that wave function itself as continuous across the boundary

$$-\frac{\partial}{\partial x}\phi_N(x)\bigg|_{-\epsilon}^{\epsilon} + c\phi_N(x=0) = 0. \tag{15.50}$$

For reasons which will become clear later, this boundary condition is more conveniently written

$$\left(\frac{\partial}{\partial x_{j+1}} - \frac{\partial}{\partial x_j} - c\right)\phi_N\left(\{x\}\right)\bigg|_{x_{j+1}=x_j+0} = 0, \qquad j = 1, 2, \ldots, N-1. \tag{15.51}$$

The notation $x_j + 0$ stands for $\lim_{\epsilon\to 0}(x_j + \epsilon)$. Note that, in order to derive (15.50) and (15.51), respectively, we used the Bose symmetry of the wave function. Note further that at this stage we could have used the wave functions in any other two sectors $F_N^{(Q')}$ and $F_N^{(Q'')}$ that coincide except for the exchange of two consecutive coordinates $x_{q'_j}$ and $x_{q'_{j+1}}$, which we than take to approach each other $x_{q'_{j+1}} \to x_{q'_j}$.

15.5 Form of the Bethe eigenfunctions

We now seek the eigenfunctions of the Schrödinger equation (15.26) by a coordinate Bethe ansatz in analogy to the Bethe ansatz for the Heisenberg quantum spin chain.

This approach will again be most transparent if we consider in turn the small particle numbers $N = 2$ and $N = 3$.

For $N = 1$ particle, obviously no interaction can take place and we simply deal with a free particle described by a plane wave

$$\phi(x) \propto \exp(ikx) \tag{15.52}$$

with a continuous momentum $P_1 = k$ and energy $E_1 = k^2$.

15.5.1 Two particle case

For $N = 2$ particles, we use a similar ansatz wave function as for the Heisenberg spin chain problem. Away from the boundary $x_1 = x_2$ inside the fundamental sector $F_2^{(I)}$: $x_1 < x_2$, the solution of the Schrödinger equation (15.26) is a superposition of two plane waves, one an incoming, the other an outgoing wave

$$\phi_2^{(12)}(x_1, x_2) = A^{(12)}(12)e^{i(k_1x_1+k_2x_2)} + A^{(12)}(21)e^{i(k_2x_1+k_1x_2)} \tag{15.53}$$

where the momenta $(k_1,\ k_2)$ have been permuted between the two particles with corresponding amplitudes $A^{(12)}(12)$ and $A^{(12)}(21)$. The upper index (12) refers to the fundamental sector $F_2^{(I)}$ and indicates permutations of the ordering of coordinates in configuration space. Together with the wave function in the other sector $F_2^{(Q)}$: $x_1 > x_2$ (labelled by (21)) of the two-particle configuration space, we need to determine $2! \cdot 2! = 4$ amplitudes $A^{(12)}(12)$, $A^{(12)}(21)$ and $A^{(21)}(12)$, $A^{(21)}(21)$. The wave function (15.53) corresponds to a total momentum $P_2 = k_1 + k_2$ and total energy $E_2 = k_1^2 + k_2^2$ of the two particles.

Let us check for the two-particle system whether the required symmetry of the wave function for Bosons indeed implies that we can restrict our analysis to the fundamental sector $F_2^{(I)}$. The two-particle wave function in the sector $F_2^{(Q)}$: $x_1 > x_2$ is

$$\phi_2^{(21)}(x_1, x_2) = A^{(21)}(12)e^{i(k_1x_1+k_2x_2)} + A^{(21)}(21)e^{i(k_2x_1+k_1x_2)}. \tag{15.54}$$

We can combine the wave functions in the two sectors by using appropriate step functions θ

$$\phi_2(x_1, x_2) = \phi_2^{(12)}(x_1, x_2)\theta(x_2 - x_1) + \phi_2^{(21)}(x_1, x_2)\theta(x_1 - x_2). \tag{15.55}$$

This two-particle wave function is now formally valid in the whole configuration space.

Bose symmetry now implies

$$\phi_2(x_1, x_2) = \phi_2^{(12)}(x_1, x_2)\theta(x_2 - x_1) + \phi_2^{(21)}(x_1, x_2)\theta(x_1 - x_2) \tag{15.56}$$

$$= \phi_2(x_2, x_1) = \phi_2^{(12)}(x_2, x_1)\theta(x_1 - x_2) + \phi_2^{(21)}(x_2, x_1)\theta(x_2 - x_1) \tag{15.57}$$

from which we can infer that

$$\phi_2^{(12)}(x_1, x_2) = \phi_2^{(21)}(x_2, x_1) \tag{15.58}$$

and, hence, we obtain for the amplitudes in the two sectors

$$A^{(12)}(12) = A^{(21)}(21), \tag{15.59}$$

$$A^{(12)}(21) = A^{(21)}(12). \tag{15.60}$$

Thus, for two particles, to consider the amplitudes in one sector alone, usually chosen to be the fundamental sector $F_2^{(I)}$, suffices. Of the $2! \cdot 2! = 4$ amplitudes we had initially, we only need to consider $2! = 2$.

Two particles can interact via the δ-potential and we need to take the boundary condition (15.51) into account, which gives us a condition on the amplitudes $A^{(12)}(12)$ and $A^{(12)}(21)$ in the fundamental sector $F_2^{(I)}$ that, until now, have been arbitrary constants. The boundary condition (15.51) implies that the amplitudes satisfy

$$\frac{A^{(12)}(21)}{A^{(12)}(12)} = \frac{k_2 - k_1 + ic}{k_2 - k_1 - ic} \equiv -e^{-i\Theta(k_2 - k_1)} \tag{15.61}$$

where we have defined the phase shift

$$\Theta(k) = -i \ln\left(\frac{1 + \frac{ik}{c}}{1 - \frac{ik}{c}}\right) = 2\tan^{-1}\left(\frac{k}{c}\right), \qquad -\pi < \Theta(k) < \pi, \tag{15.62}$$

which encapsulates the interaction potential and can be interpreted as a scattering phase shift. The limits $k \to \pm\infty$, taken with the correct choice of branch for the logarithm, i.e. its phase $(-1) = e^{\pm i\pi}$ for $k \to \pm 1$, produce the asymptotic values $\Theta \to \pm\pi$. Furthermore, the scattering phase shift is antisymmetric in its argument

$$\Theta(k) = -\Theta(-k). \tag{15.63}$$

We have thus reduced the number of independent amplitudes from $2! \cdot 2! = 4$ to one, namely $A^{(12)}(12)$, by using the Bose symmetry of the wave function and the boundary condition derived from the δ-function potential of the Schrödinger equation.

15.5.2 Three-particle case

For three particles, there are $3! = 6$ sectors of the configuration space distinct by the ordering by magnitude of the coordinates. Again, the fundamental sector may be chosen as $F_3^{(I)} : x_1 < x_2 < x_3$ with other sectors reached by permutations of the order relation of the fundamental sector.

Moreover, there are $3! = 6$ arrangements of the plane waves

$$\exp i(k_{p_1}x_1 + k_{p_2}x_2 + k_{p_3}x_3) \tag{15.64}$$

from which we build the wave function with appropriate amplitudes. Hence, we have to determine $3! \cdot 3! = 36$ amplitudes in the three-particle case.

EXERCISE 15.2 Amplitudes for three particles 1 Convince yourself that the conclusions from Bose symmetry we found for two Bose particles also hold for three Bose particles. In particular, show that the amplitudes are related by

$$A^{(ijk)}(abc) = A^{(P(ijk))}(P(abc)) \tag{15.65}$$

where the upper index (ijk) refers to one of the 3! sectors (the sector (123) is the fundamental sector $F_3^{(I)}$ for three particles) while abc labels the 3! amplitudes within a given sector. Note that the permutation on the right-hand side of the equation is the same for sector and amplitude within the sector.

For example, (15.65) means that

$$A^{(123)}(123) = A^{(213)}(213) = A^{(231)}(231) = A^{(132)}(132) = A^{(312)}(312) = A^{(321)}(321), \tag{15.66}$$

$$A^{(213)}(123) = A^{(123)}(213) = \ldots, \tag{15.67}$$

$$\ldots.$$

Hence, all amplitudes of sectors different from the fundamental sector $F_3^{(I)} = (123) : x_1 < x_2 < x_3$ can be related to the amplitudes $A^{(123)}(P(123))$ of the fundamental sector (123). Of the initially $3! \cdot 3!$ amplitudes only 3! independent ones remain.

As we have already indicated, the wave function $\phi_3(x_1, x_2, x_3)$ in the fundamental sector $F_3^{(I)} : x_1 < x_2 < x_3$ will again be assumed by the Bethe ansatz hypothesis to be a superposition of plane waves where the momenta are permuted in the individual plane waves

$$\phi_3^{(123)}(x_1, x_2, x_3) = \sum_P A^{(123)}(P) \exp\left(i \sum_{n=1}^{3} k_{p_n} x_n\right). \tag{15.68}$$

The sum is over the $3! = 6$ permutations $(p_1 p_2 p_3)$ corresponding to the $3! = 6$ ways of assigning the three-momenta (k_1, k_2, k_3) to the three particles in the superposition of plane waves with respective amplitudes $A^{(123)}(P)$. This gives the energy and momentum of the three particles in terms of the momenta (k_1, k_2, k_3)

$$E_3 = k_1^2 + k_2^2 + k_3^2, \qquad P_3 = k_1 + k_2 + k_3. \tag{15.69}$$

EXERCISE 15.3 Amplitudes for three particles 2 Calculate the ratios of the amplitudes of the three–particle wave function (15.68) using the boundary condition (15.51).

Show especially that the ratio of the amplitudes for the two permutations $P \equiv I = (123)$ and $P' = (321)$ are given by

$$\frac{A^{(123)}(P')}{A^{(123)}(I)} = \frac{A^{(123)}(321)}{A^{(123)}(123)} = (-1)^3 \exp\left(-i\Theta(k_2 - k_1) - i\Theta(k_3 - k_2) - i\Theta(k_3 - k_1)\right). \tag{15.70}$$

Moreover, show that there are two distinct possibilities to reach this result corresponding to the two possible paths to go from permutation $P' = (321)$ to $I = (123)$.

The foregoing exercise tells us that the $3! = 6$ amplitudes $A^{(123)}(P(123))$ are not completely fixed by the six equations derived from the δ-function boundary conditions. There will always be one amplitude that can be freely chosen. The choice of this amplitude then fixes the remaining five amplitudes. In other words, as the exercise also shows, the sequence of transpositions to go from (321) to (123) does not matter.

If, in the next step, we considered four Bose particles, we would find that the $4! = 24$ amplitudes, which remain to be considered after the Bose symmetry has been exploited, are connected by $4! \cdot (4-1)/2 = 36$ equations from the δ-potential boundary conditions. This overdetermination of the amplitudes is, however, only apparent. Again, the combined scattering phase in the amplitudes will be independent of the sequence of transpositions of neighbouring indices required to go from one plane wave permutation to another one, i.e which of the several possible distinct paths we choose. Therefore, the 36 equations fix the amplitudes only up to one amplitude, which remains to be freely chosen. This observation, which is at the heart of Bethe ansatz solvable models, has been spelled out for the first time for one-dimensional N-particle problems by McGuire (1964).

So far in this section, the conclusions derived from Bose symmetry and the Schrodinger equation for the form of the Bethe ansatz wave function also hold for a general particle number N; note that for the Schrodiner equations, they take the form of boundary conditions in the configuration space divided into sectors by the δ-interaction potential between the Bose particles.

It is also interesting to note that the posited Bose symmetry of the problem relates the amplitudes in different sectors of the configuration space, while the Schrödinger equation relates the amplitudes within a given sector.

15.5.3 N particle case

The direct generalization to the case of N Bose particles in the fundamental sector $F_N^{(I)}: x_1 < x_2 < \ldots < x_N$ is the Bethe ansatz wave function

$$\phi_N^{(12\ldots N)}(x_1, x_2, \ldots, x_N) = \sum_P A^{(12\ldots N)}(P) \exp\left(i \sum_{n=1}^{N} k_{p_n} x_n\right). \tag{15.71}$$

As in the cases $N = 2$ and $N = 3$, the amplitudes $A^{(Q)}(P)$ for sectors other than the fundamental sector $F_N^{(I)}$ can be related to the amplitudes $A^{(I)=(12...N)}(P(12...N))$ in the fundamental sector by exploiting the Bose symmetry. Moreover, also as in the cases $N = 2$ and $N = 3$, the boundary condition (15.51) can be used to derive conditions the amplitude ratios have to satisfy. Considering the boundary $x_{j+1} \rightarrow x_j$, we obtain a ratio between the amplitudes $A^{(12...N)}(P)$ and $A^{(12...N)}(P')$ where the corresponding permutations only differ by the exchange of the two entries p_j and p_{j+1}

$$P = (p_1 \ldots, p_j, p_{j+1}, \ldots, p_N) \rightarrow P' = (p_1 \ldots, p_{j+1}, p_j, \ldots, p_N) = P_{j,j+1}P \qquad (15.72)$$

where $P_{j,j+1}$ is a transposition of the entries at positions j and $j + 1$ in the permutation. As in the two particle case (15.61), the corresponding amplitude ratio is

$$\frac{A^{(12...N)}(P')}{A^{(12...N)}(P)} = \frac{k_{p_{j+1}} - k_{p_j} + ic}{k_{p_{j+1}} - k_{p_j} - ic} = -\exp\left(-i\Theta(k_{p_{j+1}} - k_{p_j})\right). \qquad (15.73)$$

Since $A^{(12...N)}(P')$ and $A^{(12...N)}(P)$ only differ by the exchange of one pair of momenta with successive labels p_j and p_{j+1}, it is suggestive to assume the form of $A^{(12...N)}(P)$ itself as the product of terms of the form

$$k_{p_{j+1}} - k_{p_j} + ic = \left(k_{p_{j+1}} - k_{p_j}\right)\left(1 + \frac{ic}{k_{p_{j+1}} - k_{p_j}}\right). \qquad (15.74)$$

Indeed, if we write for some permutation Q

$$A^{(12...N)}(Q) = C(-1)^Q \prod_{\substack{m,n=1 \\ m<n}}^{N} \left(1 + \frac{ic}{k_{p_m} - k_{p_n}}\right), \qquad (15.75)$$

with an arbitrary constant C, then for the amplitudes with the two permutations $Q = P$ and $Q = P' = P_{j,j+1}P$ the amplitude ratio (15.73) will be reproduced.

Again, evaluating the boundary conditions (15.51) for all $N - 1$ boundaries of successive coordinates $x_{j+1} \rightarrow x_j$ $(j = 1, \ldots, N - 1)$ will lead to $N!(N - 1)/2$ equations for the $N!$ amplitudes $A^{(I)}(P)$. Also in the general case, the overdetermination of the amplitudes is only apparent since the combined scattering phases are path-independent, i.e. independent of the sequence of transpositions between different permutations.

15.6 Construction of the Bethe ansatz wave function

Recalling our experience in section 15.3 with the impenetrable Bose gas, i.e. with the $c \rightarrow \infty$ case we now demonstrate a constructive way to arrive at the Bethe ansatz wave function. This procedure will prove very instructive and reveal the details of the Bethe ansatz wave function in a transparent way.

The starting point of the construction of the Bethe ansatz wave function for N interacting Bosons is again, as in the case of impenetrable Bosons, the Slater determinant (15.33)

$$\phi_N^{(0)}(\{k\}|\{x\}) = \frac{C}{\sqrt{N!}} \det\left(\exp\left[ik_m x_n\right]\right) = \frac{C}{\sqrt{N!}} \sum_{\mathcal{P}} (-1)^{\mathcal{P}} \exp\left[i \sum_{m=1}^{N} k_{p_m} x_m\right], \quad (15.76)$$

where C is a normalization constant that we will determine as we go along. The momenta $\{k\}$ may be complex numbers.

The Slater determinant satisfies the free Schrödinger equation in the fundamental sector $F_N^{(I)}$ or in any other sector reached by permuting the ordering of the coordinates

$$\mathcal{H}_N^{(0)} \phi_N^{(0)} = E_N \phi_N^{(0)}, \qquad E_N = \sum_{n=1}^{N} k_n^2 \quad (15.77)$$

as long as we are away from the boundaries where coordinates coincide. However, the Slater determinant is antisymmetric under permutations of the coordinates and it does not satisfy the boundary condition (15.51). In fact, the Slater determinant and all its derivatives with respect to the coordinates $\{x\}$ are continuous functions of these coordinates.

In order to satisfy the boundary condition (15.51), we modify the Slater determinant by a product of differential operators

$$\phi_N^{(1)}(\{k\}|\{x\}) = \left(\prod_{\substack{m,n=1 \\ N \geq m > n \geq 1}}^{N} \left(\frac{\partial}{\partial x_m} - \frac{\partial}{\partial x_n} + c \right) \right) \phi_N^{(0)}(\{k\}|\{x\}). \quad (15.78)$$

The partial derivatives only produce extra factors that depend on the momenta $\{k\}$. Thus, this modified Slater determinant also satisfies the free Schrödinger equation with the same eigenvalue.

In order to check the boundary condition (15.51), we apply the differential operator, which defines the boundary condition to the modified Slater determinant

$$\left(\frac{\partial}{\partial x_{j+1}} - \frac{\partial}{\partial x_j} - c \right) \phi_N^{(1)} = \left(\left[\frac{\partial}{\partial x_{j+1}} - \frac{\partial}{\partial x_j} \right]^2 - c^2 \right) \bar{\phi}_N^{(1)} \quad (15.79)$$

where, in order to produce the differential operator on the right-hand side, we have singled out the appropriate differential operator from the product of differential operators defining the modified Slater determinant. The remaining wave function can be written as

$$\bar{\phi}_N^{(1)} = \left(\prod_{\substack{m,n=1 \\ N\geq m>n\geq 1}}^{N} \left(\frac{\partial}{\partial x_m} - \frac{\partial}{\partial x_n} + c \right) \right)' \phi_N^{(0)} \tag{15.80}$$

where the prime indicates that the particular differential operator $\left(\frac{\partial}{\partial x_{j+1}} - \frac{\partial}{\partial x_j} + c\right)$ is not included in the product. However, there are still factors in the primed product that contain either the derivative $\partial/\partial x_j$ or $\partial/\partial x_{j+1}$. These derivatives always appear together in the first or second position of the differential operators of which the product of the operators is composed. This observation implies that the primed product of differential operators is symmetrical under an exchange of x_j and x_{j+1}. Since the Slater determinant wave function $\phi_N^{(0)}$ is an antisymmetric wave function under $x_j \leftrightarrow x_{j+1}$, the function $\bar{\phi}_N^{(0)}$ will also be antisymmetric, as will the right-hand side of (15.79). Because of this antisymmetry, the right-hand side of (15.79) vanishes for $x_j = x_{j+1}$, which shows that $\phi_N^{(1)}$ satisfies the boundary condition (15.51) imposed by the δ-potential.

The wave function $\phi_N^{(1)}$ is thus the solution ϕ_N of the Schrödinger equation in the fundamental sector $F_N^{(I)}$

$$\phi_N(\{x\}|\{k\}) = \phi_N^{(1)}((\{x\}|\{k\})). \tag{15.81}$$

Finally, we must symmetrize $\phi_N^{(1)}$ such that it becomes a symmetrical wave function solving the Schrödinger equation in all sectors reached from $F_N^{(I)}$ by arbitrary permutations of the ordering of the positions $\{x\} = (x_1, \ldots, x_N)$.

Executing the product of differential operators in (15.78) explicitly gives in the fundamental sector $F_N^{(I)}$

$$\phi_N = \frac{i^{N(N-1)/2} C}{\sqrt{N!}} \sum_{\mathcal{P}} (-1)^{\mathcal{P}} \exp\left[i \sum_{m=1}^{N} k_{\mathcal{P}m} x_m \right] \prod_{m>n} \left(k_{\mathcal{P}_m} - k_{\mathcal{P}_n} - ic \right). \tag{15.82}$$

We again take a cue from the impenetrable case, $c \to \infty$, where we symmetrized the wave function with the help of a product of sign functions (cf. (15.36)). In order to extend the wave function of the present case outside of the fundamental sector $F_N^{(I)}$ to any other sector, we introduce the sign function into the product of differential operators in (15.79) which, omitting the constant factor for the moment, explicitly leads to

$$\phi_N = \sum_{\mathcal{P}} (-1)^{\mathcal{P}} \exp\left[i \sum_{m=1}^{N} k_{\mathcal{P}m} x_m \right] \prod_{m>n} \left(k_{\mathcal{P}_m} - k_{\mathcal{P}_n} - ic\, \mathrm{sign}(x_m - x_n) \right). \tag{15.83}$$

In order to show that this wave function is indeed symmetric under an arbitrary permutation of the positions, we need to verify that

$$\phi_N(\{\mathcal{Q}x\}|\{k\}) = \phi_N(\{x\}|\{k\}) \tag{15.84}$$

for some arbitrary permutation $\mathcal{Q}x = x_{q_1}, \ldots, x_{q_N}$ of the positions. The wave function, again disregarding the constant factor, then becomes by definition of the permutation $\mathcal{Q}$

$$\sum_{\mathcal{P}}(-1)^{\mathcal{P}} \exp\left[i\sum_{m=1}^{N} k_{p_m}x_{q_m}\right]\prod_{m>n}\left(k_{p_m} - k_{p_n} - ic\,\mathrm{sign}(x_{q_m} - x_{q_n})\right). \tag{15.85}$$

A permutation can always be written as the product of two or more other permutations. Using this property, we introduce a permutation $\mathcal{P}'$ such that we obtain for the permutation of the momenta $\mathcal{P} = \mathcal{Q}\mathcal{P}'$. The summation in the wave function over all permutations $\mathcal{P}$ is then equivalent to the summation over all permutations $\mathcal{P}'$. Furthermore, we have

$$(-1)^{\mathcal{P}} = (-1)^{\mathcal{Q}+\mathcal{P}'}, \qquad \sum_{m=1}^{N} k_{p_m}x_{q_m} = \sum_{m=1}^{N} k_{p'_m}x_m, \tag{15.86}$$

and

$$\prod_{m>n}\left(k_{p_m} - k_{p_n} - ic\,\mathrm{sign}(x_{q_m} - x_{q_n})\right)$$
$$= (-1)^{\mathcal{Q}}\prod_{m>n}\left(k_{p'_m} - k_{p'_n} - ic\,\mathrm{sign}(x_m - x_n)\right), \tag{15.87}$$

which together prove the symmetry of the wave function under an arbitrary permutation of the positions of the particles.

The last step in the construction of the wave function is the determination of the normalization constant C which may, and indeed does depend on the momenta $\{k\}$ and the coupling constant c. The constant C consists of two factors, the product

$$\left(\prod_{m<n}\left[(k_{p_m} - k_{p_n})^2 + c^2\right]\right)^{\frac{1}{2}} = \left(\prod_{m<n}\left[k_{p_m} - k_{p_n} + ic\right]\left[k_{p_m} - k_{p_n} - ic\right]\right)^{\frac{1}{2}} \tag{15.88}$$

and a pure phase $(-1)^{N(N-1)/2}$. The latter phase factor cancels the phase factor in (15.82) together with the phase factor acquired after inserting (15.88) into the wave function (15.82). We thus arrive at the Bethe ansatz wave function valid in all sectors of the configuration space

$$\phi_N(\{x\}|\{k\}) = \frac{1}{\sqrt{N!}} \sum_{\mathcal{P}} (-1)^{\mathcal{P}} \exp\left[\sum_{m=1}^{N} k_{\mathcal{P}m} x_m\right]$$
$$\cdot \exp\left[\frac{i}{2} \sum_{m>n} \text{sign}(x_m - x_n)\Theta(k_{\mathcal{P}m} - k_{\mathcal{P}n})\right] \tag{15.89}$$

where $\Theta(k)$ is given by (15.62). This wave function is called the Bethe eigenfunction or the Bethe wave function. The corresponding eigenstate can be viewed as consisting of N Bose particles with momenta $p_0(k_n) = k_n$ and eigenenergies $\varepsilon_0(k_n) = k_n^2$. These quantities are also often called bare momenta and bare energy of the particles.

EXERCISE 15.4 Bethe wave function Convince yourself that our choice of the constant C leads to the final form of the Bethe wave function (15.89).

Although the Bethe wave function for the Bose gas with δ-interaction is symmetric under an arbitrary permutation of the particles, i.e. under an arbitrary permutation of the positions of the particles, it is antisymmetric under an arbitrary permutation of the momenta

$$\phi_N(\{x\}|\{\mathcal{Q}k\}) = (-1)^{\mathcal{Q}} \phi_N(\{x\}|\{k\}). \tag{15.90}$$

As a continuous function of all momenta $\{k\}$, it thus vanishes if two of the momenta coincide. This reflects a kind of Pauli principle for interacting Bosons in one space dimension that requires that all momenta must be mutually different $k_m \neq k_n$ for all pairs of mutually distinct particles $m \neq n$.

The following sections investigate the Bethe ansatz wave function on the whole configuration space $-\infty < x_n < \infty$ and for periodic boundary conditions where the configuration space is restricted to $0 < x_n < L$. These considerations will especially restrict the allowed values of the momenta $\{k\}$ and, hence, the energy spectrum E_N of the Bose system. We focus on the energy spectrum in these cases and under what conditions bound states, characterized by complex momenta, are possible.

15.7 Unrestricted configuration space

If the particles are free to move in the whole configuration space $-\infty < x_n < \infty$, the physically natural boundary condition is the requirement that the wave function remains finite at infinity. This restriction will determine the allowed values of the momenta and therefore the energy $E_N = \sum_{n=1}^{N} k_n^2$. Let us, in turn, look at the cases of $N = 1$ and $N = 2$ particles.

15.7.1 Case N=1

The wave function for $N = 1$ is just a simple plane wave

$$\phi_1(x|k) = e^{ikx}. \tag{15.91}$$

This wave function would grow without bounds for a non-vanishing imaginary part of the momentum. For $\Im k > 0$, we would have an unbounded wave function as $x \to +\infty$, while for $\Im k < 0$ the wave function would grow without limit as $x \to -\infty$. Hence, for a physical wave function, k must be real. For momentum and energy spectrum, we have

$$-\infty < P_1 = k < \infty, \qquad 0 \le E_1 = k^2 < \infty. \tag{15.92}$$

This is the spectrum we expect of a single free particle.

More interesting is the case of two particles, which can indeed interact.

15.7.2 Case N=2

For two particles, $N = 2$, the symmetrized wave function (15.84), disregarding an unimportant constant, becomes

$$\begin{aligned}\phi_2(x_1, x_2|k_1, k_2) = {} & e^{ik_1x_1+ik_2x_2}\,(k_2 - k_1 - ic\,\mathrm{sign}(x_2 - x_1)) \\ & - e^{ik_2x_1+ik_1x_2}\,(k_1 - k_2 - ic\,\mathrm{sign}(x_2 - x_1)).\end{aligned} \tag{15.93}$$

This wave function can be rewritten using centre of mass and relative coordinates $2X = x_1 + x_2$ and $x = x_2 - x_1$

$$\begin{aligned}\phi_2(x_1, x_2|k_1, k_2) = e^{i(k_1+k_2)X}\Big[& e^{i(k_2-k_1)x/2}\,(k_2 - k_1 - ic\,\mathrm{sign}(x)) \\ & + e^{-i(k_2-k_1)x/2}\,(k_2 - k_1 + ic\,\mathrm{sign}(x))\Big].\end{aligned} \tag{15.94}$$

For real momenta, $\Im k_1 > \Im k_2 = 0$, the wave function is bounded and we obtain a continuous spectrum

$$-\infty < P_2 = k_1 + k_2 < \infty, \qquad 0 \le E_2 = k_1^2 + k_2^2 < \infty. \tag{15.95}$$

Furthermore, to keep the factor $e^{i(k_1+k_2)X}$ and thus the wave function bounded, the total momentum must be real, i.e. $\Im(k_1 + k_2) = 0$. The only possibility left for complex momenta is thus that they are complex conjugates

$$k_1 = k + i\kappa, \qquad k_2 = k - i\kappa \tag{15.96}$$

where the imaginary part may be chosen $\kappa > 0$ without loss of generality. The wave function then becomes

$$\phi_2 = e^{2ikX}\left[e^{\kappa x}\left(-2i\kappa - ic\,\mathrm{sign}(x)\right) + e^{-\kappa x}\left(-2i\kappa + ic\,\mathrm{sign}(x)\right)\right]. \tag{15.97}$$

The condition for this wave function to remain bounded is that $\kappa = -c/2$ for both sectors, $x > 0$ and $x < 0$. Since we have $\kappa > 0$, this implies $c < 0$, which corresponds to an attractive interaction potential. For a repulsive interaction potential, the condition $\kappa = -c/2$ cannot be satisfied.

The wave function for the bound state is given by a plane wave of the centre of mass coordinate and an exponential damping with damping length proportional to $|c|^{-1}$

$$\phi_2 = 2i|c|e^{2ikX}e^{-|c|x/2}. \tag{15.98}$$

Its momentum and energy are

$$P_2 = k_1 + k_2 = 2k, \qquad E_2 = k_1^2 + k_2^2 = 2k^2 - 2\kappa^2 = \frac{P_2^2}{2} - \frac{|c|^2}{2}. \tag{15.99}$$

The first term in the energy E_2 represents the centre of mass energy of the two-particle bound state, the second is the binding energy, which lowers the energy.

We have thus learned that, as expected, the repulsive Bose gas with δ-interaction supports only particles while the attractive Bose gas also supports bound states of particles. For N particles, the picture is similar to the two-particle case. In particular, the bound states formed in the attractive case $c < 0$ can be described by complex momenta forming strings.

15.7.3 Scattering matrix

Especially in the repulsive case $c > 0$ of the Bose gas, where the particles do not form bound states, we can usefully describe the gas by scattering events between particles. Let us first consider again the case of two particles.

In the distant past $t \to -\infty$, if we consider the fundamental sector $F_2^{(I)} : x_1 < x_2$, we may choose momentum k_1 for the particle with coordinate x_1 and momentum k_2 for the particle with coordinate x_2, which are, furthermore, both real momenta satisfying $k_1 > k_2$. This situation is described by the first part of the wave function (15.94), an incoming wave, with no interaction at $t \to -\infty$. The two particles will eventually scatter and after the scattering event, in the distant future $t \to \infty$, the momenta k_1 and k_2 of the particles will have been exchanged in the outgoing wave when there will again be no interaction. For this scattering event we define the scattering matrix as the ratio of the amplitude of the outgoing wave to the amplitude of the incoming wave, i.e. according to (15.94)

$$S(k_1, k_2) = S(k_1 - k_2) = \frac{k_1 - k_2 - ic}{k_1 - k_2 + ic} = -e^{i\Theta(k_1 - k_2)}. \tag{15.100}$$

This result justifies the interpretation of the phase $\Theta(k)$ as scattering phase shift. Due to the Bose symmetry of the wave function (15.94), we obtain the same result also for the sector $F_N^{(Q)}: x_1 > x_2$ while keeping the ordering of the momenta as $k_1 > k_2$.

Similarly, we can use the wave function (15.84) or (15.90) to describe the scattering of N particles with real momenta $k_1 > k_2 > \ldots > k_N$. We proceed from an incoming wave function in the fundamental sector $F_N^{(I)}: x_1 < x_2 < \ldots < x_N$ where the particle with coordinate x_j has momentum k_j and is described by one term in (15.84) with all sign functions positive, which has no interaction in the distant past at $t \to -\infty$. The outgoing wave is also described by one term in the fundamental sector $F_N^{(I)}$ but with the momenta exchanged in (15.84). The total scattering matrix is then a product of mutual two-particle scattering matrices

$$S(k_1, k_2, \ldots, k_N) = \prod_{1 \le m < n \le N} S(k_m, k_n). \tag{15.101}$$

The number of pair scattering events between the N particles is $N(N-1)/2$, which is reflected by the number of two-particle scattering matrices in the product of the total scattering matrix.

Section 15.5 demonstrated that the decisive property for the Bethe ansatz solvability of the one-dimensional Bose gas with δ-function potential was the property that the amplitude relations were independent of the sequence of transpositions that lead from permutation (321) to (123) (see especially exercise 15.3)

$$A^{(I)}(321) = -e^{-i\Theta_{32}-i\Theta_{31}-i\Theta_{21}}A^{(I)}(123) = -e^{-i\Theta_{21}-i\Theta_{31}-i\Theta_{32}}A^{(I)}(123) \tag{15.102}$$

with $\Theta_{ij} = \Theta(k_i - k_j)$ and its generalization for N Bose particles. This translates into a form of Yang–Baxter relations for the scattering matrix (15.101)

$$S(k_3, k_2)S(k_3, k_1)S(k_2, k_1) = S(k_2, k_1)S(k_3, k_1)S(k_3, k_2), \tag{15.103}$$

which is trivially satisfied for the Bose gas since the scattering matrix is a scalar. The property was nevertheless important and will become more so for models where the corresponding quantity is indeed a matrix and (15.104) is a highly non-trivial statement. The Heisenberg quantum spin chain is a case in point as are the electronic models we discuss later.

15.8 Periodic boundary conditions: Bethe ansatz equations

For applications of the δ-Bose gas in condensed matter physics, our discussion so far of particles that may scatter but are otherwise unrestricted on the infinite line $-\infty < x_n < \infty$ is not the most useful. Rather, what is needed is a situation where the gas is considered with a fixed density $\rho = N/L$.

The standard way to obtain this in condensed matter physics, as for the Tonks–Girardeau gas in section 15.3, is to put the system in a finite interval of length L with a finite number of particles. For such a finite system all quantities, e.g. the ground state energy, are well defined in the sense that no infinities can occur. We then need to impose boundary conditions at the end of the interval. There are different choices for these boundary conditions. For instance, we may choose hard walls, i.e. infinite potentials, at the ends of the interval.

The allowed values of the momenta will depend on the chosen boundary conditions. The most convenient choice of boundary conditions are periodic boundary conditions, which we have already used for other systems too. We choose this kind of boundary condition now.

After adapting the solution of the δ-Bose gas to periodic boundary conditions, we shall study the ground state of the system in the thermodynamic limit where we let $N \to \infty$ and $L \to \infty$ in such a way that the particle density $\rho = N/L$ is kept fixed at a finite value.

Periodic boundary conditions require that the N particles are restricted to a box of length L with $0 \leq x_n \leq L$ for $n = 1, 2, \ldots, N$ and that the wave function is periodic in each coordinate x_n in turn, while all other coordinates x_m with $m \neq n$ are kept fixed, i.e. for $n = 1, \ldots, N$ the periodic boundary conditions are

$$\phi_N(x_1, \ldots, x_n = L, \ldots, x_N|\{k\}) = \phi_N(x_1, \ldots, x_n = 0, \ldots, x_N|\{k\}). \tag{15.104}$$

These conditions will now be applied to the Bethe ansatz wave functions (15.84) or, equivalently, (15.90). It will result in a system of equations for the allowed values of the momenta $\{k\}$. These equations are called Bethe ansatz equations of just Bethe equations.

Again, it is useful to start with only one and two particles to better understand how the Bethe ansatz equations come about and work.

15.8.1 Case N=1

For $N = 1$, the wave function is essentially given by $\phi_1(x, k) = e^{ikx}$ and thus the periodicity condition (15.105) is simply $e^{ikL} = 1$. This is, as expected, the well-known quantization condition for a free particle in a box

$$k_\ell = \frac{2\pi \ell}{L}, \qquad \ell = 0, \pm 1, \pm 2, \ldots. \tag{15.105}$$

More interesting will be again the case $N = 2$.

15.8.2 Case N=2

Now the wave function is given by (15.94)

$$\begin{aligned} \phi_2(x_1, x_2|k_1, k_2) = {} & e^{ik_1x_1 + ik_2x_2} \left(k_2 - k_1 - ic\, \mathrm{sign}(x_2 - x_1)\right) \\ & - e^{ik_2x_1 + ik_1x_2} \left(k_1 - k_2 - ic\, \mathrm{sign}(x_2 - x_1)\right). \end{aligned} \tag{15.106}$$

The periodicity conditions (15.104) furnish two relations for the two momenta k_1 and k_2, which are

$$e^{ik_1L} = \frac{k_1 - k_2 + ic}{k_1 - k_2 - ic} = -e^{-i\Theta(k_1-k_2)}, \tag{15.107}$$

$$e^{ik_2L} = \frac{k_2 - k_1 + ic}{k_2 - k_1 - ic} = -e^{-i\Theta(k_2-k_1)}. \tag{15.108}$$

We note that multiplying these two equations, because of the antisymmetry of the scattering phase shift $\Theta(k)$, gives

$$e^{i(k_1+k_2)L} = e^{iP_2L} = 1. \tag{15.109}$$

This equation implies that $k_1 + k_2$ is real, a conclusion at which we have arrived earlier in section 15.7.2 via a different argument.

15.8.3 Case of general N

For general particle number N, we obtain from the periodicity conditions (15.105) a set of N equations for the momenta $k_1, k_2, \ldots, k_N$

$$e^{ik_mL} = \prod_{\substack{n=1 \\ n\neq m}}^{N} \frac{k_m - k_n + ic}{k_m - k_n - ic} = -\prod_{n=1}^{N} \frac{k_m - k_n + ic}{k_m - k_n - ic}, \qquad m = 1, 2, \ldots, N. \tag{15.110}$$

Using the antisymmetric scattering phase Θ, these equations can be rewritten as

$$\exp\left(i\left[k_mL + \sum_{n=1}^{N} \Theta(k_m - k_n)\right]\right) = (-1)^{N+1}, \qquad m = 1, 2, \ldots, N, \tag{15.111}$$

the Bethe ansatz equations. Moreover, using this antisymmetry again, these equations multiplied together for all $m = 1, \ldots, N$ give

$$\exp\left[iL\sum_{m=1}^{N} k_m\right] = \exp\left[iP_NL\right] = 1. \tag{15.112}$$

Replacing $p_0(k_m) = k_m$ in the Bethe ansatz equations (15.112), these equations appear in this form for many quantum integrable models. The details of the model under consideration are encapsulated in the functions $p(k)$ and $\Theta(k)$.

15.9 Ground state of the δ-Bose gas in the thermodynamic limit

The essential importance of the Bethe ansatz equations (15.110) or (15.111) is the fact that any solution of these equations determines a state of the the Bose gas. Differently put, this means that prescribing a set of momenta $k_1, \ldots, k_N$ compatible with the Bethe ansatz equations corresponds to a state of the system. The task is therefore to solve the equations or, since this may prove quite a difficult task, find reasonable criteria that the solutions ought to satisfy in order to describe particular states of the system.

As a side remark, we note that the equations for sufficiently small numbers of particles N may also be used for numerical purposes. In such a numerical approach, the equations would, for instance, be solved by iteration starting from a suitable initial assumption for the values of the momenta.

15.9.1 Logarithmic form of the Bethe ansatz equations

In order to further discuss the Bethe ansatz equations, it is useful to take the logarithm of (15.112). In their logarithmic form, the Bethe ansatz equations are

$$Lk_m + \sum_{n=1}^{N} \Theta(k_m - k_n) = 2\pi n_m, \qquad m = 1, 2, \ldots, N \tag{15.113}$$

where the numbers n_m are integers $n_m = 0, \pm 1, \pm 2, \ldots$ if the number of particles N is odd or half-odd integers $n_m = \pm 1/2, \pm 3/2, \ldots$ if the number of particles N is even. The numbers n_m are often called quantum numbers.

From here on we focus on the repulsive case $c > 0$ where, demonstrated in section 15.7.2, all solutions k_m are real.

Before we use the Bethe ansatz equations (15.113) to extract information about the physical states of the Bose gas and their energies, we demonstrate that the solutions k_m of the Bethe ansatz equations are uniquely determined by a given set of integer or half-odd integer numbers $\{n_m\}$. For this purpose we define a function of the momenta $\{k\}$, the so-called Yang action

$$Y(\{k\}) \equiv \frac{L}{2} \sum_{n=1}^{N} k_n^2 - 2\pi \sum_{n=1}^{N} n_n k_n + \frac{1}{2} \sum_{n,m=1}^{N} \tilde{\Theta}(k_n - k_m) \tag{15.114}$$

where

$$\tilde{\Theta}(k) = \int_0^k dk' \, \Theta(k'). \tag{15.115}$$

The stationary points of this function

$$\frac{\partial Y}{\partial k_n} = 0, \qquad n = 1, 2, \ldots, N \tag{15.116}$$

are equivalent to the Bethe ansatz equations (15.114). Moreover, from the matrix of second derivatives of the Yang action

$$\frac{\partial^2 Y}{\partial k_n \partial k_m} = \delta_{nm} \left(L + \sum_{l=1}^{N} K(k_n - k_l) \right) - K(k_n - k_m) \tag{15.117}$$

where, for $c > 0$

$$K(k) = \frac{d\Theta(k)}{dk} = \frac{2c}{k^2 + c^2} > 0 \tag{15.118}$$

we can build a quadratic form for arbitrary numbers $u_n \neq 0$

$$\sum_{n,m=1}^{N} \frac{\partial^2 Y}{\partial k_n \partial k_m} u_n u_m = \sum_{n=1}^{N} L u_n^2 + \frac{1}{2} \sum_{n,l=1}^{N} K(k_n - k_l) \left(u_n - u_l \right)^2 > 0. \tag{15.119}$$

This is equivalent to the statement that the matrix (15.118) is positive definite. Therefore, (15.117) corresponds to a unique minimum for any given set of numbers $\{n_m\}$. This implies that the momenta $\{k\}$, and hence the state of the Bose gas, are uniquely determined for any given set of numbers $\{n_m\}$.

As shown section 15.6, all the momenta k_m in the Bethe ansatz equations are mutually different (the Pauli principle for one-dimensional interacting Bosons). Different sets of integer of half-odd integer numbers $\{n_m\} : n_i \neq n_j$, parametrize the solutions uniquely $\{k_m\} : k_i \neq k_j$ of the Bethe ansatz equations. Moreover, we have $n_i > n_j$ for $k_i > k_j$. Thus, integers (for N odd) or half-odd integers (for N even) $\{n_j\}$ can indeed be considered as quantum numbers parametrizing the solutions uniquely.

15.9.2 Ground state

The ground state for a system with a variable number of particles is the Fock vacuum $|0\rangle$ for which $\mathcal{H}|0\rangle = 0$ and $N = 0$, as discussed in section 15.2 (cp. especially (15.19)).

More interesting is the problem to determine the ground state for a given fixed number N of particles. Physically, we expect that the ground state for a fixed number of particles N corresponds to a choice of integers $\{n_n^{(0)}\}$ that are distributed in a symmetrical interval around zero with no vacancies or holes left between consecutive quantum numbers. That is, the ground state quantum numbers are

$$\left\{ n_n^{(0)} : \quad n_n^{(0)} = \frac{N+1}{2} - n, \quad n = 1, 2, \ldots, N \right\}. \tag{15.120}$$

The quantum numbers are thus equally spaced with spacing $\Delta n \equiv n_n^{(0)} - n_{n+1}^{(0)} = 1$ between $n^{(0)\max} = (N-1)/2$ and $n^{(0)\min} = -(N-1)/2$. Note that $n_n^{(0)} > n_{n+1}^{(0)}$. Relaxing these conditions on the quantum numbers will later lead us to a description of the excited states of the Bose gas.

We immediately obtain from the Bethe ansatz equations (15.113) that the total momentum of the ground state vanishes

$$P_N^{(0)} = \sum_{m=1}^{N} k_m^{(0)} = \frac{2\pi}{L} \sum_{m=1}^{N} n_m^{(0)} = 0. \tag{15.121}$$

The Bethe ansatz equations themselves become for the ground state

$$Lk_m^{(0)} + \sum_{n=1}^{N} \Theta(k_m^{(0)} - k_n^{(0)}) = 2\pi \left(\frac{N+1}{2} - m \right), \qquad m = 1, 2, \ldots, N. \tag{15.122}$$

15.9.3 Thermodynamic limit of the ground state

We now consider the thermodynamic limit of the Bethe ansatz equations (15.122) for the ground state of the Bose gas, which is defined as the limit of both number of particles $N \to \infty$ and length of the interval $L \to \infty$, while the particle density $\rho = N/L$ is kept fixed at a finite value. We shall, where no confusion can arise, suppress the superscript (0) for ground state quantities in the following.

Let us write the number of particles in the following way

$$N = \sum_{m=1}^{N} 1 = \sum_{m=1}^{N} \frac{k_{m+1} - k_m}{k_{m+1} - k_m} \equiv L \sum_{m=1}^{N} \rho(k_m)\,(k_{m+1} - k_m) \tag{15.123}$$

where we defined the density of solutions or root density of the Bethe ansatz equations as

$$\rho(k_m) = \frac{1}{L\,(k_{m+1} - k_m)} \equiv \frac{1}{L\Delta k_m}. \tag{15.124}$$

In the thermodynamic limit, the solutions of the Bethe ansatz equations become dense, i.e.

$$\Delta k_m = k_{m+1} - k_m = \mathcal{O}\left(\frac{1}{L}\right), \tag{15.125}$$

and we can consider $\rho(k)$ as a function of a continuous variable k. In the thermodynamic limit of the ground state, the momenta of the particles fill the symmetric interval $[-q, q]$ where q is determined by

$$\rho = \lim_{N,L\to\infty} N/L = \int_{-q}^{q} dk\, \rho(k). \tag{15.126}$$

The momentum q plays a role for the Bose gas analogous to the Fermi momentum for electrons. In order to have a succinct terminology, we shall therefore refer to q as the Fermi momentum. The density of particles ρ and the Fermi momentum q are related by $\rho = \rho(q)$. For a fixed density, therefore, the Fermi momentum $q = q(\rho)$ must be chosen such that (15.126) is satisfied.

The sum over momenta of an arbitrary function of the discrete momenta $f(k_m)$ becomes in the thermodynamic limit

$$\sum_{m=1}^{N} f(k_m) = \sum_{m=1}^{N} \frac{f(k_m)}{\Delta k_m} \Delta k_m \approx L \sum_{m=1}^{N} f(k_m)\rho(k_m)\Delta k_m \tag{15.127}$$

$$\to L \int_{-q}^{q} dk f(k)\rho(k). \tag{15.128}$$

Combining these approximations, we can write the difference of the Bethe ansatz equations for index m and $m+1$

$$L\,(k_{m+1} - k_m) + \sum_{n=1}^{N} [\Theta(k_{m+1} - k_n) - \Theta(k_m - k_n)] = 2\pi, \tag{15.129}$$

which as $k_{m+1} \to k_m$, i.e. for $N, L \to \infty$ becomes

$$L\,(k_{m+1} - k_m) + (k_{m+1} - k_m) \sum_{n=1}^{N} K(k_m - k_n) = 2\pi \tag{15.130}$$

and finally, using the density $\rho(k)$ and replacing the sum by an integral, we obtain

$$\rho(k) - \frac{1}{2\pi} \int_{-q}^{q} dk'\, K(k - k')\rho(k') = \frac{1}{2\pi}. \tag{15.131}$$

This is an integral equation for the root density of the Bethe ansatz equations describing the ground state of the Bose gas in the thermodynamic limit. This type of integral equation is classified as a second order integral equation of Fredholm type because the unknown function appears also outside the integral and because the integration limits are constants. Furthermore, it is an inhomogeneous integral equation because of the constant term $1/2\pi$.

Physically, however, the integration limits have to be determined by the normalization condition that

$$\int_{-q}^{q} dk\, \rho(k) = \lim_{N,L\to\infty} = \rho \tag{15.132}$$

is a fixed value. We shall see shortly how to deal with this difficulty.

Before doing this, note that the energy of the Bose gas in the thermodynamic limit becomes

$$E_0 = \sum_{n=1}^{N} k_n^2 \to L \int_{-q}^{q} dk\, k^2 \rho(k). \tag{15.133}$$

Lieb and Liniger (1963) devised a set of changes of variables to rewrite the integral equation (15.131) and the condition (15.132) in a way suitable for further analysis. They introduced

$$k = qy, \qquad c = q\alpha, \qquad \rho(qy) = g(y), \qquad \text{and} \qquad \gamma\rho = c \tag{15.134}$$

such that (15.131), (15.132), and (15.133) become

$$g(y) - \frac{1}{2\pi}\int_{-1}^{1} dy' \frac{2\alpha g(y')}{(y-y')^2 + \alpha^2} = \frac{1}{2\pi}, \tag{15.135}$$

$$\gamma \int_{-1}^{1} dy\, g(y) = \alpha, \tag{15.136}$$

and

$$e(\gamma) \equiv \frac{E_0}{N\rho^2} = \frac{\gamma^3}{\alpha^3}\int_{-1}^{1} dy\, y^2 g(y). \tag{15.137}$$

The parameters $\gamma = c/\rho$ and $\alpha = c/q$ represent two ways to introduce a dimensionless coupling constant.

EXERCISE 15.5 Sound velocity of the δ-Bose gas The sound velocity of a one-dimensional particle gas is defined by

$$v_S = \left(\frac{L}{m\rho}\left(\frac{\partial^2 E_0}{\partial L^2}\right)_N\right)^{1/2} = \left(\frac{2L}{\rho}\left(\frac{\partial^2 E_0}{\partial L^2}\right)_N\right)^{1/2} \tag{15.138}$$

with the mass of the particles $m = 1/2$ in our units ($\hbar = 1$ and $\hbar^2/2m = 1$).

Express the sound velocity for the δ-Bose gas in terms of $e(\gamma)$ and its derivatives $e'(\gamma)$ and $e''(\gamma)$.

The integral equation (15.135) subject to the condition (15.136) is still not solvable analytically, but it is well suited for approximative or numerical calculations. For instance, weak and strong coupling limits, i.e. $\gamma \ll 1$ and $\gamma \gg 1$, can be considered.

EXERCISE 15.6 Strong coupling limit for the ground state of the δ-Bose gas The strong coupling limit $c \gg 1$ implies $\alpha \gg 1$ (cf. (15.134)). In this limit, the denominator in the integral of the integral equation (15.135) can effectively be replaced by α^2. This replacement implies that the right-hand side of (15.135) must be constant and, thus, $g(y)$ is constant.

Express the constant which $g(y)$ assumes in this case in terms of α. The condition (15.137) then gives a relation between α and the dimensionless coupling constant γ. Use these results to also express the energy (15.138) and the integration limit $q = c/\alpha$ in terms of γ.

15.10 Excited states

We again assume a repulsive interaction $c > 0$. The excitations can be discussed using the Bethe ansatz equations (15.113) with modified quantum numbers $\{n_j\}$ for a system consisting of N particles, $j = 1, 2, \ldots, N$. These quantum numbers provide a simple way to discuss the possible excitations. There are three ways to excite the Bose gas above the ground state given by the N ground state quantum numbers

$$\{n_j^{(0),N}\} = \left\{-\frac{N-1}{2}, -\frac{N-3}{2}, \ldots, \frac{N-3}{2}, \frac{N-1}{2}\right\}, \tag{15.139}$$

which is symmetric about the origin and consists of integers (for N odd) or half-odd integers (for N even). For later convenience we added a superscript N to emphasize the dependence on the particle number.

The three possibilities are:

- a particle with momentum $|k| \leq q$ can be moved to $|k'| \geq q$, thus leaving a vacancy or hole behind;
- a particle with momentum $k^{(p)} > q$ can be added; and
- a particle with momentum $k^{(h)} \leq q$ can be removed.

The first possibility can be regarded as a combination of the latter two possibilities and, thus, we focus on these latter two, dubbed type I and type II excitations by Lieb and Liniger (1963).

15.10.1 Particle excitation

The ground state of the N particle system is described by the the Bethe ansatz equations (15.113) with N quantum numbers $\{n_j^{(0),N}\}$ given by (15.139). The addition of a particle can be described by an additional quantum number. However, since the system now consists of $N+1$ particles, by adding a particle we go from integer to half-odd integer quantum numbers, or vice versa, and hence the ground state distribution of quantum numbers for the system with one more particle is now given by

$$\{n_j^{(0),(N+1)}\} = \left\{-\frac{N}{2}, -\frac{N}{2}+1, \ldots, \frac{N}{2}-1, \frac{N}{2}, \frac{N}{2}\right\}. \tag{15.140}$$

Instead of creating a new ground state for the system of $N+1$ particles, we can choose any other quantum number outside the ground state distribution, e.g. we can give the added particle the quantum number $n_p = \frac{N}{2} + m$ with an integer $m > 0$ (alternatively, we may choose $n_p = -\frac{N}{2} - m$). Then the quantum numbers for which we have to solve the Bethe ansatz equations are

$$\{n_j^{(p),(N+1)}\} = \left\{-\frac{N}{2}, -\frac{N}{2}+1, \ldots, \frac{N}{2}-1, \frac{N}{2}, \frac{N}{2}+m\right\}. \tag{15.141}$$

We immediately observe that, while the total momentum of the ground state vanishes, the state corresponding to (15.141) has a total momentum

$$P = \frac{2\pi}{L} m. \tag{15.142}$$

This total momentum is produced by a rearrangement of all momenta of the particles of the system with respect to the momenta of the ground state of the system of N particles. The ground state momenta of the system of N particles $\{k_1, k_2, \ldots, k_N\}$ become the momenta $\{k'_1, k'_2, \ldots, k'_N, k_p\}$ of the excited state of the system of $N+1$ particles. The momentum k_p of the added particle is called the bare momentum of the particle. However, it is important to note that adding a particle also changes the momenta of all other particles $k_j \to k'_j$. Like all the other momenta $\{k'_j\}$, the momentum k_p of the added particle has to be determined from a set of Bethe ansatz equations with (15.141) as quantum numbers. The overall effect of adding a particle thus consists of all these changes in addition to the momentum k_p of the added particle and gives rise to the momentum change P (15.142) of the system of $N+1$ particles, which is therefore called dressed momentum.

We are now mostly interested in the changes to the system enacted by adding the extra particle, e.g. the excitation energy above the ground state. This can be achieved by considering the difference between the Bethe ansatz equations for the ground state of the N particle system and those of the excited state of the $N+1$ particle system. We have to keep in mind that the former is determined by the ground state distribution (15.139) of quantum numbers, and the latter by the excited state distribution (15.142). We define the difference in momentum

$$\delta k_j = k_j' - k_j, \qquad j = 1, 2, \ldots, N, \tag{15.143}$$

which is of the order of $\mathcal{O}(1/L)$, or equivalently of $\mathcal{O}(1/N)$. Note the difference between δk_j and Δk_j as defined (15.125). Using the respective Bethe ansatz equations, this difference can be expressed as

$$L\delta k_j + \sum_{l=1}^{N} \Big[\Theta(k_j' - k_l') - \Theta(k_j - k_l)\Big] + \Theta(k_j' - k_p) = \pi \tag{15.144}$$

where we suppressed the superscript (0) we used in the ground state Bethe ansatz equation (15.122). The constant π on the right-hand side is due to the change from integer to half-odd integer quantum numbers (or vice versa) between the system of N and $N+1$ particles. If we are interested in the bare momentum of the added particle, we need to solve the additional Bethe ansatz equation

$$Lk_p + \sum_{l=1}^{N} \Theta(k_p - k_l') = 2\pi \left(\frac{N}{2} + m\right). \tag{15.145}$$

The difference δk_j is of the order of magnitude $\mathcal{O}(1/L)$. Thus, we can expand (15.144) to order $\mathcal{O}(1/L)$

$$L\delta k_j + \sum_{l=1}^{N} \left[K(k_j - k_l)\left(\delta k_j - \delta k_l\right)\right] + \Theta(k_j - k_p) = \pi. \tag{15.146}$$

Since we intend to go to the thermodynamic limit, it is useful to rearrange this equation

$$\delta k_j \left(1 + \frac{1}{L}\sum_{l=1}^{N} K(k_j - k_l)\right) = \frac{1}{L}\left(\pi - \Theta(k_j - k_p)\right) + \frac{1}{L}\sum_{l=1}^{N} K(k_j - k_l)\delta k_l. \tag{15.147}$$

The left-hand side of this equation is similar to (15.130) and, thus, can, in the thermodynamic limit, be written as $2\pi\delta k\rho(k)$, where δk is the shift in momentum in the thermodynamic limit. Together with the right-hand side we obtain in the thermodynamic limit

$$2\pi\, \delta k\, \rho(k) = \frac{1}{L}\left(\pi - \Theta(k - k_p)\right) + \int_{-q}^{q} dk'\, \delta k' K(k - k')\rho(k'). \tag{15.148}$$

In the thermodynamic limit, we now introduce a function that describes the rearrangement or back-flow in the momenta after adding a particle that depends also on the momentum k_p of this added particle

$$F(k|k_p) \equiv L\,\delta k\,\rho(k) = \lim_{N,L\to\infty} \frac{k'_j - k_j}{k_{j+1} - k_j}. \tag{15.149}$$

This function depends on the momentum variable k and remains finite in the thermodynamic limit. It plays a role for the particle excited state similar to the density $\rho(k)$ of solutions of the Bethe ansatz equations for the ground state; it has several names in the literature, including one-particle shift, back-flow, or polarization function.

Using the one-particle shift function, (15.148) becomes the integral equation, also called dressing equation

$$F(k|k_p) - \frac{1}{2\pi}\int_{-q}^{q} dk'\, K(k-k')F(k'|k_p) = \frac{1}{2\pi}\left(\pi - \Theta(k-k_p)\right). \tag{15.150}$$

With the help of the one-particle shift function $F(k|k_p)$, we can express other quantities. For example, the dressed momentum, the sum of all shifted momenta and the bare momentum of the added particle, is given by

$$P = \delta p(k_p) = \frac{2\pi}{L}m = k_p + \sum_{j=1}^{N}\left(k'_j - k_j\right) = k_p + \sum_{j=1}^{N}\delta k_j = k_p + \int_{-q}^{q} dk\, F(k|k_p). \tag{15.151}$$

The dressed energy of the system with one added particle of bare momentum k_p becomes

$$\delta e(k_p) = k_p^2 + \sum_{j=1}^{N}\left(k_j'^2 - k_j^2\right) = k_p^2 + \sum_{j=1}^{N}\left(2k_j\delta k_j - (\delta k_j)^2\right) \tag{15.152}$$

$$= k_p^2 + \int_{-q}^{q} dk\, 2k\, F(k|k_p) \tag{15.153}$$

where, in the last step, we have neglected the term $\sum_{j=1}^{N}(\delta k_j)^2$, which is of the order of $\mathcal{O}(1/L)$.

These expressions demonstrate the collective nature of this excitation, which cannot be explained in a simple way by the single added Boson. There is a marked difference between the bare and the dressed quantities. The addition of a particle with bare momentum k_p and bare energy k_p^2 resulted in a rearrangement of the whole particle system, which collectively acquires the dressed momentum (15.151) and the dressed energy (15.153).

15.10.2 Hole excitation

The other type of excitation, the type II or hole excitation, starts again from the ground state determined by the Bethe ansatz equations for the quantum numbers $\{n_j^{(0),N}\}$ given by (15.139). Removing one particle from the N particle ground state means that by removing one quantum number we consider the ground state of an $N-1$ particle system, which is to be determined by the Bethe ansatz equations for the quantum numbers $\{n_j^{(0),(N-1)}\}$

$$\{n_j^{(0),(N-1)}\} = j - \frac{(N-1)+1}{2} = \left\{-\frac{N}{2}+1, -\frac{N}{2}+2, \ldots, \frac{N}{2}-2, \frac{N}{2}-1\right\}. \quad (15.154)$$

The hole excitation can now be created by removing one quantum number from this set of quantum numbers and putting it at the left or right end in the next available slot. For a hole with a positive (dressed) momentum $P = 2\pi m/L$, the corresponding distribution of quantum numbers is given by

$$\{n_j^{(h)(N-1)}\} = \left\{-\frac{N}{2}+1, -\frac{N}{2}+1, \ldots, \frac{N}{2}-m-1, \frac{N}{2}-m+1, \ldots, \frac{N}{2}\right\}. \quad (15.155)$$

From here, the analysis of the one hole excitation proceeds in close analogy to the one-particle excitation as exercise 15.7 shows.

EXERCISE 15.7 Hole excitation Show that the one-hole shift function, defined in analogy to the one-particle shift function (15.149), is determined by the integral equation

$$F(k|k_h) - \frac{1}{2\pi}\int_{-q}^{q} dk'\, K(k-k')F(k'|k_h) = -\frac{1}{2\pi}(\pi - \Theta(k-k_h)). \quad (15.156)$$

In terms of the one-hole shift function, the dressed momentum and the dressed energy for the one-hole excitation are then given by

$$\delta p(k_h) = -\frac{2\pi}{L}m = -k_h - \int_{-q}^{q} dk\, F(k|k_h) \quad (15.157)$$

and

$$\delta e(k_h) = -k_h^2 + \int_{-q}^{q} dk\, 2k\, F(k|k_h). \quad (15.158)$$

15.10.3 Particle-hole excitation

The linearity of the dressing integral equations (15.150) for a particle excitation and (15.156) for a hole excitation allows the description of more general excitations as superpositions of these two basic types of excitations. A particle-hole excitation will thus be described in the thermodynamic limit by a shift function $F_{ph}(k|k_p, k_h)$, which is a given by the linear integral equation

$$F_{ph}(k|k_p, k_h) - \frac{1}{2\pi}\int_{-q}^{q} dk'\, K(k-k')F(k'|k_p, k_h) = -\frac{1}{2\pi}\left(\Theta(k-k_p) - \Theta(k-k_h)\right)$$
$$= \frac{1}{2\pi}\int_{k_h}^{k_p} dk'\, K(k-k'). \quad (15.159)$$

In this way, states with M_p particles and M_h holes can be described by linear integral equations generalizing (15.159).

Without solving the integral equation explicitly, we can derive some important properties of the particle-hole excitations by formally manipulating (15.159). In order to do so, we need the notion of the resolvent operator $\mathcal{R}$. The integral equation can be written using the integral operator $\mathcal{K}$ whose integral kernel $K(k, k') = K(k-k')$ we have used already

$$\left(\mathcal{I} - \frac{1}{2\pi}\mathcal{K}\right) f(k) = f(k) - \frac{1}{2\pi}\int_{-q}^{q} dk'\, K(k-k')f(k'). \quad (15.160)$$

The resolvent $\mathcal{R}$ is then defined as the operator that, at least formally, solves the integral equation

$$(\mathcal{I}+\mathcal{R})\left(\mathcal{I} - \frac{1}{2\pi}\mathcal{K}\right) = \left(\mathcal{I} - \frac{1}{2\pi}\mathcal{K}\right)(\mathcal{I}+\mathcal{R}) = \mathcal{I}. \quad (15.161)$$

This operator equation implies that

$$\left(\mathcal{I} - \frac{1}{2\pi}\mathcal{K}\right)\mathcal{R} = \frac{1}{2\pi}\mathcal{K}, \qquad (\mathcal{I}+\mathcal{R})\,\mathcal{K} = 2\pi\mathcal{R}. \quad (15.162)$$

We now consider the excitation energy of the excitation, which consists of one hole and one particle. The presence of the particle and the hole changes all momenta of the ground state particles $k_j \to k_j'$. The energy δE_{ph} of the excitation is the energy of the new state minus the ground state energy. We have calculated the separate contributions of the added particle and the hole in (15.153) and (15.158), respectively. Due to the linearity of the integral equations that determine the shift functions, we obtain for the energy of the particle-hole excitation

$$\delta E_{ph} = \delta E_{ph}(k_p, k_h) = \varepsilon_0(k_p) - \varepsilon_0(k_h) + \int_{-q}^{q} dk\, F_{ph}(k|k_p, k_h)\frac{\partial \varepsilon_0(k)}{\partial k}. \quad (15.163)$$

We now use the resolvent operator $\mathcal{R}$ to obtain a formal solution of the integral equation (15.159). Employing the operator identities (15.161) and (15.162), we obtain for the particle-hole shift function from (15.159)

$$F_{ph}(k|k_p, k_h) = \int_{k_h}^{k_p} dk' \, R(k, k'), \tag{15.164}$$

where $R(k, k') = R(k', k)$ is the kernel function corresponding to the resolvent operator $\mathcal{R}$. The symmetry follows from the definition (15.161) of the resolvent $\mathcal{R}$ and from (15.162). The energy of the particle-hole excitation can now be written as

$$\delta E_{ph}(k_p, k_h) = \varepsilon_0(k_p) - \varepsilon_0(k_h) + \int_{-q}^{q} dk \int_{k_h}^{k_p} dk' \, R(k, k') \frac{\partial \varepsilon_0(k)}{\partial k} \tag{15.165}$$

$$= \varepsilon_0(k_p) - \varepsilon_0(k_h) + \int_{k_h}^{k_p} dk' \int_{-q}^{q} dk \, R(k, k') \frac{\partial \varepsilon_0(k)}{\partial k} \tag{15.166}$$

$$\equiv \varepsilon_0(k_p) - \varepsilon_0(k_h) + \int_{k_h}^{k_p} dk' \left[\mathcal{R} \frac{\partial \varepsilon_0}{\partial k} \right] (k'). \tag{15.167}$$

We now demonstrate that an expression for $\delta E_{ph}(k_p, k_h)$ can be obtained directly from an integral equation for an appropriate quantity. We do not have to solve the integral equation for the particle-hole shift function whose formal solution we shall need only for some intermediate steps in the derivation. The full motivation for this direct approach will become clear only in chapter 17 on the thermodynamics of the δ-Bose gas at finite temperature.

We consider the solution $E(k)$ of the integral equation

$$E(k) - \frac{1}{2\pi} \int_{-q}^{q} dk' \, K(k - k') E(k') = E_0(k) \tag{15.168}$$

where the inhomogeneity is given by

$$E_0(k) = \varepsilon_0(k) - \mu = k^2 - \mu. \tag{15.169}$$

The constant $\mu = \mu(q)$, which will turn out to play the role of a chemical potential in chapter 17, is determined by the condition

$$E(q) = E(-q) = 0. \tag{15.170}$$

For the present consideration, the chemical potential μ will not be of importance. It has been introduced here for later convenience.

The integral equation (15.168) is of the same type as the integral equation (15.131) for the density $\rho(k)$ of solutions of the Bethe ansatz equations or the integral equations for the shift functions, e.g. for the particle-hole shift function (15.159). These equations differ only by their respective inhomogeneity terms.

We now want to make contact between the integral equation (15.168) and the energy expression (15.166) or (15.167), respectively, identifying (15.168) as the integral equation for an energy, the dressed energy of a particle–hole pair above the ground state energy.

We achieve this by first differentiating (15.168) with respect to k, taking into account the symmetry of the kernel function $K(k) = K(-k)$ and that, due to the condition (15.170), boundary terms vanish after an integration by parts. We arrive at

$$\frac{\partial E(k)}{\partial k} - \frac{1}{2\pi}\int_{-q}^{q} dk' \, K(k-k')\frac{\partial E(k')}{\partial k'} = \frac{\partial E_0(k)}{\partial k} \tag{15.171}$$

or, written in terms of the integral operator $\mathcal{K}$

$$\left[\left(\mathcal{I} - \frac{1}{2\pi}\mathcal{K}\right)\frac{\partial E}{\partial k}\right](k) = \frac{\partial E_0(k)}{\partial k} \tag{15.172}$$

with the formal solution, according to (15.161) and (15.162),

$$\frac{\partial E(k)}{\partial k} = \frac{\partial E_0(k)}{\partial k} + \left[\mathcal{R}\frac{\partial E_0}{\partial k}\right](k). \tag{15.173}$$

Recalling that $\partial E_0(k)/\partial k = \partial \varepsilon_0(k)/\partial k = 2k$, we can use this result in (15.166) or (15.167), respectively. After integrating from k_h to k_p, we finally find the important result

$$\delta E_{ph}(k_p, k_h) = E(k_p) - E(k_h). \tag{15.174}$$

In a similar manner, we can show that the momentum of the particle–hole excitation is given by

$$\delta P_{ph} = p_0(k_p) - p_0(k_h) - \int_{-q}^{q} dk \, \big(\Theta(k-k_p) - \Theta(k-k_h)\big)\, \rho(k) \tag{15.175}$$

$$= 2\pi \int_{-k_h}^{k_p} dk \, \rho(k). \tag{15.176}$$

Due to the linearity of the integral equation for the excitation energy of the particle-hole pair (15.168) and the other integral equations of the same type for the shift functions $F(k)$ and the density $\rho(k)$, all eigenstates of the δ-Bose gas Hamiltonian in the thermodynamic limit can be constructed with a fixed number of particles. These states contain equal numbers of particles and holes, and the energy and momentum of a composite excitation are equal to the sums of the energies and momenta of individual particles and holes.

Let us summarize this section. We considered the composite two-particle excitation consisting of a one-particle and a one-hole excitation, which conserves the total number of particles. The bare momenta of these particles obey $|k_p| > q$ and $|k_h| < q$ but are otherwise undetermined. The integral equations are linear and, thus, the two-particle composite excitation can indeed be considered as composed of two one-particle excitations, one particle with bare momentum k_p and one hole with bare momentum $-k_h$. The energy of the composite excitation is positive but vanishes for $|k_p| = |k_h| = q$ which means that we have returned to the ground state. The spectrum is, thus, gapless at the Fermi points $\pm\, q$.

We shall return to the δ-Bose gas in part V, chapter 17, when we discuss the thermodynamics of Bethe ansatz solvable models at finite temperature. We shall begin our discussion there by considering the δ-Bose gas with a large macroscopic number of excitations.

Part 4

Electronic Systems: Nested Bethe Ansatz

The bird a nest, the spider a web, man friendship.
William Blake (1757–1827)

Chapter 8 on strongly correlated or interacting quantum matter discussed electronic models whose prime examples were the Hubbard model and models for magnetic impurities in metallic hosts, i.e. the Anderson and Kondo models. These models describe the interaction of electrons where both degrees of freedom of the electrons, i.e. two components of the fundamental particles, participate in the interaction, the charge, as well as the magnetic moment or spin of the electrons. These models were in contrast to models like the Heisenberg model where only one degree of freedom, the magnetic moment or spin degree of freedom, was involved in the interaction of the constituents of the system. Thus, two or more interacting components in a model provide a more realistic picture of the strong interacting effects in quantum matter but form also an additional complication to the theoretical description.

Up to now, we have demonstrated how the Bethe ansatz technique works for models of strongly interacting quantum matter where only one component interacts. These quantum models have all been one-dimensional models. Part IV, we shall extend the Bethe ansatz technique to models with more than one interacting component. In fact, in order to be definite, we shall concentrate our attention on two-component models. The models will, however, remain one-dimensional.

The motivation to study these models will be taken from condensed matter physics. Therefore the models will all be electronic models where charge and spin degrees of freedom are the two strongly interacting components.

However, we shall attempt to keep the discussion of the Bethe ansatz technique to two-component electronic models general. Thus, a whole class of electronic models will be amenable to the this analysis, in particular the one-dimensional Fermi gas, the one-dimensional Hubbard, and Kondo models.

The generalization of the Bethe ansatz technique to multicomponent systems will consist in several, in our case two, Bethe ansätze which are interleaved or nested into each other, hence the name *nested* Bethe ansatz.

In chapter 16, we first treat the one-dimensional electron or Fermi gas with interactions between the particles given by δ functions which is conceptually simpler than both, the one-dimensional Hubbard and the Kondo model. We will pursue the Bethe ansatz analysis of the one-dimensional Fermi gas to a point where we identify a particular eigenvalue problem. Then, after arguing that this same eigenvalue problem formally also appears for the Hubbard model, we shall treat the particularities of the Kondo model until again we are able to identify the formally same eigenvalue problem as for the Fermi gas and the Hubbard model. Finally, we solve this eigenvalue problem by exploiting its similarity to the transfer matrix problems of the algebraic Bethe ansatz method. This last step leads to a coupled set of algebraic equations describing the nested Bethe ansatz solution, aptly called the nested Bethe ansatz equations.

As an application of the nested Bethe ansatz, we focus again on the Kondo model and discuss some physical results that can be derived from the model's Bethe ansatz solution, more precisely the Bethe ansatz equations directly.

For a detailed treatment of the one-dimensional δ-Fermi gas, see Gaudin (1983) or its translation Gaudin (2014), especially chapter 10. Moreover, a recent review on Fermi gases in one dimension by Guan *et al.* (2013) covers not only the nested Bethe ansatz solutions on the theoretical side but also the recent experimental situation on Fermi gases in one-dimensional optical lattices.

The Hubbard model in one dimension is comprehensively treated in a recent monograph by Essler *et al.* (2005).

Classic review articles on magnetic impurity models are Andrei *et al.* (1981) and Tsvelik and Wiegmann (1983). A more recent review, which also covers the Hubbard model, can be found in Andrei (1994). Recent monographs devoted to magnetic quantum impurities are by Hewson (1993) and Kondo (2012).

16

Electronic Systems

The electron: may it never be of any use to anybody!
Joseph John Thomson (1846–1940)

So far, we have considered the Bethe ansatz solutions of many-particle systems composed of constituents that represent only one quantum degree of freedom, a particle degree of freedom in the one-dimensional Bose gas or a spin degree of freedom in the Heisenberg quantum spin chain. In quantum mechanics, however, we often encounter objects with one or more internal degrees of freedom in addition to their particle degrees of freedom, which are often called charge degrees of freedom. The prime example is the electron, which, in addition to being a point-particle, has an internal spin degree of freedom.

The Bethe ansatz can be generalized to many-particle problems where the particles have internal degrees of freedom. The generalized method can be viewed as two Bethe ansätze executed one after the other. This gave the method its name: *nested* Bethe ansatz.

Typically, we consider electronic systems that are the most relevant for quantum condensed matter physics. The following describes the most prominent electronic many-particle systems in one dimension that have been solved by the generalization of the Bethe ansatz method: the one-dimensional δ-Fermi gas, the one-dimensional Hubbard model, and the Kondo model.

The Bethe ansatz method is powerful enough to extend the solution of, e.g. the Kondo problem, also to finite temperature, to the case where an external magnetic field is applied, to higher impurity spin $S > 1/2$, and beyond the Kondo problem directly to the Anderson model (cf. section 8.9.2) from which we started in section 8.9.3 to derive the Kondo model. However, this chapter is not an exhaustive exposition of these advanced topics, but rather the analysis of the basic Bethe ansatz solution of the models.

The three systems whose Bethe ansatz solution we present in the following, the δ-Fermi gas in section 16.1, the one-dimensional Hubbard model in section 16.2, and the Kondo model in section 16.3, are sufficiently similar with respect to their Bethe ansatz solution to allow a common treatment to some extent. The major difference to the Bethe ansatz for one component systems that we will find in all three cases is a second eigenvalue equation, called spin eigenvalue equation. This spin eigenvalue equation will have the same form in all cases we consider and can be solved with a second Bethe ansatz. Section 16.4 solves the spin eigenvalue equation by an algebraic Bethe ansatz. Section 16.5 concludes this chapter and demonstrates a direct use of the two coupled sets of

Models of Quantum Matter: A First Course on Integrability and the Bethe Ansatz. Hans-Peter Eckle,
Oxford University Press (2019). © Hans-Peter Eckle. DOI: 10.1093/oso/9780199678839.001.0001

Bethe ansatz equations that we shall have obtained by the two nested Bethe ansätze to a current problem in the physics of the Kondo problem: a quantum dot tuned to Kondo resonance (cp. section 8.9.4) and side-coupled to an isolated ring.

16.1 Fermi gas in one dimension

Historically, the first example of an exact solution of an electronic system based on the Bethe ansatz method has been the one-dimensional gas of N Fermions with pair interactions given by δ-functions in early work by McGuire (1965, 1966) and then, more comprehensively, by Gaudin (1967) and independently, Yang (1967) (see also Flicker and Lieb, 1967 and Michel Gaudin's thesis, reproduced in Gaudin, (1995). The exposition in the original publications are fairly condensed, which later occasioned a more detailed presentation of the validity of the nested Bethe ansatz for this model by Fung (1981).

This model system of interacting Fermions is a direct generalization of the one-dimensional gas of Bosons, the Lieb–Liniger model, whose Bethe ansatz solution we discussed in chapter 15 and whose thermodynamic Bethe ansatz solution are discussed in chapter 17. The restriction to Bose symmetry has, however, been lifted and replaced by requiring Fermi antisymmetry.

The Hamiltonian of the Fermi system in first quantization is formally the same as for the one-dimensional Bose gas with δ-potential interaction

$$\mathcal{H} = -\sum_{i=1}^{N} \frac{\partial^2}{\partial x_j^2} + 2c \sum_{\substack{i,\, j=1 \\ i<j}}^{N} \delta(x_i - x_j) \tag{16.1}$$

where the coupling constant $c > 0$.

In particular, the Hamiltonian (16.1) exhibits no spin dependence. We assume, however, a different symmetry of the wave function, the symmetry of spin-1/2 Fermi particles, i.e. for most purposes, at least in a condensed matter context, electrons.

The steps in the analysis of the Fermi gas model will therefore be similar to those for the Bose gas. Accounting for the Fermi antisymmetry of the wave function will again relate the amplitudes in different sectors of the configuration space distinguished by different orderings of the coordinates by their magnitude. The Schrödinger equation will, via the δ-function potential, furnish relations between the amplitudes in one specific sector. Finally, we shall impose periodic (or, in principle, other) boundary conditions on the wave function as a whole. However, unlike in the Bethe ansatz solvable models discussed so far, the periodic boundary conditions will not only provide a quantization condition for the momenta, but will give rise to a second eigenvalue problem as a consequence of the internal spin degrees of freedom of the particles, which is therefore often called the spin eigenvalue problem. The solution of this second eigenvalue problem will require a second Bethe ansatz. We shall find that the relevant quantities of this eigenvalue problem are very similar in structure to those of the eigenvalue problem we found for the transfer matrix of the six-vertex model or, equivalently, the Heisenberg quantum spin chain. Thus, we shall use our experience from part II on the algebraic Bethe ansatz to tackle this second eigenvalue problem.

We shall eventually find coupled sets of equations for the momenta of the electrons or charge degrees of freedom and the parameters of the second Bethe ansatz that describe the spin degrees of freedom. These equations are the *nested* Bethe ansatz equations.

As a warm-up, we start with a simple exercise that supplements our discussion of one-dimensional Fermionic models.

EXERCISE 16.1 Spinless Fermions Convince yourself that Fermi antisymmetry means that spinless Fermions with the Hamiltonian (16.1) are always non–interacting.

Let us return to the more interesting case of electrons where the wave function is spin dependent and carries spin indices $\sigma_j = \uparrow, \downarrow$ for each electron labelled by the particle index j.

Our strategy is to consider the cases of $N = 2$ and $N = 3$ electrons in some detail and generalize from the intuition we gain in these cases to the case of an arbitrary number N of electrons. For detailed demonstrations of the case of general N, we refer to the literature. As the δ-Fermi gas can be treated to a large extent analogous to the one-dimensional Hubbard model, we especially refer to the very detailed demonstration of the validity of the Bethe ansatz for the Hubbard model in chapter 3 of Essler *et al.* (2005).

16.1.1 Two-electron case

The Bethe ansatz wave function for two electrons ϕ_2 is assumed to be

$$\phi_{2,\sigma_1\sigma_2}(x_1, x_2) = \phi^{(12)}_{2,\sigma_1\sigma_2}(x_1, x_2)\theta(x_2 - x_1) + \phi^{(21)}_{2,\sigma_1\sigma_2}(x_1, x_2)\theta(x_1 - x_2) \qquad (16.2)$$

where, as in the Bose case, the upper indices (12) and (21) signify the fundamental sector $F_2^{(12)} = F_2^{(I)} : x_1 < x_2$ and the sector $F_2^{(21)} : x_1 > x_2$. The step functions $\theta(x)$ explicitly restrict the wave function to either of these sectors. In contrast to the Bose system, we now must take into account the spin dependence of the electrons in the amplitudes of the Bethe ansatz wave function, whose two parts are

$$\phi^{(12)}_{2,\sigma_1\sigma_2}(x_1, x_2) = A^{(12)}_{\sigma_1\sigma_2}(12)e^{i(k_1x_1+k_2x_2)} + A^{(12)}_{\sigma_1\sigma_2}(21)e^{i(k_2x_1+k_1x_2)}, \qquad (16.3)$$

$$\phi^{(21)}_{2,\sigma_1\sigma_2}(x_1, x_2) = A^{(21)}_{\sigma_1\sigma_2}(12)e^{i(k_1x_1+k_2x_2)} + A^{(21)}_{\sigma_1\sigma_2}(21)e^{i(k_2x_1+k_1x_2)}. \qquad (16.4)$$

The amplitudes $A^{(Q)}_{\sigma_1\sigma_2}$ have three pairs (for general particle number N, three N-tuples) of labels: the spin labels σ_1 and σ_2, the label Q (with $Q = I = (12)$ or $Q = (21)$) referring to the sector as an upper index and the label P with $P = I = (12)$ or $P = (21)$ signifying the permutation of the momenta k_j in the plane waves. Disregarding the spin labels for the moment, we have again, as in the Bose case, $2! \cdot 2! = 4$ amplitudes for the two-electron case.

We shall not pursue the solution of the δ-Fermi gas in all detail here. However, the following exercises and subsequent considerations will be useful to gain some intuition for our discussion of the Kondo model in section 16.3.

EXERCISE 16.2 Fermi gas of two electrons: antisymmetry Determine relations between the amplitudes $A^{(Q)}_{\sigma_1\sigma_2}(P)$ from the posited Fermi antisymmetry of the wave function (16.2) together with (16.3) and (16.4). In particular, show that all amplitudes can be reduced to the amplitudes in the fundamental sector $F_2^{(I)} = F_2^{(12)}$.
Note that the spin indices of the amplitudes are exchanged by the permutation operator or matrix

$$\Pi \equiv \left(\Pi^{\sigma_1'\sigma_2'}_{\sigma_1\sigma_2}\right) \equiv \left(\Pi^{\sigma_2'}_{\sigma_2}(\sigma_1,\sigma_1')\right), \tag{16.5}$$

which is defined by (suppressing the other labels of the amplitudes for the moment)

$$A_{\sigma_2\sigma_1} = \Pi A_{\sigma_1\sigma_2} = \sum_{\sigma_1'\sigma_2'} \Pi^{\sigma_1'\sigma_2'}_{\sigma_1\sigma_2} A_{\sigma_1'\sigma_2'} = \sum_{\sigma_1'\sigma_2'} \delta^{\sigma_1'}_{\sigma_2}\delta^{\sigma_2'}_{\sigma_1} A_{\sigma_1'\sigma_2'}. \tag{16.6}$$

We introduced the second notation for the permutation matrix Π in (16.5) because we want to make contact later with the notation we used in part II on the algebraic Bethe ansatz.

EXERCISE 16.3 Fermi gas of two electrons: Schrödinger equation Use the boundary condition at $x_1 = x_2$ implied by the Schrödinger equation between the two sectors of the two-electron Fermi gas to show that the amplitudes in the fundamental sector are related by

$$A^{(12)}_{\sigma_1\sigma_2}(21) = -\frac{ic + (k_1 - k_2)\Pi}{ic + (k_1 - k_2)} A^{(12)}_{\sigma_1\sigma_2}(12) \equiv -Y(k_1 - k_2) A^{(12)}_{\sigma_1\sigma_2}(12). \tag{16.7}$$

The derivation of the boundary condition at $x_1 = x_2$ due to the δ-function in the Hamiltonian of the Schrödinger equation is similar to the corresponding derivation for the Bose gas case. However, care has to be expended because of the different symmetries of the wave functions.

More explicitly, (16.7) may be written as

$$A^{(12)}_{\sigma_1\sigma_2}(21) = -\sum_{\sigma_1'\sigma_2'} Y^{\sigma_2'}_{\sigma_2}(\sigma_1,\sigma_1')(k_1 - k_2) A^{(12)}_{\sigma_1'\sigma_2'}(12). \tag{16.8}$$

The foregoing two exercises demonstrate that the $2! \cdot 2! = 4$ amplitudes introduced to specify the Bethe ansatz wave function (16.2) can be reduced to only one amplitude and the wave function thus written as

$$\phi_{2\sigma_1\sigma_2}(x_1, x_2) = \left(e^{i(k_1x_1+k_2x_2)} - e^{i(k_2x_1+k_1x_2)}Y(k_1-k_2)\right)A^{(12)}_{\sigma_1\sigma_2}(12)\theta(x_2-x_1)+ \left(e^{i(k_1x_1+k_2x_2)}Y(k_1-k_2) - e^{i(k_2x_1+k_1x_2)}\right)\Pi A^{(12)}_{\sigma_1\sigma_2}(12)\theta(x_1-x_2). \quad (16.9)$$

The corresponding energy eigenvalue is $E = k_1^2 + k_2^2$ and the total momentum $P_2 = k_1 + k_2$.

The reduction of the amplitudes to only a single independent one in the $N = 2$ electron case was thus achieved, similar to the Bose case in chapter 15, by the $N! = 2!$ conditions imposed by the Fermi antisymmetry (instead of the Bose symmetry) of the wave function and the $N - 1 = 1$ condition derived from the δ-potential of the Schrödinger equation. Again, as in the Bose case, a similar reduction is possible for $N > 2$.

The remaining amplitude $A^{(12)}_{\sigma_1\sigma_2}(12)$ is a still undetermined function of the spin variables σ_1 and σ_2. It can be freely chosen since the Hamiltonian (16.1) has no spin dependence. The space of two spins 1/2 is spanned by an antisymmetric spin singlet state

$$\varphi_a = |\uparrow\downarrow\rangle - |\downarrow\uparrow\rangle, \quad (16.10)$$

corresponding to total spin $S = 0$ and total z-component $S^z = 0$ and the three symmetric spin triplet states

$$\varphi_s = \begin{cases} |\uparrow\uparrow\rangle \\ |\uparrow\downarrow\rangle + |\downarrow\uparrow\rangle \\ |\downarrow\downarrow\rangle \end{cases} \quad (16.11)$$

corresponding to total spin $S = 1$ and total z-components $S^z = -1, 0, 1$. The amplitude $A^{(12)}_{\sigma_1\sigma_2}(12)$ will be a linear combination of these spin states that satisfy

$$\Pi\varphi_a = -\varphi_a, \qquad \Pi\varphi_s = \varphi_s. \quad (16.12)$$

It is interesting to note that in the pure spin triplet solution

$$\phi_{2\sigma_1\sigma_2}(x_1, x_2) = \varphi_s\left(e^{i(k_1x_1+k_2x_2)} - e^{i(k_2x_1+k_1x_2)}\right), \quad (16.13)$$

which is symmetric in the spin variables and antisymmetric in the position coordinates of the two electrons, the wave function vanishes at $x_1 = x_2$ and, thus, the electrons can never occupy the same position. Hence, the electrons do not notice the local interaction via the δ-potential. As a consequence, the triplet wave function is indistinguishable from the wave function of free electrons.

We can also use the results of the foregoing two exercises to address the question of the domain in which the two electrons are supposed to exist. As in the Bose case, we may consider an infinite domain $-\infty < x < \infty$ or put the electrons in a box of finite length $0 < x < L$ with, e.g. periodic boundary conditions. Another possible choice would be hard wall conditions, i.e. an infinite potential barrier at $x = 0$ and $x = L$, cf. Woynarovich (1985) or, for the corresponding boundary condition problem in the case of the one-dimensional Hubbard model cf. Schulz (1985). We shall again choose the former, i.e. the periodic, boundary conditions, i.e.

$$\phi_{2,\sigma_1\sigma_2}(0, x_2) = \phi_{2,\sigma_1\sigma_2}(L, x_2), \tag{16.14}$$

$$\phi_{2,\sigma_1\sigma_2}(x_1, 0) = \phi_{2,\sigma_1\sigma_2}(x_1, L). \tag{16.15}$$

We can restrict our attention to the first periodic boundary condition (16.14) because the second (16.15) follows from the first as a consequence of the antisymmetry of the wave function required by the Pauli exclusion principle

$$\phi_{2,\sigma_1\sigma_2}(x_1, 0) = -\phi_{2,\sigma_2\sigma_1}(0, x_1) = -\phi_{2,\sigma_2\sigma_1}(L, x_1) = \phi_{2,\sigma_1\sigma_2}(x_1, L). \tag{16.16}$$

Since both electrons are confined to the interval $(0, L)$, the first boundary condition (16.14) means that the wave function of the left-hand side resides in the fundamental sector $F_2^{(I)} = F_2^{(12)}$ where $x_1 = 0 < x_2$, while the wave function of the right-hand side resides in the sector $F_2^{(Q)} = F_2^{(21)}$ where $x_1 = L > x_2$. Thus, the periodic boundary condition relates the wave functions (16.3) and (16.4) of the two sectors. Explicitly, we obtain for the amplitudes

$$A^{(12)}_{\sigma_1\sigma_2}(12) = A^{(21)}_{\sigma_1\sigma_2}(12)e^{ik_1L}, \tag{16.17}$$

$$A^{(12)}_{\sigma_1\sigma_2}(21) = A^{(21)}_{\sigma_1\sigma_2}(21)e^{ik_2L}. \tag{16.18}$$

These amplitude relations are often also referred to as periodic boundary conditions.

Let us initially focus on the first of these equations. In exercise 16.2, we obtained from the Fermi antisymmetry of the two-electron wave function $A^{(21)}_{\sigma_1\sigma_2}(12) = -A^{(12)}_{\sigma_2\sigma_1}(21) = -\Pi A^{(12)}_{\sigma_1\sigma_2}(21)$ and, hence,

$$A^{(12)}_{\sigma_1\sigma_2}(12) = -\Pi A^{(12)}_{\sigma_1\sigma_2}(21)e^{ik_1L}. \tag{16.19}$$

The Schrödinger equation, as in exercise 16.3, supplies a further relation between the amplitudes such that we arrive at

$$e^{-ik_1L}A^{(12)}_{\sigma_1\sigma_2}(12) = \Pi\frac{ic + (k_1 - k_2)\Pi}{ic + (k_1 - k_2)}A^{(12)}_{\sigma_1\sigma_2}(12) \tag{16.20}$$

$$= \frac{(k_1 - k_2) + ic\Pi}{(k_1 - k_2) + ic} A^{(12)}_{\sigma_1\sigma_2}(12) \tag{16.21}$$

$$\equiv \Pi Y(k_1 - k_2) A^{(12)}_{\sigma_1\sigma_2}(12) \equiv S(k_1 - k_2) A^{(12)}_{\sigma_1\sigma_2}(12). \tag{16.22}$$

This is an eigenvalue equation for the eigenvector $A^{(12)}_{\sigma_1\sigma_2}(12)$, i.e. the two-electron amplitude, with eigenvalue $\mu_1 = \exp(-ik_1L)$ of the operator

$$Y(k_1 - k_2)\Pi = \Pi Y(k_1 - k_2) \equiv S(k_1 - k_2), \tag{16.23}$$

which we identify as the two-particle scattering operator or scattering matrix. It is instructive to write (16.22) one more time giving all indices explicitly

$$e^{-ik_1L} A^{(12)}_{\sigma_1\sigma_2}(12) = \sum_{\sigma_1'\sigma_2'} S^{\sigma_2'}_{\sigma_2}(\sigma_1, \sigma_1')(k_1 - k_2) A^{(12)}_{\sigma_1'\sigma_2'}(12). \tag{16.24}$$

The other periodic boundary condition, (16.18), leads to

$$e^{-ik_2L} A^{(12)}_{\sigma_1\sigma_2}(12) = Y(k_2 - k_1)\Pi A^{(12)}_{\sigma_1\sigma_2}(12) = S(k_2 - k_1) A^{(12)}_{\sigma_1\sigma_2}(12) \tag{16.25}$$

where, in the derivation, we used that $Y(k)Y(-k) = I$.

The periodic boundary conditions (16.22) and (16.25) demonstrate the essential complication of the Fermi case over the Bose case. While in the Bose case, by writing down the periodic boundary conditions, the problem was (in principle) solved at this stage, the periodic boundary conditions in the Fermi case reveal a new, second eigenvalue problem. This is the hallmark of models with internal degrees of freedom of which electronic problems are the prime example, at least from the point of view of condensed matter theory.

The operators we introduced, especially $S(k_1 - k_2)$, play a decisive role for the construction of the Bethe ansatz wave function of the many-electron problem of the δ-Fermi gas. Analogous scattering operators can be derived for the Hubbard chain, the Kondo problem, and other models with internal degrees of freedom amenable to the Bethe ansatz method.

In close analogy to the R-matrix introduced in connection with the algebraic Bethe ansatz in section 10.4, and its graphical representation in figure 10.2, we may use a graphical representation of the scattering matrix S as given in figure 16.1.

16.1.2 The cases of three and more electrons

The case of $N = 2$ electrons can be generalized to an arbitrary number N of electrons. A careful analysis of the case $N = 3$, for instance, reveals that this case can be reduced to the case of $N = 2$ in the sense that only two-particle scattering matrices $S^{ij} = S(k_i - k_j)$ are involved. Moreover, as we shall discuss in detail later in the case of the Kondo

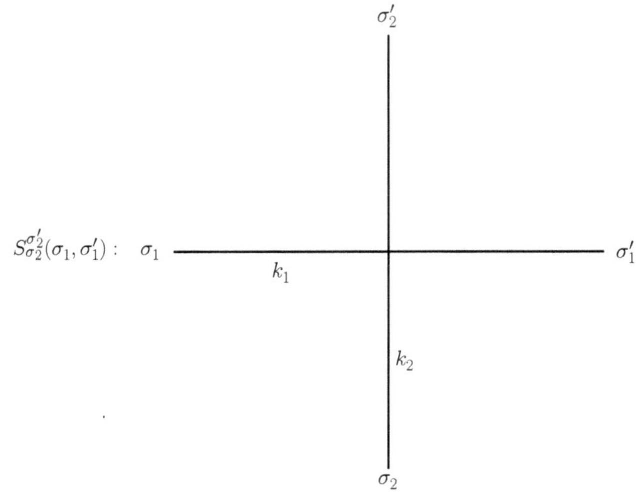

Figure 16.1 *Scattering matrix element* $S_{\sigma_2}^{\sigma_2'}(\sigma_1, \sigma_1')(k_1 - k_2)$. *The arrangement of the spin indices is given to highlight the similarity to the index arrangement of the R-matrix in section 10.4 (see figure 10.2). However, the scattering matrix also depends on the difference in the momentum variable,* $k_1 - k_2$, *of the electrons.*

model, consistency relations, i.e. the Yang–Baxter relations for the scattering matrices, are satisfied, and which we also encountered in the discussion of the algebraic Bethe ansatz of, e.g. the six-vertex model and the corresponding Heisenberg quantum spin chain in part II and for the Bose gas case in chapter 15 where, in the latter case, they were trivially satisfied.

The results for the general case of N electrons will therefore be similar to the case $N = 2$. The antisymmetry of the wave function relates all amplitudes of the wave function to those in the fundamental sector $F_N^{(I)} : x_1 < x_2 < \ldots < x_N$. Within the fundamental sector, the amplitudes are related by the Schrödinger equation, or more directly, by the boundary conditions the δ-interactions impose. Finally, periodic boundary conditions imply a second eigenvalue problem for a product of two-particle scattering matrices. However, some features of the structure of the second eigenvalue problem of the nested Bethe ansatz are not yet fully displayed in the two particle case. Moreover, the case $N = 2$ does not show the problem of the consistency of the Bethe ansatz wave function or, as it is sometimes called, the Bethe ansatz hypothesis (for this wave function, in order to be more precise).

Therefore, in order to gain some familiarity and intuition on how the consistency problem of the Bethe ansatz wave function arises, as well as with the structure of the calculations leading to the second eigenvalue problem, it is quite useful to go through some steps of the solution of the three-electron problem in some detail, before we state the general result for an arbitrary number N of electrons without delving into the detailed and highly technical derivation.

The three-electron Bethe ansatz wave function

$$\phi_{3\sigma_1\sigma_2\sigma_3}(x_1, x_2, x_3) = \sum_Q \theta(x_{(Q)})\phi^{(Q)}_{3\sigma_1\sigma_2\sigma_3}(x_1, x_2, x_3) \tag{16.26}$$

now has $3! = 6$ terms $\phi^{(Q)}_{3\sigma_1\sigma_2\sigma_3}(x_1, x_2, x_3)$ corresponding to the 3! sectors of the configuration space, which are constrained by the step functions $\theta(x_{(Q)})$, e.g. $\theta(x_{(123)}) \equiv \theta(x_2 - x_1)\theta(x_3 - x_2)$. The $3! \cdot 3! = 36$ amplitudes $A^{(Q)}_{\sigma_1\sigma_2\sigma_3}(P)$ of the 3! terms

$$\phi^{(Q)}_{3\sigma_1\sigma_2\sigma_3}(x_1, x_2, x_3) = A^{(Q)}_{\sigma_1\sigma_2\sigma_3}(P)e^{i(k_{p_1}x_1 + k_{p_2}x_2 + k_{p_3}x_3)} \tag{16.27}$$

of this wave function can again be reduced by Fermi antisymmetry to the 3! amplitudes $A^{(I)}_{\sigma_1\sigma_2\sigma_3}(P)$ of just one sector, which we conveniently choose to be the fundamental sector $F_3^{(I)} = F_3^{(123)}$.

EXERCISE 16.4 Three electrons: Fermi antisymmetry Check that the Fermi antisymmetry indeed yields relations between the amplitudes so that only the amplitudes in, e.g. the fundamental sector need be considered together with appropriate products of spin exchange operators Π.

The $3! = 6$ amplitudes $A^{(I)}_{\sigma_1\sigma_2\sigma_3}(P) = A^{(123)}_{\sigma_1\sigma_2\sigma_3}(P)$ corresponding to different momenta permutations of different plane wave terms are connected through the boundary condition imposed by the δ-function in the Schrödinger equation with Hamiltonian (16.1). There are $3 - 1 = 2$ boundaries $x_1 = x_2$ and $x_2 = x_3$ each giving equations connecting pairs of two amplitudes. Altogether, thus, we have $3!(3 - 1)/2 = 6$ equations for the $3! = 6$ amplitudes.

EXERCISE 16.5 Three electrons: boundary condition from δ-interaction Derive the relations between the amplitudes in the fundamental sector at the boundaries $x_1 = x_2$ and $x_2 = x_3$. In particular, the Yang–Baxter relations

$$S^{23}S^{13}S^{12} = S^{12}S^{13}S^{23} \tag{16.28}$$

or, written giving the spin indices explicitly

$$\begin{aligned}
&\sum_{\sigma_1''\sigma_2''\sigma_3''} \left(S^{23}\right)^{\sigma_2''}_{\sigma_2}(\sigma_3'', \sigma_3) \left(S^{13}\right)^{\sigma_3'}_{\sigma_3''}(\sigma_1, \sigma_1'') \left(S^{12}\right)^{\sigma_2'}_{\sigma_2''}(\sigma_1'', \sigma_1') \\
&= \sum_{\sigma_1''\sigma_2''\sigma_3''} \left(S^{12}\right)^{\sigma_2''}_{\sigma_2}(\sigma_1, \sigma_1'') \left(S^{13}\right)^{\sigma_3''}_{\sigma_3}(\sigma_1'', \sigma_1') \left(S^{23}\right)^{\sigma_3'}_{\sigma_3''}(\sigma_2'', \sigma_2'),
\end{aligned} \tag{16.29}$$

for the scattering matrices

$$S^{ij} = \frac{(k_i - k_j) + ic\Pi^{ij}}{(k_i - k_j) + ic} \tag{16.30}$$

will be interesting since for the three-electron case these can indeed be shown to be satisfied (as opposed to the two-electron case, where they do not yet make an appearance).

Find also the graphical representation of the Yang–Baxter relations and compare with the graphical representation of the Yang–Baxter relations for the R-matrices in figure 10.5.

Furthermore, the matrices must also satisfy

$$S^{ji}S^{ij} = I, \qquad \text{and, more generally,} \qquad S^{ij}S^{mn} = S^{mn}S^{ij} \tag{16.31}$$

and the inversion relation

$$S^{ij}(k)S^{ij}(-k) = I \tag{16.32}$$

with i, j, m, n all mutually distinct.

....

For an arbitrary number N of electrons, the above argument generalizes to $N - 1$ boundaries $x_1 = x_2, \ldots, x_{N-1} = x_N$ again connecting pairs of amplitudes from the fundamental sector. Thus, there are $N!(N - 1)/2$ equations for $N!$ amplitudes which are, however, consistent. The general proof of consistency, using some results from group theory we have not developed in this book, can be found in the literature, (see Zinn-Justin and Brézin (1966) or chapter 10 of the book by Gaudin (1983) or its English translation, Gaudin (2014)).

More interesting are the periodic boundary conditions. Let us, for instance, consider the condition on the Bethe ansatz wave function

$$\phi_{3\sigma_1\sigma_2\sigma_3}(x_1 = 0, x_2, x_3) = \phi_{3\sigma_1\sigma_2\sigma_3}(x_1 = L, x_2, x_3). \tag{16.33}$$

Imagining the three Fermi particles sitting on a ring of circumference L, we have to choose an orientation for the propagation of the particles. Thus, the periodic boundary condition connects either the sector $F_3^{(123)} : x_1 < x_2 < x_3$ with the sector $F_3^{(231)} : x_2 < x_3 < x_1$ or the sectors $F_3^{(132)} : x_1 < x_3 < x_2$ and $F_3^{(321)} : x_3 < x_2 < x_1$. We may choose one of these cases, say the first one, i.e. the periodic boundary condition equates the wave functions $\phi^{(123)}_{3\sigma_1\sigma_2\sigma_3}(x_1, x_2, x_3)$ at $x_1 = 0$ and $\phi^{(231)}_{3\sigma_1\sigma_2\sigma_3}(x_1, x_2, x_3)$ at $x_1 = L$.

Furthermore, in order for one particle to be able to pass the others, we need an ordering by magnitude of the momenta k_1, k_2, k_3. For our case, this ordering must be chosen to be $k_1 > k_2, k_3$. This ordering selects two conditions between the amplitudes, one of which is

$$e^{-ik_1L}A^{(123)}_{\sigma_1\sigma_2\sigma_3}(123) = A^{(231)}_{\sigma_1\sigma_2\sigma_3}(123). \tag{16.34}$$

The other condition involving the momentum k_1 is equivalent to (16.34) due to the amplitude relations derived from the Fermi antisymmetry and the Schrödinger equation. It suffices thus to consider the periodic boundary condition (16.34).

Now, again with the help of the relations between amplitudes, the right-hand side of (16.34) can be expressed through the amplitude on the left-hand side, i.e. it can be written as

$$\begin{aligned} A^{(231)}_{\sigma_1\sigma_2\sigma_3}(123) &= Y(k_1-k_3)\Pi_{13}Y(k_1-k_2)\Pi_{12}A^{(123)}_{\sigma_1\sigma_2\sigma_3}(123) \\ &= S(k_1-k_3)S(k_1-k_2)A^{(123)}_{\sigma_1\sigma_2\sigma_3}(123). \end{aligned} \tag{16.35}$$

The product of scattering matrices with all spin indices shown explicitly

$$\big(S(k_1-k_3)S(k_1-k_2)\big)^{\sigma_1'\sigma_2'\sigma_3'}_{\sigma_1\sigma_2\sigma_3} = \sum_{\sigma_1''} S^{\sigma_3'}_{\sigma_3}(\sigma_1,\sigma_1'')(k_1-k_3)S^{\sigma_2'}_{\sigma_2}(\sigma_1'',\sigma_1')(k_1-k_2) \tag{16.36}$$

reveals a structure similar to the product (10.19) of two R-matrices. This observation, which will generalize to N electrons, will prove to be the key to the complete solution of the model. Again we note that, as in (10.19) for the R-matrices, the product (16.36) is *not* a conventional matrix product.

EXERCISE 16.6 Eigenvalue equation for three electrons 1 Convince yourself of the validity of the eigenvalue equation (16.35) for three electrons.

Combining the relations (16.34) and (16.35) leads again, as in the case of two electrons, to an eigenvalue equation for the eigenvector $A^{(123)}_{\sigma_1\sigma_2\sigma_3}(123)$, i.e. the three-electron amplitude, with eigenvalue $\mu_1 = e^{-ik_1L}$ of the product $S^{13}S^{12} = S(k_1-k_3)$ $S(k_1-k_2)$ of scattering operators.

EXERCISE 16.7 Eigenvalue equation for three electrons 2 Determine the two other eigenvalue equations for the eigenvalues $\mu_2 = e^{-ik_2L}$ and $\mu_3 = e^{-ik_3L}$ from the periodic boundary conditions $\phi_{3\sigma_1\sigma_2\sigma_3}(x_1, x_2 = 0, x_3) = \phi_{3\sigma_1\sigma_2\sigma_3}(x_1, x_2 = L, x_3)$ and $\phi_{3\sigma_1\sigma_2\sigma_3}(x_1, x_2, x_3 = 0) = \phi_{3\sigma_1\sigma_2\sigma_3}(x_1, x_2, x_3 = L)$.

Let us summarize the consequences of the periodic boundary conditions for the three relevant amplitudes, i.e. the three eigenvalue equations:

$$e^{-ik_1L}A^{(123)}_{\sigma_1\sigma_2\sigma_3}(123) = S^{13}S^{12}A^{(123)}_{\sigma_1\sigma_2\sigma_3}(123), \tag{16.37}$$

$$e^{-ik_2L}A^{(231)}_{\sigma_1\sigma_2\sigma_3}(123) = S^{21}S^{23}A^{(231)}_{\sigma_1\sigma_2\sigma_3}(123), \tag{16.38}$$

$$e^{-ik_3L}A^{(312)}_{\sigma_1\sigma_2\sigma_3}(123) = S^{32}S^{31}A^{(312)}_{\sigma_1\sigma_2\sigma_3}(123). \tag{16.39}$$

Before we interpret these three equations, we introduce a more appropriate notation, going back to Belavin (1979), for the amplitudes that helps shift the focus from the upper index relating to the sector in configuration space to the spin indices. The general definition is for N particles

$$A^{(Q)}_{\sigma_1\sigma_2\ldots\sigma_N}(P) = (-1)^P A_{\sigma_{q_1}\sigma_{q_2}\ldots\sigma_{q_N}}(P\cdot Q) \tag{16.40}$$

where $Q = (q_1, q_2, \ldots, q_N)$ is a permutation of the sectors in configuration space and $P \cdot Q$ is the product permutation, i.e. the permutation where the permutation Q is followed by the permutation P. Applying this notation to the periodic boundary conditions for the amplitudes of the three particle case, we obtain

$$e^{-ik_1L}A_{\sigma_1\sigma_2\sigma_3}(123) = S^{13}S^{12}A_{\sigma_1\sigma_2\sigma_3}(123), \tag{16.41}$$

$$e^{-ik_2L}A_{\sigma_2\sigma_3\sigma_1}(231) = S^{21}S^{23}A_{\sigma_2\sigma_3\sigma_1}(231), \tag{16.42}$$

$$e^{-ik_3L}A_{\sigma_3\sigma_1\sigma_2}(312) = S^{32}S^{31}A_{\sigma_3\sigma_1\sigma_2}(312). \tag{16.43}$$

These equations reveal that the operators $\mathcal{Z}_j$ ($j = 1, 2, 3$), defined as

$$\mathcal{Z}_1 = S^{13}S^{12}, \qquad \mathcal{Z}_2 = S^{21}S^{23}, \qquad \text{and} \qquad \mathcal{Z}_3 = S^{32}S^{31}, \tag{16.44}$$

act in the same way on the spin indices σ_j and the indices formed from the sector indices Q relating to the configuration space and the permutation indices P relating to the momenta of the plane wave in the Bethe ansatz wave function. In this sense, the operators $\mathcal{Z}_j$ only act on the spin degrees of freedom of the electron system.

The three eigenvalue equations of the periodic boundary conditions can now be interpreted: we take, e.g. particle 1 and carry it around the ring that represents the periodicity of the system, first past particle 2, then past particle 3, until particle 1 is back at its original position. While being carried around the ring, the amplitude has been modified by two scattering events with a product of scattering matrices $S^{13}S^{12}$ that results in an overall phase factor $\mu_1 = e^{-ik_1L}$ picked up by the amplitude. The eigenvalue equation (16.41) encodes this process, while (16.42) and (16.43) encode the corresponding processes for particles 2 and 3, respectively.

This interpretation suggests the general definition, valid for an arbitrary number N of electrons, of an operator that is the product of scattering operators

$$\mathcal{Z}_j = S^{jj-1}S^{jj-2}\ldots S^{j1}S^{jN}S^{jN-1}\ldots S^{jj+1}. \tag{16.45}$$

Noting again that the product of scattering matrices resulting in the operator $\mathcal{Z}_j$ is not a conventional matrix product but a product where 'inner' spins are summed over as in (16.36), we state the dependence on the spin variables of the N electrons explicitly

$$\mathcal{Z}_j = \left(\mathcal{Z}_j\right)^{\sigma'_1 \dots \sigma'_N}_{\sigma_1 \dots \sigma_N}. \tag{16.46}$$

Here, particle j is carried around the ring, first past particle $j + 1$ and so on until it is carried past particle N. Because of the periodicity of the particle system on the ring, the next particle after particle N will be particle 1, past which particle j is now carried, then past 2, and so on until particle j reaches its original position after it has finally been carried past particle $j - 1$. The corresponding eigenvalue equation with $\mu_j = e^{-ik_jL}$ is thus

$$\begin{aligned}&\mu_j A_{\sigma_j\sigma_{j+1}\dots\sigma_N\sigma_1\sigma_2\dots\sigma_{j-1}}(jj+1\dots N12\dots j-1) = \\ &\quad \mathcal{Z}_j A_{\sigma_j\sigma_{j+1}\dots\sigma_N\sigma_1\sigma_2\dots\sigma_{j-1}}(jj+1\dots N12\dots j-1), \\ &\quad j = 1, \dots, N.\end{aligned} \tag{16.47}$$

Once we have generalized the definition of the scattering operators S^{ij} to the situation when i and j are *not* nearest neighbours, a somewhat lengthy but direct algebraic calculation shows that the operators $\{\mathcal{Z}_j\}$ mutually commute

$$\mathcal{Z}_i \mathcal{Z}_j = \mathcal{Z}_j \mathcal{Z}_i \tag{16.48}$$

and therefore can be all diagonalized simultaneously.

The operators $\mathcal{Z}_j$ exhibit a similar mathematical structure as the objects encountered in the discussion of the algebraic Bethe ansatz in part II, namely, the monodromy and the transfer matrices for vertex models which we introduced and examined in section 10.7. Below, after identifying analogous objects also for the one-dimensional Hubbard and the Kondo model, we shall make these similarities more precise and explicit, which will allow us to transfer the machinery of the algebraic Bethe ansatz from part II to the present and similar problems.

Comparing the Bose case (cf. chapter 15, section 15.8) with our present case of the Fermi gas of electrons, we observe that the quantity corresponding to the operator $\mathcal{Z}_j$ is given in the Bose case by the scalar function

$$\exp i\left(2\pi n_j - \sum_{n=1}^{N} \Theta(k_j - k_n)\right) \tag{16.49}$$

(cf. section 15.9.1). In other words, for the one-dimensional Bose gas, the problem of diagonalizing the model Hamiltonian was solved in the sense that it was reduced to a problem of finding the solutions of a set of algebraic equations. This problem, in itself not trivial, was then solved, e.g. for the ground state of the Bose gas in the thermodynamic limit (see section 15.9).

The Fermi case, however, presents a whole new and fully fledged eigenvalue problem at this point of the analysis. The solution of this new eigenvalue problem proceeds first of all by the insight that we are dealing with a spin problem that has some similarity with the Heisenberg quantum spin chain. The original Bethe ansatz solution of the δ-Fermi gas used this insight to formulate a second ansatz composed of single-particle spin functions, which has the character of a second coordinate Bethe ansatz (see Fung, 1981 for this approach).

We shall not follow this strategy here. The analogy with the spin chain problem can be put to even more fruitful use when we consider the algebraic Bethe ansatz instead of only the coordinate Bethe ansatz. The structure of the operators $\mathcal{Z}_j$ and also their commutation property (16.48) suggest that the algebraic Bethe ansatz will be very well adapted to solve this second eigenvalue problem.

Before we move on to solve the second eigenvalue problem, we end our discussion of the Bethe ansatz solution (specifically of the one-dimensional Fermi gas) and, after briefly looking at the one-dimensional Hubbard model, continue with a treatment of the specifics of the Kondo model. As we shall see, all three models will lead to a formally identical second eigenvalue problem, i.e. to the eigenvalue problem (16.47) for suitably defined operators $\mathcal{Z}_j$, which are, however, different in detail for the three models. We then return to the second eigenvalue problem by bringing it into a form best suited for the methods of the algebraic Bethe ansatz.

16.2 One-dimensional Hubbard model

The Hubbard model first appeared in chapter 8 with increasing frequency as one of the fundamental models discussed in condensed matter theory. We then fully introduced the model in section 8.7 and explained some of its properties. In its one-dimensional manifestation it may be considered a discrete version of the one-dimensional δ-Fermi gas introduced in section 16.1. But, vice versa, the latter model may be construed as a continuum version of the former.

Shortly after the solution by an extended version of the Bethe ansatz of the δ-Fermi gas, Lieb and Wu (1968) extended the solution method to solve the one-dimensional version of the Hubbard model, whose Hamiltonian for a lattice of L sites is

$$\mathcal{H} = -t \sum_{i=1\sigma}^{L} \left(c_{i\sigma}^{\dagger} c_{i+1\sigma} + c_{i+1\sigma}^{\dagger} c_{i\sigma} \right) + U \sum_{i=1}^{L} n_{i\uparrow} n_{i\downarrow} \equiv \mathcal{H}_t + \mathcal{H}_U. \tag{16.50}$$

Furthermore, the original publication by Lieb and Wu (1968) was very dense, and the authors considered it worth their while to return to the subject many years later (Lieb and Wu, 2003). Since there is a comprehensive monograph by Essler *et al.* (2005) devoted to the one-dimensional Hubbard model, we shall not go into any details of the solution of this model. Instead, this section outlines how the equations pertaining to the solution of the δ-Fermi gas, have to be adapted to the case of the Hubbard model.

The states spanning the Hilbert space $\mathscr{H}_N$ of $N \leq 2L$ electrons with respect to the vacuum state $|0\rangle$ defined by

$$c_{i\sigma}|0\rangle = 0 \tag{16.51}$$

are given by

$$|\phi\rangle = \sum_{\sigma_1\ldots\sigma_N}\sum_{n_1\ldots n_N} \phi_{\sigma_1\ldots\sigma_N}(n_1\ldots n_N)\prod_{i=1}^{N} c_{n_i\sigma_i}|0\rangle \tag{16.52}$$

where $\sigma_j = \pm 1$ is the spin-projection and n_j the position of the jth electron.

The eigenvalue equation $\mathcal{H}|\phi\rangle = E|\phi\rangle$ then becomes in first quantized form

$$h\phi_{\sigma_1\ldots\sigma_N}(n_1\ldots n_N) = E\phi_{\sigma_1\ldots\sigma_N}(n_1\ldots n_N) \tag{16.53}$$

with the first quantized Hamiltonian

$$h = -t\sum_{j=1}^{N}\Delta_j + U\sum_{\substack{j,l=1\\ j<l}}^{N}\delta_{n_j,n_l} \tag{16.54}$$

where Δ_j is the hopping operator, a discrete version of the Lapacian whose action on the position n_j of the jth electron is

$$\begin{aligned}\Delta_j\phi_{\sigma_1\ldots\sigma_N}(n_1\ldots n_j\ldots n_N) =&\phi_{\sigma_1\ldots\sigma_N}(n_1\ldots n_j+1\ldots n_N)\\ &+\phi_{\sigma_1\ldots\sigma_N}(n_1\ldots n_j-1\ldots n_N).\end{aligned} \tag{16.55}$$

EXERCISE 16.8 Hubbard Hamiltonian in first quantization Derive the first quantized Hamiltonian (16.54) with (16.55) for the state (16.52) and second quantized Hubbard Hamiltonian (16.50).

An analysis similar to the two-particle analysis for the δ-Fermi gas leads to the two-particle scattering matrix for the one-dimensional Hubbard model

$$S^{ij} \equiv S_{\sigma_j}^{\sigma_j'}(\sigma_i,\sigma_i') = \frac{(\sin k_i - \sin k_j)I_{\sigma_j}^{\sigma_j'}(\sigma_i,\sigma_i') + i\frac{u}{2}\Pi_{\sigma_j}^{\sigma_j'}(\sigma_i,\sigma_i')}{(\sin k_i - \sin k_j) + i\frac{u}{2}} \tag{16.56}$$

with $u = U/t$ (for details of this calculation, see chapter 3 of Essler *et al.*, 2005). A comparison shows that a formal replacement $k \rightarrow \sin k$ and $c \rightarrow u/2$ identifies the scattering matrices of the δ-Fermi gas and the Hubbard model. Energy and momenta are given by

$$E = -2t \sum_{j=1}^{N} \cos k_j, \qquad P = \sum_{j=1}^{N} k_j. \tag{16.57}$$

In full analogy to the discussion of periodic boundary conditions for the δ-Fermi gas, we can obtain an eigenvalue equation that is formally identical to (16.47) but where the operator $\mathcal{Z}_j$ is now formed by the scattering matrices (16.56) of the Hubbard model. Again, for a full demonstration of these calculations and much more, we recommend Essler and colleagues' (2005) comprehensive monograph on the one-dimensional Hubbard model.

We now turn our attention to the Kondo model, which, in the sense discussed in section 8.9.3.2, can be considered as a genuine one-dimensional model where an electronic system interacts with a localized magnetic moment or spin.

16.3 Kondo model

Section 8.9 investigated how magnetic impurities behave in metallic hosts. As a prototypical model for magnetic impurities, we derived the Kondo model. Section 8.9.3.2 showed that the Kondo problem of a single magnetic impurity with spin σ_0 corresponding to a spin quantum number $s = 1/2$ in an otherwise non-magnetic metallic host may be described by a one-dimensional Kondo model whose Hamiltonian is given by

$$\mathcal{H} = -i \int_{-\infty}^{\infty} dx \sum_{\sigma} \psi_{\sigma}^{\dagger}(x) \frac{\partial}{\partial x} \psi_{\sigma}(x) + \mathcal{J} \boldsymbol{\sigma}_0 \cdot \sum_{\sigma\sigma'} \psi_{\sigma}^{\dagger}(0) \boldsymbol{\sigma}_{\sigma\sigma'} \psi_{\sigma'}(0)$$
$$+ \mathcal{J}' \sum_{\sigma} \psi_{\sigma}^{\dagger}(0) \psi_{\sigma}(0). \tag{16.58}$$

In order to show that also a slightly more general model Hamiltonian is possible, we have also indicated a term that couples with strength $\mathcal{J}'$ to the charge density at the impurity position $x = 0$. This term corresponds to an additional potential scattering from the impurity site. We have also absorbed the factor $1/2\pi$ of (8.735) into a redefinition of the coupling constants $\mathcal{J}$ and $\mathcal{J}'$. In the following, however, we dispose with the potential scattering term and set accordingly $\mathcal{J}' = 0$ because keeping the term would only obfuscate the formulas without adding to the physical insight in our intended context.

Our task is now to diagonalize this Hamiltonian, using the methods of the Bethe ansatz we developed in the previous chapters. This task can formally be accomplished in a very similar way to the treatments of the δ-Fermi gas of section 16.1 and the one-dimensional Hubbard model of section 16.2. More concretely, this means that we first have to find the scattering matrix. We shall, thus, in following similar steps as for the δ-Fermi gas, only have to take proper care to include the impurity in the analysis.

The original studies of the Kondo problem were undertaken by Andrei (1980), Andrei and Lowenstein (1981), Wiegmann (1980), Wiegmann (1981), Filyov *et al.* (1981) (see also the reviews by Andrei *et al.*, 1981, and Tsvelik and Wiegmann, 1983).

16.3.1 Bethe ansatz for the Kondo model

The many-particle Hilbert space of a system consisting of N electrons and the impurity in a spin state $|s = \pm 1/2\rangle$ is spanned by the states

$$|\Psi\rangle = \int d^N x \sum_{\sigma_1 \ldots \sigma_N} \sum_s \phi_{\sigma_1,\ldots,\sigma_N,s}(x_1,\ldots,x_N)\psi^\dagger_{\sigma_1}(x_1)\ldots\psi^\dagger_{\sigma_N}(x_N)|0\rangle_{\text{el}}|s\rangle \quad (16.59)$$

where $\phi_{\sigma_1,\ldots,\sigma_N,s}(x_1,\ldots,x_N)$ is a wave function that depends on the N electron coordinates x_j and the $N+1$ spin variables $\sigma_j, j = 1,\ldots,N$ for the electrons and $s = \pm 1/2$ for the impurity. The state with zero electrons $|0\rangle_{\text{el}}$ and the impurity state $|s = \pm 1/2\rangle$ is for all σ and x annihilated by $\psi_\sigma(x)$

$$\psi_\sigma(x)|0\rangle = \psi_\sigma(x)|0\rangle_{\text{el}}|s\rangle = 0. \quad (16.60)$$

The stationary Schrödinger eigenvalue equation in second quantized form

$$\mathcal{H}|\Psi\rangle = E|\Psi\rangle \quad (16.61)$$

translates into the corresponding Schrödinger eigenvalue equation for the wave function ϕ

$$h\phi_{\sigma_1,\ldots,\sigma_N,s}(x_1,\ldots,x_N) = E\phi_{\sigma_1,\ldots,\sigma_N,s}(x_1,\ldots,x_N) \quad (16.62)$$

in the Hilbert space $\mathscr{H} = \mathscr{H}_N \otimes \mathscr{H}_{\text{imp}}$ of N electrons and the impurity where the first quantized Hamiltonian has become

$$h = -i\sum_{j=1}^{N} \frac{\partial}{\partial x_j} + \mathcal{J}\boldsymbol{\sigma}_0 \cdot \sum_{j=1}^{N} \delta(x_j)\boldsymbol{\sigma}_j. \quad (16.63)$$

EXERCISE 16.9 Impurity and one and two electrons Convince yourself that (16.63) is the correct first quantized Hamiltonian corresponding to the second quantized Hamiltonian (16.58) for $\mathcal{J}' = 0$ in the special cases of one and two electrons interacting with the impurity.

We now want to construct in a consistent way the many-particle wave function of the Hamiltonian (16.63) from single-particle wave functions or, more precisely, from the wave function for the problem of the impurity and a single electron.

16.3.1.1 Impurity and single electron

We now consider the simplest case that will reveal a large piece of the problem: a single electron j at position x_j interacting with the impurity, which we shall henceforth label 0. The Schrödinger equation of this problem has the explicit form

$$\sum_{s'}\sum_{\sigma_j'}\left(-i\frac{\partial}{\partial x_j}I_s^{s'}(\sigma_j,\sigma_j')+\mathcal{J}\delta(x_j)\,(\boldsymbol{\sigma}_0)_s^{s'}\cdot(\boldsymbol{\sigma})_{\sigma_j}^{\sigma_j'}\right)\phi_{\sigma_j's'}(x_j)=E\phi_{\sigma_js}(x_j) \tag{16.64}$$

where $I_s^{s'}(\sigma_j,\sigma_j')$ are the elements of the unit matrix.

As long as the electron is away from the impurity at $x=0$, i.e. for $x_j\neq 0$, we expect the solution of the eigenvalue equation of the first quantized Hamiltonian h, (16.63) or, more precisely, the Schrödinger equation (16.64), to be a plane wave. Let us, however, write this plane wave in the particular form

$$\phi_{\sigma_js}(x_j)=e^{ikx_j}\left(A_{\sigma_js}\theta(-x_j)+B_{\sigma_js}\theta(x_j)\right) \tag{16.65}$$

where $\theta(x)$ is the step function for which we use the convention

$$\theta(x)=\begin{cases}1 & \text{for} \quad x>0\\ \frac{1}{2} & \text{for} \quad x=0\\ 0 & \text{for} \quad x<0\end{cases}. \tag{16.66}$$

The coefficient factors A_{σ_js} and B_{σ_js} do not depend on the position of the electron, but only on the spin states of the impurity and the electron. This form of the wave function will prove to be very well adapted to our impurity problem, since we allow for two different amplitudes depending on whether the electron is right ($x_j>0$)

$$\phi_{\sigma_js}(x_j)=B_{\sigma_js}e^{ikx_j} \tag{16.67}$$

or left ($x_j<0$)

$$\phi_{\sigma_js}(x_j)=A_{\sigma_js}e^{ikx_j} \tag{16.68}$$

of the impurity.

In other words, we have introduced two sectors, an approach employed previously in the Bethe ansatz analysis of other models. The two sectors are $F^{(I)}=(10):x_j<0$, which we may choose as fundamental sector, and $F^{(Q)}=(01):0<x_j$. Accordingly, we shall later use a notation for the amplitudes

$$A_{\sigma_js}=A_{\sigma_js}(10)\,,\qquad B_{\sigma_js}=A_{\sigma_js}(01). \tag{16.69}$$

From the eigenvalue equation (16.64), we obtain for $x_j\neq 0$ the relativistic energy or momentum eigenvalue

$$E(k)=P(k)=k. \tag{16.70}$$

For $x = 0$, on the other hand, the eigenvalue equation becomes

$$(h - E)\,\phi_{\sigma_j s}(0) = \sum_{\sigma'_j s'} \Big\{ -iI_s^{s'}(\sigma_j, \sigma'_j)\left(B_{\sigma'_j s'} - A_{\sigma'_j s'}\right)$$

$$+\frac{1}{2}\mathcal{J}(\boldsymbol{\sigma}_0)_s^{s'} \cdot (\boldsymbol{\sigma}_j)_{\sigma_j}^{\sigma'_j}\left(A_{\sigma'_j s'} + B_{\sigma'_j s'}\right)\Big\}\,\delta(0) = 0 \qquad (16.71)$$

where, using (16.66), we employed the convention

$$\theta(x)\delta(x) = \theta(-x)\delta(x) = \frac{1}{2}\delta(0). \qquad (16.72)$$

From this equation, we can determine the relation between the amplitude $B_{\sigma_j s}$ on the right and the amplitude $A_{\sigma_j s}$ on the left of the impurity

$$B_{\sigma''_j s''} = \sum_{\sigma'_j s'} \sum_{\sigma_j s} \left\{M^{-1}\right\}_{s''}^{s} (\sigma_j, \sigma''_j) \cdot \left\{ iI_s^{s'}(\sigma_j, \sigma'_j) + \frac{1}{2}\mathcal{J}(\boldsymbol{\sigma}_0)_s^{s'} \cdot (\boldsymbol{\sigma}_j)_{\sigma_j}^{\sigma'_j} \right\} A_{\sigma'_j s'} \qquad (16.73)$$

$$\equiv \sum_{\sigma'_j s'} S_{s''}^{s'}(\sigma''_j, \sigma'_j) A_{\sigma'_j s'} \qquad (16.74)$$

where M^{-1} is the inverse of the matrix

$$M_s^{s'}(\sigma_j, \sigma'_j) = \left\{ iI_s^{s'}(\sigma_j, \sigma'_j) - \frac{1}{2}\mathcal{J}(\boldsymbol{\sigma}_0)_s^{s'} \cdot (\boldsymbol{\sigma}_j)_{\sigma_j}^{\sigma'_j} \right\}. \qquad (16.75)$$

Dropping the double primes from the indices, we arrive at the definition of the scattering matrix S for the scattering between an electron and the impurity

$$B_{\sigma_j s} = \sum_{\sigma'_j s'} S_s^{s'}(\sigma_j, \sigma'_j) A_{\sigma'_j s'}, \qquad (16.76)$$

which can be expressed as

$$\mathcal{S}^{j0} \equiv S_s^{s'}(\sigma_j, \sigma'_j) = e^{i\phi}\frac{I^{j0} - icP^{j0}}{1 - ic}. \qquad (16.77)$$

Here, we have introduced

$$P^{j0} \equiv \frac{1}{2}\left(I^{j0} + \boldsymbol{\sigma}_0 \cdot \boldsymbol{\sigma}_j\right) \qquad (16.78)$$

$$\equiv P_s^{s'}(\sigma_j, \sigma_j') = \frac{1}{2}\left(I_s^{s'}(\sigma_j, \sigma_j') + (\boldsymbol{\sigma}_0)_s^{s'} \cdot (\boldsymbol{\sigma}_j)_{\sigma_j}^{\sigma_j'}\right) = \delta_{\sigma_j}^{s'} \delta_s^{\sigma_j'}, \tag{16.79}$$

the permutation operator that exchanges the spin of the jth electron in the conduction band and the spin of the impurity. The identity matrix, which leaves impurity and electron spin untouched, is explicitly

$$I^{j0} \equiv I_s^{s'}(\sigma_j, \sigma_j') = \delta_{\sigma_j}^{\sigma_j'} \delta_s^{s'}. \tag{16.80}$$

Assuming, for instance, the impurity to be initially in the spin-up state ($\Uparrow$), the exchange described by the operator (16.78) or (16.79), respectively, may be depicted as

$$\uparrow\Uparrow \quad \Longrightarrow \quad \uparrow\Uparrow, \tag{16.81}$$

$$\downarrow\Uparrow \quad \Longrightarrow \quad \uparrow\Downarrow. \tag{16.82}$$

The scattering matrix contains a phase factor depending on the coupling constant $\mathcal{J}$

$$e^{i\phi} = \frac{1 - \frac{3}{4}\mathcal{J}^2 - 2i\mathcal{J}}{1 + \frac{3}{4}\mathcal{J}^2 + i\mathcal{J}} \tag{16.83}$$

and the effective coupling constant

$$c = \frac{2\mathcal{J}}{1 - \frac{3}{4}\mathcal{J}^2}. \tag{16.84}$$

In summary, the wave function for the impurity and a single electron expressed using the scattering matrix is

$$\phi_{\sigma_j s}(x_j) = e^{ikx_j}\left(I_s^{s'}(\sigma_j, \sigma_j')\theta(-x_j) + S_s^{s'}(\sigma_j, \sigma_j')\theta(x_j)\right) A_{\sigma_j' s'}. \tag{16.85}$$

The scattering matrix will again play a prominent role in our further analysis.

EXERCISE 16.10 Electron-impurity scattering matrix Convince yourself by an explicit calculation that (16.77) follows from (16.73) together with (16.75), the phase (16.83), and the effective coupling constant (16.84).

16.3.1.2 Impurity and two electrons

We now know how to describe the scattering between a single electron and the impurity through the scattering matrix S^{j0}. Next, we look at three constituents, two electrons i and j and the impurity, which raises the question of how to include the electron-electron scattering in the emerging picture. Since the electrons do not interact among themselves, it is tempting to assume that the identity matrix $S^{ij} = I^{ij}$ would correctly take the electron-electron scattering into account. However, this is not the case.

An identity matrix for the electron-electron scattering, $S^{ij} = I^{ij}$, would not satisfy the Yang–Baxter relation for the scattering matrices between electrons and between electrons and the impurity of the Kondo problem, expressed by

$$S^{ij} S^{i0} S^{j0} = S^{j0} S^{i0} S^{ij} \tag{16.86}$$

and demonstrated graphically in figure 16.2.

Moreover, $S^{ij} = I^{ij}$ would imply that (16.86) reduces simply to

$$S^{i0} S^{j0} = S^{j0} S^{i0}. \tag{16.87}$$

However, this last equation is not fulfilled (see exercise 16.11), i.e. we in fact have instead

$$S^{i0} S^{j0} \neq S^{j0} S^{i0}, \tag{16.88}$$

which demonstrates an important physical aspect of the Kondo problem. The state of the impurity with which the electrons interact depends on whether a spin-up electron scatters first and then a spin-down electron, or vice versa. This can be illustrated by considering the impurity initially in, say, a spin-up state, depicted by $\Uparrow$. If we now let a

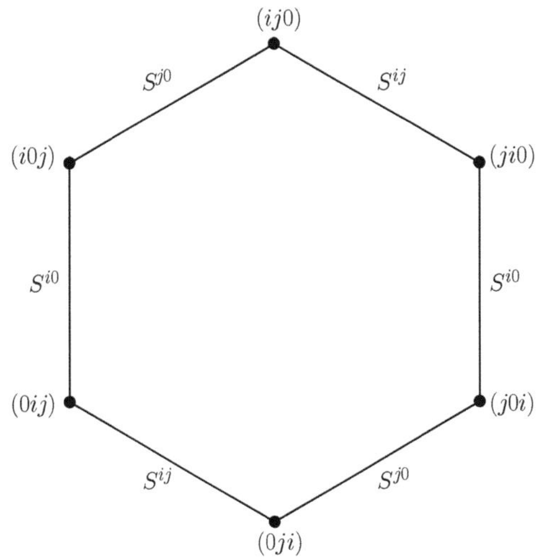

Figure 16.2 *The Yang–Baxter relation for the interaction of electrons and Kondo impurity. There are two paths from (ij0), where both electrons are considered to be to the left of the impurity, to (0ji), where electrons and impurity have changed relative position. These paths are visualized in the diagram. Since the final state is the same, the two paths must be equivalent, i.e. the Yang–Baxter relation must be satisfied for the scattering matrices:* $S^{ij} S^{i0} S^{j0} = S^{j0} S^{i0} S^{ij}$.

spin-up electron (↑) scatter first, and then a spin-down electron (↓), we get the following process and final state of impurity and electrons

$$\downarrow\ \uparrow\ \Uparrow\ \Longrightarrow\ \downarrow\ \Uparrow\ \uparrow\ \Longrightarrow\ \Downarrow\ \uparrow\ \uparrow\,. \tag{16.89}$$

Next, consider the scattering processes where first a spin-down and then a spin-up electron scatters off the spin-up impurity

$$\uparrow\ \downarrow\ \Uparrow\ \Longrightarrow\ \uparrow\ \Downarrow\ \uparrow\ \Longrightarrow\ \Uparrow\ \downarrow\ \uparrow\,. \tag{16.90}$$

We observe that both the impurity and the conduction electrons as well as the electrons among themselves become strongly correlated because of the scattering off the impurity. In particular, the conduction electrons, although they do not interact with each other directly, become strongly correlated. This system of conduction electrons, which are now strongly correlated due to the impurity scattering, is often called the Kondo screening cloud. For a recent review on the Kondo screening cloud, see Affleck (2010).

The scattering of electrons off a magnetic impurity is thus markedly different from the scattering of electrons off a potential impurity (o) where the spins of the electron do not change their state and, hence, do not become correlated:

$$\uparrow\downarrow\ o \Longrightarrow \uparrow\ o\ \downarrow \Longrightarrow\ o\ \uparrow\downarrow\,. \tag{16.91}$$

EXERCISE 16.11 Non-commutativity of scattering matrices Prove formally that $S^{i0}S^{j0} \neq S^{j0}S^{i0}$ from the definition of the electron-impurity scattering matrix (16.77), which is depicted in (16.81) and (16.82).

Since the identity $S^{ij} = I^{ij}$ cannot work as the scattering matrix for the electrons among themselves, the considerations above, i.e. the Yang–Baxter relation and the physical picture presented, suggest that the correct choice of S^{ij} will prove to be

$$S^{ij} = P^{ij} = \frac{1}{2}\left(I^{ij} + \boldsymbol{\sigma}_i \cdot \boldsymbol{\sigma}_j\right), \tag{16.92}$$

the spin exchange operator between the electron spins.

Although it is expected, it can also be shown formally that $N = 2$ electrons, which scatter off each other back and forth, satisfies

$$S^{ij}S^{ji} = I^{ij}, \tag{16.93}$$

i.e. the inverse of the scattering matrix S^{ij} is

$$\left(S^{ij}\right)^{-1} = S^{ji}. \tag{16.94}$$

The Yang–Baxter relations, together with this relation for the inverse of the scattering matrix, guarantee the path independence of the choice of amplitudes in the wave function for an arbitrary number N of electrons as has been shown by Zinn-Justin and Brézin (1966).

16.3.2 Bethe basis

This section returns to the question of the correct scattering matrix for the electrons. We emphasize that the electrons in the Kondo model we consider are non-interacting electrons. However, we have seen that the scattering matrix cannot simply be the identity matrix. Rather, we suggested that the scattering matrix of the electrons between themselves is the spin exchange operator $S^{ij} = P^{ij}$.

How can we understand this? Does it not mean that we introduced explicit interactions between the electrons?

Consider the case of two electrons whose Hamiltonian is

$$h = -i\left(\frac{\partial}{\partial x_1} + \frac{\partial}{\partial x_2}\right). \tag{16.95}$$

An eigenfunction of this non-interacting Hamiltonian is given by plane waves with arbitrary independent coefficients $A_{\sigma_1\sigma_2}$ and $A_{\sigma_2\sigma_1}$

$$\phi^q_{\sigma_1\sigma_2}(x_1, x_2) = A_{\sigma_1\sigma_2}e^{ik_1x_1+ik_2x_2} + A_{\sigma_2\sigma_1}e^{ik_2x_1+ik_1x_2} \tag{16.96}$$

The spectrum of this system is, however, given by

$$E = k_1 + k_2 = (k_1 + q) + (k_2 - q) \tag{16.97}$$

for arbitrary q. Thus, the spectrum is infinitely degenerate. We can thus use a continuously infinite number of eigenfunctions

$$\phi^q_{\sigma_1\sigma_2}(x_1, x_2) = A_{\sigma_1\sigma_2}e^{i(k_1+q)x_1+i(k_2-q)x_2} + A_{\sigma_2\sigma_1}e^{i(k_2-q)x_1+i(k_1+q)x_2} \tag{16.98}$$

to build up by appropriate summation the eigenfunction basis best adapted to our problem.

This means that the wave function

$$\phi_{\sigma_1\sigma_2}(x_1, x_2) = \mathcal{A}e^{ik_1x_1+ik_2x_2}\left(A^{12}_{\sigma_1\sigma_2}\theta(x_1 - x_2) + A^{21}_{\sigma_1\sigma_2}\theta(x_2 - x_1)\right) \tag{16.99}$$

$$= \mathcal{A}e^{ik_1x_1+ik_2x_2}\left(A^{12}_{\sigma_1\sigma_2}\theta(x_1 - x_2) + \left(S^{12}A^{12}\right)_{\sigma_1\sigma_2}\theta(x_2 - x_1)\right) \tag{16.100}$$

where $\mathcal{A}$ is the antisymmetrization operator that acts only on the electron coordinates, may be built from eigenfunctions of the type (16.98) and is thus an eigenfunction of the Hamiltonian (16.95) for any choice of scattering matrix. The choice $S^{12} = P^{12}$ corresponds to the so-called Bethe basis while the choice $S^{12} = I^{12}$ would correspond to the Fock basis. The former basis leads to a spin-charge separated wave function

$$\phi_{\sigma_1\sigma_2}(x_1, x_2) = \left(e^{ik_1x_1+ik_2x_2} - e^{ik_1x_2+ik_2x_1}\right) \\ \times \left(A_{\sigma_1\sigma_2}\theta(x_1 - x_2) + A_{\sigma_2\sigma_1}\theta(x_2 - x_1)\right). \tag{16.101}$$

Lowenstein (1982) provides a more detailed discussion of the Bethe basis.

16.3.3 General case: impurity and N electrons

The Bethe ansatz wave function for the impurity and N electrons is a direct generalization of the wave function (16.65). As usual with Bethe wave functions, we have to divide the configuration space into sectors that are defined by a particular ordering of the coordinates.

For instance, the configuration space for two electrons with coordinates x_1 and x_2 and the impurity at $x = 0$ consists of $3! = 6$ sectors. For example, the sector (120) represents $x_1 < x_2 < 0$. We label the six permutations of (120) by $Q^{(3)} = Q^{(2+1)}$.

Within a particular sector labelled by a particular $Q^{(3)}$ we can write the wave function as

$$e^{ik_1x_1+k_2x_2}A_{\sigma_1\sigma_2s}\left(Q^{(3)}\right) \tag{16.102}$$

while the total wave function is given by a summation over all permutations of the sectors

$$\phi_{\sigma_1\sigma_2s}(x_1, x_2) = \mathcal{A}\, e^{ik_1x_1+k_2x_2}\sum_{Q^{(3)}} A_{\sigma_1\sigma_2s}\left(Q^{(3)}\right)\theta\left(x_{Q^{(3)}}\right). \tag{16.103}$$

The step function $\theta(Q^{(3)})$ restricts the wave function to within the sector $Q^{(3)}$, e.g. for (120): $\theta(x_{(120)}) = \theta(x_2 - x_1)\theta(-x_2)$. If the coordinates are ordered according to the particular permutation $Q^{(3)}$, then $\theta\left(x_{Q^{(3)}}\right) = 1$, otherwise, if the coordinates are ordered according to any other permutation $Q'^{(3)}$, $\theta\left(x_{Q'^{(3)}}\right)$ vanishes.

The S-matrices introduced earlier can now be given the additional interpretation to connect the different sectors. If we start in sector (120), we can reach sector (021) by two distinct paths. Since both paths must be equivalent, we again arrive at the Yang–Baxter relation

$$S^{20}S^{10}S^{12} = S^{12}S^{10}S^{20}, \tag{16.104}$$

which we already encountered in the previous section and which is depicted in figure 16.2.

As we have seen explicitly in previous chapters for other one-dimensional models that can be solved exactly by Bethe ansatz, it can be shown also for the Kondo model that the cases of the impurity and $N = 3, 4, \ldots$ electrons present no new complications. They can all be described by using two-particle scattering matrices S^{j0} between electrons and impurity and S^{ij} between electrons. The Yang–Baxter relation suffices to decompose many-particle scattering for the general system of impurity and N electrons into two-particle scattering between impurity and electrons and among electrons. Thus, the wave function is given in a path-independent way by

$$\phi_{\sigma_1\ldots\sigma_N s}(x_1,\ldots,x_N) = \mathcal{A}\, e^{i\sum_{j=1}^N k_j x_j} \sum_{Q^{(N+1)}} A_{\sigma_1\ldots\sigma_N s}\left(Q^{(N+1)}\right)\theta\left(x_{Q^{(N+1)}}\right). \tag{16.105}$$

As in the case of the δ-Fermi gas discussed in section 16.1, rather than considering all the $(N+1)!$ sectors in this wave function, it suffices to consider one reference sector $Q^{(N+1)}$, which we label $I^{(N+1)}$ and connect to the other sectors using the requisite number of scattering matrices. The wave function then becomes

$$\phi_{\sigma_1\ldots\sigma_N s}(x_1,\ldots,x_N) = \mathcal{A} e^{i\sum_{j=1}^N k_j x_j}\cdot$$
$$\cdot \sum_{Q^{(N+1)}} S\left(Q^{(N+1)}\right) A_{\sigma_1\ldots\sigma_N s}\left(I^{(N+1)}\right)\theta\left(x_{Q^{(N+1)}}\right) \tag{16.106}$$

where all $S\left(Q^{(N+1)}\right)$ are products of two-particle scattering matrices. The set of momenta $\{k_j\}$ is the same in all sectors $Q^{(N+1)}$.

16.3.4 Periodic boundary conditions

In order to obtain a constraint on the set of momenta $\{k_j\}$, we have to impose boundary conditions. The conventional choice are periodic boundary conditions on a ring with circumference L where we put the impurity at $x = 0 \equiv L$ and restrict all electron positions to $-L/2 \le x_j \le L/2, j = 1, 2, \ldots, N$

$$\phi_{\sigma_1\ldots\sigma_N s}\left(x_1,\ldots,x_j = \frac{L}{2},\ldots,x_N\right) = \phi_{\sigma_1\ldots\sigma_N s}\left(x_1,\ldots,x_j = -\frac{L}{2},\ldots,x_N\right). \tag{16.107}$$

We use this different definition of the periodic boundary conditions compared to the definition for the δ-Fermi gas because of the special role of the position $x = 0$ as the location of the impurity spin in the Kondo model. The further results will not be significantly affected by this choice. The condition (16.107) connects different sectors of the configuration space in the wave function (16.105). In order to use it, we have to fold back the amplitudes of the different sectors to just one particular sector with the help of the scattering matrices as it is formally done in the wave function (16.106). We must solve whether such a sector exists and how to find it. We shall adopt a constructive approach to these questions.

Again, it is useful to begin with the simplest case of just the impurity and a single electron where the wave function is given by (16.65) which, written using the scattering matrix, is given in (16.85)

$$\phi_{\sigma_j s}(x_j) = e^{ikx_j} \sum_{\sigma_j' s'} \left(I_s^{s'}(\sigma_j, \sigma_j')\theta(-x_j) + S_s^{s'}(\sigma_j, \sigma_j')\theta(x_j) \right) A_{\sigma_j' s'}. \tag{16.108}$$

The periodic boundary condition $\phi_{\sigma_j s}(-L/2) = \phi_{\sigma_j s}(L/2)$ leads to the condition for the amplitude

$$A_{\sigma_j s} e^{-ikL} = \sum_{\sigma_j' s'} S_s^{s'}(\sigma_j, \sigma_j') A_{\sigma_j' s'} \equiv S^{j0} A_{\sigma_j s}. \tag{16.109}$$

This equation again has the character of an eigenvalue equation for the scattering matrix S^{j0} with $A_{\sigma_j s}$ as eigenfunction and e^{-ikL} as eigenvalue. We encountered a similar eigenvalue equation previously for the δ-Fermi gas.

Guided by this simple case, let us now proceed to the general case. The periodic boundary condition (16.107) implies that there are two sectors labelled by the permutations $Q_{\text{left}}^{(N+1)}$ and $Q_{\text{right}}^{(N+1)}$ for which the amplitudes are related by

$$A_{\sigma_1,\ldots,\sigma_N,s}\left(Q_{\text{left}}^{(N+1)}\right) e^{-ik_jL/2} = A_{\sigma_1,\ldots,\sigma_N,s}\left(Q_{\text{right}}^{(N+1)}\right) e^{ik_jL/2} \tag{16.110}$$

where $A_{\sigma_1,\ldots,\sigma_N,s}\left(Q_{\text{left}}^{(N+1)}\right)$ and $A_{\sigma_1,\ldots,\sigma_N,s}\left(Q_{\text{right}}^{(N+1)}\right)$ are the amplitudes in the two distinct sectors of the configuration space. The two permutations $Q_{\text{left}}^{(N+1)}$ and $Q_{\text{right}}^{(N+1)}$ have the properties that the leftmost element of $Q_{\text{left}}^{(N+1)}$ and the rightmost element of $Q_{\text{right}}^{(N+1)}$ are both equal to j, e.g. $Q_{\text{left}}^{(N+1)} = j, q_1, \ldots, q_{j-1}, q_{j+1}, \ldots, q_{N+1}$ and correspondingly for $Q_{\text{right}}^{(N+1)}$.

We now ask if there an amplitude $A_{\sigma_1,\ldots,\sigma_N,s}\left(Q^{(N+1)}\right)$ that connects the two amplitudes $A_{\sigma_1,\ldots,\sigma_N,s}\left(Q_{\text{left}}^{(N+1)}\right)$ and $A_{\sigma_1,\ldots,\sigma_N,s}\left(Q_{\text{right}}^{(N+1)}\right)$ when we apply the requisite number of scattering matrices to it. This amplitude initially has to satisfy only one requirement: in the set of elements of the permutation $Q^{(N+1)}$: $q_1, q_2, \ldots, q_j, \ldots, q_N, q_{N+1}$, one of which corresponds to the fixed position $x = 0$ of the impurity, the jth element must be $q_j = j$. All other elements q_k with $k \neq j$ have no specified values.

Let us construct an equation that such an amplitude would have to satisfy. If this equation, which we expect again to have the character of an eigenvalue equation, can be solved, we shall also have demonstrated that such an amplitude exists.

By acting with S^{jj+1} on the amplitude $A_{\sigma_1\ldots\sigma_N s}\left(Q^{(N+1)}\right)$ in the particular sector labelled by the permutation $Q^{(N+1)}$ just described, we exchange particle j with particle $j+1$, which is to the immediate right of j so that now particle j is to the right of particle $j+1$: $\ldots jj+1\ldots \rightarrow \ldots j+1j\ldots$. This amplitude is

$$S^{jj+1}A_{\sigma_1\ldots\sigma_N s}\left(Q^{(N+1)}\right). \tag{16.111}$$

In this new amplitude, particle j is now immediately to the left of particle $j+2$ and can be exchanged with this particle by applying S^{jj+2}. We continue this process until we reach the right end of the system to arrive at the amplitude

$$A_{\sigma_1,\ldots,\sigma_N,s}\left(Q^{(N+1)}_{\text{right}}\right) = S^{jN+1}S^{jN}S^{jN-1}\ldots S^{jj+2}S^{jj+1}A_{\sigma_1\ldots\sigma_N s}\left(Q^{(N+1)}\right). \tag{16.112}$$

This amplitude has now the required property that particle j is at the right end of the system of particles.

Next, we act with S^{j-1j} on the same initial amplitude $A_{\sigma_1\ldots\sigma_N s}\left(Q^{(N+1)}\right)$ to exchange particle j with its immediate neighbour $j-1$ on the left and continue to act with further scattering matrices $S^{j-2j}, \ldots, S^{1j}$ until the left end of the system is reached. This results in the amplitude

$$A_{\sigma_1,\ldots,\sigma_N,s}\left(Q^{(N+1)}_{\text{left}}\right) S^{1j}S^{2j}\ldots S^{j-2j}S^{j-1j}A_{\sigma_1\ldots\sigma_N s}\left(Q^{(N+1)}\right), \tag{16.113}$$

an amplitude that represents particle j at its left end. We now connect the amplitudes (16.112) and (16.113) by the periodic boundary condition (16.110)

$$\begin{aligned} &S^{1j}S^{2j}\ldots S^{j-2j}S^{j-1j}A_{\sigma_1\ldots\sigma_N s}\left(Q^{(N+1)}\right)e^{-ik_jL} \\ &\quad = S^{jN+1}S^{jN}S^{jN-1}\ldots S^{jj+2}S^{jj+1}A_{\sigma_1\ldots\sigma_N s}\left(Q^{(N+1)}\right). \end{aligned} \tag{16.114}$$

Recalling $(S^{mn})^{-1} = S^{nm}$, this becomes

$$\begin{aligned} &e^{-ik_jL}A_{\sigma_1\ldots\sigma_N s}\left(Q^{(N+1)}\right) \\ &\quad = S^{jj-1}\ldots S^{j1}S^{jN+1}\ldots S^{jj+1}A_{\sigma_1\ldots\sigma_N s}\left(Q^{(N+1)}\right). \end{aligned} \tag{16.115}$$

Again we recognize that this equation is an eigenvalue equation that connects the eigenvector $A_{\sigma_1\ldots\sigma_N s}\left(Q^{(N+1)}\right)$ and the corresponding eigenvalue $\mu_j = e^{-ik_jL}$, or rather, of course, k_j.

The eigenvector is a matrix in spin space with spin indices $\sigma_1\ldots\sigma_N s$. Written explicitly, the eigenvalue equation is

$$\sum_{\sigma'_1\ldots\sigma'_N,s'} (Z_j)^{\sigma'_1\ldots\sigma'_N,s'}_{\sigma_1\ldots\sigma_N,s}A_{\sigma'_1\ldots\sigma'_N,s'}\left(Q^{(N+1)}\right) = e^{-ik_jL}A_{\sigma_1\ldots\sigma_N,s}\left(Q^{(N+1)}\right) \tag{16.116}$$

where the operator $\mathcal{Z}_j$ is given by the product of scattering matrices

$$(\mathcal{Z}_j)_{\sigma_1\ldots\sigma_N,s}^{\sigma'_1\ldots\sigma'_N,s'} = \left(P^{jj-1}\ldots P^{j1}P^{jN+1}\ldots e^{i\phi}\frac{I - icP^{j0}}{1-ic}\ldots P^{jj+1}\right)_{\sigma_1\ldots\sigma_N s}^{\sigma'_1\ldots\sigma'_N s'}. \tag{16.117}$$

Here, we recalled that the scattering matrices are $S^{mn} = P^{mn}$ for the electron-electron scattering and $S^{m0} = e^{i\phi}\left(I - icP^{m0}\right)/(1 - ic)$ for the electron-impurity scattering. The multiplicative phase factor $e^{i\phi}$ in the electron-impurity scattering matrix only shifts all momenta k_j by the same amount, and so we can therefore discard it from here on out.

The derivation of this eigenvalue equation is equivalent, but more general, to our treatment of the corresponding eigenvalue equation of the δ-Fermi gas.

For one-component Bethe ansatz solvable systems, like the Heisenberg quantum spin chain or the one-dimensional Bose gas, the periodic boundary conditions already furnished a set of equations, the Bethe ansatz equations, whose solution solves or diagonalizes the Schrödinger eigenvalue equation because they yield, at least in principle, the momenta and, hence, the energies of ground and excited states. Again, in the case of an electronic problem like the Kondo problem, where particles have two degrees of freedom, often denoted as charge and spin degrees of freedom, we find that imposing periodic boundary conditions (or some other kind of boundary conditions) leads us to a second eigenvalue equation, the eigenvalue equation (16.116) for the operator $\mathcal{Z}_j$ given explicitly in (16.117). We have found the same kind of complication also before in the case of the δ-Fermi gas and the one-dimensional Hubbard model.

As mentioned, Gaudin (1967) and Yang (1967), for the case of the one-dimensional Fermi gas, dealt with the eigenvalue equation (16.116) by implementing a second coordinate Bethe ansatz directly. This strategy could now also be applied to the Kondo problem (see also, Andrei *et al.*, 1981).

However, with our experience of the algebraic version of the Bethe ansatz from part II, we also realize that the operator $\mathcal{Z}_j$ resembles a monodromy or transfer matrix. Section 16.4 makes this similarity explicit and uses the algebraic Bethe ansatz to solve the eigenvalue problem of the prototypical kind (16.116).

16.4 Algebraic Bethe ansatz for the spin eigenvalue problem

Our experience with the algebraic Bethe ansatz showed that, starting from the vertex models, it was natural to introduce certain conceptually important quantities, like the monodromy matrix and the R-matrix, which are connected to the transfer matrix. The largest eigenvalue of the transfer matrix then gave the partition function and the free energy of the vertex models and, hence, its equilibrium thermodynamics. Moreover, chapter 12 showed that it was possible and useful to transfer these concepts to models like the modified quantum Tavis–Cummings model without an underlying statistical mechanical model.

In a very similar way, we now proceed with the solution of the electronic models we have been treating so far in this chapter. In the process of solving these models with a coordinate Bethe ansatz for the wave function of the particles, we encountered quantities whose mathematical structure is very similar to quantities with which we are already familiar from the algebraic Bethe ansatz approach: the scattering matrix bears similarities to the R-matrix, the operators $\mathcal{Z}_j$, constructed from a string of scattering matrices, is, hence, very similar to the monodromy matrix $\mathcal{T}$ or transfer matrix T constructed in an analogous way from R-matrices. We now need to make these similarities precise and explicit. In particular, we need now to figure out whether $\mathcal{Z}_j$ corresponds to rather a monodromy matrix $\mathcal{T}$ or directly a transfer matrix T or in what senses $\mathcal{Z}_j$ may correspond to either of these objects.

Let us summarize the spin eigenvalue problem for the three electronic models we have investigated in general form. The matrix elements of the operator in whose eigenvalues and eigenfunctions we are interested is

$$\left(\mathcal{Z}_j\right)_{\sigma_1\ldots\sigma_K}^{\sigma_1'\ldots\sigma_K'} = \left(S^{jj-1}\ldots S^{j1}S^{jK}\ldots S^{jj+1}\right)_{\sigma_1\ldots\sigma_K}^{\sigma_1'\ldots\sigma_K'} \tag{16.118}$$

where, for the δ-Fermi gas and the one-dimensional Hubbard model, $K = N$, the number of electrons, and, for the Kondo model, $K = N + 1$, where N is the number of electrons and we have one impurity. For definiteness, we recall the following notations: L is the length of the system for the δ-Fermi gas and the Kondo model, while L denotes the total number of sites for the Hubbard model; the number of down-spin electrons is denoted by M, and the z-component of the total spin is $N/2 - M$.

The general form of the scattering matrix is

$$S^{ij}(p_i - p_j) = \frac{(p_i - p_j)I^{ij} + i\kappa P^{ij}}{(p_i - p_j) + i\kappa} \tag{16.119}$$

where κ represents the interaction of the model and p is a continuous parameter used to distinguish the models and plays a role similar to the spectral parameter.

For the Kondo problem, choosing $p = 1$ for an electron and $p = 0$ for the impurity and $\kappa = -c$ for the effective coupling, returns the electron-impurity

$$S^{j0}(p_j - p_0) = S^{j0}(1) = \frac{I^{j0} - icP^{j0}}{1 - ic} \tag{16.120}$$

and the electron-electron scattering matrix

$$S^{ij}(p_i - p_j) = S^{ij}(0) = P^{ij}. \tag{16.121}$$

The latter scattering matrix coincides with what we obtained from the L-operator in the isotropic limit of the six-vertex model in section 11.4.1 (although we did not use the terminology 'scattering matrix' there). The operators discussed there described the isotropic Heisenberg quantum spin chain.

The choices $p = k$ with the interaction $\kappa = c$ and $p = \sin k$ with the interaction $\kappa = u/2 = U/2t$, respectively, result in the scattering matrices for the δ-Fermi gas and the one-dimensional Hubbard model, respectively.

16.4.1 Algebraic representation of the spin eigenvalue problem

In the derivations of the spin eigenvalue problems for the δ-Fermi gas (and, by association, the one-dimensional Hubbard model) and the Kondo model, we highlighted the similarities of the respective scattering matrices and the respective operators $\mathcal{Z}_j$ built from the scattering matrices with corresponding objects of the algebraic Bethe ansatz for the vertex models. We shall now work out these similarities in precise and explicit detail to make full use of and carry over the results of the algebraic Bethe ansatz for the vertex models to the present situation of the spin eigenvalue problem of electronic models.

First, we use the embedding of the Pauli matrices $\sigma_j{}^\alpha$ ($j = 1, \ldots, K$, $\alpha = x, y, z$) for a system of K constituents, which is familiar from the Heisenberg quantum spin chain (cf. section 13.1)

$$\sigma_j^\alpha = I_2^{\otimes(j-1)} \otimes \sigma^\alpha \otimes I_2^{\otimes(K-j)} \tag{16.122}$$

for the scattering matrix (16.119). Here $I_2^{\otimes n}$ is the n-fold tensor product of the 2×2 unit matrix; the embedding was discussed in section 10.7.1.1. Generalizing the results from section 10.7.1.1, we can write the permutation operator in (16.119)

$$P^{ij} = \frac{1}{2}\left(I_2^{\otimes K} + \sum_\alpha \sigma_i^\alpha \sigma_j^\alpha\right). \tag{16.123}$$

The unit matrix in (16.119) has now also to be construed in this embedded way: $I^{ij} = I_2^{\otimes K}$.

So far, we have been vague about whether the operators $\mathcal{Z}_j$ are rather to be compared with the monodromy $\mathcal{T}$ or the transfer matrix T[1]. We recall, however, that for the vertex models these two matrices were intimately connected. The monodromy $\mathcal{T}$ being a kind of transfer matrix T where the first and last horizontal spin variables were left free or, in other words, no boundary conditions on these spin variables has been specified. The transfer matrix was then obtained as a particular trace that corresponded to identifying the first and last horizontal spin variables, i.e. imposing periodic boundary conditions in the horizontal direction.

Our present task now consists in constructing operators on the basis of the operators $\mathcal{Z}_j$ that can play the role of the monodromy matrix, i.e. can be written as 2×2 matrices with matrix entries whose trace in the 2×2 space yields the transfer matrices for the

[1] Our notational choice of the letter $\mathcal{Z}$ should therefore not be construed as to predetermine this question.

spin eigenvalue problem. Furthermore, we shall have to find explicit analogs of the R-matrices and the L-operators or -matrices so as to fully employ the machinery of the algebraic Bethe ansatz.

First, we observe that the matrix (16.118) for a system of K particles, electrons or electrons and an impurity contains $K-1$ scattering matrices S^{ij} that altogether form a $2^K \times 2^K$ matrix. We could now consider any one pair of spin indices (σ'_k, σ_k) for some $k \in (1, \ldots, K)$ as belonging to an auxiliary space. Then taking the trace over this auxiliary space would give us the transfer matrix, although of a system consisting of $K-1$ constituents, electrons or electrons and an impurity, depending on the physical model we intend to solve. It is slightly more elegant to extend the Hilbert space by an auxiliary spin $\mathscr{H}_K \to \mathscr{H}_K \times \mathscr{H}_{\text{aux}}$. This can be done by joining the scattering matrix S^{aj} of one more constituent at the right end of the system and so consider the mathematical object

$$\begin{aligned}\left(\mathcal{Z}_j^{(a)}(\lambda)\right)_{\sigma_1\ldots\sigma_K\sigma_a}^{\sigma'_1\ldots\sigma'_K\sigma'_a} = \Big(& S_{(a)}^{aj-1}(\lambda - i\frac{\kappa}{2} - p_{j-1}) S_{(a)}^{aj-2}(\lambda - i\frac{\kappa}{2} - p_{j-2}) \\ & \ldots S_{(a)}^{a1}(\lambda - i\frac{\kappa}{2} - p_1) S_{(a)}^{aK}(\lambda - i\frac{\kappa}{2} - p_K) \\ & \ldots S_{(a)}^{aj+1}(\lambda - i\frac{\kappa}{2} - p_{j+1}) S_{(a)}^{aj}(\lambda - i\frac{\kappa}{2} - p_j)\Big)_{\sigma_1\ldots\sigma_K\sigma_a}^{\sigma'_1\ldots\sigma'_K\sigma'_a}\end{aligned} \tag{16.124}$$

where we introduced the parameter λ in this specific way, the use of which will become clear in a moment. Matrices in the space $\mathscr{H}_K \times \mathscr{H}_{\text{aux}}$ are of dimension $2^{K+1} \times 2^{K+1}$. This is also indicated by the labels (a). The object (16.124) can thus act as a monodromy matrix if we write

$$\left(\mathcal{Z}_j^{(a)}(\lambda)\right)_{\sigma_1\ldots\sigma_K\sigma_a}^{\sigma'_1\ldots\sigma'_K\sigma'_a} = \left\{\left(\mathcal{Z}_j^{(a)}(\lambda)\right)_{\sigma_1\ldots\sigma_K}^{\sigma'_1\ldots\sigma'_K}\right\}_{\sigma_a}^{\sigma'_a} \tag{16.125}$$

to highlight that it is a 2×2 matrix in the auxiliary space.

Now, setting $\lambda = i\kappa/2 + p_j$, we obtain

$$\begin{aligned}\left\{\left(\mathcal{Z}_j^{(a)}(i\frac{\kappa}{2} + p_j)\right)_{\sigma_1\ldots\sigma_K}^{\sigma'_1\ldots\sigma'_K}\right\}_{\sigma_a}^{\sigma'_a} = \Big\{\Big(& S_{(a)}^{aj-1}(p_j - p_{j-1}) S_{(a)}^{aj-2}(p_j - p_{j-2}) \\ & \ldots S_{(a)}^{a1}(p_j - p_1) S_{(a)}^{aK}(p_j - p_K) \\ & \ldots S_{(a)}^{aj+1}(p_j - p_{j+1}) S_{(a)}^{aj}(0)\Big)_{\sigma_1\ldots\sigma_K}^{\sigma'_1\ldots\sigma'_K}\Big\}_{\sigma_a}^{\sigma'_a}.\end{aligned} \tag{16.126}$$

We note that

$$S_{(a)}^{aj}(0) = P_{(a)}^{aj} = \frac{1}{2}\left(I_2 \otimes I_2^{\otimes K} + \sum_\alpha \sigma^\alpha \otimes \sigma_j^\alpha\right) \tag{16.127}$$

and furthermore $P^{ak}_{(a)}P^{aj}_{(a)} = P^{aj}_{(a)}P^{jk}_{(a)}$, so that we can permute the permutation matrix $P^{aj}_{(a)}$ through the whole string of scattering matrices in (16.126), because for the scattering matrices (16.119) we have

$$S^{ak}_{(a)}(p)P^{aj}_{(a)} = P^{aj}_{(a)}S^{jk}_{(a)}(p). \tag{16.128}$$

The significance of this expression stems from the fact that $S^{jk}_{(a)}(u)$ is a scattering matrix which only acts non-trivially in the Hilbert space $\mathcal{H}_K$ while it has become trivial in the auxiliary space $\mathcal{H}_{\text{aux}}$, i.e.

$$S^{jk}_{(a)}(p) = I_2 \otimes S^{jk}(p). \tag{16.129}$$

Suppressing the indices of the spin variables, the matrix (16.126) then takes the form

$$\begin{aligned} \mathcal{Z}^{(a)}_j\left(i\frac{\kappa}{2} + p_j\right) = P^{aj}_{(a)}S^{jj-1}_{(a)}(p_j - p_{j-1})S^{jj-2}_{(a)}(p_j - p_{j-2})\ldots \\ \ldots S^{j1}_{(a)}(p_j - p_1)S^{jK}_{(a)}(p_j - p_K)\ldots S^{jj+1}_{(a)}(p_j - p_{j+1}) \end{aligned} \tag{16.130}$$

or, with (16.127) and (16.129)

$$\begin{aligned} \mathcal{Z}^{(a)}_j\left(i\frac{\kappa}{2} + p_j\right) =& \frac{1}{2}\left(I_2 \otimes I_2^{\otimes K} + \sum_\alpha \sigma^\alpha \otimes \sigma^\alpha_j\right)\cdot \\ &\cdot\Big(I_2 \otimes S^{jj-1}(p_j - p_{j-1})\ldots S^{j1}(p_j - p_1)\cdot \\ &\cdot S^{jK}(p_j - p_K)\ldots S^{jj+1}(p_j - p_{j+1})\Big). \end{aligned} \tag{16.131}$$

In this latter form of the matrix, we can take the trace in auxiliary space to obtain

$$\begin{aligned} \mathcal{Z}_j = \mathrm{Tr}_a\left(\mathcal{Z}^{(a)}_j\left(i\frac{\kappa}{2} + p_j\right)\right) = S^{jj-1}(p_j - p_{j-1})\ldots S^{j1}(p_j - p_1)\cdot \\ \cdot\ S^{jK}(p_j - p_K)\ldots S^{jj+1}(p_j - p_{j+1}). \end{aligned} \tag{16.132}$$

This is the matrix of the spin eigenvalue problem which we found for the three models we considered in the previous three sections, the δ-Fermi gas, the one-dimensional Hubbard model, and the Kondo model. Through the construction above and by comparison with the construction of the vertex models, we can now identify it as a transfer matrix for the electronic models, the matrix (16.131), as the corresponding monodromy matrix. The eigenvalue Λ of the matrix (16.132) thus satisfies $\Lambda = e^{-ik_jL}$.

In order to proceed with the identification of the objects required for the algebraic Bethe ansatz of the electronic models, we now need to extract the equivalents of the R-matrix and the L-matrices. For this purpose, let us now consider the Yang–Baxter relations (16.28) of exercise 16.5. It is easy to generalize them to the scattering of arbitrary particles i, j, and k. We only have to replace the upper indices 1, 2, and 3 by i, j, and k such that

$$S^{jk}(p_j - p_k)S^{ji}(p_j - p_i)S^{ki}(p_k - p_i) = S^{ki}(p_k - p_i)S^{ji}(p_j - p_i)S^{jk}(p_j - p_k) \quad (16.133)$$

where we have used $S^{nm}(p_n - p_m)S^{mn}(p_m - p_n) = I$ to change the order of the indices for the inner two scattering matrices on each side of the equation compared to the exercise. The differences of variables occurring in this expression can be be replaced by introducing two generic variables λ and μ, so that it takes the form

$$S^{jk}(\lambda)S^{ji}(\lambda + \mu)S^{ki}(\mu) = S^{ki}(\mu)S^{ji}(\lambda + \mu)S^{jk}(\lambda). \quad (16.134)$$

We now shift the variables one more time. First, let $\lambda \to \lambda - \mu$ and then let $\lambda \to \lambda - i\kappa/2$ and $\mu \to \mu - i\kappa/2$. We thus arrive at

$$S^{jk}(\lambda - \mu)S^{ji}\left(\lambda - i\frac{\kappa}{2}\right)S^{ki}\left(\mu - i\frac{\kappa}{2}\right) = S^{ki}\left(\mu - i\frac{\kappa}{2}\right)S^{ji}\left(\lambda - i\frac{\kappa}{2}\right)S^{jk}(\lambda - \mu). \quad (16.135)$$

After these preparations, let us rewrite these Yang–Baxter relations embedded in two auxiliary spaces $\mathcal{H}_K \to \mathcal{H}_K \times \mathcal{H}_{\text{aux}} \times \mathcal{H}_{\text{aux}}$. We indicate this in general by the subscript (aa') and by the matrix indices a and b. The Yang–Baxter relations (16.135) then become

$$S^{ab}_{(aa')}(\lambda - \mu)\, S^{ai}_{(aa')}\left(\lambda - i\frac{\kappa}{2}\right)S^{bi}_{(aa')}\left(\mu - i\frac{\kappa}{2}\right) = S^{bi}_{(aa')}\left(\mu - i\frac{\kappa}{2}\right)S^{ai}_{(aa')}\left(\lambda - i\frac{\kappa}{2}\right)S^{ab}_{(aa')}(\lambda - \mu). \quad (16.136)$$

In order to extract an equivalent to the R-matrix, we need to multiply these relations from the left by the permutation matrix $P^{ab}_{(aa')}$ in auxiliary space (aa') for which we have the three relations

$$P^{ab}_{(aa')}S^{ab}_{(aa')}(\lambda) = Y^{ab}_{(aa')}(\lambda), \quad (16.137)$$

$$P^{ab}_{(aa')}S^{bi}_{(aa')}(\lambda) = S^{ai}_{(aa')}(\lambda)P^{ab}_{(aa')}, \quad (16.138)$$

$$P^{ab}_{(aa')}S^{ai}_{(aa')}(\lambda) = S^{bi}_{(aa')}(\lambda)P^{ab}_{(aa')}. \quad (16.139)$$

It suffices to apply the first of these relations to the left-hand side of (16.136), while we need the latter two equations, similar to (16.128), for the right-hand side of (16.136). This procedure brings the Yang–Baxter relations (16.136) into the form

$$Y^{ab}_{(aa')}(\lambda-\mu)\,S^{ai}_{(aa')}\left(\lambda-i\frac{\kappa}{2}\right)S^{bi}_{(aa')}\left(\mu-i\frac{\kappa}{2}\right)=$$
$$S^{ai}_{(aa')}\left(\mu-i\frac{\kappa}{2}\right)S^{bi}_{(aa')}\left(\lambda-i\frac{\kappa}{2}\right)Y^{ab}_{(aa')}(\lambda-\mu). \quad (16.140)$$

This latter form of the Yang–Baxter relations can now be brought into a form which is familiar from the Yang–Baxter relations (11.18) of the six-vertex model. In order to make this correspondence manifest, we first write the relation in auxiliary space

$$Y^{ab}_{(aa')}(\lambda)=\frac{2i\kappa I_2\otimes I_2+\lambda\left(I_2\otimes I_2+\sum_\alpha\sigma^\alpha\otimes\sigma^\alpha\right)}{2i\kappa+2\lambda}\otimes I_2^{\otimes K}=R(\lambda)\otimes I_2^{\otimes K}, \quad (16.141)$$

which we recognize as an R-matrix, e.g. by comparison with the R-matrix of the isotropic six-vertex model (11.63). This explains why we multiplied (16.136) by $P^{ab}_{(aa')}$. Explicitly, the R-matrix may be written as

$$R(\lambda)=\begin{pmatrix}1&0&0&0\\0&b(\lambda)&c(\lambda)&0\\0&c(\lambda)&b(\lambda)&0\\0&0&0&1\end{pmatrix} \quad (16.142)$$

with

$$b(\lambda)=\frac{i\kappa}{\lambda+i\kappa}\quad\text{and}\quad c(\lambda)=\frac{\lambda}{\lambda+i\kappa}. \quad (16.143)$$

Furthermore, we can identify the L-matrix by writing (changing the index from i to n to have a more conventional notation, which may also reduce the danger of confusion in the following)

$$S^{an}_{(aa')}\left(\lambda-i\frac{\kappa}{2}\right)=L_n(\lambda)\otimes I_2\quad\text{and}\quad S^{bn}_{(aa')}\left(\mu-i\frac{\kappa}{2}\right)=I_2\otimes L_n(\mu) \quad (16.144)$$

where we have used

$$\sigma^\alpha\otimes I_2\otimes\sigma^\alpha_n=\sigma^\alpha\otimes\sigma^\alpha_n\otimes I_2, \quad (16.145)$$

which can be verified by a direct calculation.

Hence, the L-matrix is given by

$$L_n(\lambda) = \frac{\lambda I_2^{\otimes K} + i\frac{\kappa}{2}\sum_{\alpha=1}^{3}\sigma^\alpha \otimes \sigma_n^\alpha}{\lambda + i\frac{\kappa}{2}} = c(2\lambda)I_2^{\otimes K} + b(2\lambda)\sum_{\alpha=1}^{3}\sigma^\alpha \otimes \sigma_n^\alpha$$
$$= \begin{pmatrix} c(2\lambda)I_2^{\otimes(K-1)} + b(2\lambda)\sigma_n^z & 2b(2\lambda)\sigma_n^- \\ 2b(2\lambda)\sigma_n^+ & c(2\lambda)I_2^{\otimes(K-1)} - b(2\lambda)\sigma_n^z \end{pmatrix}. \tag{16.146}$$

Compare this expression with the L-matrix of the six-vertex model (11.19) or, for the isotropic case (11.61).

The Yang–Baxter relations (16.140) thus take the form

$$R(\lambda-\mu)\,(L_n(\lambda)\otimes L_n(\mu)) = (L_n(\mu)\otimes L_n(\lambda))\,R(\lambda-\mu). \tag{16.147}$$

Furthermore, we can write the matrices $\mathcal{Z}_j^{(a)}$ (16.124) and $\mathcal{Z}_j$ (16.132), the monodromy and transfer matrices of the spin eigenvalue problem, in terms of the L-matrices as

$$\mathcal{Z}_j^{(a)}(\lambda) = L_{j-1}(\lambda - p_{j-1})\ldots L_1(\lambda - p_1)L_K(\lambda - p_K)\ldots L_j(\lambda - p_j) \tag{16.148}$$

and

$$\mathcal{Z}_j = L_{j-1}(\lambda - p_{j-1})\ldots L_1(\lambda - p_1)L_K(\lambda - p_K)\ldots L_{j+1}(\lambda - p_{j+1}) \tag{16.149}$$

respectively. Since the L-matrices for different indices commute, $[L_n, L_m] = 0$ for $n \neq m$, we obtain by iteration the Yang–Baxter relations

$$R(\lambda-\mu)\left(\mathcal{Z}_j^{(a)}(\lambda)\otimes\mathcal{Z}_j^{(a)}(\mu)\right) = \left(\mathcal{Z}_j^{(a)}(\mu)\otimes\mathcal{Z}_j^{(a)}(\lambda)\right)R(\lambda-\mu). \tag{16.150}$$

Now, we have achieved our goal to reduce the spin eigenvalue problem to acquire the same structure as the eigenvalue problem of the six-vertex model. The (trigonometrically parameterized) R-matrix (11.21) of the six-vertex-model is structurally the same as the R-matrix (16.142). The Yang–Baxter relations for the L-matrix in the form (16.147) are identical in structure to those for the six-vertex model (11.18), the Yang–Baxter relations (16.150) for the monodromy matrix are comparable to the Yang–Baxter relations for the six-vertex model (11.34).

With these preparations, we are now in a position to perform the programme of the algebraic Bethe ansatz in the same way we formulated it in chapter 10 for the general case of two-dimensional vertex models and carried it through for the six-vertex model in chapter 11. In fact, taking proper care with definitions and keeping in mind a few differences, we can just copy the results from the case of the six-vertex model.

As mentioned, the monodromy matrix can again be viewed as a 2×2 matrix in auxiliary space

$$\mathcal{Z}_j^{(a)}(\lambda) = \begin{pmatrix} A(\lambda) & B(\lambda) \\ C(\lambda) & D(\lambda) \end{pmatrix} \tag{16.151}$$

whose trace in auxiliary space determines the transfer matrix

$$\mathcal{Z}_j(\lambda) = \mathrm{Tr}_a \mathcal{Z}_j^{(a)}(\lambda) = A(\lambda) + D(\lambda). \tag{16.152}$$

This transfer matrix is the matrix we want to consider in order to diagonalize the spin eigenvalue problem, cf. (16.132), with the scattering matrices for the various electronic models inserted; for the Kondo model, the scattering matrix (16.77) for the electron-impurity scattering and (16.92) for the electron-electron scattering, for the one-dimensional Hubbard model, the scattering matrix (16.56), and for the δ-Fermi gas, the scattering matrix (16.23).

16.4.2 Diagonalization of the transfer matrix

We follow the programme of the algebraic Bethe ansatz demonstrated for the vertex models and carried through especially for the six-vertex model in chapters 10 and 11. The first step consists in exploiting the fundamental relations of integrability, the Yang–Baxter relations, in the form (16.150) with $\mathcal{Z}_j^{(a)}$ written in the form (16.151). Explicitly written, these are 16 equations that are bilinear in A, B, C, and D and can be interpreted as generalized commutation relations, often called fundamental commutation relations. Again, we need only to consider in detail a subset of the 16 equations that are formally identical to (11.35), (11.36), and (11.37) of section 11.3. Since the further analysis runs exactly parallel to that of section 11.3 for the six-vertex model, we shall restrict our attention to the few differences in detail between the six-vertex model and the present electronic models.

First, we denote the (global) vacuum or reference state by

$$|0\rangle \equiv |\uparrow\uparrow \ldots \uparrow\rangle \equiv \begin{pmatrix} 1 \\ 0 \end{pmatrix}^{\otimes K}. \tag{16.153}$$

Note the different choice in (11.23) where we started from a *local* vacuum state $\omega_n = (1, 0)^T$. After we have taken a step by step approach in section 11.3, we can take a shortcut now and use the global reference state already with the L-matrix.

The result of acting with the L-matrix (16.146) on the reference state (16.153) is

$$L_n(\lambda)|0\rangle = \begin{pmatrix} (c(2\lambda) + b(2\lambda))\,|0\rangle & * \\ 0 & (c(2\lambda) - b(2\lambda))\,|0\rangle \end{pmatrix} = \begin{pmatrix} 1|0\rangle & * \\ 0 & \frac{\lambda - i\frac{\kappa}{2}}{\lambda + i\frac{\kappa}{2}}|0\rangle \end{pmatrix}. \tag{16.154}$$

Again, as in section 11.3, we do not explicitly need the entry in the upper right corner of this matrix, and therefore it is just indicted by a placeholder *.

Acting with the full monodromy matrix (16.148) on the reference state thus yields

$$\mathcal{Z}_j^{(a)}(\lambda)|0\rangle = \begin{pmatrix} \alpha(\lambda)|0\rangle & * \\ 0 & \delta(\lambda)|0\rangle \end{pmatrix} \tag{16.155}$$

where now

$$\alpha(\lambda) = 1 \qquad \text{and} \qquad \delta(\lambda) = \prod_{j=1}^{K} \frac{\lambda - p_j - i\frac{\kappa}{2}}{\lambda - p_j + i\frac{\kappa}{2}}. \tag{16.156}$$

In complete analogy to section 11.3, we define a general state by acting with a product of operators $B(\lambda_l)$ for distinct values of the parameters λ_l on the reference state

$$|\{\lambda_l\}\rangle = \prod_{l=1}^{M} B(\lambda_l)|0\rangle. \tag{16.157}$$

The number $M \leq N/2$ is the number of down-spin electrons among the N electrons of the three models. As for the case of the six-vertex model, we find that this state is an eigenstate of the transfer matrix $\mathcal{Z}_j$, (16.152),

$$\mathcal{Z}_j(\lambda)\,|\{\lambda_l\}\rangle = (A(\lambda) + D(\lambda))\,|\{\lambda_l\}\rangle = \Lambda(\lambda|\lambda_1, \ldots, \lambda_M)\,|\{\lambda_l\}\rangle \tag{16.158}$$

with eigenvalue

$$\Lambda(\lambda|\lambda_1, \ldots, \lambda_M) = \prod_{l=1}^{M} \frac{1}{c(\lambda_l - \lambda)} + \delta(\lambda) \prod_{l=1}^{M} \frac{1}{c(\lambda - \lambda_l)} \tag{16.159}$$

provided the parameters λ_l satisfy

$$\delta(\lambda_l) = \prod_{k=1,k\neq l}^{M} \frac{c(\lambda_l - \lambda_k)}{c(\lambda_k - \lambda_l)} = \prod_{k=1,k\neq l}^{M} \frac{\lambda_l - \lambda_k - i\kappa}{\lambda_l - \lambda_k + i\kappa} \qquad l = 1, \ldots, M \tag{16.160}$$

which yields using (16.156)

$$\prod_{j=1}^{K} \frac{\lambda_l - p_j - i\frac{\kappa}{2}}{\lambda_l - p_j + i\frac{\kappa}{2}} = \prod_{k=1,k\neq l}^{M} \frac{\lambda_l - \lambda_k - i\kappa}{\lambda_l - \lambda_k + i\kappa}, \qquad l = 1, \ldots, M. \tag{16.161}$$

These sets of equations are the equivalent of the Bethe ansatz equations (11.42) for the six-vertex model.

However, unlike in the six-vertex case, we now have to determine two sets of parameters, two sets of Bethe ansatz roots: λ_l, $l = 1, \ldots, M$ and p_j, $j = 1 \ldots, K$. The former are interpreted as pertaining to the spin degrees of freedom, the latter to the charge degrees of freedom of the electrons of the one-dimensional systems.

The second set of equations for the Bethe ansatz roots λ_l and p_j is provided by the periodic boundary conditions and the requirement to evaluate the eigenvalue (16.159) for $\lambda = p_j + i\kappa/2$ where $\delta(\lambda)$ vanishes. The latter requirement is motivated by our discussion in section 11.4 where we evaluated the transfer matrix for a special value of the spectral parameter. Recalling that $p_j = p(k_j)$, we obtain for the eigenvalue (16.159)

$$e^{-ik_jL} = \Lambda(p_j + i\kappa/2) = \prod_{l=1}^{M} \frac{\lambda_l - p_j + i\frac{\kappa}{2}}{\lambda_l - p_j - i\frac{\kappa}{2}}, \qquad j = 1, \ldots, N. \tag{16.162}$$

The two coupled sets of equations (16.161) and (16.162) are the *nested* Bethe ansatz equations that may now be specialized depending on the choice of the function $p(k)$ to the three models: the δ-Fermi gas for $p(k) = k$, the Hubbard model for $p(k) = \sin k$, and the Kondo model for $p(k) = 1$ for electrons and $p(k) = 0$ for the impurity.

For many calculational purposes, it is advantageous to use the nested Bethe ansatz equations (16.161) and (16.162) in their logarithmic form

$$k_jL = 2\pi I_j + \sum_{l=1}^{M} \theta\left(2(p_j - \lambda_l)\right), \qquad j = 1, \ldots, N, \tag{16.163}$$

$$\sum_{k=1, k\neq l}^{M} \theta\left(\lambda_k - \lambda_l\right) = 2\pi \mathcal{J}_l + \sum_{j=1}^{K} \theta\left(2(p_j - \lambda_l)\right), \qquad l = 1, \ldots, M, \tag{16.164}$$

where $\theta(x) = -2\tan^{-1}(x/\kappa)$. The Bethe ansatz quantum numbers I_j and $\mathcal{J}_l$ are integers or half-odd integers. For M an even integer, the numbers I_j are integers, for M an odd integer, they are half-odd integers. The numbers $\mathcal{J}_l$ are integers if $K - M + 1$ is even and half-odd integers if $K - M + 1$ is odd.

The Bethe ansatz quantum numbers I_j are usually called charge quantum numbers, the Bethe ansatz quantum numbers $\mathcal{J}_l$ spin quantum numbers. The corresponding sets of variables are called charge and spin rapidities, k_j and λ_l, respectively.

We have seen in the case of the one-component models that we had one set of Bethe ansatz quantum numbers that determined the quantum state of the systems. Now we have two sets of quantum numbers, I_j and $\mathcal{J}_l$, which again determine the quantum state of the two-component (or spinfull) electronic systems. Although mathematically these numbers come about by taking logarithms, they play a decisive role in the physics of the Bethe ansatz solvable models. By choosing them arbitrarily within the restrictions mentioned, a particular quantum state has been selected. The quantum state selected, though, takes more insight into the problem at hand.

Having now laid the foundations of the Bethe ansatz method for two-component systems, namely the nested Bethe ansatz equations exemplified for the three electronic models considered in this chapter, we could go on to explore first their ground state properties at zero temperature and their low-lying excitations as we did in chapter 14 for the anisotropic Heisenberg quantum spin chain and in chapter 15 for the one-dimensional δ-Bose gas, the Lieb–Liniger model. And then we might also become interested in the question of whether we can access their finite temperature, i.e. thermodynamic, properties within a generalization of the Bethe ansatz methodology. However, we instead return in the next two chapters of part V to the simpler one-component systems we treated earlier, the δ-Bose gas and the isotropic Heisenberg quantum spin chain, and investigate their finite temperature properties by introducing an extension to finite temperature of the Bethe ansatz for these systems.

Before turning to the thermodynamic Bethe ansatz, we shall, however, end this chapter (without going into all calculational details) at a problem connected to the Kondo effect in nanostructures (see also section 8.9.4) where the nested Bethe ansatz equations play the key role directly, i.e. where we do not consider them in the limit of a very large system or where we use them as the starting point for a finite temperature thermodynamic treatment.

In order to do so, we first write down the nested Bethe ansatz equations for the Kondo model explicitly.

16.4.2.1 Nested Bethe ansatz equations for the Kondo model

Recalling that $\kappa = -c$ and $p_j = 1$ for an electron and $p_j = 0$ for the impurity, the nested Bethe ansatz equations (16.162) and (16.161) specialized to the Kondo model become

$$e^{ik_jL} = \prod_{l=1}^{M} \frac{\lambda_l - 1 + i\frac{c}{2}}{\lambda_l - 1 - i\frac{c}{2}}, \qquad j = 1, \ldots, N, \tag{16.165}$$

$$\left(\frac{\lambda_l - 1 + i\frac{c}{2}}{\lambda_l - 1 - i\frac{c}{2}}\right)^N \frac{\lambda_l + i\frac{c}{2}}{\lambda_l - i\frac{c}{2}} = \prod_{k=1,k\neq l}^{M} \frac{\lambda_l - \lambda_k + ic}{\lambda_l - \lambda_k - ic}, \qquad l = 1, \ldots, M, \tag{16.166}$$

or, in logarithmic form,

$$k_jL = 2\pi I_j + \sum_{l=1}^{M} \theta\left(2(\lambda_l - 1)\right), \qquad j = 1, \ldots, N, \tag{16.167}$$

$$N\theta\left(2(\lambda_l - 1)\right) + \theta\left(2\lambda_l\right) = -2\pi \mathcal{J}_l + \sum_{k=1,k\neq l}^{M} \theta\left(\lambda_l - \lambda_k\right) \qquad l = 1, \ldots, M \tag{16.168}$$

where now $\theta(x) = -2\tan^{-1}(x/c)$. Each allowed choice of the integer or half-odd integer numbers $\{I_j, \mathcal{J}_l\}$ determines an eigenstate of the Kondo Hamiltonian. Thus, the numbers $\{I_j, \mathcal{J}_l\}$ are the quantum numbers of the state. Note that the sign in $-\ 2\pi\mathcal{J}_l$ of (16.168) is quite unimportant as the Bethe ansatz quantum numbers are integrers $I_j, \mathcal{J}_l \in \mathbb{Z}$ of half-odd integers $I_j, \mathcal{J}_l \in \mathbb{Z} + 1/2$. Accordingly, we drop this sign in the following when it is convenient.

We note that the nested Bethe ansatz equations, (16.167) and (16.168), for the Kondo model are coupled in a peculiar way. While the corresponding two sets of nested Bethe ansatz equations for the δ-Fermi gas and the Hubbard model (see exercise 16.12) are mutually coupled, for the Kondo model the Bethe ansatz equations for the electron momenta k_j, (16.167), also called the charge Bethe ansatz equations, are coupled to the Bethe ansatz equations (16.168) for the spin variables, the latter conventionally called spin rapidities. However, the reverse is not true for the Kondo model in the sense that the Bethe ansatz equation (16.167) has no influence back on (16.168). In other words, we can solve (16.168) independently of the solution set $\{k_j\}$ of (16.167).

The energy of this relativistic system of electrons coupled to the magnetic impurity follows directly from (16.167)

$$E = \sum_{j=1}^{N} k_j = \frac{2\pi}{L}\sum_{j=1}^{N} I_j + \rho \sum_{l=1}^{M} \theta(2\lambda_l - 2) \tag{16.169}$$

where $\rho = N/L$ is the electron density. This energy expression allows the interpretation that the impurity is reflected in the term $\theta(2\lambda_l)$ in the Bethe ansatz equations (16.168). Namely, if the term $\theta(2\lambda_l)$ were absent from (16.168), we would obtain

$$E = \frac{2\pi}{L}\sum_{j=1}^{N} I_j + \rho \sum_{l=1}^{M} \theta(2\lambda_l - 2) \tag{16.170}$$

$$= \frac{2\pi}{L}\sum_{j=1}^{N} I_j + \frac{1}{L}\sum_{l=1}^{M}\left[-2\pi\mathcal{J}_l + \sum_{k=1,k\neq l}^{M} \theta(\lambda_k - \Lambda_l)\right] \tag{16.171}$$

and, hence, because the scattering phases $\theta(x)$ are odd functions

$$E = \frac{2\pi}{L}\left(\sum_{j=1}^{N} I_j - \sum_{l=1}^{M} \mathcal{J}_l\right). \tag{16.172}$$

This is the energy for a non-interacting gas of Dirac Fermions (Dirac electrons) in a basis where the electrons' charge and spin properties are decoupled.

EXERCISE 16.12 Nested Bethe ansatz equations for the δ-Fermi gas and the Hubbard model Derive the nested Bethe ansatz equations in logarithmic form for the δ-Fermi gas and the Hubbard model.

The following section uses the nested Bethe ansatz equations (16.167) and (16.168) and investigates a question in the physics of quantum impurities in nanoscopic structures. This section demonstrates the direct usefulness of the nested Bethe ansatz equations in a concrete physical problem situation. However, we shall not give all the details of the arguments but refer to the cited literature. Compare also section 8.9.4 on the physics of the Kondo effect in quantum dots for background.

16.5 Magnetic impurities in nanostructures: Bethe ansatz results

Section 8.9.4 explored the physics of magnetic quantum impurities, especially the Kondo effect, when the magnetic moment could be produced and studied in a controllable way in a quantum dot connected to thin leads as electron reservoirs. This situation is not only an ideal setting for the study of the Kondo effect in nanodevices, but also presents a quasi-one-dimensional system by construction. The expectation is that Bethe ansatz integrable models might be useful as models for such nanoscopic systems. However, due to the electric currents involved in most experiments, we are faced with a non-equilibrium system. Attempts to extend the Bethe ansatz methodology to problems like these have been made in recent years but a discussion of these developments (which were initiated by Mehta and Andrei, 2006) is outside the scope of this book. For a broad overview of the research situation into the physics of strongly interacting quantum matter out of equilibrium, see the Les Houches summer school volume by Giamarchi *et al.* (2016).

There are, however, nanostructures, ultrasmall rings, which have been studied widely and which allow for equilibrium currents, the so-called persistent currents when the rings are threaded by a magnetic flux. These rings are made from semiconductor or metallic materials. The possibility of persistent currents was first discussed by Byers and Yang (1961) where the metallic ring is superconductive. Büttiker *et al.* (1983) then predicted the existence of persistent currents also in normal metallic rings. For a review, see e.g. Viefers *et al.* (2004).

This section considers this equilibrium response, the persistent current, of a doubly connected system, a one-dimensional mesoscopic ring of circumference L, to a magnetic Aharonov–Bohm flux (Aharonov and Bohm, 1959) piercing the system when the electrons in the ring are coupled by a spin exchange to a magnetic impurity but are otherwise considered to be free electrons. The magnetic impurity is realized as a quantum dot at Kondo resonance side-coupled to the ring such that the ring itself remains intact when the coupling to the quantum dot is switched off.

For certain privileged values of the Aharonov–Bohm fluxes, the problem can be mapped onto an integrable model that is exactly solvable by a nested Bethe ansatz.

The exact solution found under these conditions demonstrates that the presence of the quantum dot has no effect on the persistent current. In contrast, the Kondo resonance interferes with the spin-dependent Aharonov–Casher effect (Aharonov and Casher, 1984). For more details on this, see Eckle *et al.* (2000) and Eckle *et al.* (2001).

The ring geometry of the physical set-up naturally imposes periodic boundary conditions on the electron field operators $\psi_\sigma(x)$ of the one-dimensional system and which are modified through the effect of the Aharonov–Bohm flux. A gauge transformation (Byers and Yang, 1961) then leads to twisted boundary conditions

$$\psi_\sigma(L) = \psi_\sigma(0)e^{i\phi} \tag{16.173}$$

where $\phi = 2\pi\Phi/\Phi_0$ with Φ the magnetic flux enclosed the ring and $\Phi_0 = hc/e$ the elementary flux quantum.

For the problem situation outlined here, we can use the Kondo Hamiltonian of (8.731) and (8.733) in the form

$$\mathcal{H} = \mathcal{H}_0 + \mathcal{H}_{\text{imp}} \tag{16.174}$$

$$= -i\frac{v_F}{2\pi}\sum_\sigma \int_0^L dx \left(\psi^\dagger_{\sigma,L}(x)\frac{\partial}{\partial x}\psi_{\sigma,L}(x) - \psi^\dagger_{\sigma,R}(x)\frac{\partial}{\partial x}\psi_{\sigma,R}(x)\right) \tag{16.175}$$

$$+\mathcal{J}\boldsymbol{\sigma}\cdot\sum_{\sigma\sigma'}\left[\psi^\dagger_{\sigma,L}(0) + \psi^\dagger_{\sigma,R}(0)\right]\boldsymbol{\sigma}_{\sigma\sigma'}\left[\psi_{\sigma',L}(0) + \psi_{\sigma',R}(0)\right] \tag{16.176}$$

where we restricted the infinite interval of (8.731) to the finite circumference L of the ring and redefined the Kondo interaction constant $\mathcal{J}$ for our present purposes. In the following, we choose the Fermi momentum k_F that appears in (8.727) as $k_F = 2\pi n_F/L$ with an integer n_F.

We now introduce a basis with a definite parity, a so-called Weyl basis, which defines by

$$\psi^\dagger_{\sigma,\text{even}}(x) = \frac{1}{\sqrt{2}}\left(\psi^\dagger_{\sigma,R}(x) + \psi^\dagger_{\sigma,L}(-x)\right) \tag{16.177}$$

a right-moving electron field operator with even parity and by

$$\psi^\dagger_{\sigma,\text{odd}}(x) = \frac{1}{\sqrt{2}}\left(\psi^\dagger_{\sigma,R}(-x) - \psi^\dagger_{\sigma,L}(x)\right) \tag{16.178}$$

a left-moving electron field operator with odd parity.

In the definite parity basis, the Kondo Hamiltonian takes the form

$$\mathcal{H} = \mathcal{H}_0^{\text{odd}} + \mathcal{H}_0^{\text{even}} + \mathcal{H}_{\text{imp}}^{\text{even}} \tag{16.179}$$

where the the non-interacting parts of the Hamiltonian for the odd and even electron quantum fields are, respectively,

$$\mathcal{H}_0^{\text{odd}} = -i\frac{v_F}{2\pi}\sum_\sigma \int_0^L dx\, \psi^\dagger_{\sigma,\text{odd}}(x)\frac{\partial}{\partial x}\psi_{\sigma,\text{odd}}(x) \tag{16.180}$$

and

$$\mathcal{H}_0^{\text{even}} = i\frac{v_F}{2\pi}\sum_\sigma \int_0^L dx\, \psi^\dagger_{\sigma,\text{even}}(x)\frac{\partial}{\partial x}\psi_{\sigma,\text{even}}(x). \tag{16.181}$$

These two parts of the Hamiltonian describe independent relativistic electrons while the impurity in the definite parity basis now couples only the even electron quantum field

$$\mathcal{H}_{\text{imp}}^{\text{even}} = \mathcal{J}\boldsymbol{\sigma}\cdot\sum_{\sigma\sigma'}\psi^\dagger_{\sigma,\text{even}}(0)\boldsymbol{\sigma}_{\sigma\sigma'}\psi_{\sigma',\text{even}}(0). \tag{16.182}$$

The Hamiltonian for the even electron quantum field, i.e. in the even parity channel,

$$\mathcal{H}_{\text{K}}^{\text{even}} = \mathcal{H}_0^{\text{even}} + \mathcal{H}_{\text{imp}}^{\text{even}} \tag{16.183}$$

is the Hamiltonian of the spin-1/2 Kondo model. It is a *chiral* Hamiltonian because it involves only even parity, right-moving electrons coupled to the impurity.

On the level of the Hamiltonian, the odd and even parity channels are now decoupled in a free relativistic odd parity channel and a chiral even parity Kondo channel. However, the twisted boundary condition (16.173) couples the odd and even channels back together through

$$\begin{pmatrix} \psi_{\sigma,\text{even}}(L) \\ \psi_{\sigma,\text{odd}}(L) \end{pmatrix} = \begin{pmatrix} \cos\phi & i\sin\phi \\ -i\sin\phi & \cos\phi \end{pmatrix}\begin{pmatrix} \psi_{\sigma,\text{even}}(0) \\ \psi_{\sigma,\text{odd}}(0) \end{pmatrix}. \tag{16.184}$$

For the special values when the magnetic flux satisfies $\phi = f\pi$, where f is an integer, the matrix in (16.184) reduces to the unit matrix or its negative, and the even and odd parity quantum field operators again completely decouple from each other.

The Hamiltonian $\mathcal{H}_{\text{K}}^{even}$ is now the Hamiltonian of the Kondo model for which we have derived the nested Bethe ansatz equations earlier in this chapter. Our problem has thus been reduced to an exactly solvable problem for flux values for which $f \in \mathbb{Z}$. It consists of a left-moving odd-parity branch of independent relativistic electrons together with a decoupled right-moving even-parity branch defined by the one-dimensional Kondo model. This model is, moreover, by construction valid in the vicinity of the Fermi level. For generic values of the flux ϕ, it is not possible to choose a basis that renders the Hamiltonian and the boundary conditions simultaneously diagonal, suggesting that the model is not integrable in general.

We now concentrate on how to extract the persistent current from the analysis of the model of a Kondo impurity in a quantum dot side-coupled to a metallic ring that we have rendered in a form solvable by a nested Bethe ansatz. For a more detailed discussion of the physical properties of the model, we again refer to Eckle *et al.* (2000) and Eckle *et al.* (2001).

Since the effective model constructed is valid near the Fermi level, the charge persistent current of the system can only depend on properties close to the Fermi level. The first issue will therefore be the definition of the *excess* numbers

$$\Delta N_{\mathrm{R/L}}(\phi) = \frac{L}{2\pi}\left(\left|k_{\mathrm{R/L,F}}(\phi)\right| - \min\left[\left|k_{\mathrm{L,F}}(0)\right|, k_{\mathrm{R,F}}(0)\right]\right) \tag{16.185}$$

of electrons or holes on the right versus the left branches of the energy dispersion. These particles and holes carry the current due to the Aharonov–Bohm flux ϕ. The momenta $k_{\mathrm{R/L,\,F}}(\phi)$ depend on the flux and represent the highest occupied level on the respective branch of the linearized energy dispersion of the model. These numbers define the persistent current

$$I(\Phi) = -\frac{ev_F}{L}\left(\Delta N_{\mathrm{R}}(\phi) - \Delta N_{\mathrm{L}}(\phi)\right). \tag{16.186}$$

We note that this treatment of persistent currents differs from the usual one. In contrast to the one-dimensional relativistic electrons with a linearized spectrum of our present problem, usually non-relativistic electrons with a non-linear spectrum (e.g. quadratic for free electrons) are considered (cp. Büttiker *et al.* (1983) and Cheung *et al.* (1988)).

In order to determine the flux-dependent momenta $k_{\mathrm{R/L,\,F}}(\phi)$ we finally require the nested Bethe ansatz equations, which we already know to hold because the model we constructed is Bethe ansatz solvable. The N_{odd} odd-parity left movers, which are decoupled from the impurity and of which M_{odd} have spin down are characterized by the quantum numbers of free, chiral electrons. We write the corresponding quantization condition in the Bethe basis

$$Lk_{n_{\mathrm{L}}} = -2\pi n_{\mathrm{L}} + \pi f + \frac{2\pi}{N_{\mathrm{odd}}}\sum_{l=1}^{M_{\mathrm{odd}}} \mathcal{J}_l^{\mathrm{odd}}. \tag{16.187}$$

The Kondo Hamiltonian $\mathcal{H}_{\mathrm{K}}$ describing the N_{even} even-parity right movers interacting with the impurities is diagonalized by the nested Bethe ansatz equations (16.167) and (16.168) which we write in the form

$$Lk_{n_{\mathrm{R}}} = 2\pi n_{\mathrm{R}} + \pi f + \sum_{l=1}^{M_{\mathrm{even}}} \theta(2\lambda_l - 2), \tag{16.188}$$

$$N_{\rm even}\theta(2\lambda_l - 2) + \theta(2\lambda_l) = 2\pi \mathcal{J}_l^{\rm even} + \sum_{k=1,k\neq l}^{M_{\rm even}} \theta(\lambda_l - \lambda_k) \tag{16.189}$$

adapted to our present problem.

The further arguments leading to the final result that the charge persistent current in a metallic ring of the free electrons side-coupled to the magnetic impurity of a quantum dot is indeed unaffected by the presence of this Kondo impurity, i.e. that we again obtain

$$I = -\frac{ev_F}{L}\frac{2\Phi}{\Phi_0} \tag{16.190}$$

as for free electrons in a ring, can be found in Eckle *et al.* (2000) and Eckle *et al.* (2001). The question of whether this result is due to the linearization of the spectrum and thus not generic (Affleck and Simon, 2002; Eckle *et al.*, 2002) for integrable models, has been partially addressed in Nilsson *et al.* (2007) for a model with a quadratic spectrum for the electrons and answered in the negative. However, in order to keep it integrable, this model had to be fine tuned by special potential terms. Unfortunately, an experimental investigation of this system combining a quantum dot tuned to Kondo resonance and an ultra-small metallic ring seems still out of reach.

Part 5

Thermodynamic Bethe Ansatz

But if your theory is found to be against the second law of thermodynamics I can give you no hope; there is nothing for it but to collapse in deepest humiliation.

Arthur Stanley Eddington (1882–1944)

Previous chapters have discussed the excitations of quantum integrable models, notably the δ-Bose gas and the Heisenberg quantum spin chain, using the Bethe ansatz technique. Moreover, they examined electronic systems for which the Bethe ansatz needed to be extended to a two-step nested version. In each case, we considered microscopically small numbers of excitations and their properties above a macroscopic ground state, i.e. low-lying excitations. This is tantamount to considering the models at zero temperature. For finite temperature, there will develop a macroscopic number of excitations. Part V investigates the question of how to extend the Bethe ansatz to extract information about a quantum integrable model at finite temperature when there are macroscopic numbers of excitations.

Although possible also for the multicomponent electronic models, we shall restrict our attention to single-component systems. In fact, chapter 17 introduces the extension of the Bethe ansatz to finite temperature, the thermodynamic Bethe ansatz, using the simplest model: the one-dimensional gas of Bosons with repulsive δ-function interaction or Lieb–Liniger model whose ground state and low-lying excitations for vanishing temperature we obtained by Bethe ansatz in chapter 15.

Furthermore, chapter 18 investigates the thermodynamics of the isotropic antiferromagnetic spin chain. The quantum spin chain, however, provides explanation for technical issues not present in the repulsive δ-Bose gas, which make it especially worthwhile to study after the basic approach has been shown in the simpler case of the latter model.

We state first the main idea of the thermodynamic Bethe ansatz, without any of the involved technicalities, for the quantum spin chain. The basic idea for the δ-Bose gas is, however, quite similar, but it lacks some specifics of the quantum spin chain. We discuss the differences in the respective chapters.

The lowest energy states of the isotropic antiferromagnetic spin chain can be described in every sector, labelled by $S = N/2 - M$, by a 'compact' set of quantum numbers and a real distribution of rapidities. The low-lying excitations are then given by holes in the set of quantum numbers and corresponding strings of complex rapidities.

Of course, if we want to consider thermodynamics, i.e. macroscopic properties, we shall have to work with the thermodynamic limit $N \to \infty$ in mind right from the start. This will allow a few slight simplifications in what follows.

If greater numbers of these excitations become excited thermally, we eventually have to deal with a macroscopic number of excitations, i.e. their number becomes proportional to N. The idea of the thermodynamic Bethe ansatz is to introduce, in the limit $N \to \infty$, densities of holes and particles. The holes are described by the missing quantum numbers in the set of quantum numbers, the particles are represented by the (real) centres of the string solutions discussed in 14.5.3.

In order to make contact with a thermodynamic interpretation, the various densities are then used to write down expressions for the entropy and related thermodynamic quantities.

The original publication on the thermodynamics of the δ-Bose gas is Yang and Yang (1969) on which further work on the thermodynamic Bethe ansatz builds. At about the same time Michel Gaudin (1971) and Minoru Takahashi (1971*a*, 1971*b*; Takahashi and Suzuki, 1972) addressed the problem of the thermodynamics of the Heisenberg quantum spin chain based on the Bethe ansatz. While Gaudin addressed the antiferromagnetic case $\Delta \geq 1$, Takahashi focused on the isotropic case (1971*a*) and the case $|\Delta| < 1$, sometimes called the planar case (Takahashi, 1971*b*). Important precursory work by Robert Griffiths (1964*a*, 1964b) is also worthwhile.

Extensive treatments of this extension of the Bethe ansatz to finite temperatures can be found in a review (Takahashi, 1997) and a monograph (Takahashi, 1999) by one of the pioneers of the thermodynamic Bethe ansatz. We already had occasion previously in this book to mention the comprehensive monograph, until recently only available in French, written by another one of the key figures, Michel Gaudin (1983), which has recently (2014) been translated into English by Jean-Sébastien Caux. A recommendable very clear exposition reviewing the thermodynamic Bethe ansatz for the Heisenberg quantum spin chain is Mezincescu and Nepomechie (1992).

17

Thermodynamics of the Repulsive Lieb–Liniger Model

It seems probable to me that God, in the beginning, formed matter in solid, massy, hard, impenetrable, moveable particles ...

Issac Newton (1643–1727)

This chapter again explores the system whose Bethe ansatz solution we investigated in chapter 15: the one-dimensional Bose gas with repulsive δ-function interaction. After briefly recalling the properties, especially in the thermodynamic limit, in section 17.1 of the excitations of this system which consist of particle and hole excitations and in section 17.2 preparing the corresponding densities of Bethe ansatz roots, we discuss their extension to finite temperature in section 17.3, where the main ingredient will be an appropriate expression for the entropy of the system of Bose particles. This system is arguably the simplest system to study by Bethe ansatz, for zero temperature, which we treated in chapter 15, as well as for finite temperature, which we address in this chapter.

17.1 Thermodynamic limit, particles, and holes

In order to explore the equilibrium thermodynamics of the δ-Bose gas, we again work in the thermodynamic limit of N particles in an interval of length L, i.e.

$$N \to \infty, \qquad L \to \infty \tag{17.1}$$

such that the particle density

$$\rho = \frac{N}{L} \tag{17.2}$$

is kept at a fixed finite value.

The microscopic description of the Bose particles has been achieved by considering the Bethe ansatz equations (15.113). Specifying sets of quantum numbers $\{n_m\}$ corresponds to sets of momenta $\{k_m\}$ of the system that are to be determined by these

Models of Quantum Matter: A First Course on Integrability and the Bethe Ansatz. Hans-Peter Eckle, Oxford University Press (2019). © Hans-Peter Eckle. DOI: 10.1093/oso/9780199678839.001.0001

equations. In the case of repulsive interaction between the particles, we found that there are two types of excitations: particle excitations and hole excitations. There will be a macroscopic number of each of these types of excitations in the thermodynamic limit and at finite temperature. In order to describe them, we introduce densities in momentum space, $\rho_p(k)$ for the particle excitations, $\rho_h(k)$ for the hole excitations, such that $\rho_t(k) = \rho_p(k) + \rho_h(k)$ is the density of all excitations in a repulsive δ-Bose gas. These densities describe the momentum distributions of solutions of the Bethe ansatz equations in the same way as the momentum distributions of solutions of the Bethe ansatz equations for the ground state in the thermodynamic limit have been described by a density in section 15.9.

The fact that we have only two types of excitations in the repulsive δ-Bose gas will simplify the analysis. In the attractive case, or for the Heisenberg quantum spin chain and other Bethe ansatz solvable models, the presence of string-type solutions complicates the Bethe ansatz to some extent, as we shall see for the quantum spin chain case in chapter 18. Thus, it is reasonable to tackle one difficulty at a time and discuss first the repulsive δ-Bose gas where only particle and hole excitations occur. The complication of string excitations will then be taken into account, not for the case of the attractive δ-Bose gas, but for the case of the Heisenberg quantum spin chain in chapter 18.

17.2 Bethe ansatz equations for particles and holes

The Bethe ansatz equations (15.113) are our starting point to derive equations that connect the density functions $\rho_t(k)$, $\rho_p(k)$, and $\rho_h(k)$. Introducing the function, sometimes called counting function,

$$z(k) = k + \frac{1}{L}\sum_{l=1}^{N} \Theta(k - k_l) \tag{17.3}$$

the Bethe ansatz equations can be cast in the general form

$$z(\bar{k}) = \frac{2\pi}{L}\bar{n} \tag{17.4}$$

for any integer of half-odd integer $\bar{n}$ and for any given set of momenta $\{k_l\}$ in the counting function $z(k)$.

For $k = k_j$ determined by a quantum number of the ground state set of quantum numbers (15.120), we recover a solution of the Bethe ansatz equations (15.113). When discussing in section 15.10 the excited states of the Lieb–Liniger model, i.e. of the Bose gas, we saw that the permissible quantum numbers $\{\bar{n}\}$, i.e. any set of integers or half-odd integers, described excitations of the system. Those quantum numbers could lie either outside or within the ground state distribution of quantum numbers (15.120). The former described particle excitations, the latter hole excitations with bare momenta k_p or k_h, respectively.

In order to examine the thermodynamics of the gas, we need to consider macroscopic numbers of particle and hole excitations. In the thermodynamic limit, the densities of the momenta of these excitations can be defined in the same way as the density of momenta for the ground state in (15.124). The densities for particles and holes are, respectively,

$$\rho_p(k_m) = \frac{1}{L\left(k_{m+1} - k_m\right)}, \qquad \rho_h(k_m) = \frac{1}{L\left(k_{m+1} - k_m\right)}, \tag{17.5}$$

where the momenta of the particles lie outside those of the holes within the ground state distribution of momenta.

The numbers of particles and holes in a momentum interval dk, multiplied by the length L of the interval (which we shall eventually let grow in the thermodynamic limit), are

$$L\rho_p(k)dk, \qquad L\rho_h(k)dk, \tag{17.6}$$

and the total number of excitations is

$$L\rho_t(k)dk = L\left(\rho_p(k) + \rho_h(k)\right) dk. \tag{17.7}$$

We now derive an integral equation connecting these densities. To this end, we observe that the derivative of the counting function (17.3) with respect to k at some particle or hole momentum k_j can be written as

$$\begin{aligned} \frac{\partial z(k)}{\partial k}|_{k=k_j} &= \lim_{N,L\to\infty} \frac{z(k_{j+1}) - z(k_j)}{k_{j+1} - k_j} = \lim_{N,L\to\infty} \frac{2\pi}{L\left(k_{j+1} - k_j\right)} \\ &= 2\pi\rho_t(k_j) = 2\pi\left(\rho_p(k_j) + \rho_h(k_j)\right) \end{aligned} \tag{17.8}$$

because $z(k_{j+1})$ and $z(k_j)$ satisfy Bethe ansatz equations with two successive quantum numbers $\bar{n}_{j+1}$ and $\bar{n}_j$, hence $\bar{n}_{j+1} - \bar{n}_j = 1$ and $z(k_{j+1}) - z(k_j) = 2\pi/L$. We combine this result with (17.3), which becomes in the thermodynamic limit after differentiation with respect to k

$$\rho_t(k) = \rho_p(k) + \rho_h(k) = \frac{1}{2\pi} + \frac{1}{2\pi}\int_{-\infty}^{\infty} dk'\, K(k - k')\rho_p(k'). \tag{17.9}$$

Note that the integral on the right-hand side contains only the particle density $\rho_p(k)$, which is a consequence of the fact that we summed in (17.3) only over the particle momenta. Unlike in the case of the ground state where the momentum distribution was confined to the finite interval $[-q, q]$, the momenta of the particles of excited states are unrestricted and, hence, we have to extend the integration limits over the whole k-axis. The integral equation (17.9) differs in an important way from the integral equations we have encountered so far. The density of particle momenta $\rho_p(k)$ depends also on the undetermined density of hole momenta $\rho_h(k)$. Nevertheless, (17.9) will prove useful since it connects these two densities.

17.3 Entropy and thermodynamic Bethe ansatz

Chapter 4 on equilibrium statistical mechanics discussed a fundamental consideration in the thermodynamics of many-particle systems, which is the counting of the number of ways microstates lead to the same macroscopic behaviour. There are many ways the numbers $L\rho_p(k)dk$ of particles and $L\rho_h(k)dk$ of holes can be distributed to obtain the same state given by $L\rho_t(k)dk$ and which lead to the change in entropy

$$dS = \ln\left(\frac{[L\rho_t(k)dk]!}{\left[L\rho_p(k)dk\right]!\,[L\rho_h(k)dk]!}\right). \tag{17.10}$$

Using Stirling's formula

$$\ln(n!) \cong n(\ln n - 1) \tag{17.11}$$

for large n, the differential entropy becomes

$$dS = Ldk\left(\rho_t(k)\ln\rho_t(k) - \rho_p(k)\ln\rho_p(k) - \rho_h(k)\ln\rho_h(k)\right), \tag{17.12}$$

which can be integrated to give

$$\begin{aligned} S &= L\int_{-\infty}^{\infty} dk\,\left(\rho_t(k)\ln\rho_t(k) - \rho_p(k)\ln\rho_p(k) - \rho_h(k)\ln\rho_h(k)\right) \\ &= L\int_{-\infty}^{\infty} dk\,\left(\rho_p(k)\ln\left[\frac{\rho_p(k)+\rho_h(k)}{\rho_p(k)}\right] + \rho_h(k)\ln\left[\frac{\rho_p(k)+\rho_h(k)}{\rho_h(k)}\right]\right) \qquad (17.13) \\ &\equiv \ln\left(W\left[\rho_p,\rho_h\right]\right). \qquad (17.14) \end{aligned}$$

Section 4.9 introduced the partition function as the central object whose calculation reveals the statistical and thermodynamic properties of a many-particle system. The general expression for the partition function

$$Z = \mathrm{Tr}\left(e^{-H/T}\right) = e^{-F/T} \tag{17.15}$$

relates the trace over the eigenstates of the Hamiltonian of the many-particle system to the free energy F. For the Lieb–Liniger model, for a finite number of particles, the partition function can be written, using the energy eigenvalues $E_N = \sum_{j=1}^{N} k_j^2$, as

$$Z_N = \frac{1}{N!} \sum_{n_1, n_2 \ldots, n_N} e^{-E_N/T} = \sum_{n_1 < n_2 < \ldots < n_N} e^{-E_N/T} \tag{17.16}$$

where the summations are over the set of quantum numbers defining a particular state via the Bethe ansatz equations by, in principle, determining the set of momenta $\{k\}$ of the state. For a gas at rest as a whole, the total momentum $P = \sum_{j=1}^{N} k_j$ vanishes, which implies, using the Bethe ansatz equations, that the quantum numbers must be chosen such that their sum vanishes: $\sum_{j=1}^{N} n_j = 0$. This allows us to introduce $N-1$ independent variables $m_j \equiv n_{j-1} - n_j$ whose range is unrestricted and write the partition function as

$$Z_N = \sum_{m_1=1}^{\infty} \sum_{m_2=1}^{\infty} \ldots \sum_{m_{N-1}}^{\infty} e^{-E_N/T}. \tag{17.17}$$

The variables m_j can be rewritten in terms of the density functions ρ_t and ρ_p if we observe, similar to (17.8) but now with m_j unrestricted, that

$$m_j = n_{j-1} - n_j = \frac{L}{2\pi} \left(z(k_{j+1}) - z(k_j) \right) = L \int_{k_j}^{k_{j+1}} dk\, \rho_t(k) \tag{17.18}$$

$$= L \int_{k_j}^{k_j + 1/L\rho_p(k_j)} dk\, \rho_t(k) = \frac{\rho_t(k_j)}{\rho_p(k_j)}. \tag{17.19}$$

We use this relation to rewrite the partition function, which is expressed in (17.17) as a sum over the microscopic variables $\{m_j\}$, as a path integral over the microscopic density ratios $\rho_t(k_j)/\rho_p(k_j)$.

The energy eigenvalues are given in the thermodynamic limit in terms of the particle density

$$E_N = L \int_{-\infty}^{\infty} dk\, \varepsilon_0(k) \rho_p(k) = L \int_{-\infty}^{\infty} dk\, k^2 \rho_p(k) = Le, \tag{17.20}$$

which we can use to write the partition function as

$$Z_N = \int \mathcal{D}\left[\frac{\rho_t}{\rho_p}\right] \delta\left(L \left\{ \int_{-\infty}^{\infty} dk\, \rho_p(k) - \rho \right\} \right) e^{S - Le/T}. \tag{17.21}$$

In this path integral, we have also introduced the entropy (17.13) via $W[\rho_p, \rho_h] = e^{S(\rho_p, \rho_h)}$ and, via the δ-function, the constraint that the particle density $\rho = N/L = \int_{-\infty}^{\infty} dk\, \rho_p(k)$ has a fixed value in the thermodynamic limit. The constraint can be written in a more useful way if we use an integral representation of the δ-function along the imaginary axis

$$\delta\left(\frac{x}{T}\right) = \int_{-i\infty}^{i\infty} d\mu\, e^{\mu x/T}, \tag{17.22}$$

which corresponds to the introduction of a Lagrange multiplier μ/T. Physically, this Lagrange multiplier μ/T represents the chemical potential. The partition function now takes the form

$$Z_N = \int_{-i\infty}^{i\infty} d\mu \int \mathcal{D}\left[\frac{\rho_p + \rho_h}{\rho_p}\right] e^{-\mathscr{F}[\rho_p,\rho_h,\mu]/T} \tag{17.23}$$

where the free energy functional is given by

$$\begin{aligned}\mathscr{F}[\rho_p, \rho_h, \mu] = \mu L\rho + L\int_{-\infty}^{\infty} dk \,\Big\{(k^2 - \mu)\rho_p(k) \\ -T\left(\rho_p(k)\ln\left[\frac{\rho_p(k) + \rho_h(k)}{\rho_p(k)}\right] + \rho_h(k)\ln\left[\frac{\rho_p(k) + \rho_h(k)}{\rho_h(k)}\right]\right)\Big\}.\end{aligned} \tag{17.24}$$

The partition function will now be evaluated within a saddle-point approximation for which we need to calculate the stationary value of $\mathscr{F}[\rho_p, \rho_t, \mu]$ from the variation

$$\begin{aligned}\delta\mathscr{F}[\rho_p, \rho_h, \mu] = -L\delta\mu\left(\int_{-\infty}^{\infty} dk\, \rho_p(k) - \rho\right) + L\int_{-\infty}^{\infty} dk\,\Big\{(k^2 - \mu)\delta\rho_p(k) \\ - T\left(\delta\rho_p(k)\ln\left[\frac{\rho_p(k) + \rho_h(k)}{\rho_p(k)}\right] + \delta\rho_h(k)\ln\left[\frac{\rho_p(k) + \rho_h(k)}{\rho_h(k)}\right]\right)\Big\}.\end{aligned} \tag{17.25}$$

At this point, we need the connection between the densities ρ_p and ρ_h given in (17.9). The variation of this equation is

$$\delta\rho_p(k) + \delta\rho_h(k) = \frac{1}{2\pi}\int_{-\infty}^{\infty} K(k - k')\delta\rho_p(k'), \tag{17.26}$$

which allows us to eliminate the variation $\delta\rho_h$ from the variation $\delta\mathscr{F}[\rho_p, \rho_h, \mu]$ which becomes

$$\begin{aligned}\delta\mathscr{F}[\rho_p, \rho_h, \mu] = -L\delta\mu\left(\int_{-\infty}^{\infty} dk\, \rho_p(k) - \rho\right) + L\int_{-\infty}^{\infty} dk\,\Big\{(k^2 - \mu)\delta\rho_p(k) \\ - T\Bigg(\delta\rho_p(k)\ln\left[\frac{\rho_h(k)}{\rho_p(k)}\right] + \frac{1}{2\pi}\ln\left[\frac{\rho_p(k) + \rho_h(k)}{\rho_h(k)}\right], \\ \cdot\int_{-\infty}^{\infty} dk'\, K(k - k')\delta\rho_p(k')\Bigg).\end{aligned} \tag{17.27}$$

In the double integral of this expression, we can exchange the integration variables $k \leftrightarrow k'$ and use the fact that $K(k - k') = K(k' - k)$ to obtain an integral whose integrand is proportional to $\delta\rho_p(k)$. In summary, the condition that the variation $\delta\mathscr{F}[\rho_p, \rho_h, \mu]$ vanishes yields the following two equations. We get back the condition that the particle density is given by the fixed value

$$\int_{-\infty}^{\infty} dk\, \rho_p(k) = \rho \tag{17.28}$$

and an integral equation connecting the densities for particle and hole excitations

$$k^2 - \mu - T\left(\ln\left[\frac{\rho_h(k)}{\rho_p(k)}\right] + \frac{1}{2\pi}\int_{-\infty}^{\infty} dk'\, K(k - k')\ln\left[\frac{\rho_p(k) + \rho_h(k)}{\rho_h(k)}\right]\right) = 0. \tag{17.29}$$

The relative numbers of particles and holes are determined by a Boltzmann factor

$$\frac{\rho_h(k)}{\rho_p(k)} = e^{\mathscr{E}(k)/T}, \tag{17.30}$$

which allows to finally make contact with thermodynamics by introducing the temperature. We can thus write

$$\mathscr{E}(k) = k^2 - \mu - \frac{T}{2\pi}\int_{-\infty}^{\infty} dk'\, K(k - k')\ln\left(1 + e^{-\mathscr{E}(k')/T}\right). \tag{17.31}$$

This equation is a nonlinear integral equation for the dressed energy $\mathscr{E}$ per particle describing the particle excitations of the δ-Bose gas. It is named Yang–Yang equation after its discoverers (Yang and Yang, 1969).

It is interesting to note that the number of particle excitations relative to the number of totally available states, i.e. particle states and hole states, is given by the Fermi distribution function

$$\frac{\rho_p(k)}{\rho_p(k) + \rho_h(k)} = \frac{1}{1 + e^{\mathscr{E}(k)/T}}. \tag{17.32}$$

Using again (17.9), the density $\rho_p(k)$ of the particle excitations can now be expressed in terms of the dressed energy by the integral equation

$$\rho_p(k)\left(1 + e^{\mathscr{E}(k)/T}\right) = \frac{1}{2\pi} + \frac{1}{2\pi}\int_{-\infty}^{\infty} dk'\, K(k - k')\rho_p(k'). \tag{17.33}$$

EXERCISE 17.1 Entropy of the δ-Bose gas Show that within the saddle-point approximation we used, the entropy of the δ-Bose gas can be written as

$$S = L\int_{-\infty}^{\infty} dk \left(\frac{1}{2\pi}\ln\left(1+e^{-\mathcal{E}(k)/T}\right)+\frac{1}{T}\left(k^2-\mu\right)\rho_p(k)\right) \tag{17.34}$$

and, hence, the free energy

$$F = -T\ln Z_N = L\int_{-\infty}^{\infty} dk\, k^2\rho_p(k) - TS - L\mu\left(\int_{-\infty}^{\infty} dk\,\rho_p(k)-\rho\right) \tag{17.35}$$

$$= L\rho\mu - \frac{TL}{2\pi}\int_{-\infty}^{\infty} dk\,\ln\left(1+e^{-\mathcal{E}(k)/T}\right). \tag{17.36}$$

EXERCISE 17.2 Pressure of the δ-Bose gas Using the definition of pressure

$$P = -\left(\frac{\partial F}{\partial L}\right), \tag{17.37}$$

show that

$$\frac{\partial P}{\partial T} = \frac{S}{L} \quad \text{and} \quad \frac{\partial P}{\partial \mu} = 0. \tag{17.38}$$

In exercise 17.1, having obtained an expression for the partition function, and, hence, the free energy, of the system of interacting Bose particles, in terms of the dressed energy $\mathcal{E}(k)$, the thermodynamics of the system can be fully calculated (see e.g. exercise 17.2). The solution of the Yang–Yang equation (17.31), however, being a nonlinear integral equation, remains a formidable task.

In chapter 18, we extend the thermodynamic Bethe ansatz to the isotropic Heisenberg quantum spin chain. The existence of a further type of solutions, the string solutions discussed in section 14.5.3, presents a major complication in this model and requires some preparation.

18

Thermodynamics of the Isotropic Heisenberg Quantum Spin Chain

Simplicity is the ultimate sophistication.
Leonardo da Vinci (1452–1519)

This chapter presents the extension of the Bethe ansatz to finite temperature, i.e. the thermodynamic Bethe ansatz, for the antiferromagnetic isotropic Heisenberg quantum spin chain, the XXX quantum spin chain.

This model features extra complications compared to the repulsive δ-Bose gas, for which we introduced the thermodynamic Bethe ansatz in chapter 17. These extra complications stem from the more intricate structure of excitations of the quantum spin chain, the complex string excitations discussed in chapter 14, in sections 14.5 and especially 14.5.3.

To cope with this more complicated situation of adapting the Bethe ansatz in order to deal with the thermodynamics of the quantum spin chain, we summarize first the results from chapter 14 in section 18.1. After further preparation in section 18.2 concerning particle and hole excitations in the XXX quantum spin chain, we derive the integral equations of the thermodynamic Bethe ansatz for the XXX quantum spin chain in section 18.3.

Finally, in sections 18.4 and 18.5, we mention as a further result for the thermodynamics explicit formulas for the free energy of the quantum spin chain and some interesting physical quantities, especially making contact with predictions of conformal symmetry (cp. chapter 7, in particular section 7.10), which can be achieved on this basis of these fundamental integral equations of the thermodynamic Bethe ansatz.

18.1 Summary of Bethe ansatz for the isotropic spin chain

Let us first summarize some of the results we obtained in part III on the coordinate Bethe ansatz for the isotropic antiferromagnetic spin chain. In order to be able to calculate quantities like the magnetic susceptibility, we consider here the slightly more general isotropic antiferromagnetic spin chain in an external magnetic field $h \geq 0$ with Hamiltonian

Models of Quantum Matter: A First Course on Integrability and the Bethe Ansatz. Hans-Peter Eckle, Oxford University Press (2019). © Hans-Peter Eckle. DOI: 10.1093/oso/9780199678839.001.0001

$$\mathcal{H} = \sum_{n=1}^{N} \left(\mathbf{S}_n \mathbf{S}_{n+1} - \frac{1}{4} \right) - h \sum_{n=1}^{N} S_n^z. \tag{18.1}$$

EXERCISE 18.1 Energy eigenvalue of the isotropic spin chain in a magnetic field Convince yourself that the earlier result for the energy eigenvalue of the isotropic antiferromagnetic chain, equation (14.67) in section 14.3.1, can easily be generalized to the finite magnetic field case of Hamiltonian (18.1) to give

$$E_M = -\frac{1}{2} \sum_{i=1}^{M} \frac{1}{v_i^2 + \frac{1}{4}} - h \left(\frac{N}{2} - M \right). \tag{18.2}$$

The rapidities $\{v_i\}$ satisfy the Bethe ansatz equations

$$\left(\frac{v_i + \frac{i}{2}}{v_i - \frac{i}{2}} \right)^N = \prod_{j=i,\, j \neq i}^{M} \frac{v_i - v_j + i}{v_i - v_j - i}, \quad i = 1, 2, \ldots, M, \quad M \leq \frac{N}{2}. \tag{18.3}$$

These equations written in a more compact notation are

$$[V_1(v_i)]^N = \prod_{j=i,\, j \neq i}^{M} V_2(v_i - v_j), \tag{18.4}$$

where the functions

$$V_n(v) \equiv \frac{v + \frac{in}{2}}{v - \frac{in}{2}} \tag{18.5}$$

have been introduced.

We discussed in section 14.5.3 the string hypothesis for the solution of these equations. Let us recall the first few cases of string solutions:

- a string of length one:

$$v^{(1)} = x \tag{18.6}$$

  with x an arbitrary real number, called the *centre* of the string.

- a string of length two:

$$\begin{cases} v^{(1)} = x + \frac{i}{2} \\ v^{(2)} = x - \frac{i}{2} \end{cases}, \tag{18.7}$$

- a string of length three:

$$\begin{cases} v^{(1)} = x + i \\ v^{(2)} = x \\ v^{(3)} = x - i \end{cases}, \tag{18.8}$$

- and so on.

Recall the *string hypothesis*, which states that all solutions $\{v_i, i = 1, 2, \ldots, M\}$ of the Bethe ansatz equations (18.3) are organized in strings. This hypothesis, despite the fact that it is not correct (cf. the references given in section 14.5.3), is adopted in most treatments of the thermodynamics of Bethe ansatz solvable models, and we shall adopt it here, too. Non string-type solutions are assumed to be negligible in the thermodynamic limit.

The general form of the strings is as follows: generally, there are M_n strings of length n of the form

$$v_\alpha^{(n,j)} = v_\alpha^n + i\left(\frac{n+1}{2} - j\right), \quad j = 1, \ldots, n, \quad \alpha = 0, 1, \ldots, M_n, \quad n = 1, \ldots, \infty \tag{18.9}$$

with real centres v_α^n.

Section 14.5.3 discussed how to rewrite the Bethe ansatz equations for string solutions $\{v_\alpha^{(n,j)}\}$ in such a way that the resulting Bethe ansatz equations only contain the real centres, $\{v_\alpha^n\}$, of the strings. The result is

$$\left[V_n\left(v_\alpha^n\right)\right]^N = \prod_{m=1;m\neq n}^{M} \prod_{\beta=0;\beta\neq\alpha}^{\nu_m} V_{nm}\left(v_\alpha^n - v_\beta^m\right) \tag{18.10}$$

$$V_{nm}(x) = V_{|n-m|}(x)\, V_{|n-m|+2}^2(x) \ldots V_{n+m-2}^2(x)\, V_{n+m}(x). \tag{18.11}$$

Since we have the thermodynamic limit $N \to \infty$ in mind, we can let the upper limit of the first product over m grow beyond bounds $M \to \infty$.

Since the products in equation (18.10) only contain phases, taking the logarithm simplifies matters

$$Nk_n(v_\alpha^n) = 2\pi \mathcal{J}_\alpha^n + \sum_{m=1,m\neq n}^{\infty} \sum_{\beta=0,\beta\neq\alpha}^{\nu_m} \Theta_{nm}\left(v_\alpha^n - v_\beta^m\right) \tag{18.12}$$

where

$$k_n(v) \equiv i \ln V_n(v) \tag{18.13}$$

and

$$\Theta_{nm}(v) \equiv i \ln V_{nm}(v). \tag{18.14}$$

Note that we have already used the upper limit ∞ in the summation over m.

The numbers $\mathcal{J}_\alpha^n$ play the role of quantum numbers and are integers or half-integers, which are constrained by

$$-\mathcal{J}_{\max}^n \leq \mathcal{J}_\alpha^n \leq \mathcal{J}_{\max}^n. \tag{18.15}$$

Faddeev and Takhtajan (1984) have given an explicit description for extracting an expression for $\mathcal{J}_{\max}^n$ from equations (18.12). However, we shall not need such an expression for our purposes.

We only need that such a range in fact does exist and call any integer or half-integer within the range *admissible*. The reason for this naming will become clear soon. For every admissible set of mutually distinct quantum numbers $\{\mathcal{J}_\alpha^n\}$ there is a unique solution of mutually distinct values $\{v_\alpha^n\}$ of equation (18.12). Such solutions describe *particles* in the Bethe ansatz framework, and are, hence, called particle rapidities.

Yang and Yang (1969), in their treatment of the thermodynamics of the one-dimensional Bose gas with repulsive δ-function interaction, introduced the corresponding concept of *holes*. If we prescribe a set of particle quantum numbers $\{\mathcal{J}_\alpha^n\}$, there can be, and in a generic case will be, gaps that correspond to numbers $\{\tilde{\mathcal{J}}_\alpha^n\}$, which, though perfectly admissible, are *not* in the set of particle quantum numbers $\{\mathcal{J}_\alpha^n\}$.

Using these admissible quantum numbers, which we call hole quantum numbers, we define via

$$Nk_n(\tilde{v}_\alpha^n) = 2\pi \tilde{\mathcal{J}}_\alpha^n + \sum_{m=1, m\neq n}^{\infty} \sum_{\beta=0, \beta\neq\alpha}^{\nu_m} \Theta_{nm}\left(\tilde{v}_\alpha^n - v_\beta^m\right) \tag{18.16}$$

quantities $\{\tilde{v}_\alpha^n\}$, called hole rapidities.

Observe that the hole rapidities are determined by the particle rapidities $\{v_\beta^m\}$, and, thus, equations (18.12) and (18.16) are coupled.

18.2 Preparation of the thermodynamic Bethe ansatz: particles and holes

Now we are ready to take advantage of the thermodynamic limit: For $N \to \infty$, there will be a macroscopic number of particle quantum numbers $\mathcal{J}_\alpha^n$ and hole quantum numbers $\tilde{\mathcal{J}}_\alpha^n$; correspondingly, there will also be a macroscopic number of particle rapidities v_α^n and hole rapidities $\tilde{v}_\alpha^n$. Thus, we can introduce densities of particle and hole rapidities

$$N\rho_n(v)dv = \text{number of } v^n \text{ in } dv, \tag{18.17}$$

$$N\tilde{\rho}_n(v)dv = \text{number of } \tilde{v}^n \text{ in } dv. \tag{18.18}$$

From now on, these densities will be the primary objects of our interest. This means, in other words, that we reformulate the problem, i.e. the Bethe ansatz equations, in terms of these densities.

A first step in this direction is the introduction of the function

$$h^n(v) = \frac{1}{2\pi}\left\{Nk_n(v) - \sum_{m=1,m\neq n}^{\infty}\sum_{\beta=0,\beta\neq\alpha}^{\nu_m}\Theta_{nm}\left(v - v_\beta^m\right)\right\}. \tag{18.19}$$

Obviously, we have

$$h^n(v_\alpha^n) = \mathcal{J}_\alpha^n \tag{18.20}$$

$$h^n(\tilde{v}_\alpha^n) = \tilde{\mathcal{J}}_\alpha^n. \tag{18.21}$$

We assume that $h^n(v)$ is a monotonic function in v, from which follows that

$$\begin{aligned} N\left[\rho_n(v) + \tilde{\rho}_n(v)\right]dv &= \text{number of } v^n \text{ and } \tilde{v}^n \text{ in } dv \\ &= h^n(v + dv) - h^n(v) = dh^n. \end{aligned} \tag{18.22}$$

Hence

$$\rho_n(v) + \tilde{\rho}_n(v) = \frac{1}{N}\frac{dh^n}{dv}. \tag{18.23}$$

Next, we replace the sums $\sum_\beta^{M_n}$ in the function h^n, equation (18.19), by integrals, i.e. we now really assume that we have a macroscopic number ν_n of strings of given real centres v_α^n:

$$\frac{1}{N}\sum_{\beta=0}^{M_n}(\ldots) \to \int_{-\infty}^{\infty} dv'(\ldots)\rho_m(v'). \tag{18.24}$$

From equation (18.23) we then find

$$\rho_n(v) + \tilde{\rho}_n(v) = \frac{1}{2\pi}\frac{dk_n(v)}{dv} - \sum_{m=1,m\neq n}^{\infty}\int_{-\infty}^{\infty} dv\frac{1}{2\pi}\frac{d\Theta_{nm}(v-v')}{dv}\rho_m(v'). \tag{18.25}$$

Using the convolution operation $*$, this equation can be recast in the compact form

$$\tilde{\rho}_n + \sum_{m=1, m\neq n}^{\infty} A_{nm} * \rho_m = a_n, \tag{18.26}$$

where

$$\begin{aligned} A_{nm}(v) &\equiv \delta(v)\delta_{nm} + \frac{1}{2\pi}\frac{d\Theta_{nm}(v)}{dv} \\ &= \delta(v)\delta_{nm} + (1-\delta_{nm})a_{|n-m|}(v) + a_{n+m}(v) \\ &+ 2\sum_{l=1}^{\min(n,m)-1} a_{|n-m|+2l}(v), \end{aligned} \tag{18.27}$$

and

$$a_n(v) \equiv \frac{1}{2\pi}\frac{dk_n(v)}{dv} = \frac{1}{2\pi}\frac{n}{v^2 + \frac{n^2}{4}}. \tag{18.28}$$

The convolution $*$ is explicitly given by

$$(f * g)(v) \equiv \int_{-\infty}^{\infty} dv' f(v - v')g(v'). \tag{18.29}$$

At this point, we could take the view that equation (18.26) defines the hole density $\tilde{\rho}_n$ in terms of the particle density ρ_n.

We introduce some useful functions, which, in some sense, provide the 'inverse' to the functions A_{nm}, as

$$A^{-1}_{nm}(v) \equiv \delta(v)\delta_{nm} - s(v)(\delta_{n,m+1} + \delta_{n,m-1}), \tag{18.30}$$

where $s(v)$ is defined by

$$s(v) \equiv \frac{1}{2\cosh \pi v}, \tag{18.31}$$

and which satisfy

$$\sum_{n'=1}^{\infty}\left(A^{-1}_{nn'} * A_{n'm}\right)(v) = \delta(v)\delta_{nm}, \tag{18.32}$$

and

$$\sum_{m}^{\infty} \left(A_{nm}^{-1} * a_m\right)(v) = s(v)\delta_{n1}, \tag{18.33}$$

$$\sum_{m}^{\infty} \left(A_{nm}^{-1} * m\right)(v) = 0. \tag{18.34}$$

These relations can be derived using Fourier transformation with the conventions

$$\hat{f}(\omega) \equiv \int_{-\infty}^{\infty} dv e^{i\omega v} f(v), \tag{18.35}$$

$$f(v) \equiv \frac{1}{2\pi} \int_{-\infty}^{\infty} d\omega e^{-i\omega v} \hat{f}(\omega). \tag{18.36}$$

Useful in this connection are the convolution theorem

$$\int_{-\infty}^{\infty} dv e^{i\omega v} (f * g)(v) = \hat{f}(\omega)\hat{g}(\omega), \tag{18.37}$$

and the identities

$$\hat{a}_n(\omega) = \begin{cases} e^{-\frac{n|\omega|}{2}} & n > 0 \\ 0 & n = 0 \end{cases}, \tag{18.38}$$

$$\hat{s}(\omega) = \frac{1}{2\cosh\frac{\omega}{2}}, \tag{18.39}$$

$$\hat{A}_{nm}(\omega) = \left(\coth\frac{|\omega|}{2}\right)\left[e^{-\frac{|n-m||\omega|}{2}} - e^{-\frac{(n+m)|\omega|}{2}}\right], \tag{18.40}$$

$$\hat{A}_{nm}^{-1}(\omega) = \delta_{nm} - \hat{s}(\omega)(\delta_{n,m+1} + \delta_{n,m-1}). \tag{18.41}$$

EXERCISE 18.2 Details of the calculations Verify the results of the previous section by explicit calculation, especially

- equation (18.26), and
- equations (18.32-18.34)

using Fourier transformation, equations (18.35, 18.36), and the convolution theorem, equation (18.37).

Having rewritten the Bethe ansatz equations in terms of particle and hole densities in the form (18.26) leaves us with the task of rewriting the expression for the energy eigenvalue (18.2) in terms of the densities. First, we insert the string hypothesis in (18.2) to arrive at

$$E = -\frac{1}{2}\sum_{n=1}^{\infty}\sum_{\alpha=0}^{\nu_n}\sum_{j=1}^{n}\frac{1}{\left(v_\alpha^{(n,j)}\right)^2+\frac{1}{4}} - h\left(\frac{N}{2}-\sum_{n=1}^{\infty}n\nu_n\right). \tag{18.42}$$

The j-summation in the first term of this expression can be rewritten as

$$\frac{1}{2}\sum_{j=1}^{n}\frac{1}{\left(v_\alpha^{(n,j)}\right)^2+\frac{1}{4}} = \frac{i}{2}\frac{d}{dv_\alpha^n}\sum_{j=1}^{n}\ln V_1\left(v_\alpha^{(n,j)}\right) = \frac{i}{2}\frac{d}{dv_\alpha^n}\ln\prod_{j=1}^{n}V_1\left(v_\alpha^{(n,j)}\right) \tag{18.43}$$

$$= \frac{i}{2}\frac{d}{dv_\alpha^n}\ln V_n(v_\alpha^n) = \frac{1}{2}\frac{d}{dv_\alpha^n}k_n(v_\alpha^n) = \pi\, a_n(v_\alpha^n), \tag{18.44}$$

where in passing to the second line we have removed the full strings and are left with only the real centres v_α^n in the same way as in section 14.5.3.

The energy is, thus, given by

$$\begin{aligned} E &= -\pi\sum_{n=1}^{\infty}\sum_{\alpha=0}^{\nu_n}a_n(v_\alpha^n) - Nh\left(\frac{1}{2}-\frac{1}{N}\sum_{n=1}^{\infty}n\nu_n\right) \\ &= -N\pi\sum_{n=1}^{\infty}\int_{-\infty}^{\infty}dv\, a_n(v)\rho_n(v) \\ &\quad - Nh\left(\frac{1}{2}-\sum_{n=1}^{\infty}n\int_{-\infty}^{\infty}dv\,\rho(v)\right). \end{aligned} \tag{18.45}$$

Thus, we have have arrived at an expression for an energy level of the system given a set of particle densities $\{\rho_n\}$. Although this expression does not explicitly depend on the hole densities $\{\tilde{\rho}_n\}$, it does so nevertheless because of the equations (18.26), which couple particle and hole densities.

This has been a rather longish exposition of formalism. Therefore, to regain firm ground, let us reproduce a known result: the ground state energy of the antiferromagnetic chain, Hulthén's famous result of 1938.

In zero field, $h = 0$, we assume that there are *no* holes present in the ground state

$$\tilde{\rho}_n = 0. \tag{18.46}$$

Therefore, equation (18.26) yields

$$\sum_{m=1}^{\infty} A_{nm} * \rho_m = a_n, \tag{18.47}$$

which, upon inversion, gives

$$\rho_n = \sum_{m=1}^{\infty} A_{nm}^{-1} * a_n = s\delta_{n1}, \tag{18.48}$$

where $s(v)$ is given explicitly (see above) by

$$s(v) = \frac{1}{2\cosh \pi v}. \tag{18.49}$$

Equation (18.48) signifies that the antiferromagnetic ground state consists of strings of length one, i.e. of real solutions of the Bethe ansatz equations. From equation (18.45) for the energy eigenvalue we now obtain for the energy per site of the ground state

$$e_0 = \frac{E_0}{N} = -\pi \int_{-\infty}^{\infty} dv\, a_1(v) s(v) = -\ln 2, \tag{18.50}$$

which indeed is the well-known result of Hulthén (1938), and which is, therefore, reassuring.

There is, however, another important aspect to this result for the ground state energy of the antiferromagnetic spin chain we derived here for the second time. In both instances where we derived this result, here and in section 14.4.3, and also in the subsequent discussion of low-lying excitations, we started from the same assumption: that the ground state is given by a compact set of Bethe ansatz quantum numbers with, in the language of the present section, no holes.

This observation raises two important questions:

1. How do we establish that the ground state indeed has no holes?
2. How do we study low-lying excitations from there?

18.3 Thermodynamic Bethe ansatz equations

A physical answer to both questions involves the study of the system at finite temperature $T \geq 0$. Then, the equilibrium state at $T = 0$ will be the ground state. Equilibrium states at low but finite temperature $T > 0$ will contain information about the low-lying excitations.

With this in mind, the following section first derives, thermodynamic Bethe ansatz equations, i.e. Bethe ansatz equations valid at $T \geq 0$. Second, we use them to derive the free energy of the spin chain. A third natural step will be the evaluation of the thermodynamic Bethe ansatz equations and of the free energy to derive physical quantities at low temperature T and small field h. However, we shall not go this last step, though perfectly possible, in this book in order to limit the complexity of the derivations and calculations. In order to round off the presentation, we present a few results in the last two sections of this chapter for the free energy at finite temperature and its behaviour at low temperature and for a small magnetic field. The latter two connect the thermodynamic Bethe ansatz results to results from conformal field theory discussed in chapter 7, in particular section 7.10.

The heuristic argument that goes back to Yang and Yang (1969), but has been advocated by others for other models, notably the Kondo model (Andrei *et al.*, 1981; Tsvelik and Wiegmann, 1983; Andrei, 1994), runs as follows: an equilibrium state at temperature T is described by equilibrium densities of particles ρ_n^{eq} and holes $\tilde{\rho}_n^{\text{eq}}$. The partition function is then given by

$$Z = tr\left(e^{-\mathcal{H}/T}\right) = \sum_{\text{eigenstates}} e^{-E/T} = \sum_{\rho,\tilde{\rho}} \delta\left[\chi(\rho,\tilde{\rho})\right] W\left[\rho,\tilde{\rho}\right] e^{-E[\rho]/T} \tag{18.51}$$

where the δ-function enforces the constraint

$$\chi(\rho,\tilde{\rho}) = 0, \tag{18.52}$$

which in turn formally represents equation (18.26). The factor $W[\rho,\tilde{\rho}]$ is the number of states compatible with given densities ρ and $\tilde{\rho}$,[1] and allows the introduction of the entropy via

$$S = \ln W \tag{18.53}$$

and the free energy

$$F = E - TS. \tag{18.54}$$

Using this concept, we can write

$$Z = \sum_{\rho,\tilde{\rho}} \delta\left[\chi(\rho,\tilde{\rho})\right] e^{-(E[\rho]-TS[\rho,\tilde{\rho}])/T} = \sum_{\rho,\tilde{\rho}} \delta\left[\chi(\rho,\tilde{\rho})\right] e^{-F[\rho,\tilde{\rho}]/T}. \tag{18.55}$$

[1] W is what Boltzmann called complexions, i.e. the number of microstates compatible with a macrostate.

This expression will be dominated for $N, M \to \infty$ with N/M kept constant by the equilibrium value of the free energy

$$Z = e^{-F(\rho_{\mathrm{eq}}, \tilde{\rho}_{\mathrm{eq}})/T}, \tag{18.56}$$

where the equilibrium densities are determined by the condition that the variation of F vanishes

$$\delta F|_{\rho=\rho_{\mathrm{eq}}, \tilde{\rho}=\tilde{\rho}_{\mathrm{eq}}} = 0 \tag{18.57}$$

subject to (18.26).

In order to use these equations, we still need an expression for the entropy[2] in terms of the densities ρ and $\tilde{\rho}$. Here again Yang and Yang (1969) provided the decisive idea. The change in entropy is given by

$$\begin{aligned} dS_n &= \ln \{\text{number of states (i.e. strings of length } n) \text{ in } dv\} \\ &= \ln \left\{\text{number of ways of choosing } v^n \text{ and } \tilde{v}_n \text{ in } dv\right\} \\ &= \ln \left\{ \frac{[N(\rho_n + \tilde{\rho}_n)dv]!}{(N\rho_n dv)!(N\tilde{\rho}_n dv)!} \right\} \\ &\approx N \{(\rho_n + \tilde{\rho}_n) \ln(\rho_n + \tilde{\rho}_n) - \rho_n \ln \rho_n - \tilde{\rho} \ln \tilde{\rho}\} \, dv \end{aligned} \tag{18.58}$$

where we have used Stirling's approximation formula in passing to the last line. Entropy formulas like this are frequently encountered, e.g. in the context of mixing of fluids where they are aptly called mixing entropies.

We, thus, arrive at the entropy

$$S = \sum_{n=1}^{\infty} \int_{-\infty}^{\infty} dS_n = N \sum_{n=1}^{\infty} \int_{-\infty}^{\infty} dv \, \{(\rho_n + \tilde{\rho}_n) \ln(\rho_n + \tilde{\rho}_n) - \rho_n \ln \rho_n - \tilde{\rho} \ln \tilde{\rho}\}. \tag{18.59}$$

Now we see the crucial role of the concept of holes, which was introduced precisely for the purpose to find an expression for the entropy.

We have now assembled all the ingredients needed: the free energy $F = E - TS$ is now expressed in terms of the densities ρ_n and $\tilde{\rho}_n$ via the expressions for the energy (18.45) and the entropy (18.59).

The condition (18.57), i.e. the variation of the free energy with respect to ρ_n and $\tilde{\rho}_n$, together with the variation of the constraint (18.26) to eliminate $\delta\tilde{\rho}_n$, yields the thermodynamic Bethe ansatz equations

[2] Or, equivalently, for W.

$$T \ln\left(1 + e^{\epsilon_n/T}\right) = \sum_{m=1}^{\infty} A_{nm} * T \ln\left(1 + e^{-\epsilon_m/T}\right) \tag{18.60}$$

with $n = 1, \ldots, \infty$ and the functions $\epsilon_n(v)$ defined by

$$\epsilon_n(v) = T \ln\left(\frac{\tilde{\rho}_n^{\mathrm{eq}}(v)}{\rho_n^{\mathrm{eq}}(v)}\right). \tag{18.61}$$

The thermodynamic Bethe ansatz equations are an infinite set of coupled nonlinear integral equations. We assume that they have a unique solution $\{\epsilon_n(v)\}$.

Assuming we have a solution, i.e. determined the functions $\{\epsilon_n(v)\}$, we can determine the equilibrium densities $\{\rho_n^{\mathrm{eq}}(v)\}$ and $\{\tilde{\rho}_n^{\mathrm{eq}}(v)\}$ from the condition (18.26).

EXERCISE 18.3 Thermodynamic Bethe ansatz equations Perform the explicit calculation that leads to the thermodynamic Bethe ansatz equations (18.60).

18.4 Thermodynamics

This brief section states the equation for the free energy per site in equilibrium that can be obtained from the thermodynamic Bethe ansatz equations. It is

$$\frac{F}{N} = -T \sum_{n=1}^{\infty} \int_{-\infty}^{\infty} dv a_n(v) \ln\left(1 + e^{-\epsilon_n(v)/T}\right) - \frac{h}{2}. \tag{18.62}$$

For some hints as to how to arrive at this expression, the interested reader is referred to Mezincescu and Nepomechie (1992). There is also a convenient alternative expression

$$\frac{F}{N} = e_0 - T \int_{-\infty}^{\infty} dv\, s(v) \ln\left(1 + e^{\epsilon_1(v)/T}\right), \tag{18.63}$$

where $e_0 = -\ln 2$. This expression for the equilibrium free energy has the convenient feature that it depends only on one of the functions $\epsilon_n(v)$.

Of course, the parts we left out are the really difficult ones. But in order to keep the exposition within reasonable bounds, the interested reader is referred to the original literature. Especially we hope that with the details provided so far, the study of the review Takahashi (1997) and the monograph Takahashi (1999) will be doable.

18.5 Thermodynamics for small *T* and *h*

The next step is to solve the thermodynamic Bethe ansatz equations (18.60) for small values of temperature T and external magnetic field h and to evaluate the expression (18.63) for the free energy. We are not going to do this here since the techniques, in particular the Wiener–Hopf technique to solve certain integral equations, will be encountered also in the following part VI on the Bethe ansatz for finite systems, and we shall discuss them there.

We refer the reader to the literature, and especially recommend the review paper by Mezincescu and Nepomechie (1992).

However, we quote one particularly interesting result that the programme outlined above yields, namely the free energy per site in lowest order of temperature T

$$\frac{F}{N} = e_0 - \frac{1}{3}T^2. \tag{18.64}$$

(See, e.g. equation (3.82) of (Mezincescu and Nepomechie, 1992). On the other hand, conformal invariance predicts (Blöte *et al.*, 1986; Affleck, 1986)

$$\frac{F}{N} = e_0 - \frac{\pi c}{6v_s}T^2 \tag{18.65}$$

where c is the central charge and v_s is the velocity of sound. For the isotropic Heisenberg quantum spin chain $v_s = \frac{\pi}{2}$.[3] Therefore, we conclude for the central charge that

$$c = 1. \tag{18.66}$$

Thus, we have a nice connection between conformal invariance (cp. chapter 7, especially section 7.10.2) and results of the thermodynamic Bethe ansatz.

[3] We shall find this result from the finite size analysis in part VI.

Part 6

Bethe Ansatz for Finite Systems

It is hard to be finite upon an infinite subject, and all subjects are infinite.

Herman Melville (1819–1891)

The Bethe ansatz provides equations for a system of finite size. The method genuinely deals with a system of finite length, e.g. a spin chain consisting of a finite number of interacting spins, all equations of the Bethe ansatz are at first derived for a finite system. Exploiting these equations to extract physical quantities, be it for the ground state energy or energies of low-lying excitations or for finite temperature quantities like the free energy, proved rather involved in most cases and required a number of tricks: e.g. the string hypothesis and the limit of macroscopically large system sizes, i.e. the thermodynamic limit $N \to \infty$. These devices facilitated the calculations enormously but restricted the information contained in the results.

For very small system sizes, the Bethe ansatz equations are manageable, but not really necessary, since we can diagonalize the finite Hamiltonian matrix with elementary methods of linear algebra. For larger, but still modest, system sizes, however, this quickly becomes a task requiring large-scale numerical computations.

Therefore the questions arises: how can the Bethe ansatz be used analytically for rather large, yet finite, system sizes?

The basic idea of the Bethe ansatz for finite systems is to use the thermodynamic limit as the starting point and then to find systematic means to calculate corrections to this limit arising from the finite size of the system.

We have encountered a seed of this idea when we discussed the simplest excitations in the isotropic antiferromagnetic chain in section 14.6.

In this part on the Bethe ansatz for finite systems, in order to be able to dispense with the simplifications the thermodynamic limit affords us, we shall have to become acquainted with the mathematical tools from which then to build an analytic method to extract physical information from the Bethe ansatz for a finite system.

Chapter 19 introduces the two major ingredients for this purpose from the toolbox of applied mathematics, the first of which, the Euler–Maclaurin formula, is then used in chapter 20 on the Bethe ansatz for the finite Heisenberg chain, to transform the discrete set of Bethe ansatz equations into integral equations. These integral equations are more involved than those encountered in the thermodynamic limit. They retain, in a systematic way, corrections from the finiteness of the system.

As we shall see, these integral equations are of the so-called Wiener–Hopf type which can be treated by the eponymous mathematical technique, the Wiener–Hopf technique. Chapter 19 present an introduction to this technique that constitutes the second ingredient from the toolbox of applied mathematics.

19

Mathematical Tools

Read Euler: he is our master in everything.
Pierre-Simon Laplace (1749–1827)

This chapter presents the mathematical techniques we employ to extract analytic information from the Bethe ansatz equations for a Heisenberg quantum spin chain of finite length. Section 19.1 introduces, as a prerequisite, the Bernoulli numbers and then, using the Bernoulli numbers, the Euler–Maclaurin summation formula that will allow us to transform finite sums into integrals plus, in a systematic way, corrections taking into account the finite size of the system.

Applying this mathematical technique to the Bethe ansatz equations will result in linear integral equations that are more general than the ones we have encountered in the thermodynamic limits performed in chapter 14. These integral equations will be of the Wiener–Hopf type, which indicates that they can be solved by the Wiener–Hopf technique, the second mathematical technique we shall introduce in section 19.2 of this chapter.

19.1 Euler–Maclaurin formula

The Bethe ansatz equations encountered in the previous chapters contain sums over the finitely many discrete quasi-momenta or rapidities whose number grows with the system size N. In order to make progress with the solution of the Bethe ansatz equations, these sums were then transformed into integrals in the thermodynamic limit where $N \to \infty$. All contributions to these integrals that depend on the initially finite size of the system vanish in this limit.

In order to retain effects of the finite size of the system, we need to perform the transformation of the discrete sums into integrals in a more careful way. The mathematical tool we employ to achieve this end is the Euler–Maclaurin summation formula, which we treat in section 19.1.2. Since they play a prominent role in the Euler–Maclaurin summation formula, we first introduce the Bernoulli numbers .

19.1.1 Bernoulli numbers

The Bernoulli numbers B_n, $n = 0, 1, \ldots$, a sequence of rational numbers, which are of great importance in number theory, appear in many mathematical applications, for

Models of Quantum Matter: A First Course on Integrability and the Bethe Ansatz. Hans-Peter Eckle, Oxford University Press (2019). © Hans-Peter Eckle. DOI: 10.1093/oso/9780199678839.001.0001

example, in the Euler–Maclaurin formula, which is the subject matter of this section. The values of the first fifteen Bernoulli numbers are

$$B_0 = 1, B_1 = -\frac{1}{2}, B_2 = \frac{1}{6}, B_3 = 0, B_4 = -\frac{1}{30}, B_5 = 0, B_6 = \frac{1}{42}, B_7 = 0, \tag{19.1}$$

$$B_8 = -\frac{1}{30}, B_9 = 0, B_{10} = \frac{5}{66}, B_{11} = 0, B_{12} = -\frac{691}{2730}, B_{13} = 0, B_{14} = \frac{7}{6}. \tag{19.2}$$

Apart from the Euler–Maclaurin formula, the Bernoulli numbers also appear in the Taylor series expansions of the tangent and hyperbolic tangent functions, in formulas for the sum of powers of the first positive integers, and in connection with the Riemann ζ function.

There are various ways to define to Bernoulli numbers. One of these is via the generating function

$$\frac{ze^{xz}}{e^z - 1} = \sum_{n=0}^{\infty} \frac{B_n(x)}{n!} z^n, \tag{19.3}$$

introducing the functions $B_n(x)$ with the Bernoulli numbers given by $B_n = B_n(0)$. The functions $B_n(x)$ are the Bernoulli polynomials

$$B_n(x) = \sum_{\nu=0}^{n} \binom{n}{\nu} B_{n-\nu} x^{\nu}, \tag{19.4}$$

defined on the interval $0 \leq x \leq 1$ and continued periodically beyond that interval. We shall, however, not further discuss the Bernoulli polynomials here, but give one property of the derivative

$$B_n'(x) = nB_{n-1}(x) \tag{19.5}$$

and state explicitly the first two Bernoulli polynomials

$$B_0(x) = 1 \quad \text{and} \quad B_1(x) = x - [x] - 1 \tag{19.6}$$

where $[x]$ is the largest integer n with $n \leq x$.

For our limited purpose to derive a few elementary properties of the Bernoulli numbers, it suffices to consider the generating function (19.3) for the special value $x = 0$.

Then we can write

$$z = (e^z - 1)\sum_{n=0}^{\infty}\frac{B_n}{n!}z^n = \left(\sum_{m=1}^{\infty}\frac{z^m}{m!}\right)\left(\sum_{n=0}^{\infty}\frac{B_n}{n!}z^n\right) = \sum_{\mu=1}^{\infty}\frac{z^\mu}{\mu!}\sum_{n=0}^{\mu-1}\binom{\mu}{n}B_n. \tag{19.7}$$

EXERCISE 19.1 Bernoulli numbers Fill in the steps in the derivation of the identity (19.7), using the summation formula

$$\left(\sum_{m=0}^{\infty} a_m z^m\right)\left(\sum_{n=0}^{\infty} b_n z^n\right) = \sum_{\mu=0}^{\infty} z^\mu \sum_{m+n=\mu} a_m b_n. \tag{19.8}$$

This identity obviously implies

$$\sum_{n=0}^{\mu-1}\binom{\mu}{n}B_n = \begin{cases} 1 & \mu = 1 \\ 0 & \mu \geq 2 \end{cases}. \tag{19.9}$$

This recursion relation can be evaluated by setting successively $\mu = 1$, $\mu = 2$, and so on. We then obtain one by one the Bernoulli numbers as given above.

The generating function (19.3) also allows proof that, as we may have suspected already, all Bernoulli numbers with odd index $n > 1$ vanish: $B_n = 0$ for $n > 1$ odd. Set again $x = 0$ to obtain

$$\frac{z}{e^z - 1} + \frac{z}{2} = \sum_{n=0,\, n\neq 1}^{\infty}\frac{B_n}{n!}z^n, \tag{19.10}$$

where we used $B_1 = -1/2$ when shuffling the term for $n = 1$ of the sum to the left-hand side of the equation. Obviously the function on the left-hand side is an even function of z. Thus, we can conclude that $B_{2k+1} = 0$ for $k = 1, 2, \ldots$.

In the next section, the Bernoulli numbers play a central role in another piece of applied mathematics: the transformation of a finite sum into an integral plus some remainder.

19.1.2 The Euler–Maclaurin summation formula

For the derivation of this important formula, we consider the following expression for an arbitrary function $f(x)$ and arbitrary integers $m < n$

$$\int_m^n dx B_1(x) f'(x) = \sum_{\nu=m}^{n-1}\int_\nu^{\nu+1} dx\left(x - [x] - \frac{1}{2}\right)f'(x) \tag{19.11}$$

$$= \sum_{\nu=m}^{n-1} \left(\left(x - \nu - \frac{1}{2} \right) f(x) \Big|_{\nu}^{\nu+1} - \int_{\nu}^{\nu+1} dx f(x) \right) \tag{19.12}$$

$$= \sum_{\nu=m}^{n} f(\nu) - \frac{1}{2} (f(m) + f(n)) - \int_{m}^{n} dx f(x). \tag{19.13}$$

This can be rearranged to give the Euler–Maclaurin summation formula in its simplest form

$$\sum_{\nu=m}^{n} f(\nu) = \int_{m}^{n} dx f(x) + \frac{1}{2} (f(m) + f(n)) + \int_{m}^{n} dx B_1(x) f'(x). \tag{19.14}$$

The formula allows us to convert a sum into an integral and two correction terms:

$$\frac{1}{2} (f(m) + f(n)) : \qquad \text{from the summation limits and} \tag{19.15}$$

$$\int_{m}^{n} dx B_1(x) f'(x) : \qquad \text{a remainder term.} \tag{19.16}$$

The Euler–Maclaurin summation formula can be further refined by looking more carefully at the remainder term. Recalling the property (19.5) of the Bernoulli polynomials and that for integer values μ and ν of its argument

$$B_n(\mu) = B_n(\nu) = B_n(0) = B_n \qquad (\mu, \nu \geq 2), \tag{19.17}$$

we can write the remainder term as

$$\int_{m}^{n} dx B_1(x) f'(x) = \frac{B_2}{2!} \left(f'(n) - f'(m) \right) - \int_{m}^{n} dx \frac{B_2(x)}{2!} f''(x). \tag{19.18}$$

This procedure can be iterated such that we arrive at the general form of the Euler–Maclaurin summation formula

$$\sum_{\nu=m}^{n} f(\nu) = \int_{m}^{n} dx f(x) + \frac{1}{2} (f(m) + f(n)) + \sum_{j=2}^{k} (-1)^j \frac{B_j}{j!} \left(f^{(j-1)}(n) - f^{(j-1)}(m) \right)$$
$$+ \int_{m}^{n} dx (-1)^{k+1} \frac{B_k(x)}{k!} f^{(k)}(x). \tag{19.19}$$

This form of the Euler–Maclaurin summation formula can be rewritten using an estimate of the remaining integral on the right-hand side of (19.19) as

$$\sum_{\nu=m}^{n} f(\nu) = \int_m^n dx f(x) + \frac{1}{2}\left(f(m) + f(n)\right) + \sum_{j=1}^{k} \frac{B_{2j}}{(2j)!}\left(f^{(2j-1)}(n) - f^{(2j-1)}(m)\right) + \Theta \frac{B_{2(k+1)}}{(2(k+1))!}\left(f^{(2k+1)}(n) - f^{(2k+1)}(m)\right) \tag{19.20}$$

for some number $0 \leq \Theta \leq 1$.

This result of applied mathematics, the Euler–Maclaurin summation formula in the form (19.20), provides the first ingredient needed for the finite-size analysis of the Bethe ansatz equations.

We now turn to the second ingredient, which is again provided by applied mathematics, the Wiener–Hopf technique to solve functional equations.

19.2 Wiener–Hopf technique

The Wiener–Hopf technique is a mathematical technique for treating functional equations. These equations arise, for example, from the study of certain classes of integral equations, which is, in view of what we discussed at various places and, in particular in chapter 18, why we are interested in the Wiener–Hopf technique here.

This section considers integral equations of the type

$$f(x) = \phi(x) + \int_0^\infty dy K(x-y) f(y) \qquad -\infty < x < \infty, \tag{19.21}$$

where $\phi(x)$ is a given function, and the kernel K of the integral equation has the difference property

$$K(x,y) = K(x-y). \tag{19.22}$$

This type of integral equation cannot be readily solved by Fourier transformation because of the integration limits that are restricted to the positive real half line $[0, \infty)$.

The strategy to solve this type of integral equation consists in, formally, extending the range of integration to the whole real line $(-\infty, \infty)$. To do this, we split the unknown function $f(x)$ formally into a positive and a negative part by defining

$$f(x) \equiv f_+(x) + f_-(x), \tag{19.23}$$

where

$$f_+(x) \equiv \begin{cases} 0 & \text{for} \quad x > 0 \\ f(x) & \text{for} \quad x < 0 \end{cases}, \tag{19.24}$$

and

$$f_-(x) \equiv \begin{cases} f(x) & \text{for} \quad x > 0 \\ 0 & \text{for} \quad x < 0 \end{cases}. \tag{19.25}$$

The integral equation 19.21 then becomes

$$f_+(x) + f_-(x) = \phi(x) + \int_{-\infty}^{\infty} dy\, K(x-y) f_-(y). \tag{19.26}$$

Now that we formally integrate over the whole real line, we can use Fourier transformation. First, we introduce the following conventions for Fourier transforms

$$f(x) = \frac{1}{2\pi} \int_{-\infty}^{\infty} dk \hat{f}(k) e^{ikx}, \tag{19.27}$$

and

$$\hat{f}(k) = \int_{-\infty}^{\infty} dx f(x) e^{-ikx}. \tag{19.28}$$

The integral equation 19.26, upon Fourier transformation, becomes a functional equation for the Fourier transform $\hat{f}(k)$[1]

$$\hat{f}_-(k)\left(1 - \hat{K}(k)\right) + \hat{f}_+(k) = \hat{\phi}(k). \tag{19.29}$$

The problem, obviously, is how to solve this equation, which contains two unknown functions. It will turn out that this is possible if we investigate the analytic properties in the complex plane of the functions involved in the functional equation, and provided that they obey certain conditions. In other words, we invoke the powers of complex analysis.

First, we observe that $\hat{K}(k)$ is an analytic function in the strip

$$-\beta < \Im(k) < \alpha \tag{19.30}$$

of the complex plane, if

1 Recall that Wiener–Hopf is really a technique to solve functional equations.

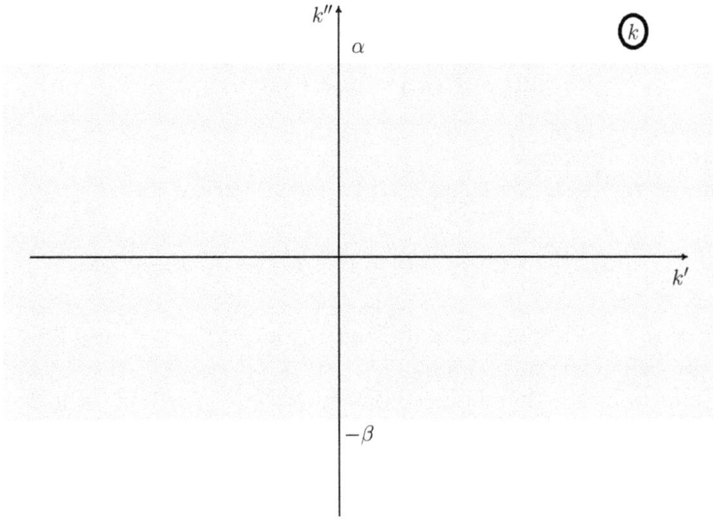

Figure 19.1 *Strip of the complex k-plane, $-\infty < k' < \infty$ and $-\beta < k'' < \alpha$, where $\hat{K}(k)$ is analytic.*

$$|K(x)| \leq \begin{cases} e^{-\alpha x} & \text{for} \quad x > 0 \\ e^{\beta x} & \text{for} \quad x < 0 \end{cases} \quad \text{with} \quad \alpha, \beta > 0. \tag{19.31}$$

The proof of this statement is left as exercise 19.2.

EXERCISE 19.2 Analytic properties of $\hat{K}(k)$ Prove the statement about the analyticity of the Fourier transformed kernel $\hat{K}(k)$ in the strip $-\beta < \Im(k) < \alpha$. Use $k = \Re(k) + i\Im(k) \equiv k' + ik''$ and split the Fourier transform of $K(x)$, i.e. $\hat{K}(k)$, into two parts, one for $0 < x < \infty$ and the other for $-\infty < x < 0$.

We now introduce the further assumption that $\hat{f}(k)$ is an analytic function in, at least, a part of the strip $-\beta < \Im(k) < \alpha$, e.g. in $-\gamma < \Im(k) < \delta$ with $\gamma \leq \beta$ and $\delta \leq \alpha$.

In order to have a more symmetrical functional equation, we define two more complex functions $A(k)$ and $B(k)$ by

$$\frac{A(k)}{B(k)} = 1 - \hat{K}(k), \tag{19.32}$$

requiring that

$A(k)$ is analytic for $\Im(k) < \beta$, and
$B(k)$ is analytic for $\Im(k) > -\gamma$.

In other words, we require that in

$$\ln(1 - \hat{K}(k)) = \ln A(k) - \ln B(k) \tag{19.33}$$

all terms are regular functions.

The functional equation now takes the more symmetrical form

$$A(k)\hat{f}_{-}(k) + B(k)\hat{f}_{+}(k) = \hat{\phi}(k)B(k) \equiv C(k) + D(k), \tag{19.34}$$

where the function $C(k)$ is analytic in the half-plane $\Im(k) > -\gamma$, and the function $D(k)$ is analytic in the half-plane $\Im(k) < \delta$..

It is important to notice that all functions in this equation have at least the common strip of analyticity $-\gamma < \Im(k) < \delta$.

Rewriting the functional equation once again, we obtain

$$B(k)\hat{f}_{+}(k) - C(k) = -A(k)\hat{f}_{-}(k) + D(k), \tag{19.35}$$

where now the left-hand side is analytic in the half-plane $\Im(k) > -\gamma$, whereas the right-hand side is analytic in the half-plane $\Im(k) < \delta$. Therefore both sides of the functional equation must be equal to an entire function. It can be shown that this entire function is a polynomial; cf. Roos (1969) for details.

We stop at this point, illustrating the Wiener–Hopf technique with an example.

19.2.1 Example of the Wiener–Hopf technique

In order to see the Wiener–Hopf technique at work, consider the integral equation

$$f(x) = e^{-|x|} + \lambda \int_0^\infty dy\, e^{|x-y|} f(y). \tag{19.36}$$

Here, the inhomogeneous function $\phi(x) = e^{-|x|}$ and the kernel $K(x - y) = \lambda e^{-|x-y|}$ are essentially the same function, whose Fourier transform is

$$\int_{-\infty}^{\infty} dx\, e^{-|x|} e^{ikx} = \int_0^\infty dx\, e^{-x(ik+1)} + \int_{-\infty}^{0} dx\, e^{-x(ik-1)} = \frac{2}{k^2+1}. \tag{19.37}$$

This leads to the functional equation

$$f_{+}(k) + f_{-}(k) = \frac{2}{k^2+1} + \frac{2\lambda}{k^2+1} f_{-}(k) \tag{19.38}$$

or

$$f_+(k) + \frac{k^2 - 2\lambda + 1}{k^2 + 1} f_-(k) = f_+(k) + \frac{k^2 - \xi^2}{k^2 + 1} f_-(k) = \frac{2}{k^2 + 1} \tag{19.39}$$

where we assume that ξ with $\xi^2 = 2\lambda - 1$ is real, i.e. $\lambda > \frac{1}{2}$.

Since

$$1 - \hat{K}(k) = \frac{k^2 - \xi^2}{k^2 + 1} \tag{19.40}$$

has singularities at $k = \pm i$, the strips where it is analytic are

$$-1 < k'' < 0 \qquad \text{or} \qquad 0 < k'' < 1. \tag{19.41}$$

Let us concentrate on the strip $-1 < k'' < 0$. The decomposition in this strip yields

$$1 - \hat{K}(k) = \frac{k^2 - \xi^2}{k^2 + 1} = \frac{A(k)}{B(k)} = \frac{k^2 - \xi^2}{k - i} \frac{1}{k + i} \tag{19.42}$$

i.e.

$$A(k) = \frac{k^2 - \xi^2}{k - i} \tag{19.43}$$

$$B(k) = k + i \tag{19.44}$$

where now $A(k)$ is analytic for $k'' < 0$ and $B(k)$ is analytic for $k'' > -1$.

Now we can write the functional equation as

$$(k + i) f_+(k) + \frac{k^2 - \xi^2}{k - i} f_-(k) = \frac{2}{k - i} \tag{19.45}$$

or

$$(k + i) f_+(k) = -\frac{k^2 - \xi^2}{k - i} f_-(k) + \frac{2}{k - i} = E(k) \tag{19.46}$$

where the left-hand side is analytic for $k'' > -1$ while the right-hand side is analytic for $k'' < 0$ and $E(k)$ is an entire function for which the simplest choice is a constant, which, for convenience, we choose to be $E(k) = ia$. Thus,

$$(k + i) f_+(k) = ia \tag{19.47}$$

$$\frac{k^2-\xi^2}{k-i}f_-(k)=\frac{2}{k-i}-ia. \qquad (19.48)$$

The functions $f_+(x)$ and $f_-(x)$ can now be calculated from (19.47) and (19.48) by Fourier transformation. The results are

$$f_+(x)=ae^x \qquad (19.49)$$

$$f_-(x)=-\frac{2}{\xi}\sin\xi x+a\left(\cos\xi x+\frac{1}{\xi}\sin\xi x\right). \qquad (19.50)$$

Besides the reference to Roos (1969), a short introduction to the Wiener–Hopf technique applied to integral equations (including the example) appear in Mathews and Walker (1970) and Wyld (1999).

We shall encounter a further, more involved, application of the Wiener–Hopf technique in chapter 20 in the context of the Heisenberg quantum spin chain.

20

Finite Heisenberg Quantum Spin Chain

If you wish to advance into the infinite, explore the finite in all directions.

Johann Wolfgang von Goethe (1749–1832)

As shown in chapter 14, in order to find solutions of the Bethe ansatz equations of the Heisenberg quantum spin chain or, by implication, any other Bethe ansatz solvable model, it proved a good strategy to perform the thermodynamic limit $N \to \infty$ of considering an infinitely large system.

However, the Bethe ansatz as such considers a finite system. Moreover, the extraction of finite-size results from the (finite) Bethe ansatz equations is of genuine interest, especially against the background of our discussions of finite-size scaling in section 5.6 and conformal symmetry in finite geometries in section 7.10. The interesting question thus arises how to come to terms with genuine finite system sizes. The mathematical techniques introduced in chapter 19 permit a systematic treatment of finite-size corrections as corrections to the thermodynamic limit of the system. Section 20.1 restates for our current purposes the Bethe ansatz for the anisotropic Heisenberg quantum spin chain. Next, in section 20.2, we perform an analysis similar to the one where we employed the thermodynamic limit, except that we keep finite-size corrections in all equations in such a way that these corrections can be systematically retained order by order in the inverse system size $1/N$. In section 20.3 this careful procedure allows the application of the Euler–Maclaurin formula transforming finite sums into integrals and finite-size corrections. The Bethe ansatz equations treated in this way will turn out to be of the special form of Wiener–Hopf integral equations with inhomogeneities representing the finite-size corrections so that we can now apply the Wiener–Hopf technique in section 20.4.

The results from a finite-size treatment of the Bethe ansatz equations will allow comparison to results for finite systems obtained from other approaches that are independent of the Bethe ansatz method.

Finally, we briefly discuss higher order corrections in section 20.5 and end with a general assessment of the finite-size method.

Models of Quantum Matter: A First Course on Integrability and the Bethe Ansatz. Hans-Peter Eckle, Oxford University Press (2019). © Hans-Peter Eckle. DOI: 10.1093/oso/9780199678839.001.0001

20.1 Bethe ansatz for the finite Heisenberg quantum spin chain

The first systematic method to calculate finite-size corrections for integrable or Bethe ansatz systems is due to de Vega and Woynarovich (1985), whose method deals with systems with a non-vanishing energy gap, e.g. the anisotropic Heisenberg quantum spin chain in the antiferromagnetic easy-axis regime.

The Hamiltonian of the anisotropic Heisenberg quantum spin chain for an even number N of spins $\frac{1}{2}$ in the form

$$\mathcal{H} = \sum_{n=1}^{N} \left(\sigma_n^x \sigma_{n+1}^x + \sigma_n^y \sigma_{n+1}^y + \Delta \sigma_n^z \sigma_{n+1}^z\right) \tag{20.1}$$

describes the easy-axis antiferromagnetic regime for anisotropy $\Delta = \cosh\gamma > 1$. In this regime considered by de Vega and Woynarovich (1985), there is a finite energy gap in the excitation spectrum even in the thermodynamic limit $N \to \infty$.

We shall not pursue the analysis for the gapped system, which involves complex solutions of the Bethe ansatz equations (see de Vega and Woynarovich, 1985 for details about gapped systems). Instead we concentrate on the Heisenberg quantum spin chain with anisotropy in the range $|\Delta| < 1$ where there is no gap, i.e. the system is critical. In this anisotropy range and for the lowest excitations in each sector labelled by the magnetization M with total spin $S = \frac{1}{2}N - M$, there are no complex solutions of the Bethe ansatz equations. Thus, the analysis will be more transparent. Moreover, there is an important motivation to study gapless models because the results can be compared to the predictions of conformal invariance for critical, i.e. gapless, systems.

The Bethe ansatz equations for anisotropy $\Delta = \cos\theta$ $(0 < \theta < \pi)$ that determine the rapidities η_α are explicitly given by

$$2N \tan^{-1}\left(\cot\frac{\theta}{2} \tanh\frac{\eta_\alpha}{2}\right) = 2\pi \mathcal{J}_\alpha + 2\sum_{\beta=1}^{M} \tan^{-1}\left(\cot\theta \tanh\frac{\eta_\alpha - \eta_\beta}{2}\right) \tag{20.2}$$

or rewritten

$$2N\phi(\eta_\alpha, \frac{1}{2}\theta) = 2\pi \mathcal{J}_\alpha + \sum_{\beta=1}^{M} \phi(\eta_\alpha - \eta_\beta, \theta), \tag{20.3}$$

introducing the useful function

$$\phi(\eta, \theta) = 2\tan^{-1}\left(cot\theta \tanh\frac{\eta}{2}\right). \tag{20.4}$$

The Bethe ansatz quantum numbers $\mathcal{J}_\alpha$

$$-\frac{1}{2}\left(\frac{1}{2}N-(S+1)\right);-\frac{1}{2}\left(\frac{1}{2}N-(S+1)\right)+1;\ldots;\frac{1}{2}\left(\frac{1}{2}N-(S+1)\right) \tag{20.5}$$

determine the lowest energy state in the sector of total spin

$$S=\frac{1}{2}N-M. \tag{20.6}$$

The ground state corresponds to the lowest energy state in the sector $S=0$.

In each sector the energy per site is given by

$$E_N^{(S)}=-\frac{1}{N}\sum_{\alpha=1}^{M}\frac{\sin^2\theta}{\cosh\eta_\alpha-\cos\theta} \tag{20.7}$$

once the rapidities have been calculated.

For a finite system, this calculation of the rapidities from the Bethe ansatz equations (20.2) or (20.3) is, however, a formidable task except for systems consisting of only very few spins. For reasonably large but still finite systems, there exists no systematic way of dealing with the Bethe ansatz equations directly.

What we can do, however, is use our knowledge of the thermodynamic limit and attempt to calculate, in a systematic way, corrections to this thermodynamic limit due to the finiteness of the system.

20.2 Finite-size corrections

From the Bethe ansatz equations (20.2) and (20.3), respectively, and from our experience with the thermodynamic limit, we are led to introduce the function

$$z_N(\eta)=\frac{1}{2\pi}\left\{2\tan^{-1}\left(\cot\frac{\theta}{2}\tanh\frac{\eta}{2}\right)\right.$$

$$\left.-\frac{1}{N}\sum_{\beta=1}^{M}2\tan^{-1}\left(\cot\theta\tanh\left(\frac{\eta-\eta_\beta}{2}\right)\right)\right\} \tag{20.8}$$

$$=\frac{1}{2\pi}\left\{\phi(\eta,\frac{1}{2}\theta)-\frac{1}{N}\sum_{\beta=1}^{M}\phi(\eta-\eta_\beta,\theta)\right\}, \tag{20.9}$$

which by definition satisfies the Bethe ansatz equations

$$z_N(\eta_\alpha)=\frac{\mathcal{J}_\alpha}{N} \tag{20.10}$$

if we insert the values of the rapidities $\eta = \eta_\alpha$. The function $z_N(\eta)$ is continuous and increases monotonically for real η.

We could now perform the thermodynamic limit $N \to \infty$ by introducing a density $\sigma_\infty(\eta)$ of Bethe ansatz roots

$$\sigma_\infty(\eta) = \frac{dz(\eta)}{d\eta} \tag{20.11}$$

$$z(\eta) \equiv z_\infty(\eta) = \lim_{N\to\infty} z_N(\eta). \tag{20.12}$$

However, instead of taking the thermodynamic limit, we keep the number of spins N finite and formally define the root density for the finite system by

$$\sigma_N(\eta) = \frac{dz_N(\eta)}{d\eta}. \tag{20.13}$$

Using

$$\phi(\eta,\theta) = 2\tan^{-1}\left(\cot\theta \tanh\frac{\eta}{2}\right), \tag{20.14}$$

we arrive at

$$\sigma_N(\eta) = \frac{1}{2\pi}\left\{\phi'(\eta,\frac{\theta}{2}) - \frac{1}{N}\sum_{\beta=1}^{M}\phi'(\eta-\eta_\beta,\theta)\right\}, \tag{20.15}$$

where

$$\phi'(\eta,\theta) = \frac{d\phi(\eta,\theta)}{d\eta} = \frac{\sin 2\theta}{\cosh\eta - \cos\theta}. \tag{20.16}$$

Next, we introduce the function

$$S_N(\eta) = \frac{1}{N}\sum_{\beta=1}^{M}\delta(\eta-\eta_\beta) - \sigma_N(\eta), \tag{20.17}$$

with which we can rewrite (20.15) as

$$\begin{aligned}\sigma_N(\eta) = \frac{1}{2\pi}\Big\{\phi'(\eta,\frac{\theta}{2}) &- \int_{-\infty}^{\infty} d\eta' S_N(\eta')\phi'(\eta-\eta',\theta)\\ &- \int_{-\infty}^{\infty} d\eta' \sigma_N(\eta')\phi'(\eta-\eta',\theta)\Big\}.\end{aligned} \tag{20.18}$$

Thus we have arrived at an equation for the root density $\sigma_N(\eta)$ that resembles the integral equation in the thermodynamic limit. We write down both equations to make comparison easier. The integral equation for the root density $\sigma_\infty(\eta)$ in the thermodynamic limit is

$$\sigma_\infty(\eta) = \frac{1}{2\pi}\frac{\sin\theta}{\cosh\eta - \cos\theta} - \int_{-\infty}^{\infty}\frac{d\eta}{2\pi}\frac{\sin 2\theta}{\cosh(\eta-\eta') - \cos 2\theta}\sigma_\infty(\eta'). \tag{20.19}$$

The solution of this equation can be found by Fourier transformation (des Cloizeaux and Gaudin, 1966) as

$$\sigma_\infty(\eta) = \frac{1}{4\theta}\frac{1}{\cosh\left(\frac{\pi\eta}{2\theta}\right)}. \tag{20.20}$$

EXERCISE 20.1 Integral equation in the thermodynamic limit Solve the integral equation (20.19) by Fourier transformation, i.e. confirm the result (20.20).

With this result for the infinite system under our belt we can now continue our attempt to find finite-size corrections. First, we bring the equation for the finite-size root density $\sigma_N(\eta)$, (20.18), in the form

$$\frac{\sin\theta}{\cosh\eta - \cos\theta} = 2\pi\sigma_N(\eta) + \int_{-\infty}^{\infty} d\eta\,\frac{\sin 2\theta}{\cosh(\eta-\eta') - \cos 2\theta}\left(\sigma_N(\eta') + S_N(\eta')\right). \tag{20.21}$$

This equation, which closely resembles an integral equation, can be treated again by Fourier transforming it back and forth. The result is

$$\sigma_N(\eta) = \frac{1}{4\theta}\frac{1}{\cosh\left(\frac{\pi\eta}{2\theta}\right)} - \int_{-\infty}^{\infty} d\eta' F(\eta-\eta')S_N(\eta') \tag{20.22}$$

where the kernel function $F(x)$ is given by

$$F(x) = \frac{1}{2\pi}\int_{-\infty}^{\infty} d\omega\frac{\exp(i\omega x)\sinh\omega(\pi - 2\theta)}{2\cosh\omega\theta\sinh\omega(\pi-\theta)}. \tag{20.23}$$

EXERCISE 20.2 Kernel function $F(x)$ Perform a Fourier transformation and back transformation of (20.21) to bring it into the form (20.22). The formulas in the book by Gradshteyn and Ryzhik (1980) will be useful for this exercise.

Compiling the results, we can write the equation for the finite-size root density as

$$\sigma_N(\eta) = \sigma_\infty(\eta) - \int_{-\infty}^{\infty} d\eta' F(\eta - \eta') S_N(\eta'). \tag{20.24}$$

This equation will be the starting point to which to apply the Euler–Maclaurin formula of section 19.1. As a result, we shall obtain a proper integral equation for the finite-size root density $\sigma_N(\eta)$ where the inhomogeneous terms describe finite-size corrections to the integral equation (20.19) for the root density in the thermodynamic limit.

Before we go on to do this, we finish this section by deriving a finite-size expression for the energy per site in the sector of total spin S, which we also express in terms of the result in the thermodynamic limit plus a contribution that contains the finite-size quantity $S_N(\eta)$. This expression will then also be evaluated with the help of the Euler–Maclaurin formula to make the finite-size corrections explicit in a systematic way.

We begin from the energy per site (20.7), which we rewite using $S_N(\eta)$

$$E_N^{(S)} = -\frac{1}{N}\sum_{\alpha=1}^{M} \frac{\sin^2\theta}{\cosh\eta_\alpha - \cos\theta} \equiv -\frac{1}{N}\sum_{\alpha=1}^{M} E(\eta_\alpha) \tag{20.25}$$

$$= -\int_{-\infty}^{\infty} d\eta\, E(\eta) S_N(\eta) - \int_{-\infty}^{\infty} d\eta\, E(\eta)\sigma_N(\eta). \tag{20.26}$$

This expression can be transformed into

$$E_N^{(S)} = E_\infty - \int_{-\infty}^{\infty} d\eta \epsilon(\eta) S_N(\eta), \tag{20.27}$$

where

$$\epsilon(\eta) = \frac{\pi}{2}\frac{\sin\theta}{\theta}\frac{1}{\cosh\left(\frac{\pi\eta}{2\theta}\right)}, \tag{20.28}$$

as we shall see in exercise 20.3.

EXERCISE 20.3 Derivation of the finite-size energy formula Insert $\sigma_N(\eta)$ from (20.22) into (20.26) to derive the energy per site (20.27) with (20.28).

Before we discuss the systematic treatment of finite-size corrections, we remark that it suffices, in order to calculate the lowest order corrections, to replace the finite-size root density $\sigma_N(\eta)$ in the function $S_N(\eta)$, (20.17), by the root density $\sigma_\infty(\eta)$ in the thermodynamic limit and the roots η_α by those of the thermodynamic limit, i.e. by those values of η_α which can be determined from

$$z_{\infty}(\eta_\alpha) = \frac{\mathcal{J}_\alpha}{N} \tag{20.29}$$

in the limit $N \to \infty$. This has been done by Avdeev and Dörfel (1986) for the isotropic case $\Delta = 1$ and by Hamer (1985, 1986) for the anisotropic case $|\Delta| < 1$.

Obviously this truncates the possibility of calculating systematically further finite-size corrections as in section 20.3 employing the Euler–Maclaurin formula.

20.3 Application of the Euler–Maclaurin formula

Our goal now is to apply the Euler–Maclaurin summation formula (19.20) to expressions of the form

$$\int_{-\infty}^{\infty} d\eta\, g(\eta) S_N(\eta) = \frac{1}{N}\sum_{\beta=1}^{M} g(\eta_\beta) - \int_{-\infty}^{\infty} d\eta\, g(\eta)\sigma_N(\eta), \tag{20.30}$$

which appear in the equation for the finite-size root density (20.22) or (20.24) and the finite-size energy (20.27). This integral can be rewritten using the definition of the finite-size root density $\sigma_N(\eta) = \frac{dz_N(\eta)}{d\eta}$

$$\int_{-\infty}^{\infty} d\eta\, g(\eta) S_N(\eta) = \frac{1}{N}\sum_{\beta=1}^{M} g(\eta_\beta) - \int_{-\infty}^{\infty} d\eta\, g(z_N)\frac{dz_N}{d\eta} \tag{20.31}$$

$$= \frac{1}{N}\sum_{\beta=1}^{M} g(\eta_\beta) - \int_{-\infty}^{\infty} dz_N\, g(z_N). \tag{20.32}$$

We now split the last integral

$$\int_{-\infty}^{\infty} dz_N\, g(z_N) = \int_{-\infty}^{-\Lambda} d\eta\, g(z_N)\frac{dz_N}{d\eta} + \int_{\Lambda}^{\infty} d\eta\, g(z_N)\frac{dz_N}{d\eta} + \int_{z_N(-\Lambda)}^{z_N(\Lambda)} dz_N\, g(z_N) \tag{20.33}$$

$$= \int_{-\infty}^{-\Lambda} d\eta\, g(\eta)\sigma_N(\eta) + \int_{\Lambda}^{\infty} d\eta\, g(\eta)\sigma_N(\eta) + \int_{z_N(-\Lambda)}^{z_N(\Lambda)} dz_N\, g(z_N) \tag{20.34}$$

where Λ is the largest root of the Bethe ansatz equations, which is implicitly given by

$$z_N(\Lambda) = \frac{1}{N}\left\{\frac{1}{2}\left(\frac{N}{2} - (S+1)\right)\right\} \tag{20.35}$$

or equivalently by

$$\int_{\Lambda}^{\infty} d\eta\, \sigma_N(\eta) = \left(1 - \frac{\theta}{\pi}\right)\frac{S}{N} + \frac{1}{2N}. \tag{20.36}$$

The property that the rapidity Λ defined in these equivalent ways is the largest root is guaranteed from the property of $z_N(\eta)$ to be a monotonically increasing function. All roots of the Bethe ansatz equations lie in the interval from $z_N(-\Lambda)$ to $z_N(\Lambda)$, which corresponds to the interval $(-\Lambda, \Lambda)$ in the rapidity variable η.

EXERCISE 20.4 Equivalence of the implicit definitions of the largest root Λ
Show the equivalence of the two definitions of Λ in (20.35) and (20.36).

It is the last integral in (20.34) to which we now apply the Euler–Maclaurin summation formula in the form

$$\begin{aligned}
&\frac{1}{2N}\left\{f\left(\frac{n_1}{N}\right) + 2\sum_{n=n_1+1}^{n_2-1} f\left(\frac{n}{N}\right) + f\left(\frac{n_2}{N}\right)\right\} - \int_{\frac{n_1}{N}}^{\frac{n_2}{N}} dx f(x) \\
&= \frac{1}{12N^2}\left\{f'\left(\frac{n_2}{N}\right) - f'\left(\frac{n_1}{N}\right)\right\} + \mathcal{O}\left(\frac{\max f'''}{N^3}\right),
\end{aligned} \tag{20.37}$$

which uses the variable transformation $x \to Nx$ with $f(x) \to f(Nx) \to f(x)$.

We obtain

$$\begin{aligned}
&\int_{-\infty}^{\infty} d\eta\, g(\eta) S_N(\eta) = \\
&\frac{1}{N}\sum_{\beta=1}^{M} g(\eta_\beta) - \left\{\int_{-\infty}^{-\Lambda} d\eta\, g(\eta)\sigma_N(\eta) + \int_{\Lambda}^{\infty} d\eta\, g(\eta)\sigma_N(\eta)\right\} \\
&\quad + \frac{1}{12N^2}\left\{\left.\frac{dg(z_N)}{dz_N}\right|_{z_N=z_N(\Lambda)} - \left.\frac{dg(z_N)}{dz_N}\right|_{z_N=z_N(-\Lambda)}\right\} \\
&\quad - \frac{1}{2N}\{g(z_N(\Lambda)) + g(z_N(-\Lambda))\} \\
&\quad - \frac{1}{N}\sum_{\beta=1}^{M\prime} g\left(z_N(\eta_\beta) = \frac{\mathcal{J}_\beta}{N}\right)
\end{aligned} \tag{20.38}$$

where in the last summation, indicated by the prime $'$, it is understood that the summation does not run over the smallest and largest roots, $-\Lambda$ and Λ, respectively. We observe that the sums in this expression almost cancel leaving only a contribution from the smallest and largest roots, $-\Lambda$ and Λ, respectively, so that we obtain

$$\begin{aligned}\int_{-\infty}^{\infty} d\eta\, g(\eta) S_N(\eta) = &-\left\{\int_{-\infty}^{-\Lambda} d\eta\, g(\eta)\sigma_N(\eta) + \int_{\Lambda}^{\infty} d\eta\, g(\eta)\sigma_N(\eta)\right\}\\ &+ \frac{1}{12N^2}\left\{\left.\frac{dg(\eta)}{d\eta}\frac{d\eta}{dz_N}\right|_{\eta=\Lambda} - \left.\frac{dg(\eta)}{d\eta}\frac{d\eta}{dz_N}\right|_{\eta=-\Lambda}\right\}\\ &- \frac{1}{2N}\{g(\Lambda) + g(-\Lambda)\}. \end{aligned} \tag{20.39}$$

Using again $\sigma_N(\eta) = \frac{dz_N(\eta)}{d\eta}$ and its symmetry $\sigma_N(\eta) = \sigma_N(-\eta)$, we can finally write

$$\begin{aligned}\int_{-\infty}^{\infty} d\eta\, g(\eta) S_N(\eta) = &-\left\{\int_{-\infty}^{-\Lambda} d\eta\, g(\eta)\sigma_N(\eta) + \int_{\Lambda}^{\infty} d\eta\, g(\eta)\sigma_N(\eta)\right\}\\ &+ \frac{1}{12N^2\sigma_N(\Lambda)}\{g'(\Lambda) - g'(-\Lambda)\}\\ &- \frac{1}{2N}\{g(\Lambda) + g(-\Lambda)\}. \end{aligned} \tag{20.40}$$

This last expression, (20.40), is in a form to which we now can apply the second mathematical technique introduced in chapter 19: The Wiener–Hopf technique. Using (20.40) in the expressions for the finite-size root density (20.22) or (20.24) finally renders them into proper integral equations which are, after some further rewriting, of the Wiener–Hopf type. From these Wiener–Hopf type integral equations we then derive functional equations whose solutions we use to evaluate the finite-size energy (20.27) which we shall also express with the help of (20.40) in a form to read off finite-size corrections.

20.4 Application of the Wiener–Hopf technique

20.4.1 Derivation of the Wiener–Hopf type integral equation

Using the approximation (20.40), the expression for the finite-size root density (20.22) becomes

$$\begin{aligned}\sigma_N(\eta) &- \int_{-\infty}^{-\Lambda} d\eta'\, F(\eta - \eta')\sigma_N(\eta') - \int_{\Lambda}^{\infty} d\eta'\, F(\eta - \eta')\sigma_N(\eta')\\ &+ \frac{1}{2N}(F(\eta - \Lambda) + F(\eta + \Lambda)) + \frac{1}{12N^2\sigma_N(\Lambda)}\left(F'(\eta - \Lambda) - F'(\eta + \Lambda)\right)\\ &= \frac{1}{4\theta}\frac{1}{\cosh\left(\frac{\pi\eta}{2\theta}\right)}. \end{aligned} \tag{20.41}$$

We now shift the integration limits $\pm\Lambda \to 0$, i.e. first we shift the integration variable $\eta' \to \eta' - \Lambda$ and then we let $\eta' \to -\eta'$

$$\int_{-\infty}^{-\Lambda} d\eta'\, F(\eta - \eta')\sigma_N(\eta') = \int_{-\infty}^{0} d\eta'\, F(\eta - \eta' + \Lambda)\sigma_N(\eta' - \Lambda) \tag{20.42}$$

$$= \int_{0}^{\infty} d\eta'\, F(\eta + \eta' + \Lambda)\sigma_N(\eta' + \Lambda) \tag{20.43}$$

where we again used the symmetry $\sigma_N(\eta) = \sigma_N(-\eta)$. Similarly, we obtain

$$\int_{\Lambda}^{\infty} d\eta'\, F(\eta - \eta')\sigma_N(\eta') = \int_{0}^{\infty} d\eta'\, F(\eta - \eta' - \Lambda)\sigma_N(\eta' + \Lambda). \tag{20.44}$$

Thus, we arrive at

$$\begin{aligned} \sigma_N(\eta) &- \int_0^\infty d\eta'\, F(\eta - \eta' - \Lambda)\sigma_N(\eta' + \Lambda) - \int_0^\infty d\eta'\, F(\eta + \eta' + \Lambda)\sigma_N(\eta' + \Lambda) \\ &+ \frac{1}{2N}\left(F(\eta - \Lambda) + F(\eta + \Lambda)\right) + \frac{1}{12N^2\sigma_N(\Lambda)}\left(F'(\eta - \Lambda) - F'(\eta + \Lambda)\right) \\ &= \frac{1}{4\theta}\frac{1}{\cosh\left(\frac{\pi\eta}{2\theta}\right)}. \end{aligned} \tag{20.45}$$

The next step of the transformation into a recognizable integral equation of Wiener–Hopf type consists of shifting the variable η, i.e. $\eta \to \eta + \Lambda$, which leads to

$$\begin{aligned} \sigma_N(\eta + \Lambda) &- \int_0^\infty d\eta'\, F(\eta - \eta')\sigma_N(\eta' + \Lambda) - \int_0^\infty d\eta'\, F(\eta + \eta' + 2\Lambda)\sigma_N(\eta' + \Lambda) \\ &+ \frac{1}{2N}\left(F(\eta) + F(\eta + 2\Lambda)\right) + \frac{1}{12N^2\sigma_N(\Lambda)}\left(F'(\eta) - F'(\eta + 2\Lambda)\right) \\ &= \frac{1}{4\theta}\frac{1}{\cosh\left(\frac{\pi(\eta+\Lambda)}{2\theta}\right)}. \end{aligned} \tag{20.46}$$

The recipe of the Wiener–Hopf technique now redefines the function $\sigma_N(\eta)$

$$\sigma_N^+(\eta) = \begin{cases} \sigma_N(\eta + \Lambda) & \eta > 0 \\ 0 & \eta < 0 \end{cases} \tag{20.47}$$

and

$$\sigma_N^-(\eta) = \begin{cases} 0 & \eta > 0 \\ \sigma_N(\eta + \Lambda) & \eta < 0 \end{cases} \tag{20.48}$$

so that we finally arrive at the desired Wiener–Hopf type integral equation

$$\begin{aligned}\sigma_N^+(\eta)+\sigma_N^-(\eta)&-\int_0^\infty d\eta'\,F(\eta-\eta')\sigma_N^+(\eta')-\int_0^\infty d\eta'\,F(\eta+\eta'+2\Lambda)\sigma_N^+(\eta')\\&+\frac{1}{2N}\left(F(\eta)+F(\eta+2\Lambda)\right)+\frac{1}{12N^2\sigma_N(\Lambda)}\left(F'(\eta)-F'(\eta+2\Lambda)\right)\\&=\frac{1}{4\theta}\frac{1}{\cosh\left(\frac{\pi(\eta+\Lambda)}{2\theta}\right)}.\end{aligned}\tag{20.49}$$

This integral equation is now in a proper form to be Fourier transformed.

20.4.2 Fourier transformation of the Wiener–Hopf type integral equation

In order to Fourier transform the Wiener–Hopf type integral equation (20.45), we recall the Fourier transform of the kernel function (20.23), which we calculated in exercise 20.2:

$$F(\eta-\eta')=\frac{1}{2\pi}\int_{-\infty}^{\infty}d\omega\frac{\exp(i\omega(\eta-\eta'))\sinh\omega(\pi-2\theta)}{2\cosh\omega\theta\sinh\omega(\pi-\theta)}.\tag{20.50}$$

Its derivative with respect to the second variable η' is given by

$$\frac{\partial}{\partial\eta'}F(\eta-\eta')=\frac{1}{2\pi}\int_{-\infty}^{\infty}d\omega(-i\omega)\frac{\exp(i\omega(\eta-\eta'))\sinh\omega(\pi-2\theta)}{2\cosh\omega\theta\sinh\omega(\pi-\theta)}.\tag{20.51}$$

These two equations for the kernel $F(\eta)$ and its derivative suffice, together with the Fourier transform of the right-hand side inhomogeneity, to write the Fourier transform of the Wiener–Hopf type integral equation (20.45) as

$$\begin{aligned}\sigma^+(\omega)+\sigma^-(\omega)&-\sigma^+(\omega)\frac{\sinh\omega(\pi-2\theta)}{2\cosh\omega\theta\sinh\omega(\pi-\theta)}\\&-\sigma^+(-\omega)\frac{\sinh\omega(\pi-2\theta)}{2\cosh\omega\theta\sinh\omega(\pi-\theta)}e^{2i\omega\Lambda}\\&+\frac{1}{2N}\frac{1}{2\pi}\frac{\sinh\omega(\pi-2\theta)}{2\cosh\omega\theta\sinh\omega(\pi-\theta)}\left[1+e^{2i\omega\Lambda}\right]\\&-\frac{1}{12N^2\sigma_N(\Lambda)}\frac{i\omega}{2\pi}\frac{\sinh\omega(\pi-2\theta)}{2\cosh\omega\theta\sinh\omega(\pi-\theta)}\left[1-e^{2i\omega\Lambda}\right]\\&=\frac{1}{4\pi}\frac{1}{\cosh\omega\theta}e^{i\omega\Lambda}.\end{aligned}\tag{20.52}$$

Let us rearrange this lengthy expression as

$$\sigma^{-}(\omega) + \frac{\sinh \omega\pi}{2\cosh \omega\theta \sinh \omega(\pi - \theta)}\sigma^{+}(\omega)$$
$$= \frac{1}{4\pi}\frac{1}{\cosh \omega\theta}e^{i\omega\Lambda}$$
$$- \frac{1}{2\pi}\frac{\sinh \omega(\pi - 2\theta)}{2\cosh \omega\theta \sinh \omega(\pi - \theta)}\left\{\left[\frac{1}{2N} - \frac{i\omega}{12N^2\sigma_N(\Lambda)}\right]\right.$$
$$\left. + \left[\frac{1}{2N} + \frac{i\omega}{12N^2\sigma_N(\Lambda)} - 2\pi\sigma^{+}(-\omega)\right]e^{2i\omega\Lambda}\right\}. \tag{20.53}$$

This equation, together with the equation determining the largest root Λ

$$\left(1 - \frac{\theta}{\pi}\right)\frac{S}{N} + \frac{1}{2N} = \int_{\Lambda}^{\infty} d\eta\, \sigma_N(\eta) = \int_{0}^{\infty} d\eta\, \sigma_N(\eta + \Lambda) = 2\pi\sigma^{+}(0), \tag{20.54}$$

an equation determining $\sigma_N(\Lambda)$

$$\sigma_N(\Lambda) = 2\int_{-\infty}^{\infty} d\omega\, \sigma^{+}(\omega), \tag{20.55}$$

and the equation for the finite-size energy per spin (20.27), together with (20.28), completely determine the problem of calculating the finite-size corrections to the energy. Of course, we still have to use (20.40), which we obtained from the application of the Euler–Maclaurin summation formula to the finite-size energy per spin (20.27). This application yields

$$E_N^{(S)} - E_\infty = \int_{-\infty}^{-\Lambda} d\eta\, \epsilon(\eta)\sigma_N(\eta) + \int_{\Lambda}^{\infty} d\eta\, \epsilon(\eta)\sigma_N(\eta)$$
$$- \frac{1}{2N}\{\epsilon(\Lambda) + \epsilon(-\Lambda)\} - \frac{1}{12N^2\sigma_N(\Lambda)}\{\epsilon'(\Lambda) - \epsilon'(-\Lambda)\}. \tag{20.56}$$

Explicitly we have for the so-called dressed energy $\epsilon(\eta)$ (see (20.28)) which we need for large values of the argument corresponding to the largest root Λ

$$\epsilon(\eta) = \frac{\pi}{2}\frac{\sin\theta}{\theta}\frac{1}{\cosh\left(\frac{\pi\eta}{2\theta}\right)} \to \pi\frac{\sin\theta}{\theta}e^{-\left(\frac{\pi|\eta|}{2\theta}\right)} \quad \text{for} \quad \eta \gg 1. \tag{20.57}$$

The derivative of this expression is

$$\epsilon'(\eta) = \left(\frac{\pi}{2}\right)^2 \frac{\sin\theta}{\theta^2}\frac{\sinh\left(\frac{\pi\eta}{2\theta}\right)}{\cosh^2\left(\frac{\pi\eta}{2\theta}\right)} \to \frac{\pi^2}{2}\frac{\sin\theta}{\theta^2}e^{-\left(\frac{\pi|\eta|}{2\theta}\right)} \quad \text{for} \quad \eta \gg 1. \tag{20.58}$$

EXERCISE 20.5 Prove (20.55) using Fourier transformation and the definition of the step function where $\Theta(0) = \frac{1}{2}$.

The various parts on the right-hand side of (20.56) have to be further manipulated. We start with

$$-\frac{1}{2N}\{\epsilon(\Lambda)+\epsilon(-\Lambda)\} - \frac{1}{12N^2\sigma_N(\Lambda)}\{\epsilon'(\Lambda)-\epsilon'(-\Lambda)\} = \\ (2\pi)^2\frac{\sin\theta}{\theta}e^{-\left(\frac{\pi\Lambda}{2\theta}\right)}\frac{(-1)}{2\pi}\left[\frac{1}{2N} - \frac{\frac{\pi}{2\theta}}{12N^2\sigma_N(\Lambda)}\right]. \tag{20.59}$$

Furthermore, we have

$$\int_{-\infty}^{-\Lambda} d\eta\, \epsilon(\eta)\sigma_N(\eta) = \int_{-\infty}^{\infty} d\eta\, \epsilon(\eta+\Lambda)\sigma^+(\eta) \tag{20.60}$$

and

$$\int_{\Lambda}^{\infty} d\eta\, \epsilon(\eta)\sigma_N(\eta) = \int_{-\infty}^{\infty} d\eta\, \epsilon(\eta+\Lambda)\sigma^+(\eta). \tag{20.61}$$

Moreover, with

$$\sigma^+(\eta) = \int_{-\infty}^{\infty} d\omega\, e^{i\omega\eta}\sigma^+(\omega) \tag{20.62}$$

the latter two contributions to the finite-size energy (20.56) become

$$\int_{-\infty}^{\infty} d\eta\, \epsilon(\eta+\Lambda)\sigma^+(\eta) = \pi\frac{\sin\theta}{\theta}e^{-\frac{\pi\Lambda}{2\theta}}\int_{-\infty}^{\infty} d\eta \int_{-\infty}^{\infty} d\omega\, e^{i\omega\eta}\sigma^+(\omega)e^{i\eta(\omega+\frac{i\pi}{2\theta})} \\ = 2\pi^2\frac{\sin\theta}{\theta}e^{-\frac{\pi\Lambda}{2\theta}}\sigma^+\left(-\frac{i\pi}{2\theta}\right). \tag{20.63}$$

We can now collect the contributions to the finite-size energy per spin

$$E_N^{(S)} - E_\infty = (2\pi)^2\frac{\sin\theta}{\theta}e^{-\frac{\pi\Lambda}{2\theta}}\left(\sigma^+\left(-\frac{i\pi}{2\theta}\right) - \frac{1}{2\pi}\left[\frac{1}{2N} - \frac{\frac{\pi}{2\theta}}{12N^2\sigma_N(\Lambda)}\right]\right). \tag{20.64}$$

This expression for the finite-size energy per spin is, of course, only an approximation. It can be shown by considering further terms in the Euler–Maclaurin summation formula that contain higher derivatives of $\epsilon(\Lambda)$, that the corrections are of order $e^{-\frac{3\pi\Lambda}{2\theta}}$.

The expressions (20.53), (20.54), (20.55), and (20.64) form a closed set of equations that we now have to further investigate by applying the Wiener–Hopf technique to the functional equation (20.53).

In order to get acquainted with the method described so far, the following exercise is recommended.

EXERCISE 20.6 Isotropic case Redo the calculations for the isotropic antiferromagnetic case $\Delta = 1$ to obtain the expressions corresponding to (20.53), (20.54), (20.55), and (20.64). The results are:

$$\begin{aligned}\sigma^-(\omega) + \frac{1}{1+e^{-2|\omega|}}\sigma^+(\omega) = \frac{1}{2\pi}\frac{e^{-2|\omega|}}{1+e^{-2|\omega|}}\cdot \\ \cdot\left\{\left(\frac{1}{2N} - \frac{i\omega}{12N^2\sigma_N(\Lambda)}\right)\right. \\ \left. + \left(\frac{1}{2N} + \frac{i\omega}{12N^2\sigma_N(\Lambda)} - 2\pi\sigma^+(-\omega)\right)e^{2\omega\Lambda}\right\}\end{aligned} \tag{20.65}$$

and

$$E_N^{(S)} - E_\infty = (2\pi)^2 e^{-\frac{\pi\Lambda}{2}}\left\{\sigma^+\left(-\frac{i\pi}{2}\right) - \frac{1}{2\pi}\left[\frac{1}{2N} - \frac{\frac{\pi}{2}}{12N^2\sigma_N(\Lambda)}\right]\right\} \tag{20.66}$$

while (20.54) and (20.55) remain unchanged.

This case generalized to an asymmetric distribution of Bethe ansatz quantum number $\mathcal{J}_\alpha$ is treated in detail in the summer school proceedings contribution of Ferenc Woynarovich (1997).

20.4.3 Functional equation

According to the prescription of the Wiener–Hopf technique, we now need to write the functional equation (20.53) in a form such that both sides of the equation decompose into two additive parts, one being analytic in the upper, the other in the lower half of the complex ω-plane.

As a first step to achieve this goal, we observe that

$$1 - \frac{\sinh\omega(\pi - 2\theta)}{2\cosh\omega\theta\,\sinh\omega(\pi-\theta)} = \frac{\sinh\omega\pi}{2\cosh\omega\theta\,\sinh\omega(\pi-\theta)} \equiv \frac{1}{G(\omega)G(-\omega)} \tag{20.67}$$

where (see Gradshteyn and Ryzhik, 1980, page 937)

$$\frac{1}{G(\omega)G(-\omega)} = \frac{1}{2(\pi-\theta)} \cdot \frac{\Gamma\left(\frac{1}{2}+i\omega\frac{\theta}{\pi}\right)\Gamma\left(1+i\omega\left(1-\frac{\theta}{\pi}\right)\right)\Gamma\left(\frac{1}{2}-i\omega\frac{\theta}{\pi}\right)\Gamma\left(1-i\omega\left(1-\frac{\theta}{\pi}\right)\right)}{\Gamma(1+i\omega)\,\Gamma(1+i\omega)}. \tag{20.68}$$

In order to factorize this expression, we have to make sure that the individual factors do not diverge for $\omega \to \infty$. Rather, the individual factors should be finite $G(\omega) \to 1$ for $\omega \to \infty$. This can be achieved by considering the asymptotic of the complex γ–function (see e.g. Abramowitz and Stegun, 1972, page 257). We obtain

$$\frac{1}{G(\omega)} = \frac{1}{\sqrt{2(\pi-\theta)}} \frac{\Gamma\left(\frac{1}{2}+i\omega\frac{\theta}{\pi}\right)\Gamma\left(1+i\omega\left(1-\frac{\theta}{\pi}\right)\right)}{\Gamma(1+i\omega)} \cdot \exp\left(-i\omega\left[\left(1-\frac{\theta}{\pi}\right)\ln\left(1-\frac{\theta}{\pi}\right)+\frac{\theta}{\pi}\ln\frac{\theta}{\pi}\right]\right). \tag{20.69}$$

The functional equation can now be written as

$$G(-\omega)\sigma^{-}(\omega) + \frac{\sigma^{+}(\omega)}{G(\omega)} = \frac{1}{4\pi}\frac{G(-\omega)}{\cosh\omega\theta}e^{i\omega\Lambda} + \frac{1}{2\pi}\left[\frac{1}{G(\omega)} - G(-\omega)\right]\left[\frac{1}{2N} - \frac{i\omega}{12N^2\sigma_N(\Lambda)}\right] \tag{20.70}$$

where, at this stage of the approximation, in the functional equation (20.53) we have neglected the term proportional to $e^{2i\omega\Lambda} = e^{2i\omega'\Lambda}e^{-2\omega''\Lambda}$.

Of the two terms on the left-hand side of the functional equation (20.70) the first can be shown to be analytic in the upper complex ω half plane, while the second is analytic in the lower complex ω half plane.

EXERCISE 20.7 Show that the statement above is correct by considering the Fourier transform of $\sigma^{\pm}(\omega)$ and the poles and zeroes of $\frac{1}{G(\pm\omega)}$.

Now we need to write also the right-hand side of (20.70) as a sum of two terms, one of which is analytic in the upper, the other in the lower half plane. In our case this can be achieved using Cauchy's integral formula (see Roos, 1969, page 169).

This leads to the expression for the part of the functional equation in the lower half plane

$$\sigma^{+}(\omega) = \lim_{\epsilon\to 0} \frac{(-G(\omega))}{(2\pi)^2 i} \int_{-\infty}^{\infty} \frac{d\omega'}{\omega' - \omega + i\epsilon} \cdot \\ \cdot \left\{ \frac{G(-\omega')}{2\cosh\omega'\theta} e^{i\omega'\Lambda} + \left[\frac{1}{G(\omega')} - G(-\omega')\right]\left[\frac{1}{2N} - \frac{i\omega'}{12N^2\sigma_N(\Lambda)}\right]\right\}. \tag{20.71}$$

We deal with the three additive terms on the right-hand side one by one. The first term can be evaluated using the residue theorem

$$\frac{-G(\omega)}{(2\pi)^2 i} \int_{-\infty}^{\infty} \frac{d\omega' e^{i\omega'\Lambda}}{\omega' - \omega + i0} \frac{G(-\omega')}{2\cosh\theta\omega'} = \frac{(-1)}{4\pi i\theta} \frac{G(-\frac{i\pi}{2\theta})G(\omega)}{\frac{i\pi}{2\theta} - \omega} e^{-\frac{\pi\Lambda}{2\theta}} + \mathcal{O}\left(e^{-\frac{3\pi\Lambda}{2\theta}}\right). \tag{20.72}$$

We only need this for $\omega = \frac{-i\pi}{2\theta}$ and for $\omega = 0$. For those two values, we obtain

$$\frac{1}{(2\pi)^2} G^2\left(-\frac{i\pi}{2\theta}\right) e^{-\frac{\pi\Lambda}{2\theta}} \tag{20.73}$$

and

$$\frac{1}{2\pi^2} G(0) G\left(-\frac{i\pi}{2\theta}\right) e^{-\frac{\pi\Lambda}{2\theta}}, \tag{20.74}$$

respectively.

The second term produces after a lengthy calculation that uses contour integration and again the residue theorem

$$\lim_{\epsilon\to 0} \frac{(-G(\omega))}{(2\pi)^2 i} \int_{-\infty}^{\infty} \frac{d\omega'}{\omega' - \omega + i\epsilon} \frac{1}{G(\omega')} \left[\frac{1}{2N} - \frac{i\omega'}{12N^2\sigma_N(\Lambda)}\right] = \\ \frac{1}{2\pi}\left[\frac{1}{2N} - \frac{\pi}{2\theta}\frac{1}{12N^2\sigma_N(\Lambda)}\right] \tag{20.75}$$

for $\omega = -\frac{i\pi}{2\theta}$ and

$$\lim_{\epsilon\to 0} \frac{(-G(\omega))}{(2\pi)^2 i} \int_{-\infty}^{\infty} \frac{d\omega'}{\omega' - \omega + i\epsilon} \frac{1}{G(\omega')} \left[\frac{1}{2N} - \frac{i\omega'}{12N^2\sigma_N(\Lambda)}\right] = \frac{1}{2\pi}\frac{1}{2N} \tag{20.76}$$

for $\omega = 0$.

For the third and last term we use the asymptotic form $G(\omega') \approx e^{\frac{\mu}{12}\frac{1}{i\omega'}} \approx 1 + \frac{\mu}{12}\frac{1}{i\omega'}$ where $\mu = \frac{\theta}{\pi-\theta} - \frac{\pi}{2\theta}$ and the expansion $\frac{1}{\left(\omega'+\frac{i\pi}{2\theta}\right)} \approx \frac{1}{\omega'} - \frac{i\pi}{2\theta\,\omega'^2}$ such that

$$\lim_{\epsilon\to 0}\frac{G(\omega)}{(2\pi)^2 i}\int_{-\infty}^{\infty}\frac{d\omega'}{\omega'-\omega+i\epsilon}G(-\omega')\left[\frac{1}{2N}-\frac{i\omega'}{12N^2\sigma_N(\Lambda)}\right]$$
$$= -\frac{G\left(-\frac{i\pi}{2\theta}\right)}{2\pi}\left[\frac{1}{2N}-\frac{\mu}{12}\frac{1}{12N^2\sigma_N(\Lambda)}-\frac{\pi}{2\theta}\frac{1}{12N^2\sigma_N(\Lambda)}\right] \tag{20.77}$$

for $\omega = -\frac{i\pi}{2\theta}$ and

$$\lim_{\epsilon\to 0}\frac{G(\omega)}{(2\pi)^2 i}\int_{-\infty}^{\infty}\frac{d\omega'}{\omega'-\omega+i\epsilon}G(-\omega')\left[\frac{1}{2N}-\frac{i\omega'}{12N^2\sigma_N(\Lambda)}\right]$$
$$= -\frac{G(0)}{2\pi}\left[\frac{1}{2N}-\frac{\mu}{12}\frac{1}{12N^2\sigma_N(\Lambda)}\right] \tag{20.78}$$

for $\omega = 0$.

Collecting terms, we arrive at the general expression

$$\sigma^+(\omega) = \frac{(-1)}{2i\theta}\frac{G\left(-\frac{i\pi}{2\theta}\right)}{2\pi}\frac{G(\omega)}{\frac{i\pi}{2\theta}-\omega}e^{-\frac{\pi\Lambda}{2\theta}} - \frac{G(\omega)}{(2\pi)^2 i}\int_{-\infty}^{\infty}\frac{d\omega'}{\omega'-\omega-i0}\cdot$$
$$\cdot\left(\frac{1}{G(\omega')}-G(-\omega')\right)\left(\frac{1}{2N}-\frac{i\omega'}{12N^2\sigma_N(\Lambda)}\right), \tag{20.79}$$

which we evaluated for $\omega = -\frac{i\pi}{2\theta}$

$$\sigma^+\left(-\frac{i\pi}{2\theta}\right) = \left(\frac{G\left(-\frac{i\pi}{2\theta}\right)}{2\pi}\right)^2 e^{-\frac{\pi\Lambda}{2\theta}} + \frac{1}{2\pi}\left(\frac{1}{2N}-\frac{\pi}{2\theta}\frac{1}{12N^2\sigma_N(\Lambda)}\right)$$
$$-\frac{G\left(-\frac{i\pi}{2\theta}\right)}{2\pi}\left(\frac{1}{2N}-\frac{\mu}{12}\frac{1}{12N^2\sigma_N(\Lambda)}-\frac{\pi}{2\theta}\frac{1}{12N^2\sigma_N(\Lambda)}\right) \tag{20.80}$$

and for $\omega = 0$, recalling (20.54),

$$2\pi\sigma^+(0) = \left(\frac{G\left(-\frac{i\pi}{2\theta}\right)G(0)}{2\pi}\right)e^{-\frac{\pi\Lambda}{2\theta}} + \frac{1}{2N} - G(0)\left(\frac{1}{2N}-\frac{\mu}{12}\frac{1}{12N^2\sigma_N(\Lambda)}\right) \tag{20.81}$$
$$= \left(1-\frac{\theta}{\pi}\right)\frac{S}{N}+\frac{1}{2N}. \tag{20.82}$$

Furthermore, we need to evaluate

$$\sigma_N(\Lambda) = 2\int_{-\infty}^{\infty} d\omega\, \sigma^+(\omega) \tag{20.83}$$

$$\begin{aligned}
&= \frac{(-1)}{i\theta}\frac{G\left(-\frac{i\pi}{2\theta}\right)}{2\pi}e^{-\frac{\pi\Lambda}{2\theta}}\int_{-\infty}^{\infty} d\omega\,\frac{G(\omega)}{\frac{i\pi}{2\theta}-\omega} \\
&- \frac{2}{(2\pi)^2 i}\int_{-\infty}^{\infty} d\omega' \left(\frac{1}{G(\omega')} - G(-\omega')\right)\cdot \\
&\cdot\left(\frac{1}{2N} - \frac{i\omega'}{12N^2\sigma_N(\Lambda)}\right)\int_{-\infty}^{\infty} d\omega\,\frac{G(\omega)}{\omega'-\omega-i0}.
\end{aligned} \tag{20.84}$$

With the help of complex integration and the asymptotic expansions for $G(\omega)$ and $(\omega-\omega')^{-1}$ we have already employed, to lowest order this becomes

$$\sigma_N(\Lambda) = \frac{1}{2\theta}G\left(-\frac{i\pi}{2\theta}\right)e^{-\frac{\pi\Lambda}{2\theta}} + \frac{\mu}{12}\left(\frac{1}{2N} - \frac{\mu}{12}\frac{1}{2}\frac{1}{12N^2\sigma_N(\Lambda)}\right). \tag{20.85}$$

The expressions (20.85), (20.82), and (20.80) suffice to determine the lowest order finite-size corrections to the energy per spin.

20.4.4 Finite-size corrections to the energy

Now employing algebraic manipulation, we can extract from (20.85), (20.82), and (20.80) the final formula for the finite-size energy per spin (20.64), which is exact to lowest order in the inverse system size $1/N$

$$E_N^{(S)} - E_\infty = (2\pi)^2\frac{\sin\theta}{\theta}\left(\frac{\left(1-\frac{\theta}{\pi}\right)S^2}{8N^2} - \frac{1}{48N^2}\right). \tag{20.86}$$

For example, the ground state finite-size energy correction in the $S=0$ sector becomes

$$N(E_N^{(0)} - E_\infty) = -\pi^2\frac{\sin\theta}{\theta}\frac{1}{12N} = -\frac{\pi\zeta}{6N} \tag{20.87}$$

where $\zeta = \frac{\pi}{2}\frac{\sin\theta}{\theta}$ whereas the finite-size correction for general S compared to the correction for $S=0$ is

$$N(E_N^{(S)} - E_N^{(0)}) = \pi^2\frac{\sin\theta}{\theta}\left(1-\frac{\theta}{\pi}\right)S^2\frac{1}{2N} = \frac{2\pi\zeta}{N}x_S \tag{20.88}$$

where $x_S = \left(1-\frac{\theta}{\pi}\right)\frac{S^2}{2}$.

These two expressions coincide precisely with the predictions of conformal invariance for critical one-dimensional systems (Blöte *et al.*, 1986; Affleck, 1986; see also Cardy, 1987).

We mention another interesting observation connected to finite-size calculations using the Bethe ansatz. The finite-size corrections of one-dimensional models in external fields show a peculiarity. At generic values of the external field, the energy spectra of the one-dimensional Bose gas and the Heisenberg quantum spin chain in an external field (a chemical potential or a magnetic field) are not of the form expected on the basis of conformal invariance. In order to recover the conformal structure of the finite-size corrections, additional conditions have to be imposed that connect the system size and the external field in a particular way. These conditions have to be satisfied when taking the continuum limit in order to arrive at a conformally invariant theory. A detailed exposition of these so-called non-analytic finite-size corrections can be found in Woynarovich *et al.* (1989) and Eckle and Hamer (1991). Quite recently, there has been renewed interest in these observations in relation to topological quantum phase transitions (Bao *et al.*, 2018).

20.5 Higher order finite-size corrections

The task to calculate higher, order corrections consists first of all in a careful analysis of the approximations made such that we can determine what contributions in powers of the inverse system size $1/N$ they will make.

Considering the degree of calculational complication we had to navigate for the lowest order finite–size corrections on the previous pages, we shall refrain from presenting a higher-order calculation in any detail. Such calculations may be found in Woynarovich and Eckle (1987*a*) and Woynarovich *et al.* (1989) for the Heisenberg quantum spin chain and the δ-Bose gas and in Woynarovich and Eckle (1987*b*) for the Hubbard model at half-filling. Another interesting reference in this regard is Hamer *et al.* (1988).

Furthermore, we have to take into account the words of Rodney Baxter (1982): 'I knew from experience how many sheets of paper go into the waste-paper basket after even a modest calculation: there was no way they could all appear in print.'

A few words of warning are also in order. The application of the Euler–Maclaurin formula (cf. (20.37) and 20.40)) to the finite-size Bethe ansatz equations of the Heisenberg quantum spin chain consists in using the leading contributions of an asymptotic expansion that does not converge. This is because the finite-size density $\sigma_N(\Lambda)$ is itself of the order of $1/N$. Therefore, all terms in the expansion, the leading ones as well as the higher ones, are of the order $1/N$. The results are nevertheless reliable as has been shown, e.g. in Woynarovich and Eckle (1987*a*), where numerical checks for long chains have been compared to the analytic results based on the Euler–Maclaurin formalism (cf. also the other literature cited above and Woynarovich (1997)) but this has to be checked in every new case when we apply this procedure to another problem. Independent methods based on nonlinear integral equations have also been developed by Klümper and Batchelor (1990), Pearce and Klümper (1991), Klümper *et al.* (1991), and Klümper and Pearce (1991), which corroborate the findings based on the Euler–Maclaurin formalism but also allow access to results that seem out of reach for this latter formalism. These issues have recently been taken up again by Kozlowski (2015) and, in the recent research literature, by Brockmann and Stéphan (2017). Recent research papers where the Euler–Maclaurin formalism has been applied to current problems in the physics of one-dimensional systems are, e.g. Nomura (1993) and Stéphan and Dubail (2011). Finally, a very recent proposal for the calculation of finite-size corrections in Bethe ansatz solvable models is due to Granet *et al.* (2018). The authors discuss the advantages and disadvantages of earlier approaches before they develop their attempt to overcome these drawbacks.

Epilogue. This book's exposition of (some of) the aspects of (some of) the models describing quantum matter, especially using Bethe ansatz methods, ends here. We hope that the reader's appetite will have been whetted to further pursue these fascinating areas. For this purpose, we hope that the directions to the literature will supply convenient entry points.

References

Abramowitz, Milton and Stegun, Irene A (1972). *Handbook of mathematical functions*. Dover, New York.

Abrikosov, Alexei A, Gorkov, Lev P, and Dzyaloshinski, Igor E (1975). *Methods of quantum field theory in statistical physics*. Dover, New York.

Affleck, Ian (1986). Universal term in the free energy at a critical point and the conformal anomaly. *Phys Rev Lett*, **56**, 746–9.

Affleck, Ian (2010). The Kondo screening cloud: What it is and how to observe it. In A Aharony and O Entin-Wohlman (Eds)., *Perspectives of mesoscopic physics: Dedicated to Yoseph Imry's 70th birthday*, pp. 1–44. World Scientific, Singapore.

Affleck, Ian and Ludwig, Andreas W W (1991). Critical theory of overscreened Kondo fixed points. *Nucl Phys*, **B360**, 641–96.

Affleck, Ian and Simon, Pascal (2002). Comment on 'Kondo resonance in a mesoscopic ring coupled to a quantum dot: Exact results for the Aharono–Bohm–Casher effects'. *Phys Rev Lett*, **88**, 139701.

Aharonov, Yakir and Bohm, David (1959). Significance of electromagnetic potentials in the quantum theory. *Phys Rev*, **115**, 485–91.

Aharonov, Yakir and Casher, Aharon (1984). Topological quantum effects for neutral particles. *Phys Rev Lett*, **53**, 319–21.

Alcaraz, Francisco, Barber, Michael, Batchelor, Murray, Baxter, Rodney, and Quispel, Reinout (1987). Surface exponents of the quantum XXZ, Ashkin-Teller, and Potts models. *J Phys A: Math Gen*, **20**, 6397.

Alcaraz, Francisco C, Barber, Michael N, and Batchelor, Murray T (1988). Conformal invariance, the XXZ chain and the operator content of two-dimensional critical systems. *Ann Phys*, **182**, 280.

Altland, Alexander and Simons, Ben D (2010). *Condensed matter field theory* (2nd edn). Cambridge University Press, Cambridge.

Amico, Luigi, Frahm, Holger, Osterloh, Andreas, and Ribeiro, Giuliano A P (2007). Integrable spin-boson models descending from rational six-vertex models. *Nucl Phys B* [*FS*], **787**, 283–300.

Anderson, Philip W (1952). An approximate quantum theory of the antiferromagnetic ground state. *Phys Rev*, **86**, 694.

Anderson, Philip W (1961). Localized magnetic states in metals. *Phys Rev*, **124**, 41–53.

Anderson, Philip W (1972). More is different. *Science*, **177**, 393.

Anderson, Philip W (1978). Local moments and localized states. *Rev Mod Phys*, **50**, 191–201.

Anderson, Philip W (1987). The resonating valence bond state in La_2CuO_4 and superconductivity. *Science*, **235**, 1196–8.

Andrei, Natan (1980). Diagonalization of the Kondo Hamiltonian. *Phys Rev Lett*, **45**, 379–82.

Andrei, Natan (1994). *Integrable models in condensed matter physics*, vol. 6. Series on Modern Condensed Matter Physics, p. 458. World Scientific, Singapore.

Andrei, Natan and Johannesson, Henrik (1984). Heisenberg chain with impurities (an integrable model). *Phys Lett*, **100A**, 108–12.

Andrei, Natan, Furuya Keiichiro and Lowenstein, John H (1981). Solution of the Kondo problem. *Rev Mod Phys*, **55**, 331–402.

Andrei, Natan and Lowenstein, John H (1981). Scales and scaling in the Kondo problem. *Phys Rev Lett*, **46**, 356–60.

Ashcroft, Neil W and Mermin, N David (1976). *Solid state physics*. Holt, Rinehart, and Winston, New York.

Audretsch, Jürgen (2007). *Entangled systems—New directions in quantum physics*. Wiley–VCH, Weinheim.

Auerbach, Assa (1994). *Interacting electrons and magnetism*. Springer, New York.

Auslaender, Ophir M, Yacoby, Amir, de Picciotto, Rafi, Baldwin, Kirk W, Pfeiffer, Loren N, and West, Ken W (2002). Tunneling spectroscopy of the elementary excitations in a one-dimensional wire. *Science*, **295**, 825–8.

Avdeev, L V and Dörfel, Bernd-D (1986). Finite-size corrections for the XXX antiferromagnet. *J Phys A: Math Gen*, **19**, L13–17.

Balser, Werner (2000). *Formal power series and linear systems of meromorphic ordinary differential equations*. Springer, New York.

Bao, Chenfeng, Yang, Shuo, Wang, Chenjie, and Gu, Zheng-Cheng (2018). Lattice model constructions for gapless domain walls between topological phases. arXiv:1801.00719 [cond-mat.str-el].

Barber, Michael N (1983). *Finite-size scaling*. In Cyril Domb and Joel L Lebowitz (Eds), *Phase transitions and critical phenomena*, vol. 8. pp. 146–475. Academic Press, London.

Bares, Pierre-Antoine (1994). Exact results for a one-dimensional *t–J* model with impurity. https://arxiv.org/abs/cond-mat/9412011.

Batchelor, Murray and Hamer, Chris (1990). Surface energy of integrable quantum spin chains. *J Phys A: Math Gen*, **23**, 761.

Batchelor, Murray T (2007). The Bethe ansatz after 75 years. *Phys Today*, **60**(1), 36.

Batchelor, Murray T and Blöte, Henk W J (1989). Conformal invariance and critical behavior of the $O(n)$ model on a honeycomb lattice. *Phys Rev B*, **39**, 2391.

Batchelor, Murray T and Foerster, Angela (2016). Yang–Baxter integrable models in experiments: From condensed matter to ultracold atoms. *J Phys A: Math Theor*, **49**, 173001.

Batchelor, Murray T, Nienhuis, Bernard, and Warnaar, Ole S (1989). Bethe ansatz results for a solvable $O(n)$ model on a square lattice. *Phys Rev Lett*, **62**, 2425.

Batchelor, Murray T and Zhou, Huan-Qiang (2015). Integrability versus exact solvability in the quantum Rabi and Dicke models. *Phys Rev A*, **91**, 053808.

Bateman, Harry (1910). The transformation of the electrodynamical equations. *P Lond Math Soc*, **8**, 223–64.

Batista, Cristian D and Ortiz, Gerardo (2001). Generalized Jordan–Wigner transformations. *Phys Rev Lett*, **86**, 1082–5.

Baxter, Rodney (1982). *Exactly solved models in statistical mechanics*. Academic Press, London. Reprinted by Dover Publications (2007) with a new chapter on *Subsequent developments* added by the author. The original 1982 edition can be freely downloaded at the web site https://physics.anu.edu.au/theophys/_files/Exactly.pdf.

Baxter, Rodney J (1971). Eight-vertex model in lattice statistics. *Phys Rev Lett*, **26**, 832–3.

Baym, Gordon and Pethick, Christopher (2004). *Landau Fermi liquid theory: Concepts and applications*. Wiley–VCH, Weinheim.

Belavin, Alexander A (1979). Exact solution of the two-dimensional model with asymptotic freedom. *Phys Lett B*, **87**, 117–21.

Belavin, Alexander A, Polyakov, Alexander M, and Zamolodchikov, Alexander B (1984*a*). Infinite conformal symmetry in two-dimensional quantum field theory. *J Stat Phys*, **34**, 763–74.

Belavin, Alexander A, Polyakov, Alexander M, and Zamolodchikov, Alexander B (1984*b*). Infinite conformal symmetry in two-dimensional quantum field theory. *Nucl Phys*, **B241**, 333–80.

Berezin, Felix A (1966). *The method of second quantization*. Academic Press, New York and London.

Bernal, John D and Fowler, Ralph H (1933). A theory of water and ionic solution, with particular reference to hydrogen and hydroxyl ions. *J Chem Phys*, **1**, 515.

Bethe, Hans (1931). Zur Theorie der Metalle. I. Eigenwerte und Eigenfunktionen der linearen Atomkette. *Z Physik*, **71**, 205.

Bethe, Hans A (1997). *Selected works of Hans A Bethe—With commentary*. World Scientific, Singapore.

Binney, James J, Dowrick, Nigel J, Fisher, Andrew J, and Newman, Mark E J (1992). *The theory of critical phenomena*. Oxford University Press, Oxford.

Bitko, D, Rosenbaum, Thomas F, and Aeppli, Gabriel (1996). Quantum critical behavior for a model magnet. *Phys Rev Lett*, 77, 940.

Bloch, Felix (1930). Zur Theorie des Ferromagnetismus. *Z Physik*, **61**, 206–19.

Bloch, Felix (1934). Inkohärente Röntgenstreuung und Dichteschwankungen eines entarteten Fermigases. *Helv Phys Acta*, 7, 385.

Bloch, Immanuel (2005). Ultracold quantum gases in optical lattices. *Nat Phys*, **1**, 30–30.

Bloch, Immanuel, Dalibard, Jean, and Zwerger, Wilhelm (2008). Many-body physics with ultracold gases. *Rev Mod Phys*, **80**, 885–964.

Blöte, Henk W, Cardy, John L, and Nightingale, M Peter (1986). Conformal invariance, the central charge, and universal finite-size amplitudes at criticality. *Phys Rev Lett*, **56**, 742–5.

Blum, Thomas and Shapir, Yonathan (1990). A soluble seven-vertex model for clusters with interfacial bending energy. *J Phys A*, **23**, L511.

Blumenhagen, Ralph and Plauschinn, Erik (2009). *Introduction to conformal field theory—With applications to string theory*, vol. 779. Lecture Notes in Physics. Springer, Berlin Heidelberg.

Blundell, Stephen (2001). *Magnetism in condensed matter*. Oxford University Press, Oxford.

Bockrath, Marc, Cobden, David H, Lu, Jia, Rinzler, Andrew G, Smalley, Richard E, Balents, Leon, and McEuen, Paul L (1999). Luttinger-liquid behaviour in carbon nanotubes. *Nature*, **397**, 598–601.

Bogoliubov, Nicholay M (1996). Exactly solvable models of quantum nonlinear optics. University of Helsinki, Research Institute of Theoretical Physics. Report Series in Theoretical Physics HU–TFT–IR–96–3.

Bogoliubov, Nicholay M, Bullough, Robin K, and Timonen, Jussi (1996). Exact solution of generalized Tavis–Cummings models in quantum optics. *J Phys A*, **29**, 6305–12.

Bohr, Niels (1911). *Studier over Metallernes Elektrontheori*. Ph.D. thesis, University of Copenhagen.

Bordag, Michael, Klimchitskaya, Galina L, Mohideen, Umar, and Mostepanenko, Vladimir M (2009). *Advances in the Casimir effect*. Oxford University Press, Oxford.

Born, Max and Oppenheimer, J Robert (1927). Zur Quantentheorie der Molekeln. *Ann Physik*, **84**, 457–84.

Bose, Satyendra Nath (1924). Plancks Gesetz und Lichtquantenhypothese. *Z Phys*, **26**, 178.

Botet, Robert and Jullien, Rémy (1983). Ground-state properties of a spin-1 antiferromagnetic chain. *Phys Rev B*, **27**, 613–15.

Braak, Daniel (2011). Integrability of the Rabi Model. *Phys Rev Lett*, **107**, 100401 (see also the online supplement).
Braak, Daniel (2013). A generalized G-function for the quantum Rabi model. *Ann Phys (Berlin)*, **525**, L23–8.
Brascamp, Herm Jan, Kunz, H, and Wu, Fa-Yueh (1973). Some rigorous results for the vertex model in statistical mechanics. *J Math Phys*, **14**, 1927–32.
Brézin, Edouard (1982). An investigation of finite size scaling. *J Physique*, **43**, 15.
Brézin, Edouard (1986). The onset of phase transitions in finite systems. *Phys Bl*, **42**, 182.
Brockmann, Michael and Stéphan, Jean-Marie (2017). Universal terms in the overlap of the ground state of the spin-1/2 XXZ chain with the Néel state. *J Phys A: Math Theor*, **50**, 354001.
Büttiker, Markus, Imry, Yoseph, and Landauer, Rolf (1983). Josephson behavior in small normal one-dimensional rings. *Phys Lett*, **96A**, 365–7.
Byers, Nina and Yang, Chen Ning (1961). Theoretical considerations concerning quantized magnetic flux in superconducting cylinders. *Phys Rev Lett*, 7, 46–9.
Callen, Herbert B (1960). *Thermodynamics*. John Wiley, New York.
Callen, Herbert B (1985). *Thermodynamics and an Introduction to Thermostatics* (second edn). John Wiley, New York.
Cardy, John L (1984). Conformal invariance and universality in finite-size scaling. *J Phys A: Math Gen*, **17**, L385–7.
Cardy, John L (1987). Conformal invariance. In Cyril Domb and Joel L Lebowitz (Eds)., *Phase transitions and critical phenomena*, vol. 11, pp. 55–126. Academic Press, London.
Cardy, John L (1990). Conformal invariance and statistical mechanics. In E Brézin and J Zinn-Justin (Eds)., *Fields, strings and critical phenomena, Proceedings of the Les Houches Ecole d'été de physique théorique Session XLIX*, pp. 169–245. North–Holland, Amsterdam.
Cardy, John L (1996). *Scaling and renormalization in statistical physics*. Cambridge University Press, Cambridge.
Cardy, John L (2010). Conformal field theory and statistical mechanics. In J Jacobson, S Ouvry, V Pasquier, D Serban, and L Cugliandolo (Eds)., *Exact methods in low-dimensional statistical physics and quantum computing, Proceedings of the Les Houches Ecole d'été de physique théorique Session LXXXIX*, pp. 65–98. Oxford University Press, Oxford.
Carr, Lincoln D (2011). *Understanding quantum phase transitions*. CRC Press, Boca Raton.
Casimir, Hendrik B G (1948). On the attraction between two perfectly conducting plates. *Proc K Ned Akad Wet*, **B51**, 793–5.
Caux, Jean-Sébastien and Mossel, Jorn (2011). Remarks on the notion of quantum integrability. *J Stat Mech*, P02023.
Cazalilla, Miguel A, Citro, Roberta, Giamarchi, Thierry, Orignac, Edmond, and Rigol, Marcos (2011). One dimensional Bosons: From condensed matter systems to ultracold gases. *Rev Mod Phys*, **83**, 1405–66.
Chandler, David (1987). *Introduction to modern statistical mechanics*. Oxford University Press.
Cheung, Ho-Fai, Gefen, Yuval, Riedel, Eberhard K, and Shih, Wei-Heng (1988). Persistent currents in small one-dimensional metal rings. *Phys Rev B*, 37, 6050–62.
Cohen-Tannoudji, Claude, Diu, Bernard, and Laloë, Franck (1977). *Quantum mechanics*, vol. 1. John Wiley, New York.
Coldea, Radu, Tennant, D Alan, Wheeler, Elisa M, Wawrzynska, Ewa, Prabhakaran, Dharmalingam, Telling, Mark, Habicht, Klaus, Smeibidl, Peter, and Kiefer, Klaus (2010). Quantum criticality in an ising chain: Experimental evidence for emergent *E*8 symmetry. *Science*, **327**, 177.

Coleman, Piers (2015). *Introduction to many-body physics*. Cambridge University Press, Cambridge.

Cronenwett, Sara M, Oosterkamp, Tjerk H, and Kouwenhoven, Leo P (1998). A tunable Kondo effect in quantum dots. *Science*, **391**, 540–4.

Cunningham, Ebenezer (1910). The principle of relativity in electrodynamics and an extension thereof. *P Lond Math Soc*, **8**, 77–98.

van Delft, Dirk (2008). Little cup of helium, big science. *Phys Today*, **61**, 36–42.

von Delft, Jan and Schoeller, Herbert (1998). Bosonization for beginners: Refermionization for experts. *Annalen Phys*, 7, 225–305.

Derzhko, Oleg (2008). Jordan–Wigner Fermionization and the theory of low-dimensional quantum spin models. Dynamic properties. In Janusz Jedrzejewski (Ed)., *Condensed matter physics in the prime of the 21st century—Phenomena, materials, ideas, methods*, pp. 35–88. World Scientific, Singapore.

des Cloizeaux, Jacques and Gaudin, Michel (1966). Anisotropic linear magnetic chain. *J Math Phys*, 7, 1384.

des Cloizeaux, Jacques and Pearson, J J (1962). Spin-wave spectrum of the antiferromagnetic linear chain. *Phys Rev*, **128**, 2131.

Deshpande, Vikram V, Bockrath, Marc, Glazman, Leonid I., and Yacoby, Amir (2010). Electron liquids and solids in one dimension. *Nature*, **464**, 209–16.

Di Francesco, Philippe, Mathieu, Pierre, and Sénéchal, David (1997). *Conformal field theory* (Corrected edn). Graduate Texts in Contemporary Physics. Springer, New York.

Dicke, Robert (1954). Coherence in spontaneous radiation processes. *Phys Rev*, **93**, 99–110.

Dotsenko, Vladimir S (1984). Critical behavior and associated conformal algebra of the Z_3 potts model. *Nucl Phys*, **B235**, 54–74.

Douçot, Benoit and Zinn-Justin, Jean (ed.) (1995). *Strongly interacting Fermions and high T_c superconductivity*. Elsevier, Amsterdam.

Dutta, Amit, Aeppli, Gabriel, Chakrabarti, Bikas K, Divakaran, Uma, Rosenbaum, Thomas F, and Sen, Diptiman (2015). *Quantum phase transitions in transverse field spin models—From statistical physics to quantum information*. Cambridge University Press, Cambridge.

Eckle, Hans-Peter (1998). Integrable quantum impurity models. In D Neilson and R F Bishop (Eds)., *Recent progress in many–body theories*, pp. 477–480. World Scientific, Singapore.

Eckle, Hans-Peter and Hamer, Chris (1991). Non-analytic finite-size corrections for the Heisenberg chain in a magnetic field with free and twisted boundary conditions. *J Phys A: Math Gen*, **24**, 191–202.

Eckle, Hans-Peter and Johannesson, Henrik (2017). A generalization of the quantum Rabi model: Exact solution and spectral structure. *J Phys A: Math Theor*, **50**, 294004.

Eckle, Hans-Peter, Johannesson, Henrik, and Stafford, Charles A. (2000). Kondo impurity in a mesoscopic ring: Charge persistent current. *J Low Temp Phys*, **118**, 475–83.

Eckle, Hans-Peter, Johannesson, Henrik, and Stafford, Charles A. (2001). Kondo resonance in a mesoscopic ring coupled to a quantum dot: Exact results for the Aharonov–Bohm–Casher effects. *Phys Rev Lett*, 87, 016602.

Eckle, Hans-Peter, Johannesson, Henrik, and Stafford, Charles A. (2002). Eckle et al. reply. *Phys Rev Lett*, **88**, 139702.

Eckle, Hans-Peter, Punnoose, Alexander, and Römer, Rudolf A (1997). Absence of backscattering at integrable impurities in one-dimensional quantum many-body systems. *Europhys Lett*, **39**, 293–8.

Editorial (2013). The Hubbard model at half a century. *Nature Phys*, **9**, 523.

Eggert, Sebastian (2007). One-dimensional quantum wires: A pedestrian approach to bosonization. arXiv:0708.0003v3[cond-mat.str-el].

Einstein, Albert (1924). Quantentheorie des einatomigen idealen Gases. *Sitzungsberichte der Preussischen Akademie der Wissenschaften, Physikalisch–mathematische Klasse*, 261–7.

Einstein, Albert (1925). Quantentheorie des einatomigen idealen Gases. 2. Abhandlung. *Sitzungsberichte der Preussischen Akademie der Wissenschaften, Physikalisch–mathematische Klasse*, 3–14.

Einstein, Albert, Podolsky, Boris, and Rosen, Nathan (1935). Can quantum-mechanical description of reality be considered complete? *Phys Rev*, **47**, 777–80.

Essler, Fabian H L, Frahm, Holger, Göhmann, Frank, Klümper, Andreas, and Korepin, Vladimir E (2005). *The one-dimensional Hubbard model.* Cambridge University Press Cambridge.

Essler, Fabian H L, van den Berg, Rianne, and Gritsev, Vladimir (2018). Integrable spin chains with random interactions. *Phys Rev B*, **98**, 024203.

Faddeev, Ludwig D and Takhtajan, Leon A (1981). What is the spin of a spin wave? *Phys Lett A*, **85**, 375.

Faddeev, Ludwig D and Takhtajan, Leon A (1984). Spectrum and scattering of excitations in the one-dimensional isotropic Heisenberg model. *J Math Sciences*, **24**, 241. Translated from the original Russian paper in: *Zapiski Nauchnykh Seminarov Leningradskogo Otdeleniya Matematicheskogo Instituta im V A Steklova* AN SSSR **109** (1981) 134.

Fazekas, Patrik (1999). *Lecture notes on electron correlation and magnetism.* World Scientific, Singapore.

Ferdinand, Arthur E and Fisher, Michael E (1969). Bounded and inhomogeneous Ising models. I. Specific–heat anomaly of a finite lattice. *Phys Rev*, **185**, 832–46.

Fetter, Alexander L and Walecka, John Dirk (2003). *Quantum theory of many-particle systems.* Dover, Mineola.

Feynman, Richard P (1965). Nobel lecture: The development of the space-time view of quantum electrodynamics. http://www.nobelprize.org/nobel_prizes/physics/laureates/1965/feynman-lecture.html. Nobelprize.org. Nobel Media AB 2014. Web. 30 Mar 2017.

Feynman, Richard P (1972). *Statistical mechanics.* Benjamin/Cummings, Reading, MA.

Feynman, Richard P (1988). *Asia-Pacific physics news*, **3**, 22.

Filyov, V M, Tsvelik, Alexei M, and Wiegmann, Pavel B (1981). Thermodynamics of the *s-d* exchange model (Kondo problem). *Phys Lett A*, **81**, 175–8.

Fisher, Michael E (1971). The theory of critical point singularities. In M S Green (Ed)., *Critical phenomena: Proceedings of the International School of Physics Enrico Fermi Course* 51, pp. 1–99. Academic Press, New York.

Fisher, Michael E and Barber, Michael N (1972). Scaling theory for finite-size effects in the critical region. *Phys Rev Lett*, **28**, 1516.

Flicker, M and Lieb, Elliott H (1967). Delta function Fermi gas with two spin deviates. *Phys Rev*, **161**, 179–88.

Flügge, Siegfried (1979). *Practical quantum mechanics* (International student edn). Springer, Berlin.

Foerster, Angela, Links, Jon, and Tonel, Arlei P (1999). Algebraic properties of an integrable *t–J* model with impurities. *Nucl Phys B*, **552**, 707–26.

Ford, George W, Kac, Mark, and Mazur, Peter (1965). Statistical mechanics of assemblies of coupled oscillators. *J Math Phys*, **6**, 504.

Fox, Mark (2006). *Quantum optics—An introduction.* Oxford University Press, Oxford.

Fradkin, Eduardo (1989). Jordan–Wigner transformation for quantum-spin systems in two dimensions and fractional statistics. *Phys Rev Lett*, **63**, 322.

Friedan, Daniel, Qiu, Zongan, and Shenker, Stephen (1984). Conformal invariance, unitarity, and critical exponents in two dimensions. *Phys Rev Lett*, **52**, 1575–8.

Friedel, Jacques (1956). On some electrical and magnetic properties of magnetic solid solutions. *Can J Phys*, **34**, 1190.

Friedel, Jacques (1958*a*). Metallic alloys. *Nuovo Cimento Suppl*, **VII**, 287 – 311.

Friedel, Jacques (1958*b*). Sur la structure électronique des métaux et alliages de transition et des métaux lourds. *J Phys Radium*, **19**, 573 – 81.

Fulde, Peter (2002). *Electron correlations in molecules and solids* (2nd edn). Springer, Berlin.

Fung, M K (1981). Validity of the Bethe–Yang hypothesis in the delta-function interaction problem. *J Math Phys*, **22**, 2017–19.

Garrison, John C and Chiao, Raymond Y (2008). *Quantum optics*. Oxford University Press, Oxford.

Gaudin, Michel (1967). Un système a une dimension des Fermions en interaction. *Phys Lett A*, **24**, 55–6.

Gaudin, Michel (1971). Thermodynamics of the Heisenberg–Ising ring for $\Delta \geq 1$. *Phys Rev Lett*, **26**, 1301.

Gaudin, Michel (1983). *La fonction d'onde de Bethe*. Masson, Paris.

Gaudin, Michel (1995). *Modèles exactement résolus—Traveaux des Michel Gaudin*. Les Èditions de Physique, Les Ulis cedex.

Gaudin, Michel (2014). *The Bethe wavefunction*. Cambridge University Press, Cambridge.

Geller, Michael R and Cleland, Andrew N (2005). Superconducting qubits coupled to nano-electromechanical resonators: An architecture for solid-state quantum-information processing. *Phys Rev A*, **71**, 032311.

Gerry, Christopher C and Knight, Peter L (2005). *Introductory quantum optics*. Cambridge University Press, Cambridge.

Giamarchi, Thierry (2004). *Quantum physics in one dimension*. Oxford University Press, Oxford.

Giamarchi, Thierry, Millis, Andrew J, Parcollet, Olivier, Saleur, Hubert, and Cugliandolo, Leticia F (Eds) (2016). Strongly interacting quantum systems out of equilibrium. *Proceedings of the Les Houches Ecole d'été de physique théorique Session XCIX*. Oxford University Press, Oxford.

Giauque, W F and Stout, J W (1936). The entropy of water and the third law of thermodynamics: The heat capacity of ice from 15 to 273°K. *J Am Chem Soc*, **58**(7), 1144–50.

Ginsparg, Paul (1990). Applied conformal field theory. In E Brézin and J Zinn-Justin (Eds)., Fields, strings and critical phenomena, Proceedings of the Les Houches Ecole d'été de physique théorique Session XLIX. North–Holland, Amsterdam.

Girardeau, Marvin (1960). Relationship between systems of impenetrable bosons and fermions in one dimension. *J Math Phys*, **1**, 516–23.

Glazman, Leonid I and Raikh, Mikhail E (1988). Resonant Kondo transparency of a barrier with quasilocal impurity states. *JETP Lett*, 47, 452 – 5.

Gogolin, Alexander O, Nersesyan, Alexander A, and Tsvelik, Alexei M (1998). *Bosonization and strongly correlated systems*. Cambridge University Press, Cambridge.

Goldenfeld, Nigel (1992). *Lectures on phase transitions and the renormalization group*. Addison-Wesley, Reading, MA.

Goldhaber-Gordon, David, Shtrikman, Hadas, Mahalu, Diana, Abusch-Magder, David, Meirav, Udi, and Kastner, Marc A (1998). Kondo effect in a single-electron transistor. *Nature*, **391**, 156–9.

Goodstein, David L (1985). *States of matter*. Dover Publications, New York.

Gradshteyn, I S and Ryzhik, I M (1980). *Table of integrals, series, and products*. Academic Press, San Diego.

Grandati, Yves (1992). Élements d'introduction à l'invariance conforme. *Ann Phys France*, **17**, 159–300.

Granet, Etienne, Jacobsen, Jesper Lykke, and Saleur, Hubert (2018). A distribution approach to finite-size corrections in Bethe ansatz solvable models. *Nucl Phys B*, **934**, 96–117.

Greiner, Markus, Mandel, Olaf, Esslinger, Tilman, Haensch, Theodor W, and Bloch, Immanuel (2002). Quantum phase transition from a superfluid to a Mott insulator in a gas of ultracold atoms. *Nature*, **415**, 39–44.

Griffiths, Robert B (1964*a*). Free energy of the antiferromagnetic linear chain. *Phys Rev*, **136**, A751.

Griffiths, Robert B (1964*b*). Magnetization curve at zero temperature for the antiferromagnetic Heisenberg linear chain. *Phys Rev*, **133**, A768.

Grobis, Michael, Rau, Ileana G, Potok, Ronald M, and Goldhaber-Gordon, David (2007). The Kondo effect in mesoscopic quantum dots. In Helmut Kronmuller and Stuart Parkin (Eds)., *Handbook of magnetism and advanced magnetic materials* vol. 5. John Wiley, Oxford.

Guan, Xi-Wen, Batchelor, Murray T, and Lee, Chaohong (2013). Fermi gases in one dimension: From Bethe ansatz to experiments. *Rev Mod Phys*, **85**, 1633.

Guan, Xi-Wen, Grimm, Uwe, Römer, Rudolf A, and Schreiber, Michael (2000). Integrable impurities for an open fermion chain. *J Phys A: Math Gen*, **33**, 3863–79.

Gunn, Michael (1988). Spin strings and superconductivity. *Phys World*, **1**, 31.

Gutzwiller, Martin C (1963). The effect of correlations on the ferromagnetism of transition metals. *Phys Rev Lett*, **10**, 159.

de Haas, Wander Johannes, de Boer, Jan Hendrik, and van den Berg, G J (1934). The electrical resistance of gold, copper, and lead at low temperatures. *Physica*, **1**, 1115.

Haken, Hermann and Wolf, Hans Christoph (1984). *Atomic and quantum physics—An introduction to the fundamentals of experiment and theory*. Springer, Berlin.

Haldane, F Duncan M (1980). General relation of correlation exponents and spectral properties of one-dimensional Fermi systems: Application to an anisotropic $S = \frac{1}{2}$ Heisenberg chain. *Phys Rev Lett*, **45**, 1358–62.

Haldane, F Duncan M (1981*a*). Demonstration of the 'Luttinger liquid' character of the Bethe-ansatz-soluble models of 1-*D* quantum liquids. *Physics Letters*, **81A**, 153–5.

Haldane, F Duncan M (1981*b*). 'Luttinger liquid theory' of one-dimensional quantum fluids: I. Properties of the Luttinger model and their extension to the general 1*D* interacting spinless Fermi gas. *J Phys C: Solid State Phys*, **14**, 2585–2609.

Haldane, F Duncan M (1983*a*). Continuum dynamics of the 1–*D* Heisenberg antiferromagnet: Identification with the $O(3)$ nonlinear sigma model. *Phys Lett A*, **93**, 464–8.

Haldane, F Duncan M (1983*b*). Nonlinear field theory of large-spin Heisenberg antiferromagnets: Semiclassically quantized solitons of the one-dimensional easy-axis Néel state. *Phys Rev Lett*, **50**, 1153–6.

Haldane, F Duncan M (1988). Exact Jastrow–Gutzwiller resonating-valence-bond ground state of the spin-$\frac{1}{2}$ antiferromagnetic Heisenberg chain with $1/r^2$ exchange. *Phys Rev Lett*, **60**, 635.

Halmos, Paul R (1978). *Measure theory*. Springer, Berlin.

Hamer, Chris J (1985). Finite-size corrections for ground states of the XXZ Heisenberg chain in the critical region. *J Phys A: Math Gen*, **18**, L1133–7.

Hamer, Chris J (1986). Finite–size corrections for ground states of the XXZ Heisenberg chain. *J Phys A: Math Gen*, **19**, 3335–51.

Hamer, Chris J, Batchelor, Murray T, and Barber, Michael N (1988). Logarithmic corrections to finite-size scaling in the four-state Potts model. *J Stat Phys*, **52**, 679–710.

Hamer, Chris, Quispel, Reinout, and Batchelor, Murray (1987). Conformal anomaly and surface energy for Potts and Ashkin–Teller quantum chains. *J Phys A: Math Gen*, **20**, 5677.

Haroche, Serge and Raimond, Jean-Michel (2006). *Exploring the quantum: Atoms, cavities, and photons*. Oxford University Press, Oxford.

Heilmann, O J and Lieb, Elliott H (1970). Violation of the non-crossing rule: the Hubbard Hamiltonian for benzene. *Trans NY Acad Sci*, **33**, 116. (Also in *Ann NY Acad Sci*, **172** (1971) 584.).

Heisenberg, Werner (1928). Zur Theorie des Ferromagnetismus. *Z Physik*, **49**, 619–36.

Heitler, Walter and London, Fritz (1927). Wechselwirkung neutraler Atome und homöopolare Bindung nach der Quantenmechanik. *Z Physik*, **44**, 455–72.

Henkel, Malte (1999). *Conformal invariance and Critical phenomena*. Texts and Monographs in Physics. Springer-Verlag, Berlin Heidelberg.

Herbut, Igor (2007). *A modern approach to critical phenomena*. Cambridge University Press, Cambridge.

Hewson, Alex C (1993). *The Kondo problem to heavy Fermions*. Cambridge University Press, Cambridge.

Holstein, Theodore and Primakoff, Henry (1940). Field dependence of the intrinsic domain magnetization of a ferromagnet. *Phys Rev*, **58**, 1098–1113.

Hubbard, John (1963). Electron correlations in narrow energy bands. *Proc Roy Soc (London)*, **A276**, 238.

Huerta, Luis and Zanelli, Jorge (1993). Bose–Fermi transformation in three-dimensional space. *Phys Rev Lett*, **71**, 3622–5.

Hulthén, Lamek (1938). Über das Austauschproblem eines Kristalles. *Ark Mat Astron Fys A*, **26**, 1.

Hurd, C M (1982). Varieties of magnetic order in solids. *Contemp Phys*, **23**, 469–93.

Ihn, Thomas (2010). *Semiconductor nanostructures—Quantum states and electronic transport*. Oxford University Press.

Imai, Takashi and Lee, Young S (2016). Do quantum spin liquids exist? *Phys Today*, **69**, 30–6.

Ising, Ernst (1925). Beitrag zur Theorie des Ferromagnetismus. *Z Physik*, **31**, 253.

Itzykson, Claude and Drouffe, Jean-Michel (1989). *Statistical field theory*, vol. 2. Cambridge University Press, Cambridge.

Izergin, Anatoli G (2000). *Introduction to the Bethe ansatz solvable models. Lectures given in 1998–1999 by Anatoli G. Izergin*. Filippo Colomo and Andrei G. Pronko (Eds).

Izykson, Claude, Saleur, Hubert, and Zuber, Jean-Bernard (Eds) (1988). *Conformal invariance and application to statistical mechanics*. World Scientific, Singapore.

Izyumov, Yurii A and Skryabin, Yurii N (1988). *Statistical mechanics of magnetically ordered systems*. Consultants Bureau, New York.

Jackson, John D (1999). *Classical electrodynamics* (3rd edn). John Wiley, New York.

Jaynes, Edwin T and Cummings, Frederick W (1963). Comparison of quantum and semi-classical radiation theories with application to the beam maser. *Proceedings of the IEEE*, **51**, 89–109.

Joerdens, Robert, Strohmaier, Niels, Guenter, Kenneth, Moritz, Henning, and Esslinger, Tilman (2008). A Mott insulator of fermionic atoms in an optical lattice. *Nature*, **455**, 204–207.

Jordan, Pascal and Wigner, Eugene (1928). Über das Paulische Äquivalenzverbot. *Z Physik*, **47**, 631.

Kadanoff, Leo P and Brown, Alan C (1979). Correlation function on the critical lines of the Baxter and Ashkin–Teller models. *Ann Phys*, **121**, 318–42.

Kaiser, Robin, Westbrook, Chris, and David, François (Eds) (1999). *Coherent atomic matter waves*. EDP Sciences, Les Ulis.

Kanamori, Junjiro (1963). Electron correlation and ferromagnetism of transition metals. *Prog Theor Phys*, **30**, 275.

Karbach, Michael, Hu, Kun, and Müller, Gerhard (1998). Introduction to the Bethe ansatz II. *Computers in Physics*, **12**, 565.

Karbach, Michael, Hu, Kun, and Müller, Gerhard (2000). Introduction to the Bethe ansatz III. arXiv:cond-mat/0008018.

Karbach, Michael and Müller, Gerhard (1997). Introduction to the Bethe ansatz I. *Computers in Physics*, **11**, 36.

Kardar, Mehran (2007*a*). *Statistical physics of fields*. Cambridge University Press, Cambridge.

Kardar, Mehran (2007*b*). *Statistical physics of particles*. Cambridge University Press, Cambridge.

Kaufman, Bruria (1949). Crystal statistics. II. Partition function evaluated by spinor analysis. *Phys Rev*, **76**, 1232.

Kaufman, Bruria and Onsager, Lars (1949). Crystal statistics. III. Short-range order in a binary Ising lattice. *Phys Rev*, **76**, 1244.

Kinoshita, Toshiya, Wenger, Trevor, and Weiss, David S (2004). Observation of a one-dimensional Tonks–Girardeau gas. *Science*, **305**, 1125–8.

Kivelson, Steven A, Rokhsar, Daniel S, and Sethna, James P (1987). Topology of the resonating valence-bond state: Solitons and high-T_c superconductivity. *Phys Rev B*, **35**, 8865.

Klimov, Andrei B and Chumakov, Sergei M (2009). *A group-theoretical approach to quantum optics—Models of atom–field interaction*. Wiley–VCH, Weinheim.

Klümper, Andreas and Batchelor, Murray T (1990). An analytic treatment of the finite-size corrections in the spin-1 antiferromagnetic XXZ chain. *J Phys A: Math Gen*, **23**, L189–95.

Klümper, Andreas, Batchelor, Murray T, and Pearce, Paul A (1991). Central charges of the 6- and 19-vertex models with twisted boundary conditions. *J Phys A: Math Gen*, **24**, 3111–33.

Klümper, Andreas and Pearce, Paul A (1991). Analytic calculation of scaling dimensions: Tricritical hard squares and critical hard hexagon. *J Stat Phys*, **64**, 13–76.

Kogut, John B (1979). An introduction to lattice gauge theory and spin systems. *Rev Mod Phys*, **51**, 659–713.

Kolb, Edward W and Turner, Michael S (1989). *The early universe*. Addisson–Wesley, Redwood City.

Kondo, Jun (2012). *The physics of dilute magnetic alloys*. Cambridge University Press, Cambridge.

Korepin, Vladimir E, Bogoliubov, Nikolay M, and Izergin, Anatoli G (1993). *Quantum inverse scattering method and correlation functions*. Cambridge University Press, Cambridge.

Kozlowski, Karol (2015). *Asymptotic analysis and quantum integrable models*. Thèse d'Habilitation. Université de Bourgogne.

Kubo, Ryogo (1988). *Statistical mechanics—An advanced course with problems and Solutions* (second edn). North–Holland, Amsterdam.

Kulish, Petr P and Sklyanin, Evgeni K (1991). The general $U_q[sl(2)]$ invariant XXZ integrable quantum spin chain. *J Phys A: Math Gen*, **24**, L435.

Lancaster, Tom and Blundell, Stephen J (2014). *Quantum field theory for the gifted amateur*. Oxford University Press, Oxford.

Landau, David P and Binder, Kurt (2015). *A Guide to Monte Carlo simulations in statistical physics* (4th edn). Cambridge University Press.

Landau, Lev and Ginzburg, Vitaly L (1950). On the theory of superconductivity. *JETP*, **20**, 1064.

Landau, Lev D (1957a). The theory of a Fermi liquid. *Soviet Phys JETP*, 3, 920–5.

Landau, Lev D (1957b). Oscillations in a Fermi liquid. *Soviet Phys JETP*, **5**, 101–8.

Landau, Lev D (1959). On the theory of the Fermi liquid. *Soviet Phys JETP*, **8**, 70–4.

Laroche, Dominique, Gervais, Guillaume, Lilly, Michael P, and Reno, John L (2014). 1D–1D Coulomb drag signature of a Luttinger Liquid. *Science*, **343**, 631.

Larson, Jonas (2013). Integrability versus quantum thermalization. *J Phys B: At Mol Opt Phys*, **46**, 224016.

Laughlin, Robert B (1998). Nobel Prize Lecture. {http://large.stanford.edu/prizes/nobel/lecture/}. Overheads from Nobel Prize Lecture, Stanford University (last accessed 14 March 2013).

Laughlin, Robert B and Pines, David (2000). The theory of everything. *Proc Nat Acad Sci*, **97**, 28–31.

Le Hur, Karyn (2015). Quantum dots and the Kondo effect. *Nature*, **526**, 203–4.

van Leeuwen, Hendrika Johanna (1921). Problème de la théorie électronique du magnétisme. *J Physique et le Radium*, **2**, 361–77.

Leggett, Anthony J (2006). *Quantum liquids—Bose condensation and Cooper pairing in condensed-matter systems*. Oxford University Press, Oxford.

Leinaas, Jon M and Myrheim, Jan (1977). On the theory of identical particles. *Nuovo Cimento Soc. Ital. Fis., B*, **37**, 1–23.

Lenz, Wilhelm (1920). Beitrag zum Verständnis der magnetischen Eigenschaften in festen Körpern. *Phys Z*, **21**, 613.

Lewenstein, Maciej, Sanpera, Anna, and Ahufinger, Verònica (2012). *Ultracold atoms in optical lattices: Simulating quantum many-body systems*. Oxford University Press, Oxford.

Lieb, Elliott H (1963). Exact analysis of an interacting Bose gas. II. The excitation spectrum. *Phys Rev*, **130**, 1616.

Lieb, Elliott H (1967*a*). Exact solution of the problem of the entropy of two-dimensional ice. *Phys Rev Lett*, **18**, 692.

Lieb, Elliott H (1967*b*). Residual entropy of square ice. *Phys Rev*, **162**, 162.

Lieb, Elliott H (1993). The Hubbard model: Some rigorous results and open problems. http://arxiv.org/abs/cond-mat/9311033.

Lieb, Elliott H and Liniger, Werner (1963). Exact analysis of an interacting Bose gas. I. The general solution and the ground state. *Phys Rev*, **130**, 1605.

Lieb, Elliott H and Mattis, Daniel C (1962). Ordering energy levels of interacting spin systems. *J Math Phys*, **3**, 749.

Lieb, Elliott H and Mattis, Daniel C (1966). *Mathematical physics in one Dimension — Exactly soluble models of interacting particles*. Academic Press, New York.

Lieb, Elliott H, Schultz, Theodore D, and Mattis, Daniel C (1961). Two soluble models of an antiferromagnetic chain. *Ann Phys*, **16**, 407.

Lieb, Elliott H and Wu, Fa-Yueh (1968). Absence of Mott transition in an exact solution of the short-range, one-band model in one dimension. *Phys Rev Lett*, **20**, 1445.

Lieb, Elliott H and Wu, Fa-Yueh (1972). Two-dimensional ferroelectric models. In Cyril Domb and M S Green (Eds)., *Phase Transitions and Critical Phenomena*, pp. 331–490. Academic, London.

Lieb, Elliott H and Wu, Fa-Yueh (2003). The one-dimensional Hubbard model: A reminiscence. *Physica A*, **321**, 1.

Lowenstein, John H (1982). *Introduction to the Bethe-ansatz approach in* (1 + 1)*–dimensional models*. In J-B Zuber and R Stora (Eds)., Recent advances in field theory and statistical mechanics. Proceedings of the Les Houches Ecole d'été de physique théorique Session XXXIX, pp. 609–81. North–Holland, Amsterdam.

Luck, Jean-Marc (1982). Finite-size scaling and the two-dimensional *xy* model. *J Phys A: Math Gen*, **15**, L169–76.

Luttinger, Joaquin M (1963). An exactly soluble Model of a many-fermion system. *J Math Phys*, **4**, 1154–62.

Ma, Shang-keng (1976). *Modern theory of critical phenomena*. Benjamin/Cummings, Reading, MA.

MacDonald, D Keith C (2006). *Introductory statistical mechanics for physicists*. Dover, Mineola, New York.

Maciejewski, Andrzej J, Przybylska, Maria, and Stachowiak, Tomasz (2014*a*). Analytical method of spectra calculations in the Bargmann representation. *Phys Lett A*, **378**, 3445–51.

Maciejewski, Andrzej J, Przybylska, Maria, and Stachowiak, Tomasz (2014*b*). Full spectrum of the Rabi model. *Phys Lett A*, **378**, 16–20.

Maciejewski, Andrzej J, Przybylska, Maria, and Stachowiak, Tomasz (2015). An exactly solvable system from quantum optics. *Phys Lett A*, **379**, 1505–9.

Madelung, Otfried (1978). *Introduction to solid-state theory*. Springer, Berlin.

Mahan, Gerald D (2000). *Many-particle physics* (3rd edn). Kluwer, New York.

Majumdar, Chanchal K (1970). Antiferromagnetic model with known ground state. *J Phys C: Solid State Phys*, **3**, 911–15.

Majumdar, Chanchal K and Ghosh, Dipan K (1969*a*). On next-nearest-neighbor interaction in linear chain. I. *J Math Phys*, **10**, 1388–98.

Majumdar, Chanchal K and Ghosh, Dipan K (1969*b*). On next-nearest-neighbor interaction in linear chain. II. *J Math Phys*, **10**, 1388–98.

Marder, Michael P (2010). *Condensed matter physics* (2nd edn). John Wiley, New York.

Maris, Humphrey J and Kadanoff, Leo P (1978). Teaching the renormalization group. *Am J Phys*, **46**, 652–7.

Marshall, W (1955). Antiferromagnetism. *Proc Roy Soc (London)*, **A232**, 48–68.

Mathews, Jon and Walker, Robert L (1970). *Mathematical Methods of Physics* (2nd edn). W.A. Benjamin, San Francisco.

Mattis, Daniel C (1993). *The many-body problem: An encyclopedia of exactly solved models in one dimension*. World Scientific, Singapore.

Mattis, Daniel C (2006). *The theory of magnetism made simple*. World Scientific, Singapore.

Mattis, Daniel C and Lieb, Elliott H (1965). Exact solution of a many-fermion system and its associated boson field. *J Math Phys*, **6**, 304–12.

McGuire, James B (1964). Study of exactly soluble one-dimensional *N*-body problems. *J Math Phys*, **5**, 622–36.

McGuire, James B (1965). Interacting Fermions in one dimension. I: Repulsive potential. *J Math Phys*, **6**, 432–9.

McGuire, James B (1966). Interacting Fermions in one dimension. II: Attractive potential. *J Math Phys*, 7, 123–32.

Mehta, Pankaj and Andrei, Natan (2006). Nonequilibrium transport in quantum impurity models: The Bethe ansatz for open systems. *Phys Rev Lett*, **96**, 216802.

Mezincescu, Luca and Nepomechie, Rafael I (1992). *Introduction to the thermodynamics of spin chains*. Proceedings of the NSERC-CAP Workshop on Quantum Groups, Integrable Models and Statistical Systems, 13–18 July, Kingston, Canada.

Mott, Neville F (1949). The basis of the electron theory of metals, with special reference to the transition metals. *Proc Royal Society (London)*, **A62**, 416.

Mott, Nevill F and Peierls, Rudolf (1937). Discussion of the paper by de Boer and Verwey. *Proc Royal Society (London)*, **49**, 72 – 3.

Mussardo, Giuseppe (2010). *Statistical field theory*. Oxford University Press.

Nagle, John F (1966). Lattice statistics of hydrogen bond crystals i. the residual entropy of ice. *J Math Phys*, 7, 1484.

Nauenberg, Michael (1975). Renormalization group solution of the one-dimensional Ising model. *J Math Phys*, **16**, 703–5.

Nazarov, Yuli V and Blanter, Yaroslav M (2009). *Quantum transport: Introduction to nanoscience*. Cambridge University Press, Cambridge.

Nazarov, Yuli V and Danon, Jeroen (2013). *Advanced quantum mechanics*. Cambridge University Press, Cambridge.

Negele, John W and Orland, Henri (1998). *Quantum many-particle systems*. Westview Press, Boulder.

Nepomechie, Raphael (1999). A spin chain primer. *Int J Mod Phys B*, **13**, 2973. {arXiv: hep-th/9810032v15Oct1998}.

Ng, Tai Kai and Lee, Patrick A (1988). On-site Coulomb repulsion and resonant tunneling. *Phys Rev Lett*, **61**, 1768 – 71.

Nielsen, Michael A and Chuang, Isaac L (2010). *Quantum computation and quantum information* (10th anniversary edn). Cambridge University Press, Cambridge.

Niemeijer, Theodorus and van Leeuwen, J M J (1976). Renormalization theory for Ising–like spin systems. In C Domb and J L Lebowitz (Eds)., *Phase Transitions and Critical Phenomena*, vol. 6. Academic Press, London.

Nightingale, M Peter (1976). Scaling theory and finite systems. *Physica*, **83A**, 561.

Nightingale, M Peter (1982). Finite-size scaling and phenomenological renormalization. *J Appl Phys*, **53**, 7927.

Nightingale, M Peter and Blöte, Henk W J (1983). The relation between amplitudes and critical exponents in finite-size scaling. *J Phys A*, **16**, L657–64.

Nilsson, Johan, Eckle, Hans-Peter, and Johannesson, Henrik (2007). Persistent currents through a quantum impurity: Protection through integrability. *Phys Rev B*, **76**, 073408.

Nishimori, Hidetoshi and Ortiz, Gerardo (2011). *Elements of phase transitions and critical phenomena*. Oxford University Press, Oxford.

Nolting, Wolfgang and Ramakanth, Anapuru (2009). *Quantum theory of magnetism*. Springer, Heidelberg.

Nomura, Kiyohide (1993). Logarithmic corrections of the one-dimensional $S = \frac{1}{2}$ Heisenberg antiferromagnet. *Phys Rev B*, **48**, 16813–17.

Onsager, Lars (1944). Crystal statistics. I. A two-dimensional model with an order-disorder transition. *Phys Rev*, **65**, 117.

Öz, Yahya and Klümper, Andreas (2018). A Hubbard model with integrable impurity. arXiv:1808.06993[cond-mat.stat-mech].

Paredes, Belén, Widera, Artur, Murg, Valentin, Mandel, Olaf, Fölling, Simon, Cirac, Ignacio, Shlyapnikov, Gora V, Hänsch, Theodor W, and Bloch, Immanuel (2004). Tonks–Girardeau gas of ultracold atoms in an optical lattice. *Nature*, **429**, 277–81.

Pariser, R and Parr, R G (1953). A semi-empirical theory of the electronic spectra and electronic structure of complex unsaturated hydrocarbons I and II. *J Chem Phys*, **21**, 466,767.

Parkinson, John B and Bonner, Jill C (1985). Spin chains in a field: Crossover from quantum to classical behavior. *Phys Rev B*, **32**, 4703–24.

Parkinson, John B and Farnell, Damian J J (2010). *An introduction to quantum spin systems*. Springer, Berlin.

Pasquier, Vincent and Saleur, Hubert (1990). Common structures between finite systems and conformal field theories through quantum groups. *Nucl Phys B*, **330**, 523.

Pauli, Wolfgang (1940). The connection between spin and statistics. *Phys Rev*, **58**, 716–22.

Pauling, Linus (1935). The structure and entropy of ice and of other crystals with some randomness of atomic arrangement. *J Am Chem Soc*, **57**(12), 2680–4.

Pauling, Linus (1960). *The nature of the chemical bond* (3rd edn). Cornell University Press, Ithaca, NY.

Paz, Robert (1989). Feynman's office: The last blackboards. *Phys. Today*, **42**(2), 88. Special issue: Richard Feynman.

Pearce, Paul A and Klümper, Andreas (1991). Finite-size corrections and scaling dimensions of solvable lattice models: An analytic method. *Phys Rev Lett*, **66**, 974–7.

Perk, Jacques H H and Au-Yang, Helen (2006). Yang–Baxter equations. In Jean-Pierre Françoise, Gregory L Naber and Tsou Sheung Tsun (Eds)., Encyclopedia of mathematical physics, vol. 5. pp. 465–73. Academic Press, New York.

Pethick, Christopher and Smith, Henrik (2008). *Bose–Einstein condensation in dilute gases* (2nd edn). Cambridge University Press, Cambridge.

Pfeuty, Pierre (1970). The one-dimensional Ising model with a transverse field. *Ann Phys*, **57**, 79.

Phillips, Philip (2012). *Advanced solid-state physics* (2nd edn). Cambridge University Press, Cambridge.

Pitaevskii, Lev and Stringari, Sandro (2008). *Bose–Einstein condensation and superfluidity* (2nd edn). Cambridge University Press, Cambridge.

Polchinski, Joseph (2005). *String theory*, vol. 1. Cambridge University Press, Cambridge.

Polyakov, Alexander M (1970). Conformal symmetry of critical fluctuations. *JETP Letters*, **12**, 381–3.

Pople, J R (1953). Electron interaction in unsaturated hydrocarbons. *Trans Faraday Soc*, **49**, 1375.

Privman, Vladimir (1985). Conformal mapping of a periodic strip at criticality. *Phys Rev B*, **32**, 6089–90.

Privman, Vladimir and Fisher, Michael E (1984). Universal critical amplitudes in finite-size scaling. *Phys Rev B*, **30**, 322–7.

Pustilnik, Michael (2006). Kondo effect in nanostructures. *Phys Stat Sol (A)*, **203**, 1137–47.

Rabi, Isidore I (1936). On the process of space quantization. *Phys Rev*, **49**, 324–8.

Rabi, Isidore I (1937). Space quantization in a gyrating magnetic field. *Phys Rev*, **51**, 652–4.

Razumov, A V and Stroganov, Yu G (2001). Spin chains and combinatorics. *J Phys A: Math Gen*, **34**, 3185.

Reichl, Linda E (2016). *A modern course in statistical physics* (4th edn). Wiley-VCH, Weinheim.

Rickayzen, G (1980). *Green's functions and condensed matter*. Academic Press, London.

Roos, Bernard W (1969). *Analytic functions and distributions in physics and engineering*. John Wiley, New York.

Sachdev, Subir (2011). *Quantum phase transitions* (2nd edn). Cambridge University Press, Cambridge.

Sachdev, Subir and Keimer, Bernhard (2011). Quantum criticality. *Phys Today*, **64**, 29. A longer version can be found at: arXiv:1102.4628v2 [cond-mat.str-el].

Saint-Aubin, Yvan (1987). Phénomènes critique en deux dimensions et invariance conforme. Note de cours de Physique Mathématique avancée, Université de Montréal.

Schmitteckert, Peter, Schwab, Peter, and Eckern, Ulrich (1995). Quantum coherence in an exactly solvable one-dimensional model with defects. *Europhys Lett*, **30**, 543–8.

Schoelkopf, Robert J and Girvin, Steven M (2008). Wiring up quantum systems. *Nature*, **451**, 664–9.

Schrieffer, J Robert and Wolff, P A (1966). Relation between the Anderson and Kondo Hamiltonians. *Phys Rev*, **149**, 491–2.

Schrödinger, Erwin (1935). Die gegenwärtige Situation in der Quantenmechanik. *Naturwissenschaften*, **23**, 807–12, 823–8, 844–9.

Schultz, Theodore D, Mattis, Daniel C, and Lieb, Elliott H (1964). Two-dimensional Ising model as a soluble problem of many fermions. *Rev Mod Phys*, **36**, 856.

Schulz, Hermann (1985). Hubbard chain with reflecting ends. *J Phys C: Solid State Phys*, **18**, 581–601.

Schwabl, Franz (2006). *Statistical mechanics* (2nd edn). Springer, Berlin.

Sethna, James P (2006). *Statistical mechanics: Entropy, order Parameters, and complexity*. Oxford University Press, Oxford.

Shastry, B Sriram (1988*a*). Decorated star-triangle relations and exact integrability of the one-dimensional Hubbard model. *J Stat Phys*, **50**, 57–79.

Shastry, B Sriram (1988*b*). Exact solution of an $S = \frac{1}{2}$ Heisenberg antiferromagnetic chain with long-ranged interactions. *Phys Rev Lett*, **60**, 639.

Shenker, Stephen H (1982). Field theories and phase transitions. In J-B Zuber and R Stora (Eds)., *Recent advances in field theories and statistical mechanics* Proceedings of the Les Houches Ecole d'été de physique théorique Session XXXIX. North–Holland, Amsterdam.

Simpson, William M R and Leonhardt, Ulf (2015). *Forces of the quantum vacuum: An introduction to Casimir physics*. World Scientific, Singapore.

Sirker, Jesko (2012*a*). The Luttinger liquid and integrable models. *Int J Mod Phys B*, **26**, 1244009.

Sirker, Jesko (2012*b*). The Luttinger liquid and integrable models: Erratum. *Int J Mod Phys B*, **26**, 1292002.

Sklyanin, Evgeny K (1988). Boundary conditions for integrable quantum systems. *J Phys A: Math Gen*, **21**, 2375.

Sklyanin, Evgeny K (1992). Quantum inverse scattering method. Selected topics. In M-L Ge (Ed)., *Quantum Group and Quantum Integrable Systems*, pp. 63–97. Nankai lectures in mathematical physics. World Scientific, Singapore.

Slavnov, Nikita A (2018). Algebraic Bethe ansatz. arXiv:1804.07350 [math-ph].

Solé, Ricard V (2011). *Phase transitions*. Princeton University Press, Princeton.

Sólyom, Jenö (2009). *Fundamentals of the physics of solids: Electronic properties*, vol. 2. Springer, Berlin.

Sólyom, Jenö (2010). *Fundamentals of the physics of solids: Normal, broken-symmetry, and correlated systems*, vol. 3. Springer, Berlin.

Sommerfeld, Arnold and Bethe, Hans (1933). Elektronentheorie der Metalle. In Hans Geiger and Karl Scheel (Eds)., *Handbuch der Physik* vol. 24, part 2. Springer, Berlin. Reprint 1967.

Sondhi, Shivashi, Girvin, Steven M, Carini, John P, and Shahar, Dan (1997). Continuous quantum phase transitions. *Rev Mod Phys*, **69**, 315–33.

Sørensen, Eric S, Eggert, Sebastian, and Affleck, Ian (1993). Integrable versus non-integrable spin chain impurity models. *J Phys A: Math Gen*, **26**, 6757–76.

Sornette, Didier (2003). *Why stock markets crash: Critical events in complex financial systems*. Princeton University Press, Princeton.

Stanley, Eugene (1971). *Introduction to phase transitions and critical phenomena*. Oxford University Press, Oxford.

Steiner, Michael, Villain, Jacques, and Windsor, Colin G (1976). Theoretical and experimental studies on one-dimensional magnetic systems. *Adv Phys*, **25**, 87–209.

Stéphan, Jean-Marie and Dubail, Jérôme (2011). Local quantum quenches in critical one-dimensional systems: Entanglement, the Loschmidt echo, and light-cone effects. *J Stat Mech*, P08019.

Stern, Ady (2008). Anyons and the quantum Hall effect: A pedagogical review. *Ann Phys*, **323**, 204–49.

Stern, Ady (2010). Non-abelian states of matter. *Nature*, **464**, 187–93.

Stone, Michael (1994). *Bosonization*. World Scientific, Singapore.

Stroganov, Yu G (2001). The importance of being odd. *J Phys A: Math Gen*, **34**, L179–85.

Sutherland, Bill (1970). Two-dimensional hydrogen bonded crystals without the ice rule. *J Math Phys*, **11**, 3183–6.

Sutherland, Bill (1985). An introduction to the Bethe ansatz. In B S Shastry, S S Jha, and V Singh (Eds)., *Exactly solvable problems in condensed matter and relativistic field theory* p. 1. Lecture Notes in Physics 242. Springer, Berlin.

Sutherland, Bill (2004). *Beautiful models: 70 years of exactly solved quantum many-body problems*. World Scientific, Singapore.

Takahashi, Minoru (1971*a*). One-dimensional Heisenberg model at finite temperature. *Prog Theo Phys*, **46**, 401.

Takahashi, Minoru (1971*b*). Thermodynamics of the Heisenberg–Ising model for $|\Delta| < 1$ in one dimension. *Phys Lett A*, **36**, 325.

Takahashi, Minoru (1997). Thermodynamical Bethe ansatz and condensed matter. In Zalán Horváth and László Palla (Eds)., *Conformal field theories and integrable models: Lectures held at the Eötvös University graduate course, Budapest, Hungary, 13–18 August 1996*, pp. 204–50. Springer, Berlin.

Takahashi, Minoru (1999). *Thermodynamics of one-dimensional solvable models*. Cambridge University Press, Cambridge.

Takahashi, Minoru and Suzuki, Masuo (1972). One-dimensional anisotropic Heisenberg model at finite temperatures. *Prog Theo Phys*, **48**, 2187.

Takhtajan, Leon A (1985). Introduction to algebraic Bethe ansatz. In B S Shastry, S S Jha, and V Singh (Eds)., *Exactly solvable problems in condensed matter and relativistic field theory*, pp. 175–219. Lecture Notes in Physics, vol. 242. Springer, Berlin.

Tavis, Michael and Cummings, Frederick (1968). Exact solution for an N-molecule-radiation-field hamiltonian. *Physical Review*, **170**, 379–84.

Tavis, Michael and Cummings, Frederick (1969). Approximate Solutions for an N-molecule-radiation-field Hamiltonian. *Physical Review*, **188**, 692.

Tennant, D Alan, Cowley, Roger A, Nagler, Stephen E, and Tsvelik, Alexei M (1995). Measurement of the spin-excitation continuum in one-dimensional $KCuF_3$ using neutron scattering. *Phys Rev B*, **52**, 13368–80.

ter Haar, Dirk (ed.) (1965). *Collected papers of L D Landau*. Gordon and Breach, New York.

Theimer, Otto and Ram, Budh (1976). The beginning of quantum statistics. *Am J Phys*, **44**, 1056–7.

Tian, Lin (2011). Cavity cooling of a mechanical resonator in the presence of a two-level-system defect. *Physical Review B*, **84**, 035417.

Tomonaga, Sin-itiro (1950). Remarks on Bloch's method of sound waves applied to many-fermion problems. *Prog Theor Phys*, **5**, 544–69.

Tonks, Lewi (1936). The complete equation of state of one, two and three-dimensional gases of hard elastic spheres. *Phys Rev*, **50**, 955–63.

Truong, Tuong T (1987). Six-vertex model. Unpublished lecture notes.

Tsvelik, Alexei M and Wiegmann, Paul B (1983). Exact results in the theory of magnetic alloys. *Adv Phys*, **32**, 453.

de Vega, Hector J and Woynarovich, Ferenc (1985). Method for calculating finite size corrections in Bethe ansatz systems: Heisenberg chain and six-vertex model. *Nucl Phys B*, **251**[**FS13**], 439–56.

Viefers, Susanne, Koskinen, Pekka, Singha Deo, Prosenjit, and Manninen, Matti (2004). Quantum rings for beginners: Energy spectra and persistent currents. *Physica E*, **21**, 1–35.

Voit, Johannes (1994). One-dimensional Fermi liquids. *Rep Prog Phys*, **57**, 977–1116.

Voit, Johannes (2005). *The statistical mechanics of financial markets* (3rd edn). Springer, Berlin.

Vojta, Matthias (2003). Quantum phase transitions. *Rep Prog Phys*, **66**, 2069–2110.

Vojta, Thomas (2000). Quantum phase transitions in electronic systems. *Ann Phys (Leipzig)*, **9**, 403.

Vojta, Thomas (2002). Quantum phase transitions. In Karl-Heinz Hoffmann and Michael Schreiber (Eds)., *Computational Statistical Physics*, pp. 211–26. Springer, Berlin.

Weigert, Stefan (1992). The problem of quantum integrability. *Physica D*, **56**, 107–9.

Wen, Xiao-Gang (2004). *Quantum field theory of many-body systems.* Oxford University Press, Oxford.

White, Steven R and Huse, David A (1993). Numerical renormalization-group study of low-lying eigenstates of the antiferromagnetic s=1 Heisenberg chain. *Phys Rev B*, **48**, 3844.

Widom, Benjamin (1965). Equation of state in the neighborhood of the critical point. *J Chem Phys*, **43**, 3898–3905.

Wiegmann, Pavel B (1980). Exact solution of the *s-d* exchange model at $T = 0$. *JETP Lett*, **31**, 364–70.

Wiegmann, Pavel B (1981). Exact solution of the *s-d* exchange model (Kondo problem). *J Phys C: Solid State Phys*, **14**, 1463–78.

Woynarovich, Ferenc (1982*a*). On the eigenstates of a Heisenberg chain with complex wavenumbers not forming strings. *J Phys C: Solid State Phys*, **15**, 6397.

Woynarovich, Ferenc (1982*b*). On the $S^z = 0$ excited states of an anisotropic Heisenberg chain. *J Phys A: Math Gen*, **15**, 2985.

Woynarovich, Ferenc (1985). Bound states and local moment in a 1-*d* interacting spin-1/2 Fermi system in the presence of an external potential. *Phys Lett A*, **108**, 401–6.

Woynarovich, Ferenc (1997). Introduction to coordinate-space Bethe ansatz. In Zalán Horváth and László Palla (Eds)., *Conformal field theories and integrable models: Lectures held at the Eötvös University graduate course, Budapest, Hungary, 13–18 August 1996*, pp. 151–203. Springer, Berlin.

Woynarovich, Ferenc and Eckle, Hans-Peter (1987*a*). Finite-size corrections and numerical calculations for long spin-1/2 Heisenberg chains in the critical region. *J Phys A: Math Gen*, **20**, L97–L104.

Woynarovich, Ferenc and Eckle, Hans-Peter (1987*b*). Finite-size corrections for the low lying states of a half-filled Hubbard chain. *J Phys A: Math Gen*, **20**, L443–9.

Woynarovich, Ferenc, Eckle, Hans-Peter, and Truong, Tuong T (1989). Non-analytic finite-size corrections in the one-dimensional Bose gas and Heisenberg chain. *J Phys A: Math Gen*, **22**, 4027–43.

Wyld, Henry W (1999). *Mathematical methods for physics* (2nd edn). Perseus Books, Reading, MA.

Yacoby, Amir, Stormer, Horst L, Wingreen, Ned S, Pfeiffer, Loren N, Baldwin, Kirk W, and West, Ken W (1996). Nonuniversal conductance quantization in quantum wires. *Phys Rev Lett*, 77, 4612–15.

Yahagi, Ryoko, Sato, Jun, and Deguchi, Tetsuo (2014). Finite-temperature behavior of an impurity in the spin-1/2 XXZ chain. *J Stat Mech*, P11020.

Yahagi, Ryoko, Sato, Jun, and Deguchi, Tetsuo (2016). Crossover temperature of the spin-1/2 XXZ chain with an impurity. *J Phys: Conference Series*, **670**, 012054.
Yamada, Tomoji (1969). Fermi-liquid theory of linear antiferromagnetic chains. *Prog Theo Phys*, **41**, 880.
Yang, Chen Ning (1967). Some exact results for the many-body problem in one dimension with repulsive delta-function interaction. *Phys Rev Lett*, **19**, 1312–15.
Yang, Chen Ning and Lee, Tsung-Dao (1952). Statistical theory of equations of state and phase transitions. I. Theory of condensation. *Phys Rev*, **87**, 404.
Yang, Chen Ning and Yang, C P (1966*a*). One-dimensional chain of anisotropic spin-spin interactions. I. Proof of Bethe's hypothesis for ground state in a finite system. *Phys Rev*, **150**, 321.
Yang, Chen Ning and Yang, C P (1966*b*). One-dimensional chain of anisotropic spin-spin interactions. II. Properties of the ground-state energy per lattice site for an infinite system. *Phys Rev*, **150**, 327.
Yang, Chen Ning and Yang, C P (1966*c*). One-dimensional chain of anisotropic spin-spin interactions. III. Applications. *Phys Rev*, **151**, 258.
Yang, Chen Ning and Yang, C P (1969). Thermodynamics of a one-dimensional system of Bosons with repulsive delta-function interaction. *J Math Phys*, **10**, 1115.
Ziman, John M (1979). *Models of disorder: The theoretical physics of homogeneously disordered systems.* Cambridge University Press.
Zinn-Justin, Jean and Brézin, Édouard (1966). Un problème à N corps soluble. *C R Acad Sci*, **263**, 670.
Zvyagin, Andrei A and Schlottmann, Pedro (1995). Aharonov–Casher effect in the Heisenberg spin chain with many impurities. *Phys Rev B*, **52**, 6569–74.

Index

The manufacturer's authorised representative in the EU for product safety is Oxford University Press España S.A. of el Parque Empresarial San Fernando de Henares, Avenida de Castilla, 2 – 28830 Madrid (www.oup.es/en or product.safety@oup.com). OUP España S.A. also acts as importer into Spain of products made by the manufacturer.

www.ingramcontent.com/pod-product-compliance
Ingram Content Group UK Ltd.
Pitfield, Milton Keynes, MK11 3LW, UK
UKHW050220160926
475595UK00015B/334

9780197919613